NEUTRONS, X-RAYS AND LIGHT

Scattering Methods Applied to Soft Condensed Matter

North-Holland
Delta Series

ELSEVIER
Amsterdam – Boston – London – New York – Oxford – Paris – San Diego – San Francisco – Singapore – Sydney - Tokyo

Neutrons, X-Rays and Light

Scattering Methods Applied to Soft Condensed Matter

Edited by

P. Lindner

Institut Laue-Langevin
Grenoble
France

Th. Zemb

DSM/DRECAM
CEA – Saclay
France

N·H

2002

ELSEVIER
Amsterdam – Boston – London – New York – Oxford – Paris – San Diego – San Francisco – Singapore – Sydney - Tokyo

ELSEVIER SCIENCE B.V.
Sara Burgerhartstraat 25
P.O. Box 211, 1000 AE Amsterdam, The Netherlands

First edition 2002

Library of Congress Cataloging-in-Publication Data
A catalog record from the Library of Congress has been applied for.

British Library Cataloguing in Publication Data
A catalogue record from the British Library has been applied for.

ISBN: 0-444-51122-9
ISSN: 0927-5029

⊗ The paper used in this publication meets the requirements of ANSI/NISO Z39.48-1992 (Permanence of Paper).

Transferred to digital printing 2006

Preface

It has been twelve years since the first European Summer School on 'Scattering Methods Applied to Soft Condensed Matter' was held in May 1990 in Bombannes, France (the lecture notes of this first 'Bombannes' school were published in 1991). Since 1990, five more events took place (1992, 1996, 1998, 2000, 2002). Over the years, these schools have provided advanced training for a total of more than 200 young European researchers at post-graduate and post-doctoral level and working in European laboratories.

Scattering experiments, using X-ray, light and neutron sources (in historical order) are key techniques for studying structure and dynamics in systems containing colloids, polymers, surfactants and biological macromolecules, summarized here as soft condensed matter. The standard of teaching in this field in Europe is very variable and frequently inadequate, which severely limits efficient use of these methods, especially at large-scale facilities. The series of 'Bombannes' schools and this completely revised and updated edition of the lecture notes are devoted to a practical approach to current methodology of static and dynamic techniques. Basic information on data interpretation, the complementarity of the different types of radiation and recent applications and developments is presented. The aim is to avoid over- as well as under-exploitation of data.

Most soft matter systems cannot be directly imaged by optical microscopy and require either fluorescent labelling or some other preparation in order to be imaged by scanning or electron microscopy techniques. For instance, X-ray microscopy in the water window is currently still in its infancy and may remain so over the next ten years. Therefore, even with the definite disadvantages of angular and time averaging, exploiting in a best possible way the results obtained by scattering studies is still of major importance.

Over the last century, the original field of 'colloids' – any heterogeneity in a fluid with a typical size from one micron to one nanometer – has split into a number of sub-categories. Polymers, surfactants, nanoparticles, clay, emulsions, mesoporous materials are all examples of colloids whose investigation has evolved into well-identified independent disciplines. For instance, polymers and mesoporous catalysts have been evaluated as the technological root of more than 20% of all industrial activity. Mineral flotation and water re-treatment, using complex fluids, is probably the most important industrial activity by mass of treated material, and not 'economic' value. Water treatment, the most crucial activity of the next century, is indeed completely dependent on colloidal processes, as is also the development of artificial photosynthesis.

Characterisations using scattering techniques are nowadays widely performed in laboratories and/or at large-scale facilities (ILL, ESRF, ...). The analytical power of the technique was discovered a long time ago, before World War II: André Guinier, who passed away recently, noticed in 1936 a quantitatively exploitable ('small angle') scattering near the edge of the beam stop of a classical X-ray camera. Shortly after, Bruno Zimm, a former radar engineer who put light scattering on an absolute scale, even in dilute solution, was able to measure masses of polymers. Otto Kratky in Graz designed in the late 40s the first widely available X-ray instrument, which is still in use today. With Günter Porod's theoretical contributions, the field started to grow rapidly in the 50s. Since these pioneering days, academic research and technological development (for instance, food formulation, pharmaceutics, cosmetics, petroleum industry, emulsification, separation of solutes, treatment of slurries, water cleaning, etc.) depend enormously on results obtained by scattering experiments. It is the aim of this book to provide direct insight into the methods to be used for any soft matter system in order to derive information on the microstructure invisible to the eye as well as on the molecular interaction and dynamics on the mesoscopic scale. The task of presenting this at a post-graduate and PhD level has been taken over by a team of 14 top-level scientists in their field from European countries, without distinction between sub-fields, ranging from polymers to nano-particles. We are indebted to the authors, each fully responsible for the contributions appearing in this book, for having accepted to focus on one particular point. The 21 chapters should allow any student or engineer to understand the peculiar and complex structures encountered in soft matter. The selected bibliography provided of similar experiments should give help to students to decide on the best experimental approach to use.

Last but not least we gratefully acknowledge the continuing support of a number of institutions and organizations: the past and present commitments of Institut Laue-Langevin Grenoble, Forschungszentrum Jülich, Risø National Laboratory, CEA and LLB Saclay, Paul Scherrer Institute Villigen, European Synchrotron Radiation Facility Grenoble, the TMR programme of the European Commission and the EU Neutron Round-Table have provided the necessary basis for organizing the six summer schools to date and for editing these completely revised 'Bombannes' lectures.

Peter Lindner and Thomas Zemb
March 2002

Cover illustration:

Small-angle neutron scattering reveals anisotropic flow enhanced concentration fluctuations in a sheared semi-dilute polymer system, above the overlap concentration. When subjected to laminar Couette shear flow, the solution shows under certain conditions (temperature, shear gradient) a pronounced scattering intensity increase in direction parallel to the flow, leading to the characteristic double-winged 'butterfly' detector pattern shown on the cover. (Morfin, I., Lindner, P. and Boué, F. (1999). *Macromolecules* **32**, 7208–7223.)

List of Contributors

O. Glatter	Department of Chemistry, Karl-Franzens Universität Graz, Heinrichstraße 28, A-8010 Graz, Austria *E-mail:* otto.glatter@uni-graz.at
R. Klein	Fakultät für Physik, Universität Konstanz, Postfach 5560, D-78434 Konstanz, Germany *E-mail:* Rudolf.Klein@uni-konstanz.de
P. Lindner	Institut Laue-Langevin, B.P. 156, F-38042 Grenoble CEDEX 9, France *E-mail:* lindner@ill.fr
R.P. May	Institut Laue-Langevin, B.P. 156, F-38042 Grenoble CEDEX 9, France *E-mail:* Roland.May@ill.fr
J.S. Pedersen	Department of Chemistry, University of Aarhus, Langelandsgade 140, DK-8000 Aarhus C, Denmark *E-mail:* jsp@chem.au.dk
G. Porte	Groupe de Dynamique des Phases Condensées, UMR 5581 CNRS/UM.II, Case 026, Université Montpellier II, F-34095 Montpellier CEDEX 5, France *E-mail:* porte@GDPC.univ-montp2.fr
P.N. Pusey	Department of Physics and Astronomy, The University of Edinburgh, Mayfield Road, Edinburgh EH9 3JZ, UK *E-mail:* p.n.pusey@ed.ac.uk
J. Rička	Institute of Applied Physics, Sidlerstraße 5, CH-3014 Bern, Switzerland *E-mail:* rick@fluor.iap.unibe.ch
J. Rieger	BASF Aktiengesellschaft, Polymer Physics/Solid State Physics, D-67056 Ludwigshafen, Germany *E-mail:* jens.rieger@basf-ag.de
A.J. Ryan	Department of Chemistry, University of Sheffield, Sheffield S3 7HF, UK *E-mail:* tony.ryan@sheffield.ac.uk

P. Schurtenberger Department of Physics, University of Fribourg, CH-1700 Fribourg, Switzerland
E-mail: Peter.Schurtenberger@unifr.ch

O. Spalla CEA Saclay, Groupe de Chimie de La Matière Ultradivisée, Service de Chimie Moléculaire, F-91191 Gif-sur-Yvette CEDEX, France
E-mail: spalla@drecam.saclay.cea.fr

Th. Zemb DSM/DRECAM, CEA Saclay, F-91191 Gif-sur-Yvette CEDEX, France
E-mail: zemb@drecam.saclay.cea.fr

R. Zorn Forschungszentrum Jülich, D-52425 Jülich, Germany
E-mail: R.Zorn@fz-juelich.de

Contents

Part I

Using general principles

Neutrons, X-Rays and Light
P. Lindner and Th. Zemb (Editors)

Chapter 1

Introduction to Scattering Experiments

Peter N. Pusey

Department of Physics and Astronomy, The University of Edinburgh, Mayfield Road, Edinburgh EH9 3JZ, UK

1. Introduction

Fig. 1a shows the top view of a typical scattering experiment. Radiation is incident on a sample (or scattering medium), say a suspension of colloidal particles or a solution of polymers or surfactants. Some of the radiation passes through the sample unaffected, some is scattered. A detector is set up at scattering angle θ, and the intensity $I(\theta,t)$ of the scattered radiation is measured. (Since, in general, the intensity fluctuates in time, or undergoes a shift in frequency, we indicate its time dependence explicitly.) Typically, the incident and scattered beams are shaped by apertures, slits, the walls of the sample's container, or by optics such as lenses. The region of sample which is both illuminated by the incident beam and 'seen' by the detector is called the 'scattering volume' V.

As we will see shortly, a (hypothetical) totally homogeneous medium does not scatter radiation away from the incident direction. Scattering is caused by *fluctuations* in the medium. Usually these fluctuations are associated with variations in the 'density of scattering material' within the medium. Thus, for example, a colloidal particle scatters light when it has a different index of refraction from the liquid in which it is dispersed.

Basically, three types of scattering experiment, each giving different information, can be performed. First, measuring the dependence on angle of the average scattered intensity – frequently called 'static' scattering – yields structural information. Thus, in a dilute system, we learn about the shapes of the individual particles and the arrangement of material within them. In a concentrated system, we gain information about positional correlations, the average spatial arrangement of the particles in the sample. Second, analysis of the time dependence of fluctuations in the scattered radiation (or analysis of frequency or energy changes) – 'dynamic', 'quasielastic', or 'inelastic' scattering – yields dynamic information. It tells us, for example, how the particles are moving in Brownian motion and how their shapes, or configurations, fluctuate in time. Third, if the equipment is fully calibrated, the absolute magnitude of the scattered intensity (averaged over time or frequency) gives information on the mass or molecular weight of the scattering objects.

The aim of this chapter is to introduce scattering experiments as applied to the study of soft condensed matter, emphasising the physical principles of simple situations rather than going into great detail. The emphasis is further restricted to static scattering. References listed at the end of this article, and later chapters in this book, can be consulted for more detail as well as for discussions of dynamic and inelastic scattering. Light (Berne and Pecora, 1976; Brown,

1996; D'Aguanno and Klein, 1996), X-rays (Guinier and Fournet, 1955; Glatter and Kratky, 1982) and neutrons (Higgins and Benoit, 1994) are the radiations most commonly used to study soft matter. While light scattering is associated with variations in dielectric properties (or refractive index), X-rays are scattered by electrons and neutrons are scattered by nuclei. For these and other reasons, each type of radiation has its own advantages and disadvantages as a probe of soft matter. We return to these briefly in Section 7. For simplicity we first outline the theory of light scattering, noting that similar principles apply to X-rays and neutrons.

2. Basic theory of light scattering

In Fig. 1b, a beam of plane-wave, monochromatic light,

$$E_{\mathrm{I}}(\boldsymbol{r},t) \equiv E_0 \exp\left[i(\boldsymbol{k}_{\mathrm{I}} \cdot \boldsymbol{r} - \omega t)\right] \tag{1}$$

with electric vector E_0 polarised perpendicular to the scattering plane (the plane of the paper), is incident on a region of scattering medium; $\boldsymbol{k}_{\mathrm{I}}$ is the propagation vector of the incident light, having magnitude $k_{\mathrm{I}} \equiv |\boldsymbol{k}_{\mathrm{I}}| = k = 2\pi/\lambda$ where λ is the wavelength of light *in the medium*; ω is its angular frequency. Light of the same polarisation is scattered at angle θ to a detector in the far field. (Thus we consider only polarised scattering, and assume that the dielectric behaviour (below) of the medium is scalar rather than tensorial. For a discussion of depolarised scattering and optically anisotropic media, see Berne and Pecora, (1976).)

We assume that the scattering is weak so that: (i) most photons pass through the sample undeviated, a few are scattered once, and the probability of double and higher-order scattering is negligible; (ii) the incident beam is not distorted significantly by the medium. (This corresponds to the first Born approximation or, in the specific context of light scattering, the so-called Rayleigh–Gans–Debye approximation.) We also assume that the scattering process is 'quasielastic', implying only a very small change in frequency. Thus the magnitude k_{S} of the propagation vector $\boldsymbol{k}_{\mathrm{S}}$ of the scattered light is also $2\pi/\lambda$.

We apply Maxwell's equations to the problem of a plane electromagnetic propagating through a medium described by a local dielectric constant $\varepsilon(\boldsymbol{r},t)$: $\varepsilon(\boldsymbol{r},t)$ is the dielectric constant of the medium at position $\boldsymbol{r}$, relative to an arbitrary origin O (Fig. 1b), and at time t. Details of this calculation are given, for example, by Berne and Pecora (1976) and D'Aguanno and Klein (1996); here we simply quote the result. The amplitude $E_{\mathrm{S}}(\boldsymbol{R},t)$ of the electric field of the radiation scattered to a point detector at position $\boldsymbol{R}$ in the far field is given by:

$$E_{\mathrm{S}}(\boldsymbol{R},t) = -\frac{k^2 E_0}{4\pi} \frac{\exp\left[i(kR - \omega t)\right]}{R} \int_V \left[\frac{\varepsilon(\boldsymbol{r},t) - \varepsilon_0}{\varepsilon_0}\right] \exp(-i\boldsymbol{q} \cdot \boldsymbol{r})\, \mathrm{d}^3 r. \tag{2}$$

Here ε_0 is the average dielectric constant of the medium, V is the scattering volume, and the 'scattering vector' $\boldsymbol{q}$ is defined as the difference between the propagation vectors of the scattered and incident light (Fig. 1b)

$$\boldsymbol{q} \equiv \boldsymbol{k}_{\mathrm{S}} - \boldsymbol{k}_{\mathrm{I}}; \qquad q \equiv |\boldsymbol{q}| = \frac{4\pi}{\lambda} \sin\frac{\theta}{2}. \tag{3}$$

Insight into the physics underlying Eq. (2) can be obtained by rewriting it as the sum of the

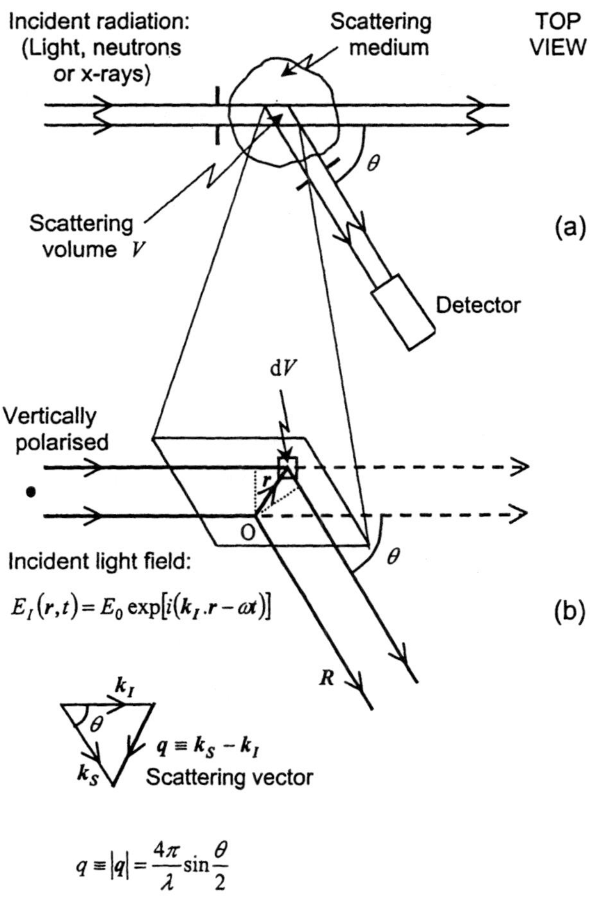

Fig. 1. (a) Top view of a typical scattering experiment. (b) Expanded view of the scattering volume, showing rays scattered at the origin O and by a volume element dV at position $\boldsymbol{r}$.

amplitudes of the fields $\mathrm{d}E_\mathrm{S}(\boldsymbol{R},t)$ scattered by volume elements $\{\mathrm{d}V \equiv \mathrm{d}^3r\}$ at positions $\{\boldsymbol{r}\}$:

$$E_\mathrm{S}(\boldsymbol{R},t) = \int_V \mathrm{d}E_\mathrm{S}(\boldsymbol{R},t), \tag{4}$$

where

$$\mathrm{d}E_\mathrm{S}(\boldsymbol{R},t) = -\frac{k^2 E_0}{4\pi}\frac{\exp\left[i(kR-\omega t)\right]}{R}\left[\frac{\varepsilon(\boldsymbol{r},t)-\varepsilon_0}{\varepsilon_0}\right]\mathrm{d}V\ \exp(-i\boldsymbol{q}\cdot\boldsymbol{r}). \tag{5}$$

Eq. (5) can be recognised as the formula describing the radiation due to an oscillating point dipole. The incident electric field, of strength E_0 and propagation vector k, induces in the volume element dV at position $\boldsymbol{r}$ a dipole moment of strength proportional to $E_0\left[\varepsilon(\boldsymbol{r},t)-\varepsilon_0\right]\mathrm{d}V$ which oscillates at angular frequency ω. This elementary dipole re-radiates, or scatters, light in all directions. The second factor in Eq. (5) describes a spherical wave of scattered radiation emanating from the origin O. The final term, $\exp(-i\boldsymbol{q}\cdot\boldsymbol{r})$, allows for the fact that the radiation scattered by the volume element at position $\boldsymbol{r}$ is shifted in phase relative to that scattered by an element at the origin O. Referring to Fig. 1b, simple geometry shows that the extra distance travelled by the light scattered at $\boldsymbol{r}$, compared to that scattered at O, is $(\boldsymbol{k}_\mathrm{I} \cdot \boldsymbol{r} - \boldsymbol{k}_\mathrm{S} \cdot \boldsymbol{r})/k$ which, with use of Eq. (3), gives a phase shift of $-\boldsymbol{q} \cdot \boldsymbol{r}$ radians.

It is immediately apparent from Eq. (2) that, as noted in Section 1, if the medium is totally homogeneous, so that $\varepsilon(\boldsymbol{r},t) = \varepsilon_0$, there is no scattering. In other words, scattering of radiation (for $\boldsymbol{q} \neq 0$) is caused by *fluctuations* in the dielectric properties of the medium.

Eq. (2) embodies the fundamental physics of light scattering. The scattered light field consists of a spherical wave emanating from the scattering volume with an angle- or q-dependent amplitude which is the spatial Fourier transform of instantaneous variations in the dielectric constant of the sample. As we will see, the intensity of the scattering, averaged over time, provides information on the sample's structure, essentially spatial correlations of the particles. Clearly any variation in time of the local dielectric constant is directly reflected in temporal variations of the amplitude of the scattered field (and its intensity). Light scattering therefore probes directly the *structure and dynamics of a sample in reciprocal space (or* $\boldsymbol{q}$*-space)*.

3. Discrete scatterers

We now specialise to the case of a sample containing discrete scattering objects suspended in a liquid. For simplicity we will call these objects 'particles', a term which could refer to a polymer molecule, a micelle, and so on, as well as to a colloidal particle. Although Eq. (6), below, is physically plausible, the development of Eq. (2) to give this result is rather subtle and is therefore relegated to the Appendix.

We consider N particles in the scattering volume V whose centres of mass at time t are described by position vectors $\{\boldsymbol{R}_j(t)\}$ (see Fig. 2). If $\boldsymbol{r}_j(t)$ is the position of volume element

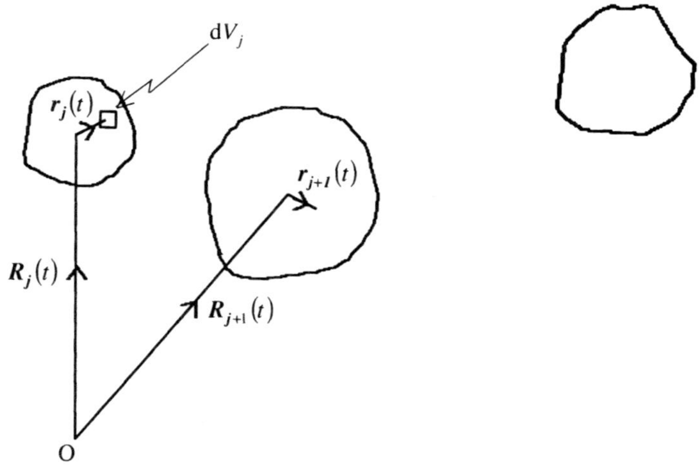

Fig. 2. Co-ordinates for discrete scatterers. Relative to an arbitrary origin O, $\boldsymbol{R}_j(t)$ is the position of the centre of mass of particle j at time t and $\boldsymbol{r}_j(t)$ is the position of volume element $\mathrm{d}V_j$ in particle j, relative to its centre of mass.

$dV_j(= d^3r_j)$ in particle j relative to its centre of mass (Fig. 2), Eq. (2) becomes (see Appendix)

$$E_S(\boldsymbol{R},t) = -\frac{k^2 E_0}{4\pi}\frac{\exp\big[i(kR-\omega t)\big]}{R} \times \sum_j \left\{ \int_{V_j} \left[\frac{\varepsilon_P(\boldsymbol{r}_j,t)-\varepsilon_L}{\varepsilon_0}\right] \exp(-i\boldsymbol{q}\cdot\boldsymbol{r}_j)\,d^3r_j \right\} \exp\big[-i\boldsymbol{q}\cdot\boldsymbol{R}_j(t)\big] \tag{6a}$$

or

$$E_S(\boldsymbol{R},t) = -E_0\frac{\exp\big[i(kR-\omega t)\big]}{R} \times \sum_j \left[\int_{V_j} \Delta\rho(\boldsymbol{r}_j,t)\exp(-i\boldsymbol{q}\cdot\boldsymbol{r}_j)\,d^3r_j \right] \exp\big[-i\boldsymbol{q}\cdot\boldsymbol{R}_j(t)\big]. \tag{6b}$$

Here V_j is the volume of particle j and $\Delta\rho(\boldsymbol{r}_j,t)$ is defined by

$$\Delta\rho(\boldsymbol{r}_j,t) = \frac{k^2}{4\pi}\left[\frac{\varepsilon_P(\boldsymbol{r}_j,t)-\varepsilon_L}{\varepsilon_0}\right], \tag{7}$$

where $\varepsilon_P(\boldsymbol{r}_j,t)$ is the local dielectric constant at position $\boldsymbol{r}_j$ in particle j, ε_L is the average dielectric constant of the liquid, and ε_0 is, as before, the average dielectric constant of the whole suspension. We see that the strength of the scattering depends on the difference between the dielectric properties of the particles and the liquid. As described in the Appendix, we assume that the 'background' scattering due to spontaneous fluctuations in the liquid itself is negligible. The quantity $\Delta\rho(\boldsymbol{r}_j,t)$ can be regarded as a measure of the local 'density of scattering material'. With the appropriate identification of $\Delta\rho(\boldsymbol{r}_j,t)$, the results to be discussed henceforth apply to X-ray and neutron scattering as well as to light scattering (see Section 7).

Eq. (6) can be rewritten as

$$E_S(\boldsymbol{R},t) = -E_0\frac{\exp\big[i(kR-\omega t)\big]}{R}\sum_{j=1}^{N} b_j(\boldsymbol{q},t)\exp\big[-i\boldsymbol{q}\cdot\boldsymbol{R}_j(t)\big] \tag{8}$$

where the 'scattering length' of particle j (see Section 6) is defined by

$$b_j(\boldsymbol{q},t) = \int_{V_j} \Delta\rho(\boldsymbol{r}_j,t)\exp(-i\boldsymbol{q}\cdot\boldsymbol{r}_j)\,d^3r_j. \tag{9}$$

From Eq. (8) we see that the total scattered electric field is the sum of the fields scattered by the individual particles. Each of these is the product of a scattering length, determined by the instantaneous distribution of material within the particle, and a phase factor determined by the instantaneous position of the particle in the sample.

Usually it is the scattered intensity that is measured directly, rather than the electric field. Intensities and fields are related by $I(\boldsymbol{q},t) = |E(\boldsymbol{q},t)|^2$. (Henceforth we regard the scattered fields and intensities as functions of the scattering vector $\boldsymbol{q}$ rather than of the position of the detector $\boldsymbol{R}$. The two are equivalent since $\boldsymbol{R}$ is in the direction of the propagation vector $\boldsymbol{k}_S$ of the scattered light, Fig. 1, and $\boldsymbol{k}_S$ and $\boldsymbol{q}$ are related by Eq. (3).) Thus from Eq. (8) we get an expression for the instantaneous scattered intensity:

$$I_S(\boldsymbol{q},t) = \frac{E_0^2}{R^2} \sum_{j=1}^{N} \sum_{k=1}^{N} b_j(\boldsymbol{q},t) b_k^*(\boldsymbol{q},t) \exp\{-i\boldsymbol{q}.[\boldsymbol{R}_j(t) - \boldsymbol{R}_k(t)]\}. \tag{10}$$

For structural information we require the ensemble average, indicated by $\langle \ldots \rangle$, of this intensity (equivalent to a time average if the scattering medium is ergodic):

$$\langle I_S(q) \rangle = \frac{E_0^2}{R^2} \left\langle \sum_{j=1}^{N} \sum_{k=1}^{N} b_j(\boldsymbol{q}) b_k^*(\boldsymbol{q}) \exp[-i\boldsymbol{q} \cdot (\boldsymbol{R}_j - \boldsymbol{R}_k)] \right\rangle. \tag{11}$$

By writing the average scattered intensity as a function of the modulus q of the scattering vector, we have assumed the sample to be spatially isotropic on average, as expected for a 'liquid-like' dispersion or solution.

Eq. (11) is the general result for the average intensity scattered by an assembly of discrete particles. We now consider some simple special cases.

3.1. Dilute systems

In a dilute system the individual particles are, on average, widely separated spatially so that their behaviours are uncorrelated. Omitting for simplicity the prefactors, Eq. (11) can then be written

$$\begin{aligned} \langle I_S(q) \rangle &= \sum_{j=1}^{N} \langle |b_j(\boldsymbol{q})|^2 \rangle + \sum_{j \neq}^{N} \sum_{k=1}^{N} \langle b_j(\boldsymbol{q}) \exp(-i\boldsymbol{q} \cdot \boldsymbol{R}_j) \rangle \langle b_k^*(\boldsymbol{q}) \exp(i\boldsymbol{q} \cdot \boldsymbol{R}_k) \rangle \\ &= \sum_{j=1}^{N} \langle |b_j(\boldsymbol{q})|^2 \rangle, \end{aligned} \tag{12}$$

where, in discarding the second term, we have exploited the fact that, over time, particle j can take up any position in the sample, unaffected by the positions of others. Thus, in the average, the exponential factors are randomly distributed about zero and $\langle b_j(\boldsymbol{q}) \exp(-i\boldsymbol{q} \cdot \boldsymbol{R}_j) \rangle = 0$. Eq. (12) shows that, in the dilute limit, the average scattered intensity is just the sum of the average intensities scattered by the individual particles. Thus a measurement of $\langle I_S(q) \rangle$ as a function of q gives information on the size, structure and shape of the particles, averaged over orientation and over any distribution of size and shape.

The average intensities scattered by identical particles are the same, so that

$$\langle I_S(q) \rangle = N \langle |b(\boldsymbol{q})|^2 \rangle, \qquad \text{(identical particles)} \tag{13}$$

which can be written

$$\langle I_S(q) \rangle = N \langle |b(0)|^2 \rangle P(q) \quad \text{with } P(q) = \frac{\langle |b(\boldsymbol{q})|^2 \rangle}{\langle |b(0)|^2 \rangle}. \tag{14}$$

$P(q)$ is called the 'form factor' of the particle. It is defined so that $P(q) \to 1$ as $q \to 0$ and provides information on the structure of the individual particles.

3.2. The form factor of a homogeneous sphere

For a homogeneous spherical particle, the distribution of scattering material is uniform throughout the sphere so that the scattering length [Eq. (9)] becomes

$$b(\boldsymbol{q}) = \Delta\rho \int_{\text{sphere}} \exp(-i\boldsymbol{q}\cdot\boldsymbol{r})\,\mathrm{d}^3r. \tag{15}$$

The integral is easily evaluated on conversion to spherical polar co-ordinates, giving

$$\begin{aligned} b(q) &= \Delta\rho\frac{4}{3}\pi R^3\frac{3}{(qR)^3}(\sin qR - qR\cos qR) \\ &\left[= \Delta\rho\frac{4}{3}\pi R^3\frac{3}{qR}j_1(qR)\right] \end{aligned} \tag{16}$$

where R is the sphere's radius (and j_1 is a spherical Bessel function of the first kind). Since $(3/x)j_1(x) \to 1$ as $x \to 0$, the form factor for a sphere is

$$P(q) = \left[\frac{3}{(qR)^3}(\sin qR - qR\cos qR)\right]^2; \tag{17}$$

this function has zeros at $\tan qR = qR$, or $qR = 4.49, 7.73, \ldots$ (see Fig. 3).

3.3. The form factor of a random-coil polymer molecule

The simplest model of a polymer molecule is the 'freely-jointed chain' in which the molecule is represented by N_P links each of length b_0, and the angle between adjacent links can take any value with equal probability (Fig. 4). For a flexible enough polymer, b_0 (roughly speaking the molecule's persistence length) is much smaller than the wavelength of light λ. Then, from the point of view of scattering, one can regard the links as point scatterers at positions $\{\boldsymbol{R}_m\}$. Thus, omitting pre-factors, the instantaneous amplitude of the field scattered by the molecule can be written

$$b_\mathrm{P}(\boldsymbol{q}) = \sum_{m=1}^{N_\mathrm{P}} \exp\left[-i\boldsymbol{q}\cdot\boldsymbol{R}_m(t)\right], \tag{18}$$

giving for the corresponding intensity averaged over all configurations of the molecule

$$\langle|b_\mathrm{P}(\boldsymbol{q})|^2\rangle = \sum_m\sum_n\left\langle\exp\left[-i\boldsymbol{q}\cdot(\boldsymbol{R}_m - \boldsymbol{R}_n)\right]\right\rangle. \tag{19}$$

For a large molecule, $N_\mathrm{P} \gg 1$, the quantity $m - n$ is also much greater than 1 for most values of m and n, so that $\boldsymbol{R}_m - \boldsymbol{R}_n$ describes the end-to-end vector of a three-dimensional random walk of many steps. Thus, by the central limit theorem, $\boldsymbol{R}_m - \boldsymbol{R}_n$ is a Gaussian-distributed random variable and, on performing the average, Eq. (19) becomes

$$\langle|b_\mathrm{P}(\boldsymbol{q})|^2\rangle = \sum_{m=1}^{N_\mathrm{P}}\sum_{n=1}^{N_\mathrm{P}}\exp\left[-\frac{q^2}{6}\langle|\boldsymbol{R}_m - \boldsymbol{R}_n|^2\rangle\right]. \tag{20}$$

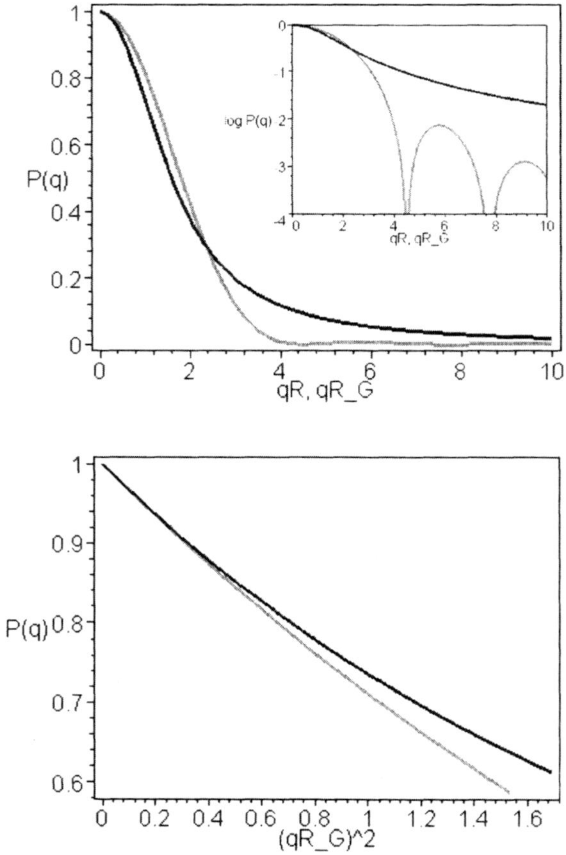

(a)

(b)

Fig. 3. (a) Form factors of a sphere (grey line, Eq. (17), plotted against qR, where R is the radius) and a Gaussian polymer coil (black line, Eq. (23), plotted against qR_G, where R_G is the coil's root-mean-square radius of gyration). The inset shows the same functions with a logarithmic vertical scale; the 'diffraction lobes' of the sphere's form factor, with zeros at $\tan qR = qR$, are clearly evident. (b) The same form factors plotted against $(qR_G)^2$, showing the validity of Guinier's expression, Eq. (30), for $(qR_G)^2$ less than about 0.25 or qR_G less than about 0.5. (For a homogeneous sphere, $R_G = \sqrt{3/5}\, R$.)

For a random walk with step length b_0,

$$\langle |\boldsymbol{R}_m - \boldsymbol{R}_n|^2 \rangle = |(n-m)| b_0^2. \tag{21}$$

Evaluation of Eqs. (20) and (21) in the limit $N_P \gg 1$ then gives

$$\langle |b_P(\boldsymbol{q})|^2 \rangle = N_P^2 P(q) \tag{22}$$

with

$$P(q) = \frac{2}{[q^2 \langle R_G^2 \rangle]^2} \left[\exp(-q^2 \langle R_G^2 \rangle) + q^2 \langle R_G^2 \rangle - 1 \right], \tag{23}$$

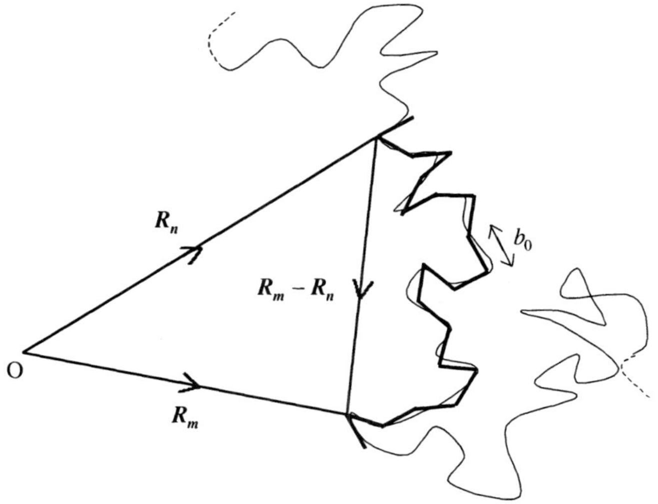

Fig. 4. A random-coil polymer molecule modelled by a freely-jointed chain. The chain contains N_p links of length b_0, at positions $\{\boldsymbol{R}_m\}$. The angles between adjacent links can take on any value. (In reality, of course, the chain occupies three dimensions.)

a result first obtained by Debye in 1947 (see Fig. 3). Here $\langle R_G^2 \rangle$ is the mean-square radius of gyration of the freely-jointed chain, given by

$$\langle R_G^2 \rangle = \frac{N_P b_0^2}{6}. \tag{24}$$

We note that, in general, the radius of gyration of N_P point masses located at positions $\{\boldsymbol{R}_m\}$ is defined by

$$R_G^2 \equiv \frac{1}{N_P} \sum_{m=1}^{N_P} |\boldsymbol{R}_m - \boldsymbol{R}_{CM}|^2 \tag{25}$$

where $\boldsymbol{R}_{CM} \equiv N_P^{-1} \sum_{m=1}^{N_P} \boldsymbol{R}_m$ is the position of the centre-of-mass of the molecule. Eq. (25) can be rewritten as (e.g. Doi and Edwards, 1986)

$$R_G^2 = \frac{1}{2N_P^2} \sum_{m=1}^{N_P} \sum_{n=1}^{N_P} |\boldsymbol{R}_m - \boldsymbol{R}_n|^2. \tag{26}$$

The result of Eq. (24) follows from averaging Eq. (26) over configurations of the random-coil molecule and using Eq. (21).

3.4. The form factor for an arbitrary particle – Guinier's expression

Any homogeneous particle can be regarded as being made up from N_P equal-sized 'beads' or volume elements at positions $\{\boldsymbol{R}_m\}$. First we assume the particle to be rigid (so that the relative positions $\boldsymbol{R}_m - \boldsymbol{R}_n$ of the beads do not vary) but allow it to take any orientation. Thus, from Eqs. (14) and (18) we can write its orientationally-averaged form factor as

$$P(q) = \frac{1}{N_{\mathrm{P}}^2} \sum_{m=1}^{N_{\mathrm{P}}} \sum_{n=1}^{N_{\mathrm{P}}} \left\langle \exp\left[-i\boldsymbol{q}.(\boldsymbol{R}_m - \boldsymbol{R}_n)\right] \right\rangle_{\mathrm{OR}} \tag{27}$$

For a vector $\boldsymbol{S}$ averaged over all orientations in three-dimensional space we have

$$\langle \exp(-i\boldsymbol{q} \cdot \boldsymbol{S}) \rangle_{\mathrm{OR}} = \frac{1}{4\pi} \int_0^{2\pi} \mathrm{d}\phi \int_0^{\pi} \sin\theta \, \mathrm{d}\theta \, \exp(-iqS\cos\theta) = \frac{\sin qS}{qS}, \tag{28}$$

giving

$$P(q) = \frac{1}{N_{\mathrm{P}}^2} \sum_{m=1}^{N_{\mathrm{P}}} \sum_{n=1}^{N_{\mathrm{P}}} \frac{\sin q|\boldsymbol{R}_m - \boldsymbol{R}_n|}{q|\boldsymbol{R}_m - \boldsymbol{R}_n|}, \tag{29}$$

another result due to Debye (1947). Expanding this expression in powers of q and using Eq. (26) gives

$$P(q) = 1 - \tfrac{1}{3}q^2 R_{\mathrm{G}}^2 + O(q^4) = 1 - \tfrac{1}{3}q^2 \langle R_{\mathrm{G}}^2 \rangle + O(q^4), \tag{30}$$

where the first expression applies to a rigid particle and the second allows for flexibility by averaging over configurations.

Eq. (30), known as Guinier's expression (Guinier and Fournet, 1955), is a very useful result since it shows that a well-defined measure of the size of any scattering object, its (mean-square) radius of gyration, can be obtained directly from its small-q scattering. (In the case of a polydisperse sample, the radius of gyration obtained from small-q scattering is further averaged over the distributions of size, shape etc.) It is straightforward to show that expanding the expressions, Eqs. (17) and (23), for the form factors of a sphere ($R_{\mathrm{G,sphere}} = \sqrt{3/5}\,R$) and a polymer molecule leads to Eq. (30), as it must (see Fig. 3b). Finally we note that the Guinier expression is sometimes written

$$P(q) = \exp\left[-(q^2 \langle R_{\mathrm{G}}^2 \rangle)/3\right]. \tag{31}$$

As can be seen by comparison with the exact result, Eq. (30), this expression is not, in general, valid beyond the $O(q^2)$ term.

4. Data analysis

We now consider briefly the analysis of experimental scattering data, $\langle I_{\mathrm{S}}(q) \rangle$ versus q, a subject which will be treated in detail later in this book by Glatter. For simplicity we limit the discussion to dilute suspensions of identical particles so that [Eqs. (13) and (14)]

$$\langle I_{\mathrm{S}}(q) \rangle = N \langle |b(\boldsymbol{q})|^2 \rangle = N \langle |b(0)|^2 \rangle P(q). \tag{32}$$

Various approaches to data analysis can be taken. First, as noted above, provided that the experimental data extend to sufficiently low values of q, one can derive the particles' radius of gyration from Eq. (30). This gives a measure of the size of the particles, but no direct information on their shape or the distribution of material within them. Second, one can try comparing data spanning a wider range of q with specific models for the form factor $P(q)$ e.g. those for a sphere, a random coil, a rod, etc. However, if a good fit of the data cannot be obtained in this way, one is forced to a more general 'inversion' approach.

From the expression [Eq.(9)]

$$b(\boldsymbol{q}) = \int_V \mathrm{d}^3 r\, \Delta\rho(\boldsymbol{r}) \exp(-i\boldsymbol{q}\cdot\boldsymbol{r}) \tag{33}$$

we get, for the intensity from a single particle,

$$|b(\boldsymbol{q})|^2 = \int_V \mathrm{d}^3 r \int_V \mathrm{d}^3 r' \Delta\rho(\boldsymbol{r})\Delta\rho(\boldsymbol{r}')\exp\left[-i\boldsymbol{q}\cdot(\boldsymbol{r}-\boldsymbol{r}')\right]. \tag{34}$$

A change of variables, $\boldsymbol{r} \to \boldsymbol{r}' - \boldsymbol{R}$ so that $\boldsymbol{r}' \to \boldsymbol{r} - \boldsymbol{R}$, gives

$$\begin{aligned} |b(\boldsymbol{q})|^2 &= \int \mathrm{d}^3 R \int \mathrm{d}^3 r\, \Delta\rho(\boldsymbol{r})\Delta\rho(\boldsymbol{r}-\boldsymbol{R})\exp(-i\boldsymbol{q}\cdot\boldsymbol{R}) \\ &= \int \mathrm{d}^3 R\, \overline{\overline{\Delta\rho^2(\boldsymbol{R})}}\exp(-i\boldsymbol{q}\cdot\boldsymbol{R}), \end{aligned} \tag{35}$$

where

$$\overline{\overline{\Delta\rho^2(\boldsymbol{R})}} = \int \mathrm{d}^3 r\, \Delta\rho(\boldsymbol{r})\Delta\rho(\boldsymbol{r}-\boldsymbol{R}). \tag{36}$$

Following a procedure similar to that used in Section 3.4, we average Eq. (35) over all orientations of $\boldsymbol{R}$, to give

$$\langle |b(\boldsymbol{q})|^2 \rangle = 4\pi \int \mathrm{d}R\, p(R) \frac{\sin qR}{qR} \tag{37}$$

where

$$p(R) = R^2\, \overline{\overline{\Delta\rho^2(\boldsymbol{R})}} \tag{38}$$

is known as the 'pair–distance distribution function'. Finally, Fourier inversion of Eq. (37) gives

$$p(R) = \frac{1}{2\pi^2} \int_0^\infty \langle |b(\boldsymbol{q})|^2 \rangle qR \sin qR\, \mathrm{d}q. \tag{39}$$

Thus the pair–distance distribution function $p(R)$ for the particles can be obtained by numerical Fourier transformation of scattering data using Eqs. (32) and (39). It can be seen from Eqs. (36) and (38), that $p(R)$ essentially describes the number of ways in which one can choose a vector of length $|\boldsymbol{R}|$ which connects scattering material within the particle. As will be discussed in detail by Glatter, it contains valuable information on the particle's structure.

5. Concentrated systems

The general expression for the average intensity scattered by an assembly of discrete particles is (Eq. (11), omitting pre-factors)

$$\langle I_S(q)\rangle = \sum_{j=1}^{N}\sum_{k=1}^{N}\langle b_j(\boldsymbol{q})b_k^*(\boldsymbol{q})\exp[-i\boldsymbol{q}.(\boldsymbol{R}_j-\boldsymbol{R}_k)]\rangle \tag{40}$$

For simplicity we first assume the particles to be identical homogeneous spheres, so that all $b_j(\boldsymbol{q}) = b(q)$. Then use of Eq. (14) gives

$$\langle I_S(q)\rangle = b^2(q)\sum_{j=1}^{N}\sum_{k=1}^{N}\langle\exp[-i\boldsymbol{q}.(\boldsymbol{R}_j-\boldsymbol{R}_k)]\rangle = Nb^2(0)P(q)S(q), \tag{41}$$

where $S(q)$ is the 'static structure factor', defined by

$$S(q) \equiv \frac{1}{N}\sum_{j=1}^{N}\sum_{k=1}^{N}\langle\exp[-i\boldsymbol{q}.(\boldsymbol{R}_j-\boldsymbol{R}_k)]\rangle. \tag{42}$$

In Eq. (41) the product $Nb^2(0)P(q)$ describes the scattering by N uncorrelated particles [compare Eq. (14)] and the structure factor represents the modification of the intensity due to the spatial correlation of the particles. Clearly, in a dilute system where particle positions are uncorrelated, cross terms, $i \neq j$, in Eq. (42) vanish and $S(q) = 1$.

The spatial correlations, or structure, in a concentrated system are usefully described in terms of the 'radial distribution function', $g(R)$, related to the probability of finding the centre of any particle at a distance R from the centre of a given particle. Specifically, for N particles in a volume V, $(N/V)g(R)\mathrm{d}V$ is the number of particles in volume element $\mathrm{d}V$ at a distance R from a given particle. It is not difficult to show that the structure factor is related to the radial distribution function by

$$S(q) = 1 + 4\pi\frac{N}{V}\int_0^\infty [g(R)-1]R^2\frac{\sin qR}{qR}\,\mathrm{d}R. \tag{43}$$

Fourier inversion of Eq. (43) gives the radial distribution function in terms of the structure factor:

$$g(R) = 1 + \frac{1}{2\pi^2}\frac{V}{N}\int_0^\infty [S(q)-1]q^2\frac{\sin qR}{qR}\,\mathrm{d}q. \tag{44}$$

The strategy for measuring the structure factor is obtained from Eq. (41). First one measures the intensity scattered by a sample dilute enough that $S(q) = 1$:

$$\langle I_S(q)\rangle_{\mathrm{dil}} = N_{\mathrm{dil}}b^2(0)P(q). \tag{45}$$

Then one measures the intensity scattered by the concentrated sample of interest:

$$\langle I_S(q)\rangle_{\mathrm{conc}} = N_{\mathrm{conc}}b^2(0)P(q)S(q). \tag{46}$$

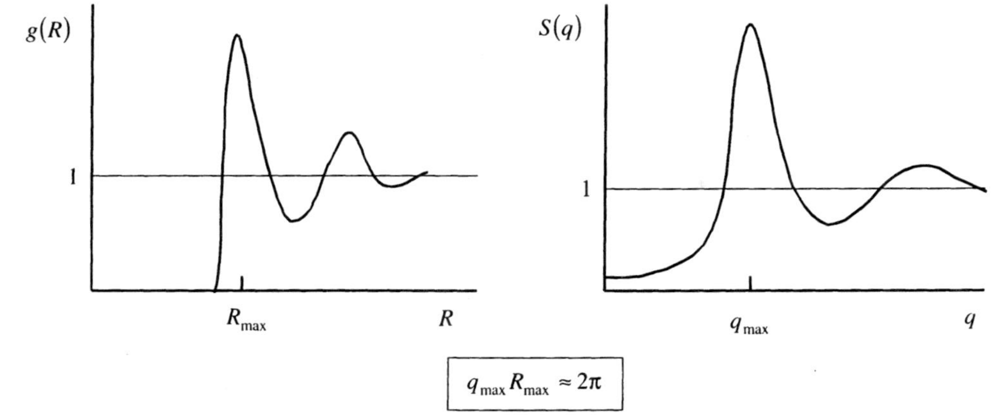

Fig. 5. Typical radial distribution function $g(R)$ and structure factor $S(q)$ for a suspension of strongly interacting colloidal particles.

The structure factor is then obtained from the ratio of these two measurements:

$$S(q) = \frac{\langle I_S(q)\rangle_{\text{conc}}}{\langle I_S(q)\rangle_{\text{dil}}} \frac{N_{\text{dil}}}{N_{\text{conc}}}. \tag{47}$$

Even if the sample is polydisperse, data can be analysed in the same way. Then one obtains the so-called 'measured structure factor' $S^M(q)$. Use of Eqs. (40), (12) and (47) shows that

$$S^M(q) = \left[\sum_{j=1}^{N} \langle |b_j(\boldsymbol{q})|^2\rangle\right]^{-1} \sum_{j=1}^{N}\sum_{k=1}^{N} \langle b_j(\boldsymbol{q}) b_k^*(\boldsymbol{q}) \exp[-i\boldsymbol{q}.(\boldsymbol{R}_j - \boldsymbol{R}_k)]\rangle, \tag{48}$$

which clearly reduces to Eq. (42) for identical particles where all $b_j(\boldsymbol{q}) = b(q)$. As will be discussed later in this book by Klein, $S^M(q)$ can be written as a generalisation of Eq. (43) in terms of 'partial' radial distribution functions which describe the spatial correlations between different species.

Fig. 5 shows sketches of a typical radial distribution function, for example for a concentrated suspension of hard-sphere colloids or a suspension of strongly interacting charged colloids, and the corresponding structure factor. Since two particles cannot occupy the same space, $g(R)$ is zero for centre-to-centre interparticle separations smaller than the particle diameter. The main peak in $g(R)$ describes the nearest-neighbour shell or 'cage' of particles around any given particle. At large interparticle separations $g(R) \to 1$ implying that spatial correlations are usually short-ranged (a crystal is an exception). The peak in $S(q)$ can be considered, approximately, to be a 'Bragg reflection' from planes of particles separated by distances equal to the mean nearest-neighbour separation. Thus one finds that the positions R_{max} and q_{max} of the main peaks in $g(R)$ and $S(q)$ are related by an approximate 'Bragg condition' $q_{\text{max}} R_{\text{max}} \approx 2\pi$. For this reason it is often said that a scattering experiment, operating at scattering vector q, 'measures structure' on a spatial scale $2\pi/q$. While this is a useful 'rule of thumb' it should not be taken too literally: it is clear from Eq. (43) that the value of $S(q)$ at a particular scattering vector q is determined by the values of $g(R)$ at all R.

6. Absolute intensities

So far we have calculated the intensity scattered to a point in the far field, and have considered only normalised q-dependent quantities such as the form and structure factors. Valuable information is also contained in the absolute magnitude of the scattered intensity. In an experiment one uses a detector of area A_d, say, and measures the flux of scattered energy by counting photons (or neutrons). Thus Eq. (11), for example, can be written as

$$\frac{\langle I_S(q)\rangle A_d}{E_0^2(A_d/R^2)} = \left\langle \sum_{j=1}^{N}\sum_{k=1}^{N} b_j(\boldsymbol{q})b_k^*(\boldsymbol{q})\exp\left[-i\boldsymbol{q}\cdot(\boldsymbol{R}_j-\boldsymbol{R}_k)\right]\right\rangle. \tag{49}$$

On the left-hand side of Eq. (49), the numerator describes the detected scattered energy per unit time and the denominator is the product of the incident flux (energy per unit area per unit time) and the detection solid angle. Thus Eq. (49) has the units of $(\text{length})^2$ and can be regarded as describing the sample's 'scattering cross-sectional area (σ) per unit solid angle (Ω)'. As noted in Section 3, the quantity $b_j(q)$, having the units of length, is called the 'scattering length'. Eq. (49) is frequently written in the form

$$\frac{\mathrm{d}\sigma(q)}{\mathrm{d}\Omega} = \left\langle \sum_{j=1}^{N}\sum_{k=1}^{N} b_j(\boldsymbol{q})b_k^*(\boldsymbol{q})\exp\left[-i\boldsymbol{q}.(\boldsymbol{R}_j-\boldsymbol{R}_k)\right]\right\rangle. \tag{50}$$

This result applies to X-ray and neutron scattering, as well as to light scattering, provided the appropriate scattering lengths (which include the coupling between the radiation and the scattering medium) are used.

For light scattering, using Eq. (7) and the relationship $\varepsilon = n^2$ between dielectric constant and refractive index, we get

$$\Delta\rho(\boldsymbol{r},t) = \frac{k^2}{4\pi}\left[\frac{n^2(\boldsymbol{r},t)-n_L^2}{n_0^2}\right], \tag{51}$$

where n_L and n_0 are respectively the average refractive indices of the liquid and the suspension. If, as is frequently the case, the difference between the refractive indices of the particles and the liquid is not too large, this becomes

$$\Delta\rho(\boldsymbol{r},t) \approx \frac{k^2}{2\pi n_0}\left[n(\boldsymbol{r},t)-n_L\right]. \tag{52}$$

Thus from Eqs. (52) and (9) we get

$$b_j(\boldsymbol{q},t) = \frac{k^2}{2\pi n_0}\int_{V_j}\left[n(\boldsymbol{r}_j,t)-n_L\right]\exp(-i\boldsymbol{q}\cdot\boldsymbol{r}_j)\,\mathrm{d}^3r_j. \tag{53}$$

For identical homogeneous particles of refractive index n_P and volume V_P, the $q \to 0$ limit of this expression becomes

$$b(q\to 0) = \frac{k^2}{2\pi n_0}(n_P-n_L)V_P. \tag{54}$$

If the refractive index of the particles is not known, the dependence of suspension refractive index on concentration c (mass of particles per unit volume of suspension) can be measured, providing dn_0/dc, the so-called 'refractive index increment'. Then, if we assume additivity by volume of refractive indices,

$$n_0 = \phi n_P + (1-\phi)n_L, \tag{55}$$

where ϕ is the volume fraction of particles, we get

$$\frac{dn_0}{dc} = \frac{dn_0}{d\phi}\frac{d\phi}{dc} = (n_P - n_L)\frac{V_P}{m_P}, \tag{56}$$

where m_P is the particles' mass and the identities $\phi = NV_P/V$ and $c = Nm_P/V$ have been used. Thus Eq. (54) becomes

$$b(q \to 0) = \frac{k^2}{2\pi n_0} m_P \left(\frac{dn_0}{dc}\right). \tag{57}$$

Then, in the limit of a dilute suspension at $q \to 0$, Eq. (50) becomes

$$\frac{1}{V}\frac{d\sigma(q)}{d\Omega} = \frac{N}{V}b^2(0) = \frac{N}{V}\left(\frac{k^2}{2\pi n_0}\right)^2 m_P^2 \left(\frac{dn_0}{dc}\right)^2. \tag{58}$$

Finally, with the further identification $m_P = M/N_A$, where M is the particle's molecular mass and N_A is Avogadro's number, we get

$$\frac{1}{V}\frac{d\sigma(q)}{d\Omega} = \left(\frac{k^2}{2\pi n_0}\right)^2 \left(\frac{dn_0}{dc}\right)^2 \frac{M}{N_A}c. \tag{59}$$

We recall that the propagation vector of the light is given by $k = 2\pi/\lambda$, where $\lambda = \lambda_0/n_0$ is the wavelength of light *in the medium* and λ_0 is its wavelength *in vacuo*. Thus Eq. (59) can be written in the frequently quoted form:

$$\frac{1}{V}\frac{d\sigma(q)}{d\Omega} = KMc, \tag{60}$$

where the 'optical constant' K is

$$K = \frac{1}{N_A}\frac{4\pi^2 n_0^2}{\lambda_0^4}\left(\frac{dn_0}{dc}\right)^2. \tag{61}$$

This result is the basis of a method to determine molecular weights from measurements of the intensity of scattered radiation, relative to the incident intensity, extrapolated to zero scattering angle and zero concentration – all the quantities in Eqs. (60) and (61) (except M) are known or can be measured.

Finally we note that, for a concentrated system studied away from the small scattering angle limit, Eq. (60) becomes

$$\frac{1}{V}\frac{d\sigma(q)}{d\Omega} = KMcP(q)S(q) \tag{62}$$

A small-q expansion of Eq. (43) gives

$$S(q) = 1 - 2\frac{N}{V}B_2 + \frac{2}{3}\pi\frac{N}{V}q^2 C + \cdots \tag{63}$$

where

$$B_2 = -2\pi \int \left[g(R) - 1\right] R^2 \, \mathrm{d}R \tag{64}$$

and

$$C = -\int \left[g(R) - 1\right] R^4 \, \mathrm{d}R. \tag{65}$$

Note that, at low concentrations, where $g(R) \approx \exp\left[-V(R)/K_B T\right]$, $V(R)$ being the potential energy of interaction of two particles, B_2 reduces to the second virial coefficient. Eq. (62), along with (63)–(65) and Guinier's expansion of $P(q)$ [Eq. (30)], form the basis of 'Zimm plots' which allow extrapolation to zero concentration and zero scattering angle of measurements made away from these limits (Zimm, 1948).

7. X-rays and neutrons

X-rays are scattered by the electrons of atoms. Except for frequencies near to an electronic transition, the scattering occurs as if the electrons were free. The scattering by a single electron (Thomson scattering) is described by a scattering length $b_0^X = 2.8 \times 10^{-15}$ m. The X-ray scattering length b_j^X of an atom or molecule containing z electrons is then $b_j^X = z b_0^X$. As with light scattering (Appendix), we must consider the contrast between the particle and the liquid in which it is dispersed. Thus we define a contrast scattering length density $\Delta\rho(\boldsymbol{r},t)$ in a small volume ΔV of the particle at position $\boldsymbol{r}$ by

$$\Delta\rho(\boldsymbol{r},t) = \frac{1}{\Delta V} \sum_j \left[b_j^{X,P} - \frac{v_P}{v_L} b^{X,L} \right], \tag{66}$$

where $b^{X,P}$ and $b^{X,L}$ are respectively the scattering lengths of the molecules comprising the particles and the liquid, v_P and v_L are the volumes of the molecules, and the sum runs over all atoms in ΔV. With this identification of $\Delta\rho(\boldsymbol{r},t)$, the formulae described above for light scattering can be used also for X-ray scattering.

Neutrons are scattered by the nuclei of atoms. Knowing the scattering lengths of various nuclei (these are tabulated), one can proceed, as for X-rays, to a contrast scattering length density.

Aside from the different mechanisms by which the radiation is scattered, a major difference between light and X-rays and neutrons is their wavelengths: ~400 nm (in the sample) for light, ~0.1 nm for X-rays and neutrons. For various scattering angles, Table 1 shows approximate values of the scattering vector q, and the spatial scale $2\pi/q$ probed by scattering. We see that, at scattering angles of 5–10°, X-rays and neutrons probe structure on a scale almost as small as that of molecules. To reach the scales of interest, >10 nm, in much of soft matter, scattering angles smaller than 1° must be used. The minimum scale, ~300 nm, probed by light scattering at large angles overlaps the maximum scale of very small-angle X-ray and neutron scattering.

While the X-ray scattering lengths of atoms are simply proportional to the atomic number, neutron scattering lengths vary irregularly with the type of nucleus and can be negative (corresponding to a phase shift of π radians in the scattering) as well as positive. Furthermore, different isotopes of the same atom (or nucleus) can have very different scattering lengths. In

Table 1

Approximate magnitudes of the scattering vector and its reciprocal for X-rays, neutrons and light at various scattering angles

X-rays, neutrons ($\lambda = 0.1$ nm)			Light ($\lambda = 400$ nm)		
θ (degrees)	q (nm^{-1})	$2\pi/q$ (nm)	θ (degrees)	q (nm^{-1})	$2\pi/q$ (nm)
0.01	0.01	600	1	3×10^{-4}	2×10^{4} (20 μm)
0.1	0.1	60	10	3×10^{-3}	2000
1	1	6	100	2.4×10^{-2}	300
10	10	0.6			

particular, the scattering length of hydrogen's nucleus (a proton) is -3.74×10^{-15} m whereas that of deuterium is 6.67×10^{-15} m. Thus partial deuteration of the components of a sample leads to marked changes in its neutron scattering properties, frequently without seriously affecting its chemical and physical properties. It is possible by this means to 'highlight' different parts of a particle and to obtain thereby detailed structural information.

The ability to vary contrast by isotopic substitution constitutes a major advantage of neutron scattering, particularly in the study of soft matter where many materials of interest contain hydrogen (see the article later in this book by Schurtenberger for more details). Limited contrast variation is possible in light scattering by using materials with different refractive indices, but this is done at the risk of changing the sample's other properties in a way which can be hard to quantify. With tunable X-rays from synchrotron sources, some contrast variation can be achieved by exploiting the fact that, close to an allowed electronic transition, an atom's scattering length varies with the frequency of the radiation.

An advantage of X-rays over neutrons, which has recently been exploited, is the ability to perform X-ray photon correlation spectroscopy, the X-ray analogue of dynamic light scattering (see, for example, Lurio et al., 2000, and Riese et al., 2000, for recent applications of this technique). Dynamic scattering of this kind requires that the incident radiation be coherent over the scattering volume of the sample. Laser light is intrinsically coherent; X-rays from intense modern synchrotron sources can be made coherent by suitable spatial and frequency filtering. (While neutron beams can also, in principle, be filtered in the same way, the resulting flux of coherent neutrons from current neutron sources is much too small to be of practical use in scattering experiments.) The advantages of X-ray dynamic scattering over dynamic light scattering are that samples opaque to light can be studied and that larger scattering vectors, and therefore smaller spatial scales, can be reached.

Appendix

It is clear from Eq. (2) that there is scattering both from the particles, dielectric constant $\varepsilon_{\mathrm{P}}(\boldsymbol{r},t)$, and the liquid, dielectric constant $\varepsilon_{\mathrm{L}}(\boldsymbol{r},t)$. Thus we can write

$$\begin{aligned} \varepsilon(\boldsymbol{r},t) - \varepsilon_0 = \varepsilon_{\mathrm{P}}(\boldsymbol{r},t) - \varepsilon_0 \quad & (\boldsymbol{r} \text{ inside any particle}) \\ + \varepsilon_{\mathrm{L}}(\boldsymbol{r},t) - \varepsilon_0 \quad & (\boldsymbol{r} \text{ outside all particles}) \end{aligned} \tag{A1}$$

where ε_0 is, as before, the average dielectric constant of the medium. If we define ε_L to be the average dielectric constant of the *liquid*, Eq. (A1) can be written in the more useful form

$$\begin{aligned}\varepsilon(\boldsymbol{r},t)-\varepsilon_0 &= [\varepsilon_P(\boldsymbol{r},t)-\varepsilon_L]+(\varepsilon_L-\varepsilon_0) && (\boldsymbol{r}\text{ inside any particle})\\ &\quad +[\varepsilon_L(\boldsymbol{r},t)-\varepsilon_L]+(\varepsilon_L-\varepsilon_0) && (\boldsymbol{r}\text{ outside all particles})\end{aligned}$$

or

$$\begin{aligned}\varepsilon(\boldsymbol{r},t)-\varepsilon_0 &= \varepsilon_P(\boldsymbol{r},t)-\varepsilon_L && (\boldsymbol{r}\text{ inside any particle})\\ &\quad +\varepsilon_L(\boldsymbol{r},t)-\varepsilon_L && (\boldsymbol{r}\text{ outside all particles})\\ &\quad +\varepsilon_L-\varepsilon_0 && (\boldsymbol{r}\text{ anywhere in } V).\end{aligned} \tag{A2}$$

When substituted into Eq. (2), Eq. (A2) leads to three contributions to the scattering. The first is the contribution of interest, essentially the scattering by the particles due to the difference between their dielectric constant (or refractive index) and that of the liquid in which they are suspended. The second corresponds to the scattering from spontaneous density fluctuations in the liquid itself (or density + concentration fluctuations if the suspension medium is a mixture of liquids). In this article we will, for simplicity, assume the intensity of this 'background' scattering to be negligible (though in practice it can be measured and subtracted from the total scattering). The third contribution in Eq. (A2) does not depend on position $\boldsymbol{r}$, so leads to a term in Eq. (2) proportional to $\int_V \exp[-i\boldsymbol{q}\cdot\boldsymbol{r}]\,\mathrm{d}^3r$. This integral describes diffraction by the whole scattering volume which, if the dimensions of V are much greater than the wavelength of the radiation, is strongly peaked around $\boldsymbol{q}=0$ and can also be neglected.

We consider N particles in the scattering volume V whose centres of mass at time t are described by position vectors $\{\boldsymbol{R}_j(t)\}$ (see Fig. 2). If $\boldsymbol{r}_j$ is the position of volume element $\mathrm{d}V_j$ ($=\mathrm{d}^3r_j$) in particle j relative to its centre of mass (Fig. 2), we can write

$$\varepsilon_P(\boldsymbol{r},t)\,\mathrm{d}^3r=\sum_{j=1}^{N}\delta\left[\boldsymbol{r}-\boldsymbol{R}_j(t)-\boldsymbol{r}_j\right]\varepsilon_P(\boldsymbol{r}_j,t)\,\mathrm{d}^3r_j \tag{A3}$$

where $\varepsilon_P(\boldsymbol{r}_j,t)$ is the local dielectric constant at position $\boldsymbol{r}_j$ in particle j. Substitution of the first term of Eq. (A2), with (A3), into (2) gives

$$\begin{aligned}E_S(\boldsymbol{R},t)= &-\frac{k^2E_0}{4\pi}\frac{\exp[i(kR-\omega t)]}{R}\\ &\times\sum_j\left\{\int_{V_j}\left[\frac{\varepsilon_P(\boldsymbol{r}_j,t)-\varepsilon_L}{\varepsilon_0}\right]\exp(-i\boldsymbol{q}\cdot\boldsymbol{r}_j)\,\mathrm{d}^3r_j\right\}\exp\left[-i\boldsymbol{q}\cdot\boldsymbol{R}_j(t)\right],\end{aligned} \tag{A4}$$

where V_j is the volume of particle j.

References

Berne, B.J. and Pecora, R. (1976). *Dynamic Light Scattering*. New York: Wiley.

Brown, W., editor. (1996). *Light Scattering: Principles and Development.* Oxford: Clarendon Press.

D'Aguanno, B. and Klein, R. (1996). *Static Properties of Colloidal Suspensions*, in *Light Scattering: Principles and Development,* edited by W. Brown. Oxford: Clarendon Press.

Debye, P. (1947). *J.Phys. and Coll. Chem.* **51**, 18.

Doi, M. and Edwards, S.F. (1986). *The Theory of Polymer Dynamics*. Oxford: Clarendon Press.
Guinier, A. and Fournet, G. (1955). *Small Angle Scattering of X-rays*. New York: Wiley Interscience.
Glatter, O. and Kratky, O. (1982). *Small-Angle X-ray Scattering*. London: Academic Press.
Higgins, J.S. and Benoît, H.C. (1994). *Polymers and Neutron Scattering*. Oxford: Clarendon Press.
Lurio, L.B., Lumma, D., Sandy, A.R., Borthwick, M.A., Falus, P., Mochrie, S.G.J., Pelletier, J.F., Sutton, M., Regan, L., Malik, A. and Stephenson, G.B. (2000). *Phys. Rev. Lett.* **84**, 785.
Riese, D.O., Wegdam, G.H., Vos, W.L., Sprik, R., Fenistein, D., Bongaerts, J.H.H and Grübel, G. (2000), *Phys. Rev. Lett.* **85**, 5460.
Zimm, B. (1948). *J. Chem. Phys.* **16**, 1093 and 1099.

Neutrons, X-Rays and Light
P. Lindner and Th. Zemb (Editors)

Chapter 2

Scattering Experiments: Experimental Aspects, Initial Data Reduction and Absolute Calibration

Peter Lindner

Institut Laue-Langevin, B.P. 156, F-38042 Grenoble CEDEX 9, France

1. Introduction and definitions

Scattering techniques are used to investigate the structure, the organization and the dynamics of matter (e.g. polymers, colloids, micellar systems, aggregates, etc.), with radiation such as neutrons, X-rays or light. In general, the specific properties of these sources and their characteristic interactions with matter allow for exploring a typical spatial scale from 1 to 20,000 nanometers and a typical dynamic range of seconds to picoseconds. This chapter resumes some basic experimental aspects for a scattering experiment, the primary steps of data reduction and absolute calibration.

1.1. Scattering setup and momentum transfer

We consider an experimental setup as sketched in Fig. 1. A well collimated beam of radiation (light, neutrons, X-rays) with wavelength λ_i and energy E_i is incident on a sample of thickness d. Most of the incident radiation is *transmitted* through the sample without any further interaction and some small parts might be *absorbed*. Only a certain fraction of the incident beam is interacting with the sample and *scattered* at an angle $\theta > 0$. In general, there is also an energy transfer between the sample and the incident beam ($\Delta E \neq 0$). A scattering experiment is thus observing the scattered intensity I as a function of scattering angle θ and energy E at a distance L away from the sample by using a convenient detection device. The measured 'scattering curve' $I = f(\theta, E)$ contains the complete information on the structure and dynamics of the sample.

In order to compare results obtained at different wavelengths, detector distances and scattering angles or with different types of radiation, a more convenient parameter is the scattering vector $\boldsymbol{Q}$, as defined in Fig. 2. Hence, the scattered intensity is discussed as a variation of the 'momentum transfer' $Q = |\boldsymbol{Q}| = (4\pi/\lambda)\sin(\theta/2)$.

In the following we focus on *static* experiments where energy transfer can be neglected ($\Delta E \cong 0$ and $\lambda \cong \lambda_i$) and experiments at *small* scattering angles ($\theta \leq 10°$), i.e. the case of '*elastic small-angle scattering*' with $|\boldsymbol{k}| \cong |\boldsymbol{k}_i|$. For learning about further aspects and more details the reader is referred to the different contributions in this book and to other standard references (Marschall and Lovesey, 1971; Berne and Pecora, 1976; Glatter and Kratky, 1982;

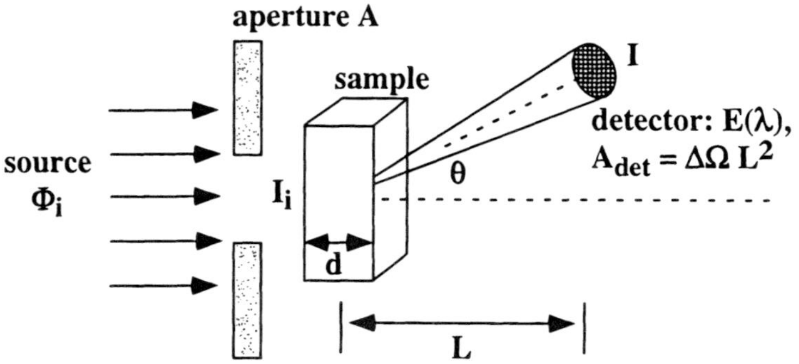

Fig. 1. General layout of a scattering experiment. Radiation source with incident flux Φ_i at wavelength λ_i, sample aperture area A, incident intensity I_i, sample thickness d, detector with efficiency $E(\lambda) < 1$ and surface area $A_{det} = \Delta\Omega L^2$ at distance L and at scattering angle θ.

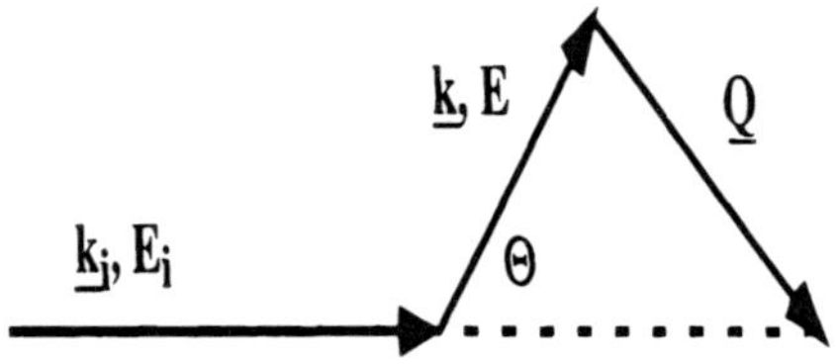

Fig. 2. Definition of the scattering vector $\boldsymbol{Q}$. The wave vector is $|\boldsymbol{k}| = 2\pi/\lambda$ and $(h/2\pi)\boldsymbol{Q} = (h/2\pi)(\boldsymbol{k} - \boldsymbol{k}_i)$, h is Planck's constant.

Alexander, 1985; Feigin and Svergun, 1987; Baruchel et al., 1993; Higgins and Benoît, 1994; Brumberger, 1995; Brown, 1996).

1.2. Differential scattering cross-section

A physical quantity needs to be defined in order to relate the intensity $I = f(\theta)$ measured in a (static) scattering experiment to the sample's structure: the experiment consists in *comparing* the incident intensity I_i with the scattered intensity I, measured at an angle θ and a distance L on a detector with area $A_{det} = \Delta\Omega L^2$ (cf. Fig. 1). We call the ratio of both intensities the *differential scattering cross-section* of the sample per unit solid angle, $d\sigma/d\Omega$:

$$I \cdot L^2/I_i = d\sigma/d\Omega\ (cm^2) \tag{1}$$

Normalization to unit sample volume V (cm^3) yields the differential scattering cross-section per unit sample volume, a quantity in units of (cm^{-1})

$$d\Sigma/d\Omega\ (cm^{-1}) = (1/V) \cdot d\sigma/d\Omega \tag{2}$$

$d\Sigma/d\Omega$ represents the probability of a particle of the incident beam being scattered out from the unit sample volume into the solid angle $\Delta\Omega$; it is identical to the 'Rayleigh ratio' used in light scattering. $d\Sigma/d\Omega$ is the quantitative representation of the interaction radiation $\leftrightarrow$ sample and contains all information on the structure of the sample. Its evaluation in absolute units of (cm^{-1}) should be the first step of data reduction, before any further interpretation of the experiment.

Since the energy transfer during a scattering process is always finite ($\Delta E \neq 0$), the scattering cross-section is to be written generally as a double differential quantity, $\delta^2\Sigma/\delta\Omega\delta E$ (see Chapter 10 by R. Zorn). For *structural* studies of soft matter systems, however, the *elastic coherent scattering process* with approximately zero energy transfer is of interest.

1.3. Differential scattering cross-section and structure

For an assembly of N atoms at individual positions r_i with fixed orientation (e.g. one polymer chain), the scattering of neutrons, X-rays and light is a manifestation of interference phenomena. The incident plane wave is interacting with the sample's atoms and leads to interference of spherically symmetric waves, scattered from each atom i at position $\boldsymbol{r}_i$ with wave functions A_i. We call b_i the fraction of radiation scattered by atom i. Differences in the techniques are related to the nature of b_i. The total scattering can be written as the sum of N wave functions A_i ('Born approximation'):

$$A(\boldsymbol{Q}) \propto \sum\nolimits_N b_i \exp[-i\,\boldsymbol{Q}\boldsymbol{r}_i] \tag{3}$$

For small-angle scattering, the observed length scale is much larger compared to inter-atomic distances and the sum over N in Eq. (3) can be replaced by an integral:

$$A(\boldsymbol{Q}) \propto \int_V \rho(\boldsymbol{r}_i)\exp[-i\,\boldsymbol{Q}\boldsymbol{r}_i]\,\mathrm{d}r \tag{4}$$

$\rho(\boldsymbol{r}_i) = \sum b_i/V$ is called the 'scattering density distribution' and $A(\boldsymbol{Q})$ is the 3rd Fourier transformation of the scattering density distribution $\rho(\boldsymbol{r}_i)$ in the sample. For neutrons, $\rho(\boldsymbol{r}_i)$ is proportional to the scattering length density, for X-rays it is proportional to the electron density and for light it is proportional to the refractive index increment.

If we now consider n polymer chains (each composed of N atoms) in a solvent, the total scattering writes as a sum for the $k = 1$ to n individual chains:

$$F(\boldsymbol{Q}) \propto \sum_k A_k(\boldsymbol{Q})\exp[-i\,\boldsymbol{Q}\boldsymbol{r}_k] \tag{5}$$

For an isotropic distribution of the scattering elements, the average over all orientations is applied and the product with the complex conjugate of Eq. (5) yields the differential scattering cross-section as a function of momentum transfer Q:

$$\frac{\mathrm{d}\Sigma}{\mathrm{d}\Omega}(Q) \propto \langle F(\boldsymbol{Q})F^*(\boldsymbol{Q})\rangle = \left\langle \sum\sum b_i b_j \exp[-i\,\boldsymbol{Q}(\boldsymbol{r}_i - \boldsymbol{r}_j)]\right\rangle \tag{6}$$

Only the relative positions of every pair of scatterers play a role in this fundamental equation; it is the basis for calculating model scattering laws for elastic scattering (static approximation). Eq. (6) can be rearranged to:

$$\frac{\mathrm{d}\Sigma}{\mathrm{d}\Omega}(Q) \propto (\rho - \langle\rho\rangle)^2 \int_V P(r)\exp[-i\,Qr]\,\mathrm{d}r \tag{7}$$

where ρ is the scattering density of the dispersed polymer and $\langle\rho\rangle$ is the average scattering density of the solution. The first term of Eq. (7), $(\rho - \langle\rho\rangle)^2$, is the square of the so-called contrast of the scatterer (e.g. the polymer chain) with respect to the reference medium (e.g. the solvent); it is radiation specific. The second integral term is a sum of the so-called 'form factor' (containing information on the conformation of the individual polymer chain) and the 'structure factor' (containing information on the interaction of the n polymer chains in solution).

1.4. Differential scattering cross-section and contrast

Let us consider for a moment only scattering from a *dilute* solution (e.g. a solution of a polymer) where inter-molecular interactions of the dissolved particles cancel each other out (second osmotic virial coefficient $A_2 = 0$, see also Chapter 11 by P. Schurtenberger). The scattering intensity I, observed on the detector as a function of the momentum transfer Q and concentration c of the solute, is proportional to the sample's differential scattering cross-section $\mathrm{d}\Sigma/\mathrm{d}\Omega(Q,c)$

$$I = f(Q,c) = C' \cdot \frac{\mathrm{d}\Sigma}{\mathrm{d}\Omega}(Q,c) \tag{8}$$

C' is a constant for a given experimental configuration (see Section 2.1) and contains instrumental parameters (sample-to-detector distance L, incident beam intensity I_i and area of sample diaphragm A), as well as sample parameters (sample transmission T and sample thickness d). The coherent differential scattering cross-section $\mathrm{d}\Sigma/\mathrm{d}\Omega(Q,c)$ can be written after appropriate background subtraction (see Section 2.3) as:

$$\frac{\mathrm{d}\Sigma}{\mathrm{d}\Omega}(Q,c) = cKMP(Q) \tag{9}$$

where c is the mass concentration of solute (g cm^{-3}), M is its (apparent) molecular mass (g mol^{-1}), K is the scattering constant (cm^2 mol g^{-2}) and $P(Q)$ is the single particle form factor. $P(Q)$ contains the complete information on the conformation (shape and size) of the dissolved particle. The scattering constant K is proportional to the square of the contrast Δb (cm g^{-1}). For *neutron scattering* with macromolecular samples, where the monomer unit is considered as the elementary scattering unit, the contrast itself is given by the differences of the scattering lengths b of 1 g of solute particles with respect to the solvent molecules in the same volume (specific scattering length difference). In the case of an incompressible two component system (solute particles + solvent) it can be written as:

$$\Delta b\ (\mathrm{cm\ g^{-1}}) = \left[(N_\mathrm{A}\sum_i b_2/M_2) - (v_2/v_1)(N_\mathrm{A}\sum_i b_1/M_1)\right] \tag{10}$$

with N_A: Avogadro's number, b_i: coherent scattering lengths in cm, v_i: partial specific volumes in cm^3 g^{-1} and M_i: molecular masses of the elementary scattering units in g mol^{-1} (index 1 is for the solvent and index 2 for the solute). For a *dilute* polymer solution the contrast Δb is given as:

$$\Delta b\ (\mathrm{cm\ g^{-1}}) = \left[(N_\mathrm{A}\sum_i b_2/M_2) - v_2^\infty \rho_1^0 (N_\mathrm{A}\sum_i b_1/M_1)\right] \tag{11}$$

with v_2^∞: partial specific volume of the solute polymer at infinite dilution in cm^3 g^{-1}, ρ_1^0: density of the pure solvent in g cm^{-3}. The neutron scattering constant K used in Eq. (9) is then:

$$K\ (\mathrm{cm^2\ mol\ g^{-2}}) = \Delta b^2/N_\mathrm{A} \tag{12}$$

For calculating Δb, the partial specific volume v_2^∞ has to be determined by densimetry. The coherent neutron scattering lengths of nuclear isotopes are tabulated in standard references (Marshall et al., 1971; Koester et al., 1981). Equivalent definitions for the contrast refer to the scattering length density difference expressed in ($\mathrm{cm^{-2}}$) as the sum of scattering lengths in a volume of 1 $\mathrm{cm^3}$ of polymer minus the corresponding sum of scattering lengths in the same volume of solvent:

$$\text{contrast}\ (\mathrm{cm^{-2}}) = \left[(N_\mathrm{A}\Sigma_i b_2/V_2) - (N_\mathrm{A}\Sigma_i b_1/V_1)\right] \tag{13}$$

with V_i: partial molar volumes ($\mathrm{cm^3\ mol^{-1}}$). Using Eq. (13) the definition of the scattering constant K is then:

$$K\ (\mathrm{cm^2\ mol\ g^{-2}}) = (\text{contrast})^2 V_2^2/(M_2^2 N_\mathrm{A}) \tag{14}$$

The large difference in scattering length b for the two hydrogen-isotopes ^{1}H and ^{2}H ($b_{^1\mathrm{H}} = -0.374 \times 10^{-12}$ cm, $b_{^2\mathrm{H}} = +0.668 \times 10^{-12}$ cm) is the reason why polymeric and colloidal systems with their large hydrogen content provide an excellent contrast in neutron scattering experiments when one of the components (either solvent or solute) is deuterated (isotopic labelling). It is generally assumed that isotopic substitution does not alter the chemical properties of the system. However, Eq. (9) defining the differential scattering cross-section is valid as well for X-ray and light scattering experiments (see also Chapter 7 by P. Schurtenberger). The *only* difference is the scattering constant K; it contains the characteristic contrast for the each type of radiation:

	neutrons	X-rays	light
contrast:	difference of scattering lengths b	difference of electron density ρ_e	refractive index increment $\delta n/\delta c$
scattering constant:	$K \propto \Delta b^2$	$K \propto \Delta\rho_\mathrm{e}^2$	$K \propto [\delta n/\delta c]^2$

1.5. General method for solving structural problems

An illustration of the general method for solving structural problems is sketched in Fig. 3. We start from the intensity I, measured as a function of angle θ between θ_min and θ_max in the scattering experiment. It would be definitely wrong to jump to structural conclusions from the observed scattering profile (left path). For instance, a decrease of the intensity with Q^{-2} in the given range of momentum transfer Q is consistent with both the model of a platelet as well as with a Gaussian polymer coil. An unambiguous distinction between these structural models is a priori impossible without further information on the sample. Consequently, a tedious procedure is necessary in order to obtain reliable results (right path in Fig. 3).

The first and crucial step in data reduction is therefore absolute intensity calibration, yielding the sample's differential scattering cross-section $\mathrm{d}\Sigma/\mathrm{d}\Omega = f(Q)$ in absolute units of ($\mathrm{cm^{-1}}$). Knowledge of $\mathrm{d}\Sigma/\mathrm{d}\Omega$ allows to evaluate structural parameters, for instance the

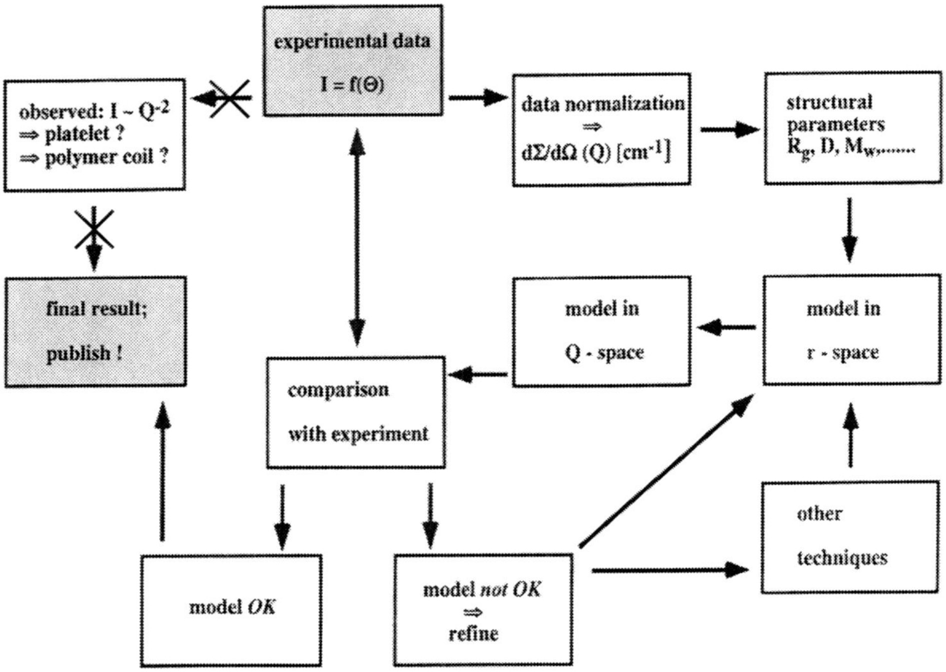

Fig. 3. General method for resolving structural problems with scattering experiments.

radius of gyration R_g, the fractal dimension D, the weight average molecular mass M_w, etc. Based on this structural information a model in the real space (r-space) can be constructed and converted by Fourier transformation techniques into a model in the experimental space (Q-space, see Chapters 4 and 5 by O. Glatter). Comparison of this basic model in Q-space with the measured experimental data then allows for a decision whether or not the model has to be refined, by modifying the model in r-space. It is extremely useful to use in this iterative process also complementary information, which is available from other techniques (e.g. NMR, microscopy, etc.). It is evident that only a careful comparison of the experimental data ('scattering curve') with the model in Q-space and its refinement will allow to correctly interpret the measurement and to yield a final result which is worth publishing.

2. The scattering experiment in practice: experimental aspects

2.1. Relating the experimental intensity to the differential scattering cross-section

We will now derive a quantitative relationship between the observed experimental intensity I (the number of neutrons, photons, . . . recorded during measuring time t) and the sample's differential scattering cross-section, $d\Sigma/d\Omega$, a quantity in units of (cm^{-1}).

We consider a sample of thickness d (cm) and with a number density ρ (cm^{-3}) of scattering elements. The radiation with wavelength λ_i (Å) (1 Å = 0.1 nm) has an incident flux Φ_i (cm^{-2} s^{-1}). The beam size is defined by an aperture with area A (cm^2) sitting in front of the sample.

At a scattering angle θ and at a distance L (cm) the scattered intensity I is observed within a solid angle $\Delta\Omega$ on the detector with efficiency $E(\lambda) < 1$ and detector area $A_{det} = \Delta\Omega L^2$ (cf. Fig. 1). The incident intensity per second I_i can be written as

$$I_i\ (\mathrm{s}^{-1}) = \Phi_i(\lambda) A E(\lambda) \tag{15}$$

The incident radiation interacts with the sample (scattering event) at position x ($0 \leq x \leq d$, d = sample thickness) with probability $d\sigma/d\Omega$ per unit solid angle. For the sake of simplicity we make the assumption that only a single scattering process occurs.

Before any interaction within the sample, the incident intensity is attenuated by a factor $\exp[-\mu x]$, μ being a sample and radiation specific linear attenuation coefficient. After the scattering process, the scattered intensity is further attenuated within the sample by a factor $\exp[-\mu(d-x)]/\cos\theta$. Integration over the sample, along x, yields the intensity I measured at scattering angle θ:

$$I(\theta) = \int_0^d \phi_x E\,\Delta\Omega \exp[-\mu x] A\rho \frac{\exp[-\mu(d-x)]}{\cos\theta} \frac{d\sigma}{d\Omega}\,dx \tag{16}$$

For neutrons and X-rays the scattering angles θ are sufficiently small so that $\cos\theta \approx 1$ and we obtain from integration the result

$$\begin{aligned} I(\theta) &= \Phi_i E \Delta\Omega A d \exp[-\mu d] \rho \frac{d\sigma}{d\Omega} \\ &= \Phi_i E \Delta\Omega A d \exp[-\mu d] \frac{d\Sigma}{d\Omega} \\ &= \Phi_i E \Delta\Omega A d T \frac{d\Sigma}{d\Omega} \end{aligned} \tag{17}$$

with the transmission of the sample being defined as

$$T = \exp[-\mu d] \tag{18}$$

Eq. (17) is basically composed of two terms. The first term is the instrumental constant $C(\lambda)$

$$C(\lambda) = \Phi_i E \Delta\Omega A = I_i \Delta\Omega \tag{19}$$

$C(\lambda)$ is defined by the configuration of the experimental setup itself, i.e. the wavelength, the detector and collimation distances, the type of detector and the size of the sample aperture. The second term of Eq. (17), $dT\,d\Sigma/d\Omega$, is specific to the sample only.

2.2. Transmission

The incident intensity is attenuated on its way through the sample (thickness d) due to scattering and absorption. The sample's transmission T_S is then defined as the ratio of the outgoing to the incoming intensity at scattering angle $\theta = 0$ (cf. Fig. 4a):

$$T_S = I(\theta = 0)/I_i = \exp[-\mu d] = \exp[-d/\Lambda] \tag{20}$$

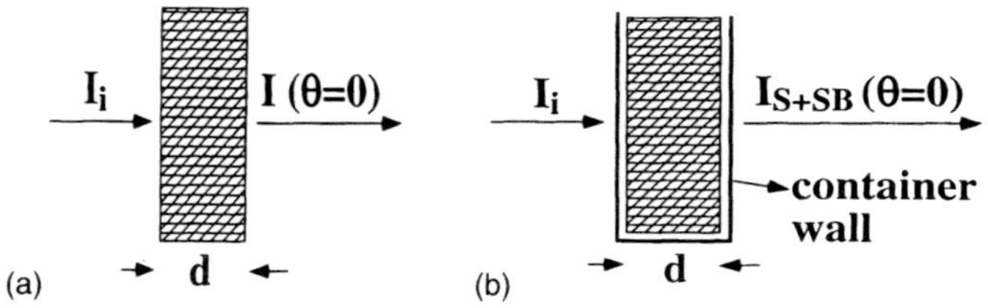

Fig. 4. (a) Schematic layout of a transmission measurement (pure sample). (b) Transmission measurement with a sample in a container.

I and I_i have to be measured under identical conditions. μ (cm^{-1}) is the linear attenuation coefficient and Λ (cm) $= 1/\mu$ is defined as the mean free path of the radiation before any collision with the sample.

The origin of the linear attenuation coefficient μ is different for different types of radiation. For light, the attenuation in the sample is due to coherent scattering into the full solid angle 4π. The attenuation coefficient μ_L is identical to the turbidity and is proportional to λ^2. The mean free path Λ_L for light ranges from about 1 m to 1000 m.

For X-rays, absorption is the essential part of attenuation and μ is identical to the absorption coefficient μ_{abs}. For X-rays, the mean free path Λ_X ranges from about 0.01 mm to 1 mm. For neutrons, the attenuation coefficient is the sum of absorption, coherent and incoherent scattering. 'Coherent' means scattering for which an incident neutron wave interacts with all the nuclei in the sample in a coordinated fashion; the scattered waves from all the nuclei have definite relative phases and can thus interfere with each other. Scattering is 'incoherent' when incident neutron waves interact independently with each nucleus in the sample. The scattered waves from different nuclei have random or indeterminate relative phases and thus cannot interfere with each other. The attenuation coefficient in neutron scattering is also called total scattering cross-section Σ (cm^{-1}). The mean free path Λ_n for neutrons ranges from about 0.5 mm to 10 mm.

As an example, the wavelength dependence of the transmission T of neutrons for light water (H_2O) for various thicknesses d is shown in Fig. 5. The total scattering cross-section $\Sigma = \ln T/d_{H_2O}$ of H_2O is shown in Fig. 6 (temperature $T = 25°C$).

From Eq. (17) we see that the scattering intensity is proportional to dT. Hence, a maximum is reached for $d = 1/\mu = \Lambda$, when the radiation mean free path is identical to the sample thickness d. On the other hand, multiple scattering is minimized for $d \ll \Lambda$. For optimization of the scattering intensity a compromise for the sample thickness has to be found. It is recommended to measure samples with a thickness as thin as possible.

Transmission measurements are essential for data reduction, e.g. for subtracting the solvent scattering (reference sample) from that of the solution (sample), as will be seen later. Their precision of is, however, often restricted, with a typical standard error of 1–3%.

The typical experimental aspects for transmission measurements are as follows:

- Attenuation of the incident beam and removal of the beam-stop (which protects for scattering measurements the central detector area). 'Attenuators' are absorbents such as perforated cadmium sheets or a piece of PMMA for neutrons. Nickel foils are used for X-rays. Due to inelastic effects (see below), the incident wavelength is changed when using ^{1}H containing absorbents, like PMMA for neutron scattering.

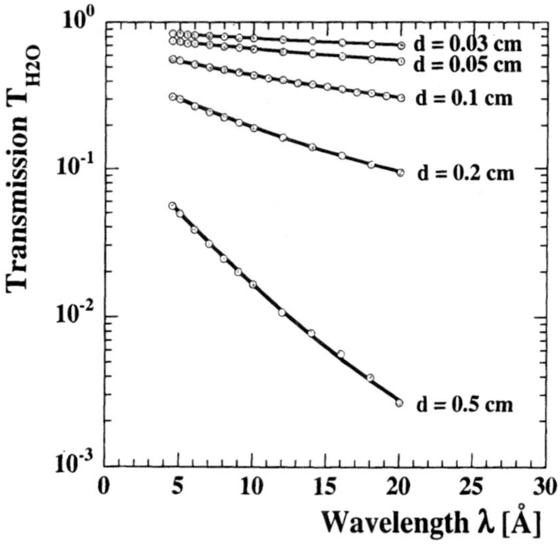

Fig. 5. Measured neutron transmission T_{H_2O} for water as a function of wavelength λ at various thicknesses d_{H_2O} and at ambient temperature $T = 25°C$ (Lindner, 2000).

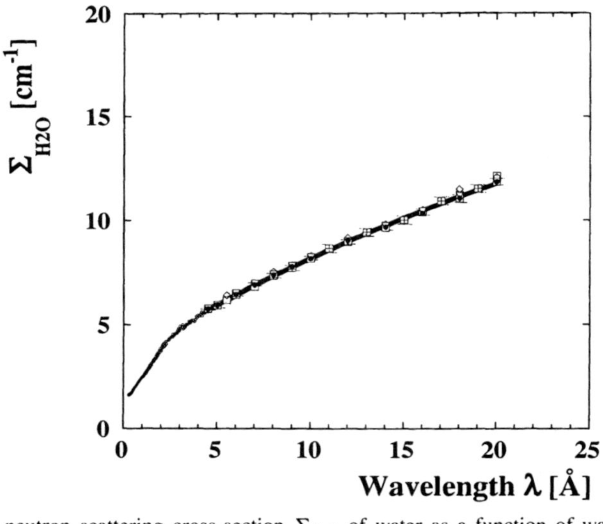

Fig. 6. Measured total neutron scattering cross-section Σ_{H_2O} of water as a function of wavelength λ at ambient temperature $T = 25°$ (Lindner, 2000). Small dots in the range 1–6 Å are results from Granada (1987).

- The attenuation of the incident beam should be strong enough to avoid detector saturation effects (dead-time losses, see Section 2.7).
- Data acquisition for the sample and for the direct beam (without sample) must be performed under identical conditions (time, collimation distance).
- For two-dimensional area detectors, the attenuated intensity is summed up in a finite but narrow central window. The 1st moment of the intensity distribution in this window does

also yield the central coordinates X_0, Y_0 for subsequent data reduction (circular averaging of the scattering data, see Section 3).

2.3. Background in scattering experiments

Several sources of background have to be considered in a scattering experiment. They can be classified into two categories, external and internal backgrounds. External backgrounds are room background, electronic noise, tails of the direct beam, scattering from the sample container and from the solvent (if the particles to be investigated are in solution). Internal background is defined as the background from the sample itself.

2.3.1. External background and its correction

The finite collimation of the incident beam and parasitic scattering from the edges of the sample aperture and of the beam-stop lead to some residual scattering (tails) of the direct beam. It is often neglected or removed from the data by an appropriate larger masking of the central region of the detector.

'Room' background originates from natural radiation. Electronic noise from the detection device does also contribute to the external background. Both can be measured by putting an absorbent at the sample position (i.e. a sheet of cadmium or B_4C for neutron scattering). We name this contribution $I_{\mathrm{Cd}}^{\mathrm{m}}$.

In scattering experiments the sample is often confined in a container (e.g. a liquid in a glass cell). The coherent container window scattering gives therefore another background contribution: the front window scattering intensity is reduced by attenuation on its way through the sample whereas the back window is scattering from an attenuated incident beam. For such a case where the sample is confined in a container, a separate scattering measurement of the empty container is necessary for background correction. We name these measured intensities $I_{\mathrm{S+SB}}^{\mathrm{m}}$ (scattering from sample + container) and $I_{\mathrm{SB}}^{\mathrm{m}}$ (empty container).

As a consequence, the transmission for a sample confined in a container is defined differently (cf. Fig. 4b) than that for a pure sample, measured without container [cf. Fig. 4a and Eq. (20)]:

$$T_{\mathrm{S}} = \frac{T_{\mathrm{S+SB}}}{T_{\mathrm{SB}}} = \frac{I_{\mathrm{S+SB}}(\theta = 0)}{I_{\mathrm{SB}}(\theta = 0)} \tag{21}$$

In order to evaluate the background corrected scattering intensity I [cf. Eq. (17)] an appropriate data treatment of the measured intensities $I_{\mathrm{S+SB}}^{\mathrm{m}}$, $I_{\mathrm{SB}}^{\mathrm{m}}$ and $I_{\mathrm{Cd}}^{\mathrm{m}}$ has to be applied. As an example we consider here the case of a sample (a solution of particles dispersed in a solvent) confined in a container (e.g. glass cell). In a first step, all measurements have to be scaled to the incident beam monitor. For the total scattering of the sample inside the container we write:

$$I_{\mathrm{S+SB}}^{\mathrm{m}} = T_{\mathrm{S}} T_{\mathrm{SB}} I_{\mathrm{S}} + T_{\mathrm{S}} T_{\mathrm{SB}} I_{\mathrm{SB}} + I_{\mathrm{Cd}} \tag{22}$$

The empty container scattering is given by:

$$I_{\mathrm{SB}}^{\mathrm{m}} = T_{\mathrm{SB}} I_{\mathrm{SB}} + I_{\mathrm{Cd}} \tag{23}$$

The room background and electronic noise measurement yields:

$$I_{\mathrm{SB}}^{\mathrm{m}} \equiv I_{\mathrm{Cd}} \tag{24}$$

From Eq. (23) we obtain:

$$I_{\mathrm{SB}} = \frac{1}{T_{\mathrm{SB}}}(I_{\mathrm{SB}}^{\mathrm{m}} - I_{\mathrm{Cd}}) \tag{25}$$

Inserting this result into Eq. (22) yields

$$I_{\mathrm{S+SB}}^{\mathrm{m}} = T_{\mathrm{S}} T_{\mathrm{SB}} I_{\mathrm{S}} + (T_{\mathrm{S}} T_{\mathrm{SB}}/T_{\mathrm{SB}})(I_{\mathrm{SB}}^{\mathrm{m}} - I_{\mathrm{Cd}}) + I_{\mathrm{Cd}} \tag{26}$$

From Eq. (26) and the definition for the transmission of a sample in a container Eq. (21)] follows the scattering intensity I_{S} of the sample (solution of solute particles in a solvent), corrected for empty container scattering, room background and electronic noise:

$$I_{\mathrm{S}} = \frac{I_{\mathrm{S+SB}}^{\mathrm{m}} - I_{\mathrm{Cd}}}{T_{\mathrm{S+SB}}} - \frac{I_{\mathrm{SB}}^{\mathrm{m}} - I_{\mathrm{Cd}}}{T_{\mathrm{SB}}} \tag{27}$$

The same treatment is applied to the solvent scattering. In a last step, the net scattering of the dispersed particles in solution is obtained by subtracting the background corrected solvent from the solution scattering (see Section 2.3.2).

2.3.2. *Internal background and its correction*

'Internal' background is defined as the background from the sample itself. Especially for neutron scattering with hydrogenous (^{1}H containing) species, the incoherent scattering cross-section gives rise to a sample inherent background. It can be shown that the total measured differential scattering cross-section of a sample (for instance a solution of polymers in a solvent) can be decomposed into two terms [see Section 1.3, Eq. (7)]:

$$\begin{aligned} \frac{\mathrm{d}\Sigma}{\mathrm{d}\Omega} &= \langle b^2 \rangle \sum \langle \exp[-i Q(r_i - r_j)] \rangle + N[\langle b^2 \rangle - \langle b \rangle^2] \\ &= \left(\frac{\mathrm{d}\Sigma}{\mathrm{d}\Omega}\right)_{\text{coherent}} + \left(\frac{\mathrm{d}\Sigma}{\mathrm{d}\Omega}\right)_{\text{incoherent}} \end{aligned} \tag{28}$$

The first term, $(\mathrm{d}\Sigma/\mathrm{d}\Omega)_{\text{coherent}}$, is Q-dependent and represents the interference of scattered waves at different nuclei $i \neq j$ with scattering length $\langle b \rangle$. It thus contains the information on the structure of the dissolved polymer particles. The second term $(\mathrm{d}\Sigma/\mathrm{d}\Omega)_{\text{incoherent}}$ does not depend on Q and is considered as a (flat) background originating from the solvent and from the incoherent scattering of the polymer. The part of the incoherent contribution from the polymer (flat internal sample background) must be measured for some cases and subtracted before the coherent differential scattering cross-section (structure) is further analyzed (Rawiso et al., 1987; Cotton, 1991).

For the case of a dilute mixture of a deuterated species in a hydrogenous solvent, the small amount of incoherent scattering from ^{2}H is negligible. It is sufficient to subtract the solvent scattering. Attention must be paid, however, when the concentration of polymer becomes high.

For the inverse case of a dilute mixture of protonated polymer in a deuterated solvent, we can distinguish several methods:

(a) The level of incoherent scattering can be estimated by measuring the sample at highest possible scattering angles, because the coherent sample scattering becomes negligible at highest momentum transfer Q. If the scattering curve levels off to a constant asymptotic value this can be used as a good approximation for the sample's internal background.

(b) The measurement of a reference sample with no structure is another method to determine the internal background. The reference sample (e.g. a H_2O/D_2O mixture) should contain a mixture of ^1H and ^2H corresponding to the volume fractions $\Phi_{^1H}$ and $\Phi_{^2H}$ in the sample. This method requires precise chemical skills.

(c) If the differential scattering cross-section has a *known* Q-dependence ($\mathrm{d}\Sigma/\mathrm{d}\Omega \sim Q^{-D}$) we can write:

$$\mathrm{d}\Sigma/\mathrm{d}\Omega = AQ^{-D} + B \qquad \text{with } B \equiv \text{ internal background} \tag{29}$$

A plot of $\mathrm{d}\Sigma/\mathrm{d}\Omega \cdot Q^D$ versus Q^D yields

$$\mathrm{d}\Sigma/\mathrm{d}\Omega \cdot Q^D = A + BQ^D \tag{30}$$

with the background B to be determined from the slope of the plot.

(d) *Polarized* neutrons can be used to distinguish between the coherent and the incoherent scattering process. This method is not widely applied because it requires a specialized instrument.

(e) The level of incoherent scattering could be estimated from table values of cross-sections. Attention must be paid here because these tables do only contain *bound* atom cross-sections, which are smaller than the real ones (due to inelastic scattering contributions, see Section 2.6).

2.4. Instrumental resolution

The fundamental question whether the sample's differential scattering cross-section can be determined with precision from the observed intensity $I = f(Q)$ for the given experimental setup is also related to the problem of limited instrumental resolution.

In general the observed scattering curve is subject to 'smearing' because of an uncertainty in the momentum transfer Q. Reasons for this uncertainty are:

- each instrument has a finite wavelength distribution $\Delta\lambda/\lambda$ of the incident beam,
- the incident beam has a finite divergence, depending on the collimation distance,
- the beam defining aperture in front of the sample has a finite size,
- the detection device has a finite spatial resolution.

Smearing of data is most pronounced for profiled scattering curves (e.g. spherical colloidal particles showing sharp Bragg peaks) but it is less critical for smooth polymer scattering curves.

Each of the contributions limiting the instrumental resolution can be treated separately and the measured data can be analyzed either by model-independent approaches or by direct modelling. For a detailed discussion the reader is referred to the contributions of O. Glatter (Chapters 4 and 5) and J.S. Pedersen (Chapters 6 and 16) in this book.

2.5. Multiple scattering

The usual approximation for structure analysis with scattering experiments is a single scattering process: the radiation interacts only once on its way through the sample. This approximation is justified for a sample thickness d being much smaller than the radiation mean free path Λ: $d \ll \Lambda$ (see Section 2.2). If this conditions is fulfilled the structure can be deduced relatively straightforward from the observed scattering intensity.

For $d \geq \Lambda$, multiple scattering occurs, i.e. the incident radiation interacts twice and more with the sample. The resulting intensity corresponds to a sum of intensities due to single, double, triple, . . . scattering. For this case, the scattering angles (or scattering vectors) are unknown and the interpretation of the observed scattering curve is quite difficult. Methods exist to correct for the effects of multiple scattering in data analysis (Berk and Hardman Rhyne, 1988). They are, however, difficult to apply for strong multiple scattering. For weak multiple scattering, the consequences are less dramatic (Goyal et al., 1983). It is recommended to further reduce the effect by a decrease of the sample thickness and to measure with the shortest possible radiation wavelength.

Fig. 7a,b show as an example the distortion of the fractal scattering of Urupunga hydrocarbon source rock, measured with SANS at different wavelengths and for different thickness of the sample (Radlinski, 2000a). The fractal structure of these systems is of great interest to oil industry because it could indicate a rock's oil-producing potential (Radlinski et al., 1999; 2000b). For a thick sample the scattering curves show a marked deviation with increasing neutron wavelength λ due to multiple scattering (Fig. 7a). The fractal structure of the sample is clearly visible at all wavelengths only for the thinnest sample (Fig. 7b). For a further discussion the reader is referred to the literature and to Chapter 20 by J. Rička which describes the recent development of multiple scattering experiments (diffusing wave spectroscopy with light).

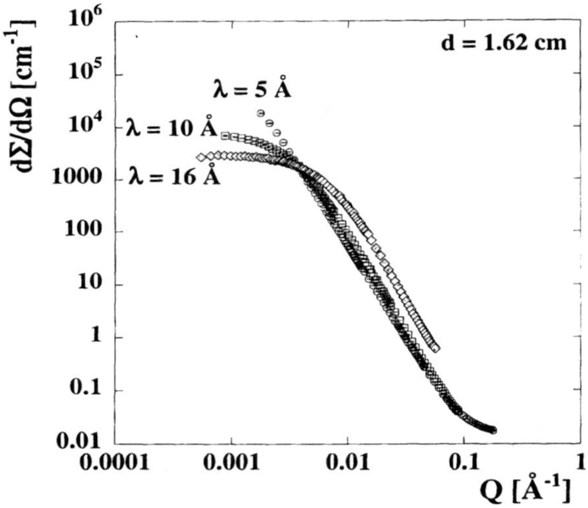

(a)

Fig. 7a. SANS of Urupunga hydrocarbon source rock measured at D11 for three wavelengths $\lambda = 5$, 10 and 16 Å. Sample thickness is $d = 1.62$ cm (Radlinksi et al., 2000a).

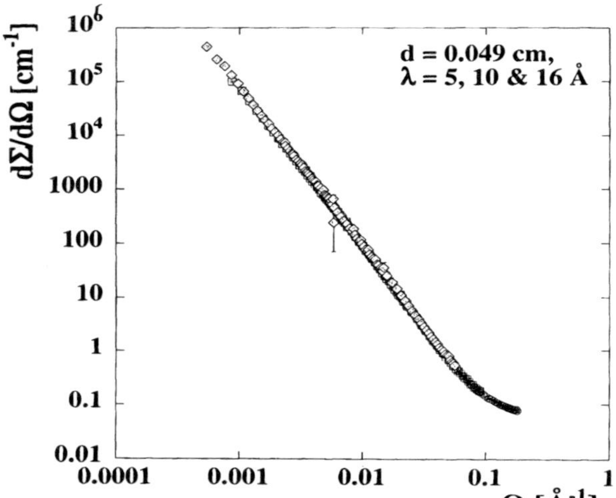

(b)

Fig. 7b. Same measuring conditions as in Fig. 7a, but the sample has a thickness of 0.049 cm.

2.6. Inelastic effects

Inelastic scattering is due a partial thermalization of the incident beam. The effect is important in small-angle neutron scattering with hydrogenous samples, using 'cold' neutrons in the wavelength range 4.5 to 20 Å. For instance, the thermal energy of a water molecule is at room temperature much larger than the kinetic energy of an incident neutron. Energy is thus transferred during the scattering process and the neutrons scattered from H_2O have shorter wavelengths than those of the incident beam (Ghosh and Rennie, 1999; Lindner et al., 2000).

Fig. 8 illustrates the effect of the inelastic scattering from water, measured with the time-of flight technique at the instrument IN5 of ILL by Ghosh and Rennie (1999). For incident wavelengths λ_i of 2, 5, 8 and 13 Å an increasing contribution of inelastic neutrons is seen in the range of 1–2 Å. The problem is relevant as in general the efficiency of the detection

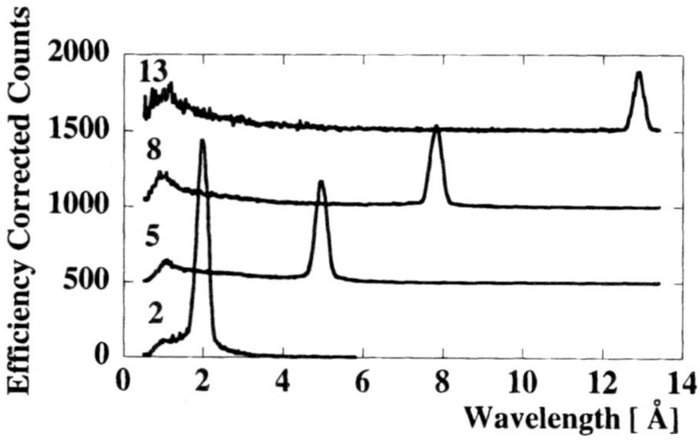

Fig. 8. Inelastic scattering of H_2O as measured with TOF at IN5 of ILL (Ghosh and Rennie, 1999).

device depends on the wavelength λ, as well as transmission measurements. It is important to take into account of these effects when using water as a calibration standard for SANS (see Section 4.3).

2.7. Detector dead time

Nearly each counting device needs a characteristic time to separate two events in order to be recorded as two separate pulses. This effect is known as the detector dead time τ and it gives an important experimental restriction, especially at high counting rates:

dead time τ (s) = minimum time for which further counting is inhibited.

It concerns the detection device itself but also the associated data acquisition electronics.

For a measured count rate n_m during a total time T, the detector does not respond for a time t and n_m is different from the true countrate n. The relationship between n_m and n can be written as (Lindner, 1998a):

$$n_m = n/[1 + n\tau] \tag{31}$$

Eq. (31) is called the 'non paralysable' model. Another model is the 'paralysable' model:

$$n_m = n \cdot \exp(-n\tau) \tag{32}$$

which differs from the 'non paralysable' model only for very high count rates.

Fig. 9 shows an illustration of both models for the measured count rate n_m as a function of true count rate n with a typical detector dead time $\tau = 2.33$ μs (BF_3 detector at D11/ILL: Lindner, 1998a).

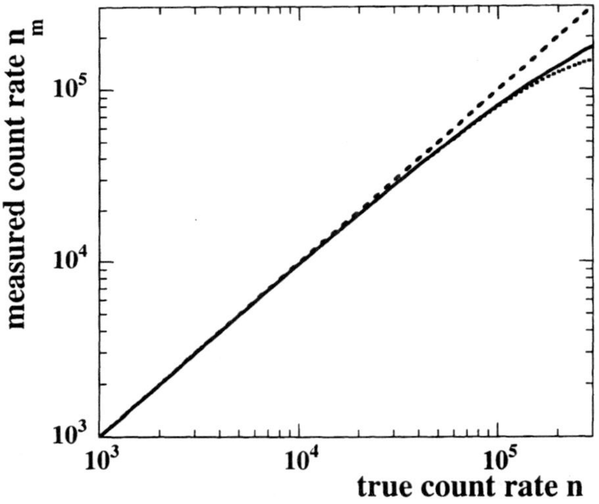

Fig. 9. Measured count rate n_m as a function of the true count rate n. Dashed line: ideal behaviour (no dead time losses). Full line: 'non-paralysable model' with a detector dead time constant $\tau = 2.33$ μs. Dotted line: 'paralysable model' ($\tau = 2.33$ μs).

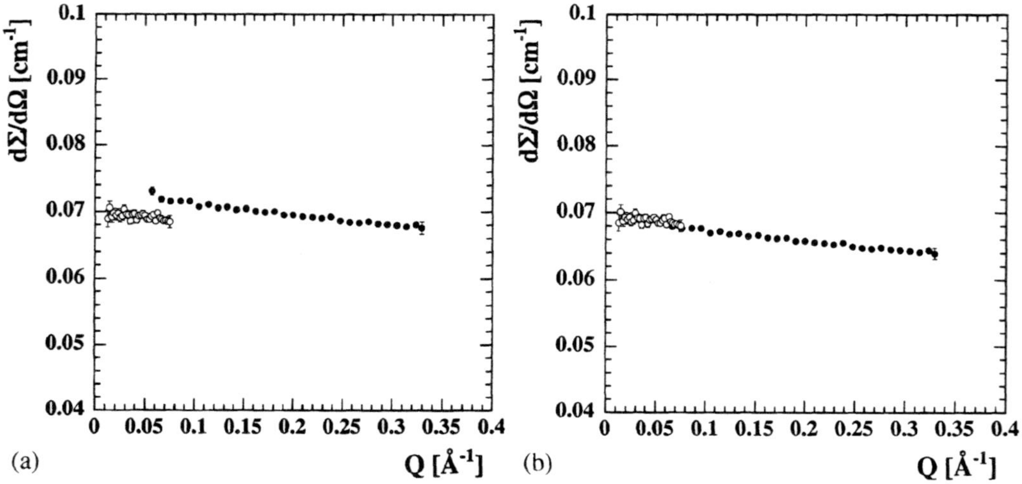

Fig. 10. (a) Differential scattering cross-section of toluene-d_8 as measured with SANS (D11) at detector distances 5 m and 1.1 m. Data were not corrected for detector dead time losses (Lindner, 1998a). (b) Same sample as (a), but including the correction for dead time losses.

The dead time problem is a possible source of error for

- strongly scattering samples with different count rates,
- experiments where the water calibration (cf. Section 4.3) and the sample measurements have different count rates,
- experiments which are performed at different detector (and collimation) distances.

Fig. 10a,b illustrate the effect of dead time losses for a SANS measurement of a deuterated sample, measured at two different detector positions with a very high count rate at the shorter detector distance. A good overlap of the background corrected and calibrated scattering curves (Fig. 10b) is only obtained after correction for dead time losses (see Section 3.1)

In practice, the gas detection devices used for instance in SANS show significant dead time losses above count rates of about 50 kHz. In the case of doubt, the incident flux should always be reduced (longer collimation distance or smaller sample aperture size). It is always recommended to correct dead time losses during the primary data reduction (see Section 3), using the characteristic dead time constant τ of the detection device.

2.8. Scaling of scattering data measured under different configurations

In particular for small-angle neutron scattering experiments at instruments with a large detector distance range (e.g. at the ILL Grenoble, D11: 1.1–37 m and D22: 1.5–18 m) intensities are frequently measured at different detector positions L over the totally available Q-range. Attention has to be paid to obtain a sufficiently large overlap region for the measurements at the different distances. For very long detector distances (e.g. 37 m detector position at D11/ILL) data cannot be directly calibrated with an incoherent water standard sample (see Section 4.3): water is a very weak scatterer for distances larger than 10 m. It is therefore necessary to calibrate those data with water scattering measured with good statistics at short detector distance. A scaling factor has then to be determined separately by one of several methods:

- Evaluation of the instrumental constants $C(\lambda)$ (Eq. (17), see Section 4.1) from an attenuated direct beam measurement for the different experimental configurations. The ratio of two values measured at two distances corresponds to the scaling factor.
- Two measurements of a strong scatterer (e.g. Teflon, sintered graphite) at long and at short distance and evaluation of the ratio of the two scattering curves. This method might be affected by multiple scattering effects (see Section 2.5).
- Two measurements of a water sample at short detector distance but with two collimation distances: a short one and the same (long) collimation distance as used for the scattering measurement of the sample at long detector distance (the water intensity scales with the detector distance $1/L^2$). Depending on the long collimation distance, this method might be quite time consuming in order to obtain good statistics! It has nevertheless an advantage as the water reference is measured under identical collimation conditions (beam divergence) as the sample.

For all these methods it is necessary to apply corrections for dead time losses (see Sections 2.7 and 3.1).

3. Initial data reduction: the example of SANS

As a prerequisite for evaluation of a scattering experiment we assume the following:

- the sample thickness is known
- the composition of the sample and the partial molar volumes are known in order to calculate the contrast (see Section 1.4)
- a reference sample (for instance the pure solvent for a solution of particles) has been measured
- the range of momentum transfer Q of the scattering experiment is well defined.

A complete experiment includes measurements of the *scattering intensities* of the sample, of its external background (for instance the empty sample container if the sample is a liquid, see Section 2.3.1) and of a standard sample used for absolute calibration (see Section 4). All scattering measurement have to be scaled to the incident beam monitor preset. Furthermore, the complete set of *transmission* measurements has to performed (attenuated incident beam, see Section 2.2), corresponding to the sample, the reference sample, the background (empty cell), the standard sample and the direct beam without any sample.

3.1. Data reduction from SANS experiments: isotropic spectra

The scattering intensity is at SANS instruments usually recorded with a position-sensitive (two-dimensional) multi detector (e.g. a ^{3}He detector with 64 × 64 'cells' of resolution 1 cm^2 at the instrument D11 of ILL Grenoble; for an overview of available software for SANS data reduction see Ghosh et al., 2000 and Dewhurst, 2002).

The result of the experiment is an intensity distribution $I^{\mathrm{m}}(x, y)$ where x, y denote the detector coordinates. Standard treatment of *isotropic* SANS data (i.e. data where the intensity distribution does not depend on the azimuthal direction on the detector) invokes a routine procedure, starting with the radial regrouping (circular averaging) of the intensity in fixed steps Δr around the centre of the primary beam (coordinates x_0, y_0). The beam centre coordinates are determined from the 1st moment of the intensity distribution of the primary beam, following separately from a direct (attenuated) beam measurement:

$$x_0 = \sum (w_i x_i) \Big/ \sum w_i \quad \text{and} \quad y_0 = \sum (w_i y_i) \Big/ \sum w_i \tag{33}$$

with x_i, y_i: position of cell i in x, y direction on the detector and w_i: content of cell i. According to the experimental conditions (wavelength λ and sample-to-detector distance L) an average value of momentum transfer Q can be attributed to each radius vector $\boldsymbol{r}$ on the detector:

$$Q = (4\pi/\lambda)\sin[0.5\,\mathrm{arctg}(|\boldsymbol{r}|/L)] \tag{34}$$

The resulting intensity distribution $I^{\mathrm{m}} = f(Q)$ is to be corrected for absorption, sample thickness d and electronic or guide hall background $I_{\mathrm{Cd}} = f(Q)$. The scattering from the empty sample container $I_{\mathrm{SB}}(Q)$ (e.g. empty cell) is treated in the same way and is then subtracted from the sample scattering (see Section 2.3). Normalization to the unit incident flux, geometrical factors and detector cell efficiency corrections are performed by using the measurement of a standard sample $I(Q)_{\mathrm{st}}$, for instance 0.1 cm thick H_2O, measured at ambient temperature. Data can be put on an absolute scale by using tabulated values for the effective differential scattering cross-section per unit volume of H_2O, determined from cross-calibration measurements (see Section 4; Lindner, 2000). For both, the sample and the (water) standard measurement, a correction for dead time losses has to be applied (see Section 2.7). As a final result the differential scattering cross-section per unit volume of the sample [cf. Eqs. (17) and (27)] can be expressed as:

$$\left(\frac{\mathrm{d}\Sigma}{\mathrm{d}\Omega}\right)_{\mathrm{sample}} (Q)\ (\mathrm{cm}^{-1}) = \frac{(I^{\mathrm{m}}_{\mathrm{S+SB}} - I_{\mathrm{Cd}}) - A_2(I^{\mathrm{m}}_{\mathrm{SB}} - I_{\mathrm{Cd}})}{(I^{\mathrm{m}}_{\mathrm{H_2O+ec}} - I_{\mathrm{Cd}}) - A_1(I^{\mathrm{m}}_{\mathrm{ec}} - I_{\mathrm{Cd}})} \cdot \frac{A_3}{A_4} \tag{35}$$

The factors $A_1 \ldots A_4$ in Eq. (35) are defined as follows:

$$A_1 = (T_{\mathrm{H_2O}}/T_{\mathrm{ec}})(1 - n_{\mathrm{m}}^{\mathrm{H_2O}}\tau)/(1 - n_{\mathrm{m}}^{\mathrm{ec}}\tau) \tag{36a}$$

$$A_2 = T_{\mathrm{S}}/T_{\mathrm{SB}}(1 - n_{\mathrm{m}}^{\mathrm{S}}\tau)/(1 - n_{\mathrm{m}}^{\mathrm{SB}}\tau) \tag{36b}$$

$$A_3 = T_{\mathrm{H_2O}} d_{\mathrm{H_2O}}(1 - n_{\mathrm{m}}^{\mathrm{H_2O}}\tau)\left(\frac{\mathrm{d}\Sigma}{\mathrm{d}\Omega}\right)_{\mathrm{H_2O,eff}} \tag{36c}$$

$$A_4 = T_{\mathrm{S}} d_{\mathrm{S}}(1 - n_{\mathrm{m}}^{\mathrm{S}}\tau) \tag{36d}$$

d_{S} and $\mathrm{d}_{\mathrm{H_2O}}$ are the thickness of the sample and the water standard. $T_{\mathrm{H_2O}}$, T_{ec}, T_{S} and T_{SB} are the transmission values of water, the empty water cell, the sample and the sample background (cf. Eq. (21), Section 2.3.1). The terms $(1 - n_{\mathrm{m}}^{\mathrm{H_2O,S}}\tau)$ and $(1 - n_{\mathrm{m}}^{\mathrm{ec,SB}}\tau)$ in $A_1 \ldots A_4$ do take into account the correction for dead time losses, with $n_{\mathrm{m}}^{\mathrm{H_2O,ec,S,SB}}$ being the integral count rates of each measurement and τ being the dead time constant of the detector. A discussion of absolute calibration with a standard water sample (using the *effective* differential scattering cross-section $(\mathrm{d}\Sigma/\mathrm{d}\Omega)_{\mathrm{H_2O,eff}}$) is to be found in Section 4.3.

The same treatment according to Eqs. (35) and (36) is applied to the reference sample (e.g. the solvent for a solution of particles). The coherent differential scattering cross-section of the solute is finally obtained by subtracting the calibrated internal background (reference sample, see Section 2.3.2) from the solution scattering.

3.2. Data reduction from anisotropic SANS experiments

As a result of a SANS experiment with an external field acting on the sample (for instance experiments under non-equilibrium conditions, cf. Chapter 17), an intensity distribution $I = f(x, y)$ with azimuthal anisotropy is usually obtained on the two-dimensional multidetector (with 'cell' coordinates x, y). One way of treating those data is a two-dimensional analysis: cell-by-cell correction and calibration yields the two-dimensional scattering cross-section $\mathrm{d}\Sigma/\mathrm{d}\Omega = f(Q_x, Q_y)$. Another way of data reduction consists of successive sector masking of detector cells with variation of the azimuthal sector opening angle $\Phi_{\perp,\parallel}$ in the preferential directions parallel and perpendicular with respect to the imposed constraint (the flow direction in a shear experiment for instance). For each sector, the intensity is treated as described in Section 3.1, yielding the differential scattering cross-section as a function of $\Phi_{\perp,\parallel}$, $\mathrm{d}\Sigma/\mathrm{d}\Omega = f(Q, \Phi_{\perp,\parallel})$. The apparent conformation parameters, such as the radius of gyration, can then be determined directly for each sector mask, followed by interpolation to $\Phi_{\perp,\parallel} \to 0$.

If the scattering object is characterized in real space by an eccentricity ε, the resulting scattering pattern in reciprocal space will also have elliptical symmetry. For instance, a deformed polymer coil with ellipsoidal shape is defined by the ratio of the radii of gyration $R_\perp/R_\parallel$ in directions perpendicular and parallel with respect to the deformation axis. In this case ($qR_\parallel < 1$) an elliptically scaled momentum transfer q^* can be defined which allows elliptical averaging of the two dimensional intensity along $q^* =$ constant with the eccentricity ε and the radius of gyration $R_\parallel$ as fitting parameters (Mildner, 1983; Reynolds and Mildner, 1984). A further possibility of analysing 2-D data sets is a Fourier analysis allowing reconstruction of the 2-D pattern for any symmetry by Fourier series expansion of the intensity (Lindner and Hess, 1989).

4. Absolute calibration

Absolute calibration is necessary for subsequent interpretation of the data, using the fundamental relationship Eq. (17) (Wignall and Bates, 1987; Cotton, 1991). Absolute calibration relates the measured intensities $I = f(\theta) = f(Q)$ to the differential scattering cross-section of the sample, $\mathrm{d}\Sigma/\mathrm{d}\Omega(Q)$. The importance of absolute calibration is evident: (i) The molecular mass, the mass per unit length or the specific area of particles can be obtained from the differential scattering cross-section. (ii) Scattering curves obtained by different scattering techniques or in different Q-ranges (measured at different sample-to-detector distances or wavelengths) can be joined together. (iii) It serves as a self-consistency test of the experiment since multiple scattering or unexpected aggregation can be detected.

Absolute calibration can be performed in two ways, either by a direct determination of the instrumental constant $C(\lambda)$ [cf. Eq. (19)] or by using standard samples with known differential scattering cross-section $(\mathrm{d}\Sigma/\mathrm{d}\Omega)_{\mathrm{st}}$.

4.1. Direct determination of $C(\lambda)$

According to Eqs. (17) and (19), the differential scattering cross-section of the sample is given by:

$$\frac{d\Sigma}{d\Omega} = \frac{I}{C(\lambda)dT} \tag{37}$$

with I being the measured intensity after background correction.

The instrumental constant $C(\lambda)$ [cf. Eq. (19)] is the product of the incident intensity I_i

$$I_i\ (s^{-1}) = \Phi_i E A \tag{38}$$

and the solid angle $\Delta\Omega$. The solid angle is known from the experimental geometry, $\Delta\Omega = A_{det}/L^2$, with the detector distance L and the detector surface area A_{det}. The incident intensity, I_i, can be evaluated in the following way from one scattering and one transmission measurement:

(a) Measuring the scattering intensity of a strong scatterer (such as a teflon sample or a sintered graphite sample for neutron scattering) with ($I_{att}(Q)$) and without ($I(Q)$) an absorbent (attenuator). The attenuation factor T_a of the absorbent is then determined from the ratio of the two scattering curves:

$$T_a = [I_{att}/I] \tag{39}$$

(b) Measurement of the attenuated direct beam (attenuator IN, beamstop OUT, no sample) for a time t under the configuration as used for the sample scattering experiment (i.e. the same detector and collimation distance as for the scattering measurement). The sum of the attenuated intensity $\sum I$ in the central window around the position of the primary beam yields the count rate $n = \sum I/t$, from which follows the incident intensity I_i as:

$$I_i = n/[T_a(1 - n\tau)] \tag{40}$$

From Eq. (40) together with the solid angle $\Delta\Omega$ the instrumental constant can be calculated:

$$C(\lambda) = n/[T_a(1 - n\tau)]\ \Delta\Omega \tag{41}$$

The term $(1 - n\tau)$ in Eq. (41) takes into account for dead time losses of the detector (see Sections 2.7 and 3.1).

Some problems are associated, however, with the described method for determining $C(\lambda)$ (Lindner, 1998b). The attenuation factor T_a is difficult to evaluate precisely: the scattering measurement with a strong scatterer (step a) might be affected by multiple scattering. Furthermore, there might be a wavelength change when using a protonated absorbent (e.g. PMMA in neutron scattering), due to inelastic effects. The absorbent (attenuator) might also increase the intensity due to higher harmonics of the wavelength λ, relative to the 1st order. Inherent mechanical imprecision (moving attenuator and collimating guides IN and OUT) might affect the quality of the measurement. In any case, a correction for dead-time losses is crucial.

In summary, $C(\lambda)$ only can be measured with a relative precision of typically 5–10% . But its determination does not 'cost' much time and it is a rapid quality check of the instrument when repeating an experiment for a given configuration (i.e. wavelength, detector distance, collimation distance, sample aperture area). Moreover, the knowledge of $C(\lambda)$ allows an estimate of the measuring time required for obtaining sufficiently good statistics.

4.2. Use of standard samples for calibration

For a standard sample with well known differential scattering cross-section $(\mathrm{d}\Sigma/\mathrm{d}\Omega)_{\mathrm{st}}$ (cm^{-1}), the instrumental constant $C(\lambda)$ is

$$C(\lambda) = I_{\mathrm{st}}/\left[(\mathrm{d}\Sigma/\mathrm{d}\Omega)_{\mathrm{st}} d_{\mathrm{st}} T_{\mathrm{st}}\right] \tag{42}$$

Various substances and their characteristic properties might be used as calibration standards: they are briefly reviewed in the following sections.

4.2.1. The compressibility χ_T of pure liquids

The measurement of the scattering intensity of a pure liquid is related to the isothermal compressibility χ_T. With an extrapolation to momentum transfer $Q \to 0$ the differential scattering cross-section can be written as

$$\left(\frac{\mathrm{d}\Sigma}{\mathrm{d}\Omega}\right)_{\mathrm{st}} = \rho b^2 kT \chi_T \quad (Q \to 0) \tag{43}$$

with the sample's number density ρ (cm^{-3}), the coherent scattering length b (cm), the Boltzmann constant k and temperature T (K). This method is convenient for light and X-ray scattering. For neutron scattering the method is less convenient because the measured intensity is generally weak and the separation of coherent from incoherent scattering is delicate.

4.2.2. Specific surface S of a porous medium

For a two-phase system with a sharp interface S applies Porod's law (see Chapter 2 by O. Spalla). The differential scattering cross-section can be written as

$$\left(\frac{\mathrm{d}\Sigma}{\mathrm{d}\Omega}\right)_{\mathrm{st}} = 2\pi\left[(N_{\mathrm{A}}/V_1)\Sigma b_1 - (N_{\mathrm{A}}/V_2)\Sigma b_2\right]^2\left[S/V\right]Q^{-4} \tag{44}$$

where $(N_{\mathrm{A}}/V_i)\Sigma b_i$ is the scattering length density of phase i, N_{A} is Avogadro's number and V_i is the molar volume of phase i.

The inconvenience of this method is: (i) the weak intensity at large momentum transfer Q due to the Q^{-4} power law, and (ii) a risk of multiple scattering at small Q-values.

4.2.3. Molecular mass M_{w} of particles (e.g. polymers)

For a solution of well characterized particles (i.e. polymer molecules) dispersed in a medium (solvent) at (sufficiently) low concentration c ($\mathrm{g\ cm}^{-3}$), the extrapolation of the measured intensity to momentum transfer $Q \to 0$ yields the apparent molecular mass M_{app}.

The differential scattering cross-section can be written as

$$\left(\frac{d\Sigma}{d\Omega}\right)_{st} = K c M_{app} \quad (\text{for } Q \to 0) \tag{45}$$

K is the contrast constant (see Section 1.4). The apparent molecular mass M_{app} is related to the weight average molecular mass M_w by

$$M_{app} = M_w/[1 + 2A_2 c M_w + \cdots] \tag{46}$$

A_2 is the second virial coefficient of the osmotic pressure series expansion. For polymer solutions in a good solvent ($A_2 > 0$) it is necessary to measure a dilution series at various concentrations and extrapolate the results to zero concentration ($c \to 0$). For ideal (so-called 'theta') solutions or for polymer mixtures composed of deuterated and protonated species ($A_2 = 0$) the calibration can be obtained with only one sample. Polymer samples can be used for the calibration of neutron, X-ray or light scattering experiments.

4.2.4. *Incoherent scattering*

Especially for neutron scattering, the measurement of samples with predominant incoherent scattering is another possibility for calibration. The differential scattering cross-section can be written as

$$\left(\frac{d\Sigma}{d\Omega}\right)_{st} = \rho(b_{incoherent})^2 = \text{constant} \tag{47}$$

with the sample's number density ρ (cm^{-3}) and the incoherent scattering length $b_{incoherent}$. An incoherent standard sample for neutron scattering is vanadium. The inconvenience of vanadium is its weak scattering intensity (about a factor 30 less compared to water, see next section), a low Q scattering from texture and parasitic incoherent scattering of absorbed Hydrogen, diffusing into the metal.

The incoherent scattering of water (H_2O) *cannot* be used directly for calibration because Eq. (47), which contains the *bound* incoherent scattering length $b_{incoherent}$, is not valid: inelastic and multiple scattering effects (see Section 2.6) are particularly important in neutron scattering and lead to a wavelength-dependent scattering cross-section. The use of H_2O as a secondary standard is described in more details in the following paragraph.

4.3. Water as a (secondary) calibration standard for SANS

A frequently used calibration standard for small-angle neutron scattering (SANS) experiments is a sample of 0.1 cm thick water (H_2O), measured at room temperature. The reason for its common use as SANS standard is that fresh, de-mineralized water is readily available, easy to handle and shows with an ideal detector a strong 'flat' scattering, i.e. independent of the scattering angle. However, for a non-ideal gas detector with a deformed inner membrane (separating the 'active' counting volume from the 'dead' compensation gas volume) a parabolic scattering profile is observed at some instruments (Lindner et al., 2000). Water is a homogeneous liquid and there are no texture or mesoscopic effects as observed for instance with solid vanadium. Its scattering intensity is predominantly incoherent; furthermore, at small angles, the remaining structured coherent part nearly vanishes by cancellation between the negative scattering lengths of the two protons and the positive length of the oxygen atom – the incoherent part of the scattering is raised to 99.7%. Compared to vanadium, the incoherent scattering of water is by about a factor of 30 larger. It is also used to normalise individual 'cells' of the large area detectors (used in SANS) for local variations in efficiency. The main disadvantages of water are inelastic effects (due to an energy exchange between the incoming radiation and the water molecules) and multiple scattering effects. As a consequence, the water cross-section does depend on wavelength and on temperature.

For the ideal (but non-realistic) case of an isotropic scattering we could assume that the incident radiation (e.g. neutrons) which is not being transmitted through the water sample is scattered uniformly in 4π. With T being the fraction of transmitted intensity and $(1-T)$ being the scattered fraction, we can write:

incident intensity × fraction scattered
= scattered intensity per unit solid angle × full solid angle

or

$$I_{\rm i}(1-T) = (I/\Delta\Omega)4\pi \tag{48}$$

From Eqs. (48) and (19) follows:

$$I\ ({\rm s}^{-1}) = C(\lambda)(1-T)/(4\pi), \qquad \text{with } C(\lambda) = I_{\rm i}\Delta\Omega \tag{49}$$

and from Eq. (17)

$$\left(\frac{{\rm d}\Sigma}{{\rm d}\Omega}\right)_{\rm H_2O,\,ideal} ({\rm cm}^{-1}) = (1-T)/[4\pi dT] \tag{50}$$

with d being the thickness and T the transmission of the water sample.

This expression is a *very crude* approximation. It should not at all be used for neutron scattering because of inelastic effects (due to the energy exchange between the incoming radiation and the water molecules) and multiple scattering effects. We repeat that cold neutrons are thermalized in water through collisions with hydrogen atoms and gain energy. These inelastically scattered neutrons have a distribution of effective wavelengths $\lambda_{\rm eff} < \lambda_{\rm i}$ ($\lambda_{\rm i} =$ incident wavelength). Because of the wavelength-dependence of the detector efficiency, they are detected with a different efficiency than the neutrons scattered elastically from the sample at $\lambda_{\rm i}$. Due to these effects the scattering of water is not totally isotropic. The assumption that those neutrons not being transmitted are scattered uniformly in the full solid angle of 4π steradian [see Eq. (50)] is therefore not valid (May et al., 1982).

The direct use of Eq. (37) for determining the differential scattering cross-section of water from the measured intensity $I_{\rm H_2O}$ at $Q \to 0$ and from the instrumental constant $C(\lambda_{\rm i})$ (as obtained from a measurement of the attenuated direct beam, see Section 4.1)

$$\left(\frac{{\rm d}\Sigma}{{\rm d}\Omega}\right)_{\rm H_2O,\,app} = I_{\rm H_2O}/[C(\lambda_{\rm i})dT] \tag{51}$$

is not recommended as well (Lindner, 1998c). This apparent value does not either represent the 'true' differential scattering cross-section of water because of the inelastic effects mentioned above (wavelength change due to thermalization): the scattered intensity of the water sample,

I_{H_2O}, is recorded with wavelengths different than the intensity of the attenuated direct beam measurement (at λ_i), from which $C(\lambda_i)$ is calculated.

Consequently, water can only be used as *secondary* standard, after cross-calibration against a primary standard, such as a particle (e.g. a polymer) with known molecular mass (see Section 4.2.3). By introducing an empirical wavelength-dependent correction factor $g(\lambda)$ (Jacrot and Zaccai, 1981; Ragnetti et al., 1985; Wignall et al., 1987, Cotton, 1991; Lindner, 2000) we can write for the *effective* differential scattering cross-section of water with thickness d_{H_2O}:

$$\left(\frac{d\Sigma}{d\Omega}\right)_{H_2O,\mathrm{eff}} = g(\lambda)(1-T)/(4\pi dT) \tag{52}$$

A method for calibration of the water sample against polymer standards (known molecular mass M_w^{real}) consists of comparing M_w^{real} with the molecular mass M_w^{exp}, as obtained from a SANS experiment:

$$\left(\frac{d\Sigma}{d\Omega}\right)_{H_2O,\mathrm{eff}} = \frac{(d\Sigma/d\Omega)_{\mathrm{polymer}}}{(I_{\mathrm{polymer}}/I_{H_2O})} = \frac{M_w^{\mathrm{real}}}{M_w^{\mathrm{exp}}} \quad (\text{for } Q \to 0) \tag{53}$$

with I_{polymer} and I_{H_2O} being the background corrected intensities of the polymer sample and the water, respectively [cf. Eqs. (27) and (35)].

It must be stressed that values for $g(\lambda)$ and $(d\Sigma/d\Omega)_{H_2O,\mathrm{eff}}$ following from cross-calibration [Eq. (53)] are only valid under defined *conditions* (i.e. those of the calibration against a primary standards with typical water thickness $d_{H_2O} = 0.1$ cm and at a temperature of $T = 25°\mathrm{C}$). Furthermore, $g(\lambda)$ and $(d\Sigma/d\Omega)_{H_2O,\mathrm{eff}}$ are instrument specific quantities and valid only for the *instrument* at which the cross-calibration has been performed. The reason for this is the characteristic detector efficiency of each instrument (given for SANS gas-detectors by the type of detection gas and pressure).

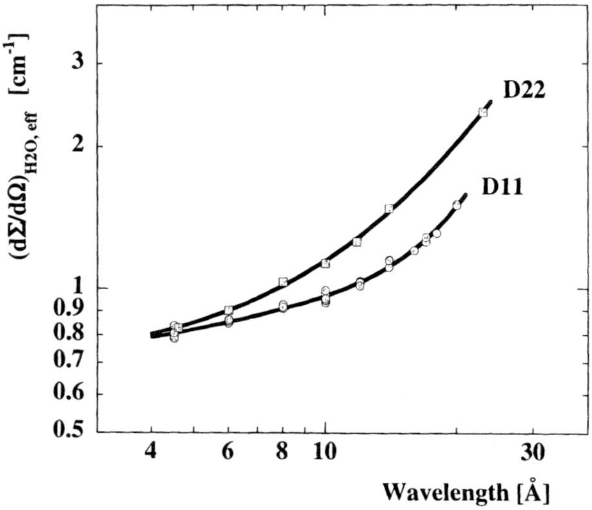

Fig. 11. Effective differential scattering cross-section of water $(d\Sigma/d\Omega)_{H_2O,\mathrm{eff}}$ from cross-calibration with polymer standards as measured with SANS at D11 and D22 at ILL (Lindner, 2000).

Fig. 11 shows the result of water calibration measurements with polymer standard samples, performed at the SANS instruments of the ILL, Grenoble (Lindner, 2000). The two instruments D11 and D22 have two different detectors (D11, with a BF_3 detector in use until May 1999 at BF_3 gas pressure 1.2 bar and D22, equipped since 1995 with a ^{3}He detector at ^{3}He gas pressure 1.85 bar). It can be clearly seen from Fig. 11 that the effective differential scattering cross-section of water is considerably different at different instruments.

5. Summary

When investigating a sample's structure by scattering methods the first important step is to evaluate the differential scattering cross-section, $d\Sigma/d\Omega$, prior to any subsequent data interpretation. Precise stages before and during the data acquisition, as well as during the primary data reduction are absolutely essential for obtaining results of good quality. Some experimental aspects have been discussed in the preceding sections with a slight bias towards elastic neutron scattering. However, the principles are valid for X-rays and light as well and they can be summarized as follows:

- One has to assume a calibrated instrument, i.e. the detector distance L and the incident wavelength λ (and the wavelength distribution $\Delta\lambda/\lambda$) are exactly known to the experimentalist.
- The experimentalist has to make a reasonable choice of the experimental configuration (λ, L) in order to perform the experiment in the range of momentum transfer Q which is appropriate to the size of the scattering elements within the sample.
- A precise knowledge of the thickness d of the sample is essential for data evaluation.
- Depending on whether the sample is in the bulk state or a liquid, the (external) background scattering has to be considered, i.e. the scattering of the empty container for the latter case.
- A reference sample, for instance the solvent for a solution of solute particles, has always to be measured and subtracted as (internal) background.
- The intensity is attenuated within the sample and careful transmission measurements are required in order to subtract the background correctly.
- Scattering curves measured at different configurations (e.g. large and small detector distances, different wavelengths) have to be matched together correctly for a final analysis.
- Some inherent problems are related to scattering experiments and need attention, for instance, multiple scattering processes, inelastic effects and instrumental resolution effects.
- Data reduction has to take into consideration the geometrically inhomogeneous detector efficiency when using area detectors, dead time losses and statistical accuracy.
- The evaluation of the sample's differential scattering cross-section in units of (cm^{-1}) requires absolute calibration of the observed scattering intensity with a calibration standard.

References

Alexander, L.E. (1985). *X-ray Diffraction Methods in Polymer Science*, Malabar: Robert E. Krieger.

Baruchel, J., Hodeau, J.L., Lehmann, M., Regnard, J.R. and Schlenker, C. (1993). *Neutron and Synchrotron Radiation for Condensed Matter Studies*. Berlin: Springer.

Berk, N.F. and Hardman Rhyne, K.A. (1988). *J. Appl. Cryst.* **21**, 645–651.

Berne, B.J. and Pecora, R. (1976). *Dynamic Light Scattering*. New York: Wiley.

Brown, W. (1996). *Light Scattering: Principles and Development*. Oxford: Clarendon Press.

Brumberger, H. (1995). *Modern Aspects of Small Angle Scattering*. Dordrecht: Kluwer Academic Publishers.
Cotton, J.P. (1991). In *Neutron, X-ray and light scattering – Introduction to an investigative tool for colloidal and polymeric systems*, edited by P. Lindner and Th. Zemb. Amsterdam: North Holland.
Damay, P., Leclercq, F., Magli, R., Formisano, F. and Lindner, P. (1998). *Phys. Rev. B* **58**, 12038–12043.
Dewhurst, C. (2002). http://www.ill.fr/LSS/Grasp/
Feigin, L.A. and Svergun, D.I. (1987). *Structure Analysis by Small Angle X-Ray and Neutron Scattering*. New York: Plenum Press.
Gawrisch, W., Brereton, M.G. and Fischer, E.W. (1981). *Polym. Bull.* **4**, 687.
Ghosh, R.E. and Rennie, A.R. (1999). *J. Appl. Cryst.* **32**, 1157.
Ghosh, R.E., Egelhaaf, S.U. and Rennie, A.R. (2000). *ILL Technical Report* ILL98GH14T. A computing guide for small angle scattering experiments.
Glatter, O. and Kratky, O. (1982). *Small Angle X-ray Scattering*. London: Academic Press.
Goyal, P.S., King, J.S. and Summerfield, G.C. (1983). *Polymer* **24**, 131.
Granada, J.R., Gillette, V.H. and Mayer, R.E. (1987). *Phys. Rev. A* **36**, 5594–5605.
Guinier, A. and Fournet, G. (1955). *Small Angle Scattering of X-rays*, New York: Wiley Interscience.
Higgins, J.S. and Benoît, H.C. (1994). *Polymers and Neutron Scattering*. Oxford: Clarendon Press.
Jacrot, B. and Zaccai, G. (1981). *Biopolymers* **20**, 2413–2426.
Koester, L. and Rauch, H. (1981). *Summary of Neutron Scattering Lengths* IAEA Contract 2517/RB.
Lindner, P. and Hess, S. (1989). *Physica B* **156&157**, 512–514.
Lindner, P. (1998a). Deadtime of the D11-BF_3 detector, Technical Report ILL98/LI12T.
Lindner, P. (1998b). Atttenuation factors at D11, Technical Report ILL98/LI13T.
Lindner P. (1998c). Standard water calibration at D11, Technical Report ILL98/LI10T.
Lindner, P. (2000). *J. Appl. Cryst.* **33**, 807–811. Water calibration at D11 verified with polymer samples.
Lindner, P., Leclercq, F. and Damay, P. (2000). *Physica B* **291**, 152–158.
Marshall, W. and Lovesey, S.W. (1971). *Theory of Thermal Neutron Scattering*. Cambridge: Oxford Clarendon.
May, R.P., Ibel, K. and Haas, J. (1982). *J. Appl. Cryst.* **15**, 15.
Mildner, D.R.F. (1983). *Macromolecules* **16**, 1760–1763.
Radlinski, A.P., Radlinska, A.Z., Agamalian, M., Wignall, G.D., Lindner, P. and Randl, O. (1999). *Phys. Rev. Lett.* **82**, 3078–3081.
Radlinski, A.P., et al. (2000a). unpublished results.
Radlinski, A.P., Boreham, C.J., Lindner, P., Randl, O., Wignall, G.D., Hinde, A. and Hope, J.M. (2000b). *Organic Geochemistry* **31**, 1–14.
Ragnetti, M., Geiser, D., Höcker, H. and Oberthür, R. (1985). *Makromol. Chem.* **186**, 1701.
Rawiso, M., Duplessix, R. and Picot, C. (1987). *Macromolecules* **20**, 630–648.
Reynolds, L.E. and Mildner, D.R.F. (1984). *J. Appl. Cryst.* **17**, 411–416.
Wignall, G.D. and Bates, F. (1987). *J. Appl. Cryst.* **20**, 28–40.

Neutrons, X-Rays and Light
P. Lindner and Th. Zemb (Editors)

Chapter 3

General Theorems in Small-Angle Scattering

Olivier Spalla

CEA Saclay, Groupe de Chimie de La Matière Ultradivisée, Service de Chimie Moléculaire, F-91191 Gif-sur-Yvette CEDEX, France

1. Introduction

The use of scattering techniques (light, neutrons and X-rays) to probe the structure of materials at the mesoscopic (1 nm to 1 μm) scale is very popular. This is due to different properties, an important one being that these techniques are non-invasive. A second advantage is that they yield physical quantities averaged over the whole sample. The last and most important attraction is that these quantities (such as volume or specific surface) can be extracted with almost no approximation or model. This is due to powerful and general theorems that will be presented in this lecture. Analysis beyond these integral theorems, i.e. the complete fitting of scattering diagrams, constitutes the every day life of many researchers using thermodynamic and geometrical models and is far beyond the scope of this text.

2. Definition of the absolute scattered intensity

When a coherent beam (either light, neutrons or X-rays) of flux N_0 (counts/s) illuminates a sample of volume V and thickness e_s, a given fraction of the flux ΔN is elastically scattered in the direction $\boldsymbol{u}$ within a solid angle $\Delta\Omega$:

$$\Delta N = N_0 T \frac{e_s}{V}\frac{\mathrm{d}\sigma}{\mathrm{d}\Omega}(\boldsymbol{u})\Delta\Omega \tag{1}$$

T being the transmission of the sample, and $(\mathrm{d}\sigma/\mathrm{d}\Omega)(\boldsymbol{u})$ the differential scattering cross-section which is characteristic of the elastic interaction of the material with the beam. The intensity I scattered by a unit volume can then be defined by:

$$I\ (m^{-1}) = \frac{1}{V}\frac{\mathrm{d}\sigma}{\mathrm{d}\Omega}(\boldsymbol{u}) = \frac{\Delta N}{N_0}\frac{1}{T e_s \Delta\Omega} \tag{2}$$

This quantity is accessible experimentally providing the thickness of the sample e_s and its transmission T are measured, together with the scattering properties $(\Delta N/N_0)(1/\Delta\Omega)$.

The experimental set-up can be calibrated when the geometry of the experiment is well defined ($\Delta\Omega$) and the direct beam measured with the same efficiency than the scattered beam $\Delta N/N_0$, which imposes to develop experimental tricks as the scattered beam is in general much weaker than the direct beam. Another solution is to use secondary standards. For X-rays, pure water is an common choice with an absolute value of 0.016 cm^{-1}.

3. General expression of the scattered intensity

The scattering intensity I (m^{-1}) is directly related to the structure of the sample as it is shown below. The local interaction with the beam depends on the kind of beam and element i illuminated. For any couple, it can be characterised by a scattering length b_i, and a density of scattering length can thus be written $\rho(\boldsymbol{r}) = \sum \rho_i(\boldsymbol{r}) b_i$, with $\rho_i(\boldsymbol{r})$ (m^{-3}) being the local density of scatterers of type i. In the case of X-rays, the photons can interact with every electron present in the sample. For the associated electric field, the scattering length is the Thomson scattering length $b = e^2/(4\pi\varepsilon_0 mc^2)$ equal to 0.282×10^{-14} m. Therefore, in X-ray scattering, a material can simply be described by its local density of electrons ρ_e. When the energy of the photons is lower (light scattering experiment), the photons are scattered only by the outer part of the electronic cloud of an atom and the scattering length density is then proportional to the polarisability of the material. It is worth noting at this point, that the scattering length density of a given material (far from an absorption edge) is therefore higher for X-ray than for light. In neutron scattering, the interaction is nuclear and the value of ρ depends on the nature of the nuclei of atoms only.

For most cases, a common approximation is to consider that the interaction with a scatterer at point the r does not depend on the scattering by others scatterers (which is called the Born approximation in X-rays and Rayleigh approximation in light). This approximation is in general valid for X-rays and neutrons scattering even when the density of scatterers is large. On the other hand, in the case of light scattering, the range of validity of the Rayleigh approximation is reduced to very dilute systems containing structures small compared the wavelength. Nevertheless, within the framework of the approximation, the local scattered amplitudes simply add (Guinier and Fournet, 1955; Porod, 1982):

$$A(\boldsymbol{q}) = \int_V \rho(\boldsymbol{r})\, \mathrm{e}^{-i\boldsymbol{q}\cdot\boldsymbol{r}}\, \mathrm{d}\boldsymbol{r} \tag{3}$$

where the amplitude of the scattering vector $\boldsymbol{q} = \boldsymbol{k}_d - \boldsymbol{k}_i$ ($\boldsymbol{k}_i$ and $\boldsymbol{k}_d$ being the wave vectors of the incident and scattered beam) is related to the angle of scattering θ by $q = (4\pi n/\lambda)\sin(\theta/2)$.

The scattered intensity per unit volume V is then expressed as:

$$I(\boldsymbol{q}) = \frac{A(\boldsymbol{q})A^*(\boldsymbol{q})}{V} \tag{4}$$

The above expression does not mean that an average is made on time or over the different configurations of the sample; it simply expresses the scattered intensity per unit volume of a sample represented in space by the distribution of scattering lengths density $\rho(\boldsymbol{r})$. Using Eq. (3):

$$I(\boldsymbol{q}) = \frac{1}{V} \int_V\int_V \rho(\boldsymbol{r})\rho(\boldsymbol{r}')\, \mathrm{e}^{-i\boldsymbol{q}\cdot(\boldsymbol{r}-\boldsymbol{r}')}\, \mathrm{d}\boldsymbol{r}\, \mathrm{d}\boldsymbol{r}' \tag{5}$$

The correlation function introduced by Debye and Bueche (1949) $\gamma(\boldsymbol{r})$ can be defined as follows:

$$\gamma(\boldsymbol{r}) = \frac{1}{V} \int_V \rho(\boldsymbol{r}')\rho(\boldsymbol{r}+\boldsymbol{r}')\, \mathrm{d}\boldsymbol{r}' = \frac{1}{V} P(\boldsymbol{r}) \tag{6}$$

$P(\boldsymbol{r})$ is classically called the Patterson function by the crystallographic community. Therefore, Eq. 5 becomes:

$$I(\boldsymbol{q}) = \int_V \gamma(\boldsymbol{r})\, \mathrm{e}^{-i\boldsymbol{q}\cdot\boldsymbol{r}}\, \mathrm{d}\boldsymbol{r} \tag{7}$$

which shows that the scattered intensity is the Fourier transform of the spatial correlation function of the scattered length density.

The correlation function presents the following limits

$$\gamma(\boldsymbol{r}=0) = \langle\rho^2\rangle \qquad \text{and} \qquad \gamma(\boldsymbol{r}=\infty) = \langle\rho\rangle^2 \tag{8}$$

where the bracket symbol means a spatial averaging. In order to use formulas with dimensionless expressions, the local fluctuation of scattering length density $\eta(\boldsymbol{r})$ can be introduced:

$$\eta(\boldsymbol{r}) = \rho(\boldsymbol{r}) - \langle\rho\rangle \tag{9}$$

Thus, one easily gets

$$\langle\eta^2\rangle = \langle\rho^2\rangle - \langle\rho\rangle^2 \tag{10}$$

Introducing the normalised correlation function $\gamma_0(\boldsymbol{r})$

$$\langle\eta^2\rangle\gamma_0(\boldsymbol{r}) = \frac{1}{V}\int_V \eta(\boldsymbol{r}')\eta(\boldsymbol{r}+\boldsymbol{r}')\, \mathrm{d}\boldsymbol{r}' \tag{11}$$

one arrives at:

$$\gamma_0(\boldsymbol{r}) = \frac{\gamma(\boldsymbol{r}) - \langle\rho\rangle^2}{\langle\eta^2\rangle} \tag{12}$$

that now varies from 1 at $\boldsymbol{r} = \boldsymbol{0}$ to 0 at $\boldsymbol{r} = \infty$. At this point, it is useful to introduce a geometrical meaning of the correlation function $\gamma_0(\boldsymbol{r})$. For an isotropic two level system ($\rho_1 = 1$, $\rho_2 = 0$), $\gamma_0(\boldsymbol{r})$ is the normalized excess probability (as compared to an infinite separation r) $p(r)$, averaged over all points of the scattering structure and over all orientations of the vector $\boldsymbol{r}$, that if one point is in the scattering structure, a second point displaced by $\boldsymbol{r}$ from the first point is also in the scattering structure. This point will be very useful when introduced the large angle behaviour of the scattered intensity. The scattered intensity per unit volume has now the general form:

$$I(\boldsymbol{q}) = \langle\eta^2\rangle \int_V \gamma_0(\boldsymbol{r})\, \mathrm{e}^{-i\boldsymbol{q}\cdot\boldsymbol{r}}\, \mathrm{d}\boldsymbol{r} + \langle\rho\rangle^2\delta(\boldsymbol{q}) \tag{13}$$

The second term simply expresses that the mean density of scatterers produces a signal only in the forward direction at $q = 0$, and therefore that only the fluctuations of density contribute to a measurable signal. The general expression (13) does not assume that the system is made of particles nor that it is composed of only two phases ρ_1 and ρ_2 nor that it is isotropic.

Using the inverse Fourier transform:

$$\langle\eta^2\rangle\gamma_0(\boldsymbol{r}) = \frac{1}{(2\pi)^3}\left[\int [I(\boldsymbol{q}) - \langle\rho\rangle^2\delta(\boldsymbol{q})]\, \mathrm{e}^{i\boldsymbol{q}\cdot\boldsymbol{r}}\, \mathrm{d}\boldsymbol{q}\right] \tag{14}$$

For $r = 0$, one gets a very useful relation:

$$\int \left[I(\boldsymbol{q}) - \langle \rho \rangle^2 \delta(\boldsymbol{q}) \right] \mathrm{d}\boldsymbol{q} = (2\pi)^3 \langle \eta^2 \rangle \tag{15a}$$

One can subtract the intensity due to the average density of scatterers:

$$I_m(\boldsymbol{q}) = I(\boldsymbol{q}) - \langle \rho \rangle^2 \delta(\boldsymbol{q}) \tag{15b}$$

Experimentally, when the illuminated volume V is of finite size, the difference between I and I_m would be observable only at $q < (1/V)^{1/3}$, which means that for volume of a few millimeters it is not accessible on a standard instrument. It must be stressed that Eq. (15b) does not mean that $I_m = 0$ at $q = 0$, as will be discussed later. The two main equations can now be rewritten in their final form:

$$\boxed{\int I_m(\boldsymbol{q}) \, \mathrm{d}\boldsymbol{q} = (2\pi)^3 \langle \eta^2 \rangle} \tag{15c}$$

and

$$\boxed{I_m(\boldsymbol{q}) = \langle \eta^2 \rangle \int_V \gamma_0(\boldsymbol{r}) \, \mathrm{e}^{-i\boldsymbol{q}\cdot\boldsymbol{r}} \, \mathrm{d}\boldsymbol{r}} \tag{16}$$

Relation (15c) which links the integral of the intensity over the whole reciprocal space to the average of the square of the density fluctuation can be now examined in detail for several simplified situations.

3.1. Case of an isotropic system

When the system is isotropic, the normalised correlation function $\gamma_0(\boldsymbol{r})$ depends only on the amplitude of $\boldsymbol{r}$. Therefore, the scattered intensity will only depends on the amplitude of $\boldsymbol{q}$. Thus, the Eqs. (16) and (15c), yield to

$$I_m(q) = 4\pi \langle \eta^2 \rangle \int_0^\infty r^2 \gamma_0(r) \frac{\sin(qr)}{qr} \, \mathrm{d}r \tag{17}$$

and

$$\int_0^\infty I_m(q) q^2 \, \mathrm{d}q = 2\pi^2 \langle \eta^2 \rangle \tag{18}$$

3.2. Two-level system

When the system is composed of two phases only, one of density ρ_1 and volume fraction Φ, and the other of density ρ_2 and volume fraction $1 - \Phi$, important results can be obtained.

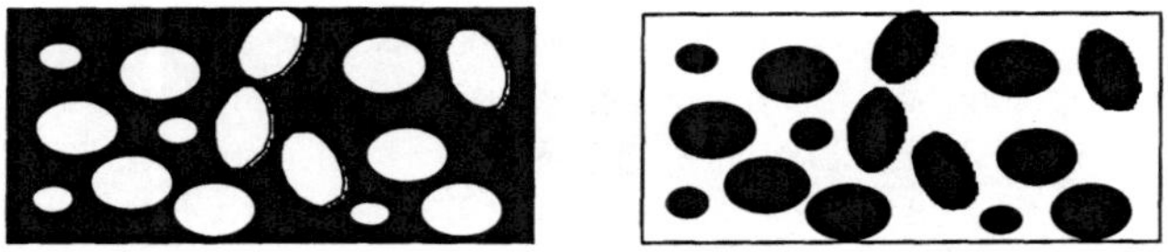

Fig. 1. (a)"White" holes in "black" matrix; (b) "Black" grains in a" white" solvent.

The first one establishes a complete symmetry between the two phases. Indeed, the amplitude is the sum of two terms:

$$A(\boldsymbol{q}) = \int_{\Phi V} \rho_1 \, \mathrm{e}^{-i\boldsymbol{q}\cdot\boldsymbol{r}} \, \mathrm{d}\boldsymbol{r} + \int_{(1-\Phi)V} \rho_2 \, \mathrm{e}^{-i\boldsymbol{q}\cdot\boldsymbol{r}} \, \mathrm{d}\boldsymbol{r} \tag{19}$$

that can be rearranged in:

$$A(\boldsymbol{q}) = \int_{\Phi V} (\rho_1 - \rho_2) \, \mathrm{e}^{-i\boldsymbol{q}\cdot\boldsymbol{r}} \, \mathrm{d}\boldsymbol{r} + \rho_2 \int_{V} \mathrm{e}^{-i\boldsymbol{q}\cdot\boldsymbol{r}} \, \mathrm{d}\boldsymbol{r} \tag{20}$$

and finally

$$A(\boldsymbol{q}) = \int_{V} \Delta\rho \, \mathrm{e}^{-i\boldsymbol{q}\cdot\boldsymbol{r}} \, \mathrm{d}\boldsymbol{r} + \rho_2 \delta(\boldsymbol{q}) \tag{21}$$

$\Delta\rho$ being equal to $\rho_1 - \rho_2$ in Φ and zero elsewhere. Using Eqs. (4) and (21) it appears that for $\boldsymbol{q} \neq \boldsymbol{0}$, the absolute scattered intensity only depends on the contrast between the two phases. This means in particular, that the two complementary structures of Fig. 1 produce exactly the same scattering. This is known as the "Babinet" Principle in optics.

The identical scattering by the two structures implies that when there are only two media the absolute sign of curvature of an interface cannot been measured by scattering techniques. Indeed, even if we precise that white is water and black is oxide, the scattering by the two media (a) and (b) remains identical meanwhile the curvature of the oxide-water interface (defined positive when concave) is changing sign from (a) to (b).

A second important theorem can be deduced concerning the volume fraction Φ. Indeed, the average square fluctuation can be easily calculated:

$$\langle \eta^2 \rangle = \Phi(1-\Phi)(\Delta\rho)^2 \tag{22}$$

yielding trough Eqs. (16) and (15c) to

$$I_m(q) = 4\pi\,\Phi(1-\Phi)(\Delta\rho)^2 \int_0^\infty r^2 \gamma_0(r) \frac{\sin(qr)}{qr} \, \mathrm{d}r \tag{23}$$

$$\boxed{Q = \int_0^\infty I_m(q) q^2 \, \mathrm{d}q = 2\pi^2 \Phi(1-\Phi)(\Delta\rho)^2} \tag{24}$$

This last integral Q is called the "invariant" (Porod, 1982) because its value does not depend on the structure of the two-phases medium but only on the volume fraction Φ and the contrast $\Delta\rho$ between the two phases. Nevertheless, the application of Eq. (24) to an experimental scattering diagram has to be done with care. Indeed, the behaviour at low and high q has to be examined in detail. The "invariant" is not measurable on an experimental diagram when (i) the intensity strongly increases at low q as it is for aggregated systems, (ii) the intensity decreases too slowly at large q (less than q^{-3}) as it is for polymeric systems.

Working at a scaled intensity is particularly interesting at this stage. Indeed, when the contrast (obtained from chemical composition of the two phases) is known an inner check of accuracy can be done between the volume fraction involved in the experiment and the invariant measure. Reciprocally, the experimental measure of the invariant allows determination of Φ, if $\Delta\rho$ is known, for example in the case of surfactants with high Critical Micellar Concentration (Warr et al., 1990).

More ambitious goals can be achieved only when every step of the experiment is closely controlled. For instance the absolute scattering by a porous glass monolith (Levitz et al., 1991) (Vycor) of thickness 3 mm has been measured on BM2 in the ESRF and is reported in Fig. 2. The invariant can be extracted (taking into account a Porod correcting factor at large angles). Taking the glass electron density equal to $\rho_e = 0.7\ \mathrm{e}/\text{Å}^3$, this leads to a scattering length density of $\rho = 2 \times 10^{11}\ \mathrm{cm}^{-2}$. Finally, the air-porosity Φ in monolith medium can be extracted and was found equal to 0.34. This value is in agreement with independent BET measurement (P. Levitz, personal communication).

4. Expressions for systems made of particles

Particulate systems are ubiquitous in chemical industry and biology. Examples of these colloidal systems are: precursors for ceramics, polymers solutions, lattices dispersions, micelles, proteins. They can be used diluted (proteins) or very concentrated (ceramics processing). The scattering techniques are very powerful tools to determine the average form of the particles and their interaction.

4.1. Form factor of a particle

When the suspension is diluted, the particles are widely separated and their positions are not correlated (with the noticeable exception of electrostatic colloidal crystals). In that situation, the intensity scattered by the different particles can be added. Therefore, let's first consider one unique particle (spherical or not) of inhomogenous density $\rho(\boldsymbol{r})$. For a given orientation of the particle, the normalised amplitude scattered by this particle is given by:

$$a(\boldsymbol{q}) = \int_{V_{\mathrm{Part}}} \rho(\boldsymbol{r})\, \mathrm{e}^{-i\boldsymbol{q}\cdot\boldsymbol{r}}\, \mathrm{d}\boldsymbol{r} = f(\boldsymbol{q}) \tag{25}$$

yielding

$$I_{\mathrm{Part}}(\boldsymbol{q}) = a(\boldsymbol{q})a^*(\boldsymbol{q}) = V_{\mathrm{Part}}^2 P(\boldsymbol{q}) \tag{26}$$

with

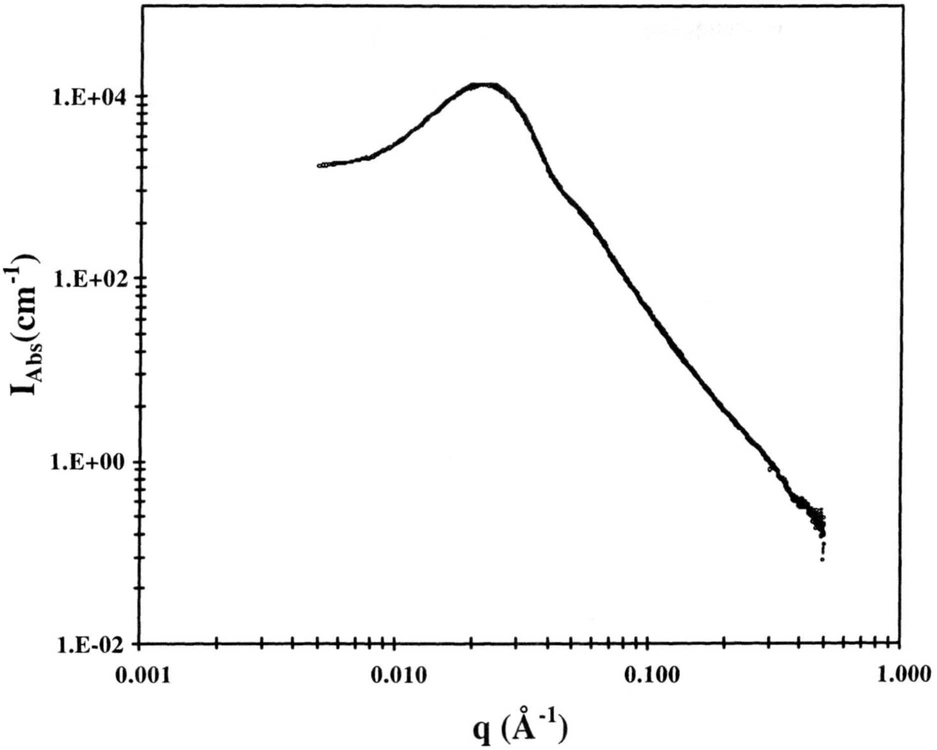

Fig. 2. Scattering diagram in log-log representation of a porous Vycor monolith (3 mm thick). At this stage, two characteristics can be extracted i.e. the correlation length by the peak $\zeta = 2\pi/Q_{\text{Peak}} = 250$ Å and the porous volume fraction $\Phi = 0.34$.

$$P(\boldsymbol{q}) = \frac{1}{V_{\text{Part}}^2} \iint\limits_{V_{\text{Part}}} \rho(\boldsymbol{u})\rho(\boldsymbol{v})\, \mathrm{e}^{-i\boldsymbol{q}\cdot(\boldsymbol{u}-\boldsymbol{v})}\, \mathrm{d}\boldsymbol{u}\, \mathrm{d}\boldsymbol{v} \tag{27}$$

being the form factor of the particle.

The intensity per unit volume of a suspension containing N identical particles in a volume V is finally:

$$I_m(\boldsymbol{q}) = \frac{N}{V} V_{\text{Part}}^2 P(\boldsymbol{q}) = \Phi V_{\text{Part}} P(\boldsymbol{q}) \tag{28}$$

The study of the form factor is very important in many colloidal systems (such as polyelectrolyte, micelles, particles) where the objects can be anisotropic or composite (core-shell particle for instance). In general, one studies the form factor in diluted suspension and the N different particles are therefore randomly oriented. Accordingly, an experimental average is done over all the orientations of the particle and the scattered intensity is isotropic. In order to fit the scattered intensity of this assembly of disoriented anisotropic objects, one first way is to calculate the scattered intensity for a given orientation using (25) and (26) and then to average it over the different available orientations.

Nevertheless, the form factor of the particle can also be linked to the correlation function $\gamma_{\text{Part}}(\boldsymbol{r})$ of the particle defined as:

$$\gamma_{\text{Part}}(\boldsymbol{r}) = \frac{1}{V_{\text{Part}}} \int_{V_{\text{Part}}} \rho(\boldsymbol{r}')\rho(\boldsymbol{r}' + \boldsymbol{r}) \, \mathrm{d}\boldsymbol{r}' = \frac{1}{V_{\text{Part}}} P_{\text{Part}}(\boldsymbol{r}) \tag{29}$$

P_{Part} being the Patterson function of the particle. Therefore it follows

$$\boxed{P(\boldsymbol{q}) = \frac{1}{V_{\text{Part}}} \int_{V_{\text{Part}}} \gamma_{\text{Part}}(\boldsymbol{r}) \, \mathrm{e}^{-i\boldsymbol{q}\cdot\boldsymbol{r}} \, \mathrm{d}\boldsymbol{r}} \tag{30}$$

Note that the integral is this time taken only over all the relative separations $\boldsymbol{r}$ between two points in the particle. Moreover, it is obvious that $\gamma_{\text{Part}}(\boldsymbol{r})$ depends on the orientation of the particle. It can be averaged over all the possible orientations of the particle:

$$\gamma_{\text{Part}}(r) = \langle \gamma_{\text{Part}}(\boldsymbol{r}) \rangle_{\Omega} \tag{31}$$

The same can be done for the form factor of the particle

$$P(q) = \langle P(\boldsymbol{q}) \rangle_{\Omega} \tag{32}$$

This can be done by, at a fixed q and r, averaging over all the orientations of the particle.

$$P(q) = \frac{1}{V_{\text{Part}}} \int_{V_{\text{Part}}} \left(\frac{1}{4\pi} \int_{\Omega} \gamma_{\text{Part}}(\boldsymbol{r}) \, \mathrm{d}\Omega \right) \mathrm{e}^{-i\boldsymbol{q}\cdot\boldsymbol{r}} \, \mathrm{d}\boldsymbol{r} \tag{33}$$

Comparing Eqs. (33) and (32) it appears that one can link the average correlation function to the average form factor, denoting by D the maximum point–point distance in the particle.

$$P(q) = \frac{1}{V_{\text{Part}}} \int_0^D 4\pi r^2 \gamma_{\text{Part}}(r) \frac{\sin(qr)}{qr} \, \mathrm{d}r \tag{34}$$

Hence, a second way to calculate the form factor of an object is to calculate its correlation function and Fourier transform it. Nevertheless, it is often much easier and efficient to calculate the amplitude then the intensity for a given orientation and averaged over all the available orientations.

4.2. Assembly of N particles

When N identical particles scatter independently Eq. (28) can be used. The more general case where the particles are identical and correlated in space is now considered. For a given set of positions and orientations of the particles, the total scattered intensity is:

$$I(\boldsymbol{q}) = \frac{A(\boldsymbol{q})A^*(\boldsymbol{q})}{V} \tag{35}$$

When the particles are diffusing in time (brownian dynamics), the intensity is changing for each configuration and the average in time is the measured quantity. For an ergodic system, it is equal to the statistical average done below:

$$I(\boldsymbol{q}) = \left\langle \frac{A(\boldsymbol{q})A^*(\boldsymbol{q})}{V} \right\rangle = \frac{1}{V}\left\langle \left\{ \int \rho(\boldsymbol{r})\, \mathrm{e}^{-i\boldsymbol{q}\cdot\boldsymbol{r}}\, \mathrm{d}\boldsymbol{r} \right\} \left\{ \int \rho(\boldsymbol{r}')\, \mathrm{e}^{i\boldsymbol{q}\cdot\boldsymbol{r}'}\, \mathrm{d}\boldsymbol{r}' \right\} \right\rangle \tag{36}$$

where the statistical average $\langle\,\rangle$ is taken over the available positions and orientations of the particles (which are indeed limited due to the presence of the others). When the system is physically static (like a porous material where the scattering is due to pores), the average means "over a volume V" large enough to include all the configurations encountered in the material. Introducing the centre position of a particle $\boldsymbol{r}_i$ one gets $\boldsymbol{r} = \boldsymbol{r}_i + \boldsymbol{u}$ and the intensity develops into:

$$I(\boldsymbol{q}) = \frac{1}{V}\left\langle \left\{ \sum_{i=1}^{N} \mathrm{e}^{-i\boldsymbol{q}\cdot\boldsymbol{r}_i} \int_{V_{\mathrm{Part}}} \rho(\boldsymbol{u})\, \mathrm{e}^{-i\boldsymbol{q}\cdot\boldsymbol{u}}\, \mathrm{d}\boldsymbol{u} \right\} \left\{ \sum_{j=1}^{N} \mathrm{e}^{i\boldsymbol{q}\cdot\boldsymbol{r}_j} \int_{V_{Part}} \rho(\boldsymbol{v})\, \mathrm{e}^{i\boldsymbol{q}\cdot\boldsymbol{v}}\, \mathrm{d}\boldsymbol{v} \right\} \right\rangle \tag{37}$$

The intensity is therefore the average of the product :

$$I(\boldsymbol{q}) = \frac{N}{V}\left\langle \left\{ \iint_{V_{\mathrm{Part}}} \rho(\boldsymbol{u})\rho(\boldsymbol{v})\, \mathrm{e}^{-i\boldsymbol{q}\cdot(\boldsymbol{u}-\boldsymbol{v})}\, \mathrm{d}\boldsymbol{u}\mathrm{d}\boldsymbol{v} \right\} \left\{ \frac{1}{N} \sum_{i=1}^{N}\sum_{j=1}^{N} \mathrm{e}^{i\boldsymbol{q}\cdot(\boldsymbol{r}_j-\boldsymbol{r}_i)} \right\} \right\rangle \tag{38}$$

A very useful case is obtained when the particles are spherical (or parallel as in a nematic or smectic structure) with identical interactions. Indeed, the average of the product is in that case equal to the product of the average. The first term is then simply the form factor $V_{\mathrm{Part}}^2 P(\boldsymbol{q})$ defined previously.

The second term is called the structure factor $S(\boldsymbol{q})$:

$$S(\boldsymbol{q}) = 1 + \frac{1}{N}\left\langle \sum_{i=1}^{N}\sum_{j\neq i}^{N} e^{i\boldsymbol{q}\cdot(\boldsymbol{r}_j-\boldsymbol{r}_i)} \right\rangle \tag{39}$$

or when going to the continuous limit

$$S(\boldsymbol{q}) = 1 + \frac{N-1}{V} \int_V g(\boldsymbol{r})\, \mathrm{e}^{-i\boldsymbol{q}\cdot\boldsymbol{r}}\, \mathrm{d}\boldsymbol{r} \tag{40}$$

with $g(\boldsymbol{r})$ being the pair correlation function of the particles centres of mass. N being much larger than 1, $N-1$ will be replaced by N in the following. Finally, a very familiar expression is obtained (Hansen and McDonald):

$$I(\boldsymbol{q}) = \Phi V_{\mathrm{Part}} P(\boldsymbol{q}) S(\boldsymbol{q}) \tag{41}$$

For $q = 0$, Eq. (39) shows that by definition $S(0) = N$ which means that all the particles are scattering in phase, independently of knowing whether they are correlated or not. Hence, as for intensity, one introduces $S_m(q)$ defined as:

$$S_m(\boldsymbol{q}) = S(\boldsymbol{q}) - \frac{N-1}{V}\delta(\boldsymbol{q}) \tag{42}$$

or when going to the continuous limit

$$S_m(\boldsymbol{q}) = 1 + \frac{N-1}{V} \int\limits_V (g(\boldsymbol{r}) - 1)\, \mathrm{e}^{-i\boldsymbol{q}\cdot\boldsymbol{r}}\, \mathrm{d}\boldsymbol{r} \tag{43}$$

Coupling Eqs. (15b)–(41)–(42), one gets

$$I_m(\boldsymbol{q}) = \Phi V_{\mathrm{Part}} P(\boldsymbol{q}) S_m(\boldsymbol{q}) \tag{44}$$

As an exercise, it is then interesting to recalculate the invariant Q:

$$\int I_m(\boldsymbol{q})\, \mathrm{d}\boldsymbol{q} = \Phi V_{\mathrm{Part}} \int P(\boldsymbol{q}) S(\boldsymbol{q}) - \Phi V_{\mathrm{Part}} \frac{N-1}{V} \int P(\boldsymbol{q}) \delta(\boldsymbol{q})\, \mathrm{d}\boldsymbol{q} \tag{45}$$

then

$$\int I_m(\boldsymbol{q})\, \mathrm{d}\boldsymbol{q} = \int I(\boldsymbol{q})\, \mathrm{d}\boldsymbol{q} - \Phi^2 (2\pi)^3 P(\boldsymbol{0}) = (2\pi)^3 \left[\langle \rho^2 \rangle - \Phi^2 \langle \rho_{\mathrm{Part}} \rangle^2\right] \tag{46}$$

and finally

$$\int I_m(\boldsymbol{q})\, \mathrm{d}\boldsymbol{q} = (2\pi)^3 \left[\langle \rho^2 \rangle - \langle \rho \rangle^2\right] \tag{47}$$

which is equal to the Eq. (15c).

As before, $S_m(q)$ and $S(q)$ only differ for $q = 0$ and $S_m(q)$ tends to 1 for uncorrelated particles and the relation (28) is recovered. The factorisation in Eq. (44) is useful because the two terms can be calculated separately. Indeed, the form factor can be calculated from the shape of a unique particle as described in the former section. On the other hand, a theoretical structure factor $S_{\mathrm{th}}(q)$ can be calculated either using the theory of liquids or crystallography. Indeed, in the liquid state, one can calculate the structure factor between colloids when the volume fraction and the interaction between the particles are known. The calculation uses statistical mechanics, with either an integral equation or simulation (Monte-Carlo, Brownian Dynamics). Finally, the calculated intensity obtained by the product of the theoretical form factor and $S_{\mathrm{th}}(q)$ can be compared with the experimental one.

Another strategy is possible: the form factor can also be measured experimentally when the system is diluted as expressed by Eq. (28). Hence, when one gets an experimental spectrum $I(q)$ such as the one in Fig. 3, an efficient way to analyse the data is to divide the intensity by an experimental form factor measured in the dilute regime ($P(q)$ in Fig. 3). In that case, an experimental structure factor $S_{\exp}(\boldsymbol{q})$ is obtained. It can finally be compared with a calculated one, trying different pair potentials $U(\boldsymbol{r})$ to recover the right $S_{\exp}(\boldsymbol{q})$.

4.3. Case of isotropic interaction

The azimuthal average is then trivial for the structure factor:

$$S_m(q) = 1 + \frac{N}{V} \int\limits_0^\infty 4\pi r^2 \left[g(r) - 1\right] \frac{\sin(qr)}{qr}\, \mathrm{d}r \tag{48}$$

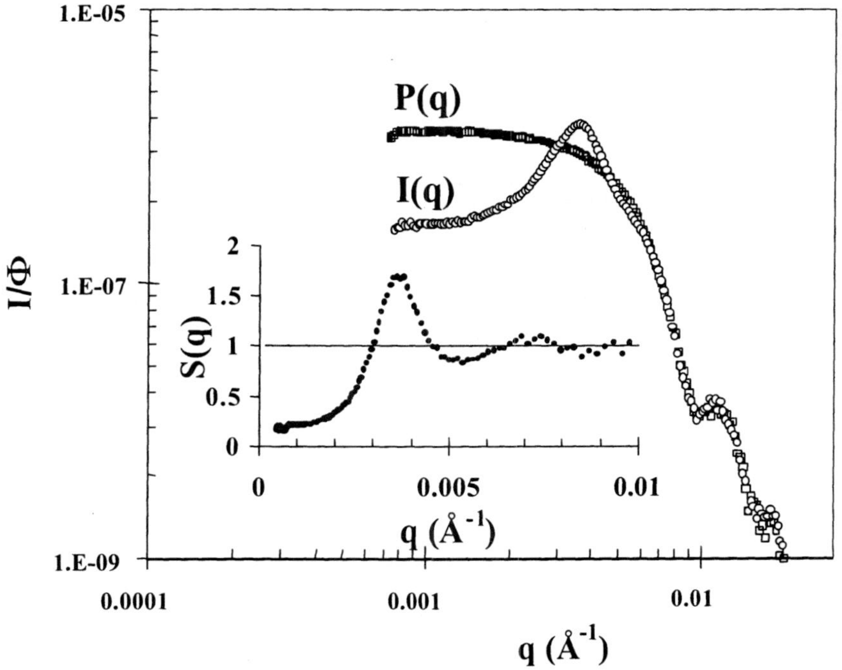

Fig. 3. Scattering diagrams by latices suspensions (per unit volume fraction but in arbitrary units). The latex radius is 45 nm. The lowest concentration ($\Phi = 1\%$ with salt) allows a measure of an experimental form factor $P(q)$, because the interactions and ordering are very weak. This is no more the case for the concentrated suspension ($\Phi = 5\%$ without salt) but a structure factor can be extracted by division.

When the particles are spherical Eq. (41) can be used. Furthermore when the particle is homogeneous, it takes the even simpler form:

$$P(q) = (\Delta\rho)^2 P_0(q) \tag{49}$$

with therefore $P_0(0) = 1$.

$$I_m(q) = \Phi V_{\text{Part}}(\Delta\rho)^2 P_0(q) S_m(q) \tag{50}$$

An interesting result can then be deduced. Indeed, for $q = 0$, $P_0(0) = 1$ and

$$I_m(0) = \Phi V_{\text{Part}}(\Delta\rho)^2 S_m(0) \tag{51}$$

and finally, using the invariant defined in Eq. (24)

$$\boxed{V_{\text{Part}} = 2\pi^2 \frac{I_m(0)}{Q} * \frac{(1-\Phi)}{S_m(0)}} \tag{52}$$

When the suspension of particles is very diluted, Φ tends to zero and $S_m(0)$ tends to one (providing the particles are not aggregated). Therefore, Eq. (52) allows a direct measure of the volume of the scattering particle. This volume is classically named the "Porod" volume. This is no more true as soon as the volume concentration is a few percent. For example, in the case of the dilute suspension of latex shown in Fig. 3, the "Porod" volume is equal to 3.7×10^5 nm^3 corresponding to a radius of 45 nm.

When the particles are polydisperse and not spherical but have all the same isotropic interaction, it can be shown from Eq. (37) that:

$$I_m(q) = \frac{N}{V}\left[\langle f^2(\boldsymbol{q})\rangle + \left(S(q) - 1\right)\langle f(\boldsymbol{q})\rangle^2\right] \tag{53}$$

$f(\boldsymbol{q})$ being defined in Eq. (25) and the average taken over the polydispersity and orientations in space. When the suspension is made of p kinds of particles of volume V_i and number N_i the application of Eq. (52) for a diluted suspension (meaning again $S_m(0) = 1$) yields an average volume:

$$\frac{2\pi^2 I_m(0)}{Q} = V_m = \frac{\sum_1^p N_i V_i^2}{\sum_1^p N_i V_i} \tag{54}$$

This average volume can be meaningless when the particles are highly polydisperse.

5. Behavior at the limit q tending to zero (forward scattering)

This limit is particularly interesting for particulate systems in the dilute regime, as will be now discussed. When q tends to 0, $P(q)$ has a general form which depends only on the radius of gyration of the particle. Indeed, a limited development of the Eq. (34) can be done:

$$P(q) = \frac{1}{V_{\text{Part}}} \int_0^D 4\pi \gamma_{\text{Part}}(r) \frac{r}{q} \left[qr - \frac{(qr)^3}{6} + \cdots\right] \mathrm{d}r \tag{55}$$

yielding

$$P(q) = \frac{1}{V_{\text{Part}}} \int_0^D 4\pi r^2 \gamma_{\text{Part}}(r)\, \mathrm{d}r \left[1 - \frac{(qR_G)^2}{3} + \cdots\right] \tag{56}$$

with

$$R_G^2 = \frac{1}{2} \frac{\int_{V_{\text{Part}}} r^2 \gamma(r)\, \mathrm{d}\boldsymbol{r}}{\int_{V_{\text{Part}}} \gamma(r)\, \mathrm{d}\boldsymbol{r}} \tag{57}$$

which is the definition of the radius of gyration of an object.

Using Eqs. (28), (56) and (57), one can show that for dilute suspension the scattered intensity decreases around the central beam as:

$$\boxed{I_m(q) \approx \Phi V_{\text{Part}} \left[1 - \frac{(qR_G)^2}{3} + \cdots\right] \approx \Phi V_{\text{Part}}\, \mathrm{e}^{-(qR_G)^2/3}} \tag{58}$$

which is called the Guinier approximation (Guinier and Fournet, 1955). As it allows to measure a geometrical characteristic of an object (its radius of gyration) only by looking at the limit behavior when q tends to zero, this is a very popular and powerful method of treating the data in small angle scattering. Nevertheless, one has always to keep in mind that it works only when the structure factor can be neglected, meaning that the dispersion has to be really diluted, in fact as much diluted as when the Porod volume is extracted. As a matter a fact, it would be absurd to try a Guinier plot on the scattering diagram of the concentrated suspension presented in Fig. 3 meanwhile it would be relevant for the scattering diagram of the diluted suspension. It is also important to note that the Guinier approximation is only valid in the so-called Guinier regime defined as $qR_G < 1$. Accordingly, using an experimental diagram on can plot $\ln(I_m)$ vs. q^2, and extract a radius of gyration from the slope $(R_G^2/3)$, but the consistency of the approximation $(qR_G < 1)$ has to be checked at the end.

6. Behavior in the large q limit

This limit is always interesting as it will be now discussed. As there is no difference between $I(q)$ and $I_m(q)$ when q is not zero, $I(q)$ will be used in the following notations.

6.1. Porod limit in a two-phase system

Let us start with the limit in the case of a dilute system of monodisperse objects of maximum size D:

$$I(q) = \frac{4\pi}{q}\Phi(\Delta\rho)^2 \int_0^D r\gamma_0(r)\sin(qr)\,\mathrm{d}r \tag{59}$$

where $\gamma_0(r) = \gamma_{\mathrm{Part}}(r)/(\Delta\rho)^2$ is the normalised correlation function of the particle. After two integrations by parts together with the conditions at limits $\gamma_0(D) = \gamma_0'(D) = 0$, we arrive at:

$$I(q) = -\frac{4\pi}{q^3}\Phi(\Delta\rho)^2 \int_0^D \sin(qr)\big[r\gamma_0(r)\big]''\,\mathrm{d}r \tag{60}$$

And a last integration by parts can be performed:

$$I(q) = \frac{4\pi}{q^4}\Phi(\Delta\rho)^2 \left\{ -2\gamma_0'(0) + \cos(qD)D\gamma_0''(D) - \int_0^D \cos(qr)\big[r\gamma_0(r)\big]'''\,\mathrm{d}r \right\} \tag{61}$$

the third term is tending to zero in the large q limit. Most of the time the second term is zero because $\gamma_0''(D)$ is zero, but this is not always true. This is false in particular for homogeneous spheres and cubes (Ciccariello, 1985).

The first term, called the "Porod" (Porod, 1951) limit, shows that for a sharp interface (in the normal direction but smooth in any parallel direction) the intensity is decreasing as q^{-4} at large q. The first derivative of the correlation function can be calculated using the development of $\gamma_0(r)$ for small r.

As stated in Eqs. (6) and (12) the normalised correlation function $\gamma_0(r)$ is proportional to the probability $p(r)$, averaged over all points of the scattering structure and over all orientations of the vector $\boldsymbol{r}$ so that if one point is in the scattering structure, a second point displaced by $\boldsymbol{r}$ from the first point is also in the scattering structure. This development is obtained as follows. When the object has a uniform density and is bounded by a smooth surface of area S, the electron density changes discontinuously at the interface. The surface can thus be considered as plane for very small r.

Let us consider the surface layer of thickness r and volume $V_L = Sr$. For every point in the volume but the layer, $V_{\text{Core}} = V - Sr$, the probability $p(r)$ is equal to one. For a point in the layer at a distance z from the surface, the probability $p(r)$ is equal to (Schmidt, 1995):

$$p(r) = \frac{z + r}{2r} \tag{62}$$

Hence, the correlation function can be developed at small r:

$$\gamma_0(r) = \frac{1}{V_L + V_{\text{Core}}}\left[\int_{V_{\text{Layer}}} p_L(r)\,\mathrm{d}z + V_{\text{Core}}\right] \tag{63}$$

which gives

$$\gamma_0(r) = 1 - \frac{S}{4V} r \tag{64}$$

Finally, combining Eqs. (61) and (64) we arrive to the Porod's law in its final form (Porod, 1951):

$$\boxed{\operatorname*{Lim}_{q\to\infty} I(q) = \frac{2\pi(\Delta\rho)^2}{q^4}\frac{S}{V}} \tag{65}$$

This allows measurement of the surface of objects. When the system is concentrated or when it is no more made of separated particles but bicontinuous, the same result holds. Indeed, it deals only with the existence of an interface. Hence one can always (at least mathematically) go far enough in q to consider a piece of surface as isolated and decorrelated from any other part of the surface. There should not be a correction concerning how the surface is distributed over the sample. From the formalistic point of view the exact demonstration can be found in different ways in many reference books, the most complete being the one of Kjems and Schofield (1985).

Finally, from the experimental point of view, when a Porod limit is observed, it helps to calculate the invariant value Q using:

$$Q = \int_0^\infty I_m(q) q^2\,\mathrm{d}q = \int_0^{q_{\exp}} I_m(q) q^2\,\mathrm{d}q + \frac{[I(q)q^4]_\infty}{q_{\max}} \tag{66}$$

providing the value $q_{\max}$ is taken in the Porod regime. On the other hand, the experimental determination of the "Porod" limit can be made easier using a plot $I(q)q^4$ versus q^4. Indeed, when a constant background A is present (due to the incoherent contribution in a neutron scattering experiment for instance), this plot yields $I(q)q^4 = B + Aq^4$ and the background A can be easily separated from the Porod limit B.

6.2. Correction due to curvature

The first correction to the Porod limit is obtained when one considers that the surface is not completely flat in the q range of the measure (qR is not large compared to one). In that case, Kirste and Porod (1962) have established that the signal is sensitive to the curvature of the interface in the following way:

$$I_{q\to\infty} = \frac{2\pi(\Delta\rho)^2}{q^4}\frac{S}{V}\left[1+\frac{1}{q^2}\left(\frac{\langle C_1+C_2\rangle^2}{4}+\frac{\langle C_1-C_2\rangle^2}{8}\right)\right] \tag{67}$$

C_1 and C_2 being the two principal curvatures of the surface. Again, it has to be noted that the signal is not sensitive to the sign of the curvature when only two media are present (which is a consequence of the "Babinet" theorem). The correction can be applied for instance to a sphere where $C_1 = C_2 = R^{-1}$.

$$I_{q\to\infty} = \frac{2\pi(\Delta\rho)^2}{q^4}\frac{S}{V}\left[1+\frac{1}{(qR)^2}\right] \tag{68}$$

which shows that the Kirste–Porod correction has nothing to do with the oscillating term in Eq. (61).

6.3. Correction to the Porod limit for thick and diffuse interface

6.3.1. Homogeneous layer of thickness d

$$Iq^4 = 2\pi\frac{S}{V}\left[\rho_1^2 + 2\rho_{\text{layer}}\left(\rho_{\text{layer}}-\rho_1\right)\left(1-\cos(qd)\right)\right] \tag{69}$$

When the layer is less dense than the core of the object, the intensity tends to be smaller at large angles. The reverse is obtained when the layer is denser than the core of the particle.

6.3.2. Continuous profile

When the profile between two media is linear with no discontinuity, the asymptotic behavior changes in nature. Indeed, there is no more a q^{-4} regime but an oscillating term decreasing like q^{-6}:

$$I = 2\pi\frac{S}{V}\left[2(\Delta\rho)^2\frac{1-\cos(qd)}{d^2q^6}\right] \tag{70}$$

This can be generalised for more complicated systems where the density of scatterers varies continuously everywhere in the sample. Again, it results that the "Porod" regime is essentially linked to the presence of smooth interfaces separating two media of different composition.

6.4. Rough surfaces

When the interface between two media is rough in the q range of observation, the scattering at large angle does not present a "Porod" limit. Let us first define a "rough" surface such as the one drawn in Fig. 4.

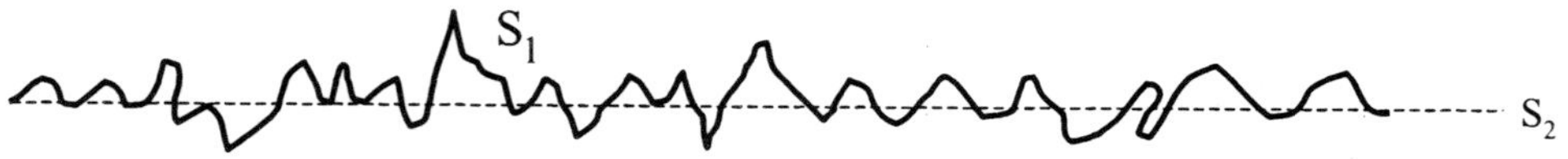

Fig. 4. Example of a "rough" surface.

At large scale an apparently smooth interface S_2 can be defined between the two media. Nevertheless, at smaller scale the interface is rough and its total extent S_1 is much greater than the "smooth" one S_2. The smallest roughness being of size ζ, a true "Porod" regime would obtained only when $q\zeta > 1$ which can be unreachable experimentally or even meaningless, when the roughness is close to the atom size. Nevertheless, one would approach at very large angle:

$$I_{q\to\infty} = \frac{2\pi(\Delta\rho)^2}{q^4}\frac{S_1}{V} \gg \frac{2\pi(\Delta\rho)^2}{q^4}\frac{S_2}{V} \tag{71}$$

which shows that at large angle the intensity must be larger than what we would expect for a smooth interface of extent S_2. Nevertheless, the full calculation (Bale and Schmidt, 1984; Wong, 1985; Sinha et al., 1988) of the scattering depends on the type of roughness and is beyond the scope of this chapter. An interesting presentation yielding the q dependence of the intensity (but not the amplitude prefactor) based on the notion of "correlation blob" introduced by T. Witten, is proposed in the lecture of L. Auvray (1991). Finally, an example of scattering by a rough surface is given in Fig. 5.

7. Application to granular and porous medium

7.1. Position of the problem

Powders formed of porous grains of micrometric size are often encountered in material science. The morphology of the grains, the total inner porosity available for ion adsorption, the size of the pores and their specific surface, have to be well characterised in view of further industrial applications. The access to the porosity through SAXS analysis requires to obtain at the absolute scale the scattering by an elementary volume of the porous grain (Schaeffer et al., 1995). In general, SAXS set-ups are designed to measure thin films of materials. Consequently, Fig. 6 shows a schematic view of a typical sample section. The granular medium does not fill the whole cell. A volume fraction (or compactness) of grains Φ^* inside the cell can be defined together with the total amount Σ_1 of surface of the envelopes of the grains. Moreover, every grain can present an inner porosity of volume fraction Φ. The true specific surface of the material is defined as σ/M where σ is the total quantity of interface and M the total mass of solid. When the grains (μm) are much bigger than the pores (nm), the amount of inner surface due to the pores is the dominating contribution to the true specific surface of the material.

The collection of data in the correct unit (cm^{-1} for the scattered intensity I) requires two mandatory characteristics of the sample, i.e. its thickness e_s and compactness Φ^*. When manipulating inorganic powders of micrometric size, this conflicts with another general requirement which is to make the sample very thin to avoid strong absorption and/or multiple scattering at small angle. For instance, when using a CuKα source (8 keV), the sample is in

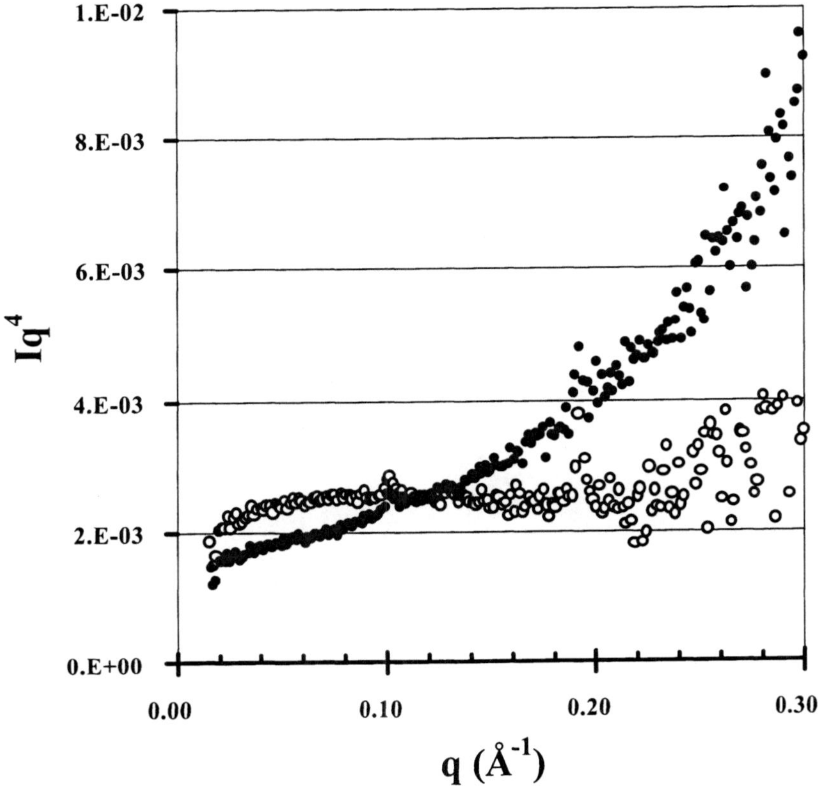

Fig. 5. Porod plot (Iq^4 vs. q) of two different porous silica. In one case, a clear plateau is obtained allowing the extraction of the specific surface (60 m^2/g). In the other case, the surface is rough and no Porod limit is accessible.

Fig. 6. Schematic drawing of a granular and porous medium.

general limited to a thin layer of grains of unknown thickness (typically 50–100 μm) and compactness. Using a Mo source (or a synchrotron) at 15 keV allows to reach a thickness of a few millimetres, but the compactness is still difficult to obtain (weighting small amount leads to strong errors). A general data analysis method has been developed in order to avoid these difficulties.

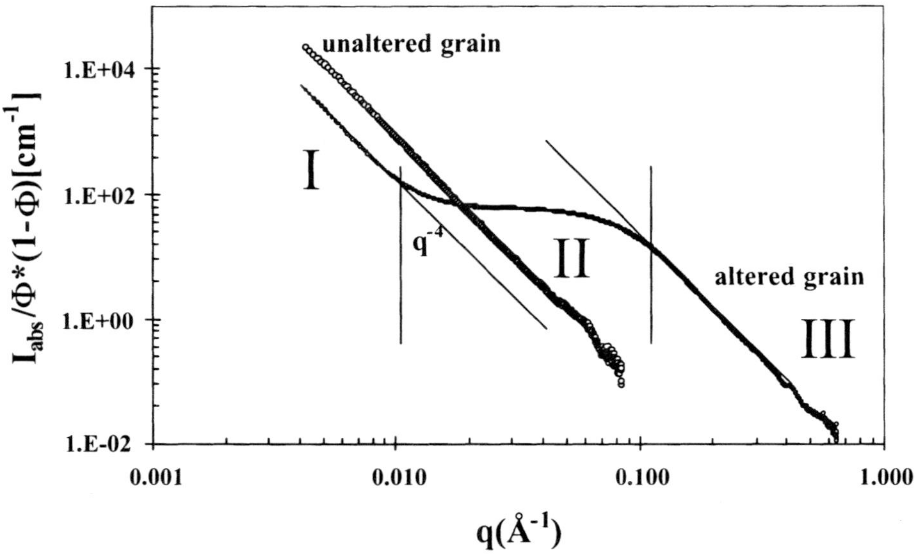

Fig. 7. Scattering diagrams of glass grains (average size 20 μm), normalised to the total solid quantity per unit volume $I_{\mathrm{abs}}/[\Phi^*(1-\Phi)]$. When the grain is unaltered, a pure Porod limit is observed, while a porous structure is revealed at large angle when the grains have been altered.

For most granular and porous systems, the two length scales: nano-pores (or particles) of few nanometers diameter on one hand, and micronic grains on the other, are largely separated. A typical signal measured is shown in Fig. 7. Three different regimes are observed on all samples: (I) a Porod's regime at low q values due to the envelope of the grains, (II) a transition regime in between, and (III) a second Porod's regime at high q values due to the inner pores of the grains.

The analyse of this type of scattering intensities to determine Σ_{I}, Φ and σ/V is based on the introduction of a so-called "apparent intensity" when the inter grain porosity $1-\Phi^*$ and/or the thickness of the sample are unknown. Due to the presence or not of solvent in the inter-grain porosity and inner pores, three different situations can be distinguished:

(i) *Solvent in the inner pores and inter-grain porosity.* When the sample can be made thick enough by dilution (for instance in a solvent containing a slight amount of xanthan gum to suspend the grains), the thickness can therefore be measured correctly. The mass fraction of grains can be measured by precisely weighting a large amount of the paste from which the sample is extracted before and after drying at 100°C. Then, knowing the density of the materials, the volume fraction of solid $\Phi_V = \Phi^*(1-\Phi)$ can be deduced and a scattered intensity per solid unit volume $I_1 = I_{\mathrm{Abs}}/\Phi_V = I_{\mathrm{Abs}}/\Phi^*(1-\Phi)$ can be obtained. Nevertheless, this works only when the "paste" is homogeneous on the size of the beam. When the incident beam is very small (a few 100 μm), another way is to deduce the volume fraction of solid Φ_V from the transmission of the sample. Indeed, the transmission of the wet paste is: $\mu_{\mathrm{Paste}} = \Phi_V \mu_{\mathrm{Solid}} + [1-\Phi_V]\mu_{\mathrm{Solvent}}$ where μ_{Solid} is calculated from the composition of the solid. In that case, μ_{Paste}, μ_{Solid} and μ_{Solvent} have to be known in order to calculate Φ_V.

(ii) *Air in the inter-grain porosity and solvent in the inner pores.* The thickness is in general not easily measured. Nevertheless, an apparent thickness e_H can be calculated from transmission T using $e_H = -\ln(T)/\mu_{\mathrm{Sample}}$, provided the absorption coefficient is correctly measured. The absorption coefficient μ_{Sample} is calculated from the composition of the powder: $\mu_{\mathrm{Sample}} = \mu_{\mathrm{Solid}}(1-\Phi) + \mu_{\mathrm{Solvent}}\Phi$. As the calculation of μ_{Sample} uses the volume fraction Φ of pores, a consistency test is obtained by comparing the assumed initial Φ used to calculate μ_{Sample} with the final determination of Φ from the invariant (see below). The procedure can be cycled until consistency is obtained. The apparent intensity that can be calculated from the experimental data is the one scattered by an equivalent compact material layer ($\Phi^* = 1$) with an inner porosity equal to Φ:

$$I_2\ (\mathrm{m}^{-1}) = \frac{\Delta N\ (\mathrm{counts/s})}{N_0 \Delta\Omega T e_H} = \frac{I_{\mathrm{abs}}}{\Phi^*} \qquad \text{with } e_H = e_s \Phi^* \tag{72}$$

(iii) *Air in the inter-grain porosity and inner pores.* The macroscopic thickness is still not accessible. Nevertheless, an apparent solid thickness e_b can be calculated from transmission. $e_b = -\ln(T)/\mu$. The absorption coefficient $\mu(\lambda)$ is calculated from the composition of the dry powder. The useful apparent intensity that can be extracted from the experimental data is therefore the same as in the case (i):

$$I_1\ (\mathrm{m}^{-1}) = \frac{C\ (\mathrm{counts/s})}{C_0 \Delta\Omega T e_b} = \frac{I_{\mathrm{abs}}}{(1-\Phi)\Phi^*} \qquad \text{with } e_b = e_s \Phi^*(1-\Phi) \tag{73}$$

7.2. Determination of the porosity Φ by the invariant Q

The analysis starts by rearranging the granular layer in a double layer one containing the material, of thickness e_H and inner porosity Φ, and one of thickness $e_s - e_H$ containing only solvent or air. During this step, the only scattering which disappears is the one which corresponds to the outer envelope of the grain, i.e., the q^{-4} regime at low q. Hence, we define the corrected intensity:

$$I_{\mathrm{Abs}}^{\mathrm{Corr}}\ (\mathrm{m}^{-1}) = I_{\mathrm{Abs}} - I_{\mathrm{Abs}}(q_*)\frac{q_*^4}{q^4} \tag{74}$$

where q_* being any value of q in the region (I). It is easy to show that the absolute intensity of the "material" layer is equal to:

$$I_{\mathrm{Abs,\,mat}}\ (\mathrm{m}^{-1}) = \frac{I_{\mathrm{Abs}}^{\mathrm{Corr}}}{\Phi^*} \tag{75}$$

Eq. (24) is applicable for the material layer and coupling with Eq. (75) yields

$$\int_0^\infty I_{\mathrm{Abs}}^{\mathrm{Corr}} q^2\,\mathrm{d}q = 2\pi^2 \Phi\Phi^*(1-\Phi)(\Delta\rho)^2 \tag{76}$$

For the three different cases it yields

(i) and (iii): $$\int_0^\infty I_1^{\mathrm{Corr}} q^2 \, \mathrm{d}q = 2\pi^2 \Phi (\Delta\rho)^2 \qquad (77)$$

(ii): $$\int_0^\infty I_2^{\mathrm{Corr}} q^2 \, \mathrm{d}q = 2\pi^2 \Phi (1-\Phi)(\Delta\rho)^2 \qquad (78)$$

We can thus access Φ directly from the experimental data (without knowing the compactness Φ^*) and the real thickness of the sample in case (ii) and (iii), provided that the electronic density of the material and solvent are known, and the transmission and the "apparent" intensities scale properly.

7.3. Determination of specific surface of pores from high-q Porod's law

In the high-q domain, the true specific surface corresponds to the interface between pores and bulk material. The specific surface Σ_2 is defined as the pore surface per cm^3 of sample. Introducing V_e as the total volume of the envelopes of the grains and σ being the total inner surface of the solid grains, one gets:

$$\Sigma_2 = \Phi^* \frac{\sigma}{V_\mathrm{e}} \qquad (79)$$

Using Eq. (65) one gets:

$$\frac{\sigma}{V_e} = \frac{\mathrm{Lim}\left[I_{\mathrm{Abs,\,mat}} q^4\right]}{2\pi(\Delta\rho)^2} = \frac{\lim_{q\to\infty}\left[I_{\mathrm{Abs}} q^4\right]}{2\pi(\Delta\rho)^2 \Phi^*} \qquad (80)$$

For the three different cases it yields

(i) and (iii): $$\frac{\sigma}{V}\,(\mathrm{m}^{-1}) = \frac{\lim_{q\to\infty}\left[I_1 q^4\right]}{2\pi(\Delta\rho)^2}(1-\Phi) \qquad (81)$$

When the specific surface is expressed in m^2/g using the density ρ_m (g/m^3) and M as the mass of the solid the relation is:

$$\frac{\sigma}{M} = \frac{\sigma}{V_\mathrm{e}} \frac{1}{\rho_m(1-\Phi)} = \frac{\lim_{q\to\infty}\left[I_1 q^4\right]}{2\pi \rho_m (\Delta\rho)^2} \qquad (82)$$

The upper equation means that the specific porous surface expressed in m^2/g can be measured independently of the specific porous volume Φ. For the second case, Eq. (65) leads to:

(ii): $$\frac{\sigma}{V_\mathrm{e}}\,(\mathrm{m}^{-1}) = \frac{\lim_{q\to\infty}\left[I_2 q^4\right]}{2\pi(\Delta\rho)^2} \qquad (83)$$

7.4. Determination of S_e/V_e from low-q Porod's law

The same method is applied to low-q Porod's regime, where the scattering arises from the contrast between air or solvent and grains of average density ρ^*:

$$\Sigma_1 = \frac{\lim_{q\to 0}\left[I_{\text{abs}} q^4\right]}{2\pi(\Delta\rho^*)^2} \tag{84}$$

If S_e is the total surface of the envelopes of the grains, $\Sigma_1 = \Phi^* \cdot S_e/V_e$ and consequently:

$$\frac{S_e}{V_e} = \frac{\lim_{q\to 0}\left[I_{\text{Abs}} q^4\right]}{2\pi\Phi^*(\Delta\rho^*)^2} \tag{85}$$

with $\Delta\rho^* = (1-\Phi)\Delta\rho$ for case (i) and (iii) and $\Delta\rho^* = (1-\Phi)\rho + \Phi\rho_s$ in case (ii). For the three different cases it yields

$$\text{(i) and (iii):} \quad \frac{S_e}{V_e}\,(\text{m}^{-1}) = \frac{\lim_{q\to 0}\left[I_1 q^4\right]}{2\pi(\Delta\rho)^2(1-\Phi)} \tag{86}$$

$$\text{(ii):} \quad \frac{S_e}{V_e}\,(\text{m}^{-1}) = \frac{\lim_{q\to 0}\left[I_2 q^4\right]}{2\pi(\Delta\rho^*)^2} \tag{87}$$

7.5. Examples of application

A first example was given in Fig. 5. Indeed, these two diagrams of scattering were obtained on powders. A thin film was deposited on a sticky kapton tape. The transmission was $T = 0.685$ for the silica powder presenting a Porod limit. Using a density of $d = 2.2 \pm 0.1$ g/cm^3 for the silica skeleton yields a true silica thickness e_b of 50 ± 2 μm. Eq. (82) allows to measure the specific surface even when the porosity Φ is unknown:

$$\frac{\sigma}{M} = \frac{\lim_{q\to\infty}\left[I_1 q^4\right]}{2\pi(\Delta\rho)^2\rho_{m,\text{SiO}_2}} \tag{88}$$

Finally, one obtains $\sigma/M = 60 \pm 8$ m^2/g, in very close agreement with BET measurement (R. Denoyel, personal communication).

Nevertheless, the inner porosity of a grain is a relevant characteristic in several fields of material science. We present below one example where the X-rays measurements have provided direct information on the system: water altered glasses (Deruelle et al., 2000). Glass grains can be altered in hot water, depending on the glass composition and the surrounding solution. This process is relevant for glass fibres which have to dissolve rapidly when accidentally inhaled. As an example, we have chosen a glass composition $SiO_2/Na_2O/B_2O_3/ZrO_2 = 70/14/14/2$ (molar fraction). For this composition, the surrounding water contains the whole B and Na after solely one month of alteration at 90°C with a surface of solid/volume of solution ratio of 20 cm^{-1}. It can be concluded that the glass is altered down to the center and is made of (Si and Zr and O) at the end of the alteration. The question is to determine the volume fraction Φ of pores and the specific surface σ/V of the final product.

It is important to work in situ in order to avoid a corruption of the altered morphology by the capillary forces. Hence, the grains (altered or not) were suspended in a dilute xanthan solution which formed a gel. The diagrams of scattering before and after alteration are shown together in Fig. 7. Before any alteration, a pure Porod regime is obtained in the whole range of q. Therefore, the specific surface of the initial powder can be determined using Eq. (86). The effective volume fraction of powder Φ_V is not well determined by weighting the sample because it can be slightly heterogeneous. It is more efficient to calculate Φ_V through the transmission $T = 0.556$ (at 17.7 keV). The thickness of the cell was 2 mm and the medium has therefore an absorption coefficient of $\mu(\text{sample}) = 2.9\ \text{cm}^{-1}$. The sample is made of water $(1 - \Phi_V)$ and glass grains (Φ_V). The dry powder has a measured density of $d = 2.68$, and its absorption coefficient at 17.7 keV is $\mu(\text{glass}) = 9.19\ \text{cm}^{-1}$. That of water being $\mu(\text{water}) = 1.16\ \text{cm}^{-1}$, the volume fraction of powder of Φ_V is found equal to 0.22. Finally, taking an electron density contrast $\Delta\rho_e$ between the glass and water of $0.77 - 0.33 = 0.44\ \text{e}/\text{Å}^3$, the specific surface of the "grains" amounts to $S/V = 0.32\ \text{m}^2/\text{g}$ or $8.65 \times 10^5\ \text{m}^{-1}$. This result is in good agreement with a BET measurement: $0.3\ \text{m}^2/\text{g}$.

The same procedure has been used for the diagram of altered grains. The transmission was equal to 0.446 yielding a sample absorption of $\mu(\text{sample}) = 4\ \text{cm}^{-1}$ and a volume fraction of $\Phi_V = 0.40$ (the absorption of the solid taking into account the leaching of some elements $\mu(\text{altered glass}) = 8.26\ \text{cm}^{-1}$). In order to calculate properly the invariant Q from the diagram, the corrected intensity $I_{\text{abs,corr}}$ has been extended by a Porod law for the q range above the maximum experimental q vector, and a constant value $I_{\min}$ below $q_{\min}$. The last unknown parameter is the electron density contrast between $\Delta\rho$ between the residual skeleton and the water in the pore. One option is to take equal to the one of the pure initial glass $(0.77 - 0.33 = 0.44\ \text{e}/\text{Å}^3)$. Using Eqs. (77), (82) and (86), this leads to

$$\rho_e = 0.768\ \text{e}/\text{Å}^3 \qquad \Phi = 0.175 \qquad \sigma/M = 119\ \text{m}^2/\text{g} \qquad S_e/V_e = 4.27 \times 10^5\ \text{m}^{-1}$$

The decrease of the value of S/V (external envelope of the grain) would mean that the size of the grain has increased by a factor more than two, which seems unrealistic. A more reasonable option is that the skeleton electron density has slightly decreased. It can be adjusted in order to recover the correct envelope specific surface:

$$\rho_e = 0.693\ \text{e}/\text{Å}^3 \qquad \Phi = 0.26\ \sigma/M = 158\ \text{m}^2/\text{g} \qquad S_e/V_e = 8.6 \times 10^5\ \text{m}^{-1}$$

The value for the skeleton remains close to the one of quartz silica (0.697). An interesting comparison can be done with the chemical titration. Indeed, it was found that all the B and Na atoms have been leached from the grain. If it is assumed (as classically done) that the volume left by their departure is dominated by the volume of the oxygen atoms bound to them in an oxide form, one arrives to a "chemical volume fraction of $\Phi_{\text{ch}} = 0.28$ close to the X-ray determination. Nevertheless, despite this nice and positive conclusion, it has to be keep in mind that the use of this method of "effective absolute intensity" require a very good knowledge of what elements are present in the beam.

References

Auvray, L. (1991). In: *Neutron, X-ray and Light Scattering* (ed. Zemb, P.L.T.) (North Holland).
Bale, H. D. and Schmidt, P.W. (1984). *Phys. Rev. Lett.*, **53**, 596.

Ciccariello, S. (1985). *Acta Cryst.*, **A41**, 560.
Debye P. and Bueche A.M. (1949). *J. Appl. Phys.*, **20**, 518.
Deruelle, O., Spalla, O., Barboux, P. and Lambard, J. (2000). *J. Non. Crys. Solids*, **261**, 237.
Guinier A. and Fournet G. (1955). *Small Angle Scattering of X-rays*. Wiley, New York.
Hansen, J.P. and McDonald, I.R. (1976). *Theory of simple liquids*. Academic Press, London.
Kirste, R. and Porod, G. (1962). *Koll. Z., Z. Polym.*, **1**.
Kjems, J.K. and Schofield, P. (1985). In: *Scaling Phenomena in Disordered Systems*. Plenum Press, New York and London, 1985, p. 141.
Levitz, P., Ehret, G., Sinha, S.K. and Drake, J.M. (1991). *J. Chem. Phys.*, **95**, 6151.
Porod, G. (1951). *Kolloid K.*, **124**, 83.
Porod, G. (1982) In: *Small Angle X-ray Scattering* (ed. Kratky, O.G.O.). Academic Press, London.
Schaeffer, D.W., Brow, R.K., Olivier, B.J., Rieker, T., Beaucage, G., Hrubesh, L. and Lin, J.S. (1995). In: *Modern Aspects of Small-Angle Scattering* (ed. Brumberger, H.). Kluwer Academic, Dordrecht.
Schmidt, P.W. (1995). In: *Modern Aspects of Small-Angle Scattering* (ed. Brumberger, H.). Kluwer Academic. Dordrecht.
Sinha, S.K., Sirota, E.B., Garoff, S. and Stanley, H.B. (1988). *Phys. Rev. B*, **38**, 2297.
Warr, G.G., Zemb, T. and Drifford, M. (1990). *J. Phys. Chem.*, **94**, 3086.
Wong, P.Z. (1985). *Phys. Rev. B*, **32**, 7417.

Neutrons, X-Rays and Light
P. Lindner and Th. Zemb (Editors)

Chapter 4

The Inverse Scattering Problem in Small-Angle Scattering

Otto Glatter

Department of Chemistry, Karl-Franzens Universität Graz, Heinrichstraße 28, A-8010 Graz, Austria

1. Introduction

The purpose of this chapter is to introduce small-angle scattering (*SAS*) as a method for investigation of soft condensed matter, complex fluids or non-periodic systems. It should create an understanding of the crucial points of this method especially for the problems of data evaluation and interpretation, i.e., the solution of the inverse scattering problem. The most important conceptions will be explained. For details and special applications one must refer to the original literature or to textbooks, i.e., the reference list is extensive but of course not complete (Guinier and Fournet, 1955; Brumberger, 1967; Glatter and Kratky, 1982; Feigin and Svergun, 1987; Brumberger, 1995).

The physical principles of scattering are the same for wide-angle diffraction and SAS. The incoming wave induces the generation of secondary waves at every scattering center which add at large distances (far field approach) to the overall scattering amplitude. All secondary waves have the same frequency but will have different phases caused by the different path lengths. Due to the high frequency it is only possible to detect the scattering intensity – the square of the scattering amplitude – and its dependence on the scattering angle. The situation is nearly the same for scattering of X-rays and neutrons. The interaction with matter is different, but the problem of interfering secondary waves is the same. Instead of the electron density for X-rays we have to deal with the *scattering length density* for neutrons. The essential fact in neutron scattering is the pronounced difference in the scattering length between hydrogen and deuterium, which is important for the variation of the contrast between the particles and the matrix. Another important feature of neutron scattering is the high incoherent scattering cross section of hydrogen. The situation would again be the same for static light scattering if the particles are small and/or have a very low refractive index ratio to the solvent. If this is not the case we have to deal with a nonlinear interaction between light and matter which can be described by the more sophisticated *Lorenz–Mie theory* (see also Chapter 8 by O. Glatter in this book). For the following we use the term density but we keep in mind that it can be electron density, scattering length density or refractive index for X-ray, neutron or light scattering, respectively.

The angle-dependent scattering amplitude is related to the electron-density distribution of the scatterer by a Fourier transformation (see also Chapter 1, *Introduction to Scattering Experiments* by P. Pusey). All this holds for both wide-angle diffraction and SAS. The main difference is that in the former we have a periodic arrangement of identical scattering centers (particles), i.e., the scattering medium is periodic in all three dimensions with a large number of repetitions, whereas in SAS these particles, for example proteins, are not ordered periodically. They are embedded with arbitrary orientation and with irregular distances in a matrix like water. The scattering centers are limited in size, non-oriented, and non-periodic, but the number of particles is high and they can be assumed to be identical like in crystallography. The Fourier transform of a periodic structure in crystallography (crystal diffraction) corresponds with a Fourier series, i.e., a periodic structure is expanded to a periodic function system.

The Fourier transform of a non-periodic limited structure (SAS) corresponds to a Fourier integral. In mathematical terms it is the expansion of a non-periodic function by a periodic function system. So the differences between crystallography and SAS are equivalent to the differences between a Fourier series and a Fourier integral. It may look foolish to expand a non-periodic function with a periodic function system, but this is how scattering works and we do not have any other powerful physical method to study these structures.

The essential effect of these differences is that in SAS we measure a continuous angle-dependent scattering intensity at discrete points instead of sharp, point-like spots as in crystallography. Another important point is that in SAS we have a linear increase of the signal (scattered intensity) with the number of particles in the measuring volume since intensities are adding due to the incoherence of the scattered waves from separated particles. Amplitudes are adding in crystallography, so we have there a quadratic relation between signal and number of particles.

In addition there is a loss of information in SAS experiments caused by the averaging over all orientations in space. The three-dimensional structure is represented by a one-dimensional function – the dependence of the scattered intensity on the scattering angle. This is also true for powder diffraction. To recover the structure uniquely is therefore impossible.

The computation of the scattering function for a known structure is called the solution of the *scattering problem*. This problem can be solved exactly or with good enough accuracy for many different structures. The inversion, i.e., the estimation of the structure of the scatterer from its scattering function is called the *inverse scattering problem*. This problem cannot be solved uniquely. The description and solution of the scattering problem gives information to the experimenter about the scattering functions to be expected in a special situation. In addition, this knowledge is the starting point for the evaluation and interpretation of experimental data (solution of the inverse problem).

There are methods that give a rough first-order approximation to the solution of the inverse scattering problem using only a minimum amount of *a priori* information about the system to get a first model. In order to improve this model one has to solve the scattering problem. The resulting theoretical model functions (see also Chapter 16 by J.S. Pedersen on *model fitting and simulation techniques*) are compared with the experimental data; if necessary, model modifications have to be deduced from the deviations. After some iterations one gets the final model. It should be noticed that it is possible to find different models that fit the data within their statistical accuracy. In order to reduce this ambiguity it is necessary to have additional independent information from other experiments. Wrong models, however, can be rejected when their scattering functions differ significantly from the experimental data.

What type of investigations can be performed with SAS? It is possible to study monodisperse and polydisperse systems. In the case of monodisperse systems it is possible to determine size, shape and under certain conditions the internal structure. Monodispersity cannot be deduced from SAS data and must therefore be assumed or checked by independent methods. For polydisperse systems a size distribution can be evaluated under the assumption of a certain shape of the particles (*particle sizing*).

All these statements are strictly true for highly diluted systems where the interparticle distances are much larger than the particle dimensions. In the case of semi-dilute systems the result of a SAS experiment is influenced by the structure of the particles and by their spatial arrangement (see also Chapter 14 by R. Klein on *interacting particles*). In the case of globular particles the scattering curve is the product of the particle scattering function (*form factor*) and of the interparticle interference function (*structure factor*). The recently developed *Generalized Indirect Fourier Transformation* (GIFT) method allows the determination of both factors directly from experimental data with a minimum of *a priori* information.

How are SAS experiments related to other scattering experiments? SAS uses radiation with a wavelength in the range of 10^{-1} to 10^{0} nm depending on the problem and on the source used. This range is similar for X-rays and neutrons, but neutrons interact with the nuclei of the atoms whereas X-rays interact with the electrons. The scattering efficiency increases linearly with the atomic number for X-rays. The dependence is much more complicated for neutrons and does not show a systematic trend. The essential fact in neutron-scattering is the pronounced difference in the scattering power between hydrogen and deuterium, which is important for the variation of the contrast between the particles and the matrix (see also Chapter 7 by P. Schurtenberger in this book).

The wavelength and the scattering efficiency limits the range of SAS experiments to systems in the size range of a few nanometers up to about one hundred nanometers. Special instruments (USAXS and USANS) permit the study of larger particles (Bonse and Hart, 1967; Koch, 1988). These instruments need a high intensity of the primary beam (synchrotron radiation or high flux reactors) and are not very common (Agamalian et al., 1999; Hainbuchner et al., 1999).

Particles in the size range from 100 nm up to some microns can be investigated by static light scattering techniques (van de Hulst, 1981; Kerker, 1969; Bohren and Huffmann, 1983; Brown, 1996). Particles exceeding this limit can be seen in an optical microscope or can be studied with Fraunhofer diffraction (Bayvel and Jones, 1981; Guesbet and Grehan, 1988).

Electron microscopy is a complementary method to all these scattering methods. It has the big advantage of giving real pictures with rather high resolution but it has the inherent disadvantage that the preparation may introduce artifacts. SAS on the other hand is a method to study macromolecules in solution which is a very important advantage for biological samples and for polymers. SAS can be used to study self-assembly and formation of aggregates of macromolecules as well as for the investigation of complex fluids. Self-diffusion NMR is an important method to study connectivity of structures (Lindman and Olsson, 1996).

Crystallography gives more information about the atomic arrangement in the particle and can be applied to relatively large systems. It is possible to study particles as large as proteins and viruses, if good crystals of these substances are available.

SAS has developed to a standard measuring method during the last decades, being most powerful for the investigation of sub-micron particles. The range of applicability of the different scattering techniques is summarized in Fig. 1. The static scattering techniques start with SAS and cover the size range between 1 nm and about 100 nm, depending on the

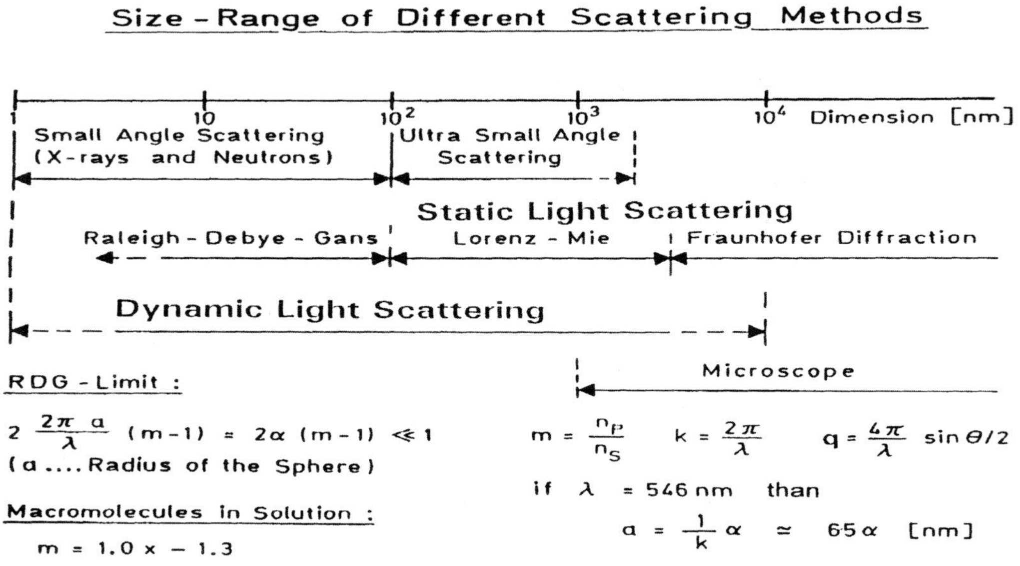

Fig. 1. Range of applicability of the scattering methods.

collimation system. Ultra-small-angle scattering with Bonse–Hart systems can be used for particles up to about one micron. In the size range from 100 nm up to some microns static light scattering can be used for the investigation of colloidal systems. Fraunhofer diffraction can be used for particle sizing in the range from a few microns up to several hundred microns. All these static scattering techniques cover a quite limited size range (up to two decades) but are high resolution techniques (Glatter et al., 1991). Dynamic light scattering (*DLS*) – or photon correlation spectroscopy (*PCS*) – can be used for particle sizing in a much wider range (about three decades) but has a lower resolution (see also Chapter 9 by P. Pusey in this book). No shape analysis is possible with this technique.

In the following we discuss the scattering problem in detail for different cases. We have to learn the scattering signature for different structures to be able to solve the inverse scattering problem. We follow the results developed in Chapter 1, using in some cases a slightly different notation.

2. Rayleigh–Debye–Gans theory

The *Rayleigh–Debye–Gans (RDG) theory* of scattering is nothing else then the rigorous application of the idea of coherent interference to scattering assuming that the incoming plane wave propagates into and through the particle regularly and homogeneously, i.e., the magnitude of the field is the same in all parts of the particle. The RDG theory describes well *small-angle X-ray scattering (SAXS)* and *small-angle neutron scattering (SANS)* as well as static light scattering *(SLS)* of small particles (size much smaller than the wavelength). The name *small angle* comes from the fact that the wavelength of X-rays and neutrons is usually much smaller than the size of the particles under investigation and so the scattered field deviates from zero essentially only at small scattering angles.

In the following we are no longer interested in the scattering of a particle in vacuum but in scattering from scattering centers embedded in a homogeneous medium like colloids or macromolecules and their aggregates in solution but also defects in glasses, pores in non-crystalline amorphous solids, i.e., the scattering centers exist in a surrounding medium of (nearly) constant electron density $\overline{\rho}$. In order to find the total scattered field we have to integrate over the whole illuminated *scattering volume* V

$$E_s(\boldsymbol{q}) = \text{const} \int_V \rho(\boldsymbol{r})\, \mathrm{e}^{-i\boldsymbol{q}\cdot\boldsymbol{r}}\, \mathrm{d}\boldsymbol{r} \tag{1}$$

(Compare with Eq. (33) in Chapter 1). We can now express the density $\rho(\boldsymbol{r})$ by its mean $\overline{\rho}$ and its fluctuations $\Delta\rho(\boldsymbol{r})$

$$\rho(\boldsymbol{r}) = \overline{\rho} + \Delta\rho(\boldsymbol{r}) \tag{2}$$

The Fourier integral in (1) is linear, so we can rewrite this equation using (2):

$$E_s(\boldsymbol{q}) = \text{const}\left[\int_V \overline{\rho}\cdot \mathrm{e}^{-i\boldsymbol{q}\cdot\boldsymbol{r}}\, \mathrm{d}\boldsymbol{r} + \int_V \Delta\rho(\boldsymbol{r})\, \mathrm{e}^{i\boldsymbol{q}\cdot\boldsymbol{r}}\, \mathrm{d}\boldsymbol{r}\right] \tag{3}$$

The scattering by the whole scattering volume V is strongly peaked at $\boldsymbol{q} = 0$ in the straight through primary beam due to the fact that the dimensions of the scattering volume are much larger than the wavelength λ. So we get for the scattering at $\boldsymbol{q} \neq 0$ only the contributions of the fluctuations $\Delta\rho(\boldsymbol{r})$ around the mean $\overline{\rho}$:

$$E_s(\boldsymbol{q}) = \text{const} \int_V \Delta\rho(\boldsymbol{r})\, \mathrm{e}^{i\boldsymbol{q}\cdot\boldsymbol{r}}\, \mathrm{d}\boldsymbol{r} \tag{4}$$

We see that the scattered field E_s is the Fourier transform of the density fluctuations $\Delta\rho$. This leads to the fact that the contrast can also be negative (whenever the density of a particle is lower than that of the solvent). Inhomogeneities of the solvent are usually very small and their contributions do hardly depend on the scattering angle and are usually taken into account experimentally by subtracting the scattering of the pure solvent from the scattering of the solution.

3. Dilute Systems

We start with the simplest possible case, a very dilute system of scatterers. This assumption will not only guarantee that the condition of single scattering is fulfilled automatically, but also leads to a situation that all scatterers are *uncorrelated*, i.e. every particle of the system can be regarded as independent in respect of translation and orientation and that there exists no long range order. This means that we assume that there exists no *interparticle interaction* at all. This condition can often be met by colloidal systems of globular particles with a low volume fraction (typically below one percent), but we have to keep in mind that high particle charges or long thin cylinders may necessitate much higher dilutions. The condition of a very dilute system leads also to the situation that $\overline{\rho}$ is equal to the density of the solvent and $\Delta\rho$ is just the difference between the density of the particle and the density of the solvent (or surrounding medium).

Monodisperse, dilute systems:

If all particles are identical we get

$$I_s(q) = N\langle|F(\boldsymbol{q})|^2\rangle = N I(q) \tag{5}$$

where we have introduced the particle scattering amplitude $F(\boldsymbol{q})$ which is the scattered field resulting from integration of Eq. (4) over the particle volume only. Now we have to have a closer look on the averaging process $\langle|F(\boldsymbol{q})|^2\rangle$ in Eq. (5).

Spatial averaging and square:

$$F(\boldsymbol{q}) = \int_V \Delta\rho(\boldsymbol{r})\,\mathrm{e}^{-i\boldsymbol{q}\cdot\boldsymbol{r}}\,\mathrm{d}\boldsymbol{r} \tag{6}$$

$$|F(\boldsymbol{q})|^2 = F(\boldsymbol{q})\cdot F^*(\boldsymbol{q}) = \iint_V \Delta\rho(\boldsymbol{r}_1)\Delta\rho(\boldsymbol{r}_2)\,\mathrm{e}^{-i\boldsymbol{q}\cdot(\boldsymbol{r}_1-\boldsymbol{r}_2)}\,\mathrm{d}\boldsymbol{r}_1\,\mathrm{d}\boldsymbol{r}_2 \tag{7}$$

We put $\boldsymbol{r}_1 - \boldsymbol{r}_2 = \boldsymbol{r}$ and use $\boldsymbol{r}_2 = \boldsymbol{r}_1 - \boldsymbol{r}$:

$$|F(\boldsymbol{q})|^2 = \iint_V \Delta\rho(\boldsymbol{r}_1)\Delta\rho(\boldsymbol{r}_1-\boldsymbol{r})\,\mathrm{e}^{-i\boldsymbol{q}\cdot\boldsymbol{r}}]\,\mathrm{d}\boldsymbol{r}_1\,\mathrm{d}\boldsymbol{r} \tag{8}$$

Now we introduce the *convolution square* $\Delta\tilde{\rho}^2(\boldsymbol{r})$ of the density fluctuations:

$$\gamma(\boldsymbol{r}) \equiv \Delta\tilde{\rho}^2(\boldsymbol{r}) = \int_V \Delta\rho(\boldsymbol{r}_1)\Delta\rho(\boldsymbol{r}_1-\boldsymbol{r})\,\mathrm{d}\boldsymbol{r}_1 \tag{9}$$

This function is also called *spatial autocorrelation function* (ACF). An excellent textbook on Fourier transformation and related topics has been written by Bracewell (1986).

Introducing Eq. (9) into (8) gives

$$|F(\boldsymbol{q})|^2 = \int_V \Delta\tilde{\rho}^2(\boldsymbol{r})\,\mathrm{e}^{-i\boldsymbol{q}\cdot\boldsymbol{r}}\,\mathrm{d}\boldsymbol{r} \tag{10}$$

So we see that the scattered intensity is essentially the Fourier transform of the spatial ACF of the density fluctuations $\Delta\rho$ of the particle.

Now we have to perform the spatial averaging process. This means that the autocorrelation function depends only on the magnitude of the distance $r = |\boldsymbol{r}|$ and the scattering intensity is a function of the magnitude q of the scattering vector $\boldsymbol{q}$. It is obvious that information about the system is lost by this average, a point that is important for the evaluation and interpretation

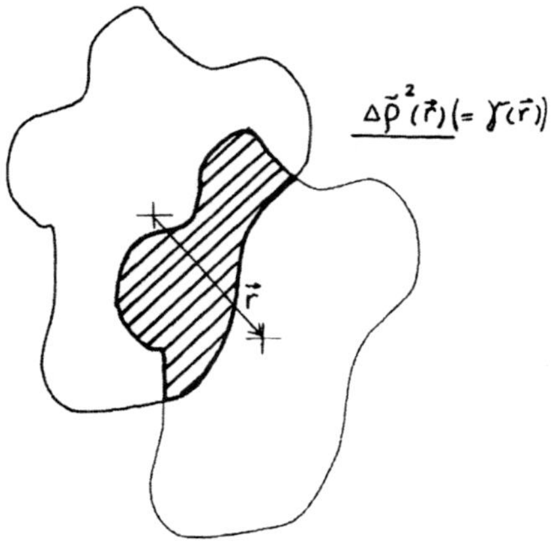

Fig. 2. The convolution square. If the electron density $\Delta\rho(\boldsymbol{r})$ is constant inside the particle $\tilde{\rho}^2(\boldsymbol{r})$ describes the overlapping volume between the particle and its ghost shifted by $\boldsymbol{r}$.

of scattering experiments. Also the three-dimensional Fourier transformation (10) reduces to the one-dimensional centro-symmetric form. The phase factor must be replaced by its average taken over all directions of $\boldsymbol{r}$:

$$\gamma(r) = \tilde{\rho}^2(r) - V(\overline{\rho})^2 = \langle \Delta\tilde{\rho}^2(\boldsymbol{r})\rangle = \left\langle \int_V \Delta\rho(\boldsymbol{r}_1)\Delta\rho(\boldsymbol{r}_1 - \boldsymbol{r})\, \mathrm{d}\boldsymbol{r}_1 \right\rangle \tag{11}$$

The convolution square is calculated as the overlap between a particle fixed in space and its identical ghost shifted by $\boldsymbol{r}$. If $\Delta\rho(\boldsymbol{r})$ is constant inside the particle $\Delta\tilde{\rho}^2$ is given by the square of this contrast $(\Delta\rho)^2$ times the overlap volume between the particle and its shifted ghost (Porod, 1948). The spatially averaged convolution square $\gamma(r) = \Delta\tilde{\rho}^2(r)$ results from the same process, the ghost is shifted by a distance $r = |\boldsymbol{r}|$, but we have to average over all possible directions in space. This correlation function $\gamma(r)$ has been introduced in the field of small-angle scattering by Debye and Bueche (1949) and in its normalized form as *characteristic function* $\gamma_0(r)$ by Porod (1951).

For homogeneous particles we find immediately that

$$\gamma(0) = (\Delta\rho)^2 \int_V \mathrm{d}\boldsymbol{r} = (\Delta\rho)^2 \cdot V \tag{12}$$

The spatial average of the phase factor in (10) can be expressed by the fundamental formula also given by Debye (1915):

$$\langle \mathrm{e}^{i\boldsymbol{q}\cdot\boldsymbol{r}} \rangle = \frac{\sin qr}{qr} \tag{13}$$

found by averaging over all possible directions of $\boldsymbol{r}$ (see also Guinier and Fournet, 1955, p. 8).

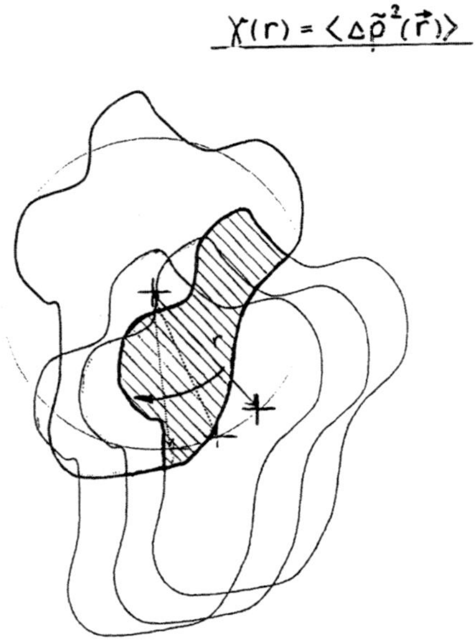

Fig. 3. Spatially averaged convolution square $\gamma(r) = \Delta\tilde{\rho}^2(r)$: here we shift the ghost again over the distance $r = |\boldsymbol{r}|$, but average over all direction in space.

Introducing Eqs. (11) and (13) into (5) and (10) we get

$$I(q) = \langle |F(\boldsymbol{q})|^2 \rangle = \left\langle \int_V \Delta\tilde{\rho}^2(\boldsymbol{r})\, \mathrm{e}^{-i\boldsymbol{q}\cdot\boldsymbol{r}}\, \mathrm{d}\boldsymbol{r} \right\rangle$$

$$= 4\pi \int_0^\infty \Delta\tilde{\rho}^2(r) \frac{\sin qr}{qr} r^2\, \mathrm{d}r$$

$$= 4\pi \int_0^\infty \gamma(r) r^2 \frac{\sin qr}{qr}\, \mathrm{d}r \tag{14}$$

and by introducing the *pair distance distribution function* (PDDF) $p(r)$ with

$$p(r) = \gamma(r) \cdot r^2 = \Delta\tilde{\rho}^2(r) \cdot r^2 \tag{15}$$

we finally get

$$I(q) = 4\pi \int_0^\infty p(r) \frac{\sin(qr)}{qr}\, \mathrm{d}r \tag{16}$$

The correlation function $\gamma(r)$ and the PDDF $p(r)$ can, in principle, be calculated from $I(q)$ by the inverse transformation

$$\gamma(r) = \frac{1}{2\pi} \int_0^\infty I(q) q^2 \frac{\sin qr}{qr}\, \mathrm{d}q \tag{17}$$

or

$$p(r) = \frac{1}{2\pi^2} \int_0^\infty I(q) qr \sin qr \, \mathrm{d}q \tag{18}$$

In practice we cannot use these equations directly for the calculation of the real space functions as $I(q)$ cannot be measured in the full regime $0 < q < \infty$.

We can summarize the above findings as follows: the scattering amplitude $F(\boldsymbol{q})$ depends on the electron density difference $\Delta\rho(\boldsymbol{r})$, they are Fourier transformation pairs and the scattering intensity $I(q)$ is the Fourier transform of the correlation function $\gamma(r)$ averaged over all directions in space. This averaging process leads to a loss of information. The whole information of the scattering experiment is contained in these two functions $I(h)$ and $\gamma(r)$. In principle, the information content of these two functions is exactly the same as they are linked to each other by the unique Fourier transformation. We see in the following that most of the special features in one space can also be found by special Fourier theorems in the corresponding reciprocal space, however the recognizability may be quite different in the two spaces. Such a problem does not exist in theoretical considerations but can be of great importance for practical applications.

For homogeneous particles the function $\gamma_0(r)$ can be interpreted as the probability to find in a distance r again a point inside the particle when starting from an arbitrary point inside the particle. For homogeneous particles this function only depends on the geometry of the particle. If we choose $r = 0$ all endpoints must be inside the particle, i.e., $\gamma_0(0) = 1$ and the function must be zero for all distances $r > D$ where D is the maximum dimension of the particle.

We now want to move on from probabilities to number of distances. This number can be found if we multiply the probability with the volume of the spherical shell with radius r and the thickness $\mathrm{d}r$. The number of distances is therefore proportional to

$$4\pi\gamma(r)r^2 \, \mathrm{d}r = 4\pi p(r) \, \mathrm{d}r$$

[see also definition (15)]. We can relate the meaning of a distance histogram to the PDDF $p(r)$ if the particles are homogeneous. The height of $p(r)$ is proportional to the number of distances that can be found inside the particle within the interval r and $r + \mathrm{d}r$ (see Fig. 4).

If we now leave the assumption of homogeneous densities in the particles we come to the (difference–density) pair–distance distribution, i.e. $p(r)$ is proportional to the difference–density length pairs in the interval r and $r + \mathrm{d}r$. Still, the orientation of all connection lines is of no importance due to the spatial averaging process. We can understand the PDDF for inhomogeneous particles as follows: we subdivide the particle into many small volume elements (small compared to the local density fluctuations). The $p(r)$ function is proportional to the product of the difference scattering lengths $n_i n_k$ $\left[n_i = \Delta\rho(\boldsymbol{r}_i)\mathrm{d}V(\boldsymbol{r}_i)\right]$ of two volume elements i and k with a center-to-center distance between r and $r + \mathrm{d}r$ and we sum over all pairs with this distance (see Fig. 5).

The difference density may be either positive or negative. Whenever we combine volume elements with opposite sign of the density we get negative contributions to $p(r)$ leading to minima and even to negative values of the PDDF at this distance in some cases. We see that

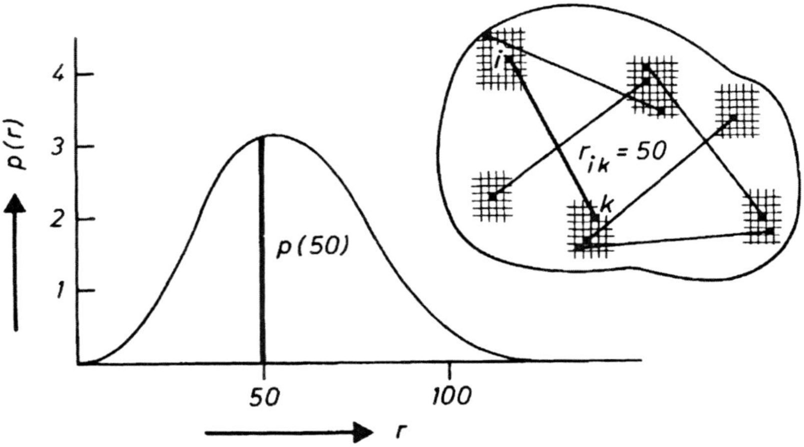

Fig. 4. Definition of the PDDF $p(r)$. The height of this function is proportional to the number of possible connection lines with a length between r and $r + \mathrm{d}r$.

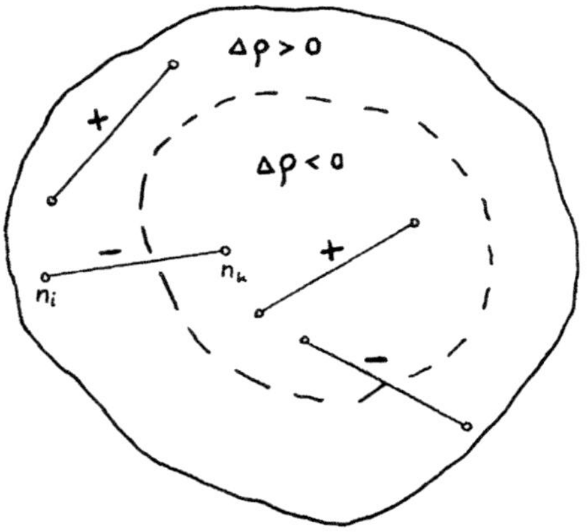

Fig. 5. Definition of the PDDF $p(r)$ of inhomogeneous particles. Each line must be weighted by $n_i n_k$.

the PDDF allows a clear interpretation for homogeneous and inhomogeneous system. This will be of great importance for the evaluation of scattering experiments.

The scattering problem and the inverse scatting problem are illustrated in Fig. 6. The calculation of the scattering functions, i.e., the calculation of the scattering curve and of the PDDF for a given structure is called the scattering problem. We have seen that the particle, after spatial averaging, is represented by its correlation function or by the PDDF [Eqs. (11) and (15)]. The Fourier transform of this PDDF $p(r)$ is the scattering function $I(q)$ [Eq. (16)].

The inverse scattering problem is the problem to find the structure of the particle from its scattering function. This task is far from trivial. The scattering curve is measured only in a limited q-regime from $q_{\min}$ to $q_{\max}$. This makes the correct computation of the PDDF difficult. In addition it is in general impossible to deconvolute the PDDF into the three-dimensional structure of the particle. In the following we shall focus on the different parts

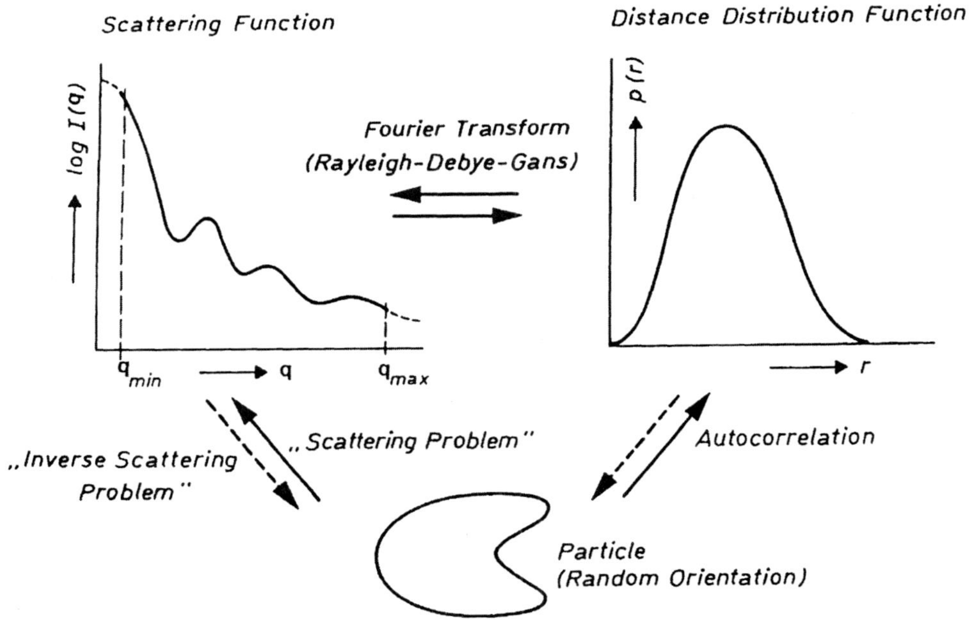

Fig. 6. The scattering problem and the inverse scattering problem .

of information we can extract from $I(q)$ and $p(r)$ in order to find some guidelines for the solution of the inverse problems. We shall learn about the importance of the PDDF for this problem. This will give the motivation to study the technical problems involved in the Fourier transformation from $I(q)$ to $p(r)$ including instrumental broadening and particle interactions (see corresponding chapters).

3.1. The radius of gyration R_g, Guinier approximation

The radius of gyration is one of the most important parameters in the field of small-angle scattering. In full analogy to the radius of interia in mechanics it is defined as

$$R_g^2 = \frac{\int \Delta\rho(r_1) r_i^2 \, \mathrm{d}V_i}{\int \Delta\rho(r_i) \, \mathrm{d}V_i} \tag{19}$$

According to the momentum theorem of Fourier transformation (Bracewell, 1986) the second moment of a function in one space is related to the second derivative (curvature) of its Fourier transform at the origin. This relation is the basis of the so-called *Guinier approximation* for the description of $I(q)$ for low$q$:

$$I(q)) = I(0)\, \mathrm{e}^{-q^2 Rg^2/3} \tag{20}$$

However, it should be kept in mind that this approximation is correct only for the q^2-term in the power series [see also discussion by Porod (1982) and Guinier and Fournet (1955)].

This approximation can be used to determine R_g with a so-called *Guinier plot*. Plotting $\ln(I(q))$ vs. q^2 one gets a straight line with a slope proportional to R_g^2.

There holds also another relation for the radius of gyration in real space:

$$R_g^2 = \frac{\int p(r) r^2 \, \mathrm{d}r}{2 \int p(r) \, \mathrm{d}r} \tag{21}$$

We can see from Eqs. (16) and (21) that the scattering intensity in forward direction ($q = 0$) is proportional to the area under the $p(r)$ function while the radius of gyration is proportional to its normalized second moment.

4. Particles with spherical symmetry

Here we discuss particles with spherical symmetry, i.e. particles with a spherical shape and $\Delta\rho(\boldsymbol{r}) = \Delta\rho(r)$. There should be no angular dependence of the density fluctuations $\Delta\rho$. With this condition we can express the scattering amplitude $F(\boldsymbol{q})$ as a function of the magnitude of the scattering vector:

$$F(q) = \langle F(\boldsymbol{q}) \rangle = 4\pi \int_0^\infty \Delta\rho(r) r^2 \frac{\sin qr}{qr} \, \mathrm{d}r. \tag{22}$$

For the homogeneous sphere with radius R we get the scattered intensity $I(q)$:

$$I(q) = (\Delta\rho)^2 V^2 P_0(q) = (\Delta\rho)^2 V^2 \left[\frac{3(\sin qR - qR \cos qR)}{(qR)^3} \right]^2 \tag{23}$$

with the normalized *form factor* $P_0(q)$

$$P_0(q) = \left[\frac{3(\sin qR - qR \cos qR}{(qR)^3} \right]^2 \tag{24}$$

This form factor is simply the square of the Bessel function of order $3/2 \, J_{3/2}(qR)$. This function has zeros for $qR = 4.493, 7.725, \ldots$ (see also Fig. 7). These zeros allow the estimation of the size of the sphere from the scattering function. But this procedure may be dangerous because we have to assume that the particles are homogeneous spheres.

There exist only a few structures where such analytical expressions exist. For homogeneous spheres it is also possible to find an analytical expression for the PDDF:

$$p(r) = \frac{3}{4\pi} \frac{r^2}{R^2} \left(2 - \frac{3}{2} \frac{r}{R} + \frac{r^3}{8R^3} \right) \qquad \text{for } r \leq 2R. \tag{25}$$

This function has its maximum at $r \approx 1.05R$, is zero for $r > 2R$ and normalized to an area of $(4\pi)^{-1}$ so that its Fourier transform according to Eq. (16) is the normalized form factor

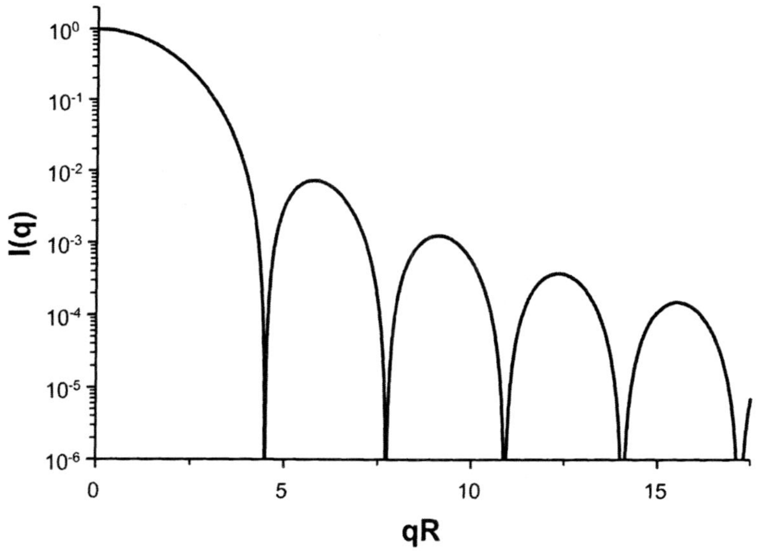

Fig. 7. Form factor $P(q)$ for homogeneous spheres in a semi-log plot: $\log(P(q))$ vs. qR.

$P_0(q)$. The characteristic function $\gamma_0(r)$ is given by

$$\gamma_0(r) = \left(1 - \frac{3}{4}\frac{r}{R} + \frac{1}{16}\frac{r^3}{R^3}\right) \qquad \text{for } r \leq 2R \tag{26}$$

In real space we can represent the particle by three functions: the PDDF $p(r)$, the correlation function $\gamma_0(r)$ and the chord distribution $G(r)$ (Tchoubar, 1991). For practical applications, especially for the solution of the inverse scattering problem, only the function $p(r)$ is really useful (Glatter, 1979).

5. Shape and structure of particles

Here we have to discuss how shape, size and structure of the scattering particle are reflected in the scattering curve $I(q)$ and in the PDDF $p(r)$. In general, it is easier to discuss the features of the PDDF, but some characteristics like symmetry give more pronounced effects in reciprocal space. This can be seen clearly in the following example where we compare two globular particles – a sphere and a cube with the same radius of gyration and scattering power $I(0)$.

It is obvious that the minima of the scattering function of the sphere are more pronounced than those of the less symmetric cube. The corresponding PDDF (Fig. 9) shows the larger maximum dimension of the cube but there are no other clear differences (see also Goodisman, 1980).

A comparison of the PDDF's for a sphere, an oblate ellipsoid of revolution (axial ratio $1:1:0.2$) and a prolate ellipsoid of revolution ($1:1:3$) is shown in Fig. 10. From this we can find the following rules: the more we deviate from the perfect sphere, the more the maximum of the PDDF $r_{\max}$ shifts to smaller r-values and the maximum dimension D increases, or

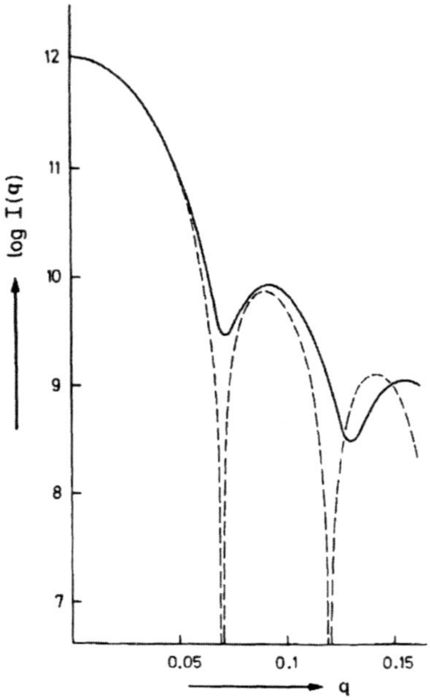

Fig. 8. Comparison of the scattering functions of a sphere (– – – –) and a cube (——) with the same radius of gyration and the same $I(0)$.

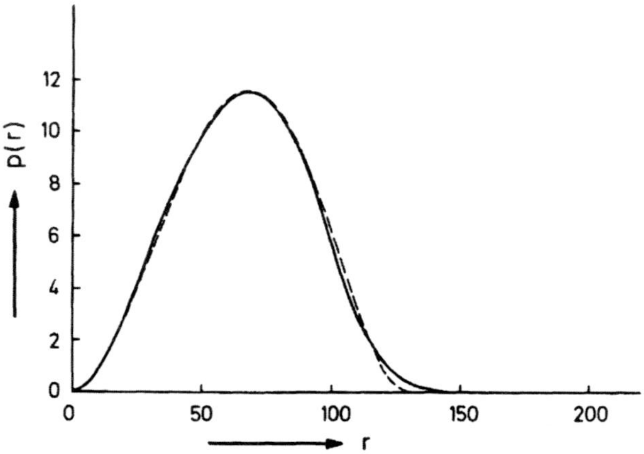

Fig. 9. PDDF of a sphere (– – – –) and a cube (——) with the same radius of gyration and the same area under the $p(r)$ function $I(0)$.

in other words, the ratio $D/r_{\max}$ increases with increasing asphericity. The effect is less pronounced for an oblate ellipsoid than it is for a prolate one. This is easy to understand as a prolate ellipsoid is extended only in one direction while the oblate ellipsoid is more or less two-dimensional and the sphere is a perfect three-dimensional structure. We see that we can start to *understand* the PDDF without a lot of exercise (at least the differences are logical).

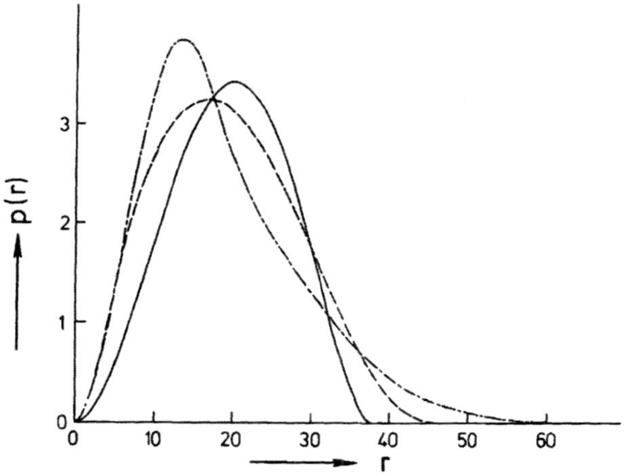

Fig. 10. PDDF of a sphere (——), of a prolate ellipsoid of revolution 1 : 1 : 3 (–·–·–) and of an oblate ellipsoid of revolution 1 : 1 : 0.2 (––––) normalized to the same R_g and $I(0)$.

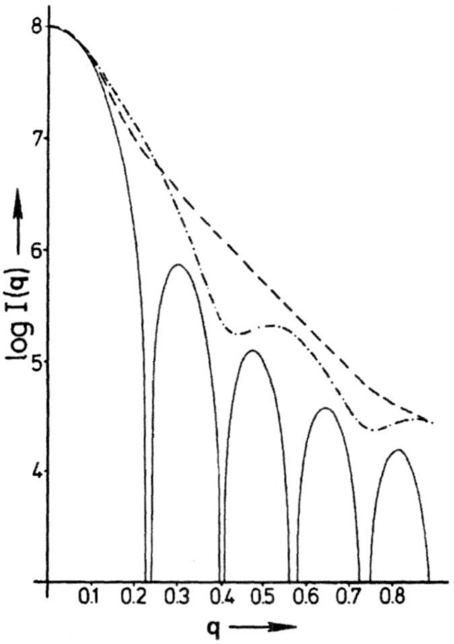

Fig. 11. Comparison of the scattering functions of a sphere, a prolate, and an oblate ellipsoid (see legend to Fig. 9).

Much more difficult would be a similar discussion of the scattering functions of these three bodies which are depicted in Fig. 11. One can clearly see differences, but it is impossible to relate them directly to structural parameters. Later we shall see that special cross-section Guinier plots may allow to recognize cylindrical and flat lamellar particles but it will remain very difficult as long as we stay in reciprocal space.

6. Rod-like particles

Let us regard a rod of length L and of cross-section $A_c = R_c^2\pi$. The cross-section A_c (with maximum dimension d) should be small in comparison to the length of the whole particle L ($d \ll L$). For $q > 1/L$ we can write

$$I(q) = \frac{L\pi}{q} \cdot I_c(q) \tag{27}$$

The cross-section scattering function $I_c(q)$ is related to the cross-section distance distribution $p_c(r)$ by (Glatter, 1982)

$$I_c(q) = 2\pi \int_0^\infty p_c(r) J_0(qr)\,\mathrm{d}r \tag{28}$$

where

$$p_c(r) = \gamma_c(r) \cdot r = 2\pi r \int_{A_c} \Delta\rho_c(r')\Delta\rho_c(r'+r)\,\mathrm{d}r \tag{30}$$

We can also define a radius of gyration of the cross-section which can be calculated from $p_c(r)$ by

$$R_c^2 = \frac{\int p_c(r) r^2\,\mathrm{d}r}{2\int p_c(r)\,\mathrm{d}r} \tag{31}$$

[see also Eq. (21)] or it can be estimated in reciprocal space form

$$I_c(q) = I_c(0)\,\mathrm{e}^{-q^2 R_c^2/2} \tag{32}$$

by a so-called cross section Guinier plot [$\log(I(q)q)$ vs. q^2]. For a homogeneous circular cylinder with radius R_c and density $\Delta\rho$ we get

$$I_c = (\Delta\rho)^2 \cdot (R_0^2\pi)^2 \left(2\frac{J_1(qR_0)}{qR_0}\right)^2 \tag{33}$$

with $R_c = R_0/\sqrt{2}$.

The different regimes can be visualized is a $\log(I)$ vs. $\log(q)$ plot of the scattering curve as demonstrated schematically in Fig.12 for a long circular cylinder.

The PDDF has to be zero for $r > D$ (or L). If we now study this function for decreasing r-values starting at $r = D$ we find a linear increase of $p(r)$ with decreasing r. We can understand this finding with the sketch of a prism given in Fig. 13.

At $(L - r) < r < L$ we can combine every point in the first slice with every point in the last slice (more precisely $p(r)$ is proportional to $\Delta\rho^2 A_c^2$). For $r < L$ we get $(L - r)$ such contributions leading to a linear regime for large r-values.

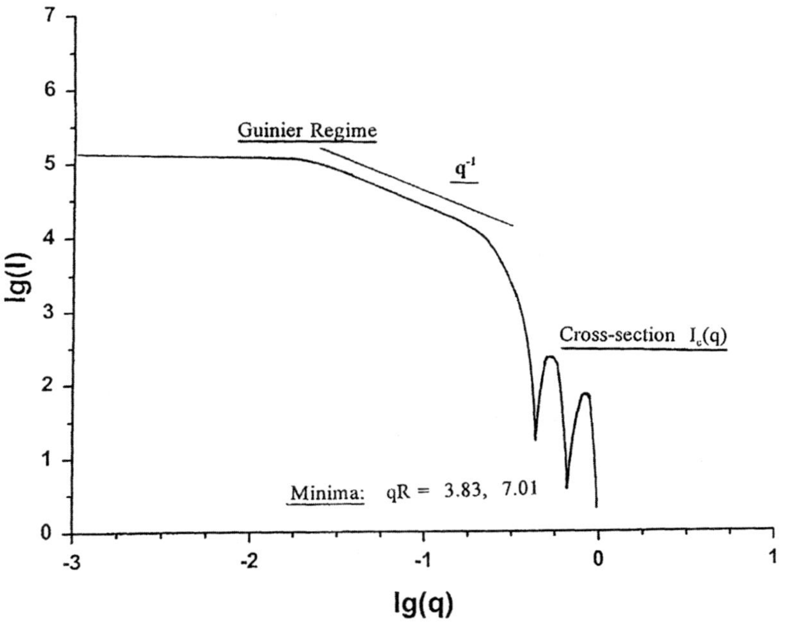

Fig. 12. Scattering curve of a long cylindrical particle (schematic representation).

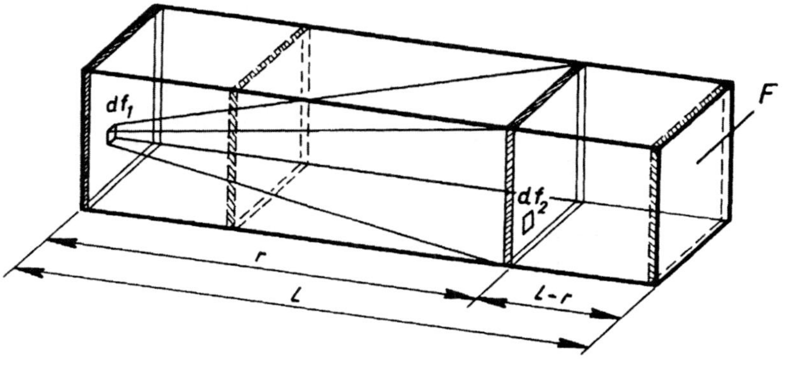

Fig. 13. Sketch of a prism for the explanation of the PDDF of a rod-like particle

The equation for this part reads

$$p(r) = \frac{2}{4\pi} \int_r^L \int_A \int_A \Delta\rho^2 \, \mathrm{d}f_1 \, \mathrm{d}f_2 \, \mathrm{d}x = \frac{1}{2\pi} \Delta\rho^2 A_c^2 (L - r), \tag{34}$$

The dependence of this slope on the square of the contrast will be important for inhomogeneous particles, where the overall contrast of the cross-section $\overline{\Delta\rho}$ can be close to zero and the long linear tail may disappear.

The maximum of the PDDF (see Fig. 14) corresponds to the cross-section ($0 \le r \le D_c$)

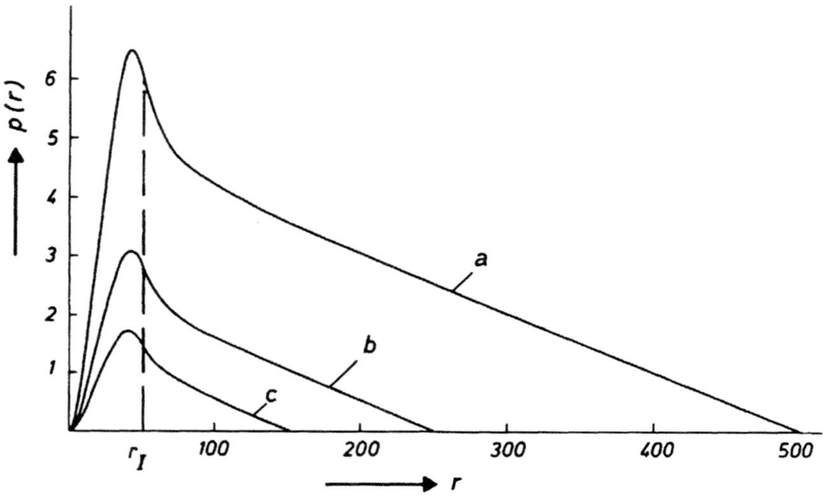

Fig. 14. PDDF from homogeneous prisms with edge lengths of: (a) 50 : 50 : 500, (b) 50 : 50 : 250, and (c) 50 : 50 : 150. The point of inflection r_I, marked with a dashed vertical line, indicates the dimension of the cross-section.

and the point of inflection r_I between the maximum and the linear region gives a rough indication for the size of the cross-section. The PDDFs of prisms with the same cross-section but different length L are shown in Fig. 14. We see that they all have the same point of inflection r_I.

The dependence of r_I on the size of the cross-section is seen clearly in Fig. 15 where three prisms with equal cross-section area A_c (same slope of the linear regime) but different cross-section dimensions are shown.

In this connection it is important to notice that the maximum of the PDDF is even for homogeneous rods not related to the maximum dimension of the cross-section, especially it is not a rough estimate for the radius of a circular cylinder.

Deviations form Eq. (34) appear as soon as the approximation $L \geq d$ breaks down, i.e., as soon as the dimension of the cross-section is no longer negligibly small. Test calculations and Figs. (14) and (15) show that an approximate limit for the occurrence of a linear part is that the relation

$$D/D_c > 2.5 \tag{35}$$

holds.

7. Flat particles

Let us now consider a flat particle, with a finite and constant thickness D_t, being extremely large in the two other dimensions with an area A. A typical example would be a circular disc with $R \geq D_t$. In full analogy to the case of the rod we can separate the scattering amplitude

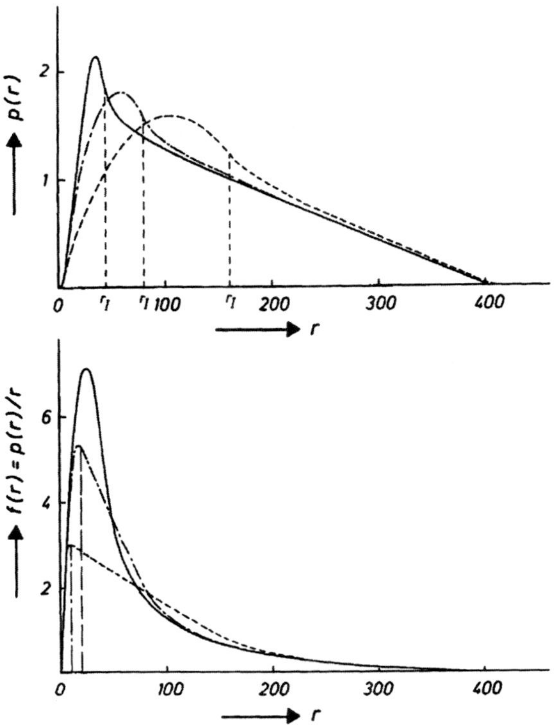

Fig. 15. PDDF and $f(r) = p(r)/r$ from three parallel epipeds with constant length L (400 Å) and a constant cross-section area A_c but varying length of the edges: ——, 40 : 40; – · – · –, 80 : 20; and – – – –, 160 : 10.

into a *planar factor* $2\pi Aq^{-2}$ and a *thickness factor* $I_t(q)$, i.e. the total intensity is given by

$$I(q) = I_{\text{plane}} \cdot I_t(q) = \frac{2\pi A}{q^2} \cdot I_t(q). \tag{36}$$

Again we can show the different regimes in a log(I) vs. log(q) plot of the scattering curve as demonstrated in Fig. 16 for a thin flat particle.

The thickness factor is related to the thickness–distance distribution $p_t(r)$ by (Glatter, 1982)

$$I_t(q) = 2\int_0^\infty p_t(r)\cos(qr)\,\mathrm{d}r \tag{37}$$

where

$$p_t(r) = \gamma_t(r) = 2\int_0^\infty \Delta\rho_t(r')\Delta\rho(r'+r)\,\mathrm{d}r. \tag{38}$$

We can also define a radius of gyration of the thickness which can be calculated from $p_t(r)$

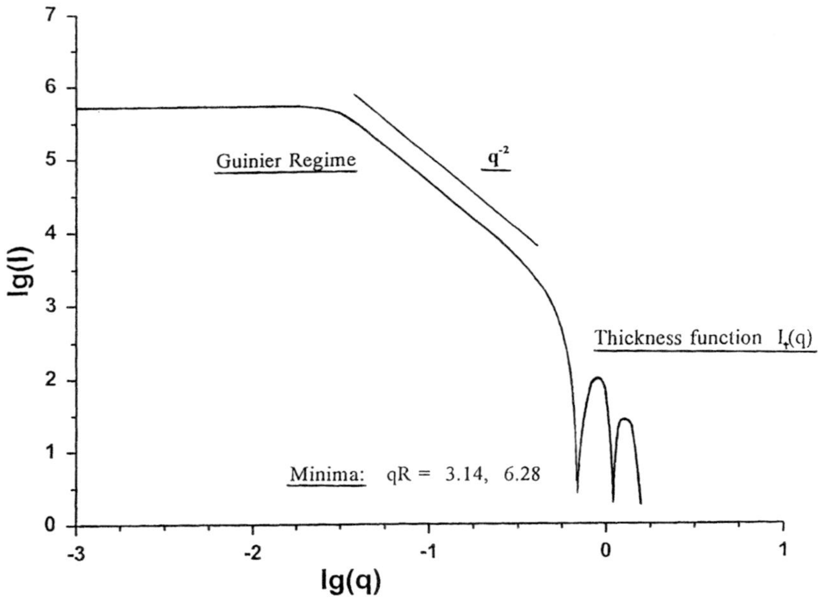

Fig. 16. Scattering curve of a thin flat particle (schematic representation).

by

$$R_t^2 = \frac{\int p_t(r) r^2 \, \mathrm{d}r}{2 \int p_t(r) \, \mathrm{d}r} \tag{39}$$

or from the scattering curve with

$$I_t(q) = I_t(0) \, \mathrm{e}^{-q^2 R_t^2} \tag{40}$$

in a so-called thickness Guinier plot $[\log(I(q)q^2)$ vs. $q^2]$. For a homogeneous flat particle of thickness D_t holds $R_t = D_t/\sqrt{12}$.

The form of the $p(r)$ function of flat particles, i.e. particles elongated in two dimensions (discs, flat prisms) is more complicated. There exists an analytic expression for the PDDF for infinitely thin circular discs (Porod, 1948).

$$p(r) = \text{const} \cdot x\left[\arccos(x) - x\sqrt{1 - x^2}\right] \qquad \text{with } x = r/D \tag{41}$$

A power series expansion shows that the first non-zero coefficient is the one for the linear term.

Intuitively, one can rationalize the main features of a flat particle, like a lamella, of finite thickness with the following consideration: let us take a point in the center of the lamella (Fig. 17). For very small values of r ($r < D_t$), $p(r)$ will increase with the second power of r, since the number of distances is proportional to the surface of a sphere with radius r (Fig. 17a). If $r > D_t$, most of the sphere lies outside the lamella and the PDDF will increase

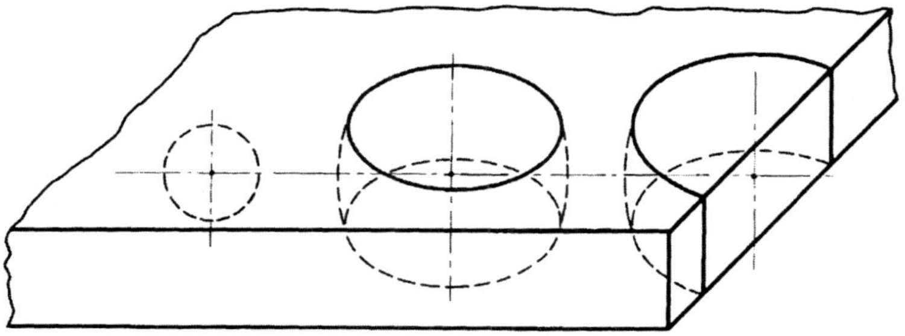

Fig. 17. Sketch for the qualitative discussion of the PDDF of a flat particle.

roughly linearly, since the intersection between the lamella and the sphere with radius r approximates a circular cylinder for $r \geq D_t$ (Fig. 17b). Fig. 17c illustrates the situation for still larger r-values; if the center of the cylinder is near the edge of the lamella, some part of the cylinder will lie outside the lamella ("boundary losses"). Evidently the boundary losses increase with increasing r, and eventually cause $p(r)$ to drop to zero for $r = D$.

Thus we expect a quadratic increase for the region $r < D_t$ (like for any confined structure), a more or less linear increase for $r > D_t$ and a drop to zero at $r = D$. This is shown in Fig. 18 for three lamellae of different thickness D_t. The transition from a parabolic to a linear increase at $r = D_t$ cannot be found easily in this plot, but a convex behavior at $r > D_t$ can be recognized distinctly. The situation becomes much clearer if we use the function

$$f(r) = p(r)/r = \gamma(r)r \tag{42}$$

This function has no physical meaning but is more useful for this purpose than $p(r)$. The $f(r)$ functions for the lamellae in Fig. 18 are plotted in Fig. 19. These functions start, as expected with a linearly increasing part and become almost flat at $r = D_t$, the slight decrease is due to boundary losses. The D_t of the particle can be read from the transition point.

So in practice the shape of the function $f(r)$ will allow the recognition of a lamellar particle and the determination of its thickness D_t. The transition from rod-like particles with compact cross-sections to more extended cross-sections has already been demonstrated in Fig. 15 for three prisms. There they have been discussed in terms of rod-like particles. Now we focus on the fact of the increasing lamellar character when changing the cross-section dimensions from 40×40 to 80×20 and 160×10 by discussing the $f(r)$ functions shown in the lower part of Fig. 15. The quadratic prism (40×40) does not indicate any lamellar structure, whereas the lamellar shape of the other prisms (80×20 and 160×10) can easily be recognized, and the thickness D_t (vertical lines) can be estimated. It should be emphasized that such a clear-cut direct classification is not possible in reciprocal space.

Summarizing this section on homogeneous particles we can say that we can clearly differentiate between globular, sphere-like particles and elongated structures using information in real space and reciprocal space. We can clearly recognize rod-like and lamellar particles without *a priori* information about the systems. However, we should not forget that we have made essential assumptions: our system has to be dilute (no particle interactions) and monodisperse.

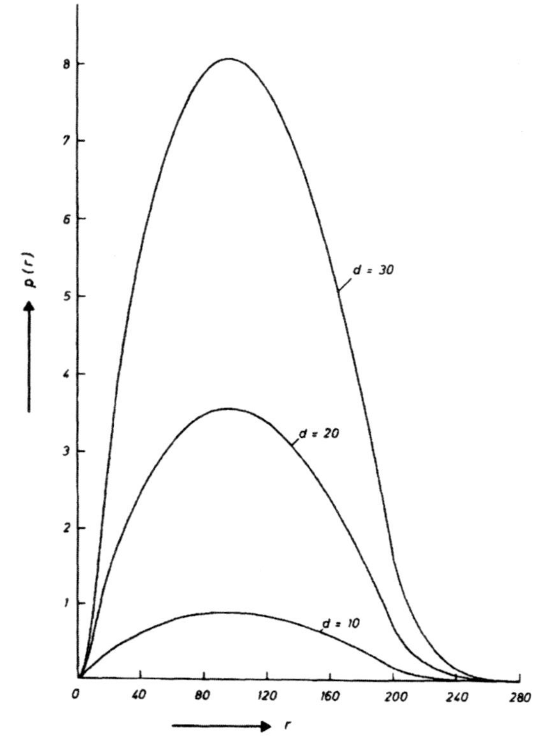

Fig. 18. PDDFs of lamellar particles with the same basal plane (100 × 100 Å) and different thickness D_t: (a) $D_t = 10$ Å, (b) $D_t = 20$ Å, and (c) $D_t = 30$ Å.

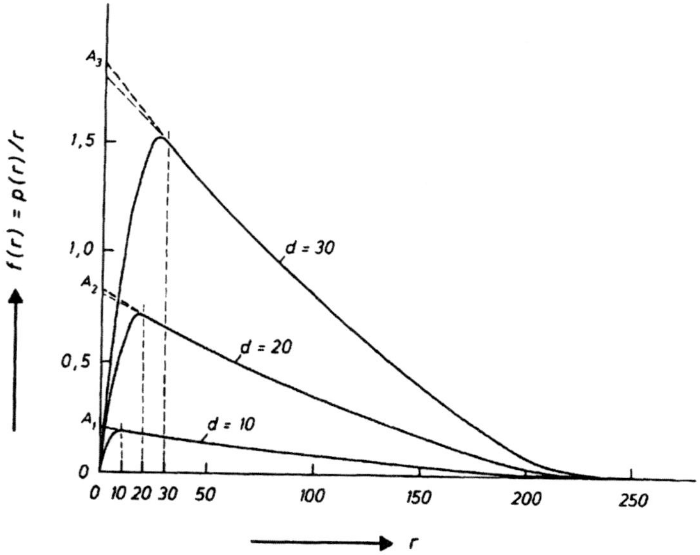

Fig. 19. Modified distance distribution $f(r) = p(r)/r$ of the particles described in Fig. 18. The transition points ($r \simeq D_t$) are indicated by the vertical dashed lines.

8. Hollow and inhomogeneous particles

We have learned to classify homogeneous particles in the previous part of this section. If we now allow internal inhomogeneities we make things more complicated and it is clear that it is impossible to get an unique reconstruction of an arbitrary inhomogeneous three-dimensional structure from its scattering function without additional *a priori* information. We restrict our considerations to special cases which are important in practical applications and which allow at least a solution in terms of a first-order approximation. In addition we have to remember the fact that the $p(r)$ function of inhomogeneous particles is a distance distribution weighted with the product of the excess densities at the endpoints, which can be negative (see also Fig. 5). Therefore a minimum in the PDDF can be caused by a small number of distances due to the particular shape, or by the addition of positive and negative contributions.

8.1. Hollow spheres

Vesicles are very important in practical applications of small angle scattering. They are hollow spheres and can be regarded as homogeneous hollow spheres in SANS experiments in D_2O while they are a prototype for inhomogeneous lamellar structure in SAXS experiments due to the differences in the scattering length densities of the surfactant head groups and tails (Iampietro et al., 1998).

A thin hollow sphere can be understood as a flat but bent particle. So we can expect a similar behavior as we found for flat, lamellar particles. Indeed, the $p(r)$ function first increases with second order of r and then runs into a linearly increasing regime at $r = d = R_o - R_i$. This linear regime ends at $r = D - d = 2R_i$ and the final part is identical to that of the full sphere. The exact linearity of the middle part is a consequence of the spherical shape. The $f(r)$ function shows this linearly increasing regime as a strictly horizontal plateau, its length increases with the hollowness of the sphere (see Fig. 20).

8.2. Inhomogeneous spherical particles

The PDDFs of spherical multilayer models with alternating signs of the excess density are illustrated in Fig. 21 together with the density profiles. The rough qualitative conclusions that can be made from the $p(r)$ functions are as follows: under the assumption of a compact structure, the accurence of minima in the $p(r)$ function signals the existence of regions or layers of density with alternating sign. The number of such regions is equal to the number of maxima in the PDDF, provided the particles are spherical (sharp minima in the scattering curve, no evidence for elongated structures in the PDDF) and the shells have similar dimensions (thickness). We shall see in the following chapter (Chapter 5) how the density profiles can be calculated from the PDDFs.

8.3. Inhomogeneous rod-like particles

Such particles can be represented by long circular cylinders. Fig. 22 shows the comparison of a full cylinder, a hollow cylinder and an inhomogeneous cylinder with a core–shell structure. The PDDFs show significant differences in the cross-section dimensions $r < D_c$.

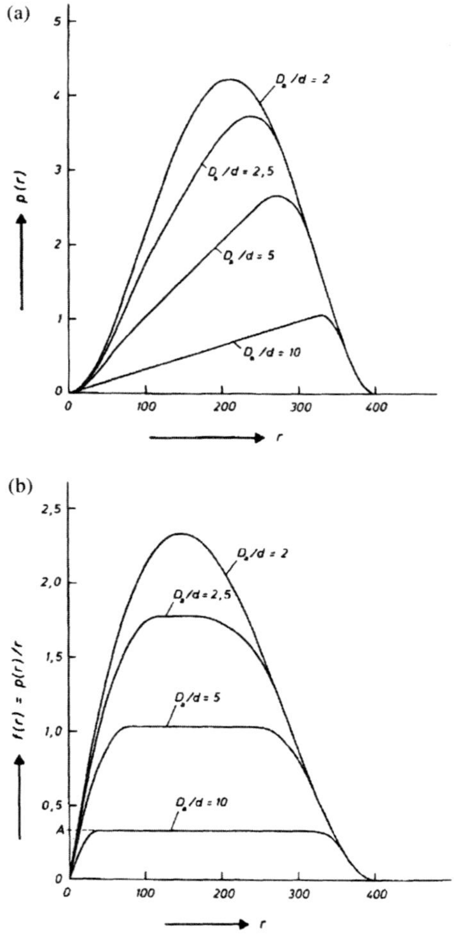

Fig. 20. (a) Hollow spheres – $p(r)$ functions: D_o is the outer diameter, d the thickness of the shell. $D_o/d = 2$ represents a full sphere, $D_o = 400$ Å. (b) Hollow spheres as in (a), $f(r)$ functions.

The linear slope for $r > D_c$ is found for all three types of structures, only the slope differs due to the different cross-section area A_c or the difference in the mean density $\overline{\rho}_c$. If we compare the homogeneous and the hollow cylinder we see that the maximum of the $p(r)$ function of the hollow cylinder is located at larger distances, since the smaller distances from the core are missing.

The PDDF of the inhomogeneous cylinder shows a negative region, i.e. there are distances which connect regions with opposite signs more frequently than regions with the same sign. The maximum is caused by the outer shell (see also hollow cylinder). The linear descent is proportional to $\overline{\rho}_c^2$, so it is always positive, even if $\overline{\rho}_c$ is negative. The linear region would coincide with the abscissa if $\overline{\rho}_c = 0$ and we would have no information at all about the length of the cylinder. We would even hardly could differentiate between an inhomogeneous cylinder and an inhomogeneous sphere.

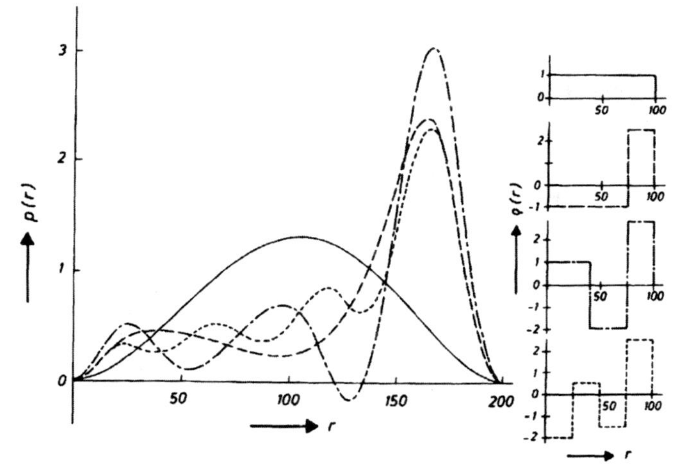

Fig. 21. Spherical multilayer models with constant outer diameter of 200 Å. PDDFs in the left part, density profiles in the right part of the figure.

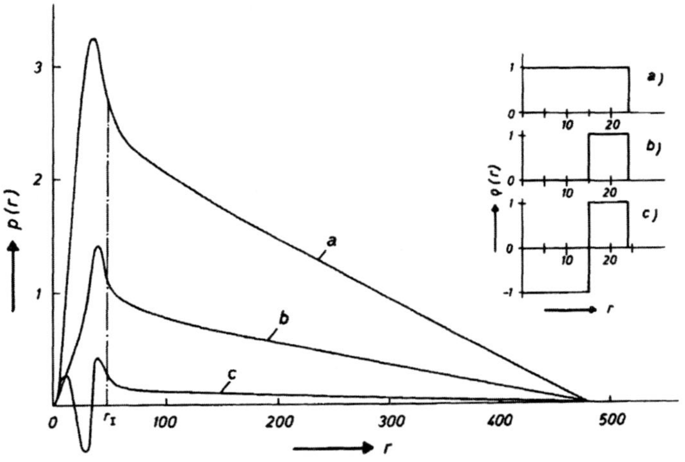

Fig. 22. Circular cylinders with a constant length of 480 Å and an outer diameter D_c of 48 Å. (a) Homogeneous cylinder, (b) hollow cylinder, (c) inhomogeneous cylinder. The PDDFs are shown on the left, the corresponding radial density distributions $\rho(r)$ on the right.

Axial Inhomogeneity:

Rod-like particles with a homogeneous cross-section but with inhomogeneities along the long axis show changes of the PDDF in the whole range of definition $0 < r < D$. Periodical axial changes of the density are assumed in the model shown in Fig. 23. The periodical changes in the density lead to the periodical ripples in $p(r)$ and the sharp maxima in $I(q)$ which could be explained as broadened reflections of odd orders. The period of the oscillations correspond with the thickness of the "alternating double layer" ($20 + 20 = 40$ Å). This double

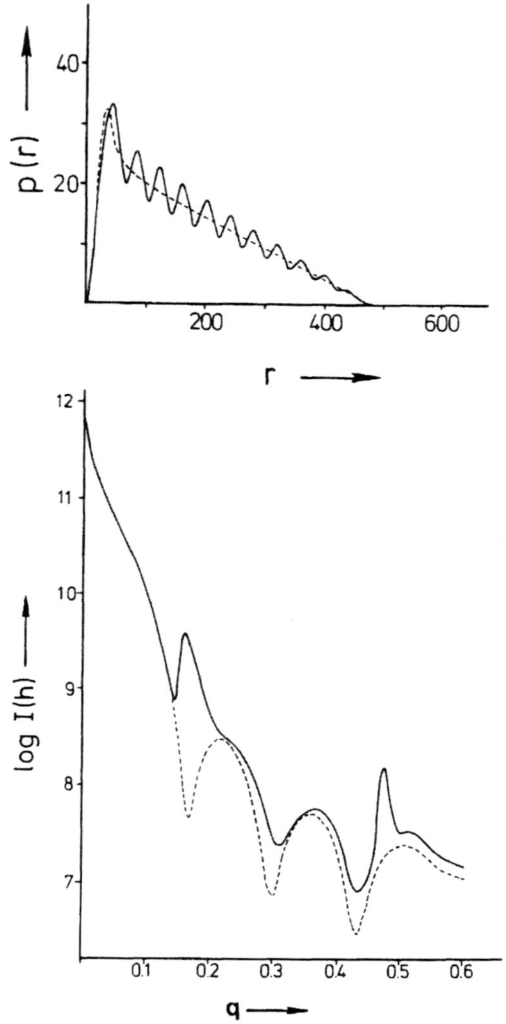

Fig. 23. Inhomogeneous circular cylinder with periodical changes of the density ($1.5\overline{\rho}$ and $0.5\overline{\rho}$) along the cylinder axis (– – – –) compared with a homogeneous cylinder (——) with the same mean density $\overline{\rho}$. Scattering function (bottom) and PDDF (top).

layer is repeated twelve times ($L = 480$ Å) and we can interpret this model as a microcrystal, where the unit cell is the double layer. The position of the sharp peaks in reciprocal space corresponds to the repeat distance ($q = 2\pi/40 \simeq 0.16$), their width is inversely proportional to the number of unit cells. The density profile is an odd function, so we get only the odd orders (1 and 3 in the figure). There is a perfect agreement of the two functions at low q-values. In the special case $\overline{\rho} = 0$ one observes a $p(r)$ function that oscillates around zero in the whole range $0 \leq r \leq D$. Oscillations about the linear descent could also occur for cylinders with homogeneous density but varying cross-section, like for a linear assembly of spheres.

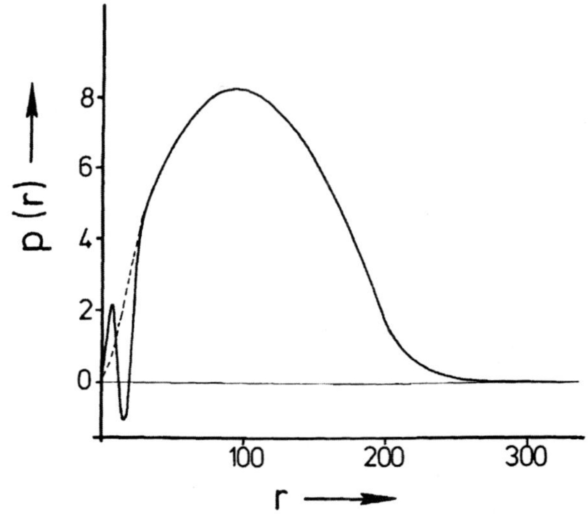

Fig. 24. PDDF of a homogeneous (– – – –) and a inhomogeneous lamella (——) with dimensions 200 × 200 × 30 Å, $\Delta\rho_{\text{hom}} = 1/3$, $\rho_t(r)$ is a three-step function (3 × 10 Å) with $\Delta\rho$-values +1, −1, +1.

Inhomogeneities which are mixtures of the axial and radial type cannot be interpreted without additional information.

8.4. Inhomogeneous lamellar particles

Inhomogeneous lamellar particles with density variations perpendicular to the basal plane, i.e., $\Delta\rho = \rho_t(r)$ where r is the distance from the central plane, show differences to a homogeneous lamella of the same size and with the same mean density in the PDDF in the range $0 < r < D_t$ where D_t is the thickness of the lamella (see Fig. 24).

There is no difference between the $p(r)$ of the homogeneous and of the inhomogeneous lamella for $r > D_t$ where the function depends only on the overall shape and size. This corresponds with the fact, that the scattering curves are identical at low q-values, but they are quite different in the high q-regime which depends on the thickness factor $I_t(q)$.

Density fluctuations along the basal plane (*in-plane inhomogeneity* like a chess-board) would lead to oscillations in the whole $p(r)$ function and would not be restricted to $r < D_t$, but they would average out with increasing distances. There are no general techniques available for the solution of the corresponding inverse problem. Extremely difficult, i.e. nearly insoluble is the problem, if a layered inhomogeneous lamella (like the one described in Fig. 24) has in addition in-plane inhomogeneities like it is the case in a real biomembrane where proteins are embedded into the lipid double-layer.

9. Polydisperse systems

Until know we have discussed all problems under the assumptions that all particles have the same size and shape and that interparticle scattering can be neglected (infinite dilution).

Especially all ideas about shape determination discussed in the previous chapter were based on the assumption of monodispersity. We now want to skip this first assumption (and still neglect interparticle contributions).

It is in principle impossible, to determine simultaneously the shape and the size distribution of a polydisperse system from its scattering data. It can be shown that the scattering curve of homogeneous ellipsoid can be exactly simulated by a corresponding distribution of homogeneous spheres (Mittelbach, 1965; Mittelbach and Porod, 1965).

The question, if the sample is polydisperse or not, must be answered by independent methods (like dynamic light scattering or ultra-centrifugation). Polydispersity is a common problem in practical applications. In these cases we have to know or to assume the shape and can then determine the size distribution or, a more theoretical problem, we could determine the shape if we would know the size distribution. In the first case we have to assume that all particles have the same shape and internal structure and that they depend only on one linear size parameter R (for example the radius of spheres or the length of cylinders with a constant cross-section, etc.). The most important case is the spherical one because many systems can be approximated by spheres (see above) and one is often interested in a rough information on the size and size distribution (monomodal, bimodal, etc.). We can define three different types of distributions, the so-called *intensity distribution* $D_i(R)$, the *volume* or *mass distribution* $D_v(R)$ and the *number distribution* $D_n(R)$. The intensity distribution gives the relative contribution of the different sizes to the total scattering intensity and is free from model assumptions. The volume or mass distribution states the relative amount of volume in the different size classes and the number distribution simply defines the relative number of particles in the different size classes. The later two distributions need the relation between volume and intensity or the size dependence of the intensity for one particle. These relations are quite simple for spheres:

$$D_i(R) = R^3 D_v(R) \tag{43}$$

and

$$D_i(R) = R^6 D_v(R) \tag{44}$$

The total scattering intensity is just the sum over the contributions from the different size classes for continuous distribution the corresponding integral

$$I(q) = c_i \int_0^\infty D_i(R) \cdot P_0(q, R)\, \mathrm{d}R$$

or

$$I(q) = c_v \int_0^\infty D_v(R) \cdot R^3 \cdot P_0(q, R)\, \mathrm{d}R \tag{45}$$

or

$$I(q) = c_n \int_0^\infty D_n(R) R^6 \cdot P_0(q, R)\, \mathrm{d}R$$

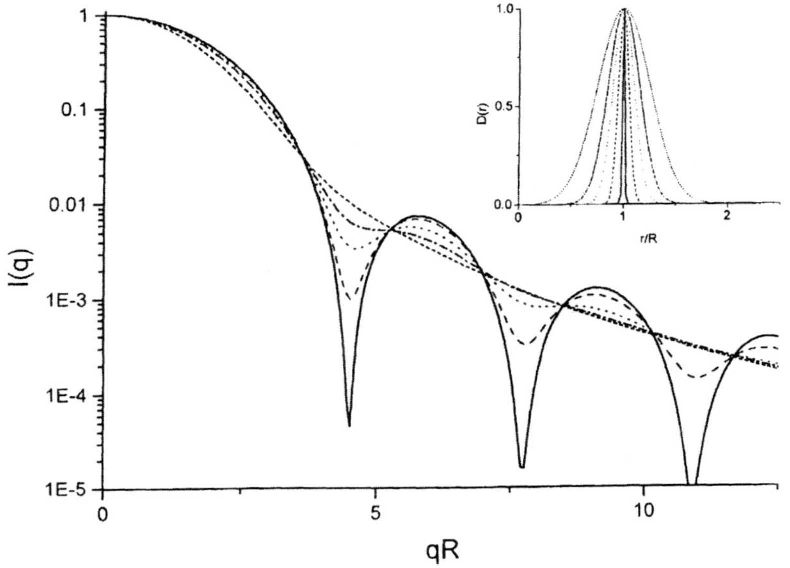

Fig. 25. Scattering curves of Gaussian size distributions of spheres with varying width.

where the coefficients c_i, c_v and c_n are normalization constants and $P_0(q, R)$ is the normalized form factor of a spheres with radius R (see Eq. (24)). Eqs. (45) are very similar to Eq. (16) for the monodisperse case, the PDDF of the particle is replaced by the size distribution $D(R)$ and the $\sin x/x$ terms are replaced by the form factor of the particles. The inversion of Eqs. (45) will be discussed in detail in Chapter 5, but we can expect that the numerical routines will be similar for the monodisperse and the polydisperse case. However, there are also some basic differences. The size distribution $D(R)$ can never become negative what can happen in the case of the PDDF. It is also possible to approximate the size distribution function by an analytical model where only a few parameters have to be estimated by a standard least squares technique. These parameters are usually the mean size or center of the distribution, its amplitude and its width (variance).

The effect of the polydispersity on the scattering curves is illustrated in Fig. 25 for the case of a Gaussian size distribution of spheres with a varying width. We can see clearly the increase of the smearing effect with increasing width and increasing q-value.

What happens if we treat a polydisperse system as a monodisperse one? Using transformation (16) or (18), respectively, we simply get the sum over all PDDFs of the different particles as the transformation is linear. Nothing would be wrong in this step. We would start to make a mistake as soon as we start to interpret this averaged $p(r)$ function as the PDDF of a monodisperse system in order to estimate the corresponding size, shape and structure.

It should be mentioned that wavelength polydispersity (like in SANS) has the same effect as size polydispersity, i.e., scattering functions like in Fig. 25 would result from monodisperse spheres if the experiment is carried out with a Gaussian wavelength distribution.

In this chapter we have discussed the influence of size, shape, internal structure, and polydispersity on the scattering functions. Some details, like symmetry, can be found easily in the scattering curves, most of the structural details can be recognized much easier in real space, especially in the PDDF $p(r)$. If we have $I(q)$ and $p(r)$ we can use this information and solve the inverse problem to a certain extend. In order to reach a final results we may have to calculate scattering functions for models, check if they fit to our results, and in most cases we shall have to improve these models iteratively (see also Chapter 6 by J.S. Pedersen). In order to receive $I(q)$ and $p(r)$ from our experimental results we have to consider resolution effects (see Chapter 6 by J.S. Pedersen) and we have to perform Fourier transformations and deconvolution processes (see Chapter 5).

References

Agamalian, M.M., Alamo, R.G., Londono, J.D., Mandelkern, L. and Wignall, G.D. (2000). *J. Appl. Cryst.*, **33**, 843–846.

Bayvel, L.P. and Jones, A.R. (1981). *Electromagnetic Scattering and Its Applications*. Applied Science Publishers, London.

Bohren, C.F. and Huffman, D.R. (1983). *Absorption and Scattering of Light by Small Particles*. Wiley, New York.

Bonse, U. and Hart, M. (1967). In: *Small Angle X-ray Scattering* (Brumberger, H., ed.). Gordon and Breach, New York, 121 pp.

Bracewell, R. (1986). *Fourier Transform and its Applications*. McGraw-Hill, New York.

Brown, W. (1996). *Light Scattering – Principles and Development*. Clarendon Press, Oxford.

Brumberger, H. (1967). *Small-Angle X-Ray Scattering*. Gordon and Breach, New York.

Brumberger, H. (1995). *Modern Aspects of Small-Angle Scattering*. Kluwer Academic Publishers, Dordrecht.

Debye, P. (1915). *Ann. Phys.*, **46**, 809.

Debye, P. and Bueche, A.M.J. (1949). *Appl. Phys.*, **20**, 518.

Feigin, L.A. and Svergun, D.I. (1987). *Structure Analysis by Small-Angle X-Ray and Neutron Scattering*. Plenum Press, New York.

Glatter, O. (1979). *J. Appl. Cryst.*, **12**, 166–175.

Glatter, O. (1982). In: *Small Angle X-ray Scattering* (Glatter, O. and Kratky, O., eds.). Academic Press, London, chapter 4.

Glatter, O. and Kratky, O. (1982). *Small Angle X-ray Scattering*. Academic Press, London.

Glatter, O., Sieberer, J. and Schnablegger, H. (1991). *Part. Part. Syst. Charact.*, **8**, 274–281.

Goodisman, J. (1980). *J. Appl. Cryst.*, **13**, 132–134.

Guesbet, G. and Grehan, G. (1988). *Optical Particle Sizing*. Plenum Press, New York.

Guinier, A. and Fournet, G. (1955). *Small Angle Scattering of X-Rays*. Wiley, New York.

Hainbuchner, M., Villa, M., Kroupa, G., Bruckner, G., Baron, M., Amenitsch, H., Seidl, E. and Rauch, H. (2000). *J. Appl. Cryst.*, **33**, 851–854.

Iampietro, D.I., Brasher, L.L., Kaler, E.W., Stradner, A. and Glatter, O. (1998). *J. Phys. Chem. B*, **102,** 3105–3113.

Kerker, M. (1969). *The Scattering of Light and Other Electromagnetic Radiation*. Academic Press, New York.

Lindman, B. and Olsson, U. (1996). *Ber. Bunsenges. Phys. Chem.*, **100**, 344–363.

Mittelbach, P. (1965). *Kolloid Z. Z. Polym.*, **206**, 152–159.

Mittelbach, P. and Porod, G. (1965). *Kolloid Z. Z. Polym.*, **202**, 40–49.

Porod, G. (1948). *Acta Phys Austriaca*, **2**, 255–292.

Porod, G. (1951). *Kolloid-Z.*, **124**, 83–114.

Porod, G. (1982). In: *Small Angle X-ray Scattering* (Glatter, O. and Kratky, O. eds.). Academic Press, London, chapter 2.

Tchoubar, D. (1991). In: *Neutron, X-Ray and Light Scattering* (Lindner, P. and Zemb, Th., eds.). North-Holland, Amsterdam, pp. 157–174.

van de Hulst, H.C. (1981). *Light Scattering by Small Particles*. Dover, New York.

Neutrons, X-Rays and Light
P. Lindner and Th. Zemb (Editors)

Chapter 5

Fourier Transformation and Deconvolution

Otto Glatter

Department of Chemistry, Karl-Franzens Universität Graz, Heinrichstraße 28, A-8010 Graz, Austria

1. Introduction

In the preceding chapter (Chapter 4) we have discussed the importance of the PDDF for the solution of the inverse scattering problem. In this chapter we discuss the most important numerical procedures for the calculation of this function from experimental small-angle scattering data. We continue here with the numbering of equations., i.e., Eqs. (1)–(45) are found in Chapter 4. Preliminary data manipulations like averaging of scattering curves, subtraction of background scattering, etc., are discussed in other contributions. Here we focus on the central problems of approximation with smoothing (*Least Squares*), transformation into real space (*Indirect Fourier Transformation*) including correction of instrumental broadening (*Desmearing*). The resulting PDDF can be deconvoluted into the radial density distribution function (*Deconvolution or Convolution Square Root Operation*) for particles with special symmetry. In the case of semi-dilute or dense systems we may take into account particle interactions via a structure factor, which is determined simultaneously with the form factor in the recently developed *Generalized Indirect Fourier Transformation* (GIFT) technique.

2. Fourier transformation including smoothing and desmearing

The main target of data processing is to obtain the two functions $I(q)$ and $p(r)$ (or $D(R)$ in the case of polydisperse systems) from the experimental scattering curve $I_{\mathrm{exp}}(q)$ represented by M data points as illustrated in Fig. 1.

In the following, we restrict ourselves to non-oriented dilute systems, for example particles in solution. In such a system, a large number of randomly oriented particles contribute to the physical scattering process, and the resulting spatial average leads to a loss of information: the three-dimensional electron density distribution $\rho(\boldsymbol{r})$ (describing the whole structure of the particle) is thereby reduced to the one-dimensional PDDF $p(r)$. We "observe" the particle indirectly by means of scattering. This physical process of scattering can be described mathematically by a Fourier transformation [transformation T_1, Eqs. (16), (27), (28), (36), (37) and (45)], which results in the unsmeared scattered intensity $I(q)$. Allowance for the various collimation and wavelength effects (transformations T_2–T_4) leads to the smeared scattered intensity; the experimental data points, $I_{\mathrm{exp}}(q)$ are affected by the statistical error $\sigma(q)$.

The target of data treatment is to obtain the functions $I(q)$ and $p(r)$ from the finite

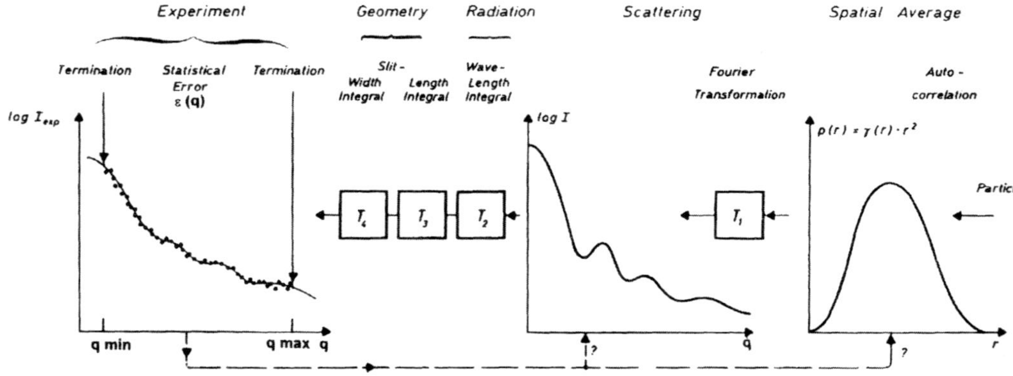

Fig. 1. Schematic representation of the correlation between a particle and its observable experimental scattering data.

number of observed intensity values, i.e. to find the inverse of the transformations T_1 to T_4. The procedure has to allow for the statistical accuracy (smoothing, least-squared approximation), has to solve the desmearing problem and has to perform the Fourier transformation with minimum termination effect. These problems could be solved in several steps (multi-step procedures) or, more preferable, in one step (single-step procedure), as the problem of error accumulation is more critical for multi-step procedures. In any case, the sequence of operations T_2–T_4 must be followed correctly (Glatter and Zipper, 1975). The essential problems of the desmearing procedure are the unavoidable smoothing of statistical fluctuations and the termination effect. The termination effect is also essential in the Fourier transformation, since the direct calculation of the inverse transformation T_1^{-1} requires in principle the knowledge of the scattering function for an infinite angular range [Eq. (18)]. Any extrapolation of the scattering function may introduce systematic errors. We shall focus in the following on a widely used single-step procedure which is known under the name *Indirect Fourier Transformation* (IFT) method.

3. The indirect Fourier transformation method

The IFT method (Glatter 1977a,b) performs the following tasks in one procedure:

- approximation (*smoothing*) of the primary data with a weighted least-squares technique;
- correction of instrumental broadening (*desmearing*) according to geometry and wavelength effects;
- transformation into real space – PDDF $p(r)$ or size distribution $D(R)$;
- minimization of termination effects during these transformations.

The basic idea of this method is that mathematics follows the experiment from real space to experimental space ("data space") in order to create first an optimized function system which allows "on the way back" the calculation of the desired solutions. This is the reason why the technique is called *indirect transformation*. A *direct transformation* would start in data space and would try to transform these data directly into real space.

A physical smoothing condition is given by the maximum intraparticle distance. If we want to use this condition, we have to find an estimate D_{max} as an upper limit of the largest particle dimension D:

$$D_{\max} \geq D \tag{46}$$

If such an estimate can be given it follows immediately that

$$p(r) = 0 \qquad \text{for } r \geq D_{\max} \tag{47}$$

for scattering curves from infinitely diluted systems (the case of non-negligible interparticle interference effects will be discussed later). For the following it is not necessary that the equality in Eq. (46) holds, i.e., $D_{\max}$ need not to be a perfect estimate of D. Essential however is the fact, that all transformations T_1–T_4 in Fig. 1 are *linear* transformations. Fourier transformation, convolution integrals and other smearing integrals are linear transformations.

We proceed as follows: we first assume that the PDDF $p(r)$ can be described by the series:

$$p_a(r) = \sum_{i=1}^{N} c_i \varphi_i(r) \qquad \text{for } 0 \leq r \leq D_{\max} \tag{48}$$

where N is the number of functions and c_i are the unknown expansion coefficients. The functions $\varphi_i(r)$ are chosen as cubic B-splines (Greville, 1969; Schelten and Hoßfeld, 1971). They are defined as multiple convolution products of a step function and they represent curves with a minimum second derivative. They are smooth, bell-shaped functions with a maximum at the position of the *knot* and they are non-zero only within five neighboring knots. They are therefore well suited for the representation of the smooth and band-limited $p(r)$ functions (see Fig. 2).

The coefficient c_i cannot be determined directly because we cannot measure the $p(r)$function. We have to follow with our "*ansatz*" the experiment by performing all necessary transformations T_1 to T_4 to reach the data space. The first step brings us into the reciprocal space:

$$I_a(q) = T_1 p_a(r) = T_1\left[\sum_{i=1}^{N} c_i \varphi_i(r)\right] = \sum_{i=1}^{N} c_i T_1 \varphi_i(r) = \sum_{i=1}^{N} c_i \psi_i(q) \tag{49}$$

$I_a(q)$ is the Fourier transform of $p_a(r)$ and is represented by the series of Fourier transformed B-splines $\psi_i(q)$. Note that due to the linearity of the Fourier transformation, we find the same coefficients c_i in Eq. (49) as in Eq. (48). The definition of the functions $\psi_i(q)$ is given by

$$\psi_i(q) = T_1 \varphi_i(r) \tag{50}$$

Now we transform according to the instrumental broadening effects T_2–T_4 (some of them may be negligible) and get

$$\tilde{I}_a(q) = T_4 T_3 T_2 I_a(q) = \sum_{i=l}^{N} c_i \chi_i(q) \tag{51}$$

where we find again the same coefficients c_i and the set of functions $\chi_i(q)$ in the experimental space

$$\chi_i(q) = T_4 T_3 T_2 \psi_i(q) = T_4 T_3 T_2 T_1 \varphi_i(r) \tag{52}$$

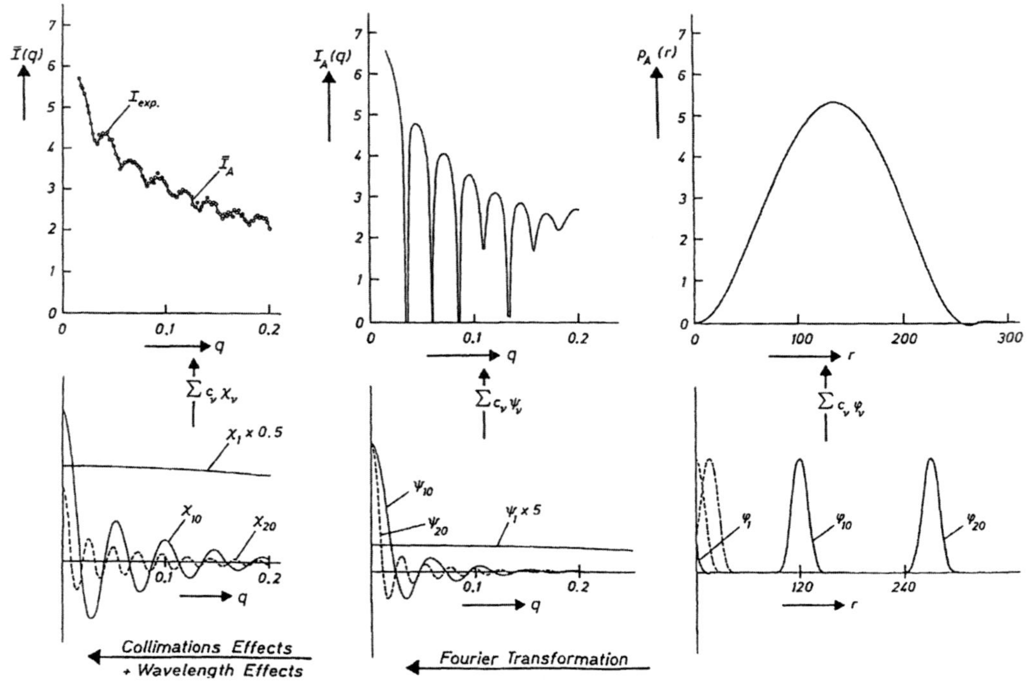

Fig. 2. Functions $\varphi_i(r)$, $\psi_i(q)$ and $\chi_i(q)$ used for the approximation of the scattering data in the indirect transformation method.

With this operation we have created the three systems of functions $\varphi_i(r)$, $\psi_i(q)$ and $\chi_i(q)$ which are optimized for the representation of the scattering functions from a particle (scattering object) with maximum dimension smaller or equal to $D_{\max}$. When the system χ_i gives a good fit to the experimental data in the *smeared experimental space* we know immediately the corresponding correct representation in *reciprocal space* (desmeared scattering curve, functions $\psi_i(q)$) and in *real space* [PDDF, functions $\varphi_i(q)$]. In order find these functions we must calculate the expansion coefficients c_i by a *weighted least-squares* operation:

$$L = \sum_{k=1}^{M} \frac{\left[I_{\exp}(q_k) - \overline{I}_a(q_k)\right]^2}{\sigma^2(q_k)} = \sum_{k=1}^{M} \frac{\left[I_{\exp}(q_k) - \sum_{i=1}^{N} c_i \chi_i(q_k)\right]^2}{\sigma^2(q_k)} \tag{53}$$

It should be kept in mind that the coefficients c_i are determined according to the condition (53) in the experimental regime $q_1 \leq q \leq q_M$ only, i.e., there is no need to extrapolate experimental data to low and high q by Guinier or Porod approximations. On the other hand, the functions $\chi_i(q)$ and $\psi_i(q)$ are defined in the whole reciprocal space and Eqs. (51) and (49) will allow to calculate solutions for all q-values of interest. This extrapolation is not restricted by an analytical expression, but depend on the function system used (splines, number of spline functions N, $D_{\max}$).

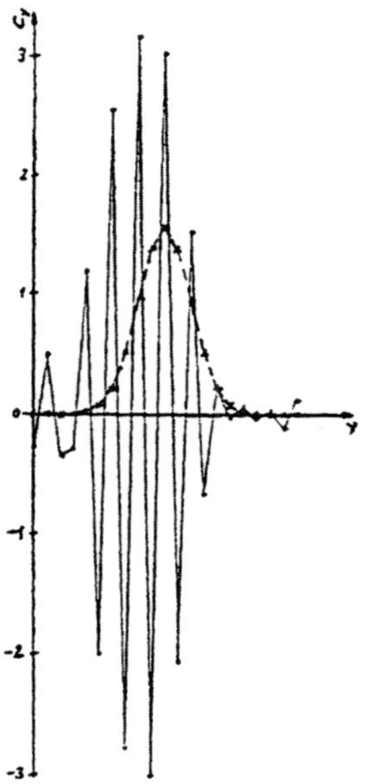

Fig. 3. −×−×−, exact expansion coefficients, ——, unstable solution.

If we would use these ideas straight forwardly we would get a good approximation of the experimental data and a reasonable desmeared curve $I_a(q)$ but the approximation $p_a(r)$ would oscillate with a very high amplitude around the correct function $p(r)$.

This phenomenon is known as *ill-posed problem* in least-squares fitting and is a result of the type of transformations T_1–T_4, the limited information in reciprocal space and the statistical error of the data points. In other words, the functions $\chi_i(q)$ are far from being orthogonal in the range $q_1 \leq q \leq q_M$ even though the function $\varphi_i(r)$ overlap only with the nearest neighbors.

An example for such an oscillating solution is shown in Fig. 3 where the solution coefficients c_i are plotted as a function of the index and are compared with the exact coefficients. These oscillations (*instabilities*) can be removed by adding a *constraint* (stabilization condition) to Eq. (53). There exist many different possible constraints. A famous one is the *maximum entropy* condition, which sounds convincing but has the disadvantage to introduce a *non-linearity* into our otherwise completely linear problem. Here we use the quite simple condition of a minimized norm of the first derivative of the solution coefficients

$$N_{c'} = \sum_{i=1}^{N-1} (c_{i+1} - c_i)^2 \tag{54}$$

This condition (54) is coupled to the least-squares condition (53) by a so-called *Lagrange-Multiplier* λ:

$$(L + \lambda N_{c'}) = \text{Min} \tag{55}$$

This Lagrange multiplier (or stabilizing parameter) defines the weight of the stabilizing condition (54) relative to the least-squares condition (53). The solution coefficients c_i are

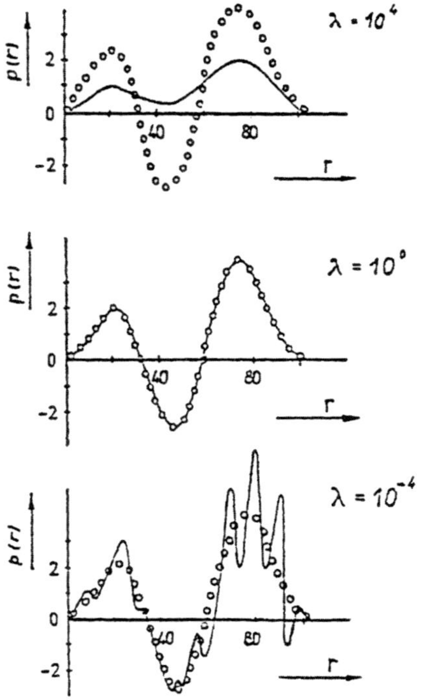

Fig. 4. Comparison of the $p(r)$ function of a given model (ooooo) with the solutions for different λ-values ($\lambda_{opt} = 10^0$).

correlated by (54), i.e., the squared difference between neighboring coefficients c_i and $c_{i\pm 1}$ is restricted proportional to $1/\lambda$. $\lambda = 0$ means that there is no restriction at all (unstabilized least squares) and $\lambda \to \infty$ means that all coefficients have the same value and the least-squares condition is not relevant in Eq. (55).

We have reduced the ill-posed problem to the non-trivial task of finding an optimum value for λ allowing a solution of the least-squares approximation (53) with a negligible increase of L while at the same time removing the artificial oscillations in $p_a(r)$, if such a value exist at all for (55). It does exist and in the following section we shall discuss how it can be determined.

In Fig. 4 we show the typical behavior of $p_a(r)$ on λ and compare it with the exact function $p(r)$. For $\lambda = 10^{-4}$ we can still see rather strong oscillations which disappear for $\lambda = 10^0$ in this example. At this optimum value $p_a(r)$ and $p(r)$ are nearly identical while a further increase of λ to 10^4 leads to a very flat solution which is no longer in agreement with the exact $p(r)$. In order to understand how λ_{opt} can be found if the exact solution is not known we create a so-called *stability plot*, where we plot the dependence of the *mean deviation* L/M and of the norm N'_c as a function of λ.

We discuss first the dependence of the mean deviation. It will be around one for low λ-values and will remain quasi constant over several decades of λ and there is a sudden increase just after λ_{opt}. The value of $N_{c'}$ decreases with increasing λ, but the slope decreases and is nearly zero (point of inflection) at λ_{opt}, i.e., the solution is nearly independent of λ

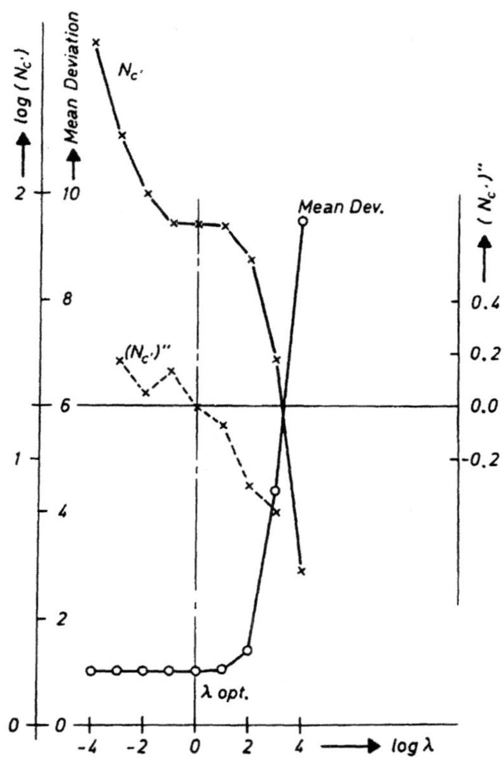

Fig. 5. *Stability plot* for the problem shown in Fig. 4: determination of the optimum Lagrange multiplier λ_{opt} by the point of inflection method.

in this regime. We find λ_{opt} in the regime where the mean deviation is still close to one and where the first derivative of $N_{c'}$ has a minimum and the second derivative $(N_{c'})''$ goes through zero (dashed line in Fig. 5).

This behavior is typical and has been found for all tests and most applications where data did not contain systematic errors and where evaluations where done within physically correct limits ($D_{max} > D$, etc.).

It was already mentioned above that the stabilization procedure introduces a correlation between the solution coefficients and thereby reduces the *effective number* of parameters which remains unknown but is in all cases smaller than N. This brings us to the important question how many spline functions N should be used. The sampling theorem of the Fourier transformation gives an estimate by the ratio $N_s = q_{max}/q_{min}$ which is typically in the regime $5 \leq N_s \leq 15$. It can be shown that the scattering functions can usually not be described by 5 to 10 splines. On the other hand, an increasing number of splines would only lead to an increase of λ_{opt}, but the solution is nearly the same for different N-values. Experimental practice leads to a standard value around 20 and limits of $10 \leq N \leq 40$, which are, however, not strict. Instrumental broadening can perfectly be corrected with this technique as can be shown by simulations.

We have discussed that it is possible to find λ_{opt} with the stability plot and have discussed the fact that the method is not very sensitive to the choice of N, the number of spline functions used for the representation of $p(r)$, but we have not yet discussed the question how to select D_{max}.

3.1. Choice of D_{max}

First of all one should not forget that there exists in most cases some *a priori* information on the system under investigation. A folded protein will hardly exceed a total length of 10 nm. Such a priori information is also necessary for the setting of the camera (size of primary beam, entrance slits, etc.) which leads to a minimum accessible q-value. The larger the objects the smaller should be the primary beam dimension. On the other hand the power of the primary beam will decrease quadratically with the reduced size of the entrance slit, so the entrance slit cannot be chosen too small. The sampling theorem of Fourier transformation (Shannon and Weaver, 1949; Bracewell, 1989) gives a clear answer to the question of largest possible particle size. If the scattering curve is sampled at increments $\Delta q \leq q_{min}$ starting at q_{min}, the scattering data contain full information for all particles with maximum dimension D_{max}

$$D_{max} = \frac{\pi}{q_{min}} \tag{56}$$

The first application of this theorem to the problem of evaluation of small-angle scattering data was given by Damaschun and Pürschel (1971a,b). In practice on will try to stay below this limit, i.e.

$$q_{min} < \frac{\pi}{D} \quad \text{and} \quad \Delta q \ll q_{min}$$

taking into account the loss of information due to counting statistics and smearing effects. In any case we can use relation (56) for a first estimation of D_{max}. If this choice of D_{max} is too small we shall not be able to fit the innermost part of the scattering curve correctly (to be seen in the residual plot and a mean deviation much greater than one). In the $p(r)$-function we usually find in this case a hump just before D_{max} because the contributions which would be at $r > D_{max}$ are shifted to $r < D_{max}$. In such a case we will also find a change in the optical appearance of the stability plot, it will look like a cross and a clear determination of λ_{opt} will not be possible. If D_{max} is chosen much bigger than π/q_{min} there will not be enough information contained in the data for a unique estimation of the PDDF for $r > \pi/q_{min}$ and the solution will start to oscillate around the abscissa axis in this regime and the solution will also be highly dependent on the actual choice of D_{max}. So a careful analysis of the results together with a variation of D_{max} can clearly solve this problem.

3.2. IFT for rod-like, lamellar or polydisperse systems

This IFT concept can also be applied for the transforms (28), (37) and (45) for rod-like, lamellar and polydisperse systems (Glatter, 1980a,b). The transform T_1 has to be defined accordingly. The situation is summarized in Fig. 6.

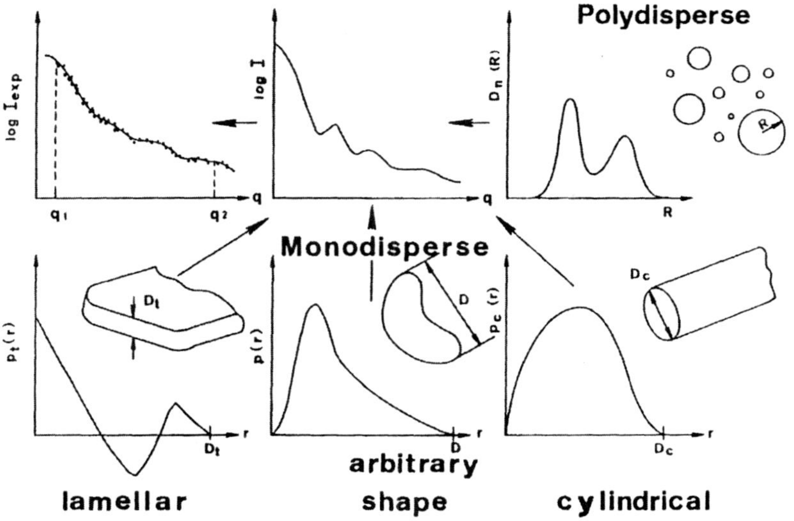

Fig. 6. Summary of the different possible applications of the indirect Fourier transformation method.

4. Deconvolution – convolution square root operation

We have seen how inhomogeneities influence the scattering functions $I(q)$ and $p(r)$, but until now we have no solution for the inverse problem, i.e., how to determine the structure of the particles from these functions. There is no general solution for this problem for arbitrary three-dimensional structures. Such methods do exist, however, for spherical symmetry, circular cylinders with centro-symmetric radial density distributions (no angular or axial dependence of the density) and for centro-symmetric lamellae without in-plane inhomogeneities.

4.1. Spherical symmetry

It is obvious that we do not lose any information by spatial averaging in the case of spherical symmetry. So we may hope that there is a chance to determine the one-dimensional information $\Delta\rho(r)$ from the measured one-dimensional function $I(q)$. We want to discuss this problem in the following section. Eqs. (14) and (15) show the correlation between the scattering intensity and the convolution square of the density distribution or the PDDF. In the case of particles with spherical symmetry we have similar equations for the scattering amplitude and the density.

We may illustrate this with the *magic square* of small-angle scattering (Fig. 7): the scattering curves in reciprocal space, the amplitude $F(q)$ and the intensity $I(q)$ are related to the density function $\Delta\rho(r)$ and the PDDF $p(r)$ in real space by a special Fourier transformation, the intensity is the square of the amplitudes and the PDDF is related by the convolution square to the density. The inversion, i.e., the calculation of $\Delta\rho(r)$ from $I(q)$, however, is still a demanding task.

The easier way seems to be the one via the amplitudes, but for the calculation of the

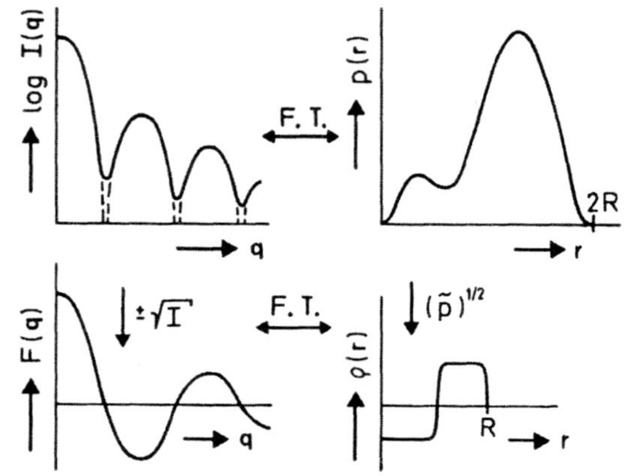

Fig. 7. The *magic square* of small-angle scattering: the correlations between the radial density $\Delta\rho(r)$ and the PDDF $p(r)$ and their Fourier transforms, the scattering amplitude $F(q)$ and scattering intensity $I(q)$ under the assumption of spherical symmetry.

scattering amplitudes from the scattering intensities one needs, beside the simple square root operation, also the right sign for the amplitudes. This is the so-called *phase problem*, a central and well-known problem in crystallography. Once this problem is solved, the radial density profile can be calculated from the amplitudes with the Fourier transformation. For some time this route via the amplitudes was the only possible way to calculate the density profile, but there exists no unique solution for the phase problem. This phase problem can be completely avoided now, since the convolution square root technique (see below) allows the deconvolution of the PDDF into the density profile. This technique avoids the phase problem [except the arbitrariness of a factor ± 1 for the whole function $\Delta\rho(r)$]. The convolution square root operation can be applied to band-limited functions, i.e., the $p(r)$ function has to be zero beyond a certain limit. This holds for particles with spherical symmetry where the limit is simply the diameter. This condition is however not fulfilled for periodical structures and can therefore not be used for ordered phases like in multi-lamellar liposomes, hexagonal phases, and in crystallography.

Similar deconvolution or square root techniques exist for rod-like and lamellar structures. Once these structures are recognized from the very shape of the PDDF of the whole particle the cross-section PDDF $p_c(r)$ or the thickness PDDF $p_t(r)$ can be calculated. These functions can then be deconvoluted into the density profiles $\Delta\rho_c(r)$ or $\Delta\rho_t(r)$ by the convolution square root technique.

Now we want to discuss shortly the principles of the deconvolution technique. Here we are facing a similar situation as in the IFT method: for a given density distribution $\rho(r)$ we can calculate the exact $p(r)$-function for all three cases (spherical, cylindrical and lamellar symmetry) by a convolution square operation but we do not have a useful description of the inverse problem, the so-called convolution square root. As an additional problem we have to keep in mind the fact, that the convolution square operation is a non-linear transformation which will not allow an inversion by the solution of a simple linear least-squares technique like in the case of the indirect Fourier transformation.

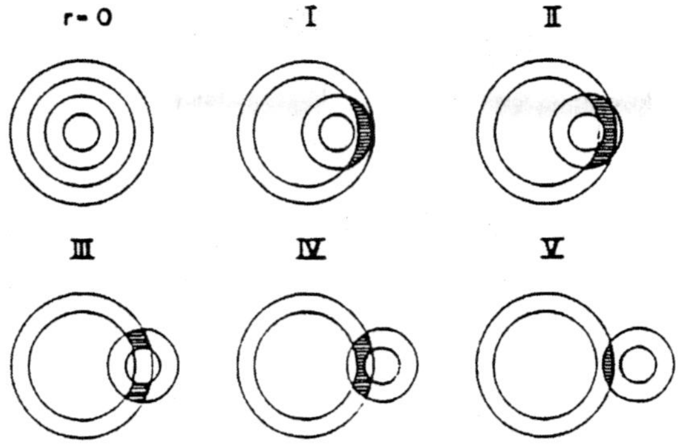

Fig. 8. Illustration of the five subregions for the calculation of the overlap integrals $V_{ik}(r)$.

We start again with a series expansion of the radial density function $\rho(r)$ in the usual way:

$$\overline{\rho}(r) = \sum_{i=1}^{N} c_i \varphi_i(r) \tag{57}$$

where we use for simplicity simple step functions for $\varphi_i(r)$. The use of B-splines would lead to severe problems in the analytical calculations of the overlap integrals. A numerical routine which allows the use of B-splines will not differ in the basic description and is developed in the meantime (Mittelbach and Glatter, 1998). The approximation $\overline{\rho}(r)$ for the density profile in Eq. (57) corresponds to an approximation $\overline{p}(r)$t o the PDDF

$$\overline{p}(r) = \sum_{i=1}^{N} V_{ii}(r) c_i^2 + 2 \sum_{i>k} V_{ik}(r) c_i c_k \tag{58}$$

The overlap integrals $V_{ik}(r)$ describe the overlapping of the ith with the kth step or shell where one function has been shifted an arbitrary distance r. These overlap or convolution integrals are very simple for the planar case (one-dimensional convolution of two step function leads simply to a triangle) but are a bit more complicated for the cylindrical and spherical case (see Fig. 8).

The analytical expressions for $V_{ik}(r)$ can be found in the original literature. For the moment it is enough to know that such expressions exist and can be used in a numerical routine to solve the convolution square root problem (Glatter, 1981; Glatter and Hainisch, 1984). Eq. (58) is non-linear in its coefficients c_i. The corresponding least-squares problem has to be linearized by a series expansion where higher-order terms are omitted. Such linearized systems must be solved iteratively as will be shown below. In addition one needs starting values $c_i^{(0)}$ for the first iteration. Here we set all coefficients equal to a constant which is estimated or normalized in such a way, that the integral over $\overline{p}^{(0)}(r)$ (calculated from $c_i^{(0)}$) is the same as the integral over the given $p(r)$ function which should be deconvoluted into $\rho(r)$.

We then calculate the difference function

$$\Delta p(r) = p(r) - \overline{p}^{(0)}(r) \tag{59}$$

which would be zero only if we would know the exact coefficients c_i. Now we calculate correction terms Δc_i in order to minimize $\Delta p(r)$ in a least-squares sense. We have to solve the set of equations:

$$\sum_{i=1}^{N} V_{ii}(r)\left[(c_i+\Delta c_i)^2\right]+2\sum_{i>k} V_{ik}(r)\left[(c_i+\Delta c_i)(c_k+\Delta c_k)-c_i c_k\right]=\Delta p(r) \tag{60}$$

which we linearize by omitting the second-order terms Δc_i^2 and $\Delta c_i \Delta c_k$ and we get

$$2\sum_{k=1}^{N}\sum_{i=1}^{N} c_i V_{ik}(r_j)\Delta c_k=\Delta p(r_j) \tag{61}$$

for $j=1,2,3,\ldots,M$ and $M>N$. These equations can be written in matrix notation

$$A_{jk}\Delta c_k=\Delta p_j \qquad \text{or} \qquad \mathbf{A}\Delta\mathbf{c}^{(\mathbf{0})}=\Delta\mathbf{p}^{(\mathbf{0})} \tag{62}$$

where the matrix elements A_{jk} are given by

$$A_{jk}=2\sum_{i=1}^{N} c_i V_{ik}(r_j) \tag{63}$$

The system (62) is solved with a weighted least-squares condition considering the standard deviations of the function $\Delta p(r)$ and we get the correction terms $\Delta\mathbf{c}$. They allow the calculation of improved coefficients $c_i^{(1)}$

$$c_i^{(1)}=c_i^{(0)}+\Delta c_i \tag{64}$$

and with these coefficients we start the next iteration at (59) and (61), get further improvements and if this iterative procedure converges we have solved the problem.

This problem is, however, again an ill-posed problem so that we have to add again a stabilization criterion and we have to solve the non-linear problem by iteration for every Lagrange multiplier. Many applications performed in the meantime have shown that the deconvolution technique works well in combination with the indirect transformation method, also in cases where the conditions of symmetry are not perfectly fulfilled (Glatter, 1988).

In Fig. 9 we see the results of a simulated example. The scattering curve for an oblate spheroid with a core–shell structure and an axial ratio of 1 : 1.2 : 1.2 is shown as data points (black dots) together with the best spherical symmetric approximation (full line). The core–shell structure has the radii 53.11 and 38.11 Å and constant densities of −0.6 in the core and +1.0 in the shell. From these data points we get a by IFT a $p(r)$ function (black dots) which is then deconvoluted into the radial profile $\rho(r)$. The $p(r)$ function for this $\rho(r)$ function is shown as a full line which fits the $p(r)$ function from the simulated data (dots) very well.

From the results we see that the core–shell structure is well recovered in the $\rho(r)$-function, the steps are smoothed due to the asymmetry of the model and the spherical approximations show sharp minima (zeros) in reciprocal space which do not exist in the original oblate model function. So the techniques work reasonably well even in such cases.

Finally it should be mentioned that it is also possible with this technique to optimize simple core–shell models by the variation of the step widths of two- or three-step models. Such calculations are not ill-posed and there is no need for stabilization for such small N-values.

The most important application of this deconvolution method for direct structure analysis is the study of complex fluids with self-assembling amphiphilic molecules in a polar or

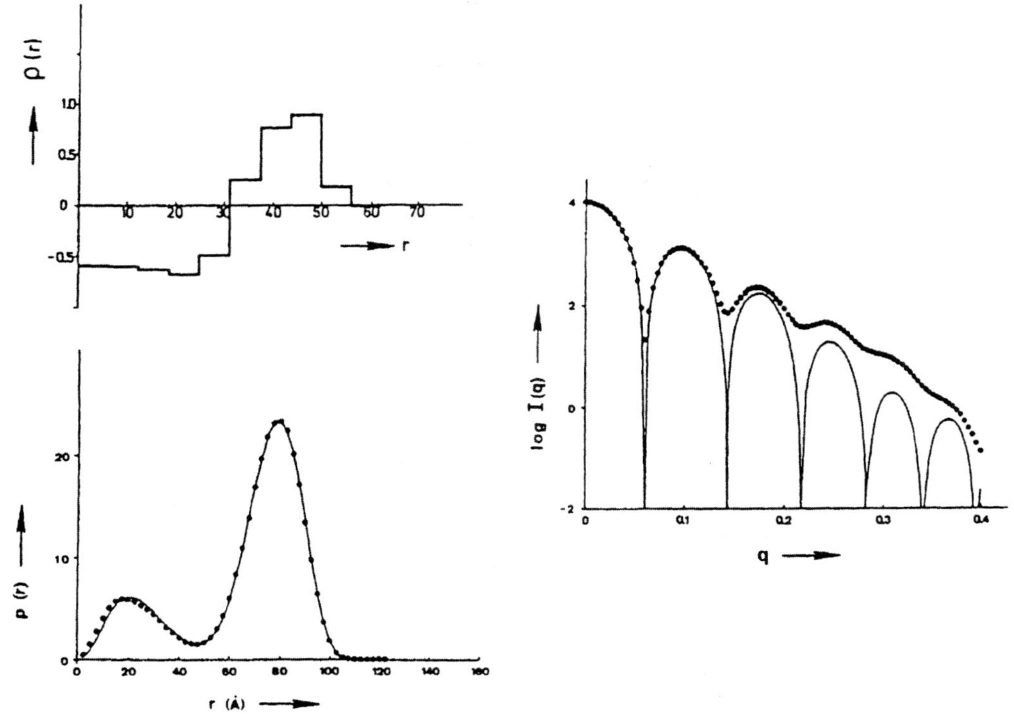

Fig. 9. Example of an inhomogeneous oblate spheroid with an axial ratio of 1 : 1.2 : 1.2.

non-polar solvent. Amphiphilic molecules consist of a hydrophilic and a hydrophobic part. The hydrophilic part – often called head group – may be ionic or non-ionic, the hydrophobic part consist in most cases of one or two alcyl chains of variable length. Depending on the chemical composition, nature of the solvent, temperature and concentration of salt many different types of aggregates may be formed: spherical and rod-like micelles, vesicles and other lamellar structures like L_3 phases (Strey et al., 1996).

In the following figures we show the general tendency of the scattering functions of such aggregates and how the methods for direct structure determination can be applied. The functions are not computed exactly according to theoretical models but illustrate the essential features.

The scattering curve of a spherical aggregate shows a series of pronounced side maxima, in theory separated by zeros (dashed-dotted line in Fig. 11); in practice, we see only minima (full line, Fig. 11, left side). From this scattering curve we can calculate the PDDF with the IFT method. If spherical symmetry can be assumed (a necessary but not sufficient condition is the appearance of pronounced side maxima in the scattering curve) we can calculate the radial density distribution $\Delta\rho(r)$ from the PDDF by the convolution square root technique. If the particle is far from spherical shape, the convolution square root technique will not be able to fit the PDDF and systematic deviations will be seen, mostly at large r-values. So the convolution square root technique can be used to test for a special symmetry like sphere, circular cylinder or symmetrical lamella.

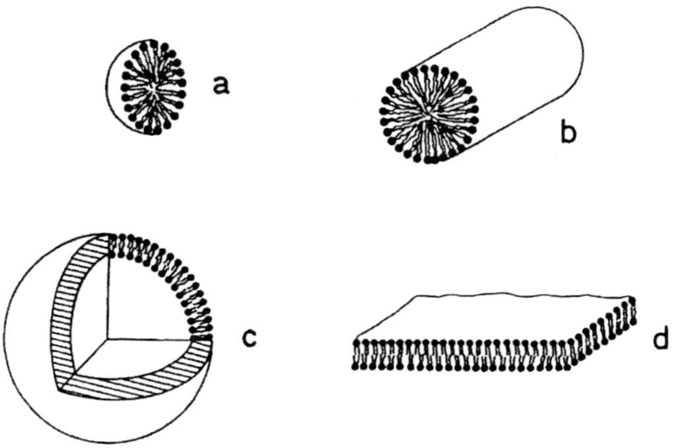

Fig. 10. Typical structures formed by amphiphilic systems: spherical micelles (a), rod-like micelles (b), vesicles (c), and lamellae (d).

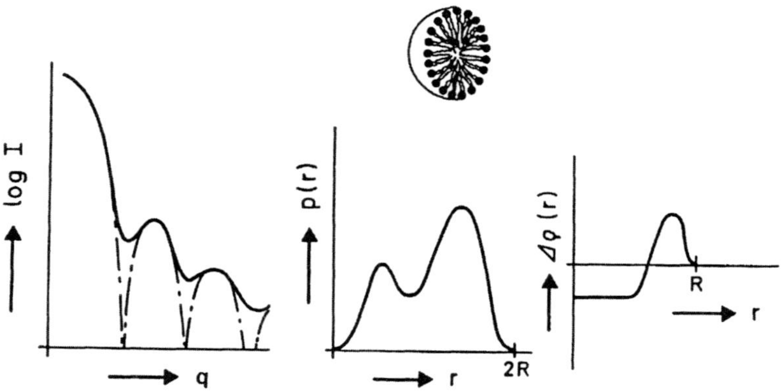

Fig. 11. Scattering pattern of spherical aggregates with radius R. The scattering curve $I(q)$ is the Fourier transform of the PDDF $p(r)$. This PDDF is the convolution square of the radial density distribution $\Delta\rho(r)$.

Rod-like (cylindrical) structures have scattering curves with a steep main maximum ($\sim q^{-1}$) and again side maxima like a sphere (Fig. 12). The position of the side maxima (and minima) is controlled by the size of the cross-section. So the difference between the spherical and the rod-like micelle is found in reciprocal space at low q-values. This difference is more pronounced in the PDDF which is again the Fourier transform of the scattering curve.

Any radial inhomogeneity of $\Delta\rho(r)$ will influence the part $0 < r < 2R$ of the PDDF and is responsible for the oscillations in the scattering curve at large q-values. The length L of the cylinder determines the length of the straight decaying line in the PDDF and is responsible for the main maximum of $I(q)$ at low q-values. If the length of the cylinder is several times larger than the diameter $2R$, one can study the radial structure of the cylindrical aggregate, i.e., the structure of the cross-section (Pedersen and Schurtenberger, 1996). We first calculate

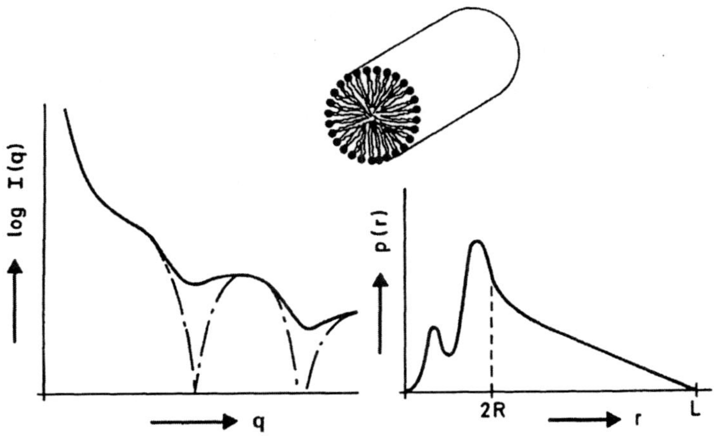

Fig. 12. Scattering function $I(q)$ and PDDF $p(r)$ for cylindrical aggregates with radius R and length L.

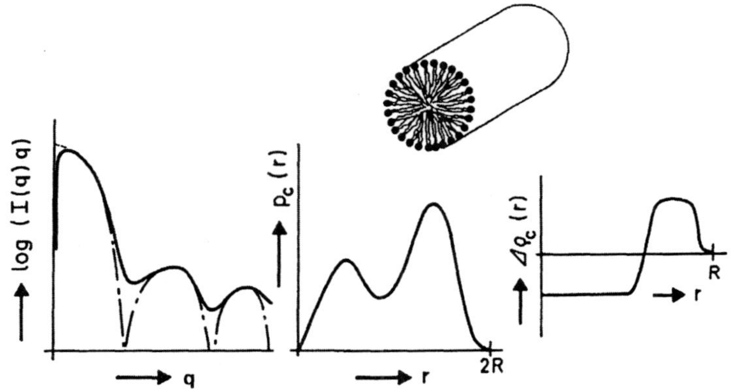

Fig. 13. Cross-section functions $I(q)q$, $p_c(r)$ and $\Delta\rho_c(r)$ for cylindrical aggregates.

the PDDF of the cross-section $p_c(r)$ from the cross-section scattering curve $I_c(q) = I(q)q$ and then by deconvolution the density profile $\Delta\rho_c(r)$ (Fig. 13).

The finite length of the cylinder forces the $I(q)q$ curve to drop to zero at $q = 0$. This part can be omitted in the indirect transformation and can therefore not cause any problems in practical applications.

There exists a similar approach for vesicles and other lamellar structures. We assume to have unilamellar systems with negligible in-plane scattering, i.e., $\Delta\rho_t$ in only a function of the distance r from the central plane of the lamella. The scattering curve shows a steep main maximum (proportional to q^{-2} at low q) but it is difficult to decide directly from the scattering curve if we have a cylindrical or a lamellar system. If the aggregates were large monodisperse unilamellar vesicles, so-called *LUVs*, the scattering function would show high-frequency oscillations (dashed line in Fig. 14). The frequency of these oscillations is a linear function of the size. In practice LUVs have a polydispersity of at least some percent up to some 10%.

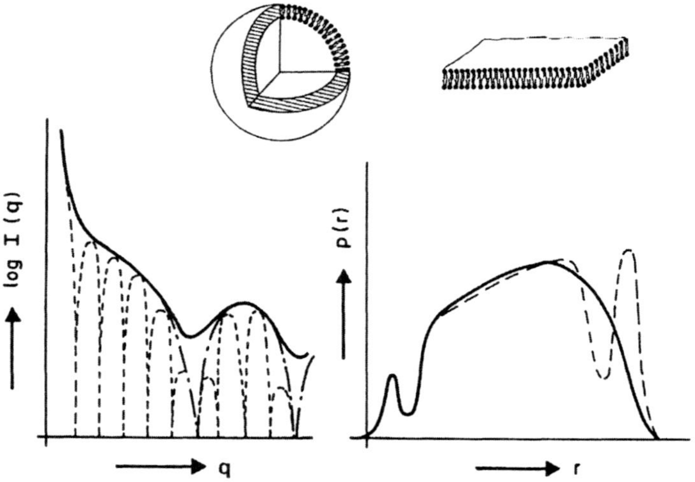

Fig. 14. Scattering function $I(q)$ and PDDF $p(r)$ for vesicles and other planar aggregates (L_3 phase). Full lines: typical results for lamellar structures; dashed lines: functions for monodisperse vesicles.

As a consequence, the oscillations are usually smeared out completely. No such oscillations occur for L_3 phases and flat lamellae. The difference between L_3 phases or flat lamellae and polydisperse LUVs is given by the innermost part of the scattering curve, which is, in most cases not accessible to the experiment.

The difference between cylindrical and lamellar aggregates can be seen from the PDDF in Figs. 12 and 14. In the case of a cylinder, we have a clear maximum within the cross-section which is followed by a monotonously decreasing part. The inhomogeneity of the particles given by $\Delta\rho_c(r)$ (Fig. 13) causes the minimum in the PDDF in the range of $0 < r < 2R$, the value of $p_c(r)$ at the minimum may be positive or negative. Such a minimum can also be found in the case of lamellae of LUVs (Figs. 14 and 15), but the PDDF shows no maximum near $r < T$ (where T is the thickness of the lamella) but runs constantly increasing into the broad, non-structured part. This part with its decay to zero would show the overall size of the aggregates if the scattering curve could be measured to small enough q-values.

Once the lamellar feature is found from the PDDF we can calculate the thickness scattering function $I_t(q) = I(q)q^2$ and by means of IFT the thickness PDDF $p_t(r)$ (Fig. 15). A deconvolution of $p_t(r)$ gives finally the density profile $\Delta\rho_t(r)$.

This calculation of the thickness function is possible as long as the overall size of the aggregates and its radius of curvature is much larger than the thickness T.

5. Interacting systems

Until now we have assumed infinite dilution which led first of all to the simplification that the mean density $\overline{\rho}$ is simply replaced by the density of the surrounding medium, i.e., the density of the solvent ρ_s. This contribution to the scattering was highly peaked at $q = 0$ and therefore omitted.

We have further assumed that all scatterers are completely uncorrelated, i.e., every particle

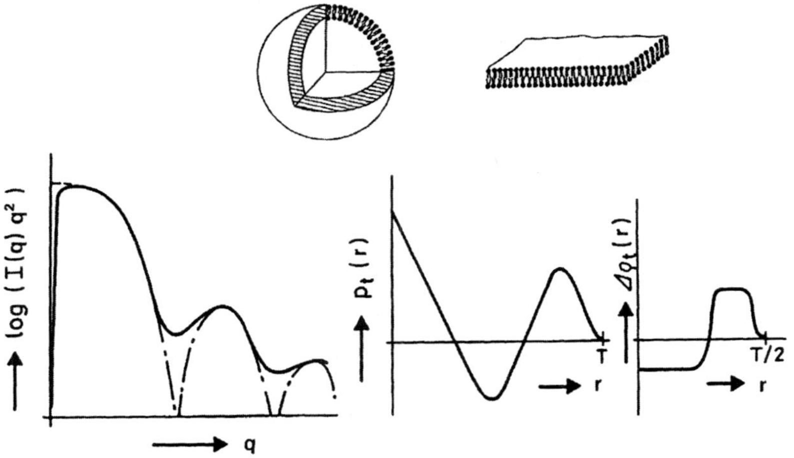

Fig. 15. Thickness functions $I(q)q^2$, $p_t(r)$ and $\Delta\rho_t(r)$ for vesicles and other planar aggregates with the thickness T.

can be regarded as independent in respect of translation and orientation, assuming no particle interaction at all. In terms of the results of this section every particle is a fluctuation from the mean of no particle at all and the resulting scattering intensity $I(q)$ is proportional to the *form factor*.

What happens if we increase the concentration? Comparing the results for different concentrations c and normalizing the intensities by a division through the concentration we find that the I/c curves fall more or less on top of each other but that they deviate with increasing concentration in the low q-part. In most cases we find a decrease of the intensity. These concentration effects are discussed in full detail in Chapter 14 by R. Klein.

For identical spheres holds

$$I(q) = N F_0^2 P(q) S(q) \tag{65}$$

the first part of this equation $N F_0^2 P(q)$ describes scattering from N single spheres and the *structure factor* $S(q)$ describes the effect of positional correlations.

We shall learn in the contribution of R. Klein that the structure factor is related to the *pair correlation function* $g(r)$ by

$$S(q) = 1 + 4\pi \frac{N}{V} \int_0^\infty [g(r) - 1] r^2 \frac{\sin qr}{qr} \, \mathrm{d}r \tag{66}$$

A detailed derivation of this relation can be found in Klein and D'Aguanno (1996) [see also Hansen and McDonald (1986), and Kaler (1995)].

The total scattering intensity $I(q)$ is with (66) given by

$$I(q) = N F_0^2 P(q) \left\{ 1 + 4\pi n \int [g(r) - 1] r^2 \frac{\sin qr}{qr} \, \mathrm{d}r \right\} \tag{67}$$

In order to show similarities and dependencies we summarize the corresponding equations in Table 1.

Table 1

Summary of the Fourier transformation equations for interacting spherical systems

$$I(q) = N F_0^2 P(q) S(q) \tag{65}$$

$$N F_0^2 P(q) = 4\pi \int p(r) \frac{\sin qr}{qr} \, \mathrm{d}r \tag{16}$$

with

$$p(r) = \gamma(r) \cdot r^2 \tag{15}$$

$$S(q) - 1 = 4\pi n \int s(r) \frac{\sin qr}{qr} \, \mathrm{d}r \tag{66}$$

with

$$s(r) = h(r) r^2 = [g(r) - 1] r^2 \tag{67}$$

and for analogy we define

$$I(q) = 4\pi \int i(r) \frac{\sin qr}{qr} \, \mathrm{d}r \tag{68}$$

For non-interacting, infinitely diluted systems we get $S(q) = 1$ and $i(r) \equiv p(r)$. We shall use the function $i(r)$ as the Fourier transform of $I(q)$ only where it is essential to differentiate between the transforms of $I(q)$ and $P(q)$. In all other cases we shall simply use $p(r)$. The reciprocal space functions $P(q)$, $S(q)$ and $I(q)$ for a system of hard spheres are shown in Fig. 16, while the corresponding real space functions $p(r)$, $h(r)$, $s(r)$ and $i(r)$ are depicted in Fig. 17. The multiplication of the form factor $P(q)$ with the structure factor $S(q)$ causes mainly changes in the low-q regime while the real space function $p(r)$ is changed in the whole r-space as can be seen from the comparison of $p(r)$ with $i(r)$. Obviously, the function $i(r)$ is no longer helpful for the determination of the shape of the particle as $p(r)$ was.

A generalized version of the IFT method, the *Generalized Indirect Fourier Transformation* (GIFT) makes use of the factorization in Eq. (65) while keeping the ideas if the IFT method. This recently developed technique (Brunner-Popela and Glatter, 1997; Weyerich et al., 1999; Brunner-Popela et al., 1999; Bergmann et al., 2000; Fritz et al., 2000) will be described shortly in the following section.

6. Generalized Indirect Fourier Transformation (GIFT)

A formal factorization like in Eq. (65) is also possible for polydisperse and non-spherical systems. In these cases the structure factor has to be replaced by a so-called effective structure factor (Weyerich et al., 1999). As can be seen in Eq. (65), $I(q)$ not only depends on $P(q)$ but also on $S(q)$, which gives rise to a highly non-linear system if both functions have to be determined simultaneously. In the following section we present a method to solve this non-linear *ill-posed* problem. The aim is the simultaneous determination of $P(q)$ and $S(q)$ with minimum a priori information.

The method used is a generalization of the IFT method, so we start in real space. The pair distance distribution function $p(r)$ is approximated by a linear combination of spline functions

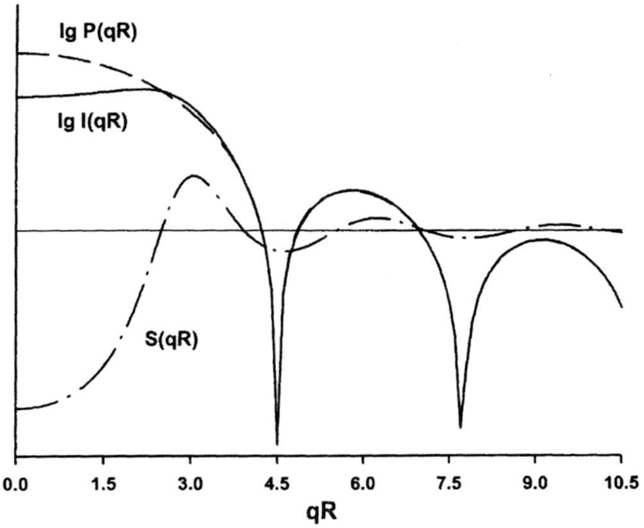

Fig. 16. Reciprocal space functions for spheres with hard sphere interaction.

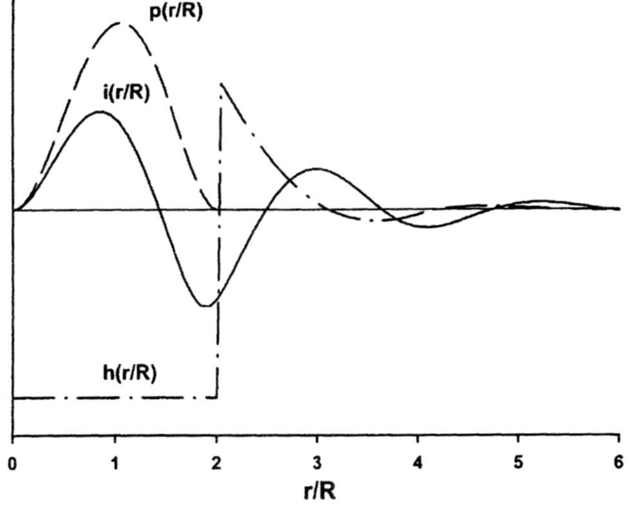

Fig. 17. Real space functions for spheres with hard sphere interaction.

(48), i.e., we make no assumptions about the shape or internal structure of the particle. The particular functions φ_ν are again transformed (T_1) into reciprocal space so that the form factor $P(q)$ can be written as the Fourier transform of $p(r)$

$$P(q) = \sum_{\nu=1}^{n} c_\nu \psi_\nu(q) \tag{49a}$$

Now the point is reached to include the structure factor according to Eq. (65):

$$I(q) = P(q)S(q,\boldsymbol{d}) = \sum_{\nu=1}^{n}\left[c_\nu \psi_\nu(q)\right]S(q,\boldsymbol{d}) \tag{69}$$

The vector $\boldsymbol{d}$ contains the coefficients d_k $(k = 1,\ldots,4)$ determining the structure factor. These coefficients correspond to the volume fraction, the size (radius), the polydispersity parameter and the charge of the particle. As can be seen from Eq. (69), the scattering intensity $I(q)$ is no longer just a function of the scattering vector q and the coefficients c_ν but from that point on also of the coefficients d_k. The scattering function can thus be represented as the product of the Fourier transformed B-splines $\psi_\nu(q)$ and the structure factor.

The functions $\psi_\nu(q)$ can be directly multiplied by $S(q,\boldsymbol{d})$ defining a new set of functions $\overline{\psi}_\nu$:

$$\overline{\psi}_\nu(q,\boldsymbol{d}) = \psi_\nu(q)S(q,\boldsymbol{d}) \tag{70}$$

With that definition we get:

$$I(q) = \sum_{\nu=1}^{n} c_\nu \psi_\nu(q)S(q,\boldsymbol{d}) = \sum_{\nu=1}^{n} c_\nu \overline{\psi}_\nu(q,\boldsymbol{d}) \tag{71}$$

$\overline{\psi}_\nu$ depends both on q and on $\boldsymbol{d}$ and therefore implicitly contains information about the structure factor.

The smearing effects in regard to wavelength distribution (transformation T_2) and geometric effects (transformations T_3 and T_4) can again be taken into account [like in (51)]:

$$\tilde{I}(q) = T_4T_3T_2I(q) = \sum_{\nu=1}^{n} c_\nu \overline{\chi}_\nu(q,\boldsymbol{d}) \tag{72}$$

where $\tilde{I}(q)$ represents the smeared scattering intensity and $\overline{\chi}_\nu$ another set of functions:

$$\overline{\chi}_\nu(q,\boldsymbol{d}) = T_4T_3T_2\overline{\psi}_\nu(q,\boldsymbol{d}) \tag{73}$$

The coefficients c_ν and d_k can be determined by a weighted least-squares approximation of the experimental data $I_{\mathrm{exp}}(q)$ by the sum over all basic functions (Bevington, 1969; Brandt, 1970)

$$L = \sum_{i=1}^{m}\left(\frac{I_{\mathrm{exp}}(q_i) - \tilde{I}(q_i,\boldsymbol{c},\boldsymbol{d})}{\sigma(q_i)}\right)^2 = \sum_{i=1}^{m}\left(\frac{I_{\mathrm{exp}}(q_i) - \sum_{\nu=1}^{n} c_\nu \overline{\chi}_\nu(q_i,\boldsymbol{d})}{\sigma(q_i)}\right)^2 = \min \tag{53a}$$

The situation is schematically presented in Fig. 18.

The main difficulty is, however, that in our case the least-squares problem leads to a highly non-linear system of equations because of the product of form factor and structure factor. This problem has first been solved be a special linearized iterative procedure (Brunner-Popela and Glatter, 1997; Weyerich et al., 1999; Brunner-Popela et al., 1999). This method worked reasonably well, but due to slow convergence and other problems in the iterative procedure it had to be replaced by a Boltzmann Simplex Simulated Annealing (BSSA), which proved

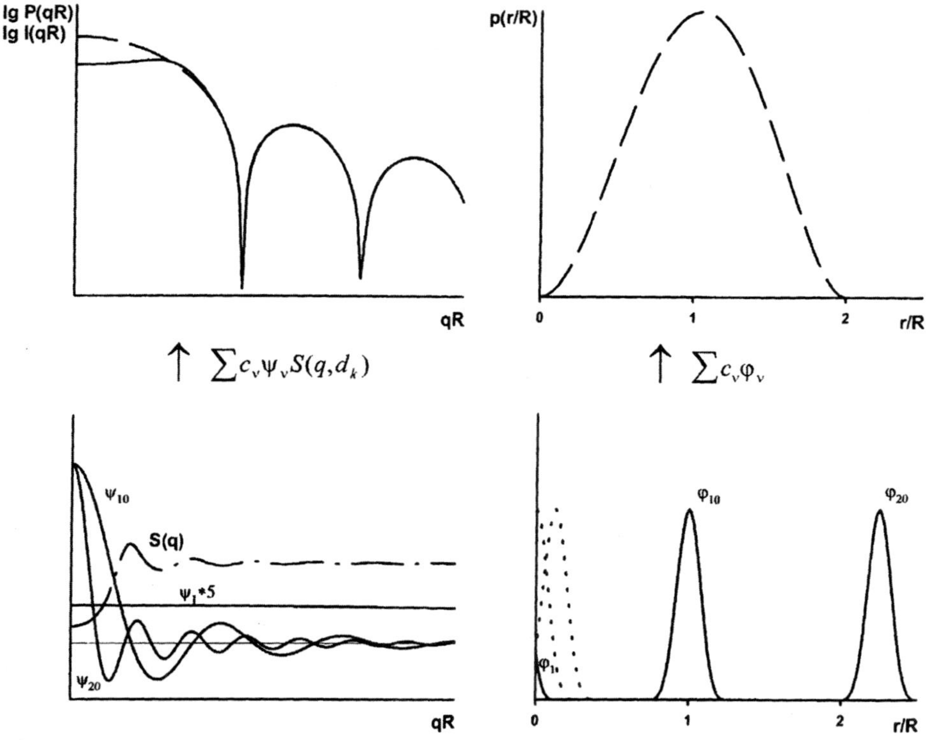

Fig. 18. Definition of the function systems used in the GIFT algorithm.

to converge to the total minimum (Bergmann et al., 2000) and had already been applied with success (Glatter et al., 2000). In the last step we have extended the method to charged systems using different closure relations for the description of particle interaction (Fritz et al., 2000; Maranzano et al., 2000).

References

Bergmann, A., Fritz, G., and Glatter, O. (2000). *J. Appl. Cryst.*, **33**, 1212–1216.

Bevington, P.R. (1969). *Data Reduction and Error Analysis for the Physical Sciences*. McGraw-Hill, New York.

Bracewell, R. (1986). *Fourier Transform and its Applications*. McGraw-Hill, New York.

Brandt, S. (1970). *Statistical and Computational Methods in Data Analysis*. North-Holland, Amsterdam.

Brunner-Popela, J. and Glatter, O. (1997). *J. Applied Cryst.*, **30**, 431–442.

Brunner-Popela, J., Mittelbach, R., Strey R., Schubert K.-V., Kaler, E.W. and Glatter, O. (1999). *J. Chem. Phys.*, **21**, 10623–10632.

Damaschun, G. and Pürschel, H.V. (1971a). *Acta Cryst.*, **A27**, 193–197.

Damaschun, G. and Pürschel, H.V. (1971b). *Mh. Chem.*, **102**, 1146–1155.

Fritz, G., Bergmann, A. and Glatter, O. (2000). *J. Chem. Phys.*, **113**, 9733–9740.

Glatter, O. (1977a). *Acta Phys. Austriaca*, **47**, 83–102.

Glatter, O. (1977b). *J. Appl. Cryst.*, **10**, 415–421.

Glatter, O. (1980a). *J. Appl. Cryst.*, **13**, 577–584.

Glatter, O. (1980b). *J. Appl. Cryst.*, **13**, 7–11.

Glatter, O. (1981). *J. Appl. Cryst.*, **14**, 101–108.

Glatter, O. (1988). *J. Appl. Cryst.*, **21**, 886–890.
Glatter, O. and Zipper, P. (1975). *Acta Phys. Austriaca*, **43**, 307–310.
Glatter, O. and Hainisch, B. (1984). *J. Appl. Cryst.*, **17**, 435–441.
Glatter, O., Fritz, G., Lindner, H., Brunner-Popela, J., Mittelbach, R., Strey, R., Stefan U. and Egelhaaf, S.U. (2000). *Langmuir*, **16**, 8692–8701.
Greville, T.N.E. (1969). *Theory and Applications of Spline Functions*. Academic Press, New York.
Hansen, J.P. and McDonald, I.R. (1986). *Theory of Simple Liquids*. Academic Press, London.
Kaler, E.W. (1995). In: *Modern Aspects of Small-Angle Scattering*, (Brumberger, H., ed.). Kluwer Academic Publishers, Dordrecht.
Klein, R. and D'Aguanno, B. (1996). In: *Light Scattering – Principles and Development* (Brown, W., ed.). Clarendon Press, Oxford, pp. 30–102.
Mittelbach, R. and Glatter, O. (1998). *J. Appl. Cryst.*, **31**, 600–608.
Pedersen, J.S. and Schurtenberger, P. (1996). *J. Appl. Cryst.*, **29**, 646–661.
Schelten, J. and Hossfeld, F. (1971). *J. Appl. Cryst.*, **4**, 210–223.
Shannon, C.E. and Weaver, W. (1949). *The Mathematical Theory of Communication*. University of Illinois Press, Urbana, IL.
Strey, R., Glatter, O., Schubert K.-V. and Kaler, E.W. (1996). *J. Chem. Phys.*, **105**, 1175–1188.
Weyerich, B., Brunner-Popela, J. and Glatter, O. (1999). *J. Appl. Cryst.*, **32**, 197–209.

Part II
Methods

Neutrons, X-Rays and Light
P. Lindner and Th. Zemb (Editors)

Chapter 6

Instrumentation for Small-Angle X-ray and Neutron Scattering and Instrumental Smearing Effects

Jan Skov Pedersen

Department of Chemistry, University of Aarhus, Langelandsgade 140, DK-8000 Aarhus C, Denmark

1. Introduction

There exist an almost infinite number of different ways of combining X-ray/neutron sources with collimators, optical devices and detector systems. Therefore the number of different instrument employed in small-angle scattering studies is very large and it is not possible in one chapter to describe all known types of instruments. In the following the most common instruments are described, however, some space is also devoted to the most recent developments which has resulted in so large improvements, that new types of science can be done with the new instruments.

A quite extensive review of small-angle scattering instrumentation was given in 1995 (Pedersen, 1995). In that paper all the most common X-ray and neutron sources as well as the most common instrument components were described in quite great detail, which gave a good basis for discussing the principles of the traditional small-angle X-rays scattering (SAXS) instruments and small-angle neutron scattering (SANS) instruments. The description of the various components will not be repeated here, however, one newly developed component, the *Göbel* mirror, will be described, as it has led to a significant improvement of laboratory SAXS instruments. For details of the more traditional components the interested reader is referred to the review from 1995.

Whether a SAXS or a SANS instrument is chosen for a particular series of measurements should depend on the actual sample to be studied. Some samples have a high contrast for X-rays whereas other have a high contrast for neutrons. The difference is due to the difference of scattering process of X-rays and neutrons (see, e.g., Glatter, 1992; May, 1992). X-rays are scattered by the electrons in the sample and the scattering power of an atom is proportional to its number of electrons (atomic number). The scattering contrast in a sample is provided by the variation in electron density in the sample, typically between particles and matrix/solvent. The largest signals are obtained from dense samples where the contrast is between air and material. The neutrons are scattered by interactions with the nuclei in the sample. The scattering ability is not proportional to the atomic number and varies in a non-systematic way with the atomic number. The interactions can also be quite different for different isotopes as it is for hydrogen and deuterium. This makes it relatively easy to control the scattering power of a solvent by mixing non-deuterated and deuterated solvent. Due to the opposite sign of the scattering length of hydrogen and deuterium one can choose the solvent scattering density to match different components of the samples. In polymer science the possibility of deuterating some of the chains or part of a chain is often exploited (see Chapter 7 by P. Schurtenberger).

2. Instrumental smearing

The requirement for having a relatively fast data acquisition leads to the use of broad wavelength band monochromators or relaxed collimation for the incident radiation as this leads to a higher flux of radiation on the sample. For X-rays, it has been a tradition to use long-slit geometries in which the beam is only well-collimated in one direction. In such a geometry, the modulus of the scattering vector $q = 4\pi \sin\theta/\lambda$, where 2θ is the scattering angle and λ is the wavelength of the radiation, has a broad distribution (see Fig. 1). This leads to a significant smearing of the scattering curve. However, the distribution of q has a sharp edge towards small q (Fig. 2) and therefore the information on the ideal scattering cross-section $\mathrm{d}\sigma(q)/\mathrm{d}\Omega$ is not lost, and it is possible in the data analysis to correct for the smearing. Figs. 1 and 2 show examples of the areas and distributions in q sampled in, respectively, a long-slit and a pinhole instrument.

For long-slit instruments, the mathematical expression for the smearing effects which is used for systems with isotropic scattering is

$$I(q) = \iiint \frac{\mathrm{d}\sigma}{\mathrm{d}\Omega}\left(\frac{\lambda_0\sqrt{(q-q_\parallel)^2+q_\perp^2}}{\lambda}\right) W_\parallel(q_\parallel)\, W_\perp(q_\perp)\, W_\lambda(\lambda)\, \mathrm{d}q_\parallel\, \mathrm{d}q_\perp\, \mathrm{d}\lambda, \qquad (1)$$

where $I(q)$ is the measured intensity, λ_0 is the nominal wavelength and $W_\lambda(\lambda)$ is the wavelength distribution. The parameters $q_\parallel$ and $q_\perp$ are the scattering vector deviations parallel and perpendicular to q, respectively, and $W_\parallel(q_\parallel)$ and $W_\perp(q_\perp)$ are the respective distributions. These two latter distributions contain contributions from both the collimation system and the spatial resolution of the detector. The weighting functions can be determined by measuring

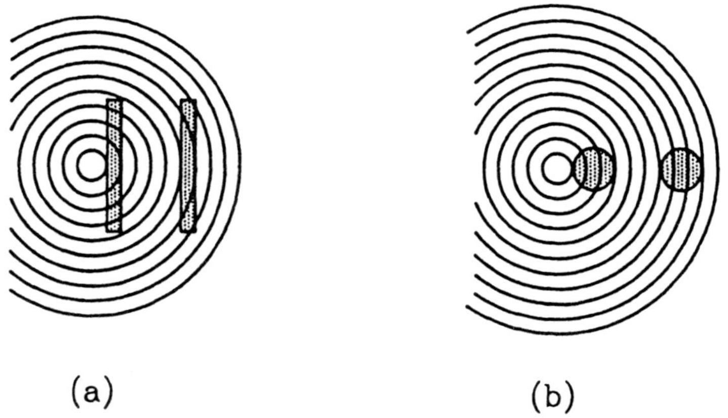

Fig. 1. The range of scattering vectors probed by (a) the long-slit geometry and (b) the pinhole geometry. The origin of the q scale is at the center of the concentric circles. The range is shown for two values of the scattering vector.

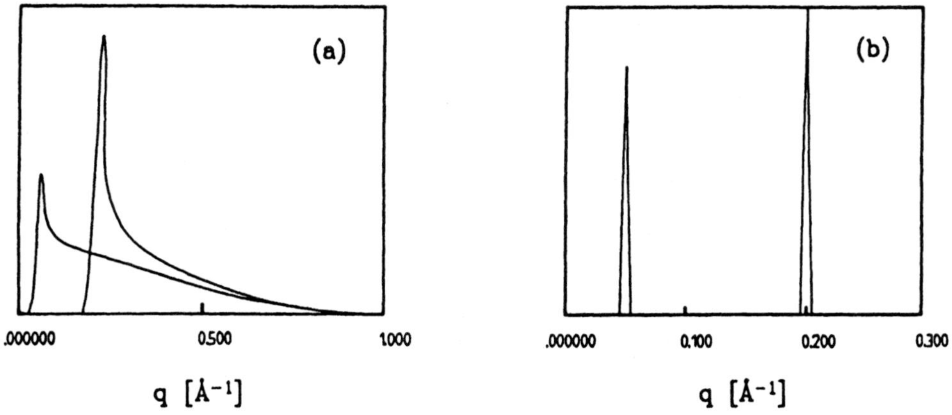

Fig. 2. The distribution of scattering vectors for two geometries: (a) long-slit geometry and (b) pinhole geometry. The distributions are shown for two values of the scattering vector.

the spectrum of the source and monochromator and by measuring the direct beam profile. The slit functions can also be calculated analytically for various long-slit geometries, as done by Hendricks and Schmidt (1967, 1973).

An alternative way of increasing the flux at the sample position is to focus the beam by means of optical elements like curved mirrors and bent monochromator crystals. If the beam is focussed on the detector (Fig. 3) there is essentially no smearing of the measured intensity for isotropic samples (see, for example, Pedersen and Riekel, 1991). In Fig. 3 one sees that for small scattering angles all rays hitting the detector at a certain point have approximately the same scattering angle and therefore there is no additional smearing. However, when colloidal single crystals are studies, the associated rocking curves, which are obtained by rotating the sample, is significantly broadened due to the variation of the angle between the incident beam and the sample across the sample. The application of the focusing geometry requires the use of larger samples, which in some situations might be a drawback or a practical limitation.

For small-angle neutron scattering (SANS) the resolutions is usually relaxed using a

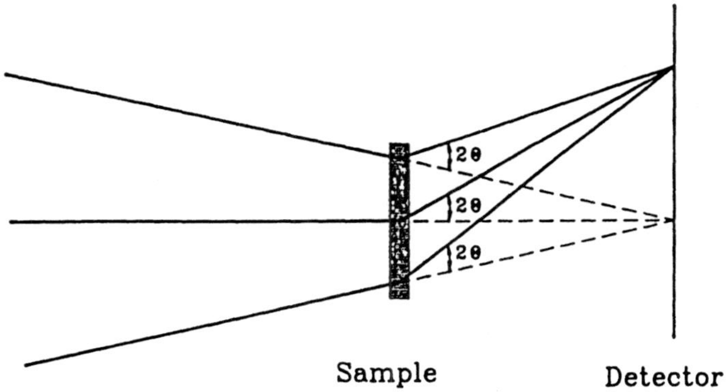

Fig. 3. Scattering angles for a setup with the beam focused at the detector.

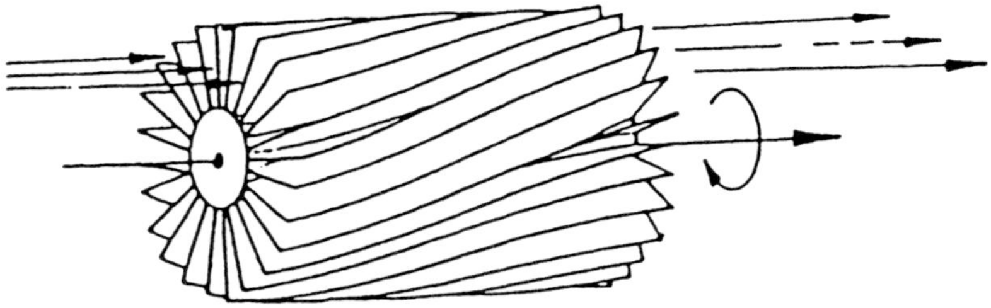

Fig. 4. A mechanical velocity selector with helical slots.

mechanical velocity selector (Holt, 1957) which allows a relatively large range of wavelengths to pass, typically 10–20% (full-width-at-half-maximum (FWHM) value). The velocity selector (Fig. 4) consists of a rotor with tilted or helical slots of neutron-absorbing material. Due to the rotation of the rotor only neutrons with a certain range of velocities are transmitted.

The effects of increasing the wavelength spread can be described by making a first-order Taylor expansion of the expression for the scattering vector with respect to the wavelength. This gives:

$$\Delta q = -q\Delta\lambda/\lambda, \tag{2}$$

and thus a spread in wavelength results in distribution of scattering vectors which increases with q. The corresponding expansion in terms of the scattering angle θ gives:

$$\Delta q = \frac{4\pi}{\lambda}\cos\theta\,\Delta\theta \approx \frac{4\pi}{\lambda}\Delta\theta, \tag{3}$$

which demonstrates that the width of distribution from the collimation effects is independent of q for small scattering angles.

The smearing effects for SANS can in principle also be described by Eq. (1), however, due to the nearly Gaussian shape of the contribution to smearing (Pedersen, 1993), it is more convenient to use a resolution function, which describes the distributions of q values probed for each instrumental setting. The use of Gaussian functions allows the combination of the various effects to be done analytically and the smearing can be included by calculating just one numerical integral (Pedersen et al., 1990; Pedersen and Riekel, 1991; Pedersen, 1993):

$$I(\langle q\rangle) = \int R(\langle q\rangle, q)\frac{\mathrm{d}\sigma(q)}{\mathrm{d}\Omega}\,\mathrm{d}q. \tag{4}$$

where $\langle q\rangle$ is the nominal scattering vector corresponding to the setting of the instrument and $R(\langle q\rangle, q)$ is the distribution of scattering vectors q around the nominal value $\langle q\rangle$. When it is assumed that the various contributions to the smearing from wavelength spread, finite collimation and finite detector resolution are independent the variance of the resolution function is given by

$$\sigma(\langle q\rangle)^2 = \langle q\rangle^2\sigma(\Delta\lambda/\lambda)^2 + (4\pi/\lambda)^2\sigma(\Delta\theta)^2, \tag{5}$$

where $\sigma(\Delta\lambda/\lambda)^2$ is the variance due to wavelength spread and $\sigma(\Delta\theta)^2$ is the variance in scattering angle due to collimation and detector resolution. The variances can be calculated or measured. The wavelength spread is commonly measured using a chopper and a time-of-flight

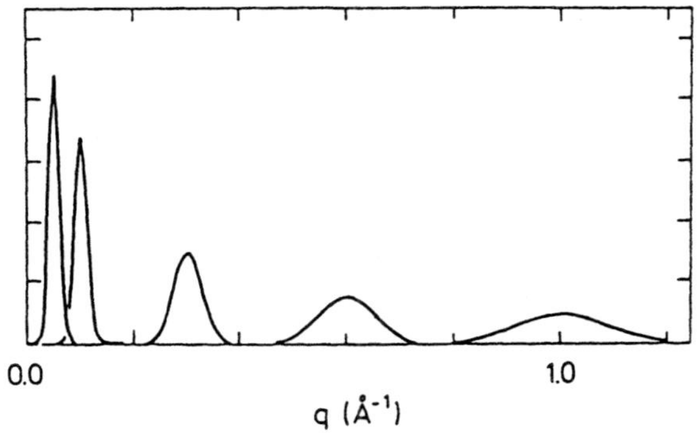

Fig. 5. Distribution of scattering vectors probed by a pinhole small-angle neutron scattering setup with the wavelength spread $\Delta\lambda/\lambda = 20\%$.

approach, whereas the scattering angle spread can be determined by measuring the profile of the direct (attenuated) beam on the detector.

Fig. 5. shows the resolution function for five different values of the nominal scattering vector. At low $\langle q \rangle$ the resolution function is dominated by the contribution from the spread in scattering angles whereas it is dominated by the contributions from the wavelength spread at high $\langle q \rangle$.

3. Small-angle X-ray scattering instruments

In SAXS, the combination of very intense incident beams and imperfections of the slits or pinholes leads to a significant parasitic scattering from the slits or pinholes. It is concentrated around the direct beam and it is a serious obstacle for obtaining scattering data with good quality at low q for weakly scattering samples. In order to reduce this undesirable parasitic scattering and extra *guard* slit or pinhole is include in the setup. Fig. 6 shows the principles. The first two slits collimate and define the beam and the third slit, the guard slit, is aligned as close as possible to the intense direct beam without touching it, as this would again give rise to parasitic scattering from this slit. The sample is placed right after the guard slit. The area of intense background is defined by the second slit and the guard slit. The scattering signal can be recorded reliably outside this region. Note that the beamstop size is chosen so that the beamstop covers the region with intense background. As large a part of the instrument as possible is evacuated in order to reduce air absorption and air scattering. One large integrated vacuum system is an advantage as the scattering from windows can be avoided.

A small-angle scattering instrument consists of a source, a monochromator, a collimating system, and a detector. Focussing mirrors, crystals and multilayers can also be incorporated. Conventional laboratory sources are sealed tubes and rotating anode sources. Usually the K_α lines of a Cu anode are employed as these give a reasonable transmission for organic material and solvents for samples with 1 mm thickness. Recently, micro-focus sources have been developed (see, for example, Gubarev et al., 2000), however, they have so far not been used in standard SAXS setups. Synchrotron X-ray sources which have a significantly higher intensity than conventional sources are also used for SAXS. Bending magnets and wigglers provide a continuous spectrum, whereas undulators have a structured spectrum with relatively

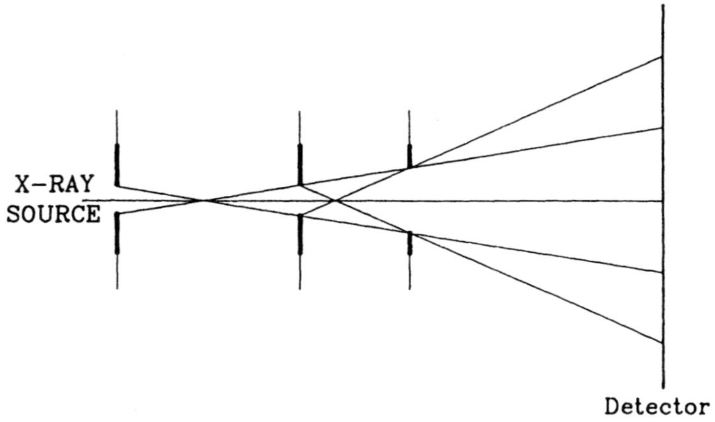

Fig. 6. The principles of a small-angle X-ray scattering setup. The two first slits define the direction of the incident beam. The third slit removes parasitic scattering from the second slit and should not touch the primary beam.

broad peaks. The intensity of wigglers is significantly higher than that of bending magnets whereas undulators has an intensity which is higher than that of wigglers. Synchrotron sources allow anomalous scattering experiments to be performed, in which the wavelength is varied in the vicinity of an absorption edge. The scattering length of atoms vary strongly close to the absorption edges and it is this effect which is exploited in contrast variation methods. However, most soft condensed matter samples do not contain elements which allows the technique to be used for wavelength in the 1–2 Å region.

In most instruments, the beam is monochromatized by Bragg reflection in a crystal. Mosaic crystals (like graphite), multilayers and Göbel mirrors (Schuster and Göbel, 1995) are usually only used in laboratory instruments. A Göbel mirror consists of a support with parabolic or elliptical shape with a length of typically 40–60 mm (see Fig. 7). The X-ray source is located in the focus of the parabola or in one of the foci of the ellipse. The shape of the support is so that the mirror is place very close to the source, typically at a 100–150 mm distance. A multilayer has been evaporated onto the support, so that a characteristic line of the source can be Bragg reflected. The period of the multilayers varies along the surface of the support, so that the beam reflected at different points for a parabola comes out parallel, whereas it for an elliptical mirror is focussed at the distant focus. Mosaic crystals and Göbel mirrors have

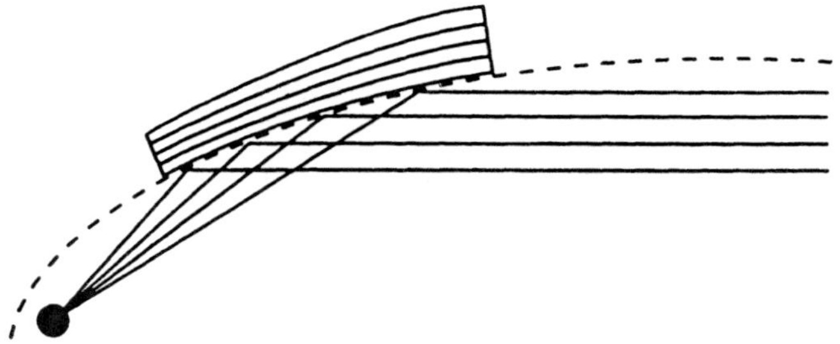

Fig. 7. The principles of a Göbel mirror. The period and scattering angle varies, so that the rays come out parallel.

relatively poor optical properties and therefore the best instrument performance is obtained when the beam is collimated after passing these components. Perfect crystals like Silicon and Germanium are optically more perfect and can be used within the collimation part without significantly deteriorating the quality of the beam.

Most detectors used in SAXS are position sensitive. Gas proportional wire detectors are still the most used detectors. More recently CCD cameras and image plates have been used. In a CCD camera the X-ray photons are converted to light which is then recorded. In image plate systems, the signal is stored on a plate as metastable colour centers, and the plate is scanned off-line by a laser, so that the stored signal can be retrieved and saved.

3.1. Kratky block camera

The most widely used 'long-slit' camera is the block camera of Kratky (see, for example, Kratky, 1982), which is traditionally used with a conventional source. The camera is commercially available. The principle of the camera is shown in Fig. 8. The long direction of the slits is in the direction perpendicular to the paper. The camera has an asymmetrical design and the scattered X-rays are recorded above the centerline C. The beamstop BS blocks the direct beam. The radiation is usually monochromatized by filters or by a graphite crystal. The blocks B_1 and B_2 provide the collimation and the sample S is situated after B_2. These blocks have to be accurately aligned with respect to the centerline and in the direction which is perpendicular to the paper.

For the Kratky camera, both a sequential and a parallel mode of data acquisition can be used. In the sequential mode a point detector and a slit are moved simultaneously in order to record the data. In the parallel mode one has a linear position-sensitive detector, usually a gas counter or an image plate (Bergmann et al., 2000). The camera typically covers the range of scattering vectors q down to 0.003 Å^{-1} and is well-suited for measurements on randomly oriented molecules, particles and aggregates in solution.

The most recent development of the camera, is to include a Göbel mirror before the collimation (Bergmann et al., 2000). This leads to such high flux and count rates that the instrument with a rotating anode source can only be used with an image plate, as all other available detectors have problems with deadtime effects. It will be interesting to see if this new development will lead to a renaissance of the Kratky camera. The main obstacle to this is probably still the fact that the data is subject to slit smearing.

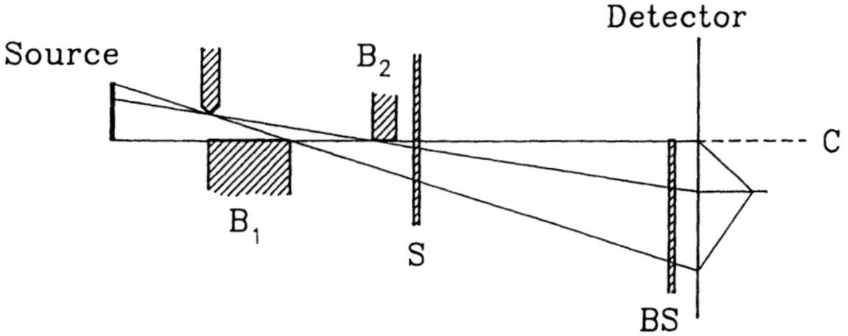

Fig. 8. The principles of the Kratky block camera.

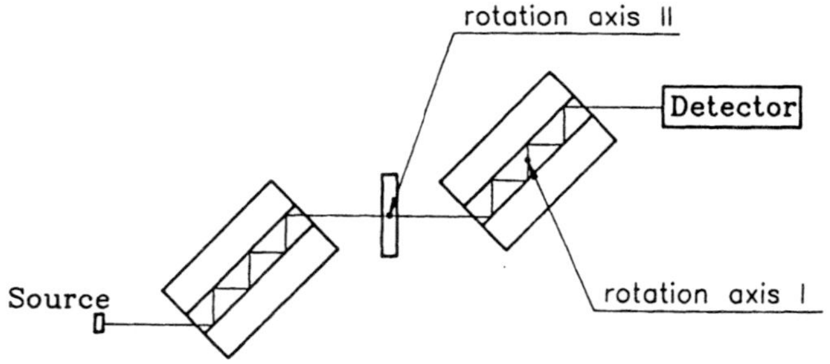

Fig. 9. The principle of the Bonse–Hart camera. In the traditional setup the second crystal rotates around its center (I). The data can also be collected by rotating a 'detector arm' around the sample center (II).

3.2. Bonse–Hart double-crystal systems

Another type of instrument is the Bonse–Hart camera (Bonse and Hart, 1965, 1966), which gives a smearing similar to that observed for a long-slit camera. In this camera multiple-bounce perfect crystals are used for monochromatizing and collimating both the incident and the diffracted beam. The beam is only collimated in the direction of the scattering plane of the crystals. Usually five bounces are used in both the monochromator and the analyzer crystal. In the traditional setup the intensity is recorded by rotating the second crystal around its center (I in Fig. 9). This gives a mechanically stable system as only a small mass has to be moved for recording the data. In a more recent version the second crystal is situated on a detector arm which rotates around the center (II) of the sample (Roux and Safinya, 1988; Lambard and Zemb, 1991; Lambard et al., 1992). Both these data recording modes are sequential. The stability of the setup is very important as one has to subtract a relatively large background from the empty camera from the data recorded with the sample.

The setup is used for ultra-small angle scattering for q values down to $\sim 10^{-4}$–10^{-3} Å^{-1}. The camera is, as the Kratky camera, best suited for samples with randomly oriented particles. Due to the very good collimation of the beam the intensity of the incident beam is usually quite low. S. Lyonnard and T. Zemb (personal communication, 2000) has recently included a mirror in front of the first crystal and in this way improved the intensity by a factor of 2–3.

A 'double' or two-dimensional Bonse–Hart camera has been introduced by Pahl et al. (1991). Multiple-bounce crystals are used for collimating in both directions, i.e., in the scattering plane and perpendicular to the scattering plane in which q is varied. With this camera there is in principle an extremely small influence of smearing of the scattering curve. However, even with a synchrotron radiation source the incident intensity is quite low and the camera can only be used for strongly scattering samples.

3.3. Simple pinhole camera

In pinhole cameras two-dimensional position-sensitive detectors can be employed. This gives a relatively effective data collection, which partly compensates for the lower incident X-ray intensity in the pinhole setup as compared to the long-slit cameras. The position-sensitive detector makes the cameras well-suited for investigating oriented or partially oriented systems.

The smallest q that can be reached is typically 0.005 $Å^{-1}$ for laboratory instruments and 0.001 $Å^{-1}$ for long instruments on synchrotron radiation sources.

Simple pinhole cameras for a conventional rotating anode source have, for example, been described by Hendricks (1978) and by Wignall et al. (1990). In such instruments the X-rays are usually monochromatized by a mosaic graphite crystal and collimation is done by three sets of slits located after the monochromator following the principle shown in Fig. 6. The scattered X-rays are recorded by a two-dimensional multiwire proportional counter. Typical instrument length are 2–10 m. The primary advantage of this type of camera as compared to the ones employing focusing optics is that it is very easy to align and that it stays aligned for long periods of time.

Simple pinhole cameras can also be installed at synchrotron radiation sources (Haubold et al., 1990). In such cameras, the beam is usually monochromatized by a set of two perfect crystal and collimated by several sets of slits. The double monochromator makes it easy to change the wavelength of the X-rays and the camera is well-suited for anomalous scattering experiments.

3.4. Huxley–Holmes and the Frank setup

The most advanced SAXS pinhole camera is the traditional Huxley–Holmes instrument (Huxley and Brown, 1967; Barrington-Leigh and Rosenbaum, 1974; Hendrix et al., 1979. Rosenbaum and Holmes, 1980; Holmes, 1982; Amemiya et al., 1983; Yoda, 1984; Koch. 1988; Riekel and Suortti, 1991), which includes a bent perfect crystal and a mirror for focusing the beam (see Fig. 10). This type of instrument is used both with conventional sources and synchrotron sources. Due to the small value of the critical angle of most mirror materials (~3–10 mrad) the mirror has to be relatively long (~1–2 m) in order to accept an appreciable part of the radiation. This of course makes it difficult to do the alignment and to keep the mirror aligned. An additional problem is that the mirror gives rise to a significantly higher background than for a simple pinhole camera (Koch, 1988). However. there are two main advantages of including the mirror: First of all it increases the flux due

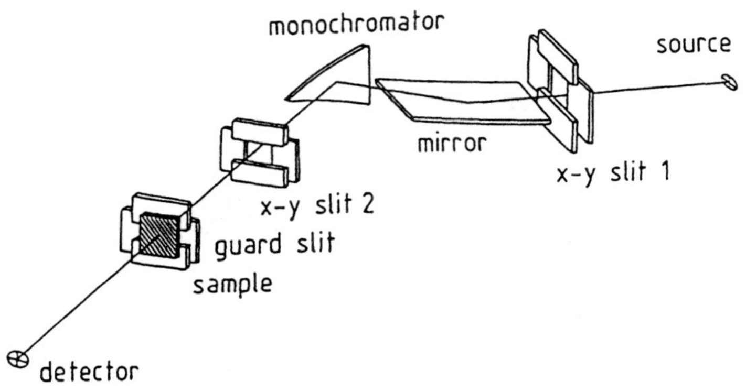

Fig. 10. The principle of the Huxley–Holmes camera with a perfect crystal monochromator scattering in the horizontal plane and an elliptically bent mirror reflecting in the vertical direction. At synchrotron sources the monochromator crystal is placed before the mirror in order to reduce the intensity and thus protect the mirror.

to the focusing and secondly it filters out higher-order contamination of the beam from the monochromator.

A somewhat similar camera is the Franks camera (Franks, 1958; Furuno et al. 1987; Hayashi et al. 1988). It uses a mirror in both the horizontal and vertical plane for focusing the beam. The setup does not include a monochromator crystal and it is usually used with a conventional source. The X-rays can be monochromatized using a filter for removing the K_β radiation and using energy discrimination for the detector for removing the high-energy part of the continuous spectrum.

3.5. High-brilliance and high-flux instrument

Due to the very high intensity of synchrotrons it is possible to have instruments with very special properties at such sources. An example is the microfocus beamline at the European Synchrotron Radiation Facility (ESRF) in Grenoble (Riekel et al., 2000). The source of this instrument is a high-brilliance undulator. The beam is monochromatized by two Si perfect crystals and focused by a double-focusing ellipsoidal mirror. The mirrors is situated much closer to the detector than to the source which results in a demagnification of the beam. A micro-collimator reduces the size of the beam so that the beam is only 20 μm at the sample position. The instrument can thus be used for performing SAXS experiments on extremely small samples or for performing position-resolved measurements by scanning the sample. Experiments have, for example, been done on single fibers from a silk spider (Riekel, 2000). The position-resolved method has been applied on wood and bone (Zizak et al., 2000).

Another setup with an ellipsoidal double-focusing mirror has been described by Stephenson (1988). The special property of this instrument is that it has a very high flux, although it is placed at a bending magnet. The instrument benefits from the fact that the spectrum is continuous so that a large gain in intensity can be obtained by increasing the bandwidth accepted by the monochromator. A synthetic multilayer with a bandwidth of about 1% is used. The flux increase allows time-resolved measurements with very short exposure time to be performed.

3.6. The NanoSTAR instrument

A pinhole laboratory camera, called the NanoSTAR instrument, has recently become commercially available from Bruker AXS [1]. The camera is described in the present context as it has already become one of the cameras which are most frequently seen in research laboratories. The NanoSTAR uses a sealed small-focus tube as source and the beam is monochromatized and made parallel by two cross-coupled Göbel mirrors placed one after the other close to the source. The collimation section is relatively short and compact. The camera was developed with the aim of performing position-resolved measurements and the beam is therefore only 0.3 mm in diameter at the sample position. The scattered beam is recorded by a position-sensitive gas detector with a high position resolution. The instrument has one integrated vacuum and the detector is mounted directly on the vacuum tube, in order to reduce

[1] http://www.bruker-axs.com/production/products/xrd/instruments/nanostar/

the number of windows and the associated background scattering.

The combination of a sealed tube and the Göbel mirrors gives a relatively high flux at the sample position. However, the flux can be significantly improved for samples where the position-resolved option is not required. This is the case in many applications to soft condensed matter, for example, in solution scattering experiments. Recent optimizations (Pedersen, 1999) by phases-space analysis methods (Pedersen and Riekel, 1990) and Monte Carlo simulations improved the flux by a factor of six by extending the length of the collimation and the size of the beam. It was furthermore suggested (Pedersen, 1999) that an extra pinhole should be included after the mirrors, so that the beam is entirely defined by slits, as this would give a lower background close to the beamstop. The geometry has been tested (A. Bergmann and O. Glatter, personal communication, 2000) and an increase in flux of a factor of 5.4 was found together with a significant reduction of the background close to the beamstop.

4. Small-angle neutron scattering instruments

The two main types of neutron sources are steady-state reactors and spallation sources. In steady-state sources neutrons are continuously produced by the fission processes. In contrast, spallation sources provide a pulsed neutron beam and therefore a time-of-flight method is used in instruments on such sources. The most common type of instrument is a pinhole camera.

The moderated neutrons at thermal energies have relatively short wavelength (<1 Å). However, the neutron absorption cross-section varies relatively slowly with the neutron wavelength ($\propto \lambda$) and neutron absorption in the sample is usually not a problem. So in contrast to the situation for X-rays for which the absorption is very strong at long wavelengths, radiation with relatively large wavelengths (2–20 Å) can be used for neutrons for going to lower scattering vectors. Cold moderators can, for example, consist of liquid hydrogen or deuterium at 20–40 K and 15 bar or liquid methane at 100 K and 4 bar (see, for example, Carpenter and Iverson, 1998).

For SANS the parasitic scattering from the pinholes is much less important compared to SAXS and therefore guard pinholes are usually not included in neutron setups. The pinholes are usually made of strongly neutron-absorbing materials (LiF) or materials which contain strongly absorbing substances like boron carbide.

The scattered neutrons are recorded by two-dimensional position-sensitive detectors. They are almost exclusively gas detectors with wire planes. A neutron-absorbing gas molecule like ^{3}He is used for converting the neutron to a charged particle which can be detected. The detectors are usually quite large, 0.5–1.0 m in diameter, with a relatively coarse resolution of 0.5-1.0 cm. In order to avoid absorption and scattering by air, the instrument is evacuated.

4.1. Steady-state instrument

Several steady-state SANS instruments are described in the literature (see, e.g., Ibel, 1976; Glinka et al., 1986; Koehler, 1986; May and Thomas, 1986; Lebech, 1990; Schwahn et al., 1991; Glinka et al. 1998; Lindner et al., 1992). The range of available scattering vectors is typically from about 10^{-3} to 0.5 Å^{-1}, which typically requires the use of three different sample–detector distances and neutron wavelengths. A particularly broad range of scattering

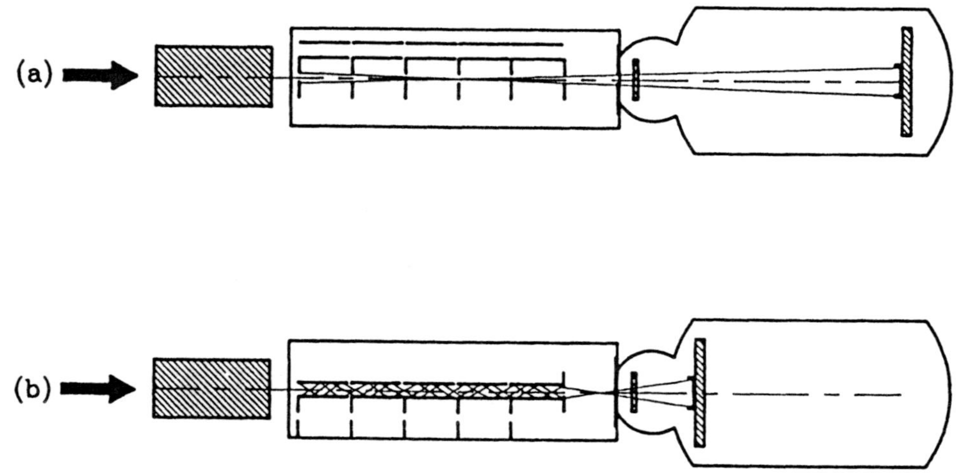

Fig. 11. The principle of a flexible steady-state SANS setup. (a) The smallest q values are reached with the longest sample–detector distance, with a long collimation, and a long neutron wavelength. (b) The largest q values are reached with the detector as close to the sample as possible and a short neutron wavelength.

vectors, from 10^{-3} to 1.5 Å^{-1} is covered by the D22 instrument[2] at ILL in Grenoble, France (May and Thomas, 1986).

Most of the available instruments are very flexible: e.g., instruments D11 and D22 at the ILL in Grenoble, France (Ibel, 1976; Lindner et al., 1992), the instruments in Jülich, Germany, (Schwahn et al., 1991) and at NIST, USA (Glinka et al. 1998). The principle of the setups is shown in Fig. 11. The sample–detector distance can be varied and the divergence of the incident beam can be chosen to match the angular range covered by the detector. This is done by varying the distance from the source pinhole to the sample. Neutron guides are inserted in front of the source pinhole in order to prevent loss of intensity. For all these instruments a mechanical velocity selector is used for monochromatizing the neutrons. The smallest q values are reached with the longest sample–detector distance for which the collimation is chosen as long as possible. The wavelength is as long as practically possible. The largest q values are reached with the detector as close to the sample as possible and with the shortest possible wavelength.

The flux at the sample position depends strongly on the wavelength and the collimation. For short wavelength (∼3 Å) the flux is typically about 10^7 neutrons per second at the sample position. For longer wavelength (∼15–20 Å) and long collimation distances (>5 m) the flux is several orders of magnitude lower. Useful measurements can be done with a flux as low as 10^3 neutrons per second, but with long acquisition times (∼4–12 hours).

For the D11 instrument at ILL (Ibel, 1976) the length of the collimator can be varied from 2.5 to 40.5 m, and the sample-to-detector distance can be varied in the same range. The velocity selector provides a wavelength resolution of 9% and wavelength in the range from 2 to 20 Å. For D22 at the ILL, the source, the neutron guide and the geometry have

[2] http://www.ill.fr/YellowBook/D22/main.html

been optimized to provide a very intense incident flux. The length of instruments collimation and flightpath is 17.6 and 18 m, respectively. With a large area sensitive detector (1.0 × 1.0 m^2) which can be offset, a very broad range of scattering vectors can be covered in a single setting. In combination with the high flux this allows time-resolved measurements to be performed (see, for example, Egelhaaf et al., 2000). The Oak Ridge instrument (Koehler, 1986) is different from these two instruments as it uses graphite monochromator crystals and has a 'fixed' wavelength of 4.75 Å and a wavelength resolution of 6%. A cooled Be filter removes higher-order contamination from the beam. The wavelength can be changed using the second-order reflection of the crystals (2.38 Å) and removing the first-order reflection by a pyrolytic graphite filter. The collimation length of the instrument can be changed from 1.5 to 7.5 m, and the sample-to-detector distance can be varied from 1.3 to 18.9 m.

4.2. Time-of-flight instruments

Spallation sources produce pulsed neutron beams by irradiation a target with heavy elements with a pulsed proton beam. The time-of-flight (TOF) technique is ideal for instruments at such sources. The pulsed nature of the source is used for determining the wavelength of the scattered neutrons. The time from a proton pulse produces the neutrons until they reach the detector is recorded, and from this and the length of the total flight path, the neutron velocity and wavelength can be determined. Therefore the beam is not monochromatized and the full 'white' beam is used. The results of a measurement are saved in a three-dimensional array with the two dimensions for the indices of the pixels of the detector and the third dimension for the time axis. The primary processing of the data provides a conversion to an intensity as a function of q or $\boldsymbol{q}$. Note that a single data set with 128 × 128 × 256 pixels takes up 12 Mbyte. Detailed descriptions and comparisons of time-of-flight (TOF) SANS instruments have been given by Hjelm (1988) and by Seeger and Hjelm (1991).

The range of scattering vectors for time-of-flight instruments is similar to the range of steady-state instruments. The main advantage of the TOF instruments is that the full range is covered with only one setting of the detector. The main drawback of the currently available instruments is that the intensity at long wavelength is quite low, which leads to poor statistics at low q. Thermalization of the neutron beam (Heenan et al., 1997), i.e. shift of neutron wavelength to shorter wavelength, makes it difficult to measure reliable transmissions, to find a good background sample, and to subtract it reliably from the sample scattering. Another problem is detector sensitivity calibration as it is difficult to find well-suited samples (Heenan et al., 1997).

Several time-of-flight instruments are described in the literature (Borso et al., 1982; Ishikawa et al., 1986; Hjelm, 1988; Seeger and Hjelm, 1991; Heenan and King, 1992; Heenan et al., 1997). A comparison of the four instruments at the Rutherford-Appleton Laboratory in UK, the Los Alamos National Laboratory in USA, the Argonne National Laboratory in USA, and the National Laboratory for High Energy Physics in Japan, has been given by Seeger and Hjelm (1991). The typical range of wavelengths is 1–15 Å, and the flux is $\sim 10^5$ neutrons per second at the sample position. The typical distance from the source to the sample is about 10 m, and the sample-to-detector distance is in the range 1–5 m.

The usual pulse frequency for pulsed neutron sources is about 50 Hz. Therefore it is difficult to use cold moderators as this would give pulse overlap and thus destroy the pulsed

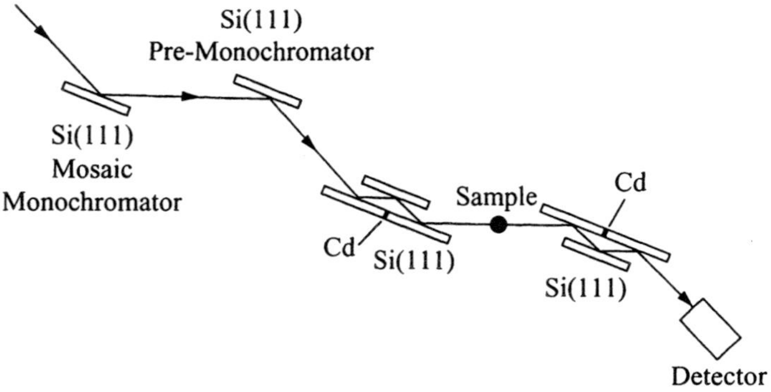

Fig. 12. The Bonse–Hart neutron scattering setup of Agamalian et al. (1997). A cadmium piece is inserted in the multi-bounce crystals in the middle of the sides with two bounces.

nature of the beam. It has therefore been suggested[3] to decrease the pulse frequency to 10 Hz and combine it with a cold moderator (see, for example, Heenan et al., 1997). When the frequency is decreased it is possible to increase the pulse power of the beam as the lower frequency gives a longer time for heat dissipation in the target.

4.3. Bonse–Hart double-crystal instrument

SANS at very small scattering vectors can be performed by use of a double-crystal Bonse–Hart setup (see, for example, Schwahn et al., 1991; Agamalian et al., 1997). Typically, the instrument is equipped with one or two premonocromator crystals, which deflect the beam so that it can enter the instrument. The instrument has typically three bounces in both the monochromator and analyzer channel-cut crystals. The instrument of Agamalian et al. uses a wavelength of 2.49 Å with a beam cross-section size of 2×4 cm^2.

Most neutron Bonse–Hart systems performs significantly worse with respect to background than theoretically predicted. In order to optimize the instrument, Agamalian et al. (1997) investigated the origins of the additional background. They found that it is due to the large penetration of neutrons in the crystals and that it could be strongly reduced by inserting a neutron absorbing cadmium piece in the middle of the side of the crystal with two bounces (see Fig. 12). This prevents a large part of the neutrons, which contributes to the tails, from passing the instrument. The instrument has successfully been applied, for example, in studies of polymer blends (Agamalian et al., 1999). For the studied samples it was possible to cover a range of scattering vectors from 3.5×10^{-5}–2×10^{-3} Å^{-1}, so that q overlap with pinhole SANS is possible. The main limitations for the application of this method is the relatively low flux of the incident neutron beam and the requirement that the magnitude of the scattering signal is so large that it can be distinguished from the background, i.e., the rocking curves of the multi-bounce crystals.

---

[3] http:/www.isis.rl.ac.uk/targetstation2/

4.4. Instrument with double-focusing mirror

The Bonse–Hart system is currently the neutron instrument which reaches the lowest q values. As the beam is only collimated in the scattering plane of the crystals, the measured intensity is subject to the strong slit smearing. An interesting alternative instrument with pinhole collimation, which covers nearly the same scattering vector regime, has been proposed and is under construction (Alefeld et al., 2000). The instrument has a mechanical velocity selector and a toroidal or ellipsoidal double-focusing mirror. The high-quality mirror is 1.2 m long and it is produced by the same technology as used for X-ray telescopes. Theoretical calculations shows that the instrument is superior to a conventional pinhole camera with respect to flux already for q values smaller than about 0.008 Å^{-1}. It is expected that q values down to 1×10^{-4} Å^{-1} can be reached. The main drawback of such instruments is that the mirror is relatively expensive.

5. Summary and conclusions

An overview of a large part of the available instrumentation for SAXS and SANS has been given with some emphasis on the most recent developments. The instruments have been described in some detail and discussed with respect to their strength and weaknesses. The information on the q ranges covered by the instrument and on the flux have not been very specific. This is due to the fact that for most instruments the parameters and specifications changes continuously due to modifications and improvements. For fully updated information the reader is encouraged to look at the various homepages or to contact the instrument responsibles directly.

The choice of instrument for a particular set of samples depends of course on the desirable range of scattering vectors. It should in principle always be as large as possible, as this allows the largest possible structures to be resolved with the best resolution. However, the minimum q does not need to be less than about a fifth of $1/R_g$, where R_g is the radius of gyration of the structures, as the intensity is constant for $q \ll 1/R_g$.

For some systems the choice between X-ray or neutron scattering is decided by the scattering contrast being sufficient for one type of radiation and absent for the other. For randomly oriented samples any instrument can be used if it covers the range of scattering vectors of interest and if the sample has sufficient scattering contrast. Oriented samples can only be investigated properly on a setup with a two-dimensional position-sensitive detector.

The contrast variation SANS method can provide a large amount of extra information for samples in which one or more of the components can be deuterated or for samples in which the components have different compositions, so that there is a natural scattering length density difference of the components. For samples consisting of aggregates in solution, the various parts of the aggregates can be highlighted in different mixtures of deuterated and hydrogenated solvents. Typical examples of applications are block copolymer micelles and microemulsions.

For biological samples with several components (proteins, DNA/RNA, or lipids) the SANS contrast variation can also be used. The contrast is varied by changing the D_2O/H_2O concentration of the solvent. The various components have scattering length densities with different characteristic dependencies on the D_2O/H_2O concentration as well as different match points. By proper analysis one can obtain information on the spatial arrangement of the components.

Contrast variation studies on macromolecular solutions can also be performed by X-rays. It can be done by adding sucrose to the solvent and thus changing the scattering contrast of the solvent. However, the changes in the scattering density of the solvent and the corresponding change in the scattering intensity are relatively small as compared to what is obtained for neutron scattering. Therefore SANS is much better suited than SAXS for performing contrast variation studies.

One of the easiest ways of performing contrast variation studies is to perform SAXS and SANS measurements on the same sample. The SANS measurements are almost always done in a deuterated solvent as this enhances the contrast and reduces the incoherent background from scattering from hydrogen in the sample. For SANS protonated particles in a deuterated solvent has a 'homogeneous' contrast, i.e., all component have a large excess scattering length density with the same sign. For SAXS the situation is often more complicated, as most components have electron densities close to that of the solvent, and therefore both positive and negative excess scattering length densities are often encountered. As an example, many surfactant molecules in water have a positive excess scattering length density of the headgroup, whereas the hydrocarbon tail has a negative excess scattering length density. Owing to the differences in the contrast for X-rays and neutron, a combination of the two techniques provides information which is very cumbersome to obtain in any other way. For SANS the same information can only be obtained by measuring on the same systems with, respectively, none of the components deuterated and with one of the components deuterated.

It is important to notice that subtle isotope effects might occur when performing SANS contrast variation. This is for most systems due to the change in mass when replacing hydrogen by deuterium and the associated effect on the hydrogen bonds. However, even for systems without hydrogen bonds, like polymers in an organic solvent, isotope effects might be present. In order to check for isotope effects it can be recommended to perform SAXS measurements on the same samples.

For neutrons the absorption is quite small in most materials and quite thick samples (1–5 mm) can often be used. This makes the sample preparation quite easy. The low absorption also allows a straightforward use of special sample environment equipment like pressure cells and shear cells (see Chapter 17 by P. Lindner in this book). For SAXS it is more complicated to construct special sample environment equipment as the absorption is in general much larger for X-rays. Therefore care has to be taken when the material for the equipment is chosen. An overview of the most recent developments has been given by Bras and Ryan (1998).

X-ray scattering has the great advantage that conventional X-ray sources are relatively cheap. It is therefore economically feasible to have an in-house X-ray instrument which can be used 24 hours a day for collecting data. Although the flux is much lower than on most synchrotron radiation based instruments, this is to a large extent compensated by the larger availability. However, there are of course experiments which require the high flux at a synchrotron, and which cannot be performed on a laboratory instrument. As an example, one can mention that the high intensity of synchrotron radiation allow time-resolved measurements to be performed with a time resolution down to milliseconds. However, due to the high flux at synchrotron radiation damage of samples, in particular of biological samples can be a problem.

6. Acknowledgements

I thank Carsten Svaneborg for corrections to the present text.

References

Agamalian, M., Wignall, G.D. and Triolo, R. (1997). *J. Appl. Cryst.*, **30**, 345–352.

Agamalian, M., Alamo, R.G., Kim, M.H., Londono, J.D., Mandelkern, L. and Wignall, G.D. (2000). *Macromolecules*, **32**, 3093–3096.

Alefeld, B., Dohmen, L., Richter, D. and Brükel, T. (2000). *Phys. B*, **276–278**, 52–54.

Amemiya, Y., Wakabayashi, K., Hamanaka, T., Wakabayashi, T., Matsushita, T. and Hashizume, H. (1983). *Nucl. Instr. Methods*, **208**, 471–477.

Barrington-Leigh, J. and Rosenbaum, G. (1974). *J. Appl. Cryst.*, **7**, 117–121.

Bergmann, A., Orthaber, D., Scherf, G. and Glatter, O. (2000). *J. Appl. Cryst.*, **33**, 869–875.

Bonse, U. and Hart, M. (1965). *Appl. Phys. Lett.*, **7**, 238–240.

Bonse, U. and Hart, M. (1966). *Z. Phys.*, **189**, 151–162.

Borso, C.S., Carpenter, J.M., Williamson, F.S., Holmblad, G.L., Mueller, M.H., Faber Jr., J., Epperson, J.E. and Danyluk, S.S. (1982). *J. Appl. Cryst.*, **15**, 443–448.

Bras, W. and Ryan, A.J. (1998). *Adv. Colloid. Interface Sci.*, **75**, 1–43.

Carpenter, J.M. and Iverson, E.B., eds. (1998). *Proceedings of the International Workshop on Cold Moderators for Pulsed Neutron Sources.* http://www.oecd.org/dsti/sti/s_t/ms/prod/proceed.htm

Egelhaaf, S.U., Olsson, U. and Schurtenberger, P. (2000). *Phys. B*, **276–278**, 326–329.

Franks, A. (1958). *Br. J. Appl. Phys.*, **9**, 349–352.

Furuno, T., Sasabe, H. and Ikegami, A. (1987). *J. Appl. Cryst.*, **20**, 16–22.

Glatter, O. (1992). In: *International Tables for Crystallography, Volume C* (Wilson, A.J.C.. ed). Kluwer Academic Publishers, Dordrecht, Chapter 2.

Glinka, C.J., Rowe, J.M. and LaRock, J.G. (1986). *J. Appl. Cryst.*, **19**, 427–439.

Glinka, C.J., Barker, J.G., Hammouda, B., Krueger, S., Moyer, J.J. and Orts, W.J. (1998). *J. Appl. Cryst.*, **31**, 430–445.

Gubarev, M., Ciszak, E., Ponomarev, I., Gibson, W., Joy, M. (2000). *J. Appl. Cryst.*, **33**, 882–887.

Haubold, H.-G., Gruenhagen, K., Wagener, M., Jungbluth, H., Heer, H., Pfeil, A., Rongen, H., Brandenberg, G., Moeller, R., Matzerath, J., Hiller, P. and Halling, H. (1990). *Rev. Sci. Instrum.*, **60**, 1943-1946.

Hayashi, H., Hamada, F., Suehiro, S., Masaki, N., Ogawa, T. and Miyaji, H. (1988). *J. Appl. Cryst.*, **21**, 330-339.

Heenan, R.K. and King S.M. (1992). *J. Non-Cryst. Solids*, **150**, 153-156.

Heenan, R.K., Penfold, J. and King, S.M. (1997). *J. Appl. Cryst.*, **30**, 1140–1147.

Hendricks, R.W. (1978). *J. Appl. Cryst.*, **11**, 15–30.

Hendricks, R.W. and Schmidt, P.W. (1967). *Acta Phys. Austriaca* **26**, 97–122.

Hendricks, R.W. and Schmidt, P.W. (1973). *Acta Phys. Austriaca*, **37**, 20–30.

Hendrix, J., Koch, M.H.J. and Bordas, J. (1979). *J. Appl. Cryst.*, **12**, 467–472.

Hjelm Jr., R.P. (1988). *J. Appl. Cryst.*, **21**, 618–628.

Holt, N. (1957). *Rev. Sci. Instrum.*, **28**, 1–3.

Huxley, H.E. and Brown, W. (1967). *J. Mol. Biol.*, **30**, 383–434.

Ibel, K. (1976). *J. Appl. Cryst.*, **9**, 296–309.

Ishikawa, Y., Furusaka, M., Niimura, N., Arai, M. and Hasegawa, K. (1986). *J. Appl. Cryst.*, **19**, 229–242.

Koch, M.H.J. (1988). *Makromol. Chem., Macromol. Symp.*, **15**, 79-90.

Koehler, W.C. (1986). *Phys. B*, **137**, 320–329.

Kratky, O. (1982). In: *Small Angle X-Ray Scattering* Glatter, O. and Kratky, O., eds.). Academic Press, London, pp. 53–83.

Lambard, J. and Zemb, T. (1991). *J. Appl. Cryst.*, **24**, 555–561.

Lambard, J., Lesieur, P. and Zemb, T. (1992). *J. Phys. I France*, **2**, 1191–1213.

Lindner, P., May, R.P. and Timmins, P.A. (1992). *Phys. B*, **180–181**, 967–972.

Lebech, B. (1990). *Neutron News*, **2**, 7–13.

May, R.P. (1992). In: *International Tables for Crystallography, Volume C* (Wilson, A.J.C., ed.). Kluwer Academic Publishers, Dordrecht, Chapter 2.

May, R.P. and Thomas, M. (1986). ILL Technical Report 86MA07T. Institute Laue-Langevin, Grenoble, France.

Pahl, R., Bonse, U., Pekala, R.W. and Kinney, J.H. (1991). *J. Appl. Cryst.*, **24**, 771–776.

Pedersen, J.S. (1993). *J. Phys. IV*, **3(C8)**, 490–498.

Pedersen, J.S. (1995). In: *Modern Aspects of Small Angle Scattering*. (Brumberger, H., ed.). Kluwer Academic Publishers, Dordrecht.

Pedersen, J.S. (1999). Optimization of pinhole SAXS cameras using phase-space analysis and Monte-Carlo simulations (unpublished).

Pedersen, J.S. and Riekel, C. (1991). *J. Appl. Cryst.* **24**, 893–909.

Pedersen, J.S., Posselt, D. and Mortensen, K. (1990). *J. Appl. Cryst.*, **23**, 321–333.

Riekel, C. (2000). *Rep. Prog. Phys.*, **63**, 233–262.

Riekel, C. and Suortti, P. (1991). *J. Appl. Cryst.*, **24**, 549–554.

Riekel, C., Burghammer, M. and Müller, M. (2000). *J. Appl. Cryst.*, **33**, 421–423.

Rosenbaum, G. and Holmes, K.C. (1980). In: *Synchrotron Radiation Research* Wininck, H. and Doniach, S., eds.). Plenum Press, New York, pp. 533–564.

Roux, D. and Safinya, C.R. (1988). *J. Phys. France*, **49**, 307–318.

Schwahn, D., Meier, G. and Springer, T. (1991). *J. Appl. Cryst.*, **24**, 568–570.

Schuster, M. and Göbel, H. (1995). *J. Phys. D*, **28**, A270–A275.

Seeger, P.A. and Hjelm Jr., R.P. (1991). *J. Appl. Cryst.*, **24**, 467-478.

Stephenson, G.B. (1988). *Nucl. Instrum. Methods*, **A266**, 447–451.

Wignall, G.D., Lin, J.S. and Spooner, S. (1990). *J. Appl. Cryst.*, **23**, 231–245.

Yoda, O. (1984). *J. Appl. Cryst.*, **17**, 337–343.

Zizak, I., Paris, O., Roschger, P., Bernstorff, S., Amenitsch, H., Klaushofer, K. and Fratzl, P. (2000). *J. Appl. Cryst.*, **33**, 820–823.

Neutrons, X-Rays and Light
P. Lindner and Th. Zemb (Editors)

Chapter 7

Contrast and Contrast Variation in Neutron, X-Ray and Light Scattering

Peter Schurtenberger

Department of Physics, University of Fribourg, CH-1700 Fribourg, Switzerland

1. Introduction

The intensity measured in a scattering experiment depends not only on the structure of the individual scattering elements and their interactions, but also on the so-called scattering contrast. This is a measure of the interaction between the probing radiation (neutrons or photons) and the investigated matter. The resulting $\boldsymbol{q}$-dependence of the intensity is then given by a combination of three different factors:

- the spatial arrangement of the scattering elements expressed by the static structure factor $S(\boldsymbol{q})$

$$S(\boldsymbol{q}) := \frac{1}{N}\left\langle \sum_{j=1}^{N}\sum_{k=1}^{N} e^{i\boldsymbol{q}\cdot\boldsymbol{r}_{jk}} \right\rangle. \tag{1}$$

- the interaction between the probing radiation and the investigated matter expressed by a scattering length b, and
- an instrumental resolution function that depends on the actual instrumental design used in the experiment (see Chapter 6 by J.S. Pedersen)

While for simple two-component systems the contrast simply defines the signal to background ratio, it is much more important for the complex multicomponent systems that one generally encounters in soft condensed matter research. Here a controlled variation of the scattering contrast provides access to detailed structural information well beyond an overall assessment of the global structure. It allows us to 'color' individual components and investigate their detailed structural properties, or to study their dynamic properties in quasielastic or inelastic scattering experiments. In this lecture we shall learn more about the quantity 'scattering contrast' for various scattering probes and see how we can use a controlled variation of this contrast in order to learn more about the structural and dynamic properties of complex multicomponent solutions.

2. Scattering length and cross-section

The starting point of this short summary of important definitions and relations is the normalized scattering intensity from an ensemble of identical point scatterers

$$\frac{d\sigma}{d\Omega}(\boldsymbol{q}) = \frac{\langle I_s(\boldsymbol{q})\rangle}{I_0} R'^2 = b^2 \sum_{j,k=1}^{N} \left\langle e^{i\boldsymbol{q}\cdot\boldsymbol{r}_{jk}} \right\rangle, \tag{2}$$

where I_0 is the intensity of the incident beam, R' is the distance between the scattering volume and the detector, b is the scattering length and $\boldsymbol{r}_{jk} := \boldsymbol{r}_j - \boldsymbol{r}_k$. The scattering length b corresponds to the amplitude of the scattered wave relative to that of the incident wave, i.e., the differential scattering cross-section of a point scatterer which can be written as

$$\frac{d\sigma}{d\Omega}(\boldsymbol{q}) = b^2. \tag{3}$$

The scattering length b can have either positive or negative values, where a negative value then signifies a change of the phase on scattering (i.e., the scattered wave is one-half wavelength shifted with respect to the incident wave).

As we generally consider the case of particles suspended in a solvent which also scatters, we have to subtract the solvent background

$$\frac{d\sigma}{d\Omega}(\boldsymbol{q}) = \frac{\langle I_s(\boldsymbol{q})\rangle - \left\langle I_{s,\text{solv}}\right\rangle}{I_0} R'^2 \simeq \Delta b^2 \sum_{j,k=1}^{N} \left\langle e^{i\boldsymbol{q}\cdot\boldsymbol{r}_{jk}} \right\rangle. \tag{4}$$

where $\Delta b = b - b_{\text{solv}}$ is the excess scattering length and b_{solv} is the scattering length of a solvent molecule. Such a simple subtraction scheme is correct if we can treat a particle suspension as an incompressible liquid (details are given, for example, in Higgins and Benoit, 1996). In particular in neutron scattering one frequently uses an average scattering length density ρ rather than the scattering length b

$$\rho = \frac{1}{V_1} \sum_j b_j, \tag{5}$$

where the b_j represent the coherent scattering lengths of all atoms (or nuclei for SANS) in a volume V_1, and where the summation runs over all atoms (or nuclei) in one 'monomer'. Instead of the excess scattering length Δb we then use the excess scattering length density $\Delta\rho$

$$\Delta\rho = \rho - \rho_{\text{Lsm}} = \frac{1}{V_1}\left(\sum_j b_j - \rho_{\text{Lsm}} V_1\right) \tag{6}$$

This leads for example to the following expression for the scattering amplitude $A(\boldsymbol{q})$ of a discrete particle:

$$A(\boldsymbol{q}) = \int_V d^3r \, e^{i\boldsymbol{q}\cdot\boldsymbol{r}} \Delta\rho(\boldsymbol{r}). \tag{7}$$

where the integration runs over the particle volume V and $\Delta\rho(\boldsymbol{r})$ contains the information about the position dependent excess scattering length density distribution in the particle.

3. Light scattering

The starting point of our short theoretical description of light scattering is schematically shown in Fig. 1. (Alternative treatments of static light scattering and the link between the scattered intensity and the relevant optical properties of the scatterers can be found in Chapter 1 by P. Pusey and Chapter 14 by R. Klein.) Here we concentrate on a generalized notation that can be used for all different types of incoming radiation and use a description based on simple relations between polarizability and index of refraction, which allows us to obtain the scattering length in a straightforward way. However, it is important to point out that the resulting expression for the scattering length and the intensity of a point scatterer agrees with the more elaborate treatment presented in the other chapters. A point scattering source is illuminated by polarized (in the z-direction) light, and we consider the case of far field detection. The electric field of the incident primary wave $\boldsymbol{E}(t,x) = \boldsymbol{E}_0 \cos(\omega t - kx)$ causes the scattering center to oscillate. The thus induced dipole oscillates with the same frequency, which results in the emission of a scattered spherical wave. Such a scattering source acts like a Hertzian dipole, and its dipole moment is given by:

$$\boldsymbol{p}(t,x) = \alpha \boldsymbol{E}(t,x) = \alpha \boldsymbol{E}_0 \cos(\omega t - kx), \tag{8}$$

where α is the polarizability (in general: we have to consider a polarizability tensor $\underline{\underline{\alpha}}$).

The scattered field $\boldsymbol{E}_\mathrm{s}$ at a distance $\boldsymbol{R}'$ in a direction which is tilted against the direction of the polarization of the incident wave by an angle ϕ can be written as:

$$\boldsymbol{E}_\mathrm{s}(\boldsymbol{R}') = \frac{1}{c^2} \frac{\partial^2 \boldsymbol{p}}{\partial t^2} \frac{\sin\phi}{R'}. \tag{9}$$

With

$$\frac{\partial^2 \boldsymbol{p}}{\partial t^2} = \alpha\omega^2 \boldsymbol{E}_0 \cos(\omega t - kx) \tag{10}$$

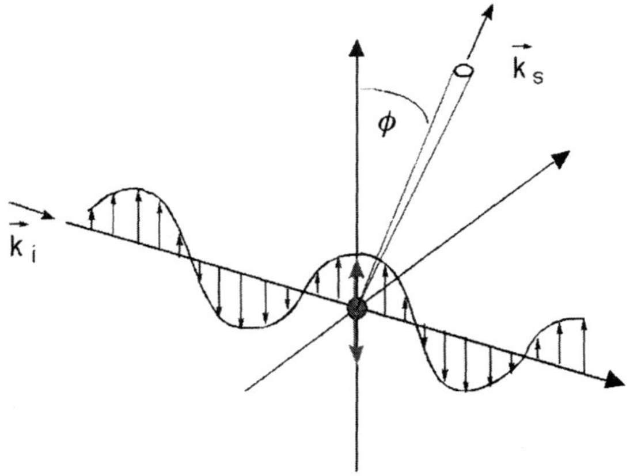

Fig. 1. The incident light wave induces an oscillating dipole moment in polarizable matter, ϕ is the angle between the direction of polarization and observation.

we then obtain

$$\boldsymbol{E}_{\mathrm{s}}(\boldsymbol{R}') = \alpha \frac{\omega^2}{c^2} \boldsymbol{E}_0 \frac{\sin\phi}{R'} \cos(\omega t - kx) \tag{11}$$

for the scattered field. As we measure intensities rather than fields, we have to use the corresponding relation

$$I = \epsilon_0 c \langle E^2 \rangle. \tag{12}$$

The ratio between the intensity of the scattered and incident light is thus

$$\frac{I_{\mathrm{s}}}{I_0} = \frac{E_{\mathrm{s}}^2}{E^2} = \frac{\omega^4}{c^4} \alpha^2 \frac{\sin^2\phi}{R'^2}, \tag{13}$$

which together with $\omega = 2\pi/\nu$ and $c = \lambda\nu$ leads to

$$\frac{I}{I_0} = \frac{16\pi^4}{\lambda^4} \alpha^2 \frac{\sin^2\phi}{R'^2}. \tag{14}$$

For the differential scattering cross-section we then obtain

$$\frac{\mathrm{d}\sigma}{\mathrm{d}\Omega} = \frac{I}{I_0} R'^2 = \frac{16\pi^4}{\lambda^4} \alpha^2 \sin^2\phi. \tag{15}$$

3.1. Scattering length, polarizability and index of refraction

We know from Eq. (15):

$$\frac{\mathrm{d}\sigma}{\mathrm{d}\Omega} \sim \alpha^2. \tag{16}$$

In a next step we now have to see how we can interpret the polarizability α, and how we can measure it. Moreover, we have to consider again the background contribution, which on the assumption of an incompressible solution allows us to rewrite Eq. (16):

$$\frac{\mathrm{d}\sigma}{\mathrm{d}\Omega} \sim (\Delta\alpha)^2. \tag{17}$$

Both the polarizability α as well as the excess polarizability $\Delta\alpha$ are not directly measurable quantities. However, we can find a link between $\Delta\alpha$ and the index of refraction n.

For an ideal gas we know the equation derived by *Clausius–Mosotti*, where:

$$\alpha = \frac{1}{4\pi} \frac{V}{N} \left(n^2 - 1\right). \tag{18}$$

For a dilute, ideal macromolecular solution we can write an analogous relation between $\Delta\alpha$ and the differences in index of refraction between solution and solvent $\left(n^2 - n_0^2\right)$:

$$\Delta\alpha = \frac{1}{4\pi} \frac{V}{N} \left(n^2 - n_0^2\right), \tag{19}$$

where N is the number of 'particles' in a solution volume V. For dilute solutions we can furthermore write:

$$n^2 - n_0^2 \simeq \left(n_0 + \frac{\mathrm{d}n}{\mathrm{d}c} c\right)^2 - n_0^2 \simeq 2n_0 \left(\frac{\mathrm{d}n}{\mathrm{d}c}\right) c. \tag{20}$$

Together with Eq. (19) this leads to:

$$\Delta\alpha = \frac{1}{2\pi}\frac{V}{N}n_{\mathrm{Lsm}}\left(\frac{\mathrm{d}n}{\mathrm{d}c}\right)c = \frac{1}{2\pi}\frac{M}{N_{\mathrm{A}}}n_{\mathrm{Lsm}}\left(\frac{\mathrm{d}n}{\mathrm{d}c}\right), \tag{21}$$

because $c = NM/(VN_{\mathrm{A}})$, where M is the molar mass of a 'particle' and N_{A} is Avogadro's number. The relation between the differential cross-section and the *scattering length b* of a point scatterer is given by Eq (3). Therefore we obtain for the differential scattering cross-section of a point scatterer:

$$\begin{aligned}\frac{\mathrm{d}\sigma}{\mathrm{d}\Omega} &= b^2 \\ &= \frac{16\pi^4}{\lambda^4}(\Delta\alpha)^2\sin^2\phi \\ &= \frac{4\pi^2}{\lambda^4}\frac{M^2}{N_{\mathrm{A}}^2}n_0^2\left(\frac{\mathrm{d}n}{\mathrm{d}c}\right)^2\sin^2\phi.\end{aligned} \tag{22}$$

In general the direction of polarization is perpendicular to the scattering plane, i.e., $\phi = 90^\circ$. This leads to:

$$b = \frac{2\pi}{\lambda^2}\frac{M}{N_{\mathrm{A}}}n_{\mathrm{Lsm}}\left(\frac{\mathrm{d}n}{\mathrm{d}c}\right). \tag{23}$$

In order to obtain an idea of the magnitude of b, we need to know the value of $\mathrm{d}n/\mathrm{d}c$. We can have, for example, a look at typical values of $\mathrm{d}n/\mathrm{d}c$ that can be found for synthetic polymers in organic solvents. For polymer solutions the values are generally of the order of $0.1\ \mathrm{cm}^3\ \mathrm{g}^{-1}$. Experimentally one finds for example ($\lambda_0 = 4880$Å, $T = 25$°C):

$$\text{polystyrene in toluene:} \qquad \frac{\mathrm{d}n}{\mathrm{d}c} = 0.11\ \frac{\mathrm{cm}^3}{\mathrm{g}}, \tag{24}$$

$$\text{polystyrene in cyclohexane:} \qquad \frac{\mathrm{d}n}{\mathrm{d}c} = 0.175\ \frac{\mathrm{cm}^3}{\mathrm{g}}. \tag{25}$$

In general one observes a relatively large dependence of the refractive index increment on temperature, but only a weak dependence upon the wave length.

4. Neutron scattering

In contrast to light scattering, the calculation of the scattering length b is difficult for neutron scattering. We therefore restrict ourselves and take the most important properties of b directly from experiment (or literature): b depends in an unsystematic way on the nucleus at which the neutrons are being scattered. Moreover, for a given nucleus we can find dramatic variations in b for an isotopic substitution, and b also depends upon the direction of the nuclear spin (magnetic scattering!). The weak interaction of neutrons with matter makes them an ideal probe for investigation of optically dense materials, as problems with multiple scattering are much less important (although existent).

4.1. Coherent and incoherent scattering

The concepts of coherent and incoherent scattering are of central importance for an understanding of neutron scattering. We write for the differential cross-section:

$$\frac{\mathrm{d}\sigma}{\mathrm{d}\Omega}(\boldsymbol{q}) = \sum_{j,k=1}^{N} \left\langle b_j b_k \, \mathrm{e}^{\mathrm{i}\boldsymbol{q}\cdot\boldsymbol{r}_{jk}} \right\rangle . \tag{26}$$

The values of b_j, b_k depend on the given isotope and spin state. Since the individual scattering lengths are independent, we can write:

$$j \neq k: \quad \langle b_j b_k \rangle = \langle b_j \rangle \langle b_k \rangle = \langle b \rangle^2 \tag{27}$$

$$j = k: \quad \langle b_j b_k \rangle = \langle b_j^2 \rangle = \langle b^2 \rangle \tag{28}$$

which leads to

$$\langle b_j b_k \rangle = \langle b \rangle^2 + \delta_{jk} \left(\langle b^2 \rangle - \langle b \rangle^2 \right) . \tag{29}$$

If one assumes that the spin state is independent of the position of the scatterer, we obtain for the differential cross-section:

$$\frac{\mathrm{d}\sigma}{\mathrm{d}\Omega}(\boldsymbol{q}) = \langle b \rangle^2 \sum_{j,k=1}^{N} \left\langle \mathrm{e}^{\mathrm{i}\boldsymbol{q}\cdot\boldsymbol{r}_{jk}} \right\rangle + N \left(\langle b^2 \rangle - \langle b \rangle^2 \right) . \tag{30}$$

The first term corresponds to the *coherent part* of the scattering intensity: This describes the scattering from nuclei which all have the same (average) value of $\langle b \rangle$. All the interference terms are included in this expression for the coherent scattering, and thus it contains the entire structural information.

The second term corresponds to the *incoherent scattering*: It has no interference terms (only $j = k$) and is caused by the 'disorder' of the isotopes. $b_{\mathrm{incoh}}^2 = \langle b^2 \rangle - \langle b \rangle^2$ is the mean square deviation of the scattering length. This expression contains no information on the structure of the system.

The distinction between coherent and incoherent neutron scattering is important as it allows to obtain different information about the system under investigation: With coherent scattering we study collective properties of a large number of scatterers such as for example their spatial arrangement. With incoherent scattering we can investigate, due to the missing phase relation, the behavior of individual scatterers such as for example their self diffusion.

The scattering cross-section for *coherent scattering* is given by

$$\sigma_{\mathrm{coh}} = 4\pi \langle b \rangle^2 \tag{31}$$

and for *incoherent scattering* by

$$\sigma_{\mathrm{incoh}} = 4\pi \left(\langle b^2 \rangle - \langle b \rangle^2 \right) . \tag{32}$$

Typical values for the coherent and incoherent scattering cross-section are given in Table 1.

The scattering lengths, which can be negative, are even more interesting than the comparison between the coherent and incoherent scattering cross-sections. This is shown in Table 2, where values of b are given for a few important nuclei. It is this enormous variation in b which

Table 1

Examples for the coherent and incoherent cross-section of several nuclei

Isotope	Nuclear spin	σ_{coh} (10^{-28} m^2)	σ_{incoh} (10^{-28} m^2)
^{1}H	1/2	1.8	79.7
^{2}H	1	5.6	2.0
^{12}C	0	5.6	–
^{14}N	1	11.6	0.3
^{16}O	0	4.2	–

Table 2

Examples for the coherent scattering length in neutron scattering experiments

Scattering length	^{1}H	^{2}H	^{12}C	^{16}O
b (10^{-14} m)	−0.38	0.66	0.66	0.58

leads to the so-called contrast variation. Particularly in the field of colloids, microemulsions and polymers this method has led to a number of important breakthroughs. The basic principle is that the substitution of a nucleus by one of its isotopes does not introduce a significant chemical perturbation of the properties of the material, but can lead to a strong variation in the scattering length. It is most fortunate that polymers, surfactants, proteins and many other constituents of soft condensed matter systems contain a large proportion of hydrogen, which can (at least in principle, and provided one has a good chemist and/or some money) easily be substituted by the deuterium isotope.

We can introduce a mean specific scattering length density as given by Eq. (5) where b_j correspond to the coherent scattering lengths of all the nuclei in a volume V_1 with linear dimensions $\geq \frac{\lambda}{\pi}$ and the summation is performed over all nuclei of a monomer. As the experiments are normally conducted with solutions or suspensions, we thus use the excess scattering length density $\Delta\rho$ given by Eq. (6).

Equation (6) serves as the basis for a short general discussion of contrast variation in neutron scattering, its possibilities and limits. We have already seen that the isotopic substitution offers a quite unique way of modifying the scattering contrast in soft condensed matter systems that are rich in hydrogen. Here we can work on the following principles:

- Increase a contrast: This allows us to increase the scattering contrast for particles such as polymers, proteins, microemulsions or micelles in solution and at the same time reduce the incoherent background by using for example deuterated solvents (see Table 3 for $\Delta\rho = \rho$(h-styrene) $-$ ρ(h-cyclohexane) and $\Delta\rho = \rho$(h-styrene) $-$ ρ(d-cyclohexane), respectively).
- Create a contrast: Individual particles can be made visible in bulk systems with no natural contrast by working with mixtures of deuterated and hydrogenated particles. This has for example allowed for a true breakthrough in our understanding of the chain conformation for polymers in the melt or in amorphous solids.
- Vary the contrast: Here we can adjust the excess scattering length density such that it becomes minimal for a given solvent by either choosing an appropriate mixture of h/d

Table 3

Examples for the scattering length density ρ in neutron scattering experiments

Substance	Formula	M_1 (g/mol)	ρ (10^{10} cm^{-2})
h-styrene	C_8H_8	104.15	1.413
d-styrene	C_8D_8	112.19	6.50
h-cyclohexane	C_6H_{12}	84.16	−0.24
d-cyclohexane	C_6D_{12}	96.23	6.01
h-benzene	C_6H_6	78.12	1.19
d-benzene	C_6D_6	84.15	5.44

solvents (such as H_2O and D_2O), i.e., through a variation of the scattering length density of the solvent, or by partial deuteration, i.e., by changing the overall scattering length density of the particle. This is particularly interesting for multicomponent systems, for composite particles or particles with an internal structure, and for the separation of inter- and intraparticle correlation effects.

When discussing the possibilities for isotopic substitution in contrast variation experiments, it is also necessary to provide some words of warning for possible problems that can arise:

- Change in solvent properties: It is known that deuteration does lead to a shift in the theta-temperature θ for polymers, which is of the order of $\theta \pm 4°$C for d-polystyrene in C_6H_{12} or h-polystyrene in C_6D_{12}. Moreover, it can cause a measurable shift in the location of a critical point.
- Variation of melting point: We find a shift in the melting point of polymers upon deuteration which is for example for polyethylene of the order of approximately 6°C.
- Induce phase separation: It is known that one observes phase separation in a 50/50 mixture of deuterated and hydrogenated polystyrene in the melt.
- Shift in the critical micellar concentration: The *cmc* of micelles is generally larger in H_2O than in D_2O.
- Exchange of hydrogen and deuterium: Hydrogens (or deuterium) from certain chemical groups such as OH or NH can be exchanged by solvent hydrogens (or deuterium). This means that the overall excess scattering length density can vary with time, and that the scattering length density of the particle may depend on the h/d ratio of the solvent.

4.2. Overall contrast and molar mass determination with light and neutron scattering

Before going to X-rays as the third important scattering probe, we make a first comparison between light and neutron scattering and consider the overall scattering contrast in the case of a simple molar mass determination for polymers in solution.

Light scattering: After having calculated the excess scattering length Δb for both light and neutron scattering, we calculate the scattering intensity of a dilute solution of noninteracting polymers. We write for the differential scattering cross-section in this case:

$$\frac{\mathrm{d}\sigma}{\mathrm{d}\Omega}(\boldsymbol{q}) = N_\mathrm{p}\left(\frac{\mathrm{d}\sigma}{\mathrm{d}\Omega}\right)_{\text{Polymer}}(\boldsymbol{q}) = N_\mathrm{p}\Delta b^2 N^2 P(\boldsymbol{q}), \tag{33}$$

where N_p is the number of polymer chains, N the number of monomers per chain and $P(\boldsymbol{q})$

the form factor of the polymer chain. With (23) this leads to:

$$\frac{\mathrm{d}\sigma}{\mathrm{d}\Omega}(\boldsymbol{q}) = N_{\mathrm{p}} \frac{4\pi^2}{\lambda^4} \frac{M_1^2}{N_{\mathrm{A}}^2} n_{\mathrm{Lsm}}^2 \left(\frac{\mathrm{d}n}{\mathrm{d}c}\right)^2 N^2 P(\boldsymbol{q}). \tag{34}$$

Here we have used the monomer mass M_1 in order to express the scattering length b. We now introduce a contrast term K_{LS} for light scattering

$$K_{\mathrm{LS}} := \frac{4\pi^2}{\lambda^4} \frac{n_{\mathrm{Lsm}}^2}{N_{\mathrm{A}}} \left(\frac{\mathrm{d}n}{\mathrm{d}c}\right)^2 \tag{35}$$

which leads to

$$\frac{\mathrm{d}\sigma}{\mathrm{d}\Omega}(\boldsymbol{q}) = K_{\mathrm{LS}} N_{\mathrm{p}} \frac{M_1^2}{N_{\mathrm{A}}} N^2 P(\boldsymbol{q}). \tag{36}$$

If we now use weight concentration c with

$$c = \frac{N_{\mathrm{p}} M_1}{N_{\mathrm{A}} V} N \tag{37}$$

in units of g cm^{-3} and normalize the measured intensity with the volume of the scattering volume V, we obtain:

$$\frac{1}{V} \frac{\mathrm{d}\sigma}{\mathrm{d}\Omega}(\boldsymbol{q}) = K_{\mathrm{LS}} c M P(\boldsymbol{q}), \tag{38}$$

where $M = N M_1$ is the molar mass of a polymer chain. We therefore obtain the molar mass of the polymer from an extrapolation of the intensity

$$\lim_{q \to 0} \frac{\frac{1}{V} \frac{\mathrm{d}\sigma}{\mathrm{d}\Omega}(\boldsymbol{q})}{K_{\mathrm{LS}} c} = M. \tag{39}$$

Neutron scattering: In analogy to Eq. (33) we can write the expression for the normalized differential scattering cross-section for a dilute polymer solution obtained with neutron scattering:

$$\frac{1}{V} \frac{\mathrm{d}\sigma}{\mathrm{d}\Omega}(\boldsymbol{q}) = \frac{N_{\mathrm{p}}}{V} \left(\sum_j b_j\right)^2 N^2 P(\boldsymbol{q}). \tag{40}$$

where we sum over all nuclei in a monomer of the polymer chain. With Eq. (5) this leads to

$$\frac{1}{V} \frac{\mathrm{d}\sigma}{\mathrm{d}\Omega}(\boldsymbol{q}) = c\rho^2 \frac{V_1^2 N_{\mathrm{A}}}{M_1} N P(\boldsymbol{q}). \tag{41}$$

We express the volume of a monomer V_1 using a more readily accessible quantity such as the density ρ_{M} or the partial volume $\overline{v} = 1/\rho_{\mathrm{M}}$, and obtain

$$V_1 \approx \frac{M_1}{\rho_{\mathrm{M}} N_{\mathrm{A}}} \approx \frac{\overline{v} M_1}{N_{\mathrm{A}}}. \tag{42}$$

This then leads to:

$$\frac{1}{V} \frac{\mathrm{d}\sigma}{\mathrm{d}\Omega}(\boldsymbol{q}) = c K_{\mathrm{SANS}} M P(\boldsymbol{q}), \tag{43}$$

Table 4

Examples for the contrast constant K_{SANS} and K_{LS} in neutron and light scattering experiments

Solution	K_{SANS} (10^{-4} g^{-2} cm^2 mol)	K_{LS} (10^{-4} g^{-2} cm^2 mol)
h-polystyrene in		
h-cyclohexane	4.3	0.00630
d-cyclohexane	33.5	0.00630

where the contrast term K_{SANS} for neutron scattering is given by

$$K_{SANS} := \overline{v}^2 \rho^2 \frac{1}{N_A}. \tag{44}$$

As we have to consider the excess scattering length density $\Delta\rho$ rather than ρ, we obtain

$$\Delta\rho = \rho - \rho_{Lsm} = \frac{1}{V_1}\left(\sum_j b_j - \rho_{Lsm} V_1\right). \tag{45}$$

This leads to

$$K_{SANS} = \overline{v}^2 (\Delta\rho)^2 \frac{1}{N_A}. \tag{46}$$

In Table 3 we find several examples for the scattering length density of styrene and two different solvents. Depending upon the choice of the isotopes selected for the polymer and the solvent (deuterated or non-deuterated) we achieve very different values for $\Delta\rho$ and K_{SANS}.

It is interesting to note that in this case we obtain a much higher scattering contrast for SANS when compared to light scattering ($K_{SANS}/K_{LS} \sim 5 \times 10^3$). This is illustrated in Table 4, where we have listed this constant for polystyrene in two different solvents, h-cyclohexane and d-cyclohexane. However, this is more than compensated by the much higher photon flux of a laser when compared to the neutron sources with the highest neutron flux currently available.

5. X-ray scattering

We shall look at X-ray scattering in a less detailed manner, as most of the important relations have already been introduced in the previous two sections. Moreover, there are only limited possibilities for contrast variation experiments, and they are not yet as routinely applied to soft condensed matter problems as with light and neutron scattering.

5.1. Scattering length

The scattering of electromagnetic radiation on matter is frequency dependent. When we cross over from low to high frequencies, we observe for the scattering cross-section ($d\sigma/d\Omega$) of small point-like scatterers a transition from the classical Rayleigh scattering typical for visible light, through a peak that is caused by resonant scattering at the resonance frequency ω_0, to the so-called Thomson scattering for X-rays. This is illustrated in Fig. 2, where we have

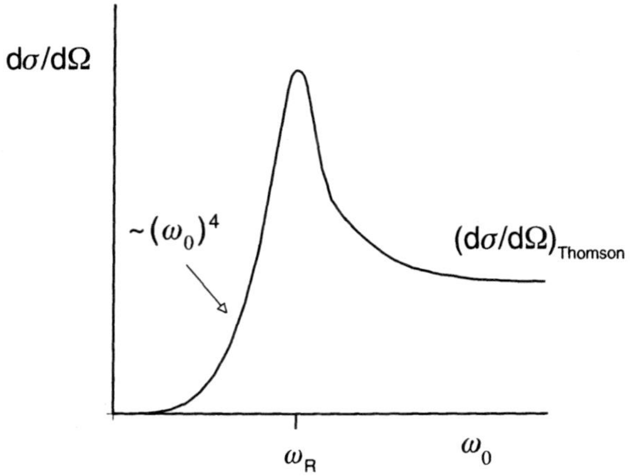

Fig. 2. The frequency dependence of the differential scattering cross-section $d\sigma/d\Omega$ for a point-like scatterer for electromagnetic radiation.

plotted a schematic view of the scattering cross-section of a point source for the scattering of electromagnetic radiation. We thus have a frequency-dependent atomic structure factor or scattering length. The cross-section for Thomson scattering for a single electron can be written as:

$$\frac{d\sigma}{d\Omega} = {r_{el}}^2 \frac{1+\cos^2\theta}{2}, \tag{47}$$

where $r_{el} = e^2/m_e c^2 = 0.28 \times 10^{-5}$ nm is the electron radius (Thomson radius) and θ the scattering angle. The $(1+\cos^2\theta)/2$ is due to the fact that we consider the case of dipole radiation with an unpolarized incident electric field. When considering small-angle X-ray scattering, with correspondingly small values of θ, on the electron cloud of an atom we obtain

$$\frac{d\sigma}{d\Omega} = {r_{el}}^2 Z^2, \tag{48}$$

where Z is the number of electrons in the atom (in principle the scattering amplitude of an atom for SAXS corresponds to the Fourier transform of the electron probability density).

Therefore we obtain the following expression for the scattering length b [see Eq. (3)]:

$$b = r_{el} Z. \tag{49}$$

Similarly we obtain the expression for the excess scattering length density $\Delta\rho$:

$$\Delta\rho = \rho - \rho_{Lsm} = \left(\frac{\sum_j Z_j}{V_1} - \frac{\sum_k Z_{k,Lsm}}{V_1} \right) r_{el}. \tag{50}$$

Values for several solvents and the surfactant SDS are given in Table 5. We see from this table that many systems in soft condensed matter may suffer from a weak scattering contrast, and a typical example where such low contrasts can be found are micellar solutions

Table 5
Examples for the scattering length density in neutron and X-ray scattering experiments

Group	ρ_{SANS} (10^{10} cm^{-2})	ρ_{SAXS} (10^{10} cm^{-2})
H_2O	−0.56	9.46
D_2O	6.39	9.46
C_2H_5OH	−0.344	7.6
C_2H_5OD	6.15	7.6
C_6H_{14}	−0.575	6.5
C_6D_{14}	6.14	6.5
$C_{12}H_{25}OSO_3Na$	0.388	10.7
$C_{12}D_{25}OSO_3Na$	6.7	10.7

of nonionic surfactants of the C_iE_j type in water. Moreover, it is not straightforward to change $\Delta\rho$, since this requires for example the use of heavy metal salts or sugars as additives, which may strongly alter the properties of the system.

5.2. Anomalous scattering and contrast variation

However, the situation changes considerably for X-rays with an energy close to the absorption edge of an atom. Under these conditions, the scattering length can be lowered by a few electron units, which allows for an energy-dependent contrast variation without having to interfere with the sample. This process is called *anomalous scattering*, and has developed into an interesting alternative in soft condensed matter research due to the availability of synchrotron radiation sources which provide monochromatic photons of high intensity over a quite broad energy range. However, the method is still hampered by the fact that the corresponding absorption edges for the light atoms important in soft condensed matter science, such as C, H, N, and O, are at very low X-ray energies and thus outside the accessible range.

For photons of energy E near the inner shell absorption edge of an atom, the coherent scattering length b can be written as composed from three terms:

$$b = r_{\mathrm{el}}\left(Z + f'(E) + f''(E)\right), \tag{51}$$

where $f'(E)$ and $f''(E)$ are the energy-dependent *anomalous scattering factors*. They have only appreciable amplitudes close to the absorption edge and can be neglected in standard experiments. The imaginary part f'' is related to the absorption cross-section σ_a

$$f''(E) = \frac{mc}{4\pi e^2}\frac{E^2}{h}\sigma_a. \tag{52}$$

The anomalous scattering factors $f'(E)$ and $f''(E)$ are related by the Kramers–Kronig relation

$$f'(\omega) = \frac{2}{\pi}\int_0^{\infty} \frac{\omega' f''(\omega')}{(\omega'^2 - \omega^2)}\,\mathrm{d}\omega', \tag{53}$$

where ω is the X-ray frequency.

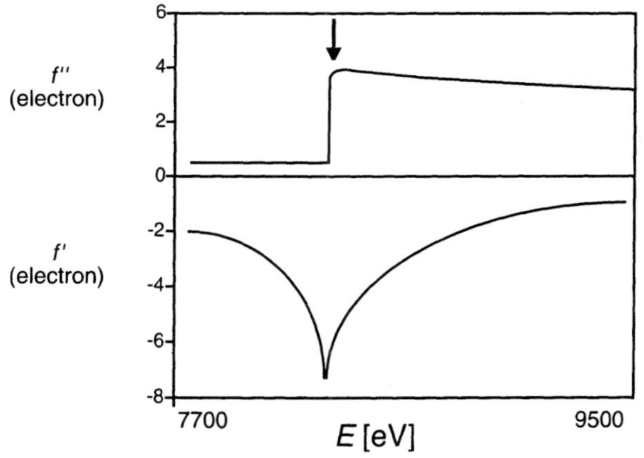

Fig. 3. Energy dependence of f' and f'' near the K-edge of nickel at 8333 eV (from Williams, 1991).

Therefore f' is negative close to the absorption edge and this results in an apparent decrease of the number of electrons. A typical energy dependence of f' and f'' is given in Fig. 3. Each element has its characteristic edges, and a subsequent variation of the photon energy near the corresponding K, L and M edges provides a selective alteration of the contribution from this particular atom to the overall intensity. In general the location of the absorption edge may be shifted away from the free atom value for mixed materials or solutions. The anomalous scattering factors are associated with scattering via absorption and reemission. Fluorescence and Raman scattering are among the most important emission processes, thus creating a parasitic background which one either has to correct for or try to avoid by an optimization of the experimental conditions.

In the design of an anomalous SAXS experiment (*ASAXS*), one has to consider several points:

- The choice of suitable elements is mainly limited by the available photon energies (currently approximately 4–15 keV). This considerably limits the experimentally accessible elements and we find Z values of 20–36 for the K edges and 50–82 for the L-edges. Therefore the most common atoms (C, O, H, N) in soft condensed matter systems can not be used as anomalous scatterers. This limits the technique for contrast variation experiments to systems such as micelles, polyelectrolytes or colloids that can contain heavy atoms.
- The effects of anomalous scattering are significant in a quite narrow range of energies close to the absorption edge only. Therefore a high resolution monochromator ($\Delta E/E < 10^{-3}$) is necessary for the ASAXS experiment.
- Whereas f'' is low and almost constant below the edge, sample absorption is large and inelastic processes give rise to a strong isotropic background. Therefore ASAXS experiments are normally conducted at energies below the edge. In order to avoid contributions from resonant Raman scattering and fluorescence one normally limits the energy range to energy differences of 5 eV and more from the edge, although this decreases the achievable contrast variation effect considerably.
- In order to perform a quantitative analysis of the anomalous scattering effects, one has to

know f' and f''. These values are tabulated for most elements and energy differences of 10 eV and more from the edge. Very close to the edge one has to measure sample absorption (f'') and calculate f' using Eq. (53).

- In general the performance of an ASAXS experiment requires very precise knowledge (or measurement) of the detector efficiency at different energies and precise subtraction of the corresponding inelastic background.

6. Selected examples

6.1. Polymer conformation and flexibility

Contrast variation in polymer solutions, melts and bulk samples has provided important information about the chain conformation in these systems. It is only the possibility to obtain the single chain form factor in a mixture of h- and d-labelled polymers that has allowed to obtain a clear proof of the random walk conformation (i.e., ideal chain behavior) of polymers in the melt. Details of these experiments can be found in the literature (see, for example, Higgins and Benoit, 1996). The theoretical basis for this experiment is given by the expression for the scattered intensity of a mixture of two different polymers with N monomers which are identical except for their coherent scattering lengths b_h and b_d. The total number of chains is n, where xn chains are deuterated and $(1-x)n$ chains hydrogenated. One can then show that the scattered intensity is given by

$$\frac{d\sigma}{d\Omega} = (b_d - b_h)^2 x(1-x) n N^2 P(q), \tag{54}$$

where $P(q)$ is the form factor of the chain. As a classical example of such a polymer we can look at polystyrene (Fig. 4). The corresponding intensity for a mixture of d- and h-polystyrene is shown in Fig. 5. We clearly observe the typical signature of a random walk chain given by the asymptotic q^{-2} behavior.

We can not only obtain information about chain conformations in the melt or bulk using mixtures of chains with different scattering contrast, but we can investigate local structural details of single chains using partially deuterated polymers. For semiflexible polymers, the SANS data should reveal a series of different regimes with a behavior characteristic of the various length scales of the chains: For very low values of q ($1/q > R_g$), i.e., in the so-called Guinier regime, the scattered intensity $I(q)$ becomes insensitive to structural details and is dominated by the finite overall length of the particles, and we can determine the radius of gyration R_g of the particles. At intermediate q (cross-section radius $R_c \ll 1/q \ll R_g$), $I(q)$ becomes much more sensitive to the local aggregate structure, and polymer theory predicts

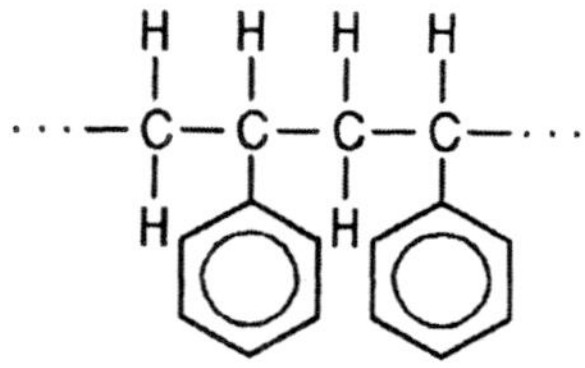

Fig. 4. Chemical structure of polystyrene.

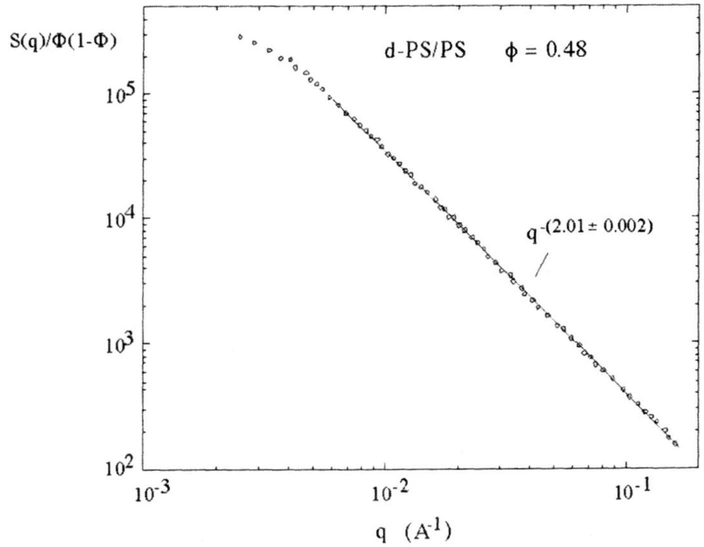

Fig. 5. q-dependence of the scattered intensity for a melt of h- and d-polystyrene (from Schwahn, 1991)

for flexible polymer coils that $I(q)$ should decay with a power law of the form $I(q) \sim Q^{-x}$, where $x = 1.66$ for a self-avoiding random walk chain and $x = 2.0$ for an ideal random walk chain. At large values of q, $I(q)$ is controlled by distances over which polymers are rod-like rather than flexible, and we expect a crossover to an asymptotic q^{-1}-dependence for $I(q)$ which is typical for locally cylindrical structures. However, it is important to point out that the experimental verification of the asymptotic behavior can be quite difficult, and in particular the crossover from a flexible to a locally stiff chain that contains the very important information on the polymer flexibility is often masked by the local cross-section structure. This can be demonstrated with data from partially deuterated polymer chains, where different parts of the monomer unit contribute differently to the overall intensity. The scattering functions of atactic polystyrene (PS) in carbon disulfide (CS_2) with different selective deuteration of the polymer have been determined by Rawiso et al. (1980) using SANS. CS_2 is a good solvent for PS, and this is reflected in the scattering function which shows a $q^{-1.7}$-behavior at intermediate q values. The data for a molar mass of $M_w = 50{,}000$ and two different selective deuterations ((i) fully deuterated, (ii) deuterated in the backbone) are shown in Fig. 6. CS_2 has a low scattering length density, C and D have a relatively large scattering length, and H has a negative scattering length. Therefore it is mainly the deuterated parts of PS which contribute to the scattering from the polymer coil. With the second labeling scheme the chains are effectively infinitely thin in the SANS experiment, and we indeed find the predicted three distinctly different regimes in the q-dependence of the measured differential cross-section: a Guinier regime, followed by an asymptotic $q^{-1.7}$ behavior for excluded volume chains which then crosses over to the q^{-1} behavior typical for rigid rods on length scales smaller than the so-called Kuhn length as a measure of the chain flexibility. For the other scheme, however, the polymers possess a cross-section with finite size that contributes in the crossover region and hides the q^{-1} behavior. Fig. 6 shows the scattering data for PS together with the

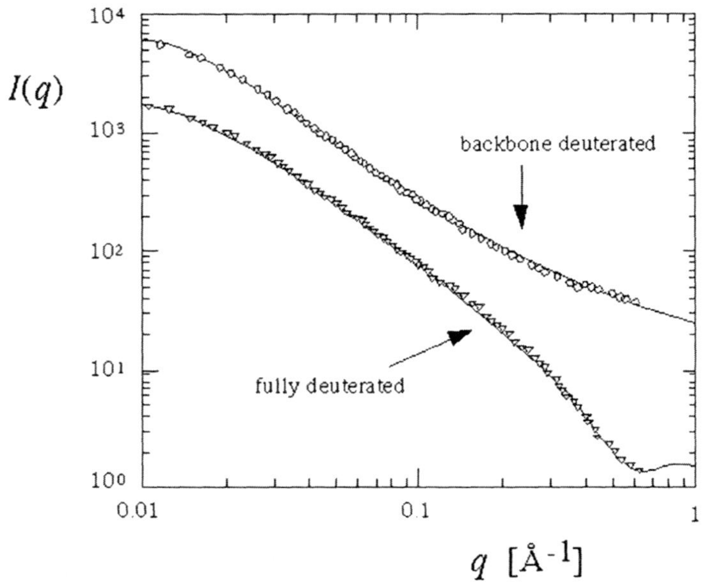

Fig. 6. Scattering functions for PS ($M_w = 50{,}000$) in CS_2 (Rawiso et al., 1980). The lower data are for fully deuterated PS, and the upper data for the backbone deuterated. The curves are fits for a model scattering function that includes excluded volume effects (Pedersen and Schurtenberger, 1996).

theoretical expression for the scattering function of semi-flexible chains with excluded volume effects obtained from a recent Monte Carlo simulation study and an additional cross-section term (Pedersen and Schurtenberger,1996). The agreement between the curve and the data is excellent and demonstrates the influence of the appropriate labeling of the polymer in the attempt to obtain information on local structure and flexibility.

6.2. Local structure of polymerlike micelles: X-rays vs. neutrons

Surfactants in solution exhibit a complex aggregation behaviour as a result of a delicate balance of opposing forces (Israelachvili, 1992; Safran, 1992; Evans and Wennerström, 1994). Micellar solutions and microemulsions represent thermodynamically stable liquid dispersions containing surfactant aggregates, which can often be found in a large region of the phase diagram of two- or multi-component surfactant systems. In micellar dispersions, the aggregates are formed by self-assembly of surfactant monomers. While micelles are frequently spherical, they can also exhibit a sphere-to-rod transition and even grow to giant flexible and polymer-like aggregates. These giant micelles normally have a high degree of flexibility, and their overall structure is generally well described by polymer theory (Cates and Candau, 1990; Jerke et al., 1997; Schurtenberger et al., 1999). In this section we shall demonstrate how detailed and quantitative knowledge on the local micellar structure can be obtained from SAXS and SANS experiments. We first use sugar-based surfactants (alkyl glucosides, APG) as an example, which have recently attracted considerable interest (Schurtenberger et al., 1999; Stradner et al., 2000). The structure and overall size of the micellar sugar

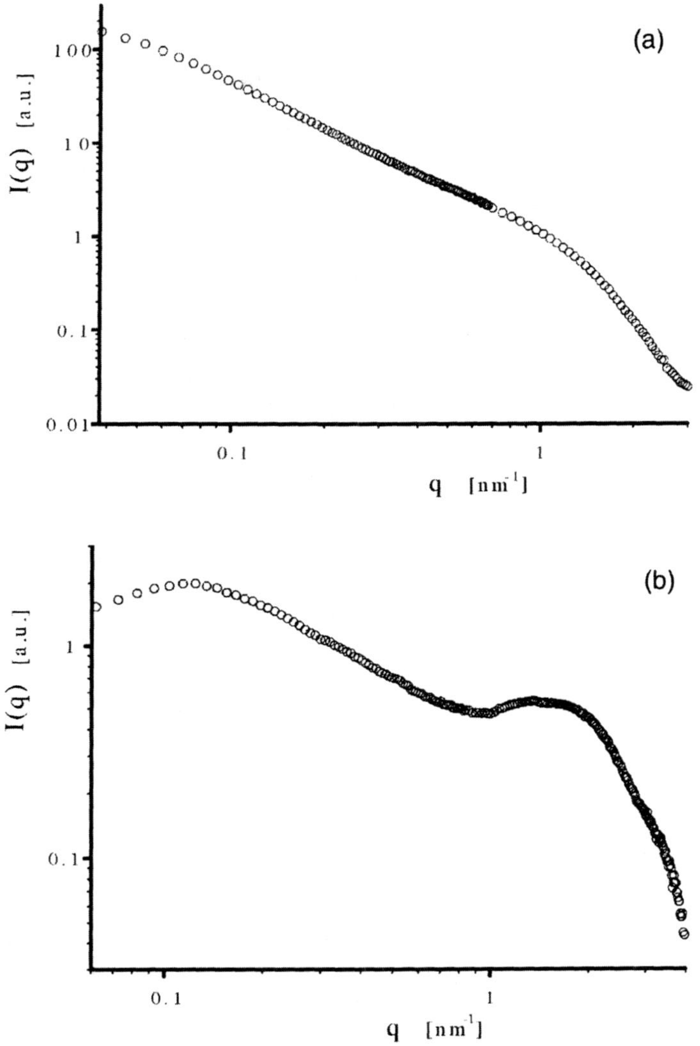

Fig. 7. q-dependence of the scattered intensity for an APG sample with $c = 5$ wt% and 0.8 wt% added hexanol. (a) from SANS; (b) from SAXS data (adapted from Stradner et al., 1999).

surfactant aggregates can be tuned by the amount of added hexanol. A qualitative explanation of the hexanol-induced micellar growth can be given based on the cosurfactant dependence of the spontaneous curvature or packing parameter of the surfactant molecule. Hexanol acts as a cosurfactant that dissolves mainly in the interfacial layer of the sugar surfactant aggregates, thereby changing its packing properties and curvature. The analogy to polymers is demonstrated in Fig. 7a, where SANS data from a dilute sample at high hexanol content is plotted. A detailed analysis of the micellar structure on all length scales is given elsewhere (Stradner et al., 2000). Here we focus on the local cylindrical cross-section structure.

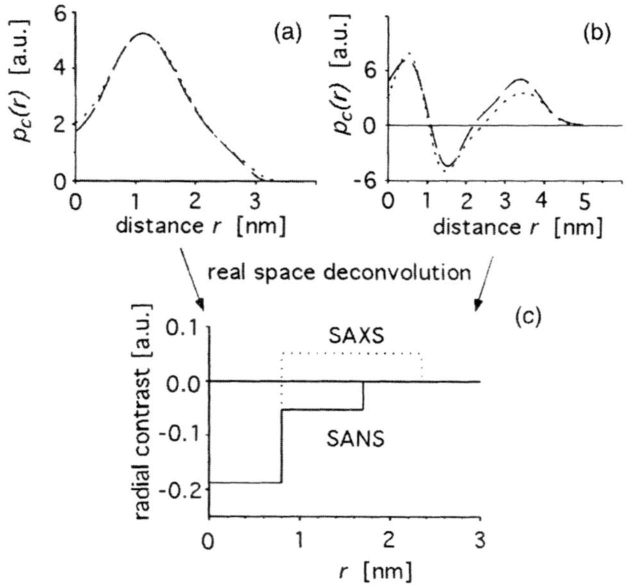

Fig. 8. Cross-section pair distance distribution functions $p_{cs}(r)$ for samples with $c = 5$ wt% and with 0.57 wt% (dashed line) and 0.8 wt % (dotted line) added hexanol. (a) from SANS; (b) from SAXS data (adapted from Stradner et al., 1999). (c) Cross-section excess scattering length density profile for 0.8 wtline) (from Schurtenberger et al., 1999).

This information can be obtained from an analysis of the scattering data using the so-called indirect Fourier transformation (IFT) technique (Glatter, 1977, 1980, 1981; Glatter and Hainisch, 1984; Pedersen et al., 1995; Schurtenberger et al., 1996). For locally cylindrical particles, IFT yields the cross-section pair distance distribution function $p_{cs}(r)$, which provides important real space information on the cross-section structure of the aggregate such as the maximum dimension, and also allows to obtain an estimate of the cross-section scattering length density distribution $\Delta\rho(r)$. The $p_{cs}(r)$ for two samples with low and high amount of added hexanol are shown in Fig. 8a. Both curves virtually overlap, thus indicating the negligible modifications of the cylindrical cross-section upon the addition of the small amounts of hexanol required to induce the formation of giant polymerlike micelles. It is quite instructive to look at the corresponding $p_{cs}(r)$ that have been obtained from an analysis of SAXS experiments with the same samples given in Fig. 8b. They exhibit a qualitatively different behavior at low values of r when compared to those obtained from the SANS data. The $p_{cs}(r)$ determined from the SAXS data shows very pronounced oscillations, and a larger maximum distance of $D_{max,cs} \sim 4.5$–5nm. The oscillations are due to the core-shell contrast with a very low overall contrast caused by the different electron densities of the surfactant tail and head group, respectively. This is also illustrated in the q-dependence of the scattered intensity obtained from SAXS shown in Fig. 7b, where the intensity at low q is considerably reduced, and where the polymerlike features are not visible. The different values of $D_{max,cs}$ obtained from SANS and SAXS are due to the fact that for SANS the head groups are almost matched due to the exchange of the hydrogens from the sugar-OH groups with deuterium.

Therefore, in the SANS experiment only the hydrocarbon chains are fully visible, and we observe mostly the micellar core. The values for the core diameter and overall diameter of the micellar cross-section are in good agreement with the known hydrocarbon chain length and the expected dimension of the sugar head group. This very different cross-section scattering length density profile $\Delta\rho_{cs}(r)$ of the micelles for SAXS and SANS is illustrated in Fig. 8c, where a deconvolution of the data in Fig. 8a,b has been made using a simplified two-shell model. This illustrates the complementarity of these two scattering techniques and demonstrates that with such a combination one can perform contrast variation experiments without having to alter the chemical composition of the sample.

6.3. Size and polydispersity of microemulsions: an optical contrast variation study

Microemulsions are formed in three- or multicomponent systems. The microemulsion aggregates have a liquid core [oil-in-water (o/w), water-in-oil (w/o) microemulsions] which is surrounded and stabilized by a surfactant monolayer. They frequently possess a droplet-like structure, but they can also grow into large tubular or sponge-like and multi-connected structures (Israelachvili, 1992; Langevin, 1992; Safran, 1992; Evans and Wennerström, 1994). The size, structure and polydispersity of microemulsions is a subject of continued study. These features of microemulsions can for example been studied with SANS experiments, where contrast variation experiments can be easily performed using combinations of deuterated and protiated solvents and surfactants. An example of such an experiment is given in Fig. 9. In this figure a schematic view of an oil-in-water microemulsion with a nonionic surfactant together with the corresponding excess scattering length density distribution $\Delta\rho(r)$ for so-called bulk (D_2O, protiated surfactant and oil) and shell (D_2O, protiated surfactant, deuterated oil) contrast is shown in a). In Fig. 9b the resulting scattered intensity for the system $D_2O/C_{12}E_5$/deuterated decane, i.e., for a microemulsion in shell contrast, is given. The use of shell contrast produces

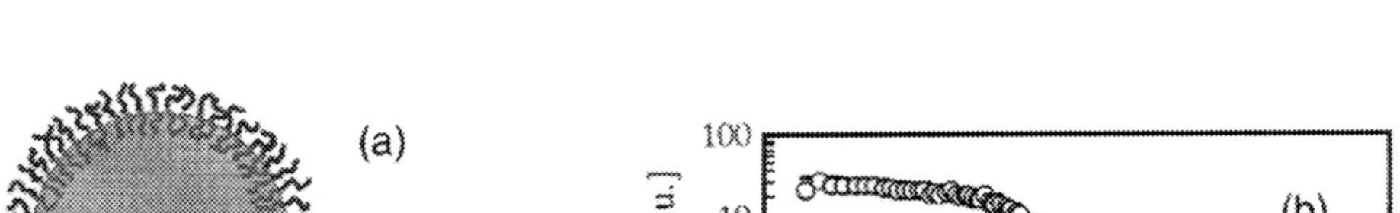

Fig. 9. (a) Schematic view of an oil-in-water microemulsion with a nonionic surfactant together with the corresponding excess scattering length density distribution $\Delta\rho(r)$ for so-called bulk (D_2O, protiated surfactant and oil, dashed line) and shell (D_2O, protiated surfactant, deuterated oil, solid line) contrast. (b) Scattered intensity from a SANS experiment with a microemulsion in shell contrast ($D_2O/C_{12}E_5$/deuterated decane). (From Olsson et al., 1997.)

a particularly pronounced form factor of the microemulsion droplet that can be used to study for example the effects of polydispersity and shape fluctuations (Olsson et al., 1997).

Contrast variation experiments with microemulsion droplets can not only be performed with neutron scattering. The so-called optical match point of w/o microemulsions offers a particularly elegant way of addressing these questions. The total average intensity at $q = 0$ for a polydisperse solution of microemulsion particles is given by (Rička et al., 1991)

$$\frac{1}{V}\frac{\mathrm{d}\sigma}{\mathrm{d}\Omega}(0) = \frac{\langle \Delta\alpha^2 \rangle \phi}{\lambda^4 \langle V \rangle}, \tag{55}$$

where $\langle V \rangle$ is the average volume and $\langle \Delta\alpha^2 \rangle$ is the mean square excess polarizability given by

$$\langle \Delta\alpha^2 \rangle \sim \int n(R)\alpha^2(R)\,\mathrm{d}R,$$

where $n(R)$ is the explicitly chosen size distribution of the microemulsion droplets. Due to the layered structure of the microemulsion droplets, their optical polarizability does not obey a simple R^3 law as found for homogeneous colloids. This is illustrated in Fig. 10, where the polarizability of a w/o microemulsion droplet is shown as a function of the water core radius R for two different solvents which differ in their index of refraction. We see that $\Delta\alpha(R)$ reaches a maximum after an initial increase and then decreases and even crosses zero at the optical match point, where the droplets become invisible in a light scattering experiment. These special optical properties of w/o microemulsions have important practical consequences when performing static and dynamic light scattering experiments. As we shall see, they permit an extremely accurate determination of the size polydispersity, and they provide us with an option to perform contrast variation experiments usually only accessible in neutron scattering. However, they also invalidate common approaches to interpret the measured structure factors. As we see from Fig. 10, the polarizability of a microemulsion droplet vanishes at a specific value of the water core radius, and size polydispersity thus results in a considerable residual scattering intensity.

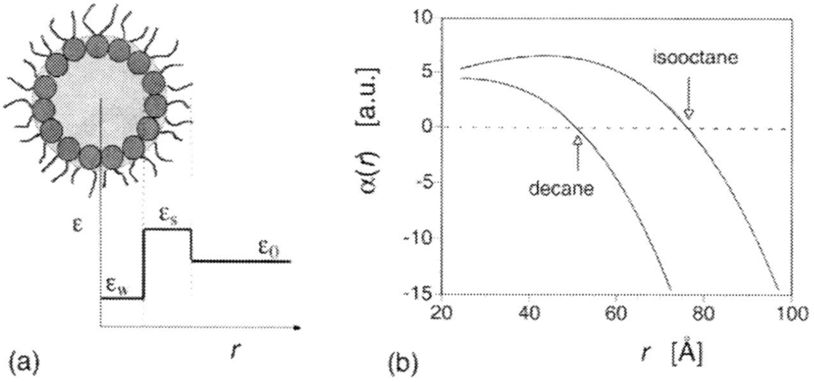

Fig. 10. (a) Shown is the dielectric constant profile for an AOT reverse micelle modelled as a layered dielectrical sphere (see text for details). (b) Dependence of the optical excess polarizability $\Delta\alpha(r)$ of an AOT reverse micelle on the water core radius r in the systems AOT–water–isooctane and AOT–water–decane, respectively (see Christ and Schurtenberger, 1994, for details).

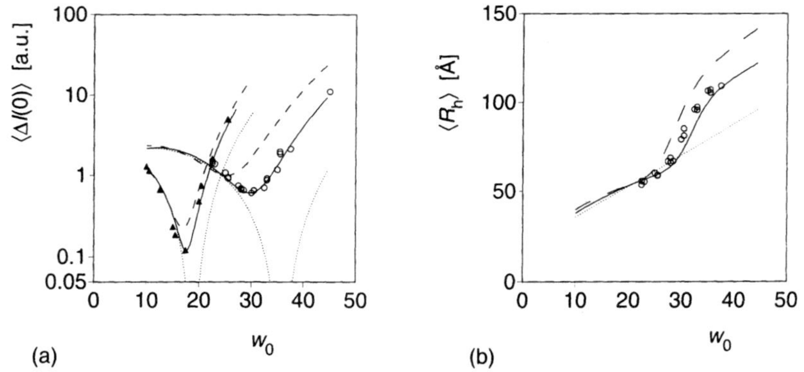

Fig. 11. (a) The normalized intensity $I(0)$ as a function of the water content w_0 for the systems AOT–water–decane (triangles) and AOT–water–isooctane (circles). Also shown are the theoretical curves for a model of monodisperse (dotted lines) and polydisperse layered spheres. For the polydisperse microemulsion droplets in decane and isooctane a Schulz distribution with a polydispersity $\gamma = (\langle R^2 \rangle - \langle R \rangle^2)/\langle R \rangle^2$ was used. The shape of the curves is completely given by the value of the polydispersity index γ. Calculations were made for decane (solid line): $\gamma = 0.01$, (dashed line): $\gamma = 0.02$, and isooctane (solid line): $\gamma = 0.035$, (dashed line): $\gamma = 0.07$ (see Christ and Schurtenberger, 1994, for details). (b) Calculated intensity-averaged hydrodynamic radius $\langle R_h \rangle$ as a function of the water content w_0 for monodisperse (solid line) and polydisperse microemulsion droplets in decane and isooctane. The parameters used are the same as in (a) (see Christ and Schurtenberger, 1994, for details).

This has first been recognized by Zulauf and Eicke, who analyzed the dependence of the scattered intensity on the water-to-surfactant molar ratio w_0 for AOT–water–isooctane microemulsions in terms of a simple shell model (Zulauf and Eicke, 1979). For monodisperse w/o microemulsion particles, the water core radius R is related to w_0 by $R = (3w_0 v_w/a_s)$, where v_w is the molecular volume of water and a_s the area per surfactant molecule at the oil-water interface. Recently it was shown that it is possible to quantitatively interpret this in the context of a polydisperse layered sphere model (Rička et al., 1990, 1991). The effect of polydispersity is illustrated in Fig. 11. We see that both the position and depth of the intensity minimum are extremely sensitive to the width of the size distribution, and that this allows an accurate determination of microemulsion droplet polydispersity. Polydispersity combined with optical inhomogeneity not only influences the scattered intensity, but has a profound influence on the results from dynamic light scattering (DLS) experiments. For polydisperse systems we measure an intensity-averaged mean diffusion coefficient or (apparent) hydrodynamic radius $\langle R_h \rangle$. This has enormous influence on the weighting of the different droplet sizes in the vicinity of the optical match point. This is also illustrated in Fig. 11, where the intensity weighted $\langle R_h \rangle$ is given for the same systems AOT w/o microemulsion droplets in isooctane and decane, respectively.

An important aspect of optically inhomogeneous particles such as w/o microemulsion droplets is the fact that, close to the optical match point, the measured structure factor $S^M(0)$ does not only reflect the thermodynamic quantity osmotic compressibility, but strongly depends upon the optical contrast also. This is based on the fact that for such particles, optical (i.e., excess polarizability $\Delta\alpha$) and size polydispersity are completely coupled, which has to be taken into account in any attempt to estimate $S^M(0)$ from scattering data. This was first pointed out by Rička and co-workers, who showed that the apparent 'virial' coefficient

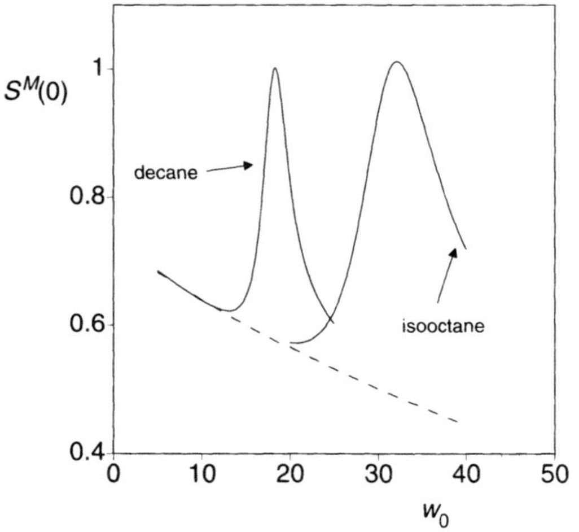

Fig. 12. Dependence of the 'measured' static structure factor $S^M(0)$ at a constant surfactant concentration of [AOT] = 0.1 M as a function of the water content w_0 for AOT w/o microemulsion droplets in decane and isooctane assuming hard sphere interactions (see Yan and Clarke, 1990, for details). Also shown is the corresponding calculation for optically homogeneous hard spheres at the same volume fractions

k_M^I as determined from an expansion of the initial slope of the concentration dependence of $S^M(0)$ vanishes in the vicinity of the optical match point (Rička et al., 1990). A more refined calculation of $S^M(0)$ for w/o microemulsion droplets was given by Yan and Clarke, who calculated $S^M(0)$ for a core and shell model of polydisperse particles interacting through a hard sphere potential (Yan and Clarke, 1990). Based on the Percus–Yevick approximation, they were able to derive the corresponding equations for a general case with complete coupling between scattering power and size polydispersity. The result of such a calculation is shown in Fig. 12, which illustrates the striking effect of the variation of the optical contrast on $S^M(0)$ for a given total surfactant concentration for two different w/o-microemulsion systems. The variation of the optical contrast is caused by an increase in water content w_0. The increase in w_0 corresponds to an increase in core radius as shown in Fig. 10, where the dependence of the optical excess polarizability $\Delta\alpha(r)$ of an AOT reverse micelle on the water core radius r for the same systems (AOT–water–isooctane and AOT–water–decane) had been given. We clearly see from Fig. 12 that $S^M(0)$ is strongly enhanced in the vicinity of the optical match point. This contrasts with the calculations for the same systems assuming polydisperse but optically homogeneous hard spheres, where $S^M(0)$ is independent of the optical contrast and depends on the volume fraction only. At constant total surfactant concentration the overall droplet volume fraction increases slightly with increasing w_0, which leads to a small but monotonic decrease of $S^M(0)$.

The optical properties of w/o microemulsions not only have to be taken into account when estimating the influence of interparticle interactions in static light scattering, but they also strongly modify the situation for dynamic light scattering when compared to classical homogeneous colloidal particles. One can in fact show that at the optical match point for these systems one obtains a perfect selection of the self-contribution to the dynamic structure factor. This property of w/o microemulsion droplets was in fact used by Rička and collaborators in order to measure the concentration dependence of the self diffusion coefficient of AOT microemulsions (Rička et al., 1990).

6.4. Nucleation and phase separation in microemulsions: SANS contrast variation experiments

In this section we look at a time resolved small-angle neutron scattering study of an Ostwald ripening process in a well characterized microemulsion system composed of water, decane (or hexadecane) and the nonionic surfactant pentaethylene glycol dodecylether ($C_{12}E_5$) (for details see Egelhaaf et al., 1999). At a surfactant-to-oil volume fraction ratio of $\phi_s/\phi_o = 0.815$ a water rich (oil-in-water) microemulsion phase is stable in the temperature range from 25 to 32°C, while below 25°C the microemulsion phase separates with a pure oil as the second phase (Olsson et al., 1997). The temperature dependence of the phase equilibria arises from a strong temperature dependence of the surfactant monolayer spontaneous curvature. At the phase boundary at 25°C, which is remarkably independent on water concentration, the solution consists of spherical microemulsion oil droplets corresponding to the maximum curvature towards oil given the constraint of the area-to-enclosed volume ratio imposed by the ratio ϕ_s/ϕ_o. The radius is approximately 7.5 nm, the polydispersity low and structural and dynamical properties follow closely those of a hard sphere fluid (Olsson et al., 1997). When the temperature is decreased below 25°C the preferred curvature towards oil increases and when equilibrium is reached the microemulsion droplets have decreased in size and an excess oil phase has appeared.

Let us now consider the dynamic process by which oil is removed from the microemulsion phase when a sample originally in the equilibrium state of spherical droplets is rapidly temperature quenched into the two-phase area. To reach the new equilibrium state the oil phase has to nucleate. From turbidity experiments we have evidence that this is a homogeneous nucleation process (Morris et al., 1997). The mechanism involves the growth of some of the droplets allowing the majority of droplets to decrease in size. This is schematically illustrated in Fig. 13. However, while turbidity measurements may report on the onset of nucleation, SANS is much better suited to follow the phase separation process in more detail and to obtain quantitative structural information. The fraction of the initial droplet population that grow in size is small and in order to experimentally detect this fraction already at early

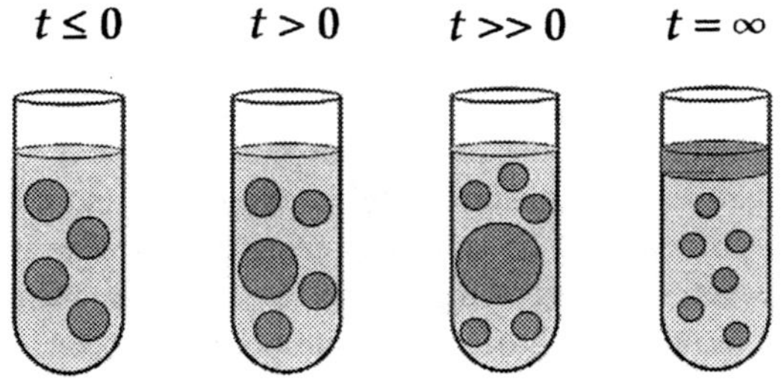

Fig. 13. At $t = 0$ an equilibrium system of oligodisperse oil-in-water microemulsion droplets is quenched into a two-phase area where, at final equilibrium ($t = \infty$), smaller droplets coexist with an excess oil phase. The oil phase nucleates at a few of the initial droplets which subsequently grow ($t > 0$), allowing the majority of droplets to decrease in size. This growth phase proceeds with a constant number of large drops ($t \gg 0$).

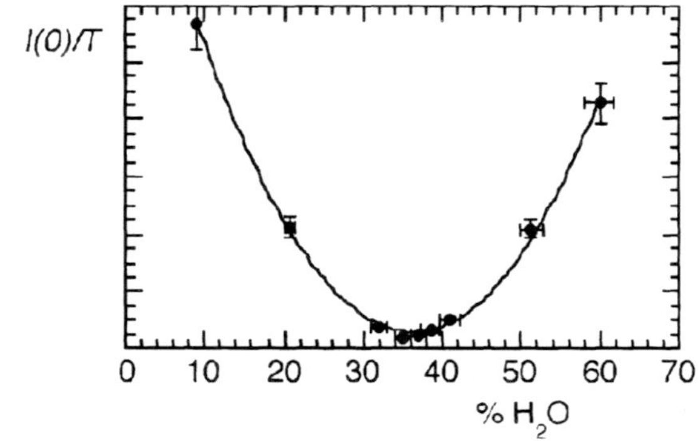

Fig. 14. Normalized scattering intensity $I(0)/T$ extrapolated to $q = 0$, where T is the transmission of the sample, for an o/w microemulsion ($C_{12}E_5$/deuterated hexadecane) in a mixture of H_2O–D_2O as a function of the H_2O volume fraction (total volume fraction $\phi_s + \phi_o = 0.1$).

stages we have utilized contrast matching to reduce the forward scattering intensity from the small microemulsion core-shell droplets. Initially they have a radius $R_1 = 6.8$ nm, are composed of fully deuterated decane or hexadecane, and covered by a shell of $C_{12}E_5$ having a thickness $d = 1.5$ nm. The average scattering length density of these droplets can be matched by dissolving them in an appropriate mixture of D_2O and H_2O. This is illustrated in Fig. 14, which shows the transmission-normalized scattering intensity $I(0)$ extrapolated to $q = 0$ for a contrast variation series with the droplet microemulsion under equilibrium conditions to determine the optimum H_2O–D_2O-ratio. We see that the microemulsion droplets have near zero average contrast at a H_2O volume fraction of 0.355, which has subsequently been used in the SANS experiments. In this case the scattered intensity from the microemulsion droplets is very small for $q < 0.2$ nm^{-1}, allowing us to monitor a small fraction of growing larger oil drops. However, the small droplets still contribute at larger scattering angles. It is therefore possible to follow the growth of the big drops (at small q) as well as the shrinkage of the small droplets ('hump' at relatively large q, see Fig. 15).

The small-angle neutron-scattering experiments were performed on the instrument D22 at the ILL, Grenoble. This instrument is particularly suited for time-resolved experiments due to the high neutron flux and the large area detector, which in addition can be offset, giving a large scattering vector coverage within a single detector setting (Egelhaaf et al., 1997). The detector was positioned so that we could follow both the evolution of growing emulsion drops at lower q, where the scattering of the small droplets was suppressed, as well as the decrease in size of the smaller droplets indicated by the form factor maximum at $q \sim 0.4$ nm^{-1}. The sample was temperature quenched into the unstable regime where the equilibrium droplet radius in the microemulsion is approximately 4.5 nm. Scattering curves were measured for periods of 30 seconds over an observation period of more than two hours as shown in Fig. 15.

Dramatic changes in the scattering occur as time progresses particularly at low q-values. Apart from the first 20 minutes or so, the scattering at low q is completely dominated by the contribution from the large drops. The concentration of the large drops is so low that their relative positions are uncorrelated, and we can also neglect interference effects between large and small droplets due to the size difference. Thus, the scattering in the low q regime can

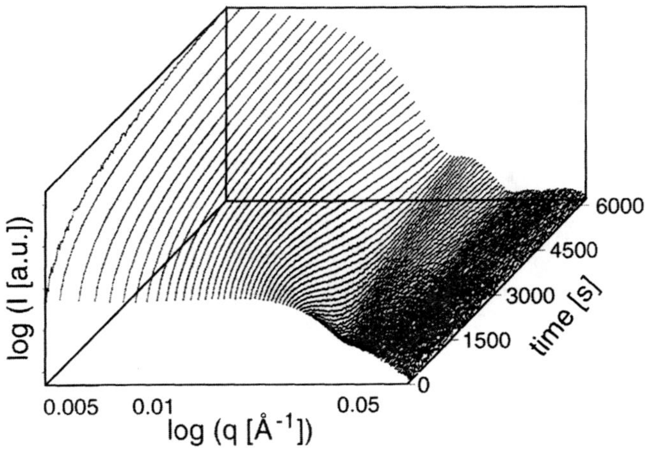

Fig. 15. Sequence of scattering curves from a time-resolved small-angle neutron-scattering experiment recorded during the nucleation and growth of an excess oil phase from oil-in-water microemulsion droplets. For clarity only every second measurement and only the first 6000 s are shown.

be described as arising from a low concentration of large drops dispersed in an essentially continuous solvent made up of water and the small droplets. The small number of large drops retain a relatively monodisperse size distribution as can be seen qualitatively from the 'hump' in the scattering curves ($q \sim 0.02$ A^{-1}), which reflects the form factor of the larger oil drops.

The time-resolved SANS experiments provide accurate data on the initial formation and subsequent growth of oil drops nucleated after a temperature quench of an oil-in-water microemulsion into a two-phase region. For the first time we have access to the simultaneous temporal evolution of the size distribution of oil drops and small microemulsion spheres, which allows for a detailed, quantitative description of the processes. Right after the temperature quench there is a fast nucleation process followed by a phase where these nuclei grow by diffusion of oil from the small droplets. After approximately 45 minutes the small droplets have reached their (near) equilibrium size and the large drops continue to evolve through Ostwald ripening at constant oil volume. From the measured intensity at $q = 0$ we not only verify that the number of drops decreases with time as t^{-1} but we also obtain a quantitative measure of the drop concentration. Independently, the Guinier analysis provides the temporal evolution of the mean drop radius which displays a $t^{1/3}$ growth law at longer times. The time resolved SANS experiments on a very well characterized model system combined with suitable contrast variation have provided us with a unique set of data on Ostwald ripening processes in microemulsions.

Acknowledgements

I gratefully acknowledge the important contributions made by Jan Skov Pedersen, Stefan U. Egelhaaf, Sylvie Christ, Ulf Olsson and Anna Stradner. Without their help this chapter would not have been possible.

References

Cates, M.E. and Candau, S.J. (1990).*J. Phys.: Condens. Matter*, **2**, 6869.

Christ, S. and Schurtenberger, P. (1994). *J. Phys. Chem.*, **98**, 12708.

Egelhaaf, S.U. and Schurtenberger, P. (1997). *Phys. B*, **234–236**, 276.

Egelhaaf, S., Olsson, U., Schurtenberger, P., Morris, J. and Wennerström, H. (1999). *Phys. Rev. E*, **60**, 5681.

Evans, D.F. and Wennerström, H. (1994). *The Colloidal Domain: Where Physics, Chemistry, Biology and Technology Meet*. VCH Publishers, New York.

Glatter, O. (1977). *J. Appl. Cryst.*, **10**, 415.

Glatter, O. (1980). *J. Appl. Cryst.*, **13**, 577.

Glatter, O. (1981). *J. Appl. Cryst.*, **14**, 101.

Glatter, O. and Hainisch, B. (1984). *J. Appl. Cryst.*, **17**, 435.

Higgins, J.S. and Benoit, H.C. (1996). *Polymers and Neutron Scattering*. Oxford University Press, Oxford.

Israelachvili, J.N. (1992). *Intermolecular and Surfaces Forces*, 2nd edition. Academic Press, London.

Jerke, G., Pedersen, J.S., Egelhaaf, S.U. and Schurtenberger, P. (1997). *Phys. Rev. E*, **56**, 5772, and references therein.

Langevin, D. (1992). *Annu. Rev. Phys. Chem.*, **43**, 341.

Morris, J., Olsson, U. and Wennerström, H. (1997). *Langmuir*, **13**, 606.

Olsson, U., Bagger-Jörgensen, H., Leaver, M., Morris, J., Mortensen, K., Strey, R., Schurtenberger, P. and Wennerström, H. (1997). *Prog. Colloid Polym. Sci.*, **106**, 6.

Pedersen, J.S. and Schurtenberger, P. (1996). *Macromolecules*, **29**, 7602.

Pedersen, J.S., Egelhaaf, S.U. and Schurtenberger, P. (1995). *J. Phys. Chem.*, **99**, 1299.

Rawiso, M., Duplessix, R. and Picot, C. (1980). *Macromolecules*, **13**, 1518.

Rička, J., Borkovec, M., Hofmeier, U. and Eicke, H.-F. (1990). *Europhys. Lett.*, **11**, 379.

Rička, J., Borkovec, M. and Hofmeier, U. (1991). *J. Chem. Phys.*, **94**, 8503.

Safran, S.A. (1992). In: *Structure and Dynamics of Strongly Interacting Colloids and Supramolecular Aggregates in Solution* (Chen, S.H., Huang, J.S. and Tartaglia, P., eds.). Kluwer Academic Publishers, Dordrecht, Vol. 369, 237 pp.

Schurtenberger, P., Jerke, G., Cavaco, C. and Pedersen, J.S. (1996). *Langmuir*, **12**, 2433.

Schurtenberger, P., Stradner, A. and Egelhaaf, S.U. (1999). In: *Frontiers of Neutron Scattering* (Furrer, A., ed.). World Scientific Publishing, pp. 107–130 .

Schwahn, D. (1991). In: *Physik der Polymere*. Forschungszentrum Julich GmbH, pp. 7.1–7.44.

Stradner, A., Mayer, B., Sottmann, T., Hermetter, A. and Glatter, O.; (1999) *J. Phys. Chem. B*, **103**, 6680.

Stradner, A., Glatter, O. and Schurtenberger, P. (2000). *Langmuir*, **16**, 5345.

Wennerström, H., Morris, J. and Olsson, U. (1997). *Langmuir*, **13**, 6972.

Williams, C.E. (1991). In: *Neutron, X-Ray and Light Scattering* Lindner, P. and Zemb, Th., eds.). Elsevier, Amsterdam, pp. 101–117.

Yan, Y.D. and Clarke, J.H.R. (1990). *J. Chem. Phys.*, **93**, 4501.

Zulauf, M. and Eicke, H.F. (1979). *J. Phys. Chem.*, **83**, 480.

Neutrons, X-Rays and Light
P. Lindner and Th. Zemb (Editors)

Chapter 8

Static Light Scattering of Large Systems

Otto Glatter

Department of Chemistry, Karl-Franzens Universität Graz, Heinrichstraße 28, A-8010 Graz, Austria

1. Introduction

Different static scattering methods can be used for sizing of colloidal systems. An overview over the size range that can be covered by these techniques is given in Fig. 1. Small-angle X-ray or neutron scattering can be used in the size range 1–100 nm (Glatter, 1980). This technique can be described by the Rayleigh–Debye–Gans (RDG) theory, which has been discussed in detail in previous chapters (see also Glatter and Kratky, 1982). Ultra-small-angle scattering could be used in principle for particles up to some micrometers but this technique needs very brilliant X-ray sources (synchrotrons) or high flux reactors and complicated instrumentation (Agamalian et al., 1999; Hainbuchner et al., 1999). This technique is important for non-transparent materials.

Static light scattering experiments are described by different theories depending on the size and the refractive index of the dissolved particles (Kerker, 1969; van de Hulst, 1981; Bohren and Huffman, 1983). Particles small compared with the wavelength of the light and/or low contrast can be described by the RDG-theory, a typical limit for colloidal systems is about 100 nm. This regime is followed by the so-called Lorenz–Mie regime which extends from about 100 nm up to several micrometers. The original work of Lorenz (1898) and Mie (1908) was related to spherical particles only, but these names are now often used for arbitrarily shaped scatterers in this size regime which is also called the resonant scattering regime, as resonances of the electric field can be found in some cases. Above this regime, i.e., from some micrometers up to about thousand micrometers, the scattering process can be described by the so-called Fraunhofer diffraction (Bayvel and Jones, 1981). It should be noted that dynamic light scattering experiments can be used for sizing of particles in a wide size regime (some nanometers up to some micrometers) but this technique has a much lower resolution than static experiments (Dahneke, 1983; Glatter et al., 1991).

This lecture will focus on static light scattering. For the following we assume negligible absorption of the particles ($n''_p \sim 0$). The RDG theory applies to very weak scatterers, the basic assumption being that the field is not altered by the presence of the particles and is identical to the incident wave. For this to be so the particle must reflect very little so that $|n_p - 1| \ll 1$, and must introduce negligible phase shift $2\alpha|m-1| \ll 1$, where $m = n'_p/n'_s$ is the ratio of the real part of the refractive indices of the particles and of the solvent, α is a size parameter defined by

$$\alpha = \frac{2\pi a}{\lambda_o} \tag{1}$$

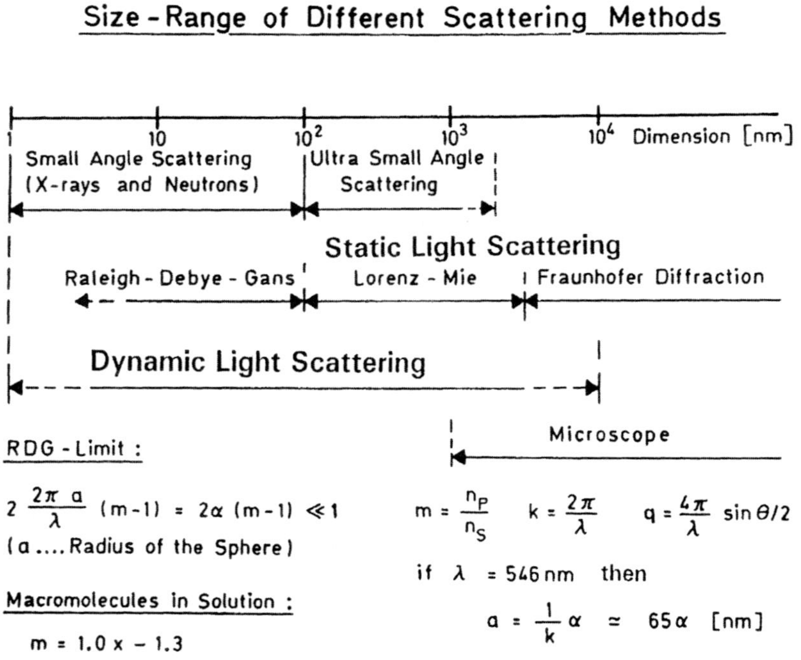

Fig. 1. Overview of the size range of different scattering methods.

where a is the radius of the sphere. In this theory we assume that each infinitesimal part of the particle acts independently as a Rayleigh scatterer and we simply add all scattered waves taking into account the phase shift due to the different path length and we find that the scattering amplitude is simply a Fourier transform of the excess refractive index profil $\alpha'(\boldsymbol{r})$ where $\boldsymbol{q}$ is the scattering vector ($q = (4\pi n/\lambda_o)\sin\theta/2$).

$$A(\boldsymbol{q}) = \text{const.} \int_v \alpha'(\boldsymbol{r})\, e^{-i\boldsymbol{qr}}\, d\boldsymbol{r} \tag{2}$$

The averaging process over all directions in space leads to the scattering intensity

$$I(q) = \text{const} \int_0^\infty p(r) \frac{\sin qr}{qr}\, dr \tag{3}$$

where $p(r)$ is the pair distance distribution of the particle. We can use this relation for an approximate shape analysis for monodisperse colloidal particles (Glatter, 1979).

The Lorenz–Mie regime is the most difficult one in terms of theoretical description. The electromagnetic radiation interacts non-linearly with the particles, the strength of the electric field depends highly on the position (close to and inside the particle). The different possible states of polarization of the wave are important. They can be described by the four Stokes parameters being components of the so-called Stokes vector. This holds for the incident and the scattered wave.

The Stokes vector of the incident wave is related to the Stokes vector of the scattered wave by the Müller or Stokes matrix (Kerker, 1969; van de Hulst, 1981; Bohren and Huffman, 1983; Hofer and Glatter, 1989). The sixteen components of this matrix depend on the scattering vector $\boldsymbol{q}$ and the orientation of the particle. The one-one element can be regarded as the classical scattering function for unpolarized experiments. These elements can be calculated by solving the Maxwell equations. New numerical methods like the T-matrix method can solve this difficult problem for regular bodies within reasonable computing time (Barber and Hill, 1990). We shall only discuss the classical scattering curves (M_{11}), because all other elements, which can be measured in principle, are difficult to interpret, i.e., there exist no solutions for the inverse scattering problem. This inverse problem is to find the structure of the scatterer from its scattering pattern. It is difficult enough for the M_{11} element.

With increasing size (above several micrometers) the field of the electromagnetic wave hardly penetrates into the particle, so the scattering process is approximated by the interaction of a planar wave with the cross-section, or aperture, of a particle (Bayvel and Jones, 1981). This process is called Fraunhofer diffraction if only the far-field solution is important (as it is for static light scattering experiments). This means that Fraunhofer diffraction cannot give information about the overall shape or about the internal structure. The same is true for the RDG approach for another reason. In the RDG regime the particles are too small to scatter an essential part of the total scattering curve into the excessible regime of an SLS instrument. If we use a HeNe laser in aqueous systems and scattering angles from 10° to 150° we get the regime $0.0023 \leq q \leq 0.0255$. This regime may be modified by a special small-angle instrument (Fraunhofer type) or by a decrease of the wavelength. These details will be discussed in the experimental section.

Here we can summarize, that for particles larger than 100 nm we cannot use RDG theory. Lorenz–Mie and Fraunhofer regime allow for particle sizing of polydisperse systems in the size regimes of about 50 nm up to a few micrometers or starting from a few micrometers up to about thousand micrometers. Only in the Lorenz–Mie regime there is a chance to determine the shape and structure of monodisperse systems.

In the following we want to discuss the scattering problem in the Lorenz–Mie regime, i.e., to see the influence of increasing size and/or contrast on the scattering data and to learn about the main features of the corresponding theory when compared to the RDG theory. All the following discussions in the next section are related to ideally diluted systems, i.e., we do not take into account any particle interaction. Here we find another important difference to SAXS and SANS: due to the high scattering power of these large particles we run much earlier into the problem of multiple scattering with increasing concentration before we see the influence of excluded volume effects. Particle interactions could only dominate for highly charged particles or in systems with vanishing contrast. But in all other cases we have to worry primarily about the problem of multiple scattering and not so much about particle interactions.

2. Monodisperse systems

Let us first summarize the general aspects of the problem. Particles in solution are randomly oriented. In a scattering experiment we get the scattering function of the particle averaged over all possible orientations in space (Fig. 2). The so-called *scattering problem* is the computation of the angle-dependent scattering function $I(q)$ for a certain particle. This

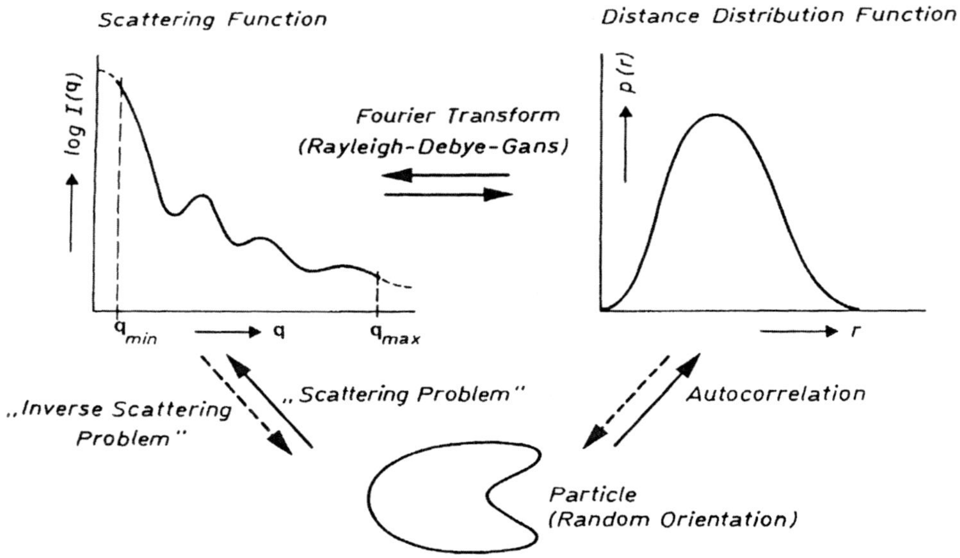

Fig. 2. Schematic representation of the correlation between a particle, its observable experimental scattering data, and its distance distribution function.

function can be measured in the limited angular range $q_{\min}$ to $q_{\max}$. This range is determined by the experimental setup ($\theta_{\min}$ and $\theta_{\max}$) and by the wavelength of the light λ_o.

The structure of the particle under investigation must be determined from this scattering function. This so-called *inverse scattering problem* has no unique solution because some information has been lost in the scattering experiment by the spatial averaging process. In addition all phases of the scattered waves have been lost by measuring the intensities of the scattered light.

The simplest way to solve such problems is to guess a model, solve the scattering problem, and compare the resulting theoretical scattering function with the experimental result. The main problem in this procedure is how to vary the model in order to improve it. The scattering functions are not only a function of the size and shape of the scattering object, but are also different for differing refractive index ratios m. This ratio varies for colloidal systems in the range $1 \leq m \leq 1.3$. From these problems we see that it is nearly impossible to solve the inverse scattering problem by a trial-and-error search for a model that is equivalent in scattering.

Our approach to the problem (Glatter et al., 1985) has been developed in the field of small-angle scattering of X-rays and neutrons. The RDG theory can be used to describe these experiments. In this theory one can show that the scattering function $I(q)$ of the particle is the Fourier transform of the PDDF (pair distance distribution function) $p(r)$. This PDDF allows a rough classification of the shape of homogeneous particles, it has characteristic shapes for globular, plate-like and rod-like structures, and it allows one to distinguish different types of aggregation. Even for inhomogeneous particles, a considerable amount of structural detail can be recognized.

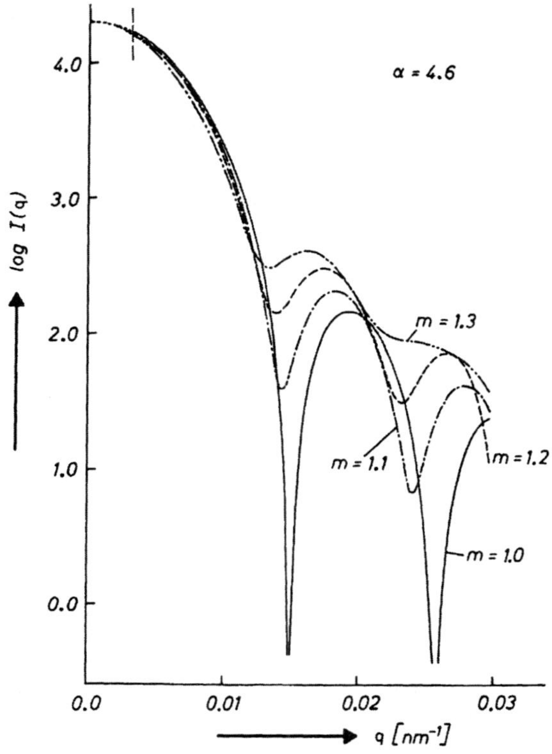

Fig. 3. Mie scattering functions for a sphere with $D = 600$ nm ($\alpha = 4.6$), $\lambda = 409.6$ nm for different m values.

2.1. Spherical objects

It can be shown that the approximate computation (Glatter, 1977) of this $p(r)$ function of spherical particles from its scattering function is possible in the ranges $1.0 \leq m \leq 1.2$ and $2.0 \leq \alpha \leq 7.0$. The results are not essentially influenced by a normally distributed statistical error of a few percent. The $p(r)$ functions calculated from Mie scattering functions show good agreement with the exact theoretical function of the particle for $0 \leq r \leq D$. Some oscillations are found beyond $r = D$. They depend on α and m and are the most significant contributions of the Mie theory in this range. Another contribution is the fact that the peak of the $p(r)$ function is slightly shifted to larger r values if $m \geq 1.2$ thus increasing a little the apparent size of the particle.

The scattering curves for a typical value of α ($\alpha = 4.6$, $D = 600$ nm) are shown in Fig. 3 for different m values ($m = 1.0$ to $m = 1.3$). Note that the value $m = 1.0$ corresponds to the ideal RDG theory. The deviations from this curve increase with increasing m values. Maxima and minima become progressively smeared out and slightly shifted toward smaller angles, corresponding to an increase in the apparent radius of gyration.

The curves are normalized to a common but arbitrary scattering intensity at zero-angle. The $p(r)$ functions computed from these scattering curves are shown in Fig. 4. For $m = 1.0$

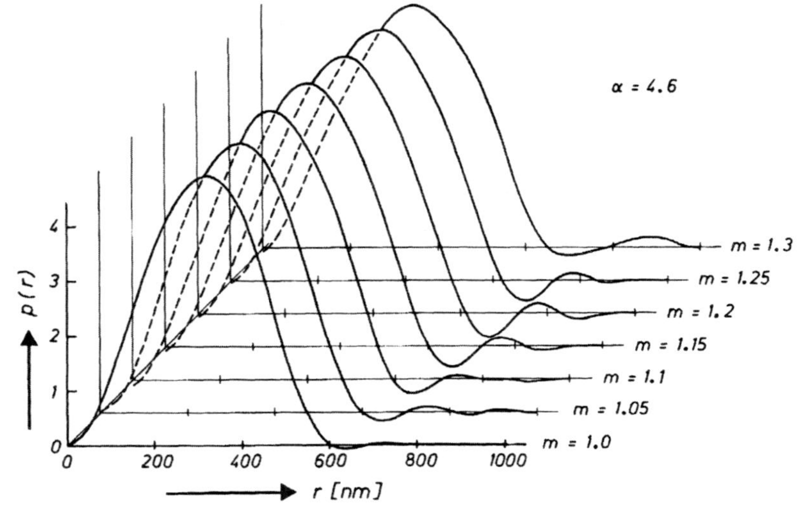

Fig. 4. The $p(r)$ functions computed from the scattering curves shown in Fig. 3. Cutoff values: $q_{\min} = 0.0032\ \mathrm{nm}^{-1}$, $q_{\max} = 0.0296\ \mathrm{nm}^{-1}$.

the curve is very close to the theoretical $p(r)$ function of a sphere, with negligible termination effects. Increasing m values introduce oscillations about the abscissa beyond the diameter D, but the shape of $p(r)$ in the range $0 \leq r \leq D$ remains nearly unchanged and the function always goes through zero close to D, even though the allowed value for $D_{\max}$ in the calculations was much larger (up to twice the value of D!).

Fig. 5 shows scattering functions for $m = 1.1$ and variable size ($\alpha = 2.2$–7.0). Only the innermost part appears for the $\alpha = 2.2$ curve, which is near its first minimum at $\theta = 150°$. The curve for $\alpha = 4.6$ already shows two side maxima, while for $\alpha = 7.0$ the scattering curve shows the onset of the fourth maximum. In the latter case, an essential portion of the central maximum appears at angles too small to be observed.

The $p(r)$ functions for these scattering curves are shown in Fig. 6. with the relative scaling r/D on the abscissa. In the range of the particle dimension ($r/D \leq 1$), the three functions agree well and they all show a shape similar to that of an ideal sphere. The functions pass through zero at $r/D = 1$ and they oscillate around the abscissa at larger r/D values. Differences in these oscillations are manifestations of different cutoff effects. They are observed for all curves with $m \neq 1.0$ and represent the unavoidable influence of the Lorenz–Mie theory. They cannot be removed by filter methods (Jenkins and Watts, 1968), such as artificial temperature factors used to smooth Fourier transformations.

3. Non-spherical objects

In order to test the applicability of the *indirect Fourier transform* (IFT) (Glatter, 1977) for solving the inverse scattering problem for non-spherical particles, one must calculate exact scattering functions of particles of different size, shape, and m values. Transforming these exact scattering curves back into real space then allows a detailed analysis of the errors

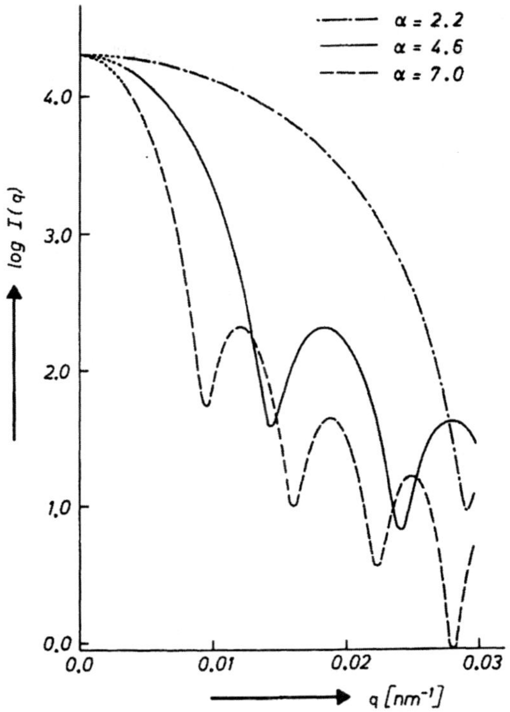

Fig. 5. Mie scattering functions for spheres with different values of α and constant refractive index $m = 1.1$.

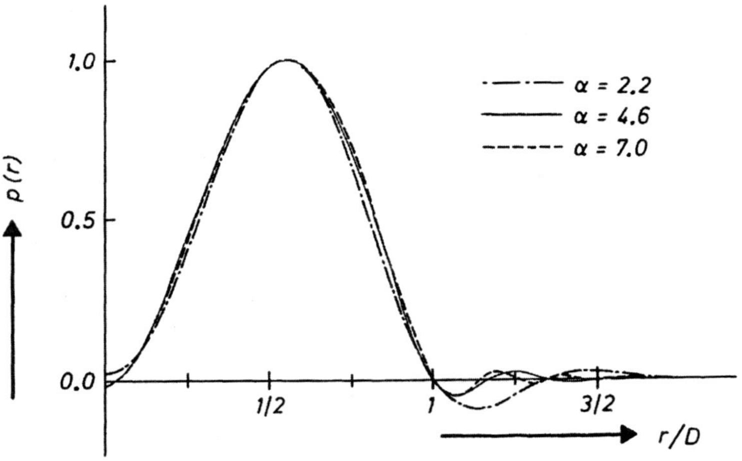

Fig. 6. The $p(r)$ functions computed from the scattering curves shown in Fig. 5, with the same cutoff values as in Figs. 3 and 4.

introduced by the IFT, which is, we should say it clearly, a RDG approximation to the inverse scattering problem.

We use the T-matrix method for the calculation of the scattering functions, referred to also under the name of *extended boundary condition method* (EBCM). P.W. Barber's FORTRAN program package, which permits angular scattering calculations of homogeneous and inhomogeneous spheres and spheroids in fixed orientation, randomly oriented in two or three dimensions, and Müller matrix calculations, is part of a book (Barber and Hill, 1990).

The programs are not suitable for spheroids with axial ratios higher than about 1 : 7. This is due to the method of numerical solution used in the calculation of Bessel and Legendre functions. It should be possible to modify the numerical solution to minimize this problem, but we have not attempted to do this. A more detailed discussion of this problem is to be found in Hofer (1987).

3.1. Prolate ellipsoids

In the next two figures we show the results for very elongated prolate ellipsoids with an axial ratio $a/c = 5$ and a size parameter $\alpha = k \cdot a = 11.5$, which corresponds to a major semi-axis of 750 nm. The scattering functions in Fig. 7 show again a slight decrease in the width of the main maximum.

The maxima of the $p(r)$ functions in Fig. 8 are slightly shifted to larger r values for increasing m, but all functions have the same overall shape. The oscillations are due to the termination effect at $q_{\min}$. The sampling theorem tells us that only scattering functions from

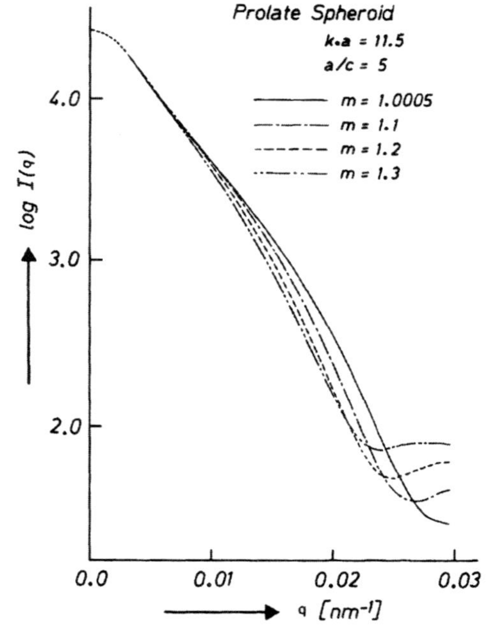

Fig. 7. Scattering functions of prolate spheroids.

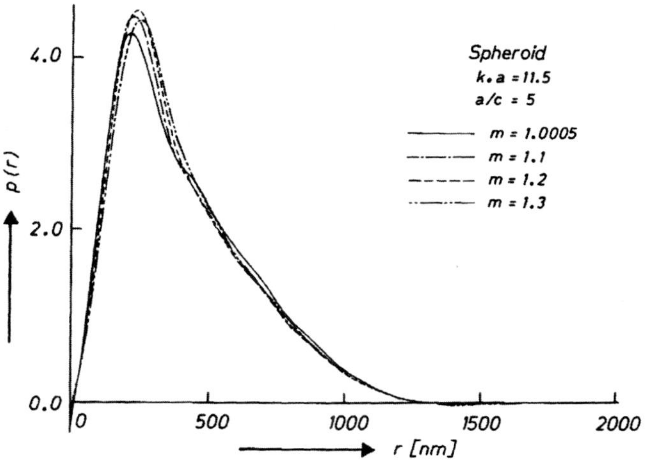

Fig. 8. Distribution functions computed from the scattering functions shown in Fig. 7.

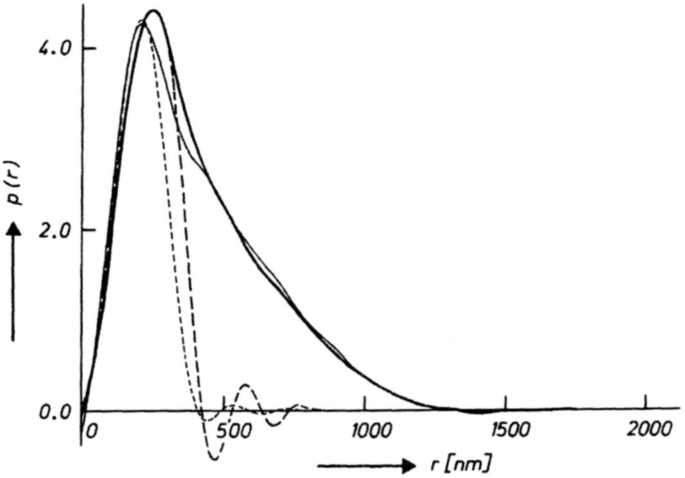

Fig. 9. Comparison of the Mie effect for a prolate ellipsoid with $\alpha = k \cdot a = 11.5$ and $a/c = 5$ (full lines) and for a sphere with $\alpha = k \cdot a = 3.068$ (dashed lines). The thin lines represent the RDG approximation ($m = 1.0005$) and the thick lines show the results for $m = 1.3$.

objects with $D_{\max} \leq \pi/q_{\min} \simeq 980$ nm can be transformed without error, but we see that the termination effects are minimized by the IFT method.

Discussing the Mie effect as a cooperative interaction between the electromagnetic field and the particle, we see from Fig. 8 that this effect is essential mainly in the cross-section, i.e., in a range, where the particle is three-dimensional and the effect is negligible at the 'one-dimensional' poles at the end of the long axis (Glatter and Hofer, 1988a).

In order to show this effect more clearly we compare the results from the prolate ellipsoid with $a/c = 5$ with the results from a spherical particle with the same m values (Fig. 9). The size parameter $\alpha = k \cdot a = 3.068$ ($R = 200$ nm) of the sphere guarantees that the position of the maximum of the PDDF for $m = 1.0005$ (thin dashed line) coincides with the maximum of the ellipsoid with the same refractive index (thin full line). The oscillations in the distance distribution function of the sphere for $r > 2R$ are caused by the termination effects. Both curves are shown in an arbitrary scaling in order to have the same height for the maximum. Increasing the m value to $m = 1.3$ shifts the peak of the $p(r)$ function of the ellipsoid to a somewhat larger r value (thick full line). The maximum of the PDDF of the sphere is shifted to exactly the same position; the oscillations at $r > D$ increase as already discussed.

The apparent increase in size can be explained roughly by the fact that the field is also changed in the close surrounding of the particle.

3.2. Oblate ellipsoids

Oblate ellipsoids are ideal particles to represent the transition form spherical to flat, two-dimensional structures. The Mie effect needs a three-dimensional structure. It increases with absolute size, m value, and axial ratio c/a. This is demonstrated with a series of $p(r)$ functions for a spheroid with $\alpha = k \cdot a = 11.2$ ($a = 730$ nm), $1.0005 \leq m \leq 1.3$ and $0.2 \leq c/a \leq 1$ in Fig. 10. The maximum dimension for all particles is the same, namely $D = 2a = 1460$ nm.

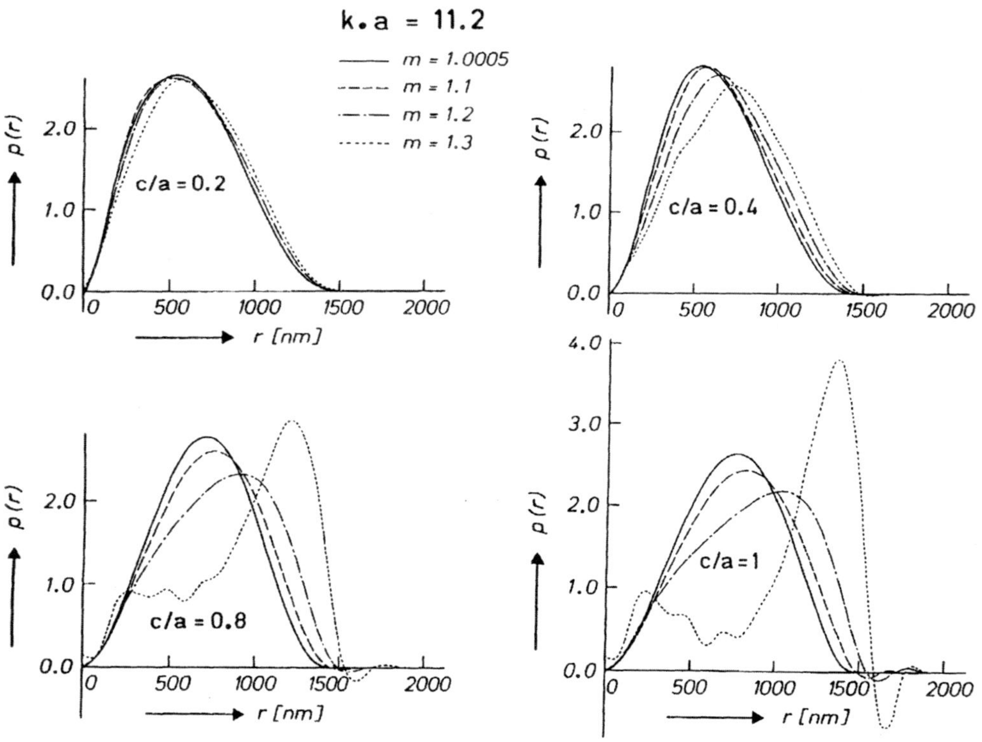

Fig. 10. Dependence of the Mie effect on the axial ratio c/a and on the m value for oblate spheroids with a constant size parameter $\alpha = k \cdot a = 11.2$. Increasing axial ratio leads to increasing effects.

The flat ellipsoid ($c/a = 0.2$) shows nearly no effects even for $m = 1.3$. Increasing axial ratio leads to increasing effects. The particles with $m = 1.3$ show the limits of the method for this α value. The PDDF is no longer comparable with the theoretical one, which is very close to the results for $m = 1.0005$. The assumption of a constant field intensity inside and close to the particle is no longer valid for globular particles with $\alpha > 10$ and $m > 1.2$.

Summarizing these results for homogeneous particles we can conclude that the method for the calculation of the $p(r)$ function works well for spheres, particles elongated in two directions (oblate ellipsoids), and particles elongated in one direction (prolate ellipsoids). The Mie effect depends on m and on the largest three-dimensional volume element; i.e., in the case of a prolate ellipsoid the effect is not controlled by the length of the long axis a, but by the size of the maximum cross-section defined by the short axis c. Any deviation from spherical symmetry decreases the Mie effect!

We do not have any data for other non-spherical structures like cylinders or aggregates at the moment, but there are no arguments why one should expect distinct differences for other elongated or flat structures. This is true also for axial ratios larger than 5 or smaller than 0.2 and for larger aggregates of chain molecules, as the Mie effect needs compact three-dimensional structures in order to show relevant contributions.

4. Inhomogeneous objects

The incoming electromagnetic field penetrates the whole particle in the pre-resonant case. Therefore it must be possible to get information on the internal structure from the scattering data. This information can be used for a direct determination of the structure only in the case of spherical, cylindrical, or lamellar symmetry. A method for the deconvolution of PDDFs to radial density profiles for RDG scatterers was developed several years ago (Glatter, 1981; Glatter and Hainisch, 1984). Suspended particles may have shells of different refractive index. Depending on the index of the solvent it is possible that they have index ratios higher and lower than one. This corresponds with a positive or negative difference polarization density (refraction index difference) α'. We have calculated scattering functions from a spherical core–shell model (Fig. 11). The size of the core is $\alpha_i = k \cdot a_i = 6.136$ ($R_i = 400$ nm) and the size of the shell is $\alpha_o = k \cdot a_o = 9.204$ ($R_o = 600$ nm). For one model we used a refractive index ratio for the core of $m_i = 0.9$ and for the shell $m_o = 1.1$ (dashed lines in Figs. 11–13) and an inverse model, i.e., $m_i = 1.1$ and $m_o = 0.9$ (full lines in Figs. 11–13). Both models would have exactly the same scattering intensity in the RDG theory. There would be only a sign inversion in the amplitudes which cannot be measured directly.

From Fig. 11 we see that the Mie theory leads to different scattering functions. The corresponding distance distribution functions are shown in Fig. 12. The main difference (apart form the different amplitudes) is that the width of the second maximum is smaller for the inverse model (full line). This second maximum is determined by the shell.

The polarization density profiles can be fitted by a small number of steps with variable height and width or with a higher number of equidistant but narrow steps. We have calculated an optimum two-step profile and a ten-step profile for both models (Fig. 13). It should be mentioned at this point that the deconvolution procedure cannot give the absolute sign of the density distribution function ; i.e., any resulting $\alpha'(r)$ function may have to be multiplied with the factor (-1). That means that the experimenter must have *a priori* information about the

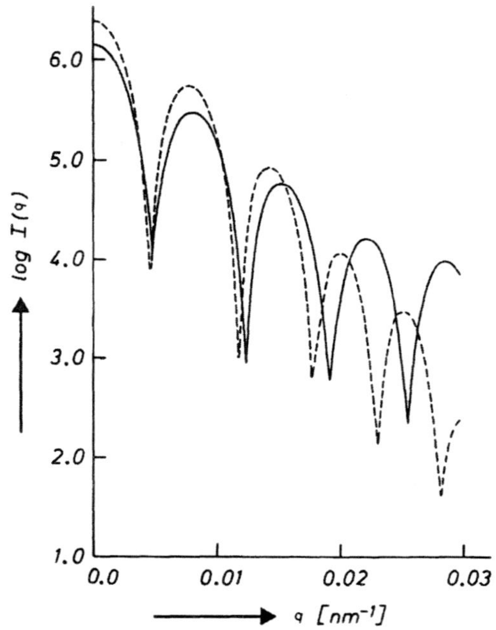

Fig. 11. Scattering functions for a spherical core–shell model with $m_i = 0.9$ and $m_o = 1.1$ (dashed line) and for the inverse model, i.e., $m_i = 1.1$ and $m_o = 0.9$ (full line).

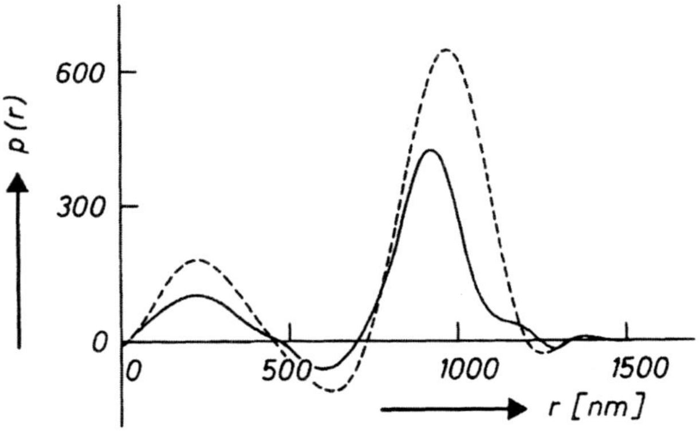

Fig. 12. The PDDFs computed from the scattering functions shown in Fig. 11.

system as to whether the region with the higher refractive index is the core or the shell. If it is the shell, we see from Fig. 13 that the two steps in the density profile are well resolved. The size of the core can be estimated with high accuracy; the outer radius of the shell is overestimated by about 4%. For the inverse model (low refractive index in the shell) we must use the phase factor (−1) and we then get again a nearly perfect estimate of the size of the core. The size of the shell is underestimated by about 5%.

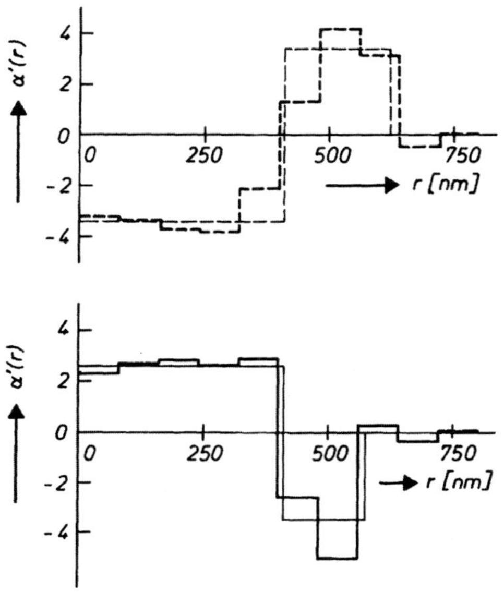

Fig. 13. Polarization density profiles $\alpha'(r)$ computed with the convolution square root technique from the $p(r)$ functions shown in Fig. 12. Upper half, the core–shell model, lower half, inverse model.

Summarizing the results we can say that we get fairly good estimates for the polarization density distribution if we have the a priori information about the sign (basic arrangement of the zones of high and low refractive index) and if the object under investigation is close to spherical symmetry.

5. Particle sizing of polydisperse systems with static light scattering

Many systems in the size range between 100 nm and several micrometers, like emulsions, show a distinct polydispersity in size. Assuming a certain shape – like spheres for emulsions – one is interested in a determination of the size distribution. Such an investigation is known as particle sizing. The most common optical methods for particle sizing are dynamic light scattering (DLS) (Dahneke, 1983) and Fraunhofer diffraction (Bayvel and Jones, 1981). There exist several commercially available instruments. In static light scattering experiments the angular dependent intensity of scattered light from non-interacting colloidal systems is influenced by the size, the shape, and the optical contrast of the particles. The most important parameter is the m value, the optical contrast (Schnablegger and Glatter, 1991).

To gain as much information as possible, we present a general approach to the problem of deducing as many parameters as possible form experimental data. In particular, we investigate the possibility of determining the relative refractive index of polydisperse colloidal particles from static light-scattering data. In the first part we deal with single-scattering curves; an extended version for application to multiple-scattering data (van de Hulst, 1980) is given in the next section.

The angular-dependent scattering curves of polydisperse systems can be regarded as a linear combination of Lorenz–Mie form factors $\Phi(q, R, m)$, where q is again the length of the scattering vector, R is the radius, and m is the relative complex refractive index of the particles in the particular solvent. If we proceed and regard the coefficients of the linear combination as a continuous function $D(R)$, we immediately obtain the Fredholm integral equation in question (Glatter and Hofer, 1988b; Schnablegger and Glatter, 1991):

$$I(q) = \int_{R_{\min}}^{R_{\max}} D(R)W(R)\Phi(q, R, m)\, \mathrm{d}R, \tag{4}$$

where $W(R) = \Phi(0, R, m)^{-1}$ for intensity distributions $D_i(R)$,
$W(R) \approx R^{-3}$ for volume or mass distributions $D_v(R)$, and
$W(R) = 1$ for number distributions $D_n(R)$.

It was shown that non-spherical particles give broadened but reasonable size distributions when evaluated with the form factors of spheres (Schnablegger and Glatter, 1993). Therefore we used the Lorenz–Mie form factors of spheres calculated by means of the S1 program of Barber and Hill (1990).

Particle sizing in the RDG regime is a special case of the Lorenz–Mie theory and can therefore be included in our considerations. The form factor for RDG spheres is

$$\Phi(q, R) = \left[3\frac{\sin(qR) - (qR)\cos(qR)}{(qR)^3}\right]^2 \tag{5}$$

The weighting factors in this case are given by

$W(R) = 1$ for intensity distribtuions $D_i(R)$,
$W(R) = R^3$ for volume or mass distributions $D_v(R)$, and
$W(R) = R^6$ for number distributions $D_n(R)$.

The integral equation in the regime of Fraunhofer diffraction reads

$$I(q) = \int_{R_{\min}}^{R_{\max}} D(R)J_1^2(2\pi nR\theta/\lambda_o)(R/\theta)^2\, \mathrm{d}R \tag{6}$$

The properties of the resulting distribution functions cannot be presumed to be different from those mentioned for the RDG regime, because in both cases the kernel is a damped oscillating Bessel function.

We see that the three regimes RDG, Lorenz–Mie and Fraunhofer diffraction have a very similar mathematical description. The most complicated kernel of the integral equation is found in the Lorenz–Mie regime. We will focus therefore on this case in the following. Only in this case we are mostly working with cylindrical cuvettes where the reflected light from the glass wall must be taken into consideration. Light scattered to an angle of 30° is partially reflected and will be detected as an additional contribution at 150°.

These contributions become apparent especially at high scattering angles (backward direction) where the scattered intensity is usually low. Fig. 14 schematically shows the scattering geometry which leads to the mathematical description of the scattering problem applied in this presentation.

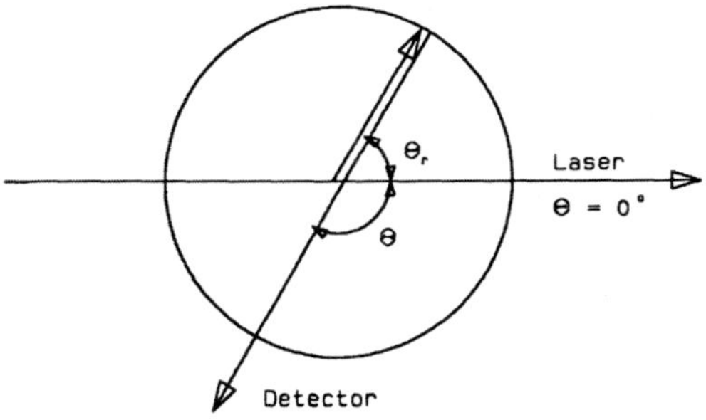

Fig. 14. The scattering geometry which leads to the mathematical description of the reflected-light contributions. In the center of the cylindrical glass cell the laser light is scattered in all directions. The intensity arriving at the detector consists of the directly scattered intensity with scattering angle θ and the reflected intensity from the glass with scattering angle $\theta_r = 180 - \theta$.

The light intensity $I'(q,m)$ arriving at the detector consists of the directly scattered light intensity $I(q,m)$ and the reflected light intensity $I(q_r,m)$ from the glass wall of the cylindrical sample holder:

$$I'(q,m) = I(q,m) + c_r R_p I(q_r,m) \tag{7}$$

q_r is the length of the scattering vector for $\theta_r = 180 - \theta$. R_p is Fresnel's reflectivity coefficient for light polarized parallel to the plane of incidence. We use the approximation for normal incidence

$$R_p = \left(\frac{n_s - n_g}{n_s + n_g}\right)^2 \tag{8}$$

where n_s and n_g are the refractive indices of the solvent and of the glass, respectively. R_p is the percentage of the reflected light for unit incident intensity and it becomes zero when the refractive indices n_s and n_g become equal. Therefore, Eq. (7) can be used even when the absolute intensity is unknown.

The coefficient c_r in Eq. (7) theoretically equals 1.0 for ideal experimental conditions. In our computational procedure we let c_r be adjustable in order to compensate for all deviations from the ideal model. These deviations may come, for example, from a not exactly known refractive index of the glass n_g, from a slight deformation of the cylindrical glass cell, and from the extinction of the reflected light in the turbid and absorbing sample. The absolute value of c_r is not of primary interest, but it must be determined in order to avoid further systematic errors. That means that neglecting the reflected light contributions cannot be falsely interpreted as a modified distribution of sizes, thus yielding a good fit. Therefore, if R_p is significantly higher than zero, the fitting procedure is sensitive enough to detect and to separate the amount of reflected light. In cases where the refractive index of the solvent is close to that of the glass ($R_p \simeq 0$), the contributions of the reflected light to the signal is negligible and therefore the value for c_r cannot and need not be determined. These cases are clearly indicated by our computational method.

That means that neglecting the reflected light contributions cannot be falsely interpreted as a modified distribution of sizes, thus yielding a good fit. Therefore, if R_p is significantly higher than zero, the fitting procedure is sensitive enough to detect and to separate the amount of reflected light. In cases where the refractive index of the solvent is close to that of the glass ($R_p \simeq 0$), the contributions of the reflected light to the signal is negligible and therefore the value for c_r cannot and need not be determined. These cases are clearly indicated by our computational method.

5.1. Computational background of the Calculations

It becomes apparent form Eq. (4) that the mathematical problem, the so-called indirect scattering problem, consists in calculating the size distribution $D(R)$ from the polysdisperse scattering function $I(q,m)$. It is clear that the resulting $D(R)$ depends on the actual value of the relative refractive index m which is being used for the calculation. First, let us assume that the value of m is known and let us extend our consideration to the case of an unknown value of m later.

Our computations begin with the assumption that the $D(R)$ can be described as a series of cubic B-splines $\varphi_i(R)$ (Greville, 1969; de Boor, 1978)

$$D(R) = \sum_{i=1}^{n} c_i \varphi_i(R) \tag{9}$$

where the linear coefficients c_i are yet unknown and must be determined. The B-spline functions $\varphi_i(R)$ are bell-shaped functions and thus they provide for a smooth distribution function $D(R)$ which, in principle, can be of any shape. This makes the technique more general because it does not assume any particular analytical form of the distribution. But it must be stated that narrow peaks in $D(R)$ can be resolved only when the widths of the B-splines are narrow enough.

Anyway, each $\varphi_i(R)$ can be transformed to the corresponding theoretical scattering function $\psi_i(q,m)$ (often called reciprocal space) since the direct scattering problem can be solved exactly:

$$\psi_i(q,m) = \int_0^{\infty} \Phi(q,R,m) W(R) \varphi_i(R)\, \mathrm{d}R \tag{10}$$

Now the Lorenz–Mie transformed B-splines $\psi_i(q,m)$ can be used to approximate the experimental scattering function $f(q)$ by a series expansion

$$f(q) \simeq \sum_{i=1}^{n} c_i \psi_i(q,m) \tag{11}$$

The coefficients c_i can be calculated by a least-squares algorithm and with these the size distribution $D(R)$ can be calculated by the use of Eq. (9). By substitution of $\psi_i(q,m)$ in Eq. (11) with the right side of Eq. (10) and by exchanging the integral with the sum, it can be easily checked that the coefficients in Eq. (11) are the same as those in Eq. (9).

Regarding the contributions of the reflected light, we have to extend Eq. (11). With Eq. (7) we can write the approximation problem:

$$f(q) \simeq \sum_{i=1}^{n} c_i \psi_i(q,m) + c_r \sum_{i=1}^{n} c_i \psi_i(q_r,m) \tag{12}$$

A further extension is the attempt to determine the relative refractive index in the case that it is not known *a priori*. Thus the least-squares problem becomes non-linear and the variables to be determined are the coefficients c_i, c_r, and m.

The least-squares problem for the determination of the coefficients is of ill-conditioned nature and has to be stabilized by regularization techniques. The employed procedure uses a stability parameter, the so-called 'Lagrange multiplier', which levels the ratio of *a priori* and *a posteriori* information about the system. The value of this Lagrange multiplier has to be determined, which is done by means of a stability plot.

In any case the values of the parameters (e.g., m or c_r) and the Lagrange multiplier have to be determined simultaneously. This cannot be done in one step but only with iterations, because of the non-linear nature of the least-squares problem. The best mathematical way to solve this problem would be the method of linearization and/or Marquardt's algorithm (Bevington, 1969) which must be modified in order to allow for the stabilization of the solution. This 'mathematically clean' way takes much time since for each iteration the most time-consuming step, the calculation of the $\psi_i(q,m)$, has to be repeated. On the other hand, there are only one or two parameters (c_r and m) which make the least-squares problem non-linear and the calculation of the gradients for all parameters, including the c_is, would be a waste of time. Therefore we considered another approach which yields the same result with fewer iterations.

This approach consists in solving the linear least-squares problem for a fixed set of values for m and/or c_r. The ill-conditioned nature of the least-squares problem demands the application of a stability parameter. Thus, for each value of the stability parameter a best fitting value of m and/or c_r can be calculated, e.g., by quadratic interpolation of the mean deviation surface. The question of finding the final best fitting value is treated in the following subsection.

The most important question arising with our computations is that of how the constraints affect the best fitting values of the additional parameters. From our experience we expect at least a range of Lagrange-multiplier values, where the additional parameters can be determined with optimum accuracy. Depending on the performance of the constraint and on the condition number of the system matrix, the width of this range can be rather small. For detecting this range of Lagrange-multiplier values we designed the so-called *sensitivity*.

It is rather obvious that a parameter which has a strong effect on the shape of the fitting function will change the mean deviation (a quality parameter) more than a parameter which has no such strong effect. Thus, the gradient of the mean deviation, i.e., the slope of the mean deviation surface, depends on the particular parameter which is being varied and on the magnitude of the stability parameter. The consequence of the fact that the mean deviation gradients have different magnitudes is that the more dominant parameter can be determined more precisely than others. We can rewrite this statement and say that the *sensitivity* of the experimental curve is not the same for all of the parameters. Hence, let us define a quality parameter which we call the sensitivity and which somehow contains the gradient, i.e., partial first derivative of the mean deviation with respect to the parameter in question.

The sensitivity S is defined by

$$S = \left\{ \sum_{i=1}^{k-1} \left[lg \left(\frac{MD_{i+1}}{MD_i} \right) \right]^2 \right\}^{1/2} \tag{13}$$

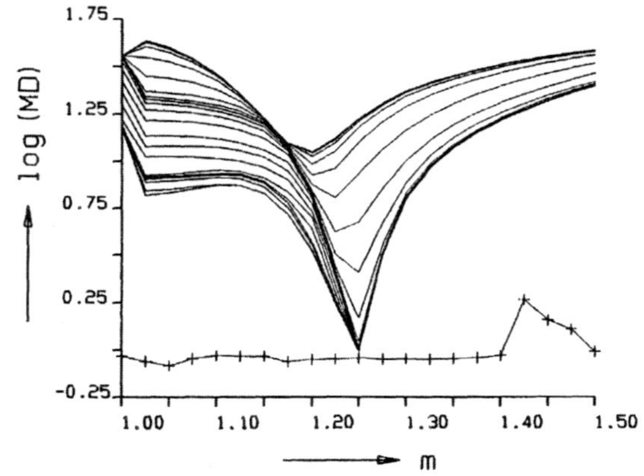

Fig. 15. Effect of the positivity constraint: the strength of this constraint is not controlled by λ, the minima in the mean deviation curves grow to a limit and then remain constant. The case of no constraints at all (pluses) is included in this graph for comparison.

where k is the number of parameter values and MD_i is the mean deviation which results when the least-squares problem is solved with the ith parameter value fixed. The mean deviation is calculated by

$$MD = \left[\frac{1}{M} \sum_{j=l}^{M} \left(\frac{f_i - g_i}{\sigma_j} \right)^2 \right]^{1/2} \tag{14}$$

where M is the number of points of the experimental scattering function f_i with standard deviation σ_j, and g_j is the sampled approximation curve defined by the right side of Eq. (12).

With the definition of the sensitivity we understand that the magnitude of the sensitivity is able to indicate the sharpness of a possible mean deviation minimum. This is not a proof for the existence of a minimum. The only unequivocal proof of a global minimum in the mean deviation surface can be drawn solely from the visualization of the whole surface in a so-called mean deviation plot (*MD* plot). Hence it follows that the sensitivity can be regarded as an additional tool only, a decision criterion to find the steepest minimum from a set of mean deviation curves, which sometimes can hardly be done by the naked eye. Thus, a sensitivity plot (*S* plot) cannot be interpreted without the picture of the corresponding mean deviation surface.

We reported extensively about the features of the positivity constraint when it is applied in a polydispersity analysis (Lawson and Hanson, 1974; Schnablegger and Glatter, 1991). One of those features is the introduction of a certain constant amount of stability, regardless of the influence of the other constraint. Thus, if λ is sufficiently small, then the total control is transferred to the positivity constraint and no dependence on λ exists anymore. Therefore the minima in the mean deviation curves (Fig. 15) grow to a limit and remain constant afterward.

The corresponding *S* plot (Fig. 16) therefore reaches a plateau when λ decreases. The case of no constraint at all (pluses) is included in Fig. 15 for comparison.

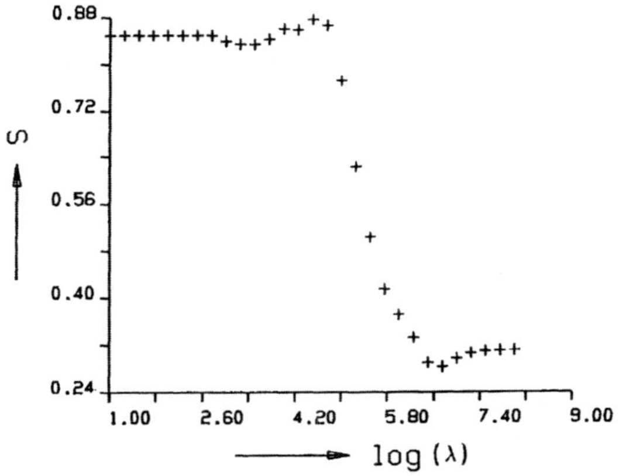

Fig. 16. Effect of the positivity constraint: the corresponding *S* plot summarizes Fig. 15. The sensitivity indicates that in this case every value below log(λ) < 5 can be used for the determination of the relative refractive index *m*. That means that below this limiting value the sensitivity for the additional parameter constantly remains at an optimum.

5.2. Application to a simulated scattering curve

In order to check the performance of the computational procedure, a size distribution of Lorenz–Mie spheres has been simulated as is depicted in Fig. 17 (full line).

It is a bimodal Gaussian volume distribution with centers at 275 nm and 450 nm, with an amplitude ratio of 2:1 and with standard deviations of 50 nm and 137.5 nm. The particles are assumed to have a relative refractive index of $m = 1.250$. The corresponding light-scattering curve which is depicted in Fig. 18 (pluses) is sampled with 71 points between the angles of

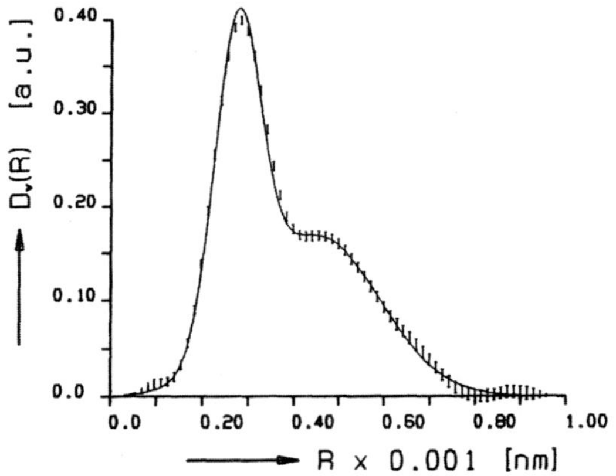

Fig. 17. Simulated size distribution of spherical particles: the reconstruction of a bimodal volume distribution $D_v(R)$ from a simulated scattering curve of Lorenz–Mie spheres with the relative refractive index of $m = 1.250$. The error bars are the reconstruction with best fitting values $m = 1.248 \pm 0.003$ and$c_r = 1.014 \pm 0.02$. The full line is the simulated distribution.

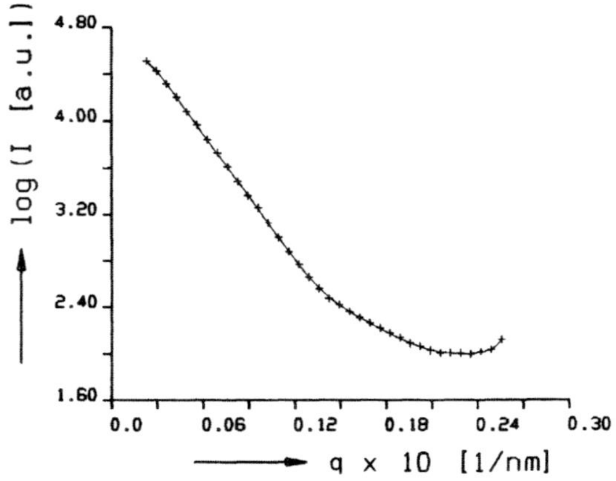

Fig. 18. Simulated size distribution of spherical particles: the scattering curve (every second data point is marked with a plus) of the simulated size distribution from Fig. 17 and the fit (full line). Reflected-light contributions with $c_r = 1.00$ and statistical noise of 1% is added.

10° and 150°. Reflected light contributions with $c_r = 1.0$ and with the refractive indices of $n_s = 1.332$ and $n_g = 1.530$ as well as 1% statistical noise has been added.

The reconstruction of the volume distribution $D_v(R)$ and of the additional parameters m and c_r has been done as described in the previous section. The resulting distribution is depicted in Fig. 17 and marked with errors bars.

The best fitting parameters were $m = 1.248$ and $c_r = 1.014$ which are within the above stated accuracy. As usual, the fit to the simulated scattering curve is very good (Fig. 18, full line). The *MD* plot and the *S* plot for this simulation are not illustrated here, since they look quite similar to Figs. 15 and 16.

5.3. Application to latex particles

We present the results of a sample which contained slightly polydisperse spheres made of Melamine-formaldehyde resin with a nominal diameter of 2900 nm. The refractive index of this polymer can be estimated applying the Lorentz–Lorenz equation (Bohn, 1975), which correlates the molar refractivity to the repeating unit of the polymer. For this estimation the density of the polymer had to be measured by matching the density of the solvent. The matching point was found by centrifugation with $d^{20} = 1.5139$ g/ml. Thus we obtained an approximate value of $n_D^{20} \simeq 1.68 \pm 0.05$ (n_D^{20}: 20°C, $D \equiv 589.3$ nm). To be sure about the size and shape of the sample particles, an electron micrograph (Fig. 19) has been made by G. Pfeiler at the Institute of Histology, University of Graz.

In order to prevent sedimentation we used an aqueous solution of 67 wt% sucrose with a refractive index of $n_{632.8}^{20} = 1.457$ and made a concentration of approximately 10^{-5} mg latex per milliliter of solution. The refractive index of the solvent was quite close to that of the glass cell.

Thus, no reflected light could be detected. The *MD* curves were flat for all values of λ and c_r, and therefore an anti-reflecting correction was neither possible nor necessary. The *MD*

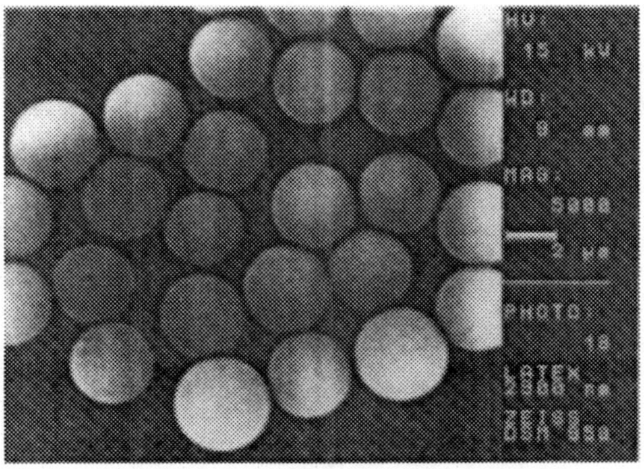

Fig. 19. This electron micrograph is taken from the investigated latex sample, which is made of Melamine-formaldehyde resin. In accordance with our results, it shows slightly polydisperse spheres, with a diameter mostly larger than the nominal diameter of 2900 nm.

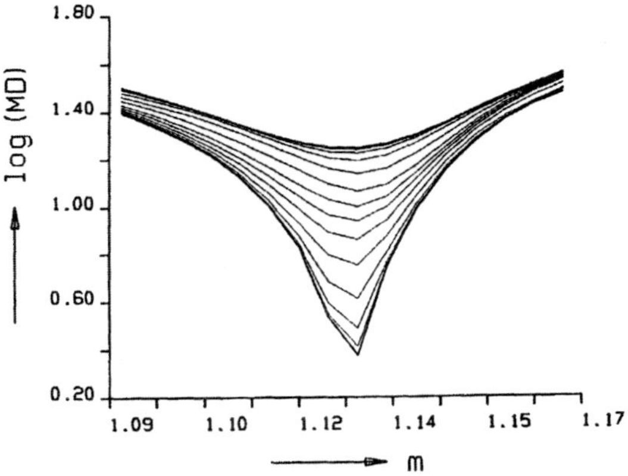

Fig. 20. Application to latex spheres: The *MD* plot for the variation of the *m* value for the experimental scattering curve of latex spheres in an aqueous solution of sucrose. A sharp *MD*-minimum indicates clearly the best fitting *m* value.

curves and the *S* plot for the variation of the *m* value are depicted in Figs. 20 and 21. The depicted *MD* plot is already the refined version (second iteration) where the range of *m* values has been reduced to allow for a better determination of the *MD*-minimum.

The first variation of the *m* value has been done for a wider range of values ($1.001 \leq m \leq 1.500$) and the corresponding *MD* plot (not depicted) did not show any other *MD*-minimum. The sharp minimum in Fig. 20 indicates clearly the best fitting *m* value of $m = 1.129 \pm 0.005$. which corresponds to a refractive index of 1.645 ± 0.007. This value is at least not in disagreement with the value of 1.68 ± 0.05 estimated via the molar refractivity. We have to state that the molar refractivity leads to a rough estimate only, but we had no possibility of obtaining the refractive index by another method.

In Fig. 22 we present the fit (full line) to the experimental data (pluses) which were sampled between 8.72° and 150.0°. From the value of $q_{\min} = 2.1996 \times 10^{-3}$ nm^{-1} we can calculate that the inversion should be performed below $R_{\max} = 714$ nm.

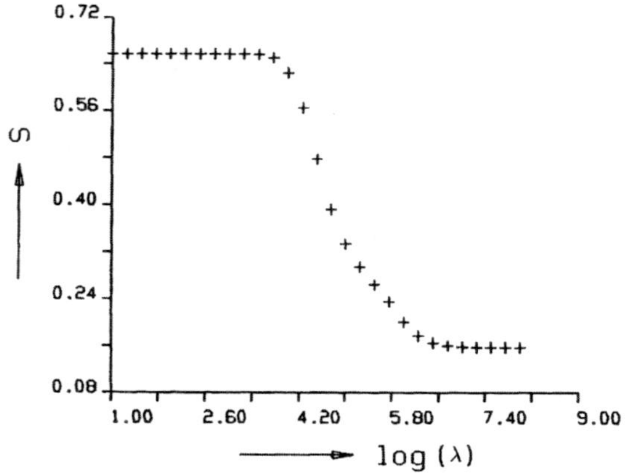

Fig. 21. Application to latex spheres: The S plot for the variation of the m value in Fig. 20. The sensitivity increases to an optimum value below $\log(\lambda) \leq 4.0$.

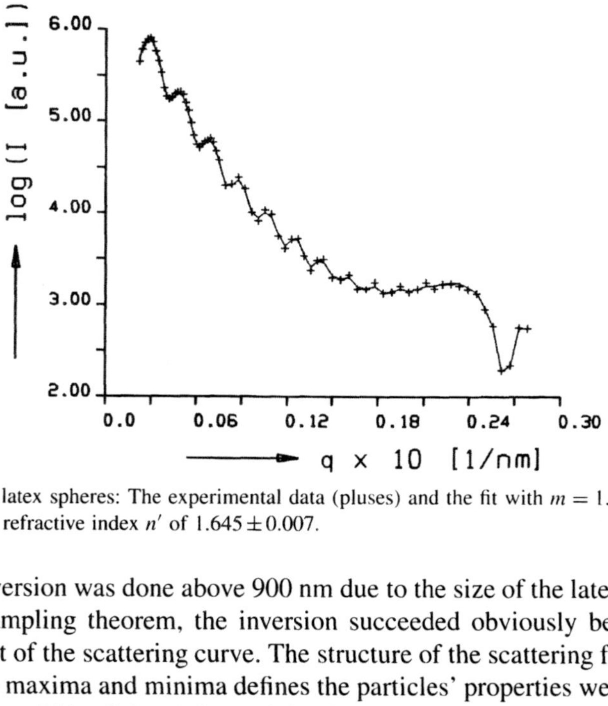

Fig. 22. Application to latex spheres: The experimental data (pluses) and the fit with $m = 1.129 \pm 0.005$ (full line), which corresponds to a refractive index n' of 1.645 ± 0.007.

Actually, the inversion was done above 900 nm due to the size of the latex. Despite a severe violation of the sampling theorem, the inversion succeeded obviously because of the high information content of the scattering curve. The structure of the scattering function that shows the positions of the maxima and minima defines the particles' properties well. That means that the position and the width of the main peak in the refractive index of the latex particles are reliable results but that the peaks at the right side are certainly misleading.

5.4. Application to an oil–water emulsion

In order to demonstrate the applicability to broad size distributions we chose an oil–water emulsion which is usually used for parenteral nutrition. The sample contained water, 20 wt%

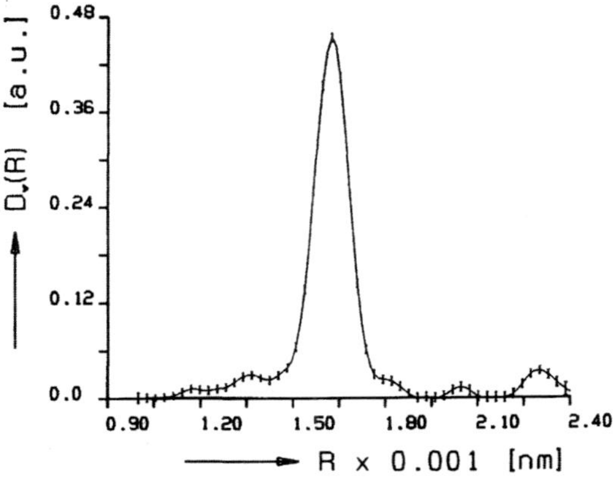

Fig. 23. Application of latex spheres: The volume distribution $D_v(R)$ simultaneously determined from the experimental data (Fig. 22). The error bars symbolize the statistical uncertainty.

soybean oil, 1.2 wt% lecithine, 2.5 wt% glycerol, 0.03 wt% oleanolic acid, and 0.003 wt% NaOH. In order to prevent multiple scattering we had to dilute the sample with water, so that the final concentration of soybean oil was 9.1×10^{-4} mg/ml. The light scattering experiment was performed within an angular range of 10.47° to 150.0° The refractive index of the solvent was $n^{20}_{632.8} = 1.3320$.

The variation of the m value (without correction for the reflected-light contributions) yielded no satisfying fit to the scattering curve. Especially in the domain of large angles the deviations from the experimental curve were exceedingly high. The substantial amount of reflected light is due to the relatively large difference of the refractive indices of the solvent and the glass cell. In order to illustrate the behavior of the mean deviation when varying c_r instead of m we present Fig. 24, where we optimized the c_r value with a constant $m = 1.07$. The minimum in Fig. 24 seems to be very flat because of the general strong decrease of the mean deviation when the Lagrange multiplier is decreased. Indeed, the optimum c_r value cannot be determined with the same accuracy that is possible for the m value owing to the relatively flat *MD*-minimum.

In this case the value of λ becomes important for the determination of c_r, but this is actually no problem, because the sensitivity in Fig. 25 indicated quite precisely the best value for λ. This λ is exactly the same value which must be used for the calculation of the true size distribution.

The final best fitting values were $m = 1.1080 \pm 0.0041$ and $c_r = 1.058 \pm 0.06$. The value of c_r is near the theoretical value of $c_r = 1$ and indicates that the approximations made for the description of the reflected-light contributions are highly applicable. The refractive index of the oil can be calculated to be $n^{20}_{632.8} = 1.4759 \pm 0.0055$. This number is comparable with the number which we obtained from the measurement of the pure oil after separation of the phases. We obtained an index of refraction of $n^{20}_{632.8} = 1.4736 \pm 0.0001$, which is in good agreement with the value determined from the light-scattering data. The fit to the data and the resulting volume distribution $D_v(R)$ are depicted in Figs. 26 and 27. The sharp increase of the intensity at large angles is due to the contribution of reflected light.

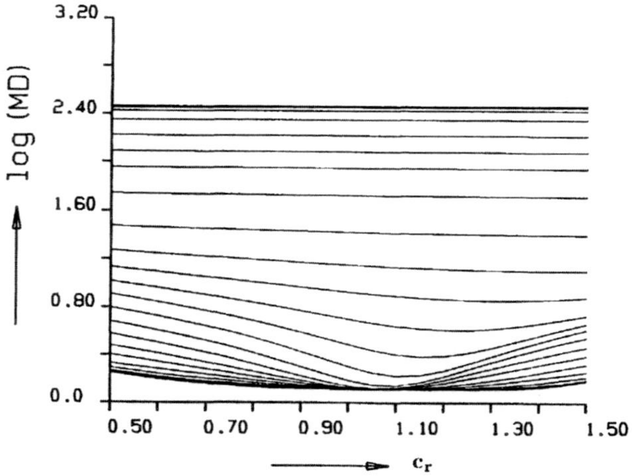

Fig. 24. Application to a polydisperse oil–water emulsion: The *MD* plot for the variation of c_r with $m = 1.07$. The flat minimum of this mean deviation surface allows for the determination of the best fitting value of $c_r = 1.058 \pm 0.06$ which is close to the theoretically predicted value of $c_r = 1$.

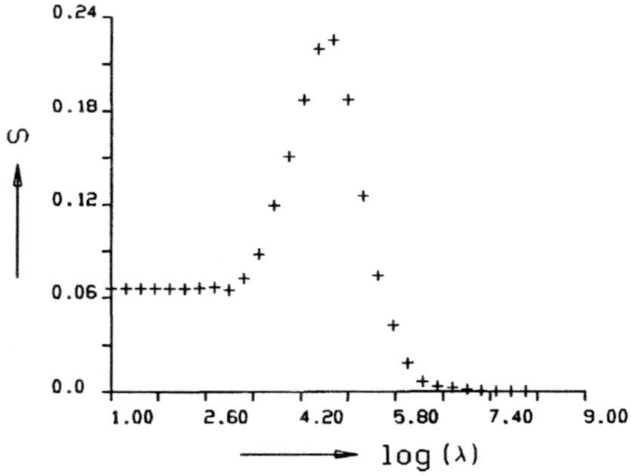

Fig. 25. Application to a polydisperse oil–water emulsion: the *S* plot for the variation of the c_r value in Fig. 24. The maximum of the sensitivity clearly indicates an optimum Lagrange multiplier.

About the consistency of the size distribution, it has to be mentioned that the shape of the distribution function does not change over a range of approximately three decades of λ values, when the optimum parameters for m and c_r are applied. This means that the solution is very stable and can be regarded as correct from the statistical point of view.

6. Multiple scattering

In order to solve the inverse scattering problem in the presence of multiple scattering, the solution of the direct problem has to be known first. Once the direct problem has been solved, the inversion can be accomplished by the 'indirect transformation method' like in the

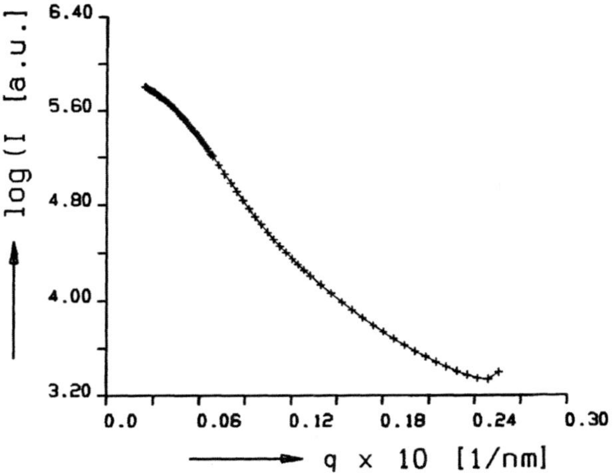

Fig. 26. Application to a polydisperse oil–water emulsion: the experimental data (pulses) and the fit with $m = 1.1080 \pm 0.0041$ and $c_r = 1.058 \pm 0.06$ (full line).

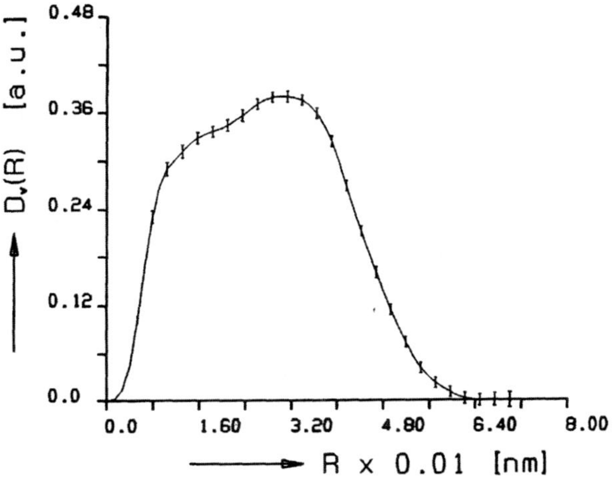

Fig. 27. Application to a polydisperse oil–water emulsion: the volume distribution $D_v(R)$ determined from the experimental data (Fig. 26). The error bars symbolize the statistical uncertainly.

single scattering case. We have developed a first-order approximation for the direct problem of multiple scattering of polydisperse systems based on Hartel's theory (Hartel, 1940). In this theory, the single scattering intensity of unpolarized monochromatic light of a sphere is expanded in a series of Legendre polynomials $P_n(\cos\theta)$

$$I_1(\theta) = \sum_{n=0}^{\infty} A_n P_n(\cos\theta) \tag{15}$$

where the expansion coefficients A_n can be calculated semi-analytically. In Hartel's simplified theory it is possible to calculate the multiple scattered light emerging from a layer with the thickness z

$$I(z,\theta) = \frac{1}{4\pi}\sum_{n=0}^{\infty}(2n+1)\{\exp[-\mu z(1-B_n)] - \exp(-\mu z)\}P_n(\cos\theta) \tag{16}$$

where $B_n = A_n/[A_0(2n+1)]$. This technique can be extended to polydisperse systems by substituting $\phi(q,R,m)$ in Eq. (4) by the right side of Eq. (15) and we finally obtain with Eq. (9) the matrix $\gamma_{n,i}$ containing the coefficients of the ith transformed B-spline function

$$\gamma_{n,i} = \int_{R=0}^{\infty} D(R)\varphi_i(R)A_n(R)\,\mathrm{d}R \tag{17}$$

with

$$A'_n = \sum_{i=1}^{N} c_i\gamma_{n,i} \tag{18}$$

These new coefficients can now be inserted in Eq. (16). This rather simple first-order approximation can be used for the inversion of experimental data on a fast PC. The essential optical parameter for these calculations is the transmittance T which can be determined experimentally. Its value should be above 0.3 (optical thickness 1.2) to give results without artifacts.

As an example we used latex suspensions with two different radii, 200 nm and 317 nm. The suspensions were mixed with a volume ratio of 2 : 1, i.e., the volume content of the 200 nm particles was twice the volume content of the 317 nm particles. The final concentrations were 2.0×10^{-4}, 5.0×10^{-3} and 2.0×10^{-2} and 5.0×10^{-2} mg/mL (the density of the polymer is 1.055 mg/mL). The transmittances of these samples are 0.981, 0.791, 0.408 and 0.104. After subtraction of the blank curve (water) we obtained the differential scattering curves which are depicted in Fig. 28, where Fig. 28a contains the intensities displayed as obtained from the

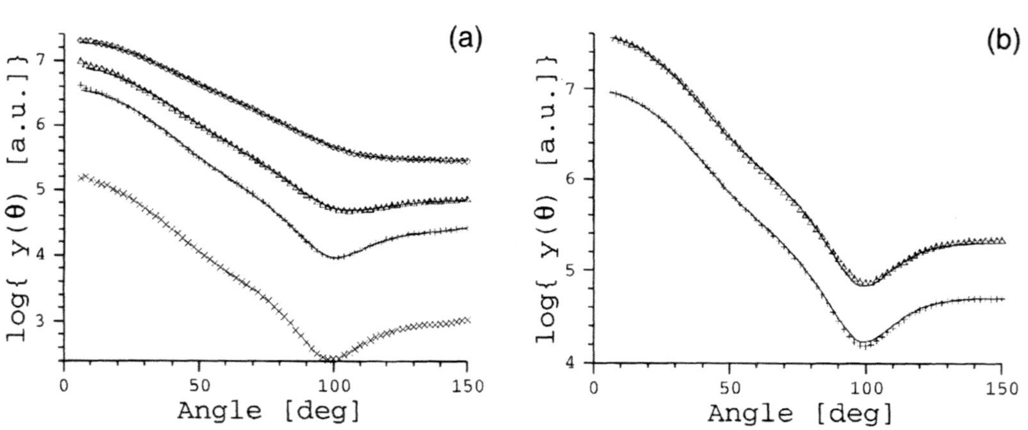

Fig. 28. (a) Differential scattering curves for a bimodal latex suspension with concentrations: 2×10^{-4} (×), 5×10^{-3} (+), 2×10^{-2} (▲) and 5×10^{-2} (◇) including fit by the inversion routine (full lines). (b) Corresponding single scattering curves to the data shown in (a). Full lines: normalized single scattering curve (lowest concentration).

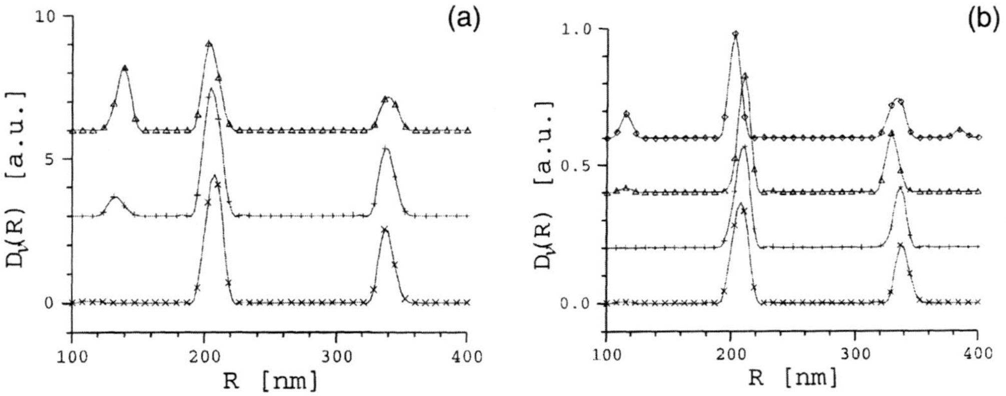

Fig. 29. (a) Size distribution from data in Fig. 28, no correction for multiple scattering. (b) Size distributions from data in Fig. 28, with multiple scattering corrections applied.

measurement. With increasing concentration the curves are marked with the symbols crosses, plusses, triangles, and diamonds. The full lines present the fit to the experimental data. The calculated coefficients c_i [see Eq. (9)] can also be used to reconstruct the corresponding single-scattering curves. In Fig. 28b we present these reconstructed functions for the second (plusses) and the third (triangles) concentration together with the experimental single scattering curve (first concentration, full lines, normalization factors about 64 and 265). The corresponding size distributions are shown in the Fig. 29. Fig. 29a shows the volume distributions (same symbols as in Fig. 28) under the assumption of single scattering, i.e., no correction for multiple scattering. An artificial peak appears at smaller radii. The results improve essentially if multiple scattering corrections are applied as can be seen in Fig. 29b. Only the highest concentration with a transmittance of about 0.1 cannot be corrected completely for multiple scattering effects (Schnablegger and Glatter, 1995). The concentration of the colloidal systems could be increased only if the sample thickness is reduced.

7. Instrumentation

7.1. Goniometer

We use a laboratory-built laser goniometer for our static light scattering experiments. The basic design is shown schematically in Fig. 30. The goniometer is mounted on a 8 cm thick slab of dark gray granite which is carried by a shock-absorbing foundation.

The axis of the goniometer runs through the granite and a half-circle hole allows for a very stable triangular swivel arm with two bearings in a distance of 30 cm. The swivel arm is driven by a stepping motor, the angular accuracy is 0.01°, the available angular range for the experiments is 4–150°. The central tower with the index-matching bath, the sample holder and the removable beam stop in the index fluid (glass rod with Brewster angle) has two cooling/heating circuits for temperature control, one in the outer housing and one inside the index bath. The tower can be brought into the central position by two orthogonal carriages. The index-matching bath can be toggled into the horizontal plane and separately moved into

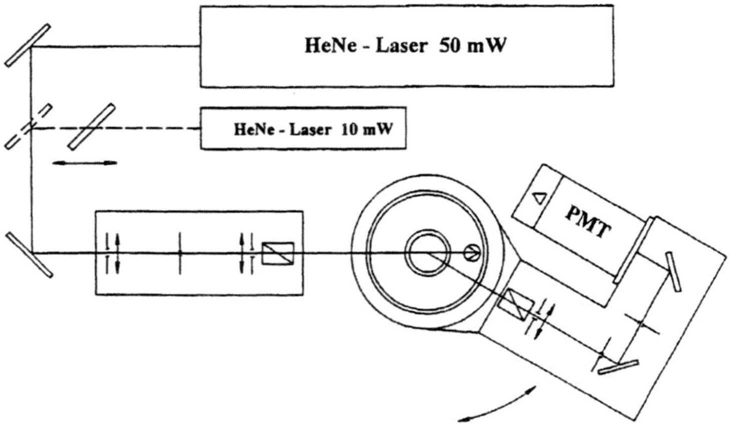

Fig. 30. Schematic drawing of the laser goniometer.

central position by two additional orthogonal carriages inside the tower. Perfect alignment of the index bath and of the tower with the sample holder is essential for the possibility to measure at very small scattering angles.

A 50 mW linearly polarized HeNe-laser or a 10 mW HeNe-laser with random polarization can be used as a light source. Two mirrors bring the primary beam into the entrance optic. A 1 : 1 Keplerian beam expander with a spatial filter and two guard pinholes are used to produce a well defined beam of 1.8 mm diameter. A high-quality polarizer can be inserted in front of and after the sample for depolarization experiments. The detection optics consists of a focusing lens in $2f$ distance from the sample followed by a spatial filter (200 μm) in $1f$ and a 1 × 3 mm rectangular definition slit in $2f$. Two mirrors bring the scattered light through an interference filter into the photomultiplier. The *PMT* (TORN EMI, delivered by ALV) is selected for a very low dark current (~10 counts/s). In this goniometer we can use cylindrical sample cells with an outer diameter of 25 mm or 10 mm (HELLMA), for high-resolution experiments one has to use the 25 mm cuvettes.

7.2. Flat-cell light scattering instrument (FCLSI)

The flat-cell small-angle light scattering instrument is similar to a Fraunhofer set-up (Fig. 31). The primary beam, emerging from a 10 mW HeNe-laser with random polarization is expanded by a 1 : 3 Galilean beam expander. The flat flow-through sample cell is of variable thickness (from a few millimeters down to about 12 μm) and is surrounded by a temperature control jacket and followed by a large plano-convex lens with a focal length of 100 mm. The intensity of the transmitted beam is measured after attenuation by a photodiode behind the detection plane. Linear photodiode arrays in the detection plane detect the scattered light in an angular regime from 1° to 60° scattering angle. The integration time of the diode elements increases with the scattering angle to take into account the decreasing intensity. The read-out time interval via the serial interface to the PC is one second. For the calibration of the diode array elements we use a dense suspension of small latex particles producing a flat, slowly

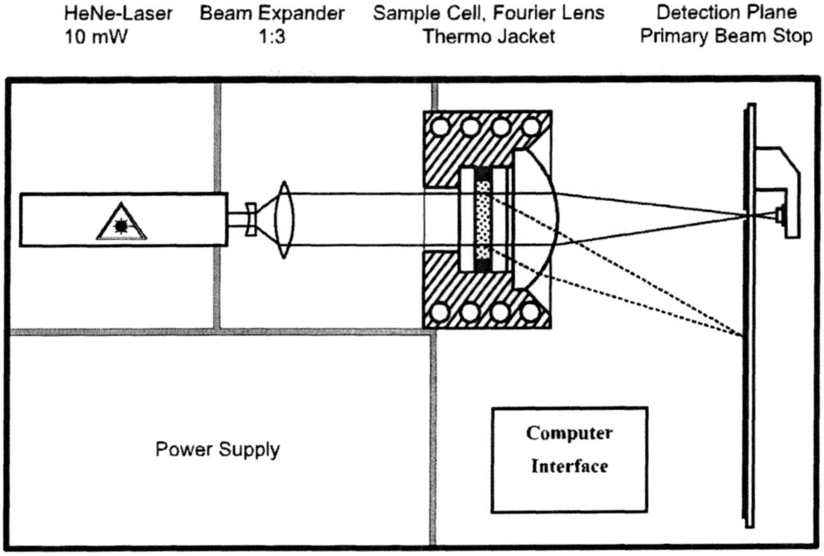

Fig. 31. Schematic drawing of the flat-cell light scattering instrument FCLSI.

decreasing scattering curve (Lehner et al., 1998). Calibration to absolute intensity could be performed by secondary standards.

7.3. Resolution

The real resolution of the goniometer (minimum scattering angle) can be found from a Guinier plot [$\log I(q)$ vs. q^2].

$$I(q)|_{q\to 0} \simeq I_0 \, e^{q^2 R_{app}^2/3} \tag{19}$$

Even in the case of Lorenz–Mie theory this plot should show a straight line for small q values. The apparent radius of gyration R_{app} is for $m > 1$ higher than $R_g^2 = 0.6R^2$ as defined in the Rayleigh–Debye–Gans theory (Glatter et al., 1985). In our system the primary beam stop blocks the primary beam up to 3.5° but some stray light is detected in the regime between 4° and 6° as can be seen from background experiments where the cell is filled with pure water. The Guinier plot for a dilute latex sample (nominal radius given by the producer, Interfacial Dynamics Corporation, is 1064 nm), after subtraction of the water file, is shown in Fig. 32. It clearly demonstrates that the data can be used down to 4.5° scattering angle (the vertical bars indicate the range of data points used for the calculation of R_{app}). This resolution is to a certain extend a function of the scattering intensity of the sample. This intensity should be at least as high as the blank scattering of the instrument. The small-angle light scattering instrument has a resolution limit of about 0.7° scattering angle.

The scattering curves of the latex sample with 1064 nm radius measured in our goniometer and in the flat-cell instrument can be seen in Fig. 33. The angular regime of the goniometer ranges from about 4.5° to 150° whereas the largest accessible angle in the FCLSI is limited to about 60°. A comparison of the two scattering curves between 1° and 60° scattering angle

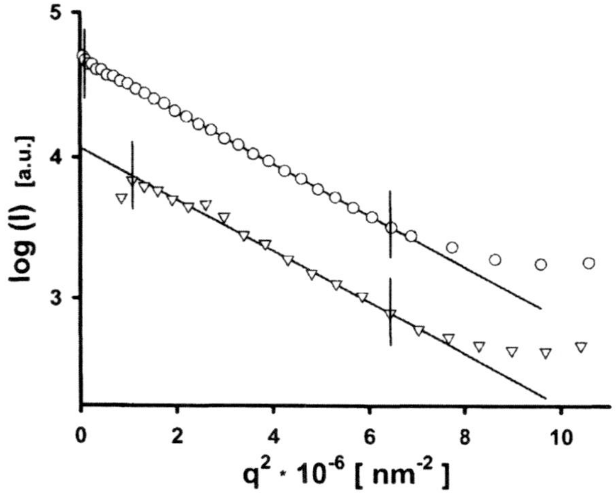

Fig. 32. *Guinier* plot for SLS data from a latex suspension with nominal radius of 1064 nm. The Guinier limits are marked by vertical lines. ∇, goniometer: $c = 0.002$ g/L, $d = 22.6$ mm, $\theta_{\min}$(Guinier) = 4.5°, $R_g = 1125$ nm; ∘, FCLSI: $c = 2$ g/L, $d = 35$ μm, $\theta_{\min}$(Guinier) = 0.7°, $R_g = 1121$ nm.

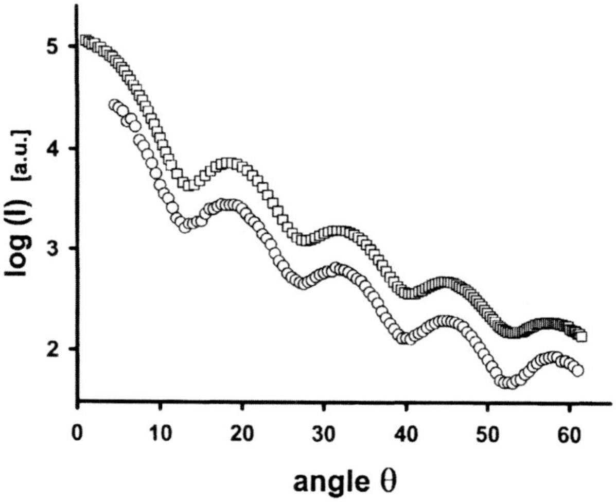

Fig. 33. Scattering curves of monodisperse 1064-nm latex from the FCLSI (□, $c = 2$ g/L, $d = 35$ μm) and from the goniometer (∘, $c = 2 \times 10^{-3}$ g/L, $d = 22.6mm$).

clearly shows the improvements by the FCLSI at small scattering angles. The agreement of the scattering curves of both instruments is satisfying despite a little broadening effect at large angles in the FCLSI.

Characterization of enzymatically induced aggregation of casein micelles was possible in natural concentration with this instrument (Lehner et al., 1999).

It should be mentioned that an efficient physical suppression of multiple scattering can be achieved with the two-color cross-correlation technique (Schätzel, 1991; Stieber and Richtering, 1995) and the three-dimensional cross-correlation technique (Urban and Schurtenberger, 1998; Sinn et al., 1999). These techniques are mostly used in the field of dynamic light scattering (see also Chapter 9 by P.N. Pusey).

References

Agamalian, M.M., Alamo, R.G., Londono, J.D., Mandelkern, L. and Wignall, G.D. (2000). *J. Appl. Cryst.*, **33**, 843–846.

Barber, P. and Hill, S.C. (1990). *Computational Light Scattering*. World Scientific, Singapore.

Bayvel, L.P. and Jones, A.R. (1981). *Electromagnetic Scattering and its Applications*. Applied Science, Barking.

Bevington, P.R. (1969). *Data Reduction and Error Analysis for the Physical Sciences*. McGraw-Hill, New York.

Bohn, L. (1975). In: *Polymer Handbook* (Brandrup, J. and Immergut, E.H., eds.). Wiley, New York, Chapter III, pp. 241–244.

Bohren, C.F. and Huffman, D.R. (1983). *Absorption and Scattering of Light by Small Particles*. Wiley, New York.

Dahneke, B.E. (1983). *Measurement of Suspended Particles by Quasi-Elastic Light Scattering*. Wiley, New York.

de Boor, C. (1978). *A Practical Guide to Splines*. Springer, New York.

Glatter, O. (1977) *J. Appl. Cryst.*, **10**, 415–421.

Glatter, O. (1979) *J. Appl. Cryst.*, **12**, 166–175.

Glatter, O. (1980). *J. Appl. Cryst.*, **13**, 7–11.

Glatter, O. (1981). *J. Appl. Cryst.*, **14**, 101–108.

Glatter, O. and Kratky, O. (1982). *Small Angle X-Ray Scattering*. Academic Press, London.

Glatter and Hainisch, B. (1984). *J. Appl. Cryst.*, **17**, 435–441.

Glatter, O. and Hofer, M. (1988a). *J. Colloid Interface Sci.*, **122**, 484–495.

Glatter, O. and Hofer, M. (1988b). *J. Colloid Interface Sci.*, **122**, 496–506. Interpretation of

Glatter, O., Hofer, M., Jorde, Ch. and Eigner, W.-D. (1985). *J. Colloid Interface Sci.*, **105**, 577–586.

Glatter, O., Sieberer, J. and Schnablegger, H. (1991). *Part. Part. Syst. Charact.*, **8**, 274–281.

Greville, T.N.E. (1969). *Theory and Application of Spline Functions*. Academic Press, New York.

Hartel, W. (1940). *Licht*, **10,** 141–143, 165, 190–191, 214–215, 232–234.

Hainbuchner, M., Villa, M., Kroupa, G., Bruckner, G., Baron, M., Amenitsch, H., Seidl, E. and Rauch, H. (2000). *J. Appl. Cryst.*, **33**, 851-854.

Hofer, M. (1987). *Analysis of Elastic Light Scattering from Pre-Resonant Mie Particles in Real Space*. Thesis, University of Graz.

Hofer, M. and Glatter, O. (1989). *Appl. Optics*, **28**, 2389–2400.

Jenkins, G.M. and Watts, D.G. (1968). *Spectral Analysis and Its Applications*. Holden-Day, Oakland.

Kerker, M. (1969). *The Scattering of Light and Other Electromagnetic Radiation*. Academic Press, New York.

Lehner, D., Kellner, G., Schnablegger, H., Glatter, O. (1998). *J. Colloid Interface Sci.*, **201**, 34–47.

Lehner, D., Worning, P., Fritz, G., Øgendal, L., Bauer, R., Glatter, O. (1999). *J. Colloid Interface Sci.*, **213**, 445–456.

Lorenz, L. (1898). *Oevres scientifiques de L. Lorenz, revues et annotées par H. Valentiner*. Libraire Lehman et Stage, Copenhague.

Mie, G. (1908). *Ann. Phys.*, **25**, 377–445.

Sinn, C., Niehüser, R., Overbeck, E., Palberg, T. (1999). *Part. Part. Syst. Charact.*, **16**, 95–101.

Schätzel, K. (1991). *J. Mod. Optics*, **38**, 1849–1865.

Schnablegger, H. and Glatter, O. (1991). *Appl. Optics*, **30**, 4889–4896.

Schnablegger, H. and Glatter, O. (1993). *J. Colloid Interface Sci.*, **158**, 228–242.

Schnablegger, H. and Glatter, O. (1995). *Appl. Optics*, **34**, 3489–3501.

Stieber, F. and Richtering, W. (1995). *Langmuir*, **11**, 4724–4727.

Urban, C. and Schurtenberger, P. (1998). *J. Colloid Interface Sci.*, **207**, 150–158.

van de Hulst, H.C. (1980). *Multiple Light Scattering*. Academic Press, New York.

van de Hulst, H.C. (1981). *Light Scattering by Small Particles*. Dover, New York.

Neutrons, X-Rays and Light
P. Lindner and Th. Zemb (Editors)

Chapter 9

Dynamic Light Scattering

Peter N. Pusey

Department of Physics and Astronomy, The University of Edinburgh, Mayfield Road, Edinburgh EH9 3JZ, UK

1. Introduction

Consider a scattering medium, such as a suspension of colloidal particles, illuminated by coherent light. At any instant, the far-field pattern of scattered light is not, as one might naively expect, a smooth envelope of intensity which varies with angle; rather it comprises a grainy random diffraction, or 'speckle', pattern (Fig. 1a). At some points in the far field the phases of the light scattered by the individual particles [the quantities $\{\boldsymbol{q} \cdot \boldsymbol{R}_j(t)\}$ in Eq. (4)] are such that the individual fields interfere largely constructively to give a large intensity; at other points destructive interference leads to a small intensity. Furthermore, as the scattering medium evolves in time – for example, the particle positions $\{\boldsymbol{R}_j(t)\}$ change due to Brownian motion – the phases change, and the speckle pattern fluctuates from one random configuration to another. Thus, as sketched in Fig. 1b, the intensity $I(\boldsymbol{q},t)$ scattered to a point in the far field fluctuates randomly in time. Clearly, information on the motions of the particles is encoded in this random signal – at the simplest level, the faster the particles move, the more rapidly the intensity fluctuates.

Normally, in dynamic light scattering (DLS), useable information is extracted from the fluctuating intensity by constructing its time correlation function, defined as:

$$\langle I(\boldsymbol{q},0)I(\boldsymbol{q},\tau)\rangle \equiv \operatorname*{Lim}_{T\to\infty} \frac{1}{T}\int_0^T \mathrm{d}t\; I(\boldsymbol{q},t)I(\boldsymbol{q},t+\tau). \tag{1}$$

As can be seen from its definition, this quantity effectively compares the signal $I(\boldsymbol{q},t)$ with a delayed version $I(\boldsymbol{q},t+\tau)$ of itself, for all starting times t and for a range of delay times τ. Typical behaviour of the intensity correlation function is sketched in Fig. 1c. At zero delay time Eq. (1) reduces to

$$\operatorname*{Lim}_{\tau\to 0}\langle I(\boldsymbol{q},0)I(\boldsymbol{q},\tau)\rangle = \langle I^2(\boldsymbol{q})\rangle. \tag{2}$$

For delay times much greater than the typical fluctuation time T_C of the intensity, fluctuations in $I(\boldsymbol{q},t)$ and in $I(\boldsymbol{q},t+\tau)$ are uncorrelated so that the average in Eq. (1) can be separated:

$$\operatorname*{Lim}_{\tau\to\infty}\langle I(\boldsymbol{q},0)I(\boldsymbol{q},\tau)\rangle = \langle I(\boldsymbol{q},0)\rangle\langle I(\boldsymbol{q},\tau)\rangle = \langle I(\boldsymbol{q})\rangle^2. \tag{3}$$

Thus, as shown in Fig. 1c, the intensity correlation function decays from the mean-square

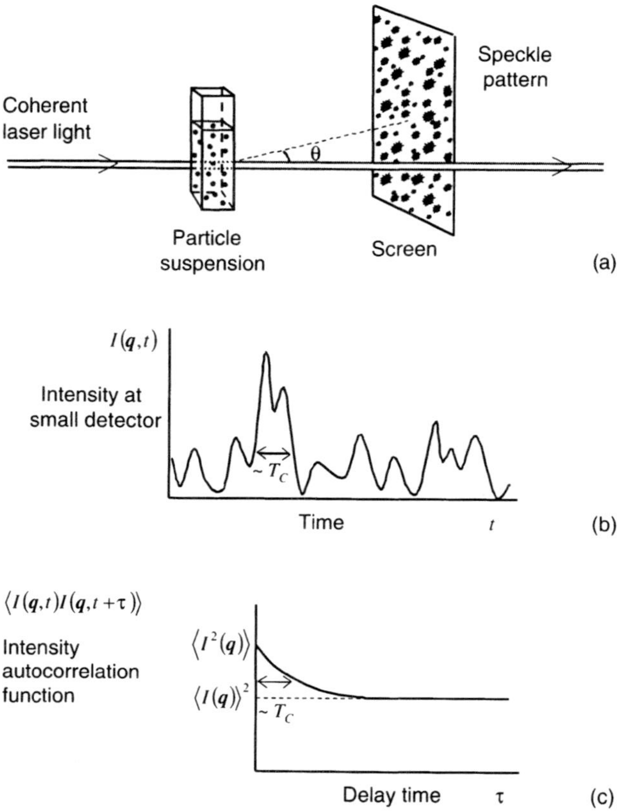

Fig. 1. (a) Coherent (laser) light scattered by a random medium such as a suspension of colloidal particles gives rise to a random diffraction, or speckle, pattern in the far field. As the particles move in Brownian motion, their positions change, as do the phases of the light that they scatter, and the speckle pattern fluctuates from one random configuration to another. (b) The fluctuating intensity observed at a detector roughly the size of one speckle. (c) The time autocorrelation function of the scattered intensity shown in (b). The time-dependent part of the correlation function decays with a time constant T_C equal to the typical fluctuation time of the scattered light.

intensity at small delay times to the square of the mean at long times; the characteristic time T_C of this decay is a measure of the typical fluctuation time of the intensity (Fig. 1b).

In the next section we consider the properties of the scattered field, speckle, and the associated correlation functions, all of which underlie the theory of dynamic light scattering. Then, as probably the simplest example of an application, we discuss particle sizing in some detail. Other topics, such as concentrated systems, non-ergodic systems, and the suppression of multiple scattering, are described in outline. The examples used are taken mainly from the field of colloids, although dynamic light scattering is widely applied in other areas, especially polymeric and amphiphilic systems. The books by Berne and Pecora (1976), Brown (1993), Chu (1991), Cummins and Pike (1974, 1977), and Pecora (1985) can be consulted for more detailed descriptions of dynamic light scattering and its applications.

2. Speckle and correlation functions

To investigate the properties of the fluctuating scattered light field we consider the simple situation of a dilute suspension of identical spherical particles. Eq. (8) of Chapter 1 gives the general expression for the instantaneous amplitude of the field $E(\boldsymbol{q},t)$ scattered by N particles to a point in the far field:

$$E(\boldsymbol{q},t) = -E_0 \frac{\exp[i(kR-\omega t)]}{R} \sum_{j=1}^{N} b_j(\boldsymbol{q},t) \exp[-i\boldsymbol{q}\cdot \boldsymbol{R}_j(t)]. \tag{4}$$

Here $\boldsymbol{q}$ is the scattering vector, $b_j(\boldsymbol{q},t)$ is the (in general time-dependent) amplitude of the field scattered by particle j, and $\boldsymbol{R}_j(t)$ is the position of the centre of mass of particle j at time t. For identical homogeneous spherical particles all $\{b_j(\boldsymbol{q},t)\}$ have the same time-independent value $b(\boldsymbol{q})$ and can therefore be taken outside the sum. As we will see below, DLS deals with normalised quantities; thus we can omit the pre-factors in Eq. (4) from the start, and can write

$$E(\boldsymbol{q},t) = \sum_{j=1}^{N} \exp[-i\boldsymbol{q}\cdot \boldsymbol{R}_j(t)]. \tag{5}$$

Each of the phase factors $\exp[-i\boldsymbol{q}\cdot\boldsymbol{R}_j(t)]$ in Eq. (5) can be represented by a unit vector in the complex plane, making an angle $\boldsymbol{q}\cdot\boldsymbol{R}_j(t)$ with the real axis (see Fig. 2). The total scattered field $E(\boldsymbol{q},t)$ is the (vectorial) sum of the individual vectors. Since, the particles are randomly distributed throughout the sample, the phase angles $\{\boldsymbol{q}\cdot\boldsymbol{R}_j(t)\}$ are randomly distributed between 0 and 2π (modulo 2π). Therefore, the field $E(\boldsymbol{q},t)$ can be represented pictorially as a two-dimensional random 'walk' of N vectors. As the particles move, for example in Brownian motion, their positions $\{\boldsymbol{R}_j(t)\}$ change, as do the phase angles $\{\boldsymbol{q}\cdot\boldsymbol{R}_j(t)\}$. Thus, the total scattered field itself fluctuates: as illustrated in Fig. 2, the field $E(\boldsymbol{q},t+\tau)$ at time $t+\tau$ is changed from its value $E(\boldsymbol{q},t)$ at time t. For small time differences τ, where the phase angles have not changed much, $E(\boldsymbol{q},t+\tau)$ and $E(\boldsymbol{q},t)$ are still correlated. However,

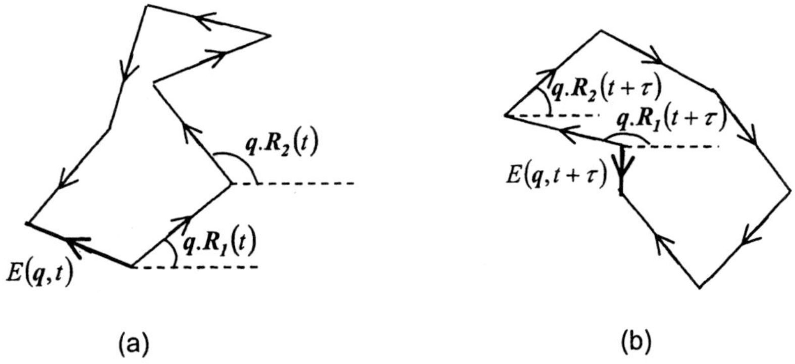

Fig. 2. (a) Representation in the complex plane of the electric field $E(\boldsymbol{q},t)$ scattered by six particles at positions $\boldsymbol{R}_1(t)$, $\boldsymbol{R}_2(t)$, etc. The field is the vectorial sum of the fields scattered by the individual particles. (b) The field $E(\boldsymbol{q},t+\tau)$ at a later time when the particles have moved to new positions $\boldsymbol{R}_1(t+\tau)$, $\boldsymbol{R}_2(t+\tau)$, etc. In the limit of a large number of particles, $N\to\infty$, $E(\boldsymbol{q},t)$ becomes a (two-dimensional) Gaussian random variable.

over times such that the phase angles change by $\sim 2\pi$ (i.e., a particle moves a distance $\boldsymbol{R}_j(t+\tau)-\boldsymbol{R}_j(t)\approx 2\pi/|\boldsymbol{q}|$), fluctuations in the scattered field become totally uncorrelated.

The average value of the scattered field is

$$\langle E(\boldsymbol{q},t)\rangle = \sum_{j=1}^{N}\big\langle \exp[-i\boldsymbol{q}\cdot\boldsymbol{R}_j(t)]\big\rangle = \sum_{j=1}^{N}\big[\langle\cos\boldsymbol{q}\cdot\boldsymbol{R}_j(t)\rangle - i\langle\sin\boldsymbol{q}\cdot\boldsymbol{R}_j(t)\rangle\big] = 0, \tag{6}$$

since the phase angles are uniformly distributed between 0 and 2π. The average of the scattered intensity is

$$\begin{aligned}\langle I(\boldsymbol{q},t)\rangle &= \langle|E(\boldsymbol{q},t)|^2\rangle = \sum_{j=1}^{N}\sum_{k=1}^{N}\big\langle\exp\{-i\boldsymbol{q}\cdot[\boldsymbol{R}_j(t)-\boldsymbol{R}_k(t)]\}\big\rangle\\ &= \sum_{j}1+\sum\sum_{j\neq k}\big\langle\exp[-i\boldsymbol{q}\cdot\boldsymbol{R}_j(t)]\big\rangle\big\langle\exp[i\boldsymbol{q}\cdot\boldsymbol{R}_k(t)]\big\rangle\\ &= N,\end{aligned} \tag{7}$$

where, in averaging the cross-terms separately, we have exploited the assumption that the suspension is dilute so that the particles' positions are uncorrelated. Eqs. (6) and (7) are well known results for the random walk: the mean is zero since the walk is, on average, symmetrical about the origin; the mean-square displacement is proportional to the number of steps. In passing we note that, if, as is usually the case, the number N of particles in the scattering volume is large, the scattered field $E(\boldsymbol{q},t)$ is a complex (two-dimensional) random variable having a Gaussian probability distribution. The more easily measured (real) intensity $I(\boldsymbol{q},t)\equiv|E(\boldsymbol{q},t)|^2$ is then distributed exponentially:

$$P[I(\boldsymbol{q},t)] = \langle I(\boldsymbol{q})\rangle^{-1}\exp\left[\frac{-I(\boldsymbol{q},t)}{\langle I(\boldsymbol{q})\rangle}\right]. \tag{8}$$

In this case, the speckle illustrated in Fig. 1a is known as 'Gaussian speckle'. At a point in the pattern (or in the trace of Fig. 1b) there is a small probability of finding a very high instantaneous intensity; the most likely value is, however, very small.

Below we will need the (normalised) time correlation function of the scattered field, defined by

$$g^{(1)}(q,\tau)\equiv\frac{\langle E(\boldsymbol{q},0)E^*(\boldsymbol{q},\tau)\rangle}{\langle I(\boldsymbol{q})\rangle}. \tag{9}$$

(As in Chapter 1, we now specialise to the case of samples which are isotropic on average so that averaged quantities such as $g^{(1)}(q,\tau)$ depend only on the modulus q of the scattering vector.) Substituting Eq. (5) into (9), and using similar manipulations as those which led to (7), gives

$$g^{(1)}(q,\tau) = N^{-1}\sum_{j}\big\langle\exp\{-i\boldsymbol{q}\cdot[\boldsymbol{R}_j(0)-\boldsymbol{R}_j(\tau)]\}\big\rangle = \big\langle\exp\{-i\boldsymbol{q}\cdot[\boldsymbol{R}(0)-\boldsymbol{R}(\tau)]\}\big\rangle. \tag{10}$$

The last step follows from the fact that the average motions of identical particles are themselves the same.

As noted in Section 1, dynamic light scattering measures directly the time correlation function of the scattered intensity. Using Eq. (5) we have

$$\begin{aligned}
&\langle I(\boldsymbol{q},0)I(\boldsymbol{q},\tau)\rangle \\
&= \left\langle \sum_{j=1}^{N}\sum_{k=1}^{N} \exp\{-i\boldsymbol{q}\cdot[\boldsymbol{R}_j(0)-\boldsymbol{R}_k(0)]\} \sum_{l=1}^{N}\sum_{m=1}^{N} \exp\{-i\boldsymbol{q}\cdot[\boldsymbol{R}_l(\tau)-\boldsymbol{R}_m(\tau)]\} \right\rangle \\
&= \sum_j\sum_k\sum_l\sum_m \langle \exp\{-i\boldsymbol{q}\cdot[\boldsymbol{R}_j(0)-\boldsymbol{R}_k(0)+\boldsymbol{R}_l(\tau)-\boldsymbol{R}_m(\tau)]\}\rangle. \qquad (11)
\end{aligned}$$

This rather intimidating quadruple sum in fact simplifies quite easily when we exploit the independence of particle positions in a dilute suspension. Consider first the terms for which one subscript, say k, is different from all the others. Thus we can factorise the average to give $\langle \exp[i\boldsymbol{q}\cdot\boldsymbol{R}_k(0)]\rangle\langle\ldots\rangle$ which, as discussed above [Eq. (6)], is zero. Next consider the terms for which $j=k=l=m$. There are N of these, each of value 1. In the large N limit, these can be neglected since, as we will see shortly, the dominant contributions to the quadruple sum are $O(N^2)$. The only remaining contributions are those for which pairs of subscripts are equal. One class of these, $j=l, k=m$, is also zero since $\langle \exp\{-i\boldsymbol{q}\cdot[\boldsymbol{R}_j(0)+\boldsymbol{R}_j(\tau)]\}\rangle = 0$. Thus non-zero contributions to Eq. (11) only arise when $j=k, l=m$ or $j=m, k=l$, giving (for $N\to\infty$)

$$\begin{aligned}
&\langle I(\boldsymbol{q},0)I(\boldsymbol{q},\tau)\rangle \\
&= \sum_j 1 \sum_l 1 + \sum_j \langle \exp\{-i\boldsymbol{q}\cdot[\boldsymbol{R}_j(0)-\boldsymbol{R}_j(\tau)]\}\rangle \sum_k \langle \exp\{i\boldsymbol{q}\cdot[\boldsymbol{R}_k(0)-\boldsymbol{R}_k(\tau)]\}\rangle \\
&= N^2 + N^2 |\langle \exp\{-i\boldsymbol{q}\cdot[\boldsymbol{R}(0)-\boldsymbol{R}(\tau)]\}\rangle|^2. \qquad (12)
\end{aligned}$$

We define the normalised time correlation function of the scattered intensity by

$$g^{(2)}(q,\tau) \equiv \frac{\langle I(\boldsymbol{q},0)I(\boldsymbol{q},\tau)\rangle}{\langle I(\boldsymbol{q})\rangle^2}. \qquad (13)$$

Use of Eqs. (7), (9), (10), (12) and (13) then gives

$$g^{(2)}(q,\tau) = 1 + \left[g^{(1)}(q,\tau)\right]^2. \qquad (14)$$

This result, sometimes called the 'Siegert relation', reflects the factorisation properties of the correlation functions of a complex Gausssian variable.

Although Eq. (14) was derived here by considering a dilute suspension of identical spheres, it is in fact applicable to any fluid system in which the range of spatial correlations is much smaller than the linear dimension of the scattering volume. Then the scattering volume contains a large number of 'correlation volumes' (volumes within which particle positions are correlated), and the derivation proceeds in much the same way with (independent) correlation volumes taking the role of the independent particles (Pusey, 1977). Thus Eq. (14) also applies, for example, to a suspension of interacting, non-identical, non-spherical and/or flexible particles, whose scattered electric field is given by Eq. (4). Use of Eqs. (9) and (4) shows that, in this case, the field correlation function $g^{(1)}(q,\tau)$ can be written

$$g^{(1)}(q,\tau) = \frac{\sum_j \sum_k \langle b_j(\boldsymbol{q},0) b_k^*(\boldsymbol{q},\tau) \exp\{-i\boldsymbol{q}\cdot[\boldsymbol{R}_j(0)-\boldsymbol{R}_k(\tau)]\}\rangle}{\sum_j \sum_k \langle b_j(\boldsymbol{q}) b_k^*(\boldsymbol{q}) \exp\{-i\boldsymbol{q}\cdot[\boldsymbol{R}_j(0)-\boldsymbol{R}_k(0)]\}\rangle}. \tag{15}$$

For reasons which will become apparent shortly, the field correlation function $g^{(1)}(q,\tau)$ is frequently called the 'measured intermediate scattering function' $f^M(q,\tau)$ i.e.

$$f^M(q,\tau) \equiv g^{(1)}(q,\tau). \tag{16}$$

For identical interacting spheres [all $\{b_j(\boldsymbol{q},t)\} = b(q)$], Eq. (15) simplifies to

$$f^M(q,\tau) \equiv f(q,\tau) \equiv \frac{F(q,\tau)}{S(q)}, \tag{17}$$

where

$$F(q,\tau) = N^{-1} \sum_j \sum_k \langle \exp\{-i\boldsymbol{q}\cdot[\boldsymbol{R}_j(0)-\boldsymbol{R}_k(\tau)]\}\rangle \quad \text{and} \quad S(q) = F(q,0). \tag{18}$$

Here $S(q)$ is the static structure factor discussed in Chapter 1. $F(q,\tau)$ is called the intermediate scattering function (and, sometimes, the dynamic structure factor, though this term is usually reserved for its Fourier transform with respect to time); $f(q,\tau)$ is the normalised intermediate scattering function.

In a DLS experiment, the intermediate scattering function $f^M(q,\tau)$ [or $g^{(1)}(q,\tau)$] is usually the quantity of interest. It can be obtained from the measured intensity correlation function $g^{(2)}(q,\tau)$ by inverting Eq. (14):

$$f^M(q,\tau) = \sqrt{g^{(2)}(q,\tau) - 1}. \tag{19}$$

In the next sections we consider the form of the measured intermediate scattering function $f^M(q,\tau)$ in a few simple cases.

For completeness we now estimate the size of a speckle in the pattern of scattered radiation. We need to answer the question: By how much must two points in the far field be separated for the fluctuations in the scattered electric fields at the two points to become uncorrelated? [If the fluctuations in the two fields (and intensities) are uncorrelated we can say that the points lie in different speckles.] Thus we consider the situation of Fig. 3 where the two points are defined by propagation vectors $\boldsymbol{k}_{S1}$ and $\boldsymbol{k}_{S2}$ and by scattering vectors $\boldsymbol{q}_1$ and $\boldsymbol{q}_2$. With use of Eq. (5) we calculate the cross-correlation function of the two fields:

$$\langle E(\boldsymbol{q}_1,t)E^*(\boldsymbol{q}_2,t)\rangle = \sum_{j=1}^{N}\sum_{k=1}^{N} \langle \exp[-i\boldsymbol{q}_1\cdot\boldsymbol{R}_j(t)] \exp[i\boldsymbol{q}_2\cdot\boldsymbol{R}_k(t)]\rangle. \tag{20}$$

For a dilute suspension, the same arguments as those used above lead to

$$\langle E(\boldsymbol{q}_1,t)E^*(\boldsymbol{q}_2,t)\rangle = \sum_{j=1}^{N} \langle \exp[i(\boldsymbol{q}_2-\boldsymbol{q}_1)\cdot\boldsymbol{R}_j(t)]\rangle = \sum_{j=1}^{N} \langle \exp[i\Delta\boldsymbol{q}\cdot\boldsymbol{R}_j(t)]\rangle, \tag{21}$$

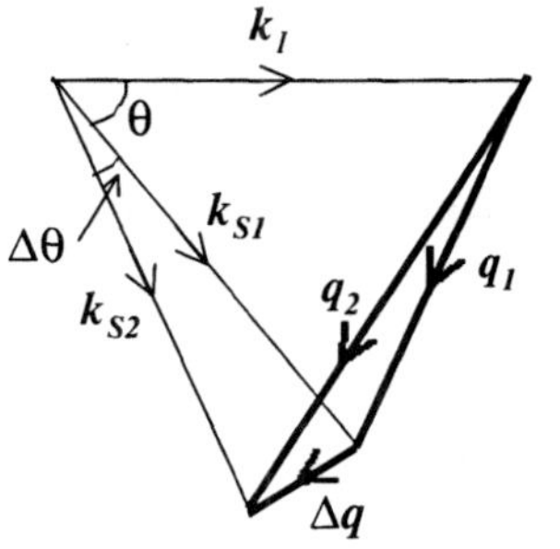

Fig. 3. Propagation and scattering vectors describing light scattered to two different points (cf. Fig. 1 of Chapter 1). When $\Delta\theta$ is small, fluctuations in the two fields are correlated; when $\Delta\theta$ is large enough that the points lie in different speckles, the fluctuations are uncorrelated. For both points to lie in the same speckle we require $\Delta\theta < \lambda/L_V$ (see text), where L_V is the linear dimension of the scattering volume in the direction of $\Delta\boldsymbol{q}$. For typical values, $\lambda = 0.5$ μm and $L_V = 200$ μm, the angle $\Delta\theta$ subtended by one speckle is about 2.5×10^{-3} radians or about 0.1°.

where $\Delta\boldsymbol{q} \equiv \boldsymbol{q}_2 - \boldsymbol{q}_1$ is the difference between the two scattering vectors (Fig. 3). If the two points coincide, $\boldsymbol{q}_1 = \boldsymbol{q}_2$, Eq. (21) simply reduces to the intensity $\langle |E(\boldsymbol{q})|^2 \rangle = N$ [cf. Eq. (7)]. If the two points are well separated, so that $\Delta\boldsymbol{q}$ is large, the arguments applied to Eq. (6) give $\langle E(\boldsymbol{q}_1,t)E^*(\boldsymbol{q}_2,t)\rangle = 0$, implying uncorrelated fluctuations. Since, in the average, the position $\boldsymbol{R}_j(t)$ of the particle can lie anywhere in the scattering volume, an estimate of the value of $\Delta\boldsymbol{q}$ at which correlation between the two fields is lost is given by

$$\Delta q L_V \approx 2\pi, \tag{22}$$

where Δq is the magnitude of $\Delta\boldsymbol{q}$ and L_V is the dimension of the scattering volume parallel to $\Delta\boldsymbol{q}$ and perpendicular to the direction of the scattering (Fig. 3). Thus the angle $\Delta\theta$ subtended by one speckle at the scattering volume is

$$\Delta\theta = \frac{\Delta q}{k_S} \approx \frac{2\pi/L_V}{2\pi/\lambda} = \frac{\lambda}{L_V}, \tag{23}$$

where $k_S = 2\pi/\lambda$ is the magnitude of the propagation vector of the light and λ is its wavelength.

We see, therefore, that the angular width of a speckle is the same as the width of the diffraction maximum that would be obtained if an aperture of the same size as the cross-section of the scattering volume were illuminated by coherent light. This is analogous to the case of a regular diffraction grating, where the overall size of the grating determines the angular width of the maxima in the diffraction pattern. Finally we note that, as pointed out in Section 2 of Chapter 1, the scattered light field can be interpreted as a spatial Fourier component, of wavevector $\boldsymbol{q}$, of variations in the 'density of scattering material' in the sample. It is known from the theory of Fourier analysis that the independent spatial Fourier components needed to describe, for example, density variations in a 'box' of linear dimension L_V are separated in reciprocal space by $\Delta q \approx 2\pi/L_V$, the result of Eq. (22) (see Pusey, 1994). Thus one speckle in the scattered light field can be identified directly with one spatial Fourier component of the sample. The trace of Figure 1b, for example, simply reflects temporal variations of the $\boldsymbol{q}$th Fourier component.

Eq. (4) gives the expression for the amplitude of the electric field scattered to a *point* in the far field; the results derived subsequently from this expression correspond to the same situation. In reality, a detector has a non-zero active area and therefore sees different scattered fields (or intensities) at different points on its surface. Then it can be shown that Eq. (14) is modified to

$$g^{(2)}(q,\tau) = 1 + \beta\left[g^{(1)}(q,\tau)\right]^2. \tag{24}$$

Here β is a factor which represents the degree of spatial coherence of the scattered light over the detector and is determined by the ratio of the detector area to the area of one speckle. When this ratio is much less than 1 (essentially a 'point detector'), $\beta \to 1$, and when it is large, corresponding to the detection of many, independently fluctuating, speckles, $\beta \to 0$. In practice the detector aperture is usually chosen to accept about one speckle [of size defined by Eq. (23)] and $\beta \approx 0.8$. Inverting Eq. (24) gives

$$\beta^{1/2} f^M(q,\tau) = \sqrt{g^{(2)}(q,\tau) - 1}; \tag{25}$$

β is usually regarded as an (uninteresting) parameter, fitted during data analysis.

3. Sizing of spherical particles by DLS

One of the main applications of DLS is to particle sizing. This is done by measuring the diffusion constants of particles undergoing Brownian motion in dilute suspension. Here we consider first the simplest case of identical spherical particles, and then the more realistic situation where spherical particles have a distribution of size (or polydispersity).

3.1. Dilute suspension of identical spheres in Brownian motion

For a dilute suspension of identical (non-interacting) spheres, the field correlation function, or intermediate scattering function, is given by Eq. (10):

$$f(q,\tau) = \left\langle \exp\left\{-i\boldsymbol{q}\cdot\left[\boldsymbol{R}(0) - \boldsymbol{R}(\tau)\right]\right\}\right\rangle = \left\langle \exp\left[i\boldsymbol{q}\cdot\Delta\boldsymbol{R}(\tau)\right]\right\rangle, \tag{26}$$

where

$$\Delta\boldsymbol{R}(\tau) \equiv \boldsymbol{R}(\tau) - \boldsymbol{R}(0) \tag{27}$$

is the displacement of the particle in time τ. We see therefore that, in this case, DLS provides information on the average motion of a single particle. Note that the result of Eq. (26) is also obtained from Eq. (15) by setting all $\{b_j(\boldsymbol{q},t)\} = b(q)$ and assuming that particle positions are uncorrelated.

The displacement of a particle in Brownian motion is a (real) three-dimensional random variable having a Gaussian probability distribution:

$$P\left[\Delta\boldsymbol{R}(\tau)\right] = \left[\frac{3}{2\pi\langle\Delta R^2(\tau)\rangle}\right]^{3/2} \exp\left[-\frac{3\Delta R^2(\tau)}{2\langle\Delta R^2(\tau)\rangle}\right], \tag{28}$$

where the particle's mean-square displacement in time τ is

$$\langle\Delta R^2(\tau)\rangle = 6D_0\tau. \tag{29}$$

The 'free-particle' diffusion constant is given by the Stokes–Einstein equation,

$$D_0 = \frac{k_B T}{6\pi\eta R}, \tag{30}$$

where k_B is Boltzmann's constant, T the temperature, η the viscosity of the liquid, and R the particles' radius. Evaluation of the average in Eq. (26) over the distribution of (28) gives

$$f(q,\tau) = \exp\left[-\frac{q^2}{6}\langle\Delta \boldsymbol{R}^2(\tau)\rangle\right] = \exp(-q^2 D_0 \tau). \tag{31}$$

In this simple case, $f(q,\tau)$ is measured by DLS [via Eq. (25)] and fitted to an exponential in delay time τ giving the decay rate $D_0 q^2$. Since the scattering vector q can be calculated from the known geometry of the experiment, laser wavelength and suspension refractive index (Eq. (3) of Chapter 1), the particles' diffusion constant can be obtained and their radius calculated from Eq. (30).

3.2. Dilute suspension of polydisperse spheres in Brownian motion

While some biological particles can be prepared to be essentially identical, synthetic colloidal and polymeric materials inevitably exhibit some distribution of size. The amplitude of the electric field of light scattered by a homogeneous spherical particle does not depend on its orientation (or on time) so that $b_j(\boldsymbol{q},t) = b_j(q)$. Thus Eq. (15) [with (16)] becomes

$$f^M(q,\tau) = \frac{\sum_j \sum_k \langle b_j(q) b_k^*(q) \exp\{-i\boldsymbol{q}\cdot[\boldsymbol{R}_j(0) - \boldsymbol{R}_k(\tau)]\}\rangle}{\sum_j \sum_k \langle b_j(q) b_k^*(q) \exp\{-i\boldsymbol{q}\cdot[\boldsymbol{R}_j(0) - \boldsymbol{R}_k(0)]\}\rangle}. \tag{32}$$

The assumption that the suspension is dilute (particle positions uncorrelated) then leads to

$$f^M(q,\tau) = \frac{\sum_j b_j^2(q) \langle \exp\{-i\boldsymbol{q}\cdot[\boldsymbol{R}_j(0) - \boldsymbol{R}_j(\tau)]\}\rangle}{\sum_j b_j^2(q)} = \frac{\sum_j b_j^2(q)\exp(-q^2 D_j \tau)}{\sum_j b_j^2(q)}, \tag{33}$$

where the second step follows from assuming that each particle performs its own Brownian motion [cf. Eqs. (26)–(31)]; D_j ($\equiv k_B T/6\pi\eta R_j$) is the diffusion constant of particle j having radius R_j. (Unfortunately, we are using 'R' for both the positions and radii of the particles; it should be clear from the context which meaning is intended.)

Eq. (33) gives the result that one might have guessed *a priori*: the intermediate scattering function for a dilute suspension of polydisperse spheres is a sum of exponentials, each corresponding to a particular particle, weighted by the intensity scattered by that particle. Eq. (33) can be written

$$f^M(q,\tau) = \int \mathrm{d}D\, P(D) \exp(-Dq^2\tau), \tag{34}$$

where $P(D)$ is the normalised intensity-weighted distribution of diffusion constants (which is determined by the distribution $P(R)$ of particle sizes). Written this way, we see that

$f^M(q,\tau)$ is the Laplace transform with respect to D of $P(D)$. Thus, in principle, inverse Laplace transformation of the measured $f^M(q,\tau)$ with respect to time yields $P(D)$, and hence information on the distribution of particle sizes. The difficulty with attempting this operation in practice is that inverse Laplace transformation is a well-known 'ill-conditioned' process. In simple language, a small, inevitable, amount of statistical or systematic error in the measured $f^M(q,\tau)$ transforms into a large error in the recovered $P(D)$. Put another way, the sum of a number of exponentials whose decay rates are not too different looks much like a single 'average' exponential (see Fig. 4). Nevertheless, there is a large literature on the application of inverse Laplace transformation and related methods to the analysis of DLS data, and a number of programs, e.g. CONTIN, are widely available (see, for example, Stepanek, 1993, for details). These methods are useful for samples, like concentrated polymer solutions, where $f^M(q,\tau)$ can often be fitted by a sum of exponentials whose decay rates differ by a factor of 10 or more.

For the sizing of particles whose distribution is relatively narrow, the method of moments [or cumulants (Koppel, 1972)] provides limited, but well-defined, information. This method recognises the insensitivity of DLS data to polydispersity by considering the deviations of the measured $f^M(q,\tau)$ from a single exponential. Thus we define moments of the intensity-weighted distribution $P(D)$ of diffusion constants by

$$\overline{D} \equiv \sum_j b_j^2(q) D_j \Big/ \sum_j b_j^2(q); \qquad \overline{D^2} \equiv \sum_j b_j^2(q) D_j^2 \Big/ \sum_j b_j^2(q), \tag{35}$$

and substitute the identity

$$D_j = \overline{D} + (D_j - \overline{D}) \tag{36}$$

into Eq. (33) to give

$$f^M(q,\tau) = \exp(-\overline{D}q^2\tau) \frac{\sum_j b_j^2(q) \exp[-(D_j - \overline{D})q^2\tau]}{\sum_j b_j^2(q)}. \tag{37}$$

Expansion of the second exponential gives

$$f^M(q,\tau) = \exp(-\overline{D}q^2\tau) \frac{\sum_j b_j^2(q) \left[1 - (D_j - \overline{D})q^2\tau + \frac{1}{2}(D_j - \overline{D})^2 q^4 \tau^2 + \cdots\right]}{\sum_j b_j^2(q)}, \tag{38}$$

or, with use of Eq. (35),

$$f^M(q,\tau) = \exp(-\overline{D}q^2\tau) \left[1 + \frac{1}{2}\left(\frac{\overline{D^2} - \overline{D}^2}{\overline{D}^2}\right)(\overline{D}q^2\tau)^2 + \cdots\right], \tag{39}$$

or equivalently,

$$\ln f^M(q,\tau) = -\overline{D}q^2\tau + \frac{1}{2}\left(\frac{\overline{D^2} - \overline{D}^2}{\overline{D}^2}\right)(\overline{D}q^2\tau)^2 + \cdots. \tag{40}$$

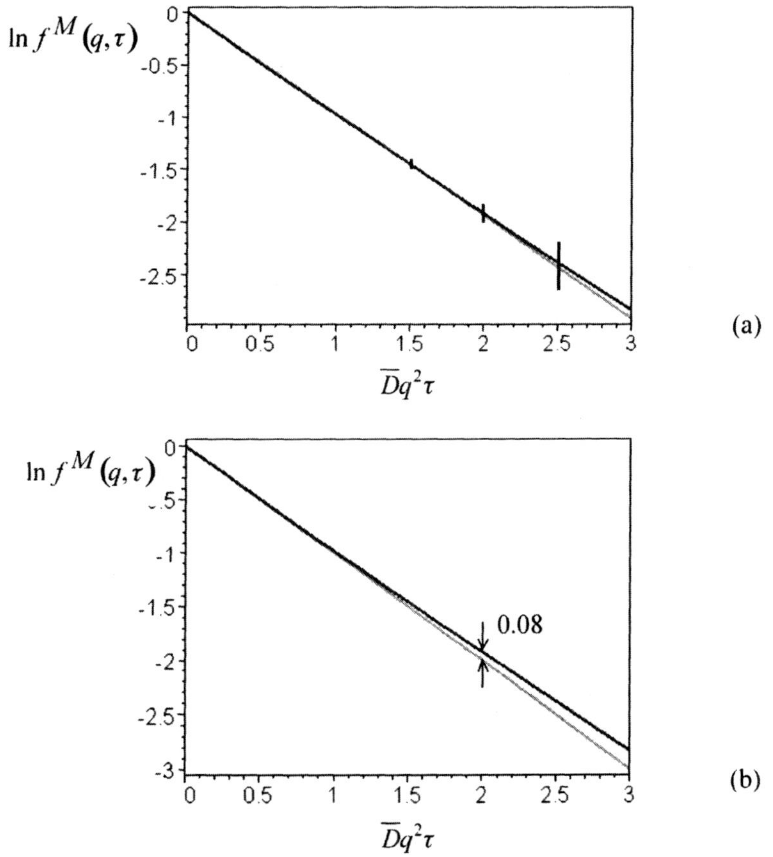

Fig. 4. (a) The black (upper) curve corresponds to $f^M(q,\tau) = 0.5[\exp(-0.8\overline{D}q^2\tau) + \exp(-1.2\overline{D}q^2\tau)]$, the situation when two species of particles, with size ration 1.5 : 1, are mixed in proportions such that each species scatters the same intensity of light. The grey curve is an approximate fit of a single exponential (a straight line in the log-linear plot). The vertical bars indicate the expected experimental uncertainty on the data for the case when the duration T of the experiment is about 10^5 T_C, where $T_C = 1/\overline{D}q^2$ is the fluctuation time of the scattered light. (The relative uncertainty in a measurement of $g^{(2)}(q,\tau)$ is approximately $(T/T_C)^{-1/2}$, where T/T_C is the number of fluctuation times taken by the measurement.) It is clear that, in this case, the expected experimental uncertainty in $f^M(q,\tau)$ is comparable to the difference between the data and the single-exponential fit. (b) The black curve is the same as in (a). The grey curve is now $-\overline{D}q^2\tau$, the first term in Eq. (40). For the situation described in (a), $(\overline{D^2} - \overline{D}^2)/\overline{D}^2 = 0.04$ so that, at a delay time such that $\overline{D}q^2\tau = 2$, the second term in Eq. (40) has a value 0.08, as indicated in the figure. Typically, with accurate data [as would be obtained by performing a long experiment with $T/T_C = 10^6$, say, so that uncertainties in the data would be $3\times$ smaller than those indicated in (a)], the experimental uncertainty in $(\overline{D^2} - \overline{D}^2)/\overline{D}^2$ is about ± 0.02. Thus, for the situation considered here, two species of particle differing in size by 50%, dynamic light scattering can (just) show that the sample is not monodisperse. In general, for polydispersities of this magnitude, fitting $\ln f^M(q,\tau)$ to Eq. (40) gives a quite accurate estimate of the mean diffusion constant, and hence the mean size, and a much less accurate estimate of the standard deviation of the size distribution.

We see, from Eqs. (39) and (40), that the effect of polydispersity can be represented by a 'correction' to single-exponential behaviour, where the correction terms constitute a power series in moments of $P(D)$. For relatively narrow particle size distributions it is frequently adequate to keep just the first two terms in Eqs. (39) or (40). Experimental data can then be analysed either by a non-linear least-squares fitting of the measured $f^M(q,\tau)$ to Eq. (39), or a linear least-squares fit of $\ln f^M(q,\tau)$ to Eq. (40). This analysis provides experimental estimates of the mean diffusion constant $\overline{D}$ and the variance $(\overline{D^2} - \overline{D}^2)/\overline{D}^2$ of the distribution of diffusion constants (see Fig. 4b). For the case of homogeneous spherical particles measured in the limit $qR \leq 1$ (i.e., particles small compared to the wavelength of light, or a measurement made at small scattering angle), these two quantities can be written in terms of moments of the particle size distribution itself. In this case we see from Eqs. (7) and (9) of Chapter 1 that the scattering length of particle j is proportional to its volume, i.e., $b_j(q \to 0) \propto R_j^3$ where R_j is its radius. Thus, from Eqs. (30) and (35) we get

$$\overline{D} = \frac{\sum_j R_j^6 D_j}{\sum_j R_j^6} = \frac{\sum_j R_j^6 \dfrac{k_B T}{6\pi\eta R_j}}{\sum_j R_j^6} = \frac{k_B T}{6\pi\eta\left(\overline{R^6}/\overline{R^5}\right)} = \frac{k_B T}{6\pi\eta R_{\text{app}}} \tag{41}$$

where the 'apparent radius' is given by

$$R_{\text{app}} = \frac{\overline{R^6}}{\overline{R^5}}, \qquad \text{with } \overline{R^n} = \frac{1}{N}\sum_{j=1}^{N} R_j^n. \tag{42}$$

Similarly we get

$$\left(\frac{\overline{D^2} - \overline{D}^2}{\overline{D}^2}\right) = \frac{\overline{R^4}\,\overline{R^6}}{\left(\overline{R^5}\right)^2} - 1. \tag{43}$$

For narrow particle size distributions, these expressions can be simplified further. We write

$$R = \overline{R}\left(1 + \frac{\Delta R}{\overline{R}}\right), \qquad \text{where } \Delta R = R - \overline{R}, \tag{44}$$

giving

$$\begin{aligned}\overline{R^n} &= \int \mathrm{d}R\, P(R) R^n = \int \mathrm{d}R\, P(R)\overline{R}^n \left(1 + \frac{\Delta R}{\overline{R}}\right)^n \\ &= \overline{R}^n \int \mathrm{d}R\, P(R)\left[1 + n\frac{\Delta R}{\overline{R}} + \frac{n(n-1)}{2}\left(\frac{\Delta R}{\overline{R}}\right)^2 + \cdots\right]\end{aligned}$$

where $P(R)$ is the (normalised) distribution of particle radii. For narrow distributions we neglect terms of higher order than the quadratic, giving

$$\overline{R^n} = \overline{R}^n\left[1 + \frac{n(n-1)}{2}\sigma^2 + \cdots\right], \qquad \text{where } \sigma \equiv \left(\frac{\overline{R^2} - \overline{R}^2}{\overline{R}^2}\right)^{1/2}. \tag{45}$$

For small standard deviation σ, use of Eq. (45) in (42) and (43) then gives

$$R_{\text{app}} = \frac{\overline{R^6}}{\overline{R^5}} = \overline{R}\left(\frac{1+15\sigma^2}{1+10\sigma^2}\right) \approx \overline{R}(1+5\sigma^2) \tag{46}$$

and

$$\left(\frac{\overline{D^2}-\overline{D}^2}{\overline{D}^2}\right) = \frac{\overline{R^4}\,\overline{R^6}}{\left(\overline{R^5}\right)^2} - 1 = \frac{(1+6\sigma^2)(1+15\sigma^2)}{(1+10\sigma^2)^2} - 1 \approx \sigma^2. \tag{47}$$

Eq. (47) is a particularly nice result. It shows that, for spherical particles measured in the limit $qR \to 0$, the parameter describing the departure of the measured intermediate scattering function from a single exponential is simply the variance of the particle size distribution. The apparent radius itself [Eq. (46)] is a well-defined function of the first two moments of the particle size distribution.

In the above we have considered mainly the case where a single DLS measurement is made, and analysed, in the limit $qR \to 0$. If the particles being studied are large enough that their form factors $b^2(q)/b^2(0)$ show significant variation over the range of accessible scattering vector q, more complete information about the particle size distribution can be obtained. In this case the exponentials in Eq. (33) are weighted differently as q is changed. Pusey and van Megen (1984) have shown that, provided the accessible range of q includes a minimum in the form factors of the particles, DLS is sensitive to quite small polydispersities. Thus the average diffusion constant $\overline{D(q)}$, Eq (35), is measured as a function of q. As q is increased towards the form factor minimum, the large particles stop scattering first, so $\overline{D(q)}$ rises. At slightly higher q, the smaller particles 'switch off' but the larger ones switch on again, and $\overline{D(q)}$ drops. Pusey and van Megen showed that the relative 'swing' in the value of $\overline{D(q)}$ depends simply on the standard deviation σ of the particle size distribution, and that standard deviations as small as 0.02 can be measured. This method can be contrasted with those based on measuring the departure of $f^M(q,\tau)$ from a single exponential. As discussed above (see also the caption to Fig. 4), the latter methods can barely detect a *variance* σ^2 of 0.02, which corresponds to a standard deviation $\sigma \approx 0.15$.

4. Concentrated systems

The information that can be obtained from dynamic light scattering experiments on concentrated suspensions is discussed in by R. Klein in Chapter 14 of this book [see also Pusey and Tough (1985) and Pusey (1991)]. Here we make a few general remarks.

In a dilute suspension each colloidal particle performs its individual random, diffusive Brownian motion due to collisions with thermally-activated liquid molecules. In a concentrated system, the particles still exchange energy and momentum predominantly with the liquid molecules, driving Brownian motion, but interparticle interactions cause the motions of different particles to become strongly coupled. Two kinds of interaction must be considered. First there are direct interactions, such as the screened Coulombic force between charged particles or the excluded-volume interaction between hard particles. Second, indirect hydrodynamic interactions result from the fact that any one particle moves in the velocity field created in the liquid by the motions of other particles. The interplay between these three influences on particle motion – Brownian motion, direct interactions, and hydrodynamic interactions – leads to a complicated many-body problem.

For identical spherical particles, DLS measures the intermediate scattering function $F(q,\tau)$ [Eqs. (17) and (18)]. As shown in Section 3.1, this quantity has a simple form for a dilute suspension: it decays with time as a single exponential with rate D_0q^2 [Eq. (31)]. In a concentrated system, $F(q,\tau)$ has a much more complicated form. First, it can depart strongly from a single-exponential decay; second, the diffusion coefficient(s) associated with the decay rate(s) can show strong dependence on scattering vector q. To obtain a qualitative understanding of the behaviour of $F(q,\tau)$, it is instructive to rewrite Eq. (18) in the form

$$F(q,\tau)=\left\langle \frac{1}{\sqrt{N}}\sum_j \exp[-i\boldsymbol{q}\cdot\boldsymbol{R}_j(0)]\frac{1}{\sqrt{N}}\sum_j \exp[i\boldsymbol{q}\cdot\boldsymbol{R}_j(\tau)]\right\rangle, \tag{48}$$

which shows that it can be viewed as the time autocorrelation function of a spatial Fourier component of wavevector $\boldsymbol{q}$ of the instantaneous arrangement of particles. Put more simply, DLS observes the decay of spontaneously created sinusoidal 'density waves' of wavelength $\Lambda = 2\pi/q$. The processes that DLS observes then depend on the value of q relative to the position q_{max} of the main peak in the static structure factor (see Fig. 5 of Chapter 1). We consider three cases: $q \ll q_{\max}$, $q \approx q_{\max}$ and $q \gg q_{\max}$.

Remembering, from Chapter 1, that $q_{\max}R_{\max} \approx 2\pi$ where $R_{\max}$ is the typical spacing between neighbouring particles, we see that $q \ll q_{\max}$ corresponds to $\Lambda \gg R_{\max}$; i.e., in this case DLS is probing fluctuations whose wavelength is much larger than the interparticle spacing. Not surprisingly, then, the diffusion coefficient describing the decay of these fluctuations is the same as that which describes the decay of a concentration gradient of macroscopic spatial extent. The magnitude of this gradient, or 'collective', diffusion coefficient is determined by competition between a 'thermodynamic drive' (the inverse of the osmotic compressibility) and a hydrodynamic drag. When $q \approx q_{\max}$, so that $\Lambda \approx R_{\max}$, light scattering observes the dominant 'structure' of the suspension, a particle and its cage of neighbours. The time constant of the (long-time) decay of $F(q = q_{\max},\tau)$ is a measure of the lifetime of a cage and can be defined as the 'structural relaxation time' τ_R of the system. Finally, when $q \gg q_{\max}$ and $\Lambda \ll R_{\max}$, DLS is mainly sensitive to the small-scale, short-time ($\tau \ll \tau_R$) motions of individual particles within their cages. (See Segrè and Pusey, 1996, for more details.)

Except at $q \gg q_{\max}$, $F(q,\tau)$ describes collective motions of the particles, gradient diffusion at small q and structural relaxation at $q \approx q_{\max}$. It is, in fact, easier to get an intuitive feel for the 'self' motions of individual particles. Self motions can be measured by the light scattering analogue of incoherent neutron scattering (achieved in neutron scattering by isotopic substitution). Thus one prepares a suspension in which some of the particles have different refractive indices from others, and therefore different scattering lengths $b(q)$, but have, so far as is possible, the same size and interactions. A nice example of the application of this method is the recent work of van Megen et al. (1998), which can be consulted for details. These authors studied the self motions of hard-sphere colloidal particles in suspensions at concentrations approaching that of the glass transition. Their results can be discussed qualitatively in terms of the 'cage picture' of particle diffusion. At times τ short compared to the cage lifetime τ_R, a particle diffuses inside its cage at a rate slowed from that of free diffusion only by hydrodynamic interactions with its neighbours. On a somewhat longer timescale, $\tau \approx \tau_R$, the particle bumps into its neighbours and its motion is slowed further. At times $\tau \gg \tau_R$, the particle diffuses through many cages at a rate slowed still further by both direct and hydrodynamic interactions. The structural relaxation time increases with increasing concentration and, at the glass transition, the particles become essentially permanently trapped in their cages.

5. Suppression of multiple scattering

The theory developed so far in this chapter has assumed that the samples under study are almost transparent, i.e., that most of the photons pass through unscattered, a few are scattered once, and the probability of multiple scattering is negligible. In fact, many soft matter samples of interest are turbid, or even opaque. Then the scattered light seen by the detector comprises not only photons which have been scattered once, but also those that have been scattered twice, three times, ..., etc. The theory of multiple scattering is much more complicated than that of single scattering and has only proved tractable in a few simple cases. Thus, in its simple form, dynamic light scattering cannot be used to obtain reliable quantitative information from turbid samples.

An important development over the last few years has been the practical implementation of light scattering schemes, based on cross correlation, which effectively suppress multiple scattering in turbid samples, allowing the interpretable single scattering to be analysed directly. The common feature of these schemes is the use of *two* illuminating beams and *two* detectors whose outputs are cross-correlated by a photon correlator. The optics are arranged so that each beam–detector pair has a different geometry but defines *exactly* the same scattering vector (in direction as well as magnitude). Thus, for single scattering the arrangement is degenerate in that each detector observes exactly the same spatial Fourier component of the sample [see Section 4, Eq. (48)] and Section 2 of Chapter 1 for discussions of the interpretation of scattering in terms of spatial Fourier components). It can be shown, however, that for double and higher-order multiple scattering this degeneracy is broken. Then each detector observes several *different* Fourier components. Different spatial Fourier components of a sample are statistically independent i.e. their temporal fluctuations are uncorrelated. Thus, when the two scattered intensities (single + multiple) are cross-correlated, non-trivial correlations are observed only between the single-scattered parts of the two intensities. The net result is that the time-dependent part of the measured time cross-correlation function of the scattered intensities reflects single scattering alone; multiple scattering contributes only to the time-independent 'background'. Thus the dynamics of a turbid sample can be analysed by the methods developed for transparent samples, studied by normal DLS. An important further development of the technique also allows interpretable *static* light scattering measurements to be performed on turbid samples. Since the amplitude of the time-dependent part of the cross-correlation function, relative to the background, depends on the ratio of the intensities of single to total (single + multiple) scattering, the average single scattered intensity can be calculated from the measured total scattered intensity.

Two schemes for multiple scattering suppression, two-colour DLS and three-dimensional DLS, both suggested by Schätzel (1991), have been developed into useful equipment. The methods and applications of multiple scattering suppression have recently been reviewed by Pusey (1999), which can be consulted for details and references.

Finally we note without further comment that, for totally opaque samples, where essentially all photons are scattered many times, useful information can still be obtained through the technique of diffusing wave spectroscopy [see Weitz and Pine (1993) and Maret (1997) for reviews].

6. Non-ergodic samples

So far we have assumed throughout this chapter that the samples under consideration are fluid-like or 'ergodic'. In effect, this means that the particles, or scattering centres, are mobile: given enough time they can diffuse throughout the sample. Thus, over time, all possible spatial configurations of the system are sampled, and time and ensemble averages are equivalent. Then, as we have tacitly assumed, one can analyse measured *time*-averaged intensity correlation functions by comparison with calculated *ensemble*-averaged functions. There is, however, an important class of samples which are non-ergodic. These are amorphous solids, such as colloidal glasses and polymer gels, in which the individual components, particles or segments of polymer, are localised and exhibit only limited Brownian excursions about fixed average positions. Such a sample is 'trapped' in a limited region of configuration space, determined by its particular configuration of fixed average positions. For an ergodic sample, the speckle pattern discussed in Section 1 undergoes 'complete' fluctuations: as the particles diffuse over large distances the intensity of each speckle loses memory of its earlier magnitude. By contrast, each speckle in the pattern corresponding to a non-ergodic sample undergoes limited fluctuations about its own persistent average value. (Scattering by a totally rigid medium, the extreme limit of non-ergodicity, results, of course, in a static, non-fluctuating speckle pattern.)

For non-ergodic samples, time and ensemble averages are no longer equivalent. Thus the measured time-averaged intensity correlation function is no longer related to the ensemble-averaged intermediate scattering function by the simple expression of Eq. (19). Several strategies have been developed to perform and interpret light scattering experiments on non-ergodic media. These are beyond the scope of this article; the reader is referred to Pusey and van Megen (1989), Joosten et al. (1990), Xue et al. (1992) and Pusey (1994) for further information.

7. Equipment for light scattering

A light scattering experiment needs a light source, a sample and optics, a detector, and signal processing electronics [see, for example, Brown and Smart (1997) for a recent review of DLS methods].

As light sources, early DLS experiments used gas lasers, e.g. helium–neon and argon or krypton ion lasers. Nowadays, improvements in the beam quality – coherence, stability, etc. – of solid-state and diode lasers make them attractive alternatives.

Frequently the sample is contained in a cylindrical scattering cell, allowing the scattering angle to be changed easily. This cell is usually placed in a cylindrical 'bath' of temperature-controlled liquid, e.g. toluene, whose refractive index is close to that of glass, thus minimising reflections and stray scattering at the cell walls. The optics are mounted on the arm of a goniometer which can rotate about the axis of the cell and bath. A common optical configuration is one in which a lens images 1 : 1 the centre of the cell onto a slit or pinhole. This arrangement defines the scattering volume and only selects light scattered from the centre of the cell, thus discriminating against stray scattering from the cell walls. A second pinhole is placed in front of the detector, its size chosen so that it subtends roughly the 'coherence angle' $\Delta\theta$ [Eq. (23)] at the first pinhole, so that the detector 'sees' about one speckle in the scattered light field. Some years ago, Brown (1987) realised that the detection optics could be simplified by using a single-mode optical fibre, pointed at the centre of the sample cell. An advantage of this method is that the detected scattered light is fully coherent so that the factor β in Eq. (24) is close to the ideal value of 1 (Brown, 1987; Ricka, 1993).

Detectors for DLS operate in the photon-counting mode: coupled to suitable electronics their output is a train of standardised pulses, typically 10–20 ns long and a few volts in amplitude, each corresponding to the detection of a single photon. Traditionally photomultiplier tubes, based on the photoelectric effect, have been used for this purpose. A disadvantage of photomultipliers is that their quantum efficiency for visible light, the fraction of photons actually converted to pulses, is low, typically a few percent. More recently detectors based on avalanche photodiodes have been developed (and are commercially available) which have efficiencies of 50% or more (Brown et al., 1987). Thus, compared to the case of photomultiplier detectors, lasers of much lower power can be used to obtain the same intensity of detected signal. The combined use of solid-state lasers and avalanche photodiode detectors has simplified DLS equipment significantly.

The process of detection converts the continuous intensity, sketched in Fig. 1b, into a digital signal, a train of pulses, or 'photocounts', which are 'bunched up' or 'spaced out' in response to fluctuations in the intensity. The final step in a DLS experiment is analysis of this signal by a digital correlator, nowadays a purpose built board plugged into a personal computer. It can be shown that the quantity constructed by the correlator, the normalised time correlation function of the photocounts, is identical to the normalised intensity correlation function of Eq. (13). The correlator operates by dividing time into a series of sample times T_S, short compared to the fluctuation time T_C of the scattered light, and counting the number of pulses that arrive in each sample time. The past history of photocounts, and therefore of intensity fluctuations, is stored and is processed to give the photocount correlation function for a range of discrete delay times τ which are integral multiples of the sample time, i.e., $\tau = nT_S$. With modern equipment [see, for example, Schätzel (1993)], accessible delay times range from less than one microsecond to many minutes, $>10^3$ s, giving the DLS technique a huge dynamic range.

References

Berne, B.J. and Pecora, R. (1976). *Dynamic Light Scattering*. Wiley, New York.

Brown, R.G.W. (1987). *Appl. Optics*, **26**, 4846.

Brown, R.G.W. and Smart, A.E. (1997). *Appl. Optics*, **36**, 7480.

Brown, R.G.W., Jones, R., Rarity, J.G. et al. (1987). *Characterization of Silicon Avalanche Photodiodes for Photon Correlation: Measurements .2. Active Quenching. Appl. Optics*, **26**, 2383.

Brown, W., ed. (1993). *Dynamic Light Scattering: The Method and Some Applications*. Clarendon Press, Oxford.

Chu, B. (1991). *Laser Light Scattering*. Academic Press, New York.

Cummins, H.Z. and Pike, E.R., eds. (1974). *Photon Correlation and Light Beating Spectroscopy*. Plenum Press, New York.

Cummins, H.Z. and Pike, E.R., eds. (1977). *Photon Correlation Spectroscopy and Velocimetry*. Plenum Press, New York.

Joosten, J.G.H., Gelade, E.T.F. and Pusey, P.N. (1990). *Phys. Rev. A*, **42**, 2161.

Koppel, D.E. (1972). *J. Chem. Phys.*, **57**, 4814.

Maret, G. (1997). *Curr. Opin. Colloid Interface Sci.*, **2**, 251.

Pecora, R., ed. (1985). *Dynamic Light Scattering: Applications of Photon Correlation Spectroscopy*. Plenum Press, New York.

Pusey, P.N. (1977). In: *Photon Correlation Spectroscopy and Velocimetry* (Cummins, H.Z. and Pike, E.R., eds). Plenum Press, New York.

Pusey, P.N. (1991). In: *Liquids, Freezing and the Glass Transition, Les Houches Session LI* (Levesque, D., Hansen, J-P. and Zinn-Justin, J., eds.). North-Holland, Amsterdam.

Pusey, P.N. (1994). *Macromol. Symp.*, **79**, 17.

Pusey, P.N. (1999). *Curr. Opin. Colloid Interface Sci.*, **4**, 177.

Pusey, P.N. and Tough, R.J.A. (1985). In: *Dynamic Light Scattering: Applications of Photon Correlation Spectroscopy* (Pecora, R., ed.). Plenum Press, New York.

Pusey, P.N. and van Megen, W. (1984). *J. Chem. Phys.*, **80**, 3513.

Pusey, P.N. and van Megen, W. (1989). *Phys. A*, **157**, 705.

Ricka, J. (1993). *Dynamic light-scattering with single-mode and multimode receivers. Appl. Optics*, **32**, 2860.

Schätzel, K. (1991). *J. Modern Optics*, **38**, 1849.

Schätzel, K. (1993). In: *Dynamic Light Scattering: The Method and Some Applications* (Brown, W., ed). Clarendon Press, Oxford.

Segrè, P.N. and Pusey, P.N. (1996). *Phys. Rev. Lett.*, **77**, 771.

Stepanek, P. (1993). In: *Dynamic Light Scattering: The Method and Some Applications* (Brown, W., ed). Clarendon Press, Oxford.

van Megen, W., Mortensen, T.C., Williams, S.R. and Muller, J. (1998). *Phys. Rev. E*, **58**, 6073.

Weitz, D.A. and Pine, D.J. (1993). In: *Dynamic Light Scattering: The Method and Some Applications* (Brown, W., ed). Clarendon Press, Oxford.

Xue, J.Z., Pine, D.J., Milner, S.T., Wu, X.L. and Chaikin, P.M. (1992). *Phys. Rev. A*, **46**, 6550.

Neutrons, X-Rays and Light
P. Lindner and Th. Zemb (Editors)

Chapter 10

Inelastic Neutron Scattering: Dynamics of Polymers

Reiner Zorn

Forschungszentrum Jülich, D-52425 Jülich, Germany

PART I. INELASTIC NEUTRON SCATTERING

In order to study the dynamical processes in glass forming materials which occur on wide time scales it is necessary to combine different inelastic neutron scattering (IENS) techniques. Fig. 1 shows the time ranges and the spatial resolutions of different IENS methods and for comparison those of other scattering (light, X-ray) or non-scattering (NMR) experiments.

It can be seen that for the observation of processes on a molecular length scale (1...10 Å) and microscopic time scale (ps...ns) IENS is an indispensable tool. Light scattering allows a much larger time range to be accessed but due to the wave length of light its spatial resolution is limited to much larger entities than molecules (e.g. colloidal particles). Recent developments of inelastic X-ray scattering techniques at intense synchrotron sources make the observation of very fast processes (<ps) possible by conventional spectroscopy and that of slow processes (>μs) by photon correlation.

The basic objective of IENS is as for all inelastic scattering experiments to measure the momentum transfer $\boldsymbol{q} = \boldsymbol{k}_\mathrm{f} - \boldsymbol{k}_\mathrm{i}$ and the energy transfer $\hbar\omega = E_\mathrm{f} - E_\mathrm{i}$. Here, $\boldsymbol{k}_\mathrm{f}$ and $\boldsymbol{k}_\mathrm{i}$ are the wave vectors ($k = 2\pi/\lambda$) and E_f, E_i the energies of the scattered (final) and incident (initial) neutrons, respectively.

This task in general requires a monochromator for the incoming neutron beam and an analyzer for the scattered neutrons. In the most straightforward setup, the triple-axis spectrometer (3AX), one uses the Bragg planes of crystals similar to the diffracting grids in an optical spectrometer (Fig. 2).

While the 3AX spectrometer is still used for purposes where a high q resolution is necessary (mainly crystals showing Bragg peaks), for 'soft matter' systems (e.g. liquids and glasses) other instruments turned out to show better performance. The most important ones will be discussed here: time-of-flight (TOF) spectrometer, backscattering (BS) spectrometer, and neutron spin echo (NSE) spectrometer.

Before doing this a short recapitulation of the fundamentals of inelastic neutron scattering is in order (Squires, 1978; Lovesey, 1984; Bée, 1988). They are similar to those in dynamic light scattering but there are some peculiarities of neutron scattering which will be pointed out. The basic quantity measured in a neutron scattering experiment is the double differential cross-section

$$\frac{\partial^2\sigma}{\partial\Omega\partial\omega} = N\,\frac{k_\mathrm{f}}{k_\mathrm{i}}\,\left(\left(\langle b^2\rangle - \langle b\rangle^2\right) S_\mathrm{inc}(\boldsymbol{q},\omega) + \langle b\rangle^2 S_\mathrm{coh}(\boldsymbol{q},\omega)\right) \tag{1}$$

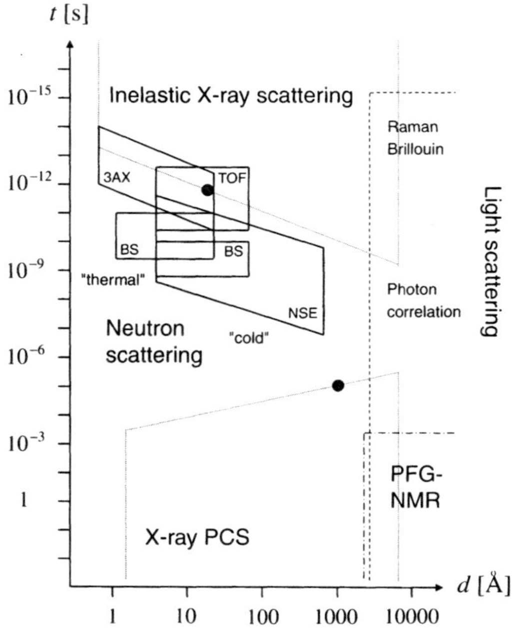

Fig. 1. Overview of dynamic scattering methods: the solid polygons indicate the space/time ranges where observation by inelastic neutron scattering is possible, the dashed lines correspond to dynamic light scattering. The (dotted) lines indicate prospective ranges for inelastic X-ray scattering extrapolated from present experiments (dots). For comparison also pulsed field gradient NMR as a non-scattering method is included (dot-dashed).

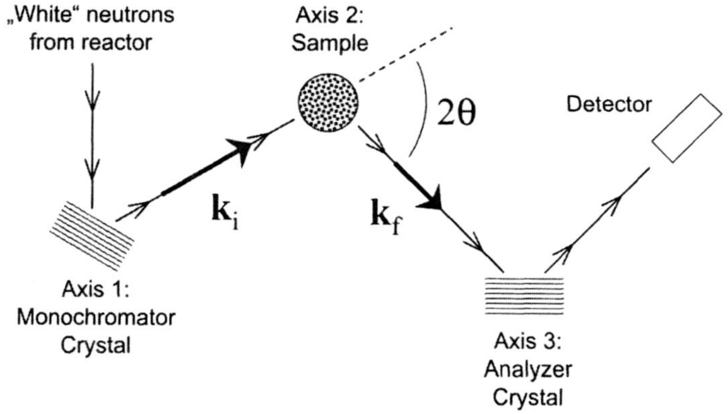

Fig. 2. Schematic setup of a triple-axis spectrometer.

which, multiplied by the flux of incoming neutrons, gives the number of neutrons scattered into a solid angle element $d\Omega$ with an energy transfer $\hbar\omega$. The variable b is the scattering length which is characteristic to each nucleus and the relative spin orientation of nucleus and incident neutron. Often several isotopes of a given element are present in the sample and even if not, the spin orientations may be randomly distributed. In both cases $\langle b^2\rangle \neq \langle b\rangle^2$ and the first term, the incoherent scattering, does not vanish. It is a major difference to light scattering that in general incoherent scattering occurs also in chemically homogeneous samples.

The scattering functions $S_{\text{coh}}(\boldsymbol{q},\omega)$ and $S_{\text{inc}}(\boldsymbol{q},\omega)$ provide a direct link to the microscopic motions of the atoms. The coherent scattering function is related by a double Fourier transform in space and time

$$S_{\text{coh}}(\boldsymbol{q},\omega) = \frac{1}{2\pi}\int \mathrm{d}^3r \int \mathrm{d}t \exp(\mathrm{i}(\boldsymbol{q}\cdot\boldsymbol{r} - \omega t))\, G(\boldsymbol{r},t) \tag{2}$$

to the density–density correlation function

$$G(\boldsymbol{r},t) = \frac{1}{\rho_0}\langle \rho(\boldsymbol{0},0)\rho(\boldsymbol{r},t)\rangle. \tag{3}$$

In the classical picture the latter can be expressed by the atom positions:

$$G(\boldsymbol{r},t) = \frac{1}{N}\left\langle \int \mathrm{d}^3\boldsymbol{r}' \sum_{i,j} \delta(\boldsymbol{r} - \boldsymbol{r}' + \boldsymbol{r}_j(t))\delta(\boldsymbol{r}' - \boldsymbol{r}_i(0))\right\rangle. \tag{4}$$

Therefore, classically $G(\boldsymbol{r},t)$ can be interpreted as the probability density for finding an atom at time t in a distance $\boldsymbol{r}$ from the position of an atom at time 0. For this correlation function it does not matter whether both atoms are the same or not. Substituting (4) into (2) leads to a direct representation of the coherent scattering function in terms of atomic movements:

$$S_{\text{coh}}(\boldsymbol{q},\omega) = \frac{1}{2\pi N}\int \mathrm{d}t \left\langle \sum_{i,j} \exp\left(\mathrm{i}\left(\boldsymbol{q}\cdot(\boldsymbol{r}_j(t) - \boldsymbol{r}_i(0)) - \omega t\right)\right)\right\rangle. \tag{5}$$

In turn, the incoherent scattering function is related by

$$S_{\text{inc}}(\boldsymbol{q},\omega) = \frac{1}{2\pi}\int \mathrm{d}^3r \int \mathrm{d}t \exp(\mathrm{i}(\boldsymbol{q}\cdot\boldsymbol{r} - \omega t))\, G_{\text{s}}(\boldsymbol{r},t) \tag{6}$$

to the self correlation function

$$G_{\text{s}}(\boldsymbol{r},t) = \frac{1}{N}\left\langle \int \mathrm{d}^3\boldsymbol{r}' \sum_{i} \delta(\boldsymbol{r} - \boldsymbol{r}' + \boldsymbol{r}_i(t))\delta(\boldsymbol{r}' - \boldsymbol{r}_i(0))\right\rangle. \tag{7}$$

This function is the probability density of finding an atom at time t in a distance $\boldsymbol{r}$ from its *own* position at time 0. From this it becomes clear that the incoherent scattering function is related to the individual motions of the atoms:

$$S_{\text{inc}}(\boldsymbol{q},\omega) = \frac{1}{2\pi N}\int \mathrm{d}t \left\langle \sum_{i} \exp(\mathrm{i}(\boldsymbol{q}\cdot\Delta\boldsymbol{r}_i(t) - \omega t))\right\rangle. \tag{8}$$

It is evident that this function can often be easier interpreted than its coherent counterpart. This is especially true when the Gaussian approximation is valid, e.g. in the limit of small q, for simple diffusion, or motion in harmonic potentials. In this case (8) reduces to

$$S_{\text{inc}}(q,\omega) = \frac{1}{2\pi}\int \mathrm{d}t \exp(-\mathrm{i}\omega t)\exp\left(-\frac{\langle \Delta r^2(t)\rangle q^2}{6}\right). \tag{9}$$

Table 1
Neutron scattering cross-sections of common elements in organic materials (1 barn = 10^{-28} m^2)

Element	$\sigma_{\text{coh}} = 4\pi \langle b \rangle^2$ (barn)	$\sigma_{\text{inc}} = 4\pi \left(\langle b^2 \rangle - \langle b \rangle^2 \right)$ (barn)
C	5.551	0.001
H	1.7568	**80.26**
D	5.592	2.05
O	4.232	0.0008
N	11.01	0.5
S	1.0186	0.007

Here the average has to be taken as an ensemble average as well as over all atoms. Thus in Gaussian approximation the incoherent scattering function depends only on the mean square displacement of the atoms, $\langle \Delta r^2(t) \rangle$.

Table 1 shows the weighting factors of the incoherent and coherent scattering function in (1), the incoherent and coherent nuclear cross-sections, for the most common elements in organic materials. It can be seen that for systems containing hydrogen usually the incoherent scattering from its nuclei is dominant by an order of magnitude. Therefore, in this case IENS observes mainly the self correlation of hydrogen motion. If a measurement of coherent scattering is desired it is necessary to replace hydrogen by deuterium. Nevertheless, also for a completely deuterated sample a considerable contribution of the incoherent scattering has to be taken into account.

An experimental separation of coherent and incoherent scattering is possible if it is due to spin disorder (not isotopic). Fortunately, this is just the case for hydrogen and deuterium. E.g. for hydrogen incoherent scattering leads to a spin flip of the scattered neutron with probability 2/3. Therefore, from a polarization analysis of the scattered neutrons the coherent and incoherent fractions can be calculated. Although some IENS instruments exist which have this facility it is seldomly used because of the low count rates obtained.

1. Time-of-flight spectrometer

The main disadvantage of the 3AX spectrometer is that it can only observe one (q, ω) point at a time. While for samples where the scattering is concentrated into Bragg peaks this may be acceptable for systems with diffuse scattering a simultaneous observation of a range of q vectors and energy transfers $\hbar\omega$ is desired. This is accomplished by surrounding the sample position with an array of detectors (Fig. 3). In addition the energy of the scattered neutrons E_{f} is here measured by their time of flight: A chopper in the incident beam defines the start time and from the time it takes a neutron to arrive at the detector its velocity and finally its energy transfer can be calculated by

$$\hbar\omega = \left(\frac{l_1{}^2}{\left(l_0 - \sqrt{2E_{\text{i}}/m_{\text{n}}} t_{\text{flight}} \right)^2} - 1 \right) E_{\text{i}}. \tag{10}$$

The monochromatization of the incoming neutron beam can either be done by Bragg reflection from a crystal or by a sequence of choppers which are phased in order to transmit a

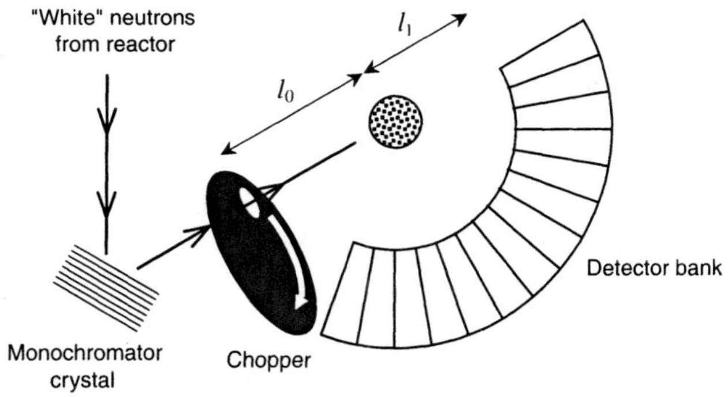

Fig. 3. Schematic setup of a time-of-flight spectrometer with crystal monochromator.

single wavelength only. The former principle usually yields higher intensities while the latter is more flexible for the selection of E_i and attains better energy resolution.

For the interpretation of TOF data it is important to know that the raw data, $n(\theta, t_{flight})$, neutron counts vs. angle and time-of-flight has a strongly different appearance than the scattering function $S(q,\omega)$. The major reason is that relation (10) is non-linear and leads to a higher weighting of the short-time channels containing the high $\hbar\omega$ part of the spectrum (Fig 4). In addition spectra of a single detector at angle θ do not correspond to a constant q but to a curved line in the (q,ω) plane described by

$$q = \frac{\sqrt{2m_n}}{\hbar}\sqrt{2E_i + \hbar\omega - 2\sqrt{E_i(E_i + \hbar\omega)}\cos 2\theta}. \tag{11}$$

Fig. 5 shows the total range covered by a typical instrument (IN6 at ILL, Grenoble) using incident neutrons with $\lambda_i = 5.1$ Å and detectors in the angular range 10°. . . 115°.

Table 2 shows some representative TOF instruments with their basic specifications. Depending on the desired incident wavelength the instruments are constructed either using neutrons directly from the reactor moderator (thermal neutrons, $\lambda_{max} \approx 1.8$ Å) or a cold source, where an additional moderation, e.g. by liquid hydrogen, takes place ($\lambda_{max} \approx 4$ Å). Thermal neutrons make a larger q range accessible while cold neutrons yield better energy resolution. Therefore, the choice of the instrument depends on the system to be observed but in general 'cold neutron' instruments are preferred for IENS in soft matter systems.

Table 2

Basic specifications of representative neutron time-of-flight spectrometers

Instrument	Type	λ_i (Å)	q_{max} (Å^{-1})	$\Delta\hbar\omega$ (meV)
IN4 (ILL)	TX	0.8–4	3–14	0.25–6
IN5 (ILL)	CC	2–15	0.8–5.7	0.01–6
IN6 (ILL)	CX	4.1–5.9	1.8–2.6	0.05–0.17
NEAT (HMI)	CC	1.8–19	0.6–6.5	0.006–5
IRIS (RAL)	I	$\lambda_f = 6.7$–20	0.6–1.9	0.001–0.015

Instrument types: TX – thermal, crystal; CX – cold, crystal; CC – cold, chopper; I – inverse. The maximal q and the energy resolution $\Delta\hbar\omega$ depend on the incident wavelength; the upper limits of their ranges correspond to the lower limit of the incident wavelength λ_i and vice versa.

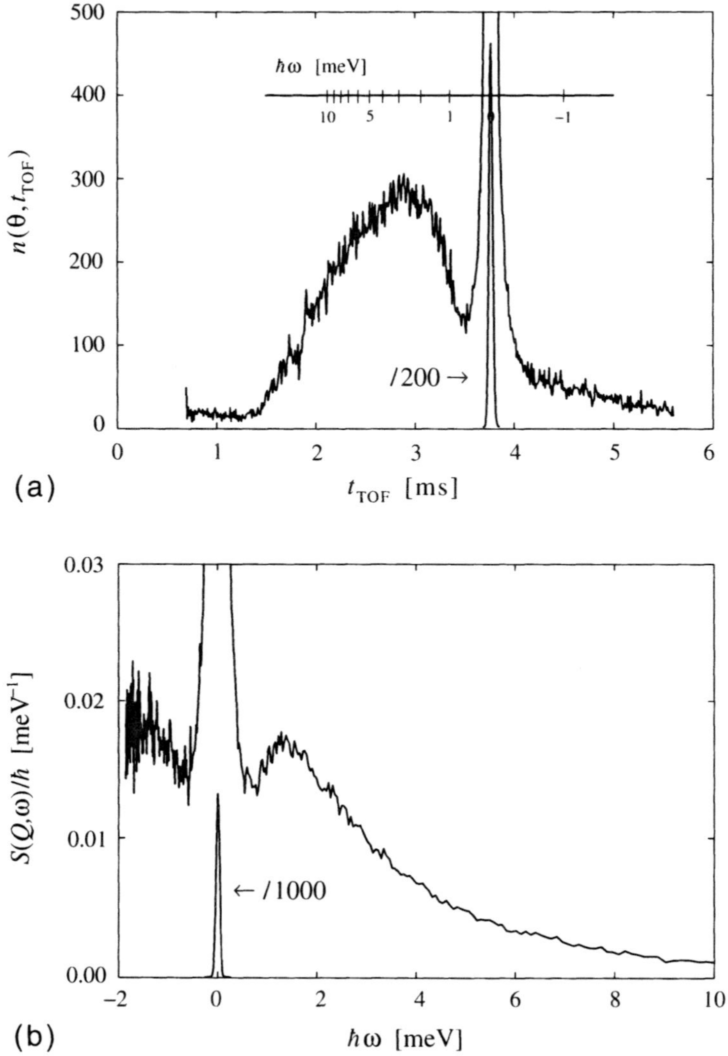

Fig. 4. (a) Raw data from TOF spectrometer: neutron counts in a time channel of 10 μs during one hour registration time. The sample is a mesoscopically confined glass-forming liquid. The floating non-linear axis indicates the energy transfers calculated by Eq. (10). Because of the strength of the elastic scattering that part of the spectrum has been reduced by a factor of 200. (b) The same TOF data converted to $S(q,\omega)$, elastic line reduced by factor 1000. The characteristic vibrational modes of the material at $\hbar\omega \approx 1.7$ meV (14 cm^{-1}) become only visible after the transformation.

A variant of the TOF spectrometer exists on spallation sources, the inverse time-of-flight spectrometer. Because the neutrons are produced in pulses by a spallation source one can use their creation time to start the TOF clock and in principle there is no need for a chopper. In this way all neutrons can be used in contrast to conventional TOF spectrometers which use only a few percent. Then usually the incident energy is measured by the time-of-flight

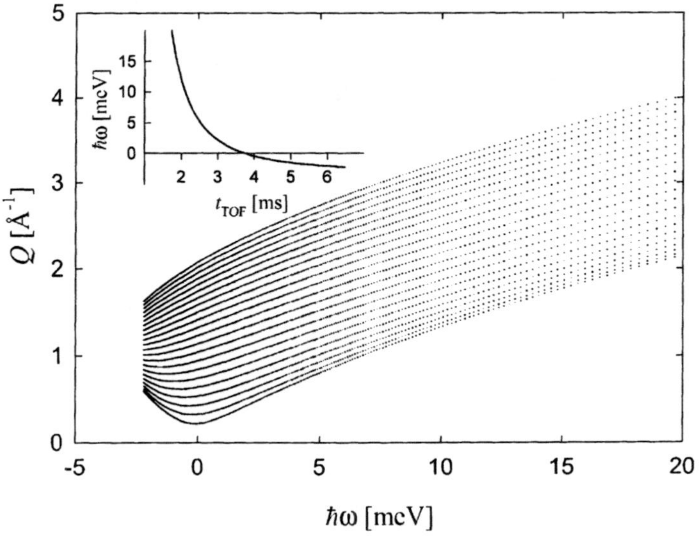

Fig. 5. Range of $S(q,\omega)$ points accessible by a typical time-of-flight spectrometer, IN6 at ILL, Grenoble: incident wavelength $\lambda_i = 5.1$ Å, secondary flight path 247 cm, detectors in the angular range 10°... 115°. The dots correspond to time channels spaced by 10 μs and detectors every 5° (increasing angle from bottom to top curve). The inset shows the energy transfer vs. TOF relation (10) for these conditions.

and the final energy is kept constant by a fixed set of analyzer crystals ('inverse geometry'). By putting those crystals into near backscattering position (see next section for details) it is possible to obtain a very good energy resolution already close to true backscattering spectrometers. Of course as a direct consequence of the good resolution function the count rates are low, especially with current spallation sources which have total neutron production rates significantly lower than reactors.

2. Backscattering spectrometer

A recurring problem of IENS investigations on soft matter systems is that processes are too slow to be observed. Without resorting to extreme setups which lead to a loss of intensity the energy resolution of TOF spectrometers is limited to about 50 μeV which corresponds to a maximal timescale of 40 ps. This is often not sufficient for e.g. the large scale motions in polymers or the glass transition related relaxation. Therefore, instruments with highest energy resolution are often needed, the backscattering (BS) and neutron spin echo (NSE) spectrometer.

The energy resolution of a TOF spectrometer is limited by the selectivity of the monochromator crystal[1]. If perfect crystals are used the spread of the selected wavelengths $\Delta\lambda/\lambda$ is determined by the angular divergence $\Delta\alpha$ of the reflected neutrons. Differentiating the Bragg

[1] For chopper spectrometers the limit is given by the pulse length which could in principle be arbitrary small. But since the counted intensity decreases quadratically with pulse length the resolution limit of an efficient experiment is in the same range.

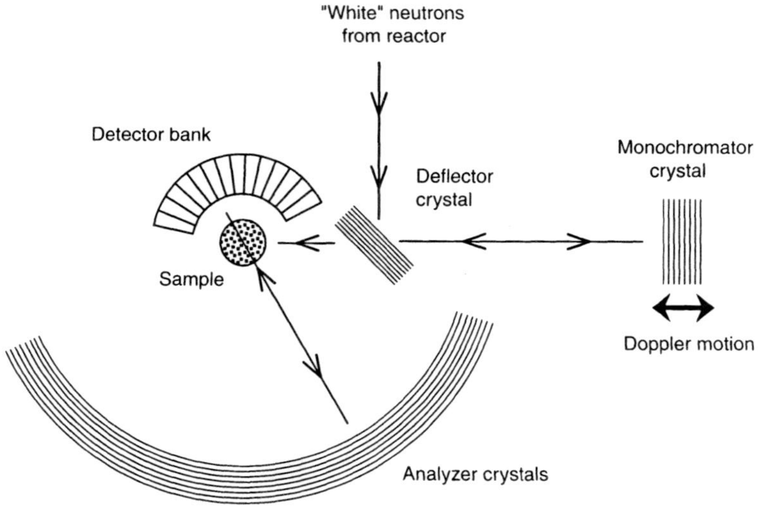

Fig. 6. Schematic setup of a backscattering spectrometer.

condition $\lambda = 2\sin\theta/d$ one obtains

$$\Delta\lambda/\lambda = \cot\theta \cdot \Delta\theta. \tag{12}$$

This expression becomes zero for $2\theta = 180°$. In practice this means that the wavelength spread becomes minimal if the neutron beam is reflected by 180°, i.e., in backscattering condition.

Fig. 6 shows schematically the instrument based on this principle. The first crystal in the beam is only a deflector with low wavelength selectivity. The actual monochromatization takes place upon the second reflection by the crystal in backscattering position. The monochromatized neutrons are then scattered by the sample which is surrounded by analyzers placed on a spherical surface. There they are again scattered under backscattering condition. The reflected neutrons pass once more through the sample and finally reach the detector.

It can be seen that the backscattering condition leads to technical problems in several places:

(1) The deflector must not accept all neutrons otherwise the monochromatized beam would be scattered back into the source. This can be solved by reducing its size deliberately below the neutron beam area or putting it on a rotating disk which removes it at the moment when the neutrons come back from the monochromator. Of course all these measures are taken at the expense of intensity.

(2) The second passage of the scattered neutrons through the sample causes additional multiple scattering and absorption. Both problems can be resolved by leaving exact backscattering condition but with the consequence that the energy resolution degrades.

So far it seems that the backscattering instrument can only observe elastic scattering ($E_i = E_f$) if the same crystals are used for monochromator and analyzer. In order to do inelastic scattering one has to change either E_i or E_f. It turns out that this is much easier for the incident energy by either using a moving monochromator (Doppler effect) or a heated monochromator (thermal expansion modifying the lattice plane distance d). The latter

Table 3
Basic specifications of representative neutron backscattering spectrometers

Instrument	Type	λ (Å)	q_{max} (Å^{-1})	$\Delta\hbar\omega$ (μeV)	$\hbar\omega_{max}$ (μeV)
IN16 (ILL)	CD	6.27	1.9	0.2–1.0	15
IN10B (ILL)	CH	6.29	2.0	1.5	120
IN13 (ILL)	TH	2.23	5.5	8	300
PI (FZJ)	CDO	6.30	1.9	2	29

Instrument types: CD – cold, Doppler monochromator (O – with offset); CH – cold, heated monochromator; TH – thermal, heated monochromator. The maximal q and the energy resolution $\Delta\hbar\omega$ depend on the incident wavelength: the upper limits of their ranges correspond to the lower limit of the incident wavelength λ_i and vice versa.

technique usually allows larger energy transfers. For very large energy transfers different crystals are used for monochromator and analyzer yielding an offset of the whole $\hbar\omega$ range. Table 3 comprises specifications of representative BS spectrometers.

3. Neutron spin echo spectrometer

In order to access even slower processes in soft matter dynamics a very high resolution technique is needed allowing to reach more than 100 ns corresponding to energy transfers in the neV range. Such a technique is provided by neutron spin echo (NSE) spectrometers (Mezei, 1980) which are able to measure directly energy changes of the neutron due to scattering.

This distinguishes NSE from conventional IENS techniques which proceed in two steps: (1) monochromatization of the incident beam to E_i, (2) analysis of the scattered beam (E_f). The energy transfer is then determined by taking the difference $E_f - E_i$. In order to achieve high energy resolutions with these conventional techniques a very narrow energy interval must be selected from the relatively low-intensity neutron spectrum of the source. Conventional high-resolution techniques therefore inevitably run into the problem of low count rates at the detector.

Unlike these methods, NSE measures the individual velocities of the incident and scattered neutrons using the Larmor precession of the neutron spin in a magnetic field. The neutron spin vector acts as the hand of an internal clock which is linked to each neutron and connects the result of the velocity measurement to the neutron itself. Thereby the velocities before and after scattering on one and the same neutron can be compared and a direct measurement of the velocity difference becomes possible. The energy resolution is thus decoupled from the monochromatization of the incident beam. Relative energy resolutions in the order of 10^{-5} can be achieved with an incident neutron spectrum of 20% bandwidth.

The motion of the neutron polarization $\boldsymbol{P}(t)$ – which is the quantum mechanical expectancy value of the neutron spin – is described by the Bloch equation

$$\frac{\mathrm{d}\boldsymbol{P}}{\mathrm{d}t} = \frac{\gamma\mu}{\hbar}(\boldsymbol{P} \times \boldsymbol{B}) \tag{13}$$

where γ is the gyromagnetic ratio ($\gamma = -3.82$) of the neutron, μ the nuclear magneton and $\boldsymbol{B}$ the magnetic field. Eq. (13) is the basis for manipulation of the neutron polarization by

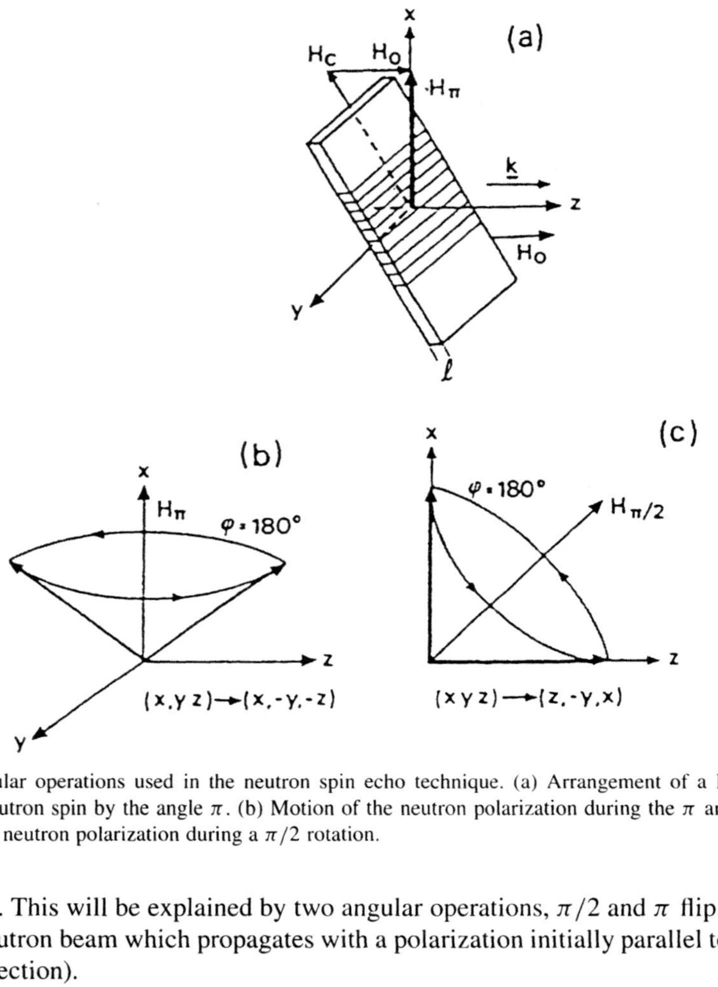

Fig. 7. Spin angular operations used in the neutron spin echo technique. (a) Arrangement of a Mezei coil for a rotation of the neutron spin by the angle π. (b) Motion of the neutron polarization during the π angular operation. (c) Motion of the neutron polarization during a $\pi/2$ rotation.

external fields. This will be explained by two angular operations, $\pi/2$ and π flip. We consider an incident neutron beam which propagates with a polarization initially parallel to its direction of flight (z direction).

$\pi/2$ flip: In order to convert the z polarization into a perpendicular (x) polarization (or vice versa) one has to turn the polarization vector by 90° ($\pi/2$). This is done by letting the neutrons pass through a magnetic field at 45° to the z-axis whose magnitude $\boldsymbol{B}_\pi$ and duration are chosen so that the neutron spin performs a Larmor precession of exactly 180° around that tilted axis (Fig. 7c). The angular velocity of the Larmor precession is $\omega_\mathrm{L} = |\gamma|\mu B_\pi/\hbar$. With the length l on which the neutrons are affected by the field and $v = h/(m_\mathrm{n}\lambda)$ being their velocity one can calculate the precession angle

$$\phi = \left(\frac{2\pi|\gamma|\mu\lambda m_\mathrm{n}}{h^2}\right) B_\pi l. \tag{14}$$

If the field is inhomogeneous $B_\pi l$ has to be replaced by the field integral $\int_0^l B\mathrm{d}z$. It is important that all neutrons are subjected to the same field integral. For this purpose the Mezei coil (Fig. 7a) is used which is wound on a slab-shaped body. In general, to conserve the polarization of the incoming beam a small 'guide field' $\boldsymbol{B}_0$ in z direction is necessary.

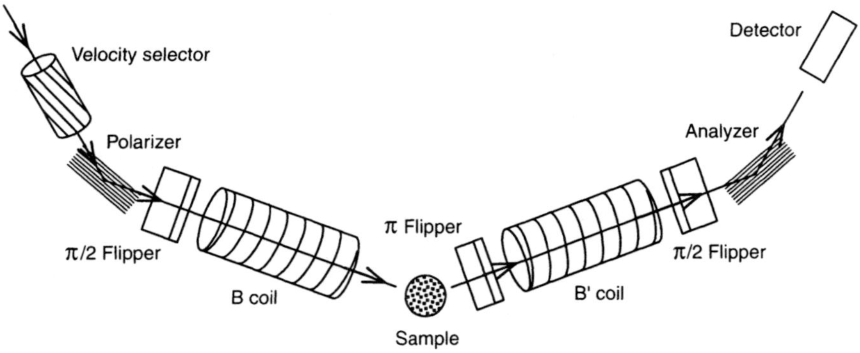

Fig. 8. Schematic setup of a neutron spin echo spectrometer.

Therefore the Mezei coil has to be placed in an oblique position so that the resulting field $\boldsymbol{B}_\pi = \boldsymbol{B}_0 + \boldsymbol{B}_c$ has the desired 45° direction.

π flip: Another spin manipulation important for the function of a NSE spectrometer is the inversion of two spatial components of the neutron spin (e.g. y and z) leaving the third one unchanged (Fig 7b). This operation is equivalent to a 180° (π) rotation in the y-z plane. It can be accomplished by a field of the same magnitude B_π as before but parallel to the spin component of the neutron to be preserved (here x).

The basic setup of an NSE spectrometer is shown in Fig. 8. A velocity selector in the primary neutron beam selects a wavelength interval of 10–20% width. In the primary and secondary flight path of the instrument precession fields B and B' parallel to the respective path are generated by cylindrical coils. Before entering the first flight path the neutron beam is polarized in forward direction[2]. A first $\pi/2$ coil rotates the polarization to the x direction perpendicular to the direction of propagation. Beginning with this well-defined initial condition the neutrons start their precession in the field B. After being scattered by the sample the neutron go through a π flipper and through the second precession field B'. Finally, the neutrons pass through another $\pi/2$ coil which under certain conditions restores their initial polarization parallel to their flight direction. In order to understand what that condition is one has to trace the changes of the spin vector (z always denoting the direction parallel to neutron propagation):

(n_x, n_y, n_z)	neutronic device
$(0,0,1)$	
	$\pi/2$ flipper
$(1,0,0)$	
	field B
$(\cos\phi, \sin\phi, 0)$	
	π flipper
$(\cos\phi, -\sin\phi, 0) = (\cos(-\phi), \sin(-\phi), 0)$	
	field B'
$(\cos(\phi' - \phi), \sin(\phi' - \phi), 0)$	
	$\pi/2$ flipper
$(0, \sin(\phi - \phi'), \cos(\phi - \phi'))$	

In total, the spin is rotated by $\phi - \phi'$ around the x-axis when a neutron passes through the spectrometer. This means that the final polarization is identical to the incident if

[2] This is done by a 'polarizing supermirror' which only reflects neutrons of that spin – similar to the Nicol prism in optics.

$\phi = \phi'$ $(+2\pi n)$, especially if $\lambda_i = \lambda_f$ (elastic scattering) and $Bl = B'l'$ (the field integrals in primary and secondary flight path are the same) as follows from (14). This condition is called 'spin echo' and is independent of the individual velocities of the neutrons because their difference alone determines $\phi - \phi'$.

Leaving spin echo condition the probability of a single neutron to reach the detector is reduced due to the polarization analyzer by $\cos(\phi' - \phi)$. If we keep the symmetry of the instrument, $Bl = B'l'$, but consider inelastic scattering the precession angle mismatch can be approximated by

$$\begin{aligned}\phi' - \phi &= \left(\frac{2\pi|\gamma|\mu m_\mathrm{n}}{h^2}\right) Bl(\lambda_\mathrm{f} - \lambda_\mathrm{i}) \\ &\approx \underbrace{\frac{|\gamma|\mu {m_\mathrm{n}}^2\lambda^3 Bl}{h^3}}_{=t_\mathrm{NSE}(B)}\,\omega \end{aligned} \tag{15}$$

for small energy transfers where $\Delta\lambda \approx \hbar\omega \big/ \frac{\mathrm{d}E}{\mathrm{d}\lambda}$ can be used. Because the energy transfer for inelastic scattering is not fixed but distributed as determined by the scattering function $S(q,\omega)$ we have to average the factor $\cos(\phi' - \phi)$ weighted by $S(q,\omega)$ to get the reduction of count rate at the detector, the effective polarization

$$P(q, t_\mathrm{NSE}) = \frac{\int_{-\infty}^{\infty} S(q,\omega)\cos(\omega t_\mathrm{NSE})\,\mathrm{d}\omega}{\int_{-\infty}^{\infty} S(q,\omega)\,\mathrm{d}\omega}. \tag{16}$$

Firstly, we note that $S(q,\omega)$ in this expression usually is the coherent scattering function. In principle, similar arguments can used for incoherent scattering because a well-defined fraction of neutrons changes its spin. This leads to a 'negative echo' because the majority of neutrons invert their polarization. But because this effect is only partial (e.g. 2/3 for hydrogen nuclei) it is much more difficult to observe. Only recently, NSE spectroscopy could be applied successfully to incoherently scattering samples.

Secondly, expression (16) reverses the temporal one of the Fourier transforms of equation (2) and therefore stands in between the microscopic function $G(\boldsymbol{r},t)$ and the observable $S(\boldsymbol{q},\omega)$. For this reason [3]

$$\begin{aligned} I(\boldsymbol{q},t) &= \int_{-\infty}^{\infty} S(\boldsymbol{q},\omega)\exp(\mathrm{i}\omega t)\,\mathrm{d}\omega \\ &= \int \mathrm{d}^3 r \exp(\mathrm{i}(\boldsymbol{q}\cdot\boldsymbol{r}))\,G(\boldsymbol{r},t) \end{aligned} \tag{17}$$

is called the intermediate scattering function and the result of the NSE experiment

$$P(q, t_\mathrm{NSE}(B)) = I(q, t_\mathrm{NSE}(B))/I(q,0) \tag{18}$$

[3] The exponential Fourier transform here and the cosine Fourier transform of (16) are identical in the classical limit when $S(\boldsymbol{q},\omega)$ is symmetric and $I(\boldsymbol{q},t)$ a real function.

Table 4

Basic specifications of representative neutron spin echo spectrometers

Instrument	λ (Å)	q_{max} (Å^{-1})	t_{max} (ns)
IN11 (ILL)	4.5–12	0.9–2.4	2–45
IN15 (ILL)	8–25	0.13–0.4	30–800
NSE (FZJ)	6–10	0.9–1.5	20–90
SPAN (HMI)	2.5–10	1.2–4.9	0.2–10

The maximal q and the maximal Fourier time t_{max} depend on the incident wavelength; the upper limit of the q range and the lower limit of t_{max} correspond to the lower limit of the incident wavelength λ and vice versa.

is the normalized intermediate scattering function. This function is often more understandable and easier to interpret than the frequency dependent scattering function.

In order to estimate typical Fourier times t_{NSE} which can be accessed by NSE we consider maximum fields of $B = B' = 500$ Gauss in precession coils of $l = l' = 2$ m length operating at $\lambda = 8$ Å. Then (15) results in a time of about 10 ns which can be reached.

From this equation it also becomes clear that the most efficient way to enlarge this time is to use longer wavelengths because λ enters in the third power. This in turn reduces the accessible q range which constitutes a drawback for studies on low molecular materials but not for the large scale properties of polymers which have to be observed at low q anyway.

Typical NSE spectrometers with their specifications are listed in Table 4. NSE spectrometers are very flexible instruments often used with different setups of which only 'typical' ones have been included. As special features have to be mentioned that IN11 and SPAN have one-dimensional detector arrays which span 60° and 150° degrees respectively, allowing the simultaneous observation of a range of q values. The instruments IN15 and NSE have two-dimensional detector arrays which can be used for studying anisotropies but cover a smaller angular range. IN15 uses a focusing mirror in order to increase neutron flux which would be otherwise very low due to its long precession coils.

PART II. DYNAMICS OF POLYMERS

Polymers exhibit unusual viscoelastic and transport properties which determine their processing characteristics and application to large extent (Ferry, 1980). These properties are also of basic interest and an understanding on the molecular level is required. Polymers are mesoscopic objects with relevant length scales from the single bond length (about 0.1 nm) to the whole-chain dimension reaching 100 nm and more. These vastly different length scales are accompanied by a large variety of time scales starting in the picosecond range and extending to macroscopic times.

A typical example of viscoelasticity of a polymer is shown in Figure 9, the shear compliance $J(t)$ of polystyrene. The shear compliance is given by the shear strain $\sigma(t) = \Delta y(t)/x$ resulting from a stress ϵ applied at $t = 0$ by $J(t) = \sigma(t)/\epsilon$. For a purely elastic medium one would expect $J(t) = \text{const}$ from Hooke's law, for a purely viscous material $J(t) = t/\eta$ from Newton's law. The polymer shows a complex behaviour, even more complex than just a superposition of both types. In the beginning the increase of $J(t)$ is sublinear. Then

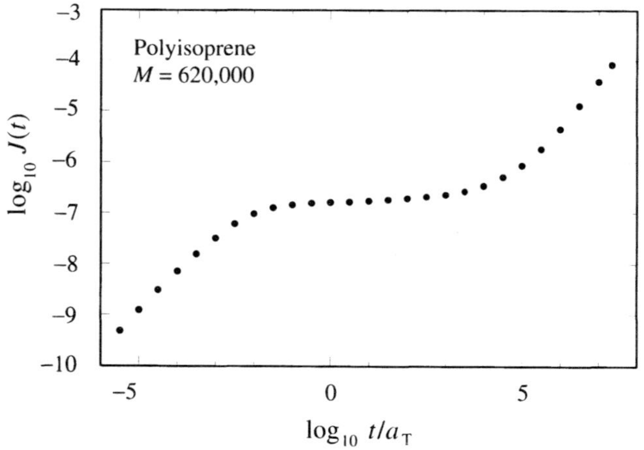

Fig. 9. Shear compliance $J(t)$ of polyisoprene (molecular weight $M_w = 6.2 \times 10^5$), double-logarithmic plot (Nemoto et al., 1971). The time axis has been constructed using the time-temperature superposition principle: Data from different temperatures were scaled by a temperature dependent factor a_T in order to obtain a master curve.

it arrests in a plateau indicating a spring- or rubber-like behaviour. Only for very long time viscous behaviour $J(t) \propto t$ is attained with a very high 'terminal' viscosity.

The length of the plateau increases with the molecular weight of the polymer shifting the viscous regime to higher values. Therefore, the viscosity increases with molecular weight M. The increase can be described by power laws, $\eta \propto M$ below a certain molecular weight M_c and typically $\eta \propto M^{3.4}$ above (Fig. 10a). A similar dependence is found for the self-diffusion constant: $D \propto M^{-1}$ for $M < M_c$, and $D \propto M^{-2}$ for $M > M_c$ (Fig. 10b).

The fact that relations as shown in Fig. 10 are applicable to nearly all polymers with the same exponents indicates that the dynamics of polymers on large scales follows universal laws. The underlying reason is that on a large length scale the individual molecular nature of the monomeric unit is not important. Rather a description by simple models (e.g. bead and spring) is possible. Such descriptions have been developed firstly for the region $M < M_c$ and later also for $M > M_c$ and will be discussed in subsections 4.1 and 4.2, respectively.

Although these concepts had an immediate success in terms of describing the rheological experiments correctly it took a long time to confirm them on a molecular basis. For this task inelastic neutron scattering, especially the Neutron Spin Echo (NSE) technique turned out to be the crucial tool to reveal the mesoscopic dynamics on its proper length and time scales.

In contrast, on short length scales the chemical nature of the polymer becomes important. The dynamic properties on that scale become similar to those of low molecular organic liquids. Especially, one finds a transition to a glassy state whose temperature T_g depends strongly on the chemistry (e.g. -120°C for polydimethylsiloxane and $+100$°C for polystyrene). Universal concepts are currently being constructed but they are by far not as developed as for the whole chain dynamics of polymers.

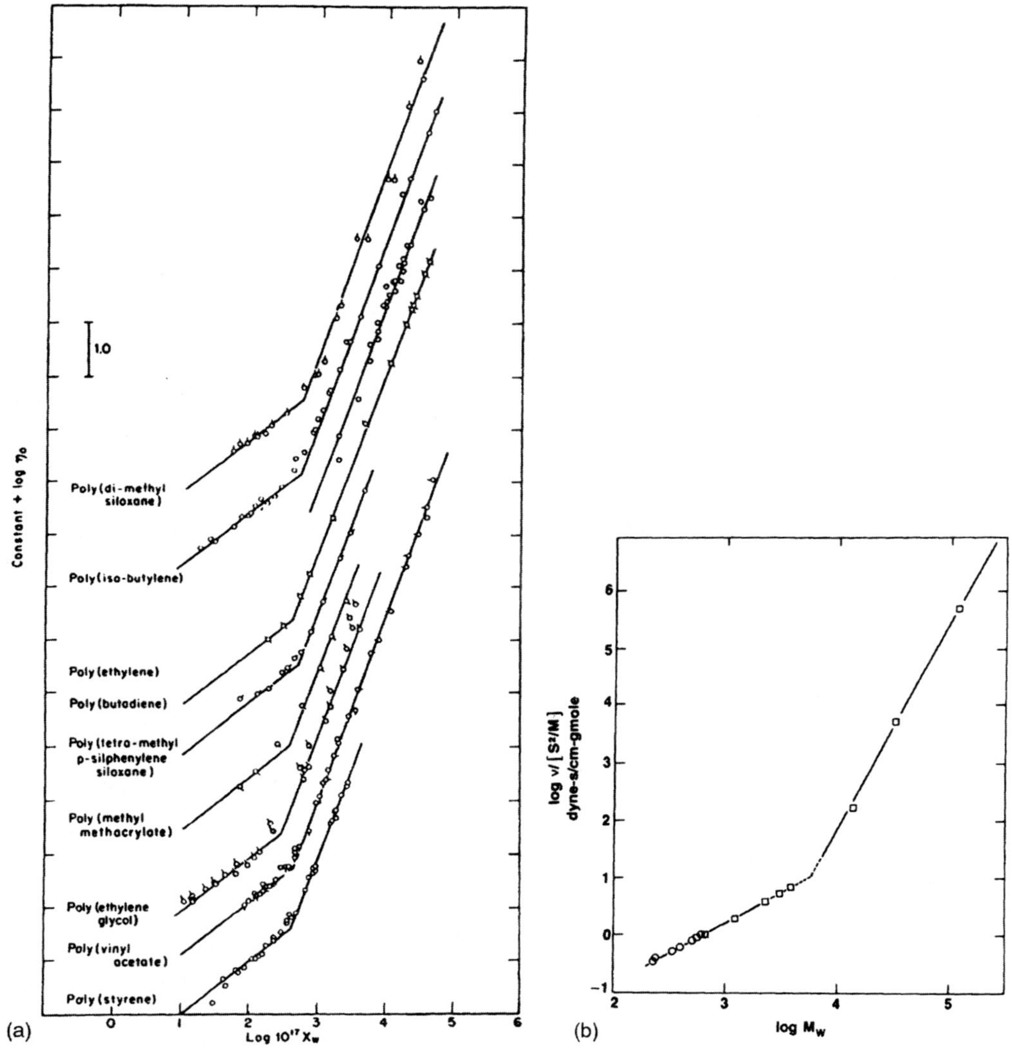

Fig. 10. Molecular weight dependence of (a) the viscosity of various polymer melts (Ferry, 1980), (b) the reciprocal (self) diffusion constant D^{-1} of polyethylene (Pearson et al., 1987).

4. Large scale dynamics

4.1. Free chain (Rouse) dynamics

Despite the simplicity of the standard model for short chain polymer dynamics, the Rouse model (Rouse, 1953), it involves some complicated calculations. The mathematical details can be found in the literature (Doi and Edwards, 1986) and the fundamental equations are summarized in Appendix A.

Here, only the basic assumptions and most important results will be presented. Firstly, the chain is subdivided into segments with a Gaussian end-to-end distance distribution similar to what is done for the calculation of the static properties. Then, the most important assumption is that there is no interaction between those segments of a polymer chain with those of other chains or the same chain except for a global friction force and a random force as assumed in the derivation of Brownian motion of a single particle.

The results for the macroscopic rheological properties, terminal viscosity and self-diffusion constant, are

$$\eta = \frac{\zeta \ell^2 \rho N}{36} \propto N \tag{19}$$

$$D = \frac{k_B T}{\zeta N} \propto N^{-1} \tag{20}$$

in perfect agreement with what is found for low molecular weight polymers. (N is the number of segments, $N \propto M$, ρ the number density of the polymer, ζ the friction coefficient, and ℓ the average length of a segment.) These relations can be motivated by the rough argument that if there is no other interaction than a friction with the environment, its total effect on a polymer molecule will simply increase with the number of rubbing units.

Despite this agreement it still has to be proven that the model is correct on a molecular level. Here, the Rouse model allows a very stringent test because it predicts the incoherent dynamic structure factor (self-correlation function), Eq. (48) of Appendix A.

Measurements of the incoherent inelastic neutron scattering are usually performed on protonated materials because of the large incoherent cross-section of hydrogen (see Table 1). But because in this case the scattering is of spin-incoherent origin which means that the polarization is partially lost NSE experiments are difficult or even impossible with samples of high hydrogen content. In order to get around this obstacle one uses a chemical trick. High molecular weight deuterated polymer is synthesized which contains short protonated sequences at random positions. In such a sample the scattering essentially originates from the contrast between the protonated sequence and the deuterated environment and is therefore coherent with respect to the individual scatterers. On the other hand, the labeled sequences are randomly distributed so that there is no constructive interference of partial waves arising from different sequences. This means that on the whole chain level the scattering is incoherent[4] and the scattering experiment measures the self-correlation function of the segments.

In Fig. 11 the scattering data obtained in this way from polydimethylsiloxane (PDMS) on the NSE spectrometer IN11 is plotted against the scaling variable of the Rouse model (50). The results for the different momentum transfers q follow a common straight line in the semilogarithmic plot which is the characteristic behaviour predicted by the Rouse model.

---

[4] At this point it becomes clear that the subdivision of (1) into coherent and incoherent scattering may be arbitrary. If we consider hydrogen and deuterium to be isotopes of the same atom we get a difference $\langle b^2 \rangle - \langle b \rangle^2$ not only from spin disorder but also from different b values of the nuclei. If we consider H and D to be different atoms (because for the samples here they are – by chemical means – put into preferential positions) we would have to extend (1) to a polyatomic expression which would then contain the same scattering in the coherent part just leaving the spin disorder part as incoherent.

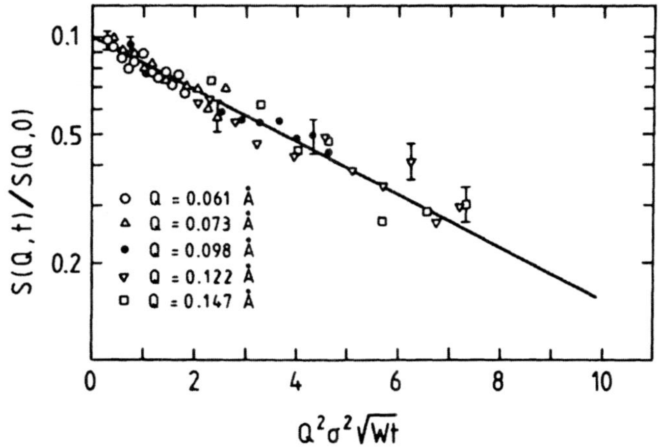

Fig. 11. NSE spectra, self-correlation function $I_{inc}(q,t)$, measured on a randomly labeled polydimethylsiloxane (molecular weight $M_w = 10^5$) sample at 100°C. Semilogarithmic plot with the time axis scaled by the Rouse variable. The solid line is a fit with the Rouse model result (48) (Richter et al., 1989).

It is also possible to test the Rouse model on a molecular level using the coherent intermediate scattering function. In this case one uses a mixture of (totally) protonated and deuterated chains as it is used for measuring the static structure factor of a chain in a small-angle neutron scattering experiment. The evaluation of the data is more complicated because there is no closed expression for the coherent intermediate scattering function. Nevertheless, because it shows the same scaling properties as the incoherent one the crucial test is the same plot versus the Rouse variable (50) (Fig. 12).

4.2. Entanglement (reptation) dynamics

The simple Rouse behaviour found for PDMS of molecular weight around 10^5 is rather the exception than the rule in polymer science. This becomes already clear from the macroscopic properties η and D which obey the predictions of the Rouse model only for $M < M_c$. In this respect, PDMS is a peculiarity because of its low friction coefficient ζ and high flexibility allowing the Rouse model to be valid in a large time range of $I(q,t)$.

Various attempts have been undertaken to understand the change of macroscopic properties for $M > M_c$. The most important path of development starts with the tube model of Edwards (1967) developed by de Gennes (1971). The basic idea is that the surrounding polymer molecules form a virtual tube (Fig. 13) which hinders the motion of the central polymer molecule. The only motion which is easily possible is that along its own contour similar to a motion of a snake. For that reason the term 'reptation'[5] has been coined for this type of motion.

It has to be noted that this virtual tube is not 'tight-fitting' but restricts the motion of a polymer chain on an intermediate length scale $d \approx 5$ nm larger than the dimension of a

[5] From the latin word reptare = to creep, to crawl; the same root as in 'reptile'.

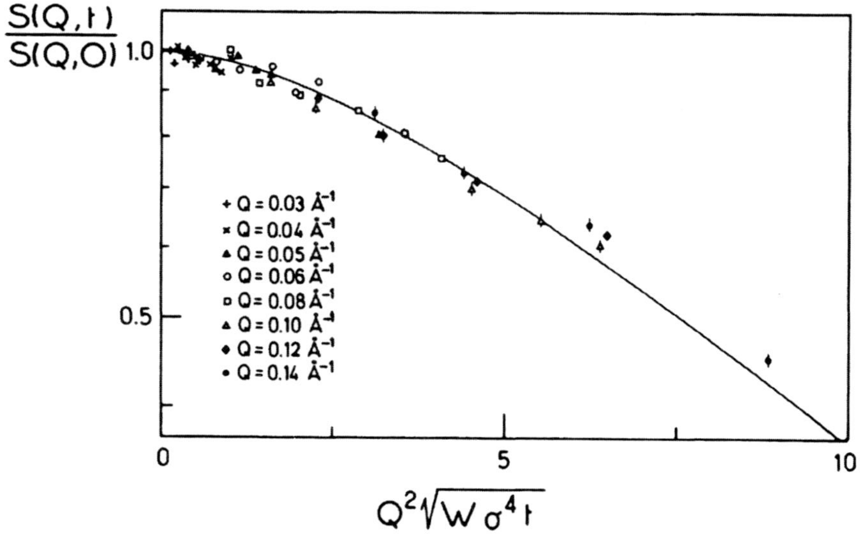

Fig. 12. NSE spectra, pair correlation function $I_{\mathrm{coh}}(q,t)$, measured on a mixture of H- and D-polydimethylsiloxane (molecular weight $M_w = 1.5 \times 10^5$) at 100°C. Semilogarithmic plot with the time axis scaled by the Rouse variable. The solid curve is a fit with the Rouse model result (51) (Richter et al., 1989).

monomer. Therefore, at short times the restriction is not active and the motion is Rouse-like. The onset of the tube interaction defines a new intermediate time scale, the entanglement time τ_e.

In this simple intuitive model the experimental results for viscosity and diffusion can be immediately understood. The viscosity of a polymer melt is determined by the longest relaxation time τ_η, in this model the time which is necessary for a polymer to leave its initial tube and constitute a new one. Within the tube the chain performs a Rouse diffusion with $D_{\mathrm{R}} \propto N^{-1}$. The chain has completely left its initial configuration when it has moved by

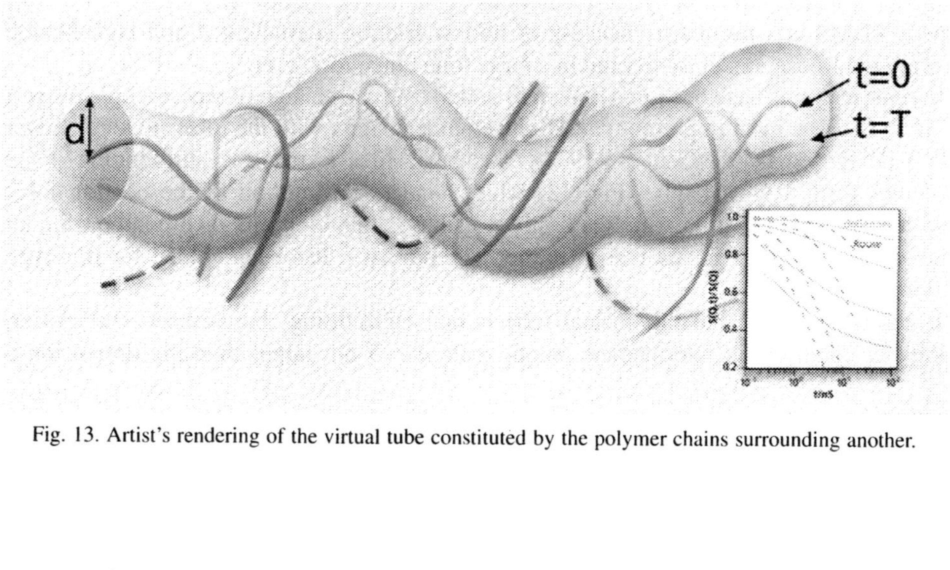

Fig. 13. Artist's rendering of the virtual tube constituted by the polymer chains surrounding another.

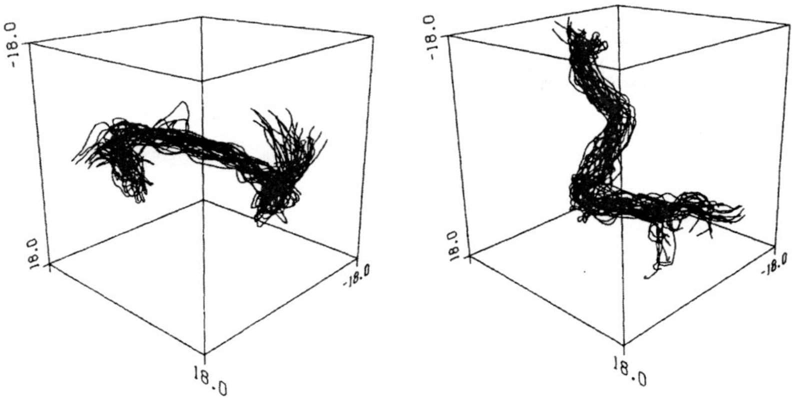

Fig. 14. Computer simulation of a bead-and-spring model of a polymer (Kremer and Grest, 1990). Shown are only two (left and right picture) of the 100 chains of length 400. 40 conformations at different times are superimposed to give an impression of the fluctuation of the polymer shape.

the contour length $L = N\ell$. (Note that here the restriction by the tube enters: It is not the end-to-end distance $\sqrt{N\ell}$ that counts but the larger curved length.) Thus one expects

$$\eta \propto \tau_\eta \propto L^2/D_R \propto N^3 \tag{21}$$

a cubic increase of viscosity with molecular weight. The actual exponent is more like 3.4 in the experiment but there are indications that for very large chain lengths the limit of 3 is attained.

In real space the contour of the chain follows a Gaussian random walk, i.e. during the time τ_η it moves by diffusion over a distance comparable to its end-to-end distance $R_E \propto \sqrt{N}$. From this one obtains the self-diffusion constant of the whole chain

$$D \propto {R_E}^2/\tau_\eta \propto N^{-2} \tag{22}$$

in perfect agreement with the experiment.

Although the explanation of the experimental results by the reptation model is quite satisfactory for a real confirmation of the model it is necessary to verify directly the molecular behaviour. One way to do this is computer simulation which has recently become able to model macromolecules. Fig. 14 shows results from a bead-and-spring model. Indeed, one can clearly see that the motion of the displayed polymer molecule is restricted to a tube except for its ends. Nevertheless, this is no true experimental test because the underlying mathematical model is a gross simplification of the complexity of a real macromolecule which can only be justified by the universality of large scale dynamics.

A test on real polymers is again possible by the NSE technique. Fig. 15 shows results of polyethylenepropylene (PEP) using a similar mixture of hydrogenated and deuterated polymer as before in Fig. 12. It can be seen that the normalized intermediate scattering function $I(q,t)/I(q,0)$ does not scale with the Rouse variable (50). Instead of decreasing exponentially for long times its value tends to go to a plateau.

This can be qualitatively understood as an effect of the restriction due to the virtual tube (Fig. 16). For short times the motion of the chain obeys Rouse diffusion. Its position becomes

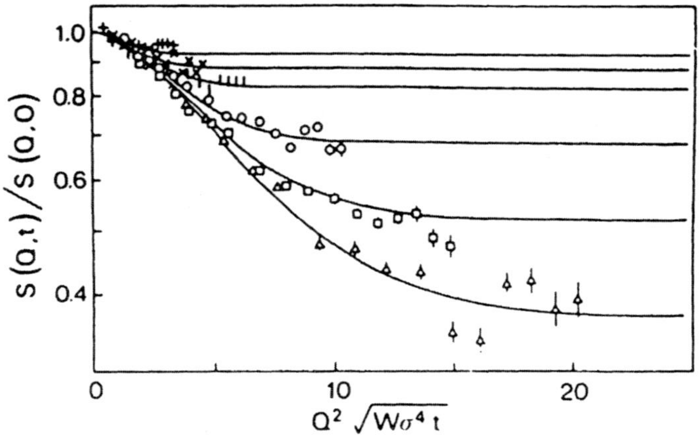

Fig. 15. NSE spectra from polyethylenepropylene at 492°C plotted semilogarithmically and scaled by the Rouse variable as in Fig. 12 (Richter et al., 1990). The solid curves are fits with the Ronca model (Ronca, 1983).

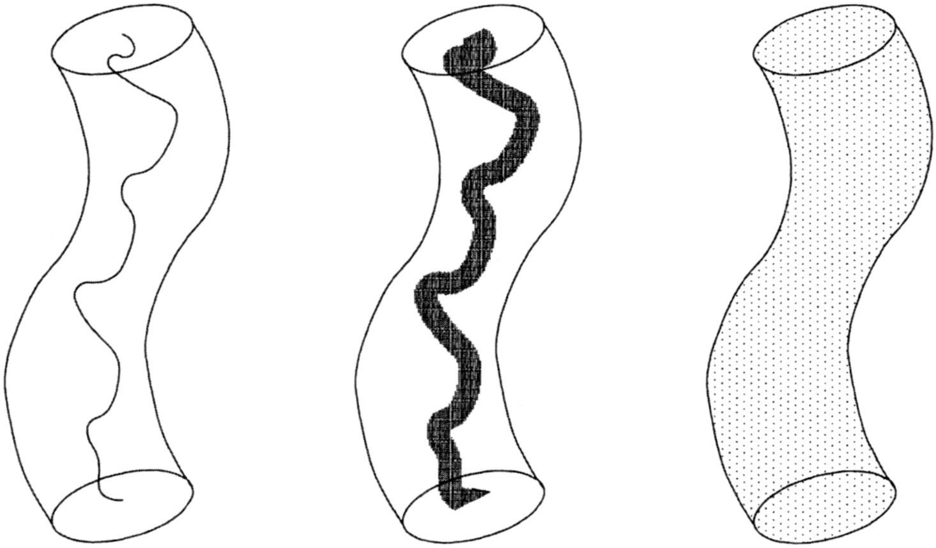

Fig. 16. Schematic visualization of the dynamics of a polymer chain in the tube of surrounding chains, left to right: 1. initial conformation of the chain; 2. distribution of conformations for short times $t < \tau_e$; 3. distribution of conformations for long times $t < \tau_e$.

on the average 'smeared out' and in consequence the correlation function decreases. As soon as the average displacement reaches the order of the tube radius ($t = \tau_e$) this process comes to a halt and the correlation function does not reduce further. Its limiting value $I(q,t \gg \tau_e)/I(q,0)$ describes the shape of the tube as $S(q)$ does for a static object. But it has to be noted that the tube 'exists' only in a certain time regime $\tau_e \ll t \ll \tau_\eta$ and therefore cannot be detected by static scattering methods.

For a quantitative comparison analytical models are required which can be compared with the data. Three different model categories shall be discussed without explaining them in detail:

(1) In so-called generalized Rouse models (Ronca, 1983; Hess, 1988) the effect of topological hindrance is described by a memory function. In the borderline case of long chains the dynamical structure factor can be explicitly calculated in models of this type in time domain which is directly comparable to the NSE experiment. The solid lines plotted in Fig. 15 correspond to a fit of the data with Ronca's generalized Rouse model (Ronca, 1983).

(2) Neglecting the initial Rouse motion which determines the common incipient part of the data in Fig. 15, in his local reptation model de Gennes explicitly calculated the collective chain motion in the localization tube (de Gennes, 1981):

$$\frac{I(q,t\gg\tau_e)}{I(q,0)} = \left(1-\exp\left(-\left(\frac{qd}{6}\right)^2\right)\right)\exp\left(\frac{k_B T\ell^2q^4t}{12}\right)\operatorname{erfc}\sqrt{\frac{k_B T\ell^2q^4t}{12}}$$
$$+\exp\left(-\left(\frac{qd}{6}\right)^2\right)\frac{8}{\pi^2}\sum_{p=1,3,\ldots}\frac{1}{p^2}\exp\left(-\frac{p^2t}{\tau_\eta}\right). \tag{23}$$

Here the first term describes the local reptation while the second represents the creep process out of the tube. Despite its complicated structure, Eq. (23) gives a simple result for the plateau

$$\frac{I(q,(qd)^4\tau_e\ll t\ll\tau_\eta)}{I(q,0)} = \exp\left(-\frac{q^2d^2}{36}\right). \tag{24}$$

where $\tau_e = d^4\zeta/3\pi^2\ell^2k_BT$.

(3) Finally, des Cloizeaux formulated a model analogous to a rubber (a polymer with fixed chemical cross-links) at intermediate times (des Cloizeaux, 1993). He assumes that the chain performs Rouse motion between fixed entanglement points.

Because all three types of models have similar underlying concepts the numerical results are not very different for shorter times in the entanglement regime $\tau_e\ldots\tau_\eta$. Therefore, it was for a long time impossible to discriminate them and decide which one is closest to the experimental truth. Only recently, a breakthrough in NSE technology could resolve this question (Schleger et al., 1998). Fig. 17 shows NSE data (IN15, ILL) up to 170 ns fitted with the models mentioned above. All fits only have one free parameter (the entanglement length d) which is listed in Table 5 together with the χ^2 quality of the fit. It is immediately clear that the full reptation expression (23) allows the best description. Next comes the 'local reptation', i.e. only the first term of (23). The slightly worse fit in this case is an indication that the reptation time scale τ_η has already an influence in the observed time window. The other two descriptions fall behind significantly what concerns the t dependence shown here as well as the wave vector q dependence (Schleger et al., 1998). It should also be mentioned that the parameter d obtained using the reptation model is the most consistent compared with that calculated from rheological measurements $d = 42$ Å (Carella et al., 1984).

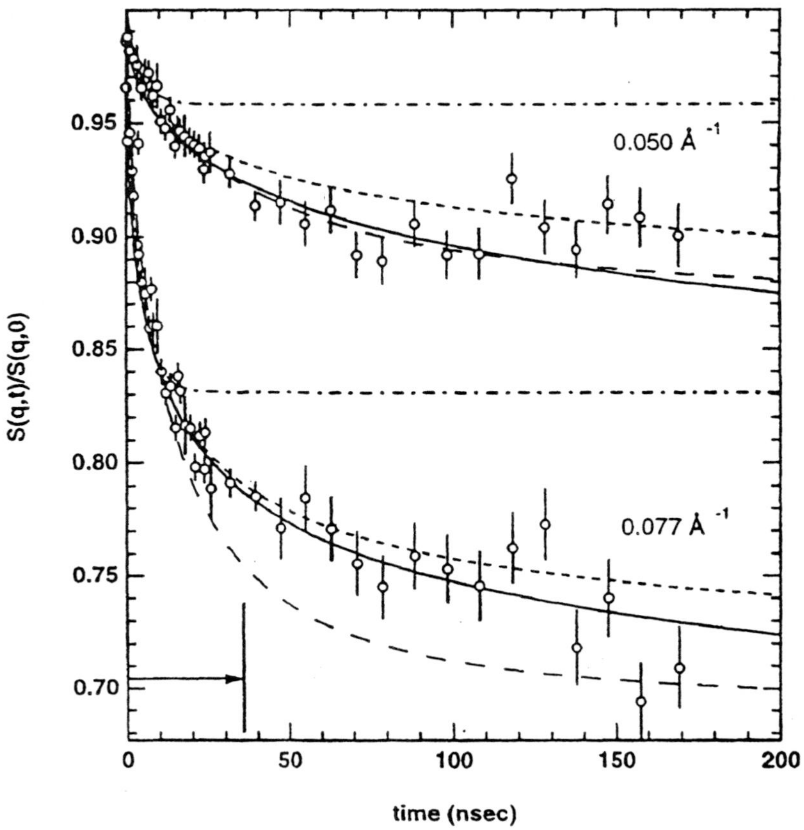

Fig. 17. NSE spectra from polyethylene at 509°C and wave vectors $q = 0.050$ and 0.077 $Å^{-1}$ with a comparison between predictions of reptation (solid curve), local reptation (short-dashed curve), the models of des Cloizeaux (1993) (dashed curve), and of Ronca (1983) (dot-dashed curve) (Schleger et al., 1998). The vertical line and the arrow indicate the upper time limit of previous experiments.

Table 5

Fit results for the entanglement distance d for the studied models (Schleger et al., 1998). The reduced χ^2 of the fits is also indicated

Model	Reference	d (Å)	Reduced χ^2
Reptation	de Gennes (1981)	46.0 ± 0.1	3.03
Local Reptation	de Gennes (1981)	46.5 ± 0.1	3.21
des Cloizeaux	des Cloizeaux (1993)	59.8 ± 0.2	7.19
Ronca	Ronca (1983)	47.4 ± 0.1	12.2

5. Small-scale dynamics

5.1. α relaxation, glass transition

It may be noticed that all calculations of Section 4 left one question open: What is the value (or the temperature dependence) of the monomeric friction coefficient ζ? All equations resulting in macroscopic quantities as viscosity or diffusion constant contain ζ as a factor. So without knowing it no absolute values can be calculated.

Fig. 18 shows that this is an important question. Obviously the temperature dependence of the viscosity $\eta(T)$ is very strong spanning several decades in a comparatively small temperature range. The other factors occurring in the formulae for $\eta(T)$ (e.g. ℓ, ρ, $k_B T$) show much weaker temperature dependence. So the basic reason for the strong temperature dependence of the viscosity (or diffusion constant, or relaxation time τ_η) must be $\zeta(T)$. Indeed, $\zeta(T)$ shows a temperature dependence which is very similar to that of the viscosity of low molecular liquids which do not crystallize but form glasses. As the 'activation style' plot of Fig. 18 shows it cannot be described by an Arrhenius law but rather by phenomenological expressions as the Vogel–Fulcher law[6]

$$\zeta(T) \propto \exp\left(\frac{B}{T - T_0}\right). \tag{25}$$

The similarity to low molecular liquids suggests that on short length scales the dynamics of polymers is not much different from the latter. For a direct observation of the short scale dynamics it is necessary to perform IENS experiments at high wave vectors q close to the molecular correlation peak of the structure factor, i.e., at about 1.4 Å^{-1}. Fig. 19 shows such data for polybutadiene (Richter et al., 1988) as obtained directly from NSE. Fig. 20 shows the same data with the time axis scaled by the viscosity and it can be seen that $I(q,t)/I(q,0)$ curves from different temperatures coincide. This shows that the microscopic dynamics on the length scale of a single monomer indeed has the same temperature dependence as the macroscopic dynamics on the whole chain level.

Nevertheless, later investigations at even higher q values destroyed the simple picture. As Fig. 21 shows, at $q = 1.88$ Å^{-1} for temperatures below 220 K the curves do not scale to a mastercurve. The reason for this deviation was discovered later and shall be discussed in Section 5.3.

5.2. 'Fast' processes

If one compares the NSE data in Figs. 19 and 21 with those of Section 4 it is striking that the curves do not extrapolate to $I(q,t)/I(q,0) = 1$ for short times. This apparent contradiction is resolved considering that we do not see the very short time (picosecond) regime on the NSE

[6] In polymer science often the mathematically equivalent Williams–Landel–Ferry law

$$\log_{10}\frac{\zeta(T)}{\zeta(T_r)} = \frac{-c_1^0(T - T_r)}{c_2^0 + T - T_r}$$

is preferred. T_r is an arbitrary reference temperature.

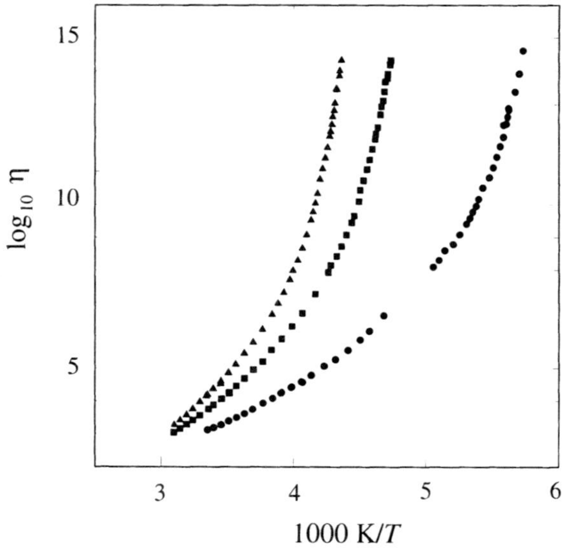

Fig. 18. Viscosity of 1,2- (•), 1,2-1,4- (■), and 1,4-polybutadiene (▲) (Zorn et al., 1995b). Note that only the lower viscosities ($< 10^5$) are actually measured on a viscosimeter, the higher values are derived using the time–temperature superposition principle.

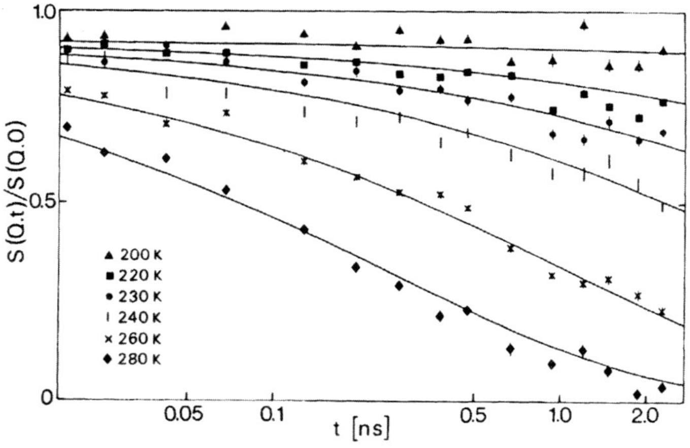

Fig. 19. NSE spectra ($I_{\mathrm{coh}}(q,t)/I(q,t)$) of polybutadiene at the wave vector $q = 1.48$ Å^{-1} which corresponds to the maximum of the structure factor $S(q)$ (Richter et al., 1988). Double-logarithmic representation, the time axis is scaled by a time τ derived from viscosity measurements ($\tau \propto \eta$). The curves are fits with the Kohlrausch expression $\exp(-(t/\tau_K)^\beta)$ with $\beta = 0.45$.

spectrometer. If another dynamic process exists at such times it leads to a decay $I(q,t)/I(q,0)$ for times not observable by NSE.

Indeed, time-of-flight spectra (Fig. 22) show the presence of inelastic scattering at energy transfers around 1 meV which corresponds roughly to 2 ps, which is just below the range detectable by NSE.

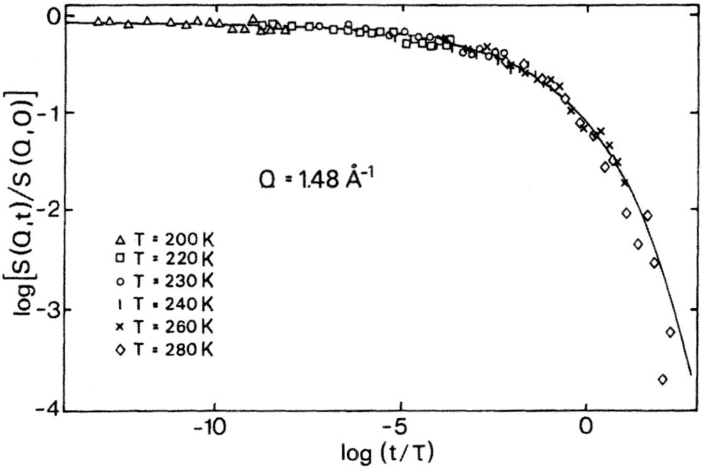

Fig. 20. Same data as in Fig. 19 but in double-logarithmic representation with the time axis scaled by a time τ derived from viscosity measurements ($\tau \propto \eta$). The curve is a fit with the Kohlrausch expression $\exp(-(t/\tau_K)^{\beta})$ with $\beta = 0.45$.

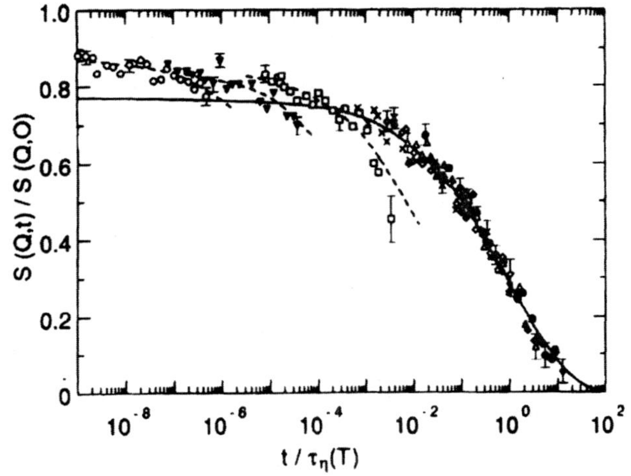

Fig. 21. NSE spectra ($I_{coh}(q,t)/I(q,t)$) of polybutadiene at the wave vector $q = 1.88$ Å^{-1} close to the minimum of the structure factor $S(q)$ (Richter et al., 1992). The logarithmic time axis is scaled by a time τ derived from viscosity measurements ($\tau \propto \eta$). The curves are fits with the Kohlrausch expression $\exp(-(t/\tau_K)^{\beta})$ with $\beta = 0.37$.

At low temperatures a peak in $S_{inc}(q,\omega)$ is visible which is very similar to what is observed in low-molecular glass-forming liquids (e.g. Fig. 4). Indeed, this so-called 'Boson peak' is characteristic for glass-forming systems. This name is derived from the fact that the peak scales with the Bose factor which indicates its origin from thermally excited vibrations. Because for incoherent scattering at low temperatures $S_{inc}(q,\omega) \propto g(\omega)/\omega^2$ one would rather expect a flat line because the density of states due to sound waves is $g(\omega) \propto \omega^2$ (Debye). But in glass-forming system there is always an excess density of states (Inoue et al., 1991).

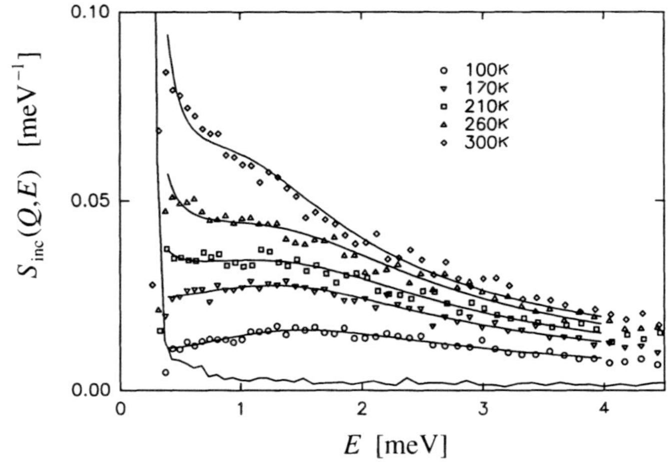

Fig. 22. Incoherent TOF spectra of 1,2-polybutadiene for the temperatures indicated and $q = 2.2$ Å^{-1}. The thick curve is the resolution function of the instrument drawn to same scale. The other curves are fits with the vibration-relaxation model (Zorn et al., 1996).

which has recently been connected to the existence of localized soft modes (Buchenau et al., 1996).

At higher temperature additional intensity closer to the elastic line appears. The discussion of this phenomenon is still controversial but as the fits indicated in Fig. 22 show a model which connects the vibrational modes and relaxation due to jump processes is successful in describing the data (Buchenau et al., 1994; Zorn et al., 1995a).

5.3. Secondary (β) relaxation

The deviation of $I(q,t)/I(q,0)$ from a mastercurve visible in Fig. 21 could recently be explained by the existence of another relaxation, the β relaxation. Using dielectric spectroscopy the such a 'secondary' relaxation has been found frequently in various glass-forming systems (Johari and Goldstein, 1970). It is faster than the α relaxation but its temperature is weaker, in general describable by an Arrhenius law $\tau \propto \exp(E_A/k_B T)$. So at a certain temperature both relaxations will merge.

If one compares dielectric and NSE data one finds that the data in Fig. 21 just cover that merging region. Therefore they have to be described by a model which takes both relaxations into account (Arbe et al., 1996). Without presenting the details of the calculation we show the resulting expression:

$$\frac{I(q,t)}{I(q,0)} = \exp\left(-\left(\frac{t}{\tau_K(q,T)}\right)^{\beta}\right)\left(1 - \frac{1}{S(q)}\int_0^{\infty} g(E_A)F_{\text{inel}}\big(q,d(E_A)\big)\,\mathrm{d}E_A \right.$$
$$\left. + \frac{1}{S(q)}\int_0^{\infty} g(E_A)\exp\left(-\frac{2t}{\tau(E_A)}\right)F_{\text{inel}}\big(q,d(E_A)\big)\,\mathrm{d}E_A\right). \tag{26}$$

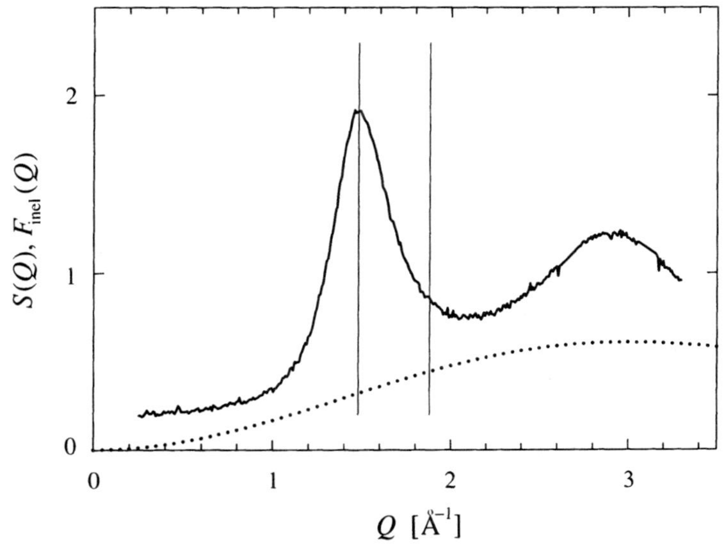

Fig. 23. Structure factor $S(q)$ (thick curve) and inelastic form factor of the two site jump model with $d = 1.5$ Å (dots). The thin lines indicate the wave vectors at which the experiments of Figs. 19 and 21 took place, $q = 1.48$ Å^{-1} and 1.88 Å^{-1}.

The first factor describes the α relaxation by the Kohlrausch function $\exp(-(t/\tau_K)^{\beta})$. It is the same functional form as resulting from the Rouse model but in general $\beta \neq 1/2$. $\tau_K(q,T)$ contains the temperature dependence expressed by (25) and a power law q dependence.

The second factor contains the β relaxation. Because α and β relaxations are assumed to be independent processes, the resulting intermediate scattering function is a product of the individual ones (Bée, 1988). $S(q)$ is the structure factor from a diffraction experiment, for coherent scattering $I(q,0) = S(q)$. $F_{\text{inel}}(q,d)$ is the inelastic form factor of the β relaxation motion. In the simplest case a jump between two positions in a distance d can be assumed (see Appendix B) but also more detailed molecular models can be conceived.

The temporal aspect of the β relaxation is included in the integrals over activation energies. As the β relaxation follows an Arrhenius law the relaxation time can be expressed as $\tau(E_A) \propto \exp(-E_A/k_B T)$. But the fact that it cannot be described by a single exponential calls for a distribution of activation energies E_A. Such a description was successful for dielectric spectroscopy data (Hofmann et al., 1996) and the $g(E_A)$ used in (26) has been taken from those data of the dielectric function. In addition a relation $d(E_A) \propto E_A^{1/4}$ has been used expressing the notion that higher energy barriers are correlated with larger distances of the potential minima.

From expression (26) it is immediately clear that at the maximum of $S(q)$ the β relaxation is of less importance. Fig. 23 shows the 'weights' of the α and β relaxation, $S(q)$ and $F_{\text{inel}}(q)$ (modeled as a two site jump with $d = 1.5$ Å), respectively. This explains the difference between Figs. 20 and 21. At a wave vector of 1.48 Å^{-1} the α relaxation is dominant while at $q = 1.88$ Å^{-1} the β relaxation is near equally important. Fig. 24 shows experiments which have been performed for many more wave vector and temperature values (Arbe et al., 1996). They can all be satisfactorily described by the model (26).

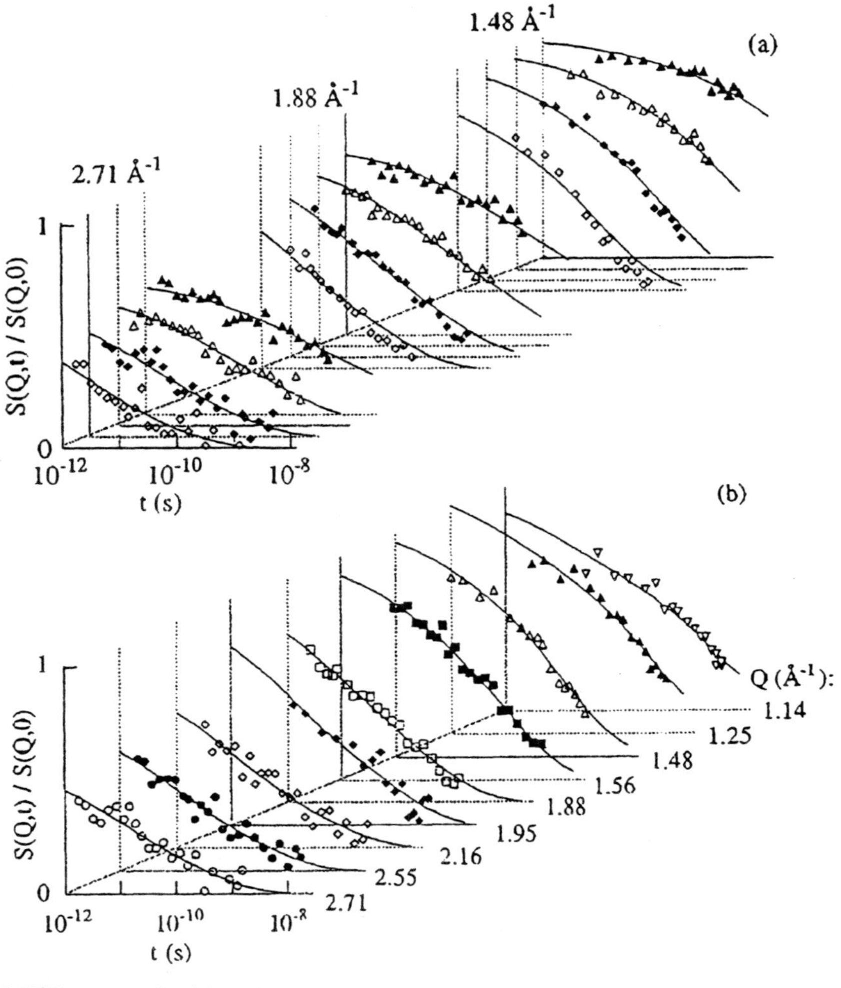

Fig. 24. (a) NSE spectra of polybutadiene for the q values indicated at 220 K (▲), 240 K (△), 260 K (♦), and 280 K (◇), (b) at 260 K for the q values indicated. Solid lines are fit curves with expression (26).

Appendix A. Rouse dynamic structure factor

The essential assumption of the Rouse model is that chains do not interfere with each other in their motion. This results in a Gaussian chain in a heat bath as the simplest model for chain relaxation. The building blocks of such a chain are N segments each consisting of sufficient monomers so that their end-to-end distance follows a Gaussian distribution with the average ℓ. We are interested in the motion of these segments on a length scale $\ell < r < R_E$ ($R_E = \sqrt{N}\ell$ is the total end-to-end distance of the polymer chain). This motion is described by a Langevin equation

$$\zeta \frac{\mathrm{d}\boldsymbol{r}_n}{\mathrm{d}t} = K\,(\boldsymbol{r}_{n+1} - 2\boldsymbol{r}_n + \boldsymbol{r}_{n-1}) + \boldsymbol{f}_n(t) \tag{27}$$

where ζ is the monomeric friction coefficient and $\boldsymbol{r}_n$ the position of the nth segment. K is a (spring) force constant and $\boldsymbol{f}_n(t)$ a stochastic force which will be determined in the following. The respective equations of motion for the first and last segment are

$$\zeta \frac{\mathrm{d}\boldsymbol{r}_1}{\mathrm{d}t} = K(\boldsymbol{r}_2 - \boldsymbol{r}_1) + \boldsymbol{f}_1(t), \qquad \zeta \frac{\mathrm{d}\boldsymbol{r}_N}{\mathrm{d}t} = K(\boldsymbol{r}_{N-1} - \boldsymbol{r}_N) + \boldsymbol{f}_N(t). \tag{28}$$

The probability of a conformation $\{\boldsymbol{r}_i\}$ is the product of the probabilities of the individual segment distances having the values $\boldsymbol{r}_{i+1} - \boldsymbol{r}_i$, in case of a Gaussian distribution with average ℓ:

$$W(\{\boldsymbol{r}_i\}) = \left(\frac{3}{2\pi\ell^2}\right)^{3N/2} \prod_{i=1}^{N-1} \exp\left(-\frac{3|\boldsymbol{r}_{i+1} - \boldsymbol{r}_i|^2}{2\ell^2}\right). \tag{29}$$

From this follows directly the entropy $S = -k_\mathrm{B} \ln W(\{\boldsymbol{r}_i\})$ and the free energy $F = E + k_\mathrm{B} T \ln W(\{\boldsymbol{r}_i\})$. Even under the assumption that there is no direct interaction between the segments ($E = 0$) there is still an entropic force between them which is given by the derivative of F with respect to a segments position $\boldsymbol{r}_n$. It is easy to carry out and results in the first right hand side term of (27) with $K = 3k_\mathrm{B}T/\ell^2$.

The stochastic force has the following properties:

$$\langle \boldsymbol{f}_n(t) \rangle = \boldsymbol{0} \tag{30}$$

$$\langle f_{m\alpha}(t) f_{n\beta}(t') \rangle = 2k_\mathrm{B} T \zeta \delta_{mn} \delta_{\alpha\beta} \delta(t - t'). \tag{31}$$

The first equation expresses the random character of the force, on average there is no net force. The second equation includes the uncorrelatedness of the force on segments m, n its vector components α, β and at times t, t'. The magnitude of $2k_\mathrm{B}T\zeta$ is related by the fluctuation dissipation theorem to the strength of the friction.

Similar to the solution of the equation of motion of an elastic string the Langevin equation (27) is solved by constructing the segment motions as a sum of normal modes $\boldsymbol{x}_p$:

$$\boldsymbol{r}_n = \sum_{p=0}^{N-1} \boldsymbol{x}_p \cos\left(\frac{p\pi\left(n - \frac{1}{2}\right)}{N}\right). \tag{32}$$

Choosing the cosine and the phase $-p\pi/2N$ has the effect that the boundary cases (28) reduce to the general case (27) by introducing fictitious segments $\boldsymbol{r}_0 = \boldsymbol{r}_1$ and $\boldsymbol{r}_{N+1} = \boldsymbol{r}_N$ [7]. Inserting (32) into the Langevin equation (27) yields

$$\zeta \sum_{p=0}^{N-1} \frac{\mathrm{d}\boldsymbol{x}_p}{\mathrm{d}t} \cos\left(\frac{p\pi\left(n - \frac{1}{2}\right)}{N}\right) = \frac{3k_\mathrm{B}T}{\ell^2} \sum_{p=0}^{N-1} \boldsymbol{x}_p \left(\cos\left(\frac{p\pi\left(n + \frac{1}{2}\right)}{N}\right) - 2\cos\left(\frac{p\pi\left(n - \frac{1}{2}\right)}{N}\right) + \cos\left(\frac{p\pi\left(n - \frac{3}{2}\right)}{N}\right)\right) + \boldsymbol{f}_n. \tag{33}$$

[7] This rather sophistical consideration of the boundary conditions loses its importance for large N. Therefore, the phase difference is often omitted in the literature and an approach chosen where n is a continuous variable instead of an integer.

Using the relation $\cos(x+y) - 2\cos x + \cos(x-y) = -2\cos x(1-\cos y)$ and introducing mode components of the stochastic force $\boldsymbol{f}_p$ such that

$$\boldsymbol{f}_n = \sum_{p=0}^{N-1} \boldsymbol{f}_p \cos\left(\frac{p\pi\left(n-\frac{1}{2}\right)}{N}\right) \tag{34}$$

we can simplify (33) to

$$\sum_{p=0}^{N-1}\left[\zeta\frac{\mathrm{d}\boldsymbol{x}_p}{\mathrm{d}t} + \frac{3k_\mathrm{B}T}{\ell^2}2\left(1-\cos\left(\frac{p\pi}{N}\right)\right)\boldsymbol{x}_p - \boldsymbol{f}_p\right]\cos\left(\frac{p\pi\left(n-\frac{1}{2}\right)}{N}\right) = \mathbf{0}. \tag{35}$$

Because of the orthogonality of the cosine functions this system of equations is only fulfilled if the term in square brackets vanishes for any p. This means that in terms of normal coordinates the equations of motions are decoupled and can be solved separately:

$$\zeta\frac{\mathrm{d}\boldsymbol{x}_p}{\mathrm{d}t} = -\frac{6k_\mathrm{B}T(1-\cos(p\pi/N))}{\ell^2}\boldsymbol{x}_p + \boldsymbol{f}_p. \tag{36}$$

To be consistent with Eqs. (30) and (31) the normal mode components of the stochastic force have to fulfill

$$\langle \boldsymbol{f}_p(t)\rangle = \mathbf{0} \tag{37}$$

$$\langle f_{p\alpha}(t) f_{q\beta}(t')\rangle = 2(2-\delta_{p0})k_\mathrm{B}T\zeta\delta_{pq}\delta_{\alpha\beta}\delta(t-t')/N. \tag{38}$$

We firstly consider the modes with $p \geq 1$: The formal solution of (36) is

$$\boldsymbol{x}_p(t) = \frac{1}{\zeta}\int_{-\infty}^{t} \exp\left(-\frac{t-t'}{\tau_p}\right)\boldsymbol{f}_p(t')\,\mathrm{d}t' \tag{39}$$

with

$$\tau_p = \frac{\zeta\ell^2}{6k_\mathrm{B}T(1-\cos(p\pi/N))} \approx \frac{\zeta N^2\ell^2}{3\pi^2 k_\mathrm{B}Tp^2} \quad \text{for } p \ll N. \tag{40}$$

Of course, this solution is undetermined because it still contains the random force $\boldsymbol{f}_p(t)$. Nevertheless, it is possible to calculate its correlators:

$$\begin{aligned}
\langle x_{p\alpha}(t)x_{q\beta}(t')\rangle &= \frac{1}{\zeta^2}\int_{-\infty}^{t}\mathrm{d}t_1\int_{-\infty}^{0}\mathrm{d}t_2\exp\left(-\frac{t-t_1-t_2}{\tau_p}\right)\langle f_{p\alpha}(t_1)f_{q\beta}(t_2)\rangle \\
&= \frac{1}{\zeta^2}\int_{-\infty}^{t}\mathrm{d}t_1\int_{-\infty}^{0}\mathrm{d}t_2\exp\left(-\frac{t-t_1-t_2}{\tau_p}\right)4\delta_{pq}\delta_{\alpha\beta}\zeta k_\mathrm{B}T\delta(t_1-t_2)/N \\
&= \frac{4\delta_{pq}\delta_{\alpha\beta}k_\mathrm{B}T}{N\zeta}\int_{-\infty}^{0}\exp\left(-\frac{t-2t_2}{\tau_p}\right)\mathrm{d}t_2 \\
&= \frac{2\delta_{pq}\delta_{\alpha\beta}k_\mathrm{B}T\tau_p}{N\zeta}\exp\left(-\frac{t}{\tau_p}\right)
\end{aligned} \tag{41}$$

On the other hand, for $p = 0$ there is no spring force term in (36) and one obtains:

$$\boldsymbol{x}_0(t) = \frac{1}{\zeta}\int_0^t \boldsymbol{f}_0(t')\,\mathrm{d}t' + \boldsymbol{x}_0(0) \tag{42}$$

and

$$\langle (x_{0\alpha}(t) - x_{0\alpha}(0))(x_{0\beta}(t) - x_{0\beta}(0))\rangle = \frac{2k_\mathrm{B}T}{N\zeta}\delta_{\alpha\beta}t. \tag{43}$$

The physical meaning of the mode zero is the motion of the center of mass:

$$\boldsymbol{r}_\mathrm{cm} = \frac{1}{N}\sum_{n=1}^{N}\sum_{p=0}^{N-1}\boldsymbol{x}_p \cos\left(\frac{p\pi\left(n-\frac{1}{2}\right)}{N}\right) = \boldsymbol{x}_0 \tag{44}$$

because the sum over n vanishes for $p \neq 0$. Therefore we can calculate the mean square displacement of the center of mass:

$$\langle \Delta r_\mathrm{cm}{}^2\rangle = \langle (\boldsymbol{r}_\mathrm{cm}(t) - \boldsymbol{r}_\mathrm{cm}(0))^2\rangle = 3\langle (x_{0\alpha}(t) - x_{0\alpha}(0))^2\rangle = \frac{6k_\mathrm{B}T}{N\zeta}t. \tag{45}$$

This is a simple diffusion law with the self-diffusion constant $D = k_\mathrm{B}T/N\zeta$ (Eq. (20)). The calculation of the viscosity is somewhat more complicated because the external shear force has to be added to the Langevin equation (27) but otherwise straightforward resulting in (19).

In order to calculate the incoherent inelastic scattering function the Gaussian approximation (9) is used. Then it is only necessary to calculate the time dependent mean square displacement of the average segment, i.e. the n average of $\langle \Delta r_n{}^2(t)\rangle = \langle (\boldsymbol{r}_n(t) - \boldsymbol{r}_n(0))^2\rangle$. Inserting (32) for $\boldsymbol{r}_n(t)$ one gets

$$\begin{aligned}\langle \Delta r_n{}^2(t)\rangle &= \sum_{p=0}^{N-1}\left\langle (\boldsymbol{x}_p(t) - \boldsymbol{x}_p(0))^2\right\rangle \cos^2\left(\frac{p\pi\left(n-\frac{1}{2}\right)}{N}\right)\\ &= \frac{6k_\mathrm{B}T}{N\zeta}t + \sum_{p=1}^{N-1}\frac{12k_\mathrm{B}T\tau_p}{N\zeta}\cos^2\left(\frac{p\pi\left(n-\frac{1}{2}\right)}{N}\right)\left(1 - \exp\left(-\frac{t}{\tau_p}\right)\right).\end{aligned} \tag{46}$$

Hitherto, the calculation of the Rouse model is exact. By inserting (46) into (9) and averaging over all segments $n = 1\ldots N$ an exact result for $S_\mathrm{inc}(q,t)$ could be obtained[8]. But in order to get a closed form expression now an approximation will be used: We assume that $t \ll \tau_\mathrm{R}$ where $\tau_\mathrm{R} \equiv \tau_1$ is the relaxation time of the fundamental mode $p = 1$, the so-called Rouse time. If this relation is fulfilled the terms of the sum in (46) decay slowly and the sum can be replaced by an integral. Furthermore, because the main contribution to the sum comes from higher p it is possible to replace $\cos^2(\ldots)$ by its average $1/2$. On the other hand the

[8] This is so because the Gaussian approximation is valid for an individual segment. The motion $\boldsymbol{r}_n(t)$ is described by a sum of integrals of random functions. Therefore its distribution is a Gaussian according to the central limit theorem. Nevertheless, for the whole chain it is approximative because the motion of the outer segments is faster than for the inner ones of the chain.

chain should be sufficiently long to ensure the validity of the approximation $\tau_p = \tau_R/p^2$ based on $p \ll N$ in (40):

$$\langle \Delta r_n{}^2(t) \rangle = \frac{4N\ell^2}{\pi^2} \int_0^\infty \frac{1}{p^2}\frac{1}{2}\left(1 - \exp\left(-\frac{tp^2}{\tau_R}\right)\right) \mathrm{d}p$$

$$= 2N\ell^2\sqrt{\frac{t}{\pi^3\tau_R}} = \sqrt{\frac{12\ell^2 k_B T}{\pi\zeta}t}. \tag{47}$$

It can be seen that – in contrast to ordinary diffusion as for the center of mass (45) – $\langle \Delta r_n{}^2(t) \rangle$ does not increase linearly with t but only with $t^{1/2}$. Inserting (47) into (9) gives the final result:

$$I_{\mathrm{inc}}(q,t) = \exp\left(-\frac{2}{\sqrt{\pi}}(\Omega_R(q)t)^{1/2}\right) \tag{48}$$

with the characteristic rate of the Rouse dynamics

$$\Omega_R(q) = \frac{k_B T \ell^2}{12\zeta} q^4. \tag{49}$$

We note that if we plot data of the incoherent intermediate scattering function versus the Rouse scaling variable

$$(\Omega_R(q)t)^{1/2} = \frac{q^2\ell^2}{6}\sqrt{Wt} \tag{50}$$

with $W \equiv 3k_B T/\ell^2\zeta$ we obtain the same exponential decay for all q values.

The calculation if the pair correlation, i.e. the coherent scattering function is much more complicated than the case before. Additional approximations are necessary which are only valid in the case $qR_E \gg 1$. The result of a lengthy but straightforward calculation (de Gennes, 1967; Doi and Edwards, 1986) is

$$I_{\mathrm{coh}}(q,t) = \frac{12}{q^2\ell^2} \int_0^\infty \mathrm{d}u \exp\left(-u - (\Omega_R(q)t)^{1/2} h\left((\Omega_R(q)t)^{-1/2}u\right)\right) \tag{51}$$

with

$$h(y) \equiv \frac{2}{\pi}\int_0^\infty \mathrm{d}x \cos(xy)\frac{1-\exp(-x^2)}{x^2}.$$

Despite its complicated form $I_{\mathrm{coh}}(q,t)$ depends (as in the incoherent case) only on the Rouse variable $(\Omega_R(q)t)^{1/2}$.

Appendix B. Inelastic form factor of two site jump model

Probably the simplest model for microscopic motions observed by IENS is the ‘two site jump model’. It describes a scattering particle which moves between two equivalent sites, e.g. in a double well potential. Although this model is in most cases too simple to describe real molecular motions it will be presented here to give an idea how inelastic form factors can be calculated in more complex situations.

Let us denote by $p(\boldsymbol{r}_1,t)$ and $p(\boldsymbol{r}_2,t)$ the probability of finding the particle at sites $\boldsymbol{r}_1$ and $\boldsymbol{r}_2$, respectively. Then $p(\boldsymbol{r}_1,t)$ and $p(\boldsymbol{r}_2,t)$ follow a set of rate equations

$$\frac{\mathrm{d}}{\mathrm{d}t}p(\boldsymbol{r}_1,t) = -\frac{1}{\tau}p(\boldsymbol{r}_1,t) + \frac{1}{\tau}p(\boldsymbol{r}_2,t) \tag{52}$$

$$\frac{\mathrm{d}}{\mathrm{d}t}p(\boldsymbol{r}_2,t) = -\frac{1}{\tau}p(\boldsymbol{r}_2,t) + \frac{1}{\tau}p(\boldsymbol{r}_1,t) \tag{53}$$

where $1/\tau$ is the jump rate from one site to the other, i.e. $\mathrm{d}t/\tau$ is the probability of a jump during a time interval of duration $\mathrm{d}t$. Assuming that the particle was initially at $\boldsymbol{r}_1$

$$p(\boldsymbol{r}_1,0) = 1 \tag{54}$$

$$p(\boldsymbol{r}_2,0) = 0 \tag{55}$$

we obtain the solution of the system of differential equations (52) and (53) by an exponential ansatz as

$$p(\boldsymbol{r}_1,t|\boldsymbol{r}_1,0) = \frac{1+\exp(-2t/\tau)}{2} \tag{56}$$

$$p(\boldsymbol{r}_2,t|\boldsymbol{r}_1,0) = \frac{1-\exp(-2t/\tau)}{2}. \tag{57}$$

Now the incoherent intermediate scattering function can be evaluated. By combining (17), (6), and (7) one gets

$$I_{\mathrm{inc}}(\boldsymbol{q},t) = \langle \exp(\mathrm{i}\boldsymbol{q}\cdot(\boldsymbol{r}(t)-\boldsymbol{r}(0)))\rangle \tag{58}$$

for a single particle. The initial probability to find the particle in one of the sites is of course

$$p(\boldsymbol{r}_1,0) = p(\boldsymbol{r}_2,0) = 1/2 \tag{59}$$

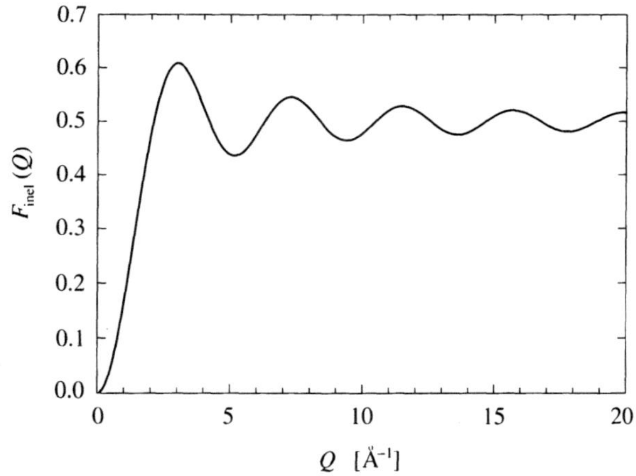

Fig. 25. Form factor of the two site jump model with $d = 1.5$ Å.

and its development in time is given by (56) and (57). With this information the average in expression (58) can be carried out resulting in

$$I_{\text{inc}}(\boldsymbol{q},t) = (1 - F_{\text{inel}}(\boldsymbol{q})) + F_{\text{inel}}(\boldsymbol{q})\exp(-2t/\tau) \tag{60}$$

where

$$F_{\text{inel}}(\boldsymbol{q}) = \frac{1 - \cos(\boldsymbol{q}\cdot(\boldsymbol{r}_2 - \boldsymbol{r}_1))}{2}. \tag{61}$$

In disordered samples or for powder specimens the orientation of $\boldsymbol{r}_2 - \boldsymbol{r}_1$ is arbitrary, only the distance is constant $|\boldsymbol{r}_2 - \boldsymbol{r}_1| \equiv d$. Therefore, an average over all orientations has to be calculated:

$$\begin{aligned} F_{\text{inel}}(q,d) &= \frac{1}{4\pi}\int_0^{2\pi} \mathrm{d}\phi \int_{-\pi}^{\pi} \sin\theta \; \mathrm{d}\theta \, F_{\text{inel}}(\boldsymbol{q}) \\ &= \frac{1}{2}\left(1 - \frac{\sin qd}{qd}\right). \end{aligned} \tag{62}$$

Fig. 25 shows the resulting function for $d = 1.5$ Å.

References

Arbe, A., Richter, D., Colmenero, J. and Farago, B. (1996). *Phys. Rev. E*, **54**, 3853–3869.

Bée, M. (1988). *Quasielastic Neutron Scattering*. Adam Hilger, Bristol.

Buchenau, U., Schönfeld, C., Richter, D., Kanaya, T., Kaji, K. and Wehrmann, R. (1994). *Phys. Rev. Lett.*, **73**, 2344–2347.

Buchenau, U., Pecharroman, C., Zorn, R. and Frick, B. (1996). *Phys. Rev. Lett.*, **77**, 659–663.

Carella, J., Graessley, W. and Fetters, L. (1984). *Macromolecules*, **17**, 2775–2786.

de Gennes, P.G. (1967). *Physics* **3**, 37.

de Gennes, P.G. (1971). *J. Chem. Phys.*, **55**, 572–579.

de Gennes, P.G. (1981). *J. Phys. (France)*, **42**, 735–740.

des Cloizeaux, J. (1993). *J. Phys. I (France)*, **3**, 1523–1539.

Doi, M. and Edwards, S.F. (1986). *The Theory of Polymer Dynamics*. Clarendon Press, Oxford.

Edwards, S.F. (1967). *Proc. Phys. Soc.*, **92**, 9–16.

Ferry, J.D. (1980). *Viscoelastic Properties of Polymers*. John Wiley and Sons, New York.

Hess, W. (1988). *Macromolecules*, **21**, 2620–2632.

Hofmann, A., Alegría, A., Colmenero, J., Willner, L., Buscaglia, E. and Hadjichristidis, N. (1996). *Macromolecules*, **29**, 129–134.

Inoue, K., Kanaya, T., Ikeda, S., Kaji, K., Shibata, K., Misawa, M. and Kiyanagi, Y. (1991). *J. Chem. Phys.* **95**, 5332–5340: "Low-energy excitations in amorphous polymers".

Johari, G.P. and Goldstein, M. (1970). *J. Chem. Phys*, **53**, 2372–2388.

Kremer, K. and Grest, G.S. (1990). *J. Chem. Phys.*, **92**, 5057–5086.

Lovesey, S.W. (1984). *Theory of Neutron Scattering from Condensed Matter*. Clarendon Press, Oxford.

Mezei, F., ed. (1980). *Neutron Spin Echo*. Springer, Heidelberg.

Nemoto, N., Moriwaki, M., Odani, H. and Kurata, M. (1971). *Macromolecules*, **4**, 215–219.

Pearson, D.S., VerStrate, G., van Merwall, E. and Schilling, F.C. (1987). *Macromolecules*, **20**, 1133–1141.

Richter, D., Frick, B. and Farago, B. (1988). *Phys. Rev. Lett.*, **61**, 2465–2468.

Richter, D., Ewen, B., Farago, B. and Wagner, T. (1989). *Phys. Rev. Lett.*, **62**, 2140–2143.

Richter, D., Farago, B., Fetters, L.J., Huang, J.S., Ewen, B. and Lartigue, C. (1990). *Phys. Rev. Lett.*, **64**, 1389–1392.

Richter, D., Zorn, R., Farago, B., Frick, B. and Fetters, L.J. (1992). *Phys. Rev. Lett.*, **68**, 71–74.
Ronca, P.F. (1983). *J. Chem. Phys.*, **79**, 1031–1044.
Rouse, P.E. (1953). *J. Chem. Phys.*, **21**, 1272–1280.
Schleger, P., Farago, B. and Lartigue, C. (1998). *Phys. Rev. Lett.*, **81**, 124–127.
Squires, G.L. (1978). *Introduction to the Theory of Thermal Neutron Scattering*. Cambridge University Press. Cambridge.
Zorn, R., Arbe, A., Colmenero, J., Frick, B., Richter, D. and Buchenau, U. (1995). *Phys. Rev. E*, **52**, 781–795.
Zorn, R., McKenna, G.B., Willner, L. and Richter, D. (1995). *Macromolecules*, **28**, 8552–8562.
Zorn, R., Kanaya, T., Kawaguchi, T., Richter, D. and Kaji, K. (1996). *J. Chem. Phys.*, **105**, 1189–1197.

Part III
Revealing microstructures of soft condensed matter

Neutrons, X-Rays and Light
P. Lindner and Th. Zemb (Editors)

Chapter 11

Static Properties of Polymers

Peter Schurtenberger

Department of Physics, University of Fribourg, CH-1700 Fribourg, Switzerland

1. Introduction

Polymers are macromolecules formed by a large number of low molecular weight building blocks, the so-called monomers. Well known examples are polyethylene (PE) consisting of N repeating groups $[-CH_2-CH_2-]_N$, polystyrene (PS) $[-CH_2-CH(C_6H_5)-]_N$, etc. The corresponding chemical structures of PE and PS are shown in Fig. 1. Polystyrene and polyethylene are so-called homopolymers, i.e. polymers made by identical monomers. Polymers that consist of different monomers are copolymers, and one uses the term statistical copolymer for chains with irregular composition and block-copolymers for polymers where the different monomers are grouped in larger blocks. Polymers can exhibit a variety of different structures such as linear and branched chains or star polymers, to name only a few. In this chapter we shall concentrate exclusively on the simplest form of a linear chain. A variety of excellent text books and reviews on all aspects of polymers exist (see for example Flory, 1953, 1969; Yamakawa, 1971; de Gennes, 1979; des Cloizeaux and Jannink, 1990; Fujita, 1990; Doi, 1995; Higgins and Benoit, 1996; Strobl, 1996). Here we shall concentrate on a few fundamental aspects of polymers in solution and the melt, and we focus on static properties and how they can be investigated with scattering experiments.

Polymers are very interesting and complex systems that are being considered in numerous areas. Modern materials sciences use the unique dependence of the properties of polymeric materials on the molar mass, composition, number of cross-links or other adjustable parameters. The number of monomers (degree of polymerization) can be very large, for example, for PE it can reach $N > 10^5$. This has enormous consequences for example for the melting temperature, the strength of a polymeric material or the viscosity of a melt. Important applications of polymer science can also be found for example in medicine, food sciences or

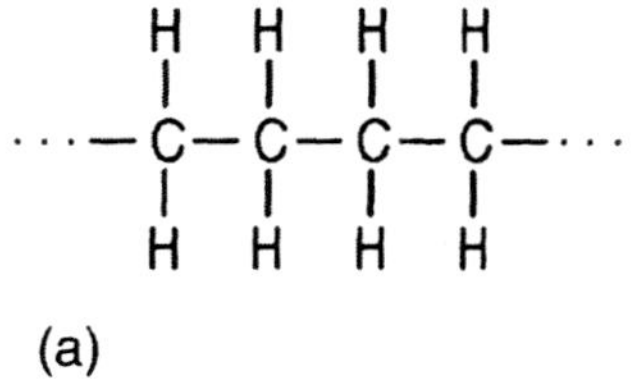

(a)

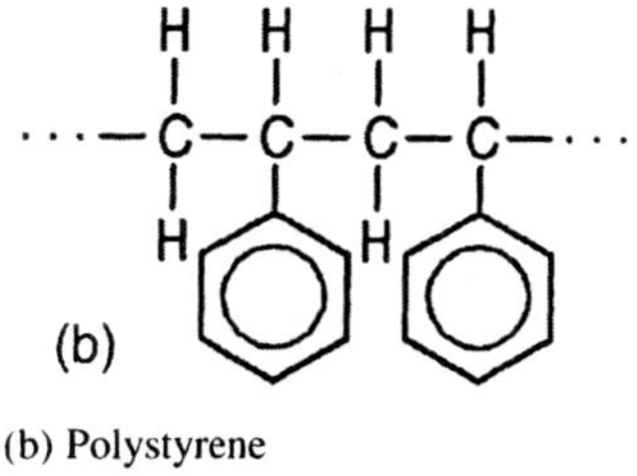

(b)

Fig. 1. (a) Polyethylene. (b) Polystyrene

pharmacology. However, polymers have not only been looked at because of their importance in 'real life', but they have attracted the attention of experimental and theoretical physicists due to the fascinating properties of polymeric fluids. The physics of polymers has been strongly influenced by the important contributions of scientists such as Staudinger (Nobel prize 1953), Debye, Kuhn and Flory (Nobel prize 1974), and more recently by Edwards and de Gennes (Nobel prize 1991). In particular the work of Pierre Gilles de Gennes, who pointed out analogies to magnetic systems and field theories, has led to a renaissance of polymer physics (de Gennes, 1979). He was instrumental in developing the important concept of universal scaling laws for the description of various properties of polymers, which we shall use in this chapter on the static properties of polymers.

An important part in the theoretical treatment of polymers is the interplay between local and global parameters, between static and dynamic properties, and between equilibrium and non-equilibrium phenomena.

static $\longleftrightarrow$ dynamic

local
$\updownarrow$ **polymer properties**
global

equilibrium $\longleftrightarrow$ non-equilibrium

The meaning of the terms *local* and *global* properties is best illustrated with Fig. 2, where two degrees of magnification of a polymer chain (PE) are shown. If we use a hypothetical magnifying glass at very high magnification, where we can resolve atomistic properties of the polymer chain, we observe chemical details such as the sequence and structure of the monomers along the chain. It is clear that these details will depend on the specific chemical nature of the chain. However, if we decrease the resolution of our magnifier, these details are lost and we see the entire polymer. At this low magnification the polymer now looks like a coiled-up line, and we can describe it using global quantities such as the end-to-end distance $\overline{R}_{\mathrm{E}}$, the (root mean-square) radius of gyration $\overline{R}_{\mathrm{G}}$ and the hydrodynamic radius $\overline{R}_{\mathrm{H}}$. On this level it is not obvious any longer whether the polymer coil that we observe is a polyethylene or a polystyrene chain. Therefore we also expect that linear macromolecules possess similar global properties independent of their molecular construction. This is for example demonstrated with the dependence of the radius of gyration on the number of monomers N of the polymer for large values of N,

$$\overline{R}_{\mathrm{G}} \approx a N^{\nu}. \tag{1}$$

The exponent ν is a universal (material independent) quantity, whereas a is a constant that depends on the local chemistry. This proportionality between $\overline{R}_{\mathrm{G}}$ and the chain length (or molar mass) is a simple example of a so-called universal scaling law (de Gennes, 1979). An important feature of scaling laws is their self similarity: At characteristic distances larger than the molecular details of the polymer it is impossible to say whether we see the entire polymer chain or only a magnified part of it because there exists no natural length scale under these conditions. Therefore the same relationship between $\overline{R}_{\mathrm{G}}$ and N exists also if we regroup the monomers into pairs and relate $\overline{R}_{\mathrm{G}}$ to the number of monomer pairs $N/2$. This property of scaling laws is the essential ingredient for the use of a so-called coarse graining model. A typical coarse graining procedure would be to regroup a number of consecutive monomers

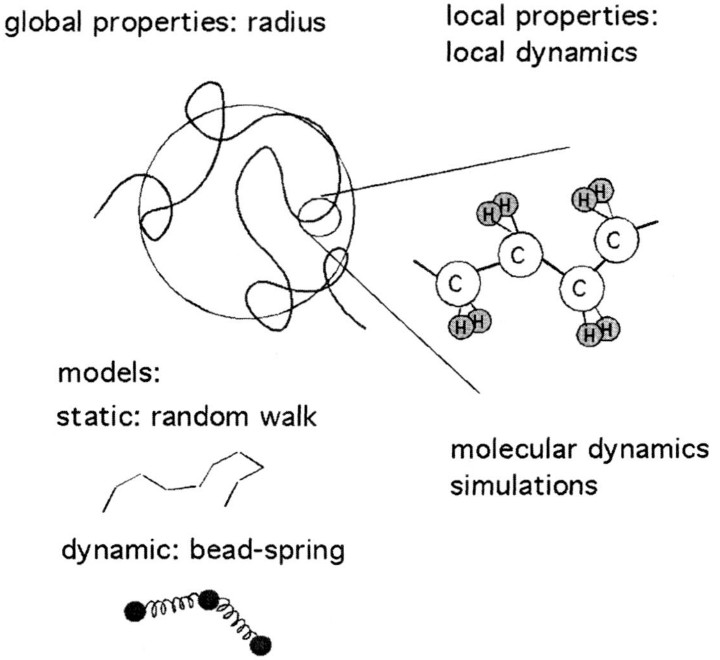

Fig. 2. Example of coarse-graining of a polymer chain (PE). Right: high degree of magnification up to the chemical structure. Left: overall structure of the same chain at a larger length scale. Also shown are the simplest theoretical models used for a treatment of static and dynamic properties of polymer solutions.

into a new subunit until we have incorporated all chemistry-specific microscopic details and recover the universal properties described by simple models such as for example a random walk.

An interesting and complicated regime can be found at intermediate length scales, where due to the influence of the chemical bonds between the molecular building blocks the polymer appears to be much stiffer. Here we have to introduce the so-called persistence of the chain as a means of incorporating molecular details. We shall see that this feature of polymers, which does have important implications for technological applications of polymers as high strength materials can in fact be modeled with a coarse graining approach, where a number of monomers is regrouped to a larger building block of the size of the persistence length, and the properties of the polymer are then looked at as a function of the number of these larger segments.

q range	Length scales	Static properties
$S(\boldsymbol{q})$, $q\overline{R}_G \ll 1$	**global**	(overall size, conformation)
$S(\boldsymbol{q})$, $q\overline{R}_G \gg 1$	**local**	(stiffness, cross-section)

In the remaining parts of this chapter we shall introduce several simple models that will enable us to understand various features of polymeric liquids. Particular emphasis will always be given to scattering experiments, and we shall show that we can use them as our magnifying glass in order to investigate polymers on various length scales. The magnification of the scattering experiment is adjusted by selecting the appropriate magnitude of the scattering vector q, and we shall use the static structure factor $S(\boldsymbol{q})$ as the main link between theory and experiment.

2. Conformation of a single coil in solution

2.1. Polymers as a random walk

The random walk is the simplest possible model that provides us with the description of the conformation of a polymer chain in solution. In the random walk model, we interpret the line that we generate by performing a random walk of step length a and number of steps $N-1$ as an individual conformation of a polymer with monomer size a and degree of polymerization N (see Fig. 3). Every new random walk then corresponds to an additional possible conformation of the same polymer chain, and various physical properties of the polymer can then be evaluated by looking at the distribution of different chain conformations. In such a random movement all the individual steps occur completely independent of each other and point in arbitrary directions. This has an important consequence for the resulting polymer chain: its monomers are completely uncorrelated. Fig. 4 shows examples of RWs with different numbers of steps. With increasing N, the walk looks qualitatively similar to a continuous polymer chain. The lattice constant defines the natural length scale similar to the monomer size for real polymers. If we go to a large number of steps, these details become unimportant and one loses the knowledge on the local details. When using such a model, the first question we ask ourselves is whether it produces meaningful results that can be compared to experimental data. In order to do so we first make the necessary definitions of our coordinate system. The lattice site j is given by the position vector r_j with $j = 1, \ldots, N$, and the connectivity vector between consecutive 'monomers' or lattice sites is given by

$$\boldsymbol{Q}_j := \boldsymbol{r}_{j+1} - \boldsymbol{r}_j, \qquad \text{with } j = 1, \ldots, N-1. \tag{2}$$

A line that connects N lattice points is thus described by $(N-1)$ connectivity vectors. Now we can try to define and calculate various quantities that characterize the chain conformation of a random walk polymer.

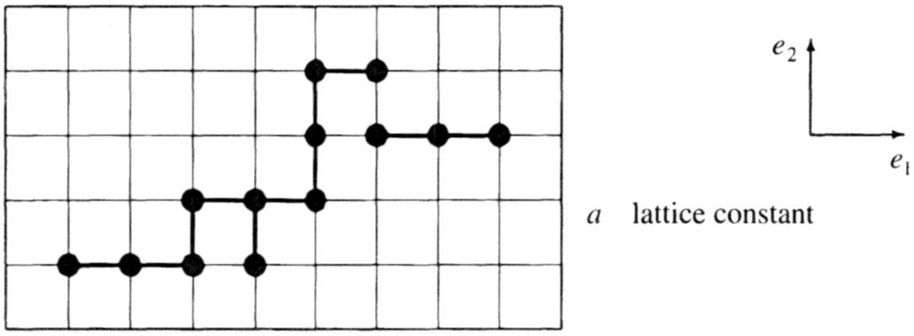

Fig. 3. Example of a random walk on a square lattice. Starting from an occupied lattice site, one chooses one out of the four neighbours with equal probability for the next step, i.e., the individual steps with step length a are completely independent of each other.

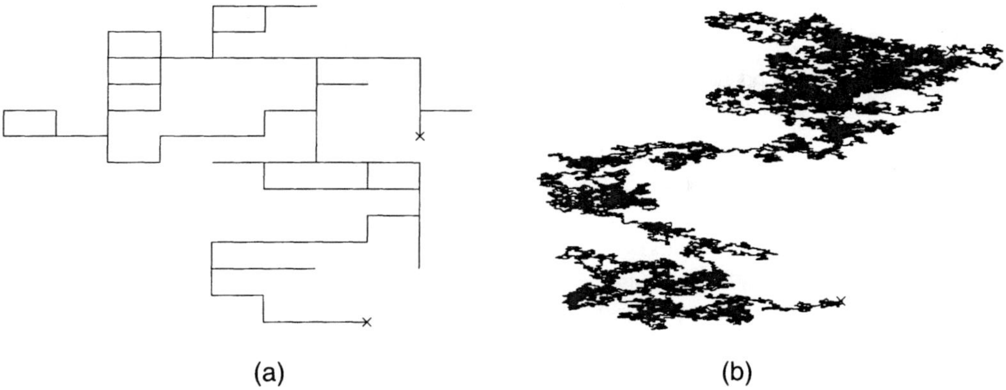

Fig. 4. Example of a linear polymer molecule generated by a random walk of N steps on a square lattice. Left: $N = 100$ steps. Right: $N = 10^5$ steps.

2.1.1. Characteristic distances and dimensions

The end-to-end vector: The end-to-end vector of a polymer conformation is defined as follows:

$$\boldsymbol{R}_{\mathrm{E}} := \boldsymbol{r}_N - \boldsymbol{r}_1 = \sum_{j=1}^{N-1} \boldsymbol{Q}_j. \tag{3}$$

The mean-square end-to-end distance then follows directly from (3), if we form the scalar product of the end-to-end vector with itself and average over all conformations:

$$\langle \boldsymbol{R}_{\mathrm{E}}^2 \rangle = \langle (\boldsymbol{r}_N - \boldsymbol{r}_1)\cdot(\boldsymbol{r}_N - \boldsymbol{r}_1) \rangle = \left\langle \sum_{j,k=1}^{N-1} \boldsymbol{Q}_j \cdot \boldsymbol{Q}_k \right\rangle = \sum_{j,k=1}^{N-1} \langle \boldsymbol{Q}_j \cdot \boldsymbol{Q}_k \rangle. \tag{4}$$

Here we have used $\langle\rangle$ in order to characterize a conformational average. Based on (4) we define the average end-to-end distance as:

$$\overline{R}_{\mathrm{E}} := \sqrt{\langle \boldsymbol{R}_{\mathrm{E}}^2 \rangle}. \tag{5}$$

We take the root mean-square distance because the average of (3) is zero due to the symmetry of the random walk: $\langle \boldsymbol{R}_{\mathrm{E}} \rangle = 0$. In the case of independent steps we have $\langle \boldsymbol{Q}_j \cdot \boldsymbol{Q}_k \rangle = \langle \boldsymbol{Q}_j \rangle \cdot \langle \boldsymbol{Q}_k \rangle$ for $j \neq k$. Therefore the scalar product of the connectivity vectors is different from zero only for $j = k$. From the definition of $\boldsymbol{Q}_j$ (2) we get:

$$\langle \boldsymbol{R}_{\mathrm{E}}^2 \rangle = \sum_{j,k=1}^{N-1} \langle \boldsymbol{Q}_j \cdot \boldsymbol{Q}_k \rangle = \sum_{j=1}^{N-1} \langle \boldsymbol{Q}_j^2 \rangle = (N-1)a^2. \tag{6}$$

In the random walk model the end-to-end distance (5) for a sufficiently large number of steps, where $(N-1) \approx N$, is thus given by:

$$\overline{R}_{\mathrm{E}} \approx a\sqrt{N} \sim N^{\nu} \quad \text{with} \quad \nu = \tfrac{1}{2}. \tag{7}$$

i.e., the scaling exponent ν for $\overline{R}_{\mathrm{E}}$ in the random walk model has a value of 0.5.

The inner distances: Similarly we can also look at the inner distances between points j and k along the polymer chain, and we directly obtain:

$$\langle (\boldsymbol{r}_j - \boldsymbol{r}_k) \cdot (\boldsymbol{r}_j - \boldsymbol{r}_k) \rangle = \langle \boldsymbol{r}_{jk} \cdot \boldsymbol{r}_{jk} \rangle = |j-k|\, a^2. \tag{8}$$

The mean-square distances are independent of the chain length and the position along the chain and follow the same scaling law as $\overline{R}_{\mathrm{E}}$.

The radius of gyration: The radius of gyration is a measure of the mass distribution of an object around its center of gravity. It is an important quantity that can be determined experimentally from scattering experiments and thus allows for a direct test of our random walk model. We start with a definition of the mean-square radius of gyration:

$$\langle \boldsymbol{R}_{\mathrm{G}}^2 \rangle := \frac{1}{N} \sum_{j=1}^{N} \left\langle (\boldsymbol{r}_j - \boldsymbol{r}_{\mathrm{CM}})^2 \right\rangle, \tag{9}$$

where $\boldsymbol{r}_{\mathrm{CM}} := \frac{1}{N} \sum_{j=1}^{N} \boldsymbol{r}_j$ is the center of gravity of the polymer. Using this relation together with Eq. (8), the mean-square radius of gyration can be rewritten as

$$\begin{aligned} \langle \boldsymbol{R}_{\mathrm{G}}^2 \rangle &= \frac{1}{2N^2} \sum_{j,k=1}^{N} \left\langle (\boldsymbol{r}_j - \boldsymbol{r}_k)^2 \right\rangle = \frac{1}{2N^2} a^2 \sum_{j,k=1}^{N} |j-k| = \frac{1}{N^2} a^2 \sum_{\substack{j,k=1 \\ j>k}}^{N} (j-k) \\ &= \frac{a^2}{N^2} \sum_{j=1}^{N} \frac{j(j-1)}{2} = \frac{a^2}{2N^2} \frac{N^3 - N}{3} = \frac{N^2-1}{6N} a^2. \end{aligned} \tag{10}$$

For a sufficiently large number of steps, where $(N+1) \approx N$, we thus find:

$$\langle \boldsymbol{R}_{\mathrm{G}}^2 \rangle = \tfrac{1}{6} \langle \boldsymbol{R}_{\mathrm{E}}^2 \rangle, \tag{11}$$

With the definition of the average radius of gyration

$$\overline{R}_{\mathrm{G}} := \sqrt{\langle \boldsymbol{R}_{\mathrm{G}}^2 \rangle}. \tag{12}$$

we see that

$$\overline{R}_{\mathrm{G}} \approx \sqrt{\tfrac{1}{6}}\, \overline{R}_{\mathrm{E}} \sim N^{\nu} \quad \text{with} \quad \nu = \tfrac{1}{2}, \tag{13}$$

i.e., that it follows the same scaling law (Eq. (7)) as the average end-to-end vector.

Distribution of inner distances: The random walk model not only provides us with the mean-square end-to-end distance, but it also allows for an estimate of the distribution of end-to-end or inner distance vectors. We can estimate the probability density $p(\boldsymbol{R}_{\mathrm{E}})\, \mathrm{d}^3 R_{\mathrm{E}}$ of finding an end-to-end vector in a volume element $\mathrm{d}^3 R_{\mathrm{E}}$ around $\boldsymbol{R}_{\mathrm{E}}$. Based on the central limit theorem we know that the sum of a large number of independent random variables (such as the steps in a random walk) should obey Gaussian statistics, i.e., we find in $d = 3$

$$p(\boldsymbol{R}_{\mathrm{E}}) = \frac{1}{\sqrt{(2\pi\theta)^3}} \exp\left(-\frac{R_{\mathrm{E}}^2}{2\theta} \right), \tag{14}$$

where $\theta = (N-1)a^2/3$. The same Gaussian statistics should also be found for the inner distances $p(\boldsymbol{r}_{jk})$ with $\boldsymbol{r}_{jk} := \boldsymbol{r}_j - \boldsymbol{r}_k$ and $|j-k| \gg 1$, as we can replace $(N-1)$ by $|j-k|$.

2.1.2. A first comparison with experimental data

We are now in a position to critically test the predictions of our simple random walk model. In this test we rely on:

- the fact that we have a quantitative prediction for the dependence of $\overline{R}_G$ on the molar mass M. This prediction should be 'chemistry-independent'.
- based on the so-called Guinier approximation ($S(\boldsymbol{q}) \simeq S(0)[1 - \frac{1}{3}q^2\overline{R}_G^2]$, see introductory chapter of P. Pusey for details) $\overline{R}_G$ can be obtained from scattering experiments (light, small-angle neutron or X-ray) without any model assumptions.

This first test of the predictions of the random walk model is summarized in Figs. 5 and 6. As our experimental model system we have chosen polystyrene in two different solvents:

- PS in cyclohexane at $T = 34.5°C$ (a so-called theta solvent) and
- PS in toluene at $T = 25°C$ (a so-called good solvent).

More details about the concept of a solvent quality (good, theta and poor solvents) will be given later. We can summarize our findings as follows:

- For PS in cyclohexane we find a very good agreement between the scaling exponent $\nu = \frac{1}{2}$ predicted by the random walk model and the experimental data for $M_w > 2 \times 10^4$.
- For PS in toluene we find a significant deviation for $M_w > 3 \times 10^5$.
- In both cases (cyclohexane, toluene) we observe a significant deviation for $M_w < 10^4$.

It is quite obvious that our simple model cannot describe the experimental data ($\overline{R}_G$ versus M_w) over the entire range of molar masses M_w investigated. There exists a need for a more refined model. Moreover, it is also clear that a characterization of the conformation beyond a

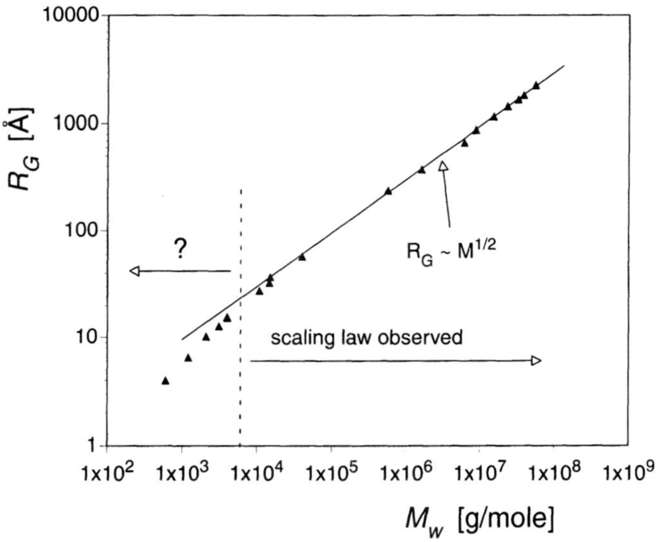

Fig. 5. Radius of gyration as a function of the molar mass for polystyrene in cyclohexane at $T = 34.5°C$ (Theta-solvent). We find good agreement between the predictions of the random walk model and the experimental data at high enough values of M_w. The slope of the straight line is given by: $\nu = \frac{1}{2}$. The data are taken from Kirste and Wild (1969), Ballard et al. (1976), Miyaki et al. (1978), Huber et al. (1985).

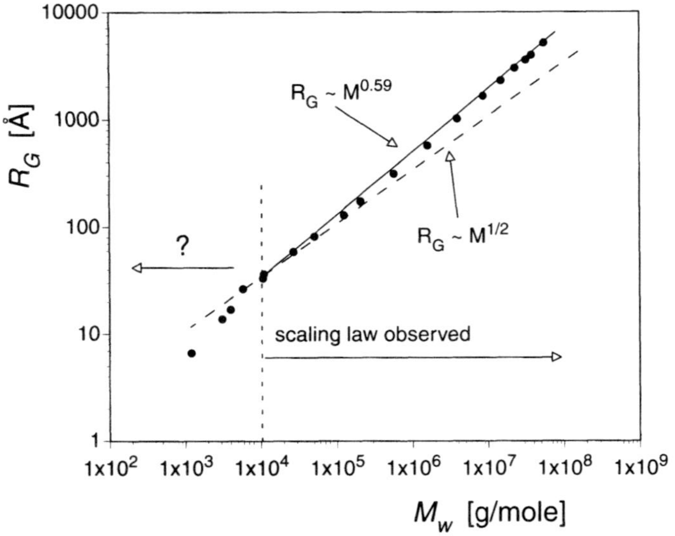

Fig. 6. Radius of gyration as a function of the molar mass for polystyrene in toluene (good solvent). The slope of the straight line is given by: $\nu = 0.59$, which can be compared with the prediction from the random walk model. The data are taken from Kirste and Wild (1969), Miyaki et al. (1978), Huber et al. (1985).

simple measure of the global coil dimensions would be highly desirable. We therefore proceed to the determination of the static structure factor $S(\boldsymbol{q})$ as the link between the theoretical description and the experimental determination of the polymer conformation at all length scales.

2.1.3. The static structure factor of a Gaussian coil

The scattering intensity of a single chain of N monomers, where each monomer is seen as a point scatterer with identical scattering length b, is given by

$$\frac{\mathrm{d}\sigma}{\mathrm{d}\Omega}(\boldsymbol{q}) = b^2 \left\langle \sum_{j=1}^{N} \sum_{k=1}^{N} \mathrm{e}^{\mathrm{i}\boldsymbol{q}\cdot\boldsymbol{r}_{jk}} \right\rangle. \tag{15}$$

If we now define the *static structure factor* $S(\boldsymbol{q})$ of the polymer chain (Higgins and Benoit, 1996)

$$S(\boldsymbol{q}) := \frac{1}{N} \left\langle \sum_{j=1}^{N} \sum_{k=1}^{N} \mathrm{e}^{\mathrm{i}\boldsymbol{q}\cdot\boldsymbol{r}_{jk}} \right\rangle, \tag{16}$$

a comparison with the scattering intensity of an ideal gas of N identical particles

$$\frac{\mathrm{d}\sigma}{\mathrm{d}\Omega}_{\mathrm{iG}} = Nb^2 \tag{17}$$

immediately reveals that

$$\frac{\frac{\mathrm{d}\sigma}{\mathrm{d}\Omega}(\boldsymbol{q})}{\frac{\mathrm{d}\sigma}{\mathrm{d}\Omega}_{\mathrm{iG}}} = S(\boldsymbol{q}). \tag{18}$$

We thus see that $S(\boldsymbol{q})$ contains important information about the structure of our system: relative positions, correlations between positions, non-ideality. The structure factor can be used for single chains ($\rightarrow$ conformation) as well as for many chain systems. Its main importance arises because of the fact that $S(\boldsymbol{q})$ is a directly measurable quantity that provides a link between theory and scattering experiment. The static structure factor can thus be used as a test quantity for the agreement between theory and experiment on all length scales, where the characteristic length scale is adjusted through the choice of $|\boldsymbol{q}|$. The next question to ask is then the relation between the structure factor as the measurable quantity in reciprocal space and the chain conformation as the quantity of interest in real space.

A first and more qualitative estimate can be made from the relation between $S(\boldsymbol{q})$ and the two-particle correlation function $g(\boldsymbol{r})$ given by

$$g(\boldsymbol{r}) = \frac{1}{N}\left\langle \sum_{j=1}^{N}\sum_{k=1}^{N} \delta(\boldsymbol{r} - \boldsymbol{r}_{jk}) \right\rangle. \tag{19}$$

$g(\boldsymbol{r})$ describes the probability of finding two particles at a directed distance $\boldsymbol{r}$, i.e., $g(\boldsymbol{r})$ is the average particle number density averaged over all particles with the normalization condition

$$\int g(\boldsymbol{r})\, \mathrm{d}^3r = N. \tag{20}$$

With the definition of the static structure factor (16) we find:

$$S(\boldsymbol{q}) = \frac{1}{N}\sum_{j,k=1}^{N} \left\langle \mathrm{e}^{\mathrm{i}\boldsymbol{q}\cdot\boldsymbol{r}_{jk}} \right\rangle \tag{21}$$

$$= \frac{1}{N}\int \mathrm{d}^3r \sum_{j,k=1}^{N} \left\langle \mathrm{e}^{\mathrm{i}\boldsymbol{q}\cdot\boldsymbol{r}}\delta(\boldsymbol{r} - \boldsymbol{r}_{jk}) \right\rangle \tag{22}$$

$$= \frac{1}{N}\int \mathrm{d}^3r\, \mathrm{e}^{\mathrm{i}\boldsymbol{q}\cdot\boldsymbol{r}} \sum_{j,k=1}^{N} \left\langle \delta(\boldsymbol{r} - \boldsymbol{r}_{jk}) \right\rangle \tag{23}$$

$$= \int \mathrm{d}^3r\, \mathrm{e}^{\mathrm{i}\boldsymbol{q}\cdot\boldsymbol{r}} g(\boldsymbol{r}), \tag{24}$$

i.e., $S(\boldsymbol{q})$ is the Fourier transform of the two-particle correlation function $g(\boldsymbol{r})$ [see also Higgins and Benoit (1996) and Chapter 14 by R. Klein for more details].

What is $g(r)$ for our polymer coil? We obtain a first estimate by drawing a sphere of radius r around each monomer and asking for the number N of monomers contained in this volume. With the relation

$$\langle \boldsymbol{r}^2 \rangle \sim Na^2 \tag{25}$$

obtained for the random walk model and the fact that $g(r)$ corresponds to the monomer density at a distance r, we immediately obtain the following estimate:

$$g(r) \sim \frac{N}{r^3} \sim \frac{1}{a^2 r}. \tag{26}$$

Since $4\pi/q^2$ is the Fourier transform of $1/r$, we thus obtain for the normalized scattering intensity:

$$\frac{\mathrm{d}\sigma}{\mathrm{d}\Omega} \sim \frac{1}{q^2}. \tag{27}$$

It is obvious that Eq. (27) describes only the asymptotic q-dependence of the scattering intensity in the regime $\overline{R}_{\mathrm{G}} \gg 1/q \gg a$. However, in the next paragraph we shall see that the random walk model indeed predicts a q^{-2}-dependence of the scattering intensity.

We can in fact calculate the static structure factor (16) for a random walk chain based on the explicit use of the Gaussian distribution of the inner distance vectors:

$$S(\boldsymbol{q}) = \frac{1}{N} \sum_{j,k=1}^{N} \left\langle \mathrm{e}^{\mathrm{i}\boldsymbol{q}\cdot\boldsymbol{r}_{jk}} \right\rangle = \frac{1}{N} \sum_{j,k=1}^{N} \mathrm{e}^{-\frac{1}{6}q^2 a^2 |j-k|}. \tag{28}$$

For long polymer chains $N \to \infty$ and in the limit $q^2 a^2 \to 0$, we then find

$$S(\boldsymbol{q}) = N \frac{2}{x^2} \left(\mathrm{e}^{-x} - 1 + x\right), \tag{29}$$

where $x = q^2 \langle \boldsymbol{R}_{\mathrm{G}}^2 \rangle$. With equation (29) we thus have a scaling law of the form

$$S(\boldsymbol{q}) = N f_{\mathrm{D}} \left(q^2 \langle \boldsymbol{R}_{\mathrm{G}}^2 \rangle\right) \tag{30}$$

where

$$f_{\mathrm{D}}(x) = \frac{2}{x^2} \left(\mathrm{e}^{-x} - 1 + x\right) \tag{31}$$

is the so-called Debye function (Debye, 1947). This scaling law is no longer a function of three variables a, N, $\boldsymbol{q}$, but depends only on the dimensionless quantity $q^2 \langle \boldsymbol{R}_{\mathrm{G}}^2 \rangle$. It is obvious that the Debye function as the exact form of the static structure factor for a random walk polymer exhibits the same asymptotic q^{-2}-dependence for large values of $q^2 \langle \boldsymbol{R}_{\mathrm{G}}^2 \rangle$, i.e., for the regime $\overline{R}_{\mathrm{G}} \gg 1/q \gg a$, that we previously obtained through a simple estimate of $g(r)$.

A comparison between the Debye function and the experimental data for polystyrene in cyclohexane at a temperature of 34.5°C obtained from light scattering is shown in Fig. 7. The data is taken from Miyaki et al. (1978), and the PS sample has a high molar mass of $M_w = 5.5 \times 10^7$ g/mol in order to ensure that the q-range covered in a light scattering experiment is sufficient to reach the asymptotic limit for $f_{\mathrm{D}} \left(q^2 \langle \boldsymbol{R}_{\mathrm{G}}^2 \rangle\right)$. We observe very good agreement between the theoretical curve derived from the random walk model and the experimental data, which indicates that this simple polymer model is capable of reproducing the polymer conformation on a large range of length scales for high enough molar masses.

However, the situation is once again very different if we change the polymer solvent and perform experiments under so-called good solvent conditions. This is illustrated in Fig. 8, where data from a small-angle neutron scattering (SANS) experiment with polystyrene in CS_2

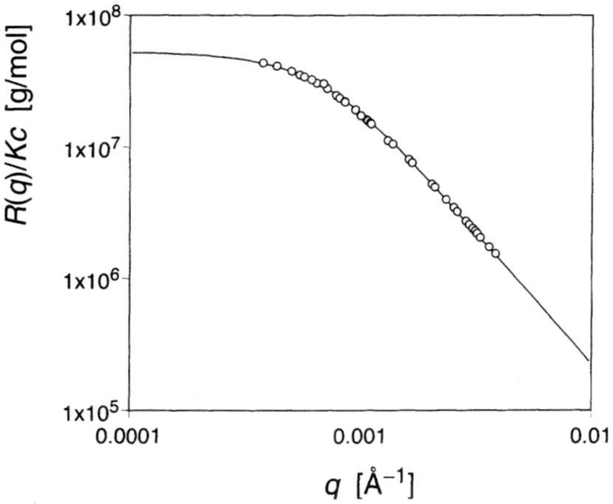

Fig. 7. q-dependence of the scattering intensity for polystyrene ($M = 55 \times 10^6$ g/mol) in cyclohexane at $T = 34.5$°C (Theta-solvent). The solid line is the Debye function. Data are taken from Miyaki et al. (1978).

are shown (Rawiso et al., 1980). We observe significant deviations between the experimental data points and the Debye function already for relatively small values of $M_w \simeq 50{,}000$ g/mol over a quite large q-range. These deviations become even more apparent if we plot the data in a so-called Kratky-plot, where $q^2 I(q)$ versus q is plotted. For a random walk polymer the data should approach a constant value at higher values of q due to the asymptotic q^{-2}-dependence of the scattering intensity. However, it is quite obvious from Fig. 8 that this is not the case for the data obtained under good solvent conditions. The data for the structure factor thus support our findings from the comparison of the experimental data with the molar mass dependence of the overall coil dimensions ($\overline{R}_G$ vs. M_w) and stress the need for a more refined polymer model that works under different solvent conditions.

2.2. Excluded volume effects

In the previous section we have used the most simple model of a random walk in order to describe the conformational properties of a polymer chain in solution. A comparison with experimental data obtained under different solvent conditions has lead to the following observations:

- Under so-called theta conditions we find good agreement between the predicted scaling law for the dependence of $\overline{R}_G$ on the molar mass for high enough values of M_w and the experimental data ($\nu = \frac{1}{2}$), and the q-dependence of the scattering intensity is well described by the Debye function
- For so-called good solvent conditions, we find a power law exponent $\nu = 0.6$ which is different from the random walk model, the resulting polymer coil appears to be swollen when compared to the theta conditions, and the q-dependence of the scattering intensity is no longer in agreement with the Debye function

A qualitative explanation for these observations can be found in the fact that every 'real' monomer possesses a finite volume, i.e., that it occupies space that is not accessible for other

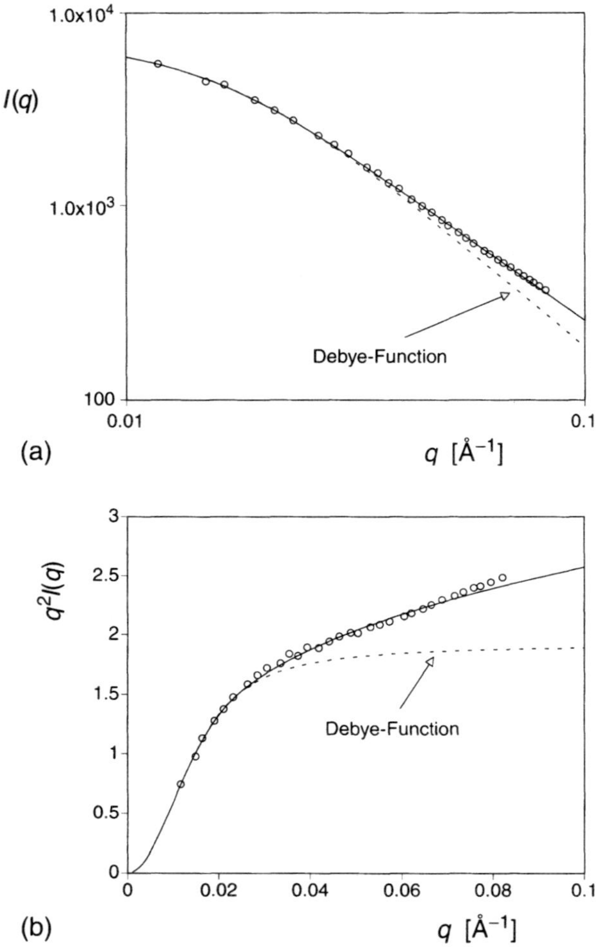

Fig. 8. q-dependence of the scattering intensity from SANS experiments with deuterated polystyrene in CS_2 with a molar mass $M_w \simeq 50{,}000$ g/mol. Also shown is the Debye function as the dashed line and the approximate scattering function for polymers with excluded volume effects introduced in Section 2.2.4. At high values of q we observe clear deviations between the Debye function and the data (A). These deviations become particularly pronounced when plotted as $q^2 I(q)$ versus q (B). The data were taken from Rawiso et al. (1980).

monomers. Excluded volume interactions between the individual monomers (belonging to the same polymer chain) are then responsible for the swelling of the polymer coil, and this has not been taken into account in the simple random walk model.

2.2.1. A simple qualitative description according to Flory

In this section we shall now try to understand the deviations between the theoretical predictions from the random walk model and the experimental data for the molar mass dependence of the radius of gyration and the q-dependence of the static structure factor.

The observations presented so far can be explained by taking into account excluded volume interactions between the individual monomers. Intuitively it is clear that the fact that every monomer has its own finite volume and occupies space that is not accessible for the other monomers will lead to a swelling of the polymer coil when compared to a simple random walk conformation. However, this swelling due to repulsive excluded volume interactions will be partly compensated by the conformational entropy that is larger for more compact conformations. It is the combination of these two opposing effects that Flory used in his mean field model in order to predict the correct dependence of $\overline{R}_{\mathrm{G}}$ on the molar mass (Flory, 1953). We now briefly summarize the model derived by Flory and show that it indeed captures the essential behavior of polymers in good solvents.

Entropy: The conformational entropy of a polymer chain is written as

$$S(\boldsymbol{R}_{\mathrm{E}}) = k_{\mathrm{B}} \ln p(\boldsymbol{R}_{\mathrm{E}}) + \text{const} = -k_{\mathrm{B}} \frac{3}{2} \frac{\overline{R_{\mathrm{E}}^2}}{Na^2} + \text{const}. \tag{32}$$

1st warning: Here we have used the result for the probability density derived from the random walk model [$p(\boldsymbol{R}_{\mathrm{E}})$, see Eq. (14)]. Interactions/correlations are thus not being considered in the calculation of the entropy.

Interaction energy: The average volume fraction occupied by N monomers of size a in a volume $\overline{R_{\mathrm{E}}^3}$ is given by

$$\phi_{\mathrm{int}} = \frac{Na^3}{\overline{R_{\mathrm{E}}^3}}. \tag{33}$$

The total average interaction energy can then be estimated as the product of the average number of interaction sites ($N\phi_{\mathrm{int}}$) multiplied by the average energy per interaction β_0:

$$E = \beta_0 N \phi_{\mathrm{int}} = \beta_0 \frac{N^2 a^3}{\overline{R_{\mathrm{E}}^3}}. \tag{34}$$

We can reformulate this and state that the interaction energy is proportional to the average number of pairs.

2nd warning: We have made a severe approximation by using an average occupational density where we have completely neglected any spatial correlations.

Flory approximation: From the entropy S and the interaction energy E we can now calculate the free energy F at a given temperature T:

$$F = E - TS = \beta_0 \frac{N^2 a^3}{\overline{R_{\mathrm{E}}^3}} + \frac{3}{2} \frac{k_{\mathrm{B}} T}{Na^2} \overline{R_{\mathrm{E}}^2}. \tag{35}$$

A minimization of F with respect to $\overline{R}_{\mathrm{E}}$ leads to:

$$\overline{R}_{\mathrm{E}} = a \left(\frac{\beta_0}{k_{\mathrm{B}} T} \right)^{1/5} N^{3/5} \sim N^{\nu}. \tag{36}$$

with $\nu = 3/5$. The thus obtaining scaling exponent ν is indeed in good agreement with predictions from renormalization group theory and computer simulations. Moreover, this value captures the experimental findings for polymers with sufficiently high molar mass in good solvents as shown in Fig. 6. However, it is important to keep in mind the two warnings stated above. The success of the Flory expression for the free energy for polymers in a good solvent is a good example for the fortuitous effect of a cancellation of errors after combining two estimates based upon rather doubtful approximations.

2.2.2. *Computer simulations*

Excluded volume effects can be studied directly and quite easily with computer simulations (for more details see, for example, des Cloizeaux and Jannink, 1990). In such a simulation we move from a pure random walk (RW) to a so-called self avoiding walk (SAW). In its simplest version we can use the following model:

- Consider all random walks with $N-1$ steps.
- Reject all random walks with multiple occupancy of a lattice site
- Give equal probability to all remaining random walks.

It is (in principle) quite easy to implement such a model on the computer, and one obtains the following results:

Coil dimensions: For large values of N we find

$$\overline{R_{\rm E}^2} = c_{\rm E} a^2 N^{2\nu}, \qquad \overline{R_{\rm G}^2} = c_{\rm G} a^2 N^{2\nu}, \tag{37}$$

with a scaling exponent of $\nu = 0.588\ (1)$ in $d = 3$ dimensions.

Distribution of end-to-end distances: Here we find

$$p(\boldsymbol{R}_{\rm E}) = \frac{1}{\theta^{3/2}} f\left(\frac{|R_{\rm E}|}{\theta^{0.5}}\right), \qquad \theta = \frac{\overline{R_{\rm E}^2}}{3}, \tag{38}$$

where the probability distribution function has the following form

$$f(x) \sim \begin{Bmatrix} x^{\kappa}\exp(-cx^{\delta}), & x \gg 1 \\ x^{g}, & x \ll 1 \end{Bmatrix}. \tag{39}$$

A comparison between the predictions for the random walk and the self avoiding walk model is given in Table 1.

We observe a distinct difference between the probability density for the random walk, which exhibits a maximum at $R = 0$, and the self-avoiding walk, which decays sharply at small values of R. This indicates that the probability of returning to the same position is vanishingly small for the SAW. Another important difference is the finding that the distribution function for inner distances depends upon the position along the chain, which is not the case for the random walk.

Table 1

A list of the exponents ν, δ, g and κ for the simple random walk model (RW) and the self avoiding walk (SAW)

	ν	δ	g	κ
RW	1/2	2	0	0
SAW	0.588(1)	2.427(6)	0.275(2)	0.249(11)

2.2.3. Asymptotic behavior of the correlation function and the structure factor

We have already obtained a simple estimate for the two-particle correlation function and the corresponding asymptotic behavior of the static structure factor for the random walk model in Section 2.1.3. We now extend this approach to excluded volume effects. We first write the correlation function $g(r)$ in a suitable form

$$g(\boldsymbol{r}) \sim \frac{N}{r^{3-(1/\nu)}} \tag{40}$$

and for the structure factor we use

$$S(\boldsymbol{q}) \sim q^{-1/\nu} \tag{41}$$

for $a \ll 1/q \ll \overline{R}_{\mathrm{G}}$. With $\nu = 3/5$ this leads to

$$g(\boldsymbol{r}) \sim r^{-4/3} \tag{42}$$

and

$$S(\boldsymbol{q}) \sim q^{-5/3}. \tag{43}$$

Note: The exponent ν will only be clearly visible in experiments for large enough molar masses M_w, for a sufficient and suitable q-range and a sufficiently good solvent. However, we shall return to this point in a forthcoming section. This is demonstrated in Fig. 9 where we have plotted the scattering intensity obtained in a SANS experiment with polystyrene in the good solvent CS_2 for two different values of the molar mass $M_w \simeq 50{,}000$ and $M_w \simeq 10^6$, respectively (Rawiso et al., 1980). We observe in both cases a good agreement of the data with the predicted asymptotic scaling behavior $I(q) \sim q^{-5/3}$ for polymers in a good solvent.

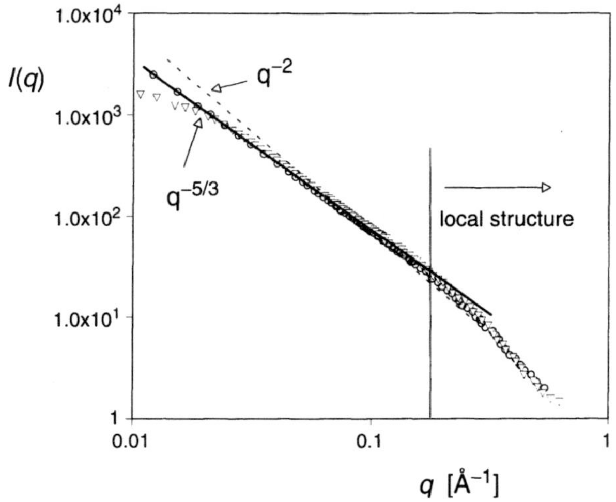

Fig. 9. SANS data obtained for deuterated polystyrene with $M_w \simeq 50{,}000$ g/mol and $M_w \simeq 10^6$ g/mol in CS_2. We see that the data clearly differ from the scaling behavior predicted for the random walk model but exhibit good agreement with the SAW model. We also observe clear deviations at high q-values, where the influence of the local structure becomes visible. The data are taken from Rawiso et al. (1980).

however the q-range where this is visible is quite limited for the shorter chains due to the fact that the Guinier regime is shifted to quite high values of q because of the relatively small radius of gyration $\overline{R}_G$.

2.2.4. *Scattering function for a single coil with excluded volume effects*

We have seen that the Debye function describes the structure factor of a random walk chain. While there exist similar calculations that take into account excluded volume effects (Duplantier, 1985; Oono, 1985), they are too complicated in the context of this introductory lecture.

However, there exist phenomenological approximations based on rather simple assumptions that do indeed yield quite good results (Sharp and Bloomfield, 1968; Utiyama et al., 1971). If we, for example, assume that the distribution function for inner distances $p_{j,k}(r)$ is independent of the position along the chain (i.e., depends on $|j-k|=n$ only) and has the same form as the distribution function $p(r)$ for the end-to-end distances R_E, the asymptotic form (39) for the self-avoiding walk then yields

$$P_n(r) \sim x_n^{\kappa} \exp(-c x_n^{\delta}), \qquad x_n = \frac{r_{jk}}{\Theta_{j,k}^{0.5}}, \qquad \Theta_{j,k} = \frac{\langle r_{jk}^2 \rangle}{3}, \qquad |j-k| = n \tag{44}$$

for large inner distances. We thus obtain a working expression for the scattering function of a single chain with excluded volume effects

$$\begin{aligned} S(q) &= \frac{1}{N} \sum_{j,k=1}^{N} \left\langle \frac{\sin(q r_{jk})}{q r_{jk}} \right\rangle \\ &\simeq 1 + \frac{2}{N} \sum_{n=1}^{N-1} (N-n) \int_0^{\infty} 4\pi r^2 P_n(r) \frac{\sin(qr)}{qr}\, \mathrm{d}r \end{aligned} \tag{45}$$

Expression (45) is indeed useful for not too high values of $q^2 \bar{R}_G^2$, i.e., under conditions where no local structural details can be resolved and where the assumption of a position-independent $p_{j,k}(r)$ would be too restrictive. This is demonstrated in Fig. 8 which shows SANS data for polystyrene in the good solvent CS_2 for $M_w \simeq 50{,}000$ g/mol.

2.2.5. *What have we achieved, and what is missing?*

We have introduced two different simple models in order to describe the conformational properties of polymer chains in solution, the random walk (RW) and the self avoiding walk (SAW), which have allowed us to predict the behavior of several measurable quantities. A comparison with the results from light and neutron scattering experiments have led to the following conclusions:

(1) Scaling law $\overline{R}_G \sim N^{\nu}$: Here the different models predict

$$\nu = \begin{Bmatrix} \frac{1}{2} & \text{(RW)} \\ 0.588\ (3/5) & \text{(SAW)} \end{Bmatrix}. \tag{46}$$

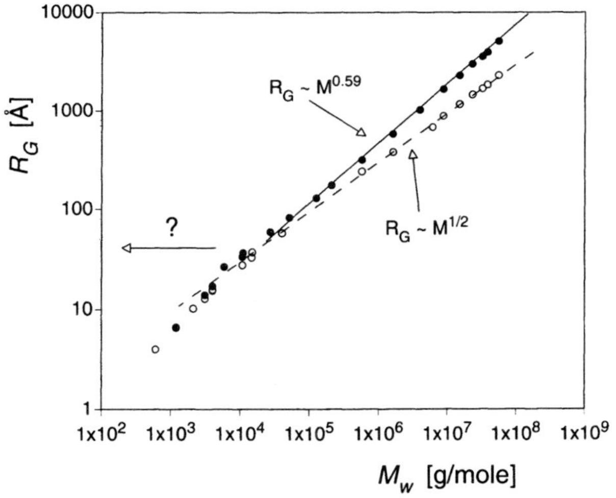

Fig. 10. Radius of gyration as a function of the molar mass for polystyrene under theta- (∘) and good (•) solvent conditions (the data is the same as shown in Figs. 5 and 6).

A comparison of the molar mass dependence of the radius of gyration obtained for polystyrene under theta and good solvent conditions shown in Fig. 10 reveals the following behavior:

- For PS in a theta solvent we find quantitative agreement with the RW model for sufficiently high values of $M_w > 10^4$.
- Under good solvent conditions and for high molar mass ($M_w > 10^5$), the polymer coils are swollen and we find $\overline{R}_G \sim M_w^{0.59}$ as predicted by the SAW model.
- At medium values of M_w ($10^4 < M_w < 10^5$) a close inspection of the data indicates that they follow the RW rather then the SAW behavior ($\bar{R}_G \sim M_w^{0.5}$). The chains are rather short under these conditions and overlap is thus unlikely to occur, which means that we can almost neglect excluded volume effects.
- At low values of M_w ($M_w < 10^4$) both data sets overlap, solvent conditions no longer matter, and we now observe the influence of the local structure (i.e., of the chemical details).

(2) Asymptotic dependence of $S(q)$ for large q:

$$S(q) \sim q^{-1/\nu} = \begin{Bmatrix} q^{-2} & \text{(RW)} \\ q^{-1.7} & \text{(SAW)} \end{Bmatrix}. \tag{47}$$

The corresponding scaling laws are indeed observed provided that we have access to high molar mass polymers in order to obtain a large enough q-range ($1/\bar{R}_G \ll q \ll 1/a$).

(3) Full scattering function or structure factor $S(q)$:
Here we have seen that the approximate expression (Eq. 45)

$$S(q) \simeq 1 + \frac{2}{N} \sum_{n=1}^{N-1} (N-n) \int_0^\infty 4\pi r^2 P_n(r) \frac{\sin(qr)}{qr}\, dr \tag{48}$$

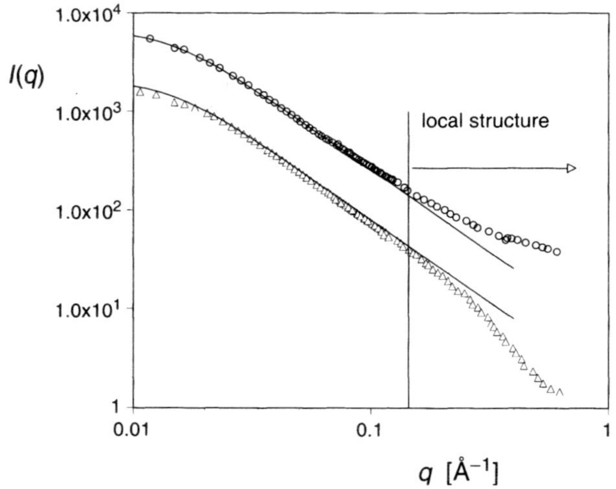

Fig. 11. Polystyrene in CS_2. SANS data from differently deuterated polystyrene ($M_w \simeq 10^6$ g/mol) in CS_2. The data are for fully deuterated PS (△) and PS with deuterated backbone only (○). The data are from Rawiso et al. (1980).

results in very good agreement with the scattering data. This is illustrated in Fig. 11 with SANS data obtained with polystyrene in the good solvent CS_2 (Rawiso et al., 1980). However, Fig. 11 also exhibits a rather unexpected q-dependence caused by the influence of local chemical details at high values of q. This effect is masked for the fully deuterated chains due to the fact that polystyrene has a relatively large cross-sectional radius (see chemical structure in Fig. 1). Therefore the high-q behavior is completely dominated by the cross-section scattering term which leads to an additional exponential decay at $q > 1/R_{G,cs}$, where $R_{G,cs}$ is the cross-sectional radius of gyration. For neutron scattering the polymer where only the backbone is deuterated resembles a thin line rather than a locally cylindrical coil. Therefore, the characteristic asymptotic behavior of a SAW chain should extend to higher values of q before the cross-section becomes once again visible. However, in reality one observes a completely different and rather unexpected behavior with a q-dependence of the scattering intensity that exhibits a cross-over from $S(q) \sim q^{-1.7}$ to $S(q) \sim q^{-1}$. We shall see that the origin for this crossover is the same that has already lead to the typical molar mass dependence of $\bar{R}_G$ at low values of M_w (Fig. 10): Stiffness of the polymer chain on short length scales due to interactions between adjacent monomers, the so-called persistence.

(4) Cross over from theta to good solvent conditions: While we have successfully designed rather simple models that describe the conformation of polymers in solution for the two limiting cases theta and good solvent (without actually attempting to understand how a theta condition can exist), we have not yet thought about intermediary conditions where we for example change the temperature and move away from the magic $T = 34.5°C$ for polystyrene in cyclohexane.

2.3. Persistence, a first step towards the incorporation of chemical details

In the previously introduced models we have always assumed that there exists no correlation between the individual steps. We can describe the correlation between the previously introduced connectivity vectors $\boldsymbol{Q}_j$ and $\boldsymbol{Q}_k$ as

$$\langle \boldsymbol{Q}_j \cdot \boldsymbol{Q}_k \rangle = \gamma_{jk}. \tag{49}$$

For a completely flexible chain (RW or SAW) we thus obtain $\gamma_{jk} = 0$ for $j \neq k$. However, for a more realistic description of a 'real' chain we expect to find $\gamma_{jk} \neq 0$ for neighbouring monomers along the polymer chain.

The origin of such correlations can be understood with the example of polyethylene ($[-CH_2 - CH_2-]_n$, PE). The backbone of this polymer is formed by carbon atoms that are also connected to two hydrogen atoms each (see Fig. 1). A schematic graphical description of a short sequence of a PE chain is given in Fig. 12, where four consecutive carbon atoms in polyethylene are shown. The valence angle between two consecutive carbon bonds is fixed and takes a value of $\Theta \approx 71°$. If we fix C_{n-3}, C_{n-2}, and C_{n-1}, this determines the position of the fourth atom C_n except for the choice of the bond rotation angle φ_n. In an idealized model of PE the rotation about this bond is free, i.e., there exists no limitation for φ. This model is frequently applied in the literature and is called the 'freely rotating chain'.

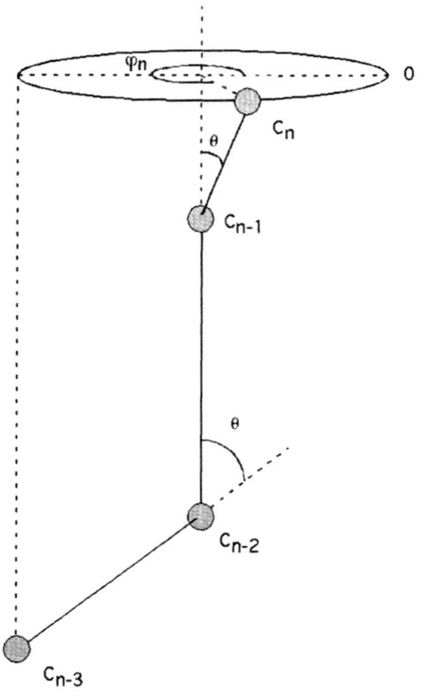

Fig. 12. Schematic description of a sequence of monomers in polyethylene. The valence angle Θ between two carbon atoms is $\Theta \approx 71°$. Due to energetic reasons the rotation angle φ is not completely random but preferentially takes values of $\varphi = 0$ (trans state), $\varphi = 120°$ (gauche state g^+) and $\varphi = -120°$ (gauche state g^-).

In a more realistic model of PE we have to take into account the fact that the angle φ_n is not completely free because of the influence of the H-sidegroups on the rotational energy. The energetic minimum is found for the so-called trans configuration, i.e., a value of $\varphi = 0°$. Two other slightly higher local minima in the rotational energy also exist for $\varphi = \pm 120°$ (the so-called gauche states). There is thus a slight preference for the configuration with the lowest rotational energy, which induces short range angular correlations between neighbouring monomers, and as a result of this the chain is locally stiff:

$$\gamma_{jk} = f(|j-k|). \tag{50}$$

A first and very important attempt to qualitatively incorporate these local correlation effects was made by Werner Kuhn in 1934 (Kuhn, 1934; Kuhn and Kuhn, 1943). His innovative approach was in fact an early version of a course graining model. He had the idea to regroup the monomers of a chain with N monomers into N_K straight segments of length b such that for large values of N/N_K the contour length L was given by

$$L = N_{\mathrm{K}} b \tag{51}$$

and the end-to-end distance of the replacement chain $\overline{R}_{\mathrm{E,K}}$ and the real chain $\overline{R}_{\mathrm{E}}$ by

$$\langle \boldsymbol{R}_{\mathrm{E,K}}^2 \rangle = \langle \boldsymbol{R}_{\mathrm{E}}^2 \rangle = N_{\mathrm{K}} b^2. \tag{52}$$

The new replacement chain is thus once again a random walk chain with a rescaled step length b. This segment length b is called the Kuhn length, and it is a measure of the chain flexibility, where for large values of b the polymer is stiff and for small values it is flexible. This remarkable idea has then been quantitatively implemented into the so-called wormlike chain model by Kratky and Porod (1949).

2.3.1. The wormlike chain model

In this model we assume that the correlations between different connectivity vectors $\boldsymbol{Q}_j$ and $\boldsymbol{Q}_k$ can be written as

$$\langle \boldsymbol{Q}_j \cdot \boldsymbol{Q}_k \rangle = a^2 \lambda^{|j-k|}. \tag{53}$$

There exists in fact a relation with the freely rotating chain with $\lambda = \langle \cos\Theta \rangle = \cos\Theta$. We can now directly calculate the end-to-end distance for a wormlike chain with N steps:

$$\langle \boldsymbol{R}_{\mathrm{E}}^2 \rangle_N = \sum_{j,k=1}^{N} \langle \boldsymbol{Q}_j \cdot \boldsymbol{Q}_k \rangle = a^2 \sum_{j,k=1}^{N} \lambda^{|j-k|} = a^2 N \left[\frac{1+\lambda}{1-\lambda} - \frac{2\lambda}{(1-\lambda)^2} \frac{1-\lambda^N}{N} \right]. \tag{54}$$

The end-to-end distance for the limiting case of a continuous chain is then obtained by allowing $N \to \infty$, $a \to 0$ and $\lambda \to 1$ while keeping the contour length $L := Na$ and the so-called persistence length $L_{\mathrm{p}} := a/(1-\lambda)$ constant. This leads to

$$\langle \boldsymbol{R}_{\mathrm{E}}^2 \rangle = 2L_{\mathrm{p}} \left[L - L_{\mathrm{p}} + L_{\mathrm{p}}\, \mathrm{e}^{-L/L_{\mathrm{p}}} \right] \tag{55}$$

with the following two limiting cases:

(1) $\langle \boldsymbol{R}_{\mathrm{E}}^2 \rangle \simeq L^2$ for $L \ll L_{\mathrm{p}}$

(2) $\langle \boldsymbol{R}_{\mathrm{E}}^2 \rangle \simeq 2LL_{\mathrm{p}}$ for $L \gg L_{\mathrm{p}}$.

The first case corresponds to a completely stiff polymer with $\overline{R}_{\mathrm{E}} = L$. For the second case we find $\langle \boldsymbol{R}_{\mathrm{E}}^2 \rangle = N_{\mathrm{K}} b^2 = Lb$ with $b = 2L_{\mathrm{p}}$ and $N_{\mathrm{K}} = L/b$. The wormlike chain model describes the conformation of a polymer as a random walk of N_K steps with step length b, i.e., it corresponds to the replacement chain of Kuhn.

A comparison of the wormlike chain with a freely rotating chain with valence angle Θ shows that in the limit $N \to \infty$ they are identical and we find

$$b := a \frac{1+\cos\Theta}{1-\cos\Theta}. \tag{56}$$

The radius of gyration can also be derived for the wormlike chain (Benoit and Doty, 1953):

$$\langle \boldsymbol{R}_{\mathrm{G}}^2 \rangle = \frac{bL}{6}\left\{1 - \frac{3}{2N_{\mathrm{K}}} + \frac{3}{2N_{\mathrm{K}}^2} - \frac{3}{4N_{\mathrm{K}}^3}(1-\exp(-2N_{\mathrm{K}}))\right\}. \tag{57}$$

In the limit $N_{\mathrm{K}} \to \infty$ we thus obtain:

$$\langle \boldsymbol{R}_{\mathrm{G}}^2 \rangle \sim \frac{bL}{6} \tag{58}$$

i.e., we recover the random walk behavior with a renormalized step length b.

We can directly test Eq. (57) by looking at the dependence of the radius of gyration $\overline{R}_{\mathrm{G}}$ of polystyrene in a theta solvent shown in Fig. 5. In order to do so we need values for the Kuhn length b as well as a relation between the molar mass M_w and the contour length L. In the literature (Pedersen and Schurtenberger, 1996) we find $b = 24.8$ Å, and $M_L = M/L = 41.2$Å for the so-called mass-per-length, which allows us to directly calculate a theoretical curve $\overline{R}_{\mathrm{G}}$ vs. M_w. Fig. 13 shows very good agreement between the wormlike chain model and the experimental data over the entire range of molar masses investigated.

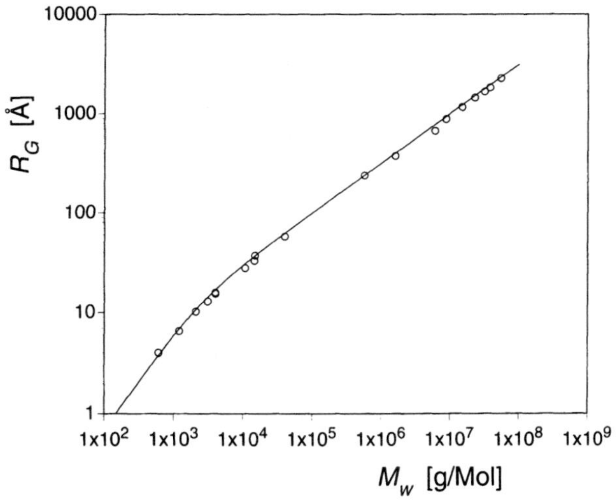

Fig. 13. Radius of gyration as a function of the molar mass for polystyrene in cyclohexane at 34.5°C (Theta-solvent). The calculated curve using the wormlike chain model (Eq. (57)) shows very good agreement. The data are identical with those shown in Fig. 5.

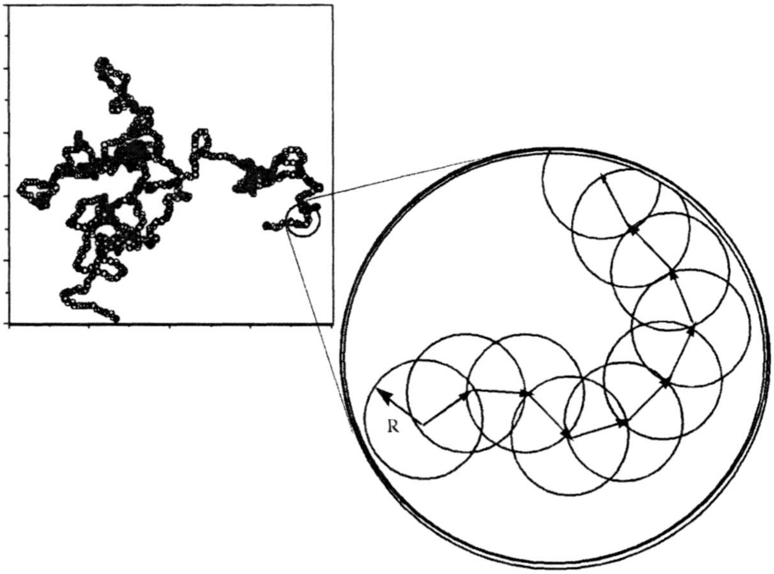

Fig. 14. Example of a chain conformation obtained from a MC simulation of a quasi-continuous wormlike chain with excluded volume effects together with a schematic representation of the model used in the simulation.

2.3.2. *Excluded volume effects*

The incorporation of excluded volume effects into the wormlike chain model is too complicated, there exists no analytical model and we thus have to resort to computer simulations. In these simulations we use a self avoiding walk simulation based on a model of a freely rotating chain with bond length a and valence angle Θ chosen such that $b = a[(1+\cos\Theta)/(1-\cos\Theta)]$. Excluded volume effects are then incorporated by placing spheres of radius R around each 'monomer' and choosing a value of $R/b = 0.1$ that matches polystyrene in good solvent conditions. The limiting case of a continuous wormlike chain with excluded volume effects is then obtained by decreasing the distance a between the centers of the spheres while leaving b and R constant. A schematic representation of the model used is shown in Fig. 14. Details of the simulation algorithm and the results can be found in the literature (Pedersen and Schurtenberger, 1996; Pedersen et al., 1996).

A comparison between the simulation results and the experimental data for the dependence of $\overline{R}_G$ on M_w for polystyrene in good solvents shows excellent agreement as demonstrated in Fig. 15. An interesting result that can be obtained from Fig. 15 is that the asymptotic limit of the predicted scaling law for excluded volume effects is only reached after about $N_K \geq 50-100$.

2.3.3. *The full scattering function of a wormlike chain*

We are now in a good position to at least qualitatively discuss the full structure factor or scattering function of wormlike chains with or without excluded volume effects (Kirste and

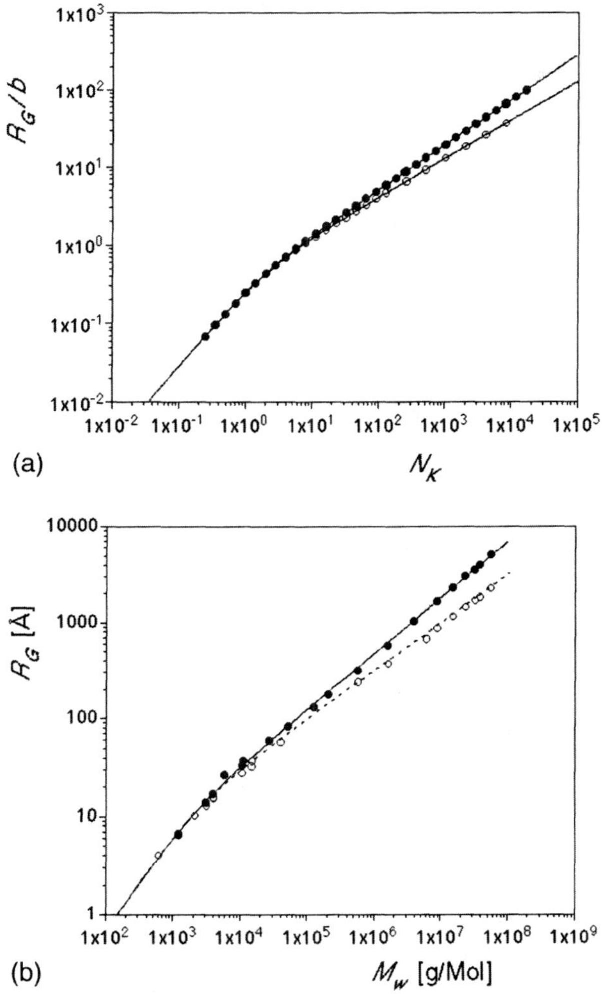

Fig. 15. (A) Radius of gyration $\overline{R}_G$ as a function of the number of Kuhn length N_K with (solid circles) and without (open circles) excluded volume interactions as obtained from Monte Carlo simulations (Pedersen et al., 1996). (B) Radius of gyration $\overline{R}_G$ as a function of the molar mass for polystyrene in toluene (good solvent; upper curve) and in cyclohexane at 34.5°C (Theta-solvent, lower curve). The theoretical curves are the results from the computer simulations of a quasicontinuous wormlike chain with and without excluded volume effects shown in A. The data are identical with those shown in Figs. 5 and 6.

Oberthür, 1982). A schematic picture of the expected q-dependence of the scattering intensity is shown in Fig. 16.

- Guinier regime ($q\overline{R}_G \ll 1$): At small values of $q\overline{R}_G$, i.e., in the so-called Guinier regime $S(\boldsymbol{q})$ is model independent and given by the approximate expression $S(\boldsymbol{q}) \approx N\left[1 - \frac{1}{3}q^2\overline{R_G^2}\right]$. This is the basis for standard applications of light, neutron or X-ray scattering experiments in polymer characterization, where the molar mass M_w and the radius of gyration $\overline{R}_G$ can be obtained.

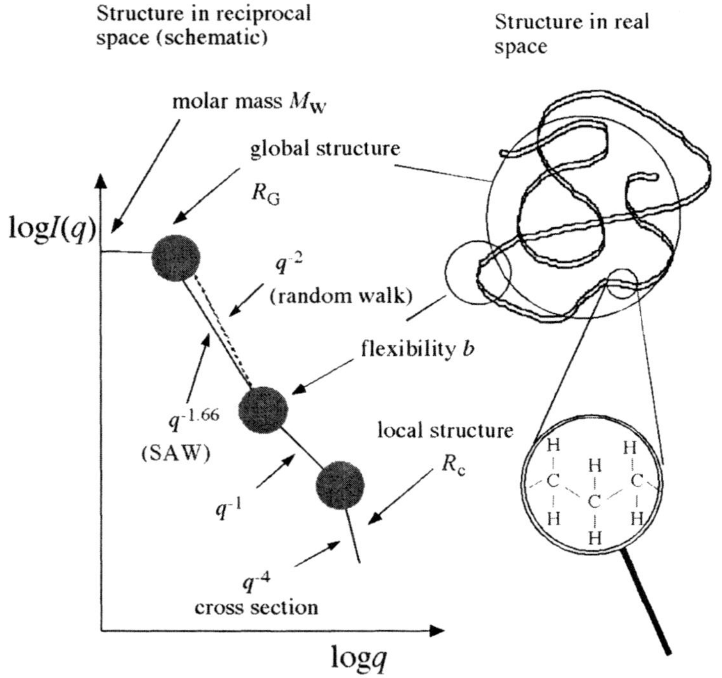

A schematic representation of the different asymptotic regions for the q-dependence of the structure factor of a wormlike chain and their link to the different structural properties on various length scales.

- Asymptotic power law regime for flexible chains ($1/\overline{R}_G \ll q \ll 1/b$): At higher values of q and for sufficiently large values of the molar mass of the polymer we then see the crossover to the characteristic power law $S(\boldsymbol{q}) \sim q^{-1/\nu}$. For a simple random walk model (theta solvent) we thus obtain q^{-2}, whereas for the self avoiding walk (good solvent) we find $q^{-5/3}$. For theta conditions the crossover is given by the Debye function, and for good solvents by the approximation in Eq. (45).
- Rigid rod regime ($1/b < q < 1/R_{G,cs}$): Under these conditions we investigate the polymer structure on length scales where the polymer resembles a rigid rod, which is characterized by an asymptotic power law of the form $S(\boldsymbol{q}) \sim q^{-1}$.
- Cross-section structure ($q R_{G,cs} \geq 1$): Here we see the local monomer structure which gives rise to a much steeper decay of the intensity due to the cross-section scattering function $S_{cs}(q)$.

A quantitative description has been derived based on an analysis of systematic Monte Carlo computer simulations of wormlike chains with and without excluded volume interactions (Pedersen and Schurtenberger, 1996). The experimental data can be reproduced by a phenomenological expression of the form:

$$S(\boldsymbol{q}) \simeq \left[(1-\chi(q,L,b))\, S_{\text{coil}}(q,L,b) + \chi(q,L,b) S_{\text{rod}}(q,L)\right] \Gamma(q,L,b) S_{cs}(q R_{cs}), \qquad (59)$$

where S_{coil} is the scattering function for a flexible polymer with or without excluded volume effects, S_{rod} the scattering function of a rigid rod, $S_{cs}(q R_{cs})$ is the cross-section scattering

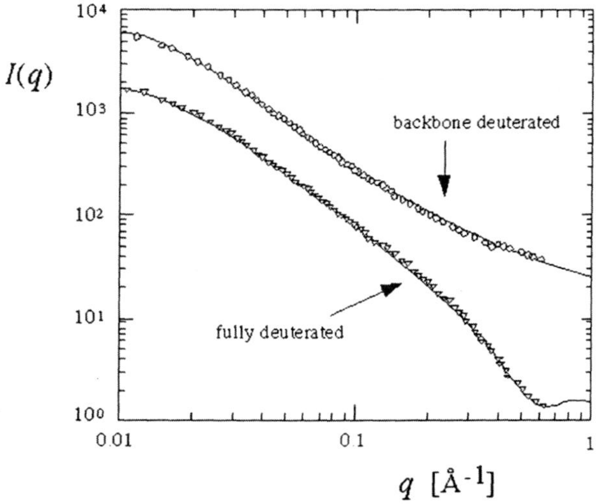

Fig. 17. SANS experiments with partially and fully deuterated polystyrene $M_w \simeq 50{,}000$ g/mol in CS_2. The data are taken from Rawiso et al. (1980), the data analysis from Pedersen and Schurtenberger (1996).

function, χ is a crossover function that describes the crossover between the two asymptotic regimes, and Γ is an additional correction term for the crossover region. Details and tabulated parameters for these functions can be found in Pedersen and Schurtenberger (1996).

The influence of the cross-section scattering function is demonstrated with data from polystyrene with different degree of deuteration in CS_2, where either the backbone only or the entire polymer with the side groups is visible (see Fig. 17). The use of an additional cross-section scattering term in Eq. (59) then results in very good agreement with the SANS data and allows to obtain additional information on the local monomer structure.

'Real' polymers exhibit a large variety of different values for the Kuhn length. Several examples are given in Table 2. It is evident from Table 2 that the experimental data are sometimes confounded with rather large error bars. This is in particular the case if b is determined through the dependence of the radius of gyration upon the molar mass via Eq. (57) as polydispersity and interchain interaction effects do play an important role. It is thus much safer to use the full scattering function in the region of the crossover from flexible coil to rigid rod scaling behavior in order to determine b from SANS or SAXS experiments.

Table 2

Typical values for the Kuhn length of polymers

Polymer	Kuhn length b
Polystyrene (PS)	22 Å
Kevlar (PPTA)	300–1,200 Å
Xanthane	~5,000 Å
fd-Virus	~39,000 Å

Table 3
A summary of the models and their most important results

	RW (theta-solvent)	? $\longleftrightarrow$	SAW (good solvent)
Global	$\overline{R}_G \sim M^{1/2} (\sim N^{1/2})$		$\overline{R}_G \sim M^{3/5} (\sim N^{3/5})$
	$S(\boldsymbol{q}) = f_D\left(q^2 \langle R_G^2 \rangle\right)$		$S(\boldsymbol{q}) = \frac{1}{N} \sum_{j,k=1}^{N} \left\langle \frac{\sin(q r_{jk})}{q r_{jk}} \right\rangle$
$\updownarrow$?			
Local	rigid $\nu = 1$		rigid $\nu = 1$

2.4. Temperature and solvent effects

We have now introduced three different models, the random walk, the self-avoiding walk, and the corresponding models with persistence. The thus achieved description of the polymer conformation on various length scales, the resulting universal scaling laws and the description of the structure factor are summarized in Table 3. We have seen that we can observe these universal features in our data obtained with scattering experiments. The main remaining question is now how the crossover between good and theta solvent or between global and local properties can be described.

The latter problem, the crossover between global and local properties, has already been described in Eq. (59) using empirical crossover functions based on MC computer simulations. However, while these crossover functions provide excellent fits to the scattering data and thus for example allow to quantitatively and precisely determine the Kuhn length of a polymer, they provide us with limited insight into the physics behind these crossover regions. A quite elegant approach that allows us to intuitively understand the origin and location of these crossover regions are so-called blob models, which is another word for a coarse graining procedure in systems that exhibit different behavior on different characteristic length scales (de Gennes, 1979). The transition from persistence to a simple random walk behavior can be taken as a simple example, and it is schematically drawn in Fig. 18A: For a random walk with local stiffness we expect to walk first along a certain direction on a lattice. After a characteristic number of steps one then expects a change of direction. If we now group these consecutive steps with the same direction into new units, the so-called blobs, we then obtain a random walk structure formed by these blobs as the new 'monomer' units. The blob size is thus a measure for the characteristic length scale for a transition from a locally stiff to a globally flexible chain.

The transition from a simple random walk to a self avoiding walk can also be described by a blob model with a typical length scale. Similar to the Kuhn length as a measure of the characteristic length scale for a transition from a locally stiff to a globally flexible chain, the blob size for the transition RW-SAW indicates that for shorter distances along the contour excluded volume effects do not matter anymore (the probability for folding back is negligible). The overall structure can then be looked at as a self avoiding walk of blobs with an internal structure of a random walk as shown in Fig. 18B. These blob models immediately yield a

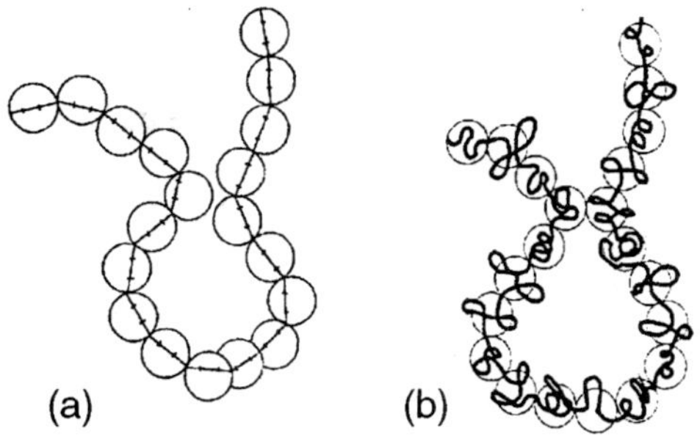

Fig. 18. Transition from persistence to random walk behavior (Part A): Consecutive steps in the same direction are regrouped to a new unit, the so-called blob. These blobs are uncorrelated and form a random walk coil. Transition from random walk to self avoiding walk behavior (Part B): There exists a blob size as a new characteristic length scale below which excluded volume effects do not play a role. Within the blobs we find RW behavior, whereas excluded volume effects influence the conformation of the blob chain which exhibits a SAW structure.

qualitative description of the structure factor in different q-regimes. For a random walk chain with persistence we expect for example a crossover from an asymptotic q^{-1}-behavior to a q^{-2}-behavior at $q \sim 1/(\text{blob size})$, while for chains in a good solvent we may even expect to find three different regimes with a crossover from q^{-1} to q^{-2} to $q^{-5/3}$ at the two different blob sizes for persistence and excluded volume effects.

After having obtained qualitative insight into these crossover phenomena with these blob models, there remain a few important questions: How can we combine and generalize these different effects? What are the necessary conditions for finding a specific type of behavior? What is the role of the solvent and the temperature (key word theta solvent)? How can we characterize these transitions quantitatively and mathematically?

2.4.1. An extended lattice model

We can for example try to incorporate flexibility, excluded volume, temperature and solvent effects into an extended lattice model for computer simulations. Starting point in this attempt is the random walk model, to which we add the following interaction effects between the individual monomers:

- $E_1 = \infty$ for doubly occupied lattice sites $\rightarrow$ excluded volume effects
- $E_2 < 0$ for each pair of occupied lattice sites on neighbouring places $\rightarrow$ attractive interactions between neighbouring lattice sites are taken into account. Their strength is determined by the solvent quality.
- $E_3 \geq 0$ for every 90° angle between consecutive steps $\rightarrow$ persistence, steps that deviate from the initial direction are energetically unfavorable.

If the interaction energies E_2 and E_3 are different from 0 and ∞, the predictions of the different types of models become temperature dependent. Such a model can be investigated by several different methods such as renormalization group theory [see, for example, Oono (1985), Freed (1987), Schäfer (1999)] or Monte Carlo computer simulations [see, for example, Meirowitch (1991) (Lattice simulation, $E_3 = 0$)].

We shall now look at the results of such a generalized polymer model, but we shall concentrate on the case $E_3 = 0$ and try to give a more quantitative description of the crossover from random walk to self avoiding walk behavior qualitatively described by the blob model. In doing so we first introduce several important quantities: $E_2/(k_B T)$ determines the strength of the attractive interactions between the monomers, θ is the theta temperature at which excluded volume effects are 'turned off', and $\tau = (T - \theta)/\theta$ is the relative deviation from the theta temperature. This model provides us with a better understanding of theta conditions: At the temperature $T = \theta$ the attractive monomer-monomer interactions fully compensate for the repulsive excluded volume interactions, and we recover the ideal random walk chain conformation.

Simulations and renormalization group theory have shown that the results for $\overline{R}_{\mathrm{G}}$ should be of the form:

$$\overline{R}_{\mathrm{G}}\left(a, N, \frac{E_2}{k_B T}\right) = \overline{R}_{\mathrm{G}}^{\theta} f\left(c\tau N^{\phi}\right) \tag{60}$$

where

$$\overline{R}_{\mathrm{G}}^{\theta} = \tilde{c} a N^{\nu_\theta} \tag{61}$$

is the radius of gyration at the theta temperature. We have thus been able to reduce a function with three unknowns to a universal dimensionless scaling function $f(c\tau N^{\phi})$ that depends on a single unknown and several parameters $\nu_\theta, \phi, c, \tilde{c}, k_B\theta/E_2$ and the crossover exponent ϕ. For $d = 3$ at the theta point we find random walk behavior, i.e., $\nu_\theta = 1/2$. The scaling function f has the following properties: (i) $f(0) = 1$; (ii) for large x, $f(x)$ has to be a power law of x (because $\overline{R}_{\mathrm{G}}$ follows a power law $\overline{R}_{\mathrm{G}} \sim N^{\nu}$). We therefore choose the constant such that $f(x) \rightarrow x^z$. For $T > \theta$ this then leads to $\nu = \nu_\theta + z\phi$ and $\phi = \frac{1}{2}$ and we find the important result:

$$\overline{R}_{\mathrm{G}}(T, N) = \overline{R}_{\mathrm{G}}^{\theta} f\left(c\tau N^{1/2}\right) \tag{62}$$

This is experimentally verified with data from polystyrene in cyclohexane with different values of the molar mass and at different temperatures shown in Fig. 19 (Arai et al., 1995). The data indeed all collapse onto a single master curve when plotted as $\overline{R}_{\mathrm{G}}/\overline{R}_{\mathrm{G}}^{\theta}$ against $\tau M^{1/2}$. We have thus achieved a simple model description that is capable of qualitatively and quantitatively describing the most important features of a single polymer chain in solution at all values of M_w and $T \geq \theta$.

3. Interacting polymer chains

In the preceding sections we have looked at the static properties of ideal solutions of noninteracting polymer chains at different length scales. In the comparison of the theoretical predictions with experimental data we have always worked in the limit of $c \rightarrow 0$. We now make a next step and increase the polymer concentration until we reach the so-called polymer melt, i.e., a dense polymer fluid without additional low molar mass solvent. As a starting point we look at SANS data for polystyrene with $M_w = 120{,}000$ g/mol in deuterated toluene as a function of the polymer concentration shown in Fig. 20 (Pedersen and Schurtenberger, 1999).

In the limit of infinite dilution we would expect the typical features of a self avoiding walk chain, i.e., a Guinier regime at low values of q followed by an asymptotic power law $I(q) \sim q^{-5/3}$. For ideal noninteracting chains the data obtained at different concentrations

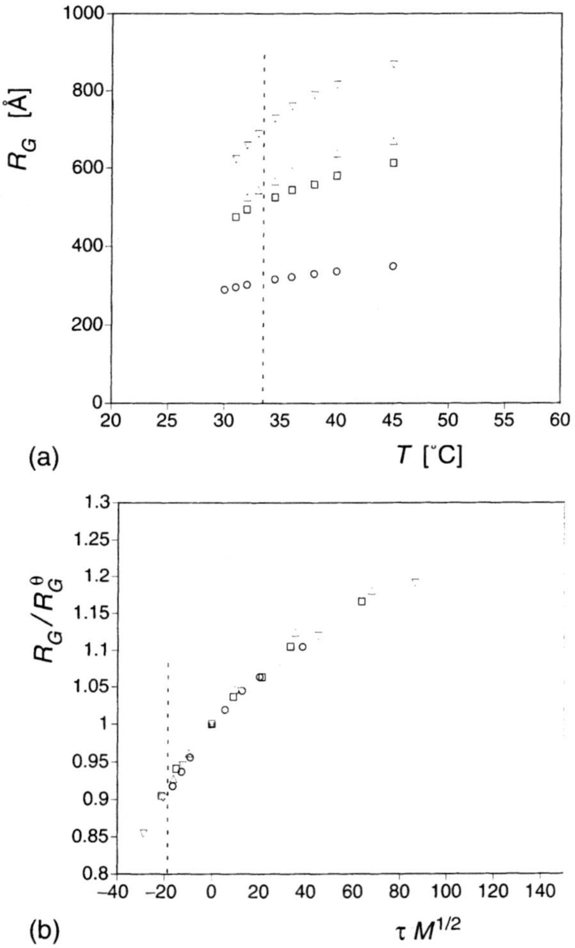

Fig. 19. (A) Radius of gyration $\overline{R}_{\mathrm{G}}$ of polystyrene in cyclohexane as a function of the temperature for four different values of the molar mass M_w (from bottom to top: 1.27, 3.47, 3.94 and 6.4 × 10^6 g/mol). (B) Plot of the values of $\overline{R}_{\mathrm{G}}$ shown in A as $\overline{R}_{\mathrm{G}}/\overline{R}_{\mathrm{G}}^{\theta}$ versus $\tau M^{1/2}$. All data points are taken from Arai et al. (1995).

should superimpose if normalized with the concentration, i.e., if plotted as $I(q)/c$ vs. q. However, at higher concentrations the experimental data significantly deviate from this pattern, and we can summarize our findings as follows:

- The normalized forward intensity $I(q \to 0)/c$ decreases with increasing concentration
- The experimental curves overlap at higher values of q and follow SAW behavior.
- We do observe deviations from the characteristic SAW behavior at very high concentrations.
- In the extreme case of a polymer melt, the single coil scattering function obtained from mixtures of deuterated and non-deuterated polystyrene chains follows in fact perfectly the prediction of a random walk chain (see Fig. 21)!

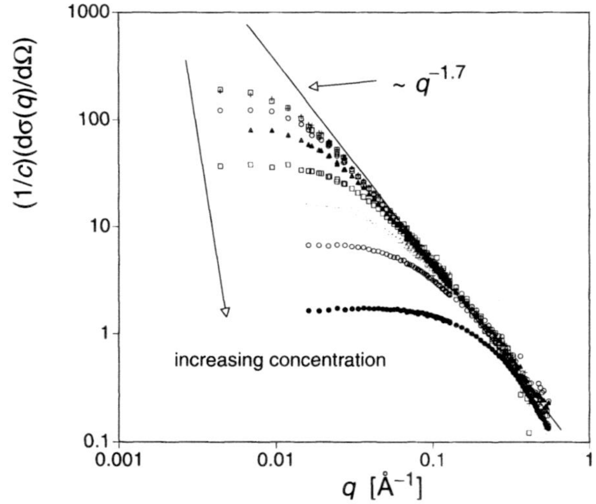

Fig. 20. Normalized scattering intensity $I(q)/c$ as a function of the scattering vector q for different polymer concentrations c for polystyrene in deuterated toluene and $M_w = 120{,}000$ g/mol. The concentrations are (from below) $c = 0.27$, 0.11, 0.054, 0.028, 0.011, 0.0052, 0.0021 and 0.0013 g/ml. The data are taken from Pedersen and Schurtenberger (1999).

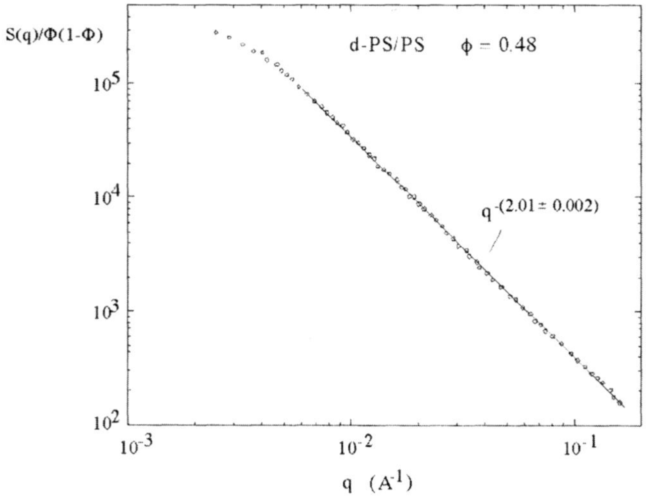

Fig. 21. SANS-experiment with a mixture of deuterated in non-deuterated polystyrene melt. The measured single chain structure factor exhibits classical random walk behavior$S(q) \sim q^{-2}$ (Schwahn, 1991).

For polymer solutions one generally distinguishes between three characteristic concentration regimes: Dilute, semi-dilute and concentrated solutions/melts. The transition from dilute to semi-dilute solutions occurs at the so-called overlap concentration c^*. At this concentration the individual polymer coils start to overlap. This is schematically shown in Fig. 22. In the remaining sections we shall now see how we can treat these concentration regimes theoretically,

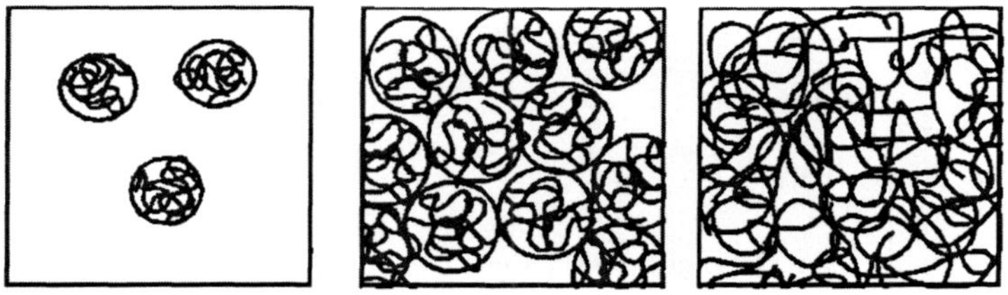

Fig. 22. A schematic representation of dilute solutions at $c < c^*$, the overlap concentration c^*, and semidilute solutions $c > c^*$.

and which experimentally accessible quantities can be used for a comparison between theory and experiment.

3.1. Scattering intensity and osmotic compressibility

For a monodisperse solution of polymer coils we know that the normalized scattering intensity or differential cross-section $\mathrm{d}\sigma/\mathrm{d}\Omega(q)$ can be expressed as

$$\lim_{q\to 0} \frac{1}{V_{sc}} \frac{\mathrm{d}\sigma}{\mathrm{d}\Omega}(q) = K c N_{\mathrm{A}} k_{\mathrm{B}} T \left(\frac{\partial \Pi}{\partial c} \right)^{-1}_{T,N}, \tag{63}$$

where V_{sc} is the scattering volume, K is the scattering contrast (see chapter on contrast and contrast variation), N_A is Avogadro's number, and $\partial\Pi/\partial c$ is the osmotic compressibility. Equation (63) is not restricted to polymer solutions but valid for any monodisperse solution of macromolecules or colloids, and it is the basis for the popular determination of molar masses of polymers with light scattering experiments, where the contrast term can be written as $K_{\mathrm{LS}} = [4\pi^2/N_{\mathrm{A}}\lambda^4] n^2_{\mathrm{Lsm}} (\mathrm{d}n/\mathrm{d}c)^2$. How can we now understand why the forward intensity decreases with increasing polymer concentration?

For a solution of ideal noninteracting chains (or more general solute particles) the osmotic pressure is given by the derivative of the free energy $\partial F/\partial V$

$$\Pi = -\frac{\partial F}{\partial V} = n_{\mathrm{p}} k_{\mathrm{B}} T = \frac{N_{\mathrm{A}} k_{\mathrm{B}} T}{M} c, \tag{64}$$

where n_{p} is the particle density, and c the mass concentration. For interacting particles, we can express the concentration dependence through a more generalized form of the noninteracting case $\Pi \sim c$. This is achieved with a so-called virial expansion such that the interaction of j particles is described by terms in jth power in the concentration c:

$$\Pi = N_{\mathrm{A}} k_{\mathrm{B}} T \left(A_1 c + A_2 c^2 + A_3 c^3 + \cdots \right). \tag{65}$$

The coefficients A_j contain the information about the interactions of j polymer chains. For noninteracting chains we recover Eq. (64), since in this case $A_1 = 1/M$ and $A_j = 0$ for $j \geq 2$. For a given polymer solution the osmotic pressure is thus a function of T, M and c, i.e.: $\Pi = \Pi(T, M, c)$. The virial coefficients A_j are functions of the temperature and the molar mass: $A_j = A_j(T, M)$.

What can we say about the virial coefficients A_j? We first look at the change in A_1 upon incorporation of two-particle interactions ($\rightarrow A_2$). This is given by the product of the number of interactions (= number of polymers N_p) and the probability of having such an interaction ($\sim$ volume fraction of polymers $\overline{R_G^3}/V$):

$$N_p \frac{\overline{R_G^d}}{V} = \frac{n_p}{n_p^*} = \frac{c}{c^*}. \tag{66}$$

Here we introduce the overlap concentration $n_p^* = 1/\overline{R_G^3}$ as the concentration where every polymer chain possesses exactly its own volume (de Gennes, 1979). When using mass concentrations this leads to $c^* = n_p^*(M/N_p)$, where N_p is the number of polymer chains. Every additional particle interaction once again contributes by this factor, i.e. the j-th particle interaction is of the order $n_p(n_p/n_p^*)^{j-1}$. Based on these arguments the osmotic pressure should be of the form

$$\Pi = \frac{RTc}{M} F\left(\frac{c}{c^*}\right) \tag{67}$$

where F is a scaling function.

For the osmotic compressibility this then leads to:

$$\frac{\partial \Pi}{\partial c} = N_A k_B T\left(\frac{1}{M} + 2A_2 c + 3A_3 c^2 + \cdots\right). \tag{68}$$

In analogy to Eq. 67 we then obtain a similar expression for the osmotic compressibility obtained from the forward scattering:

$$\frac{\partial \Pi}{\partial c} = \frac{N_A k_B T}{M} \tilde{F}\left(\frac{c}{c^*}\right) \tag{69}$$

where $\tilde{F}(x) = F(x) + xF'(x)$ is a scaling function. Together with Eq. 63 we then see that the normalized forward intensity should be a universal scaling function as a function of a reduced concentration c/c^* irrespective of the molar mass of the polymer:

$$\left[\frac{1}{Kc}\left(\lim_{q\to 0}\frac{1}{V_{sc}}\frac{d\sigma}{d\Omega}(\boldsymbol{q})\right)\right]\Big/ M_w = \left[\tilde{F}\left(\frac{c}{c^*}\right)\right]^{-1} = S(0). \tag{70}$$

A comparison between the experimentally determined values of the normalized forward scattering $S(0)$ vs. c/c^* and the predicted theoretical behavior based on Eq. (70) is made in Fig. 23. In Fig. 23 we have plotted the reduced osmotic compressibility $(N_A k_B T/M_w)/(\partial\Pi/\partial c) = S(0)$ versus c/c^* obtained from SANS experiments with polystyrene in deuterated toluene for different values of the molar mass M_w (Pedersen and Schurtenberger, 1999). We indeed find that all the experimental data collapse onto a single master curve as predicted by Eq. (70). Calculations of the scaling function $\tilde{F}(c/c^*)$ have for example been made using renormalization group theory or Monte Carlo computer simulations. The renormalization group theory expression by Ohta and Oono (1982) is also included in Fig. 23, and we find good agreement between the experimental data and the theory except for a systematic deviation at high values of c/c^*.

Note on the practical use of light scattering for polymer characterization: Eqs. (63) and (68) can be rewritten:

$$\lim_{q\to 0}\frac{1}{V}\frac{d\sigma}{d\Omega}(\boldsymbol{q}) = Kc\left[\frac{1}{M} + 2A_2 c + \cdots\right]^{-1}. \tag{71}$$

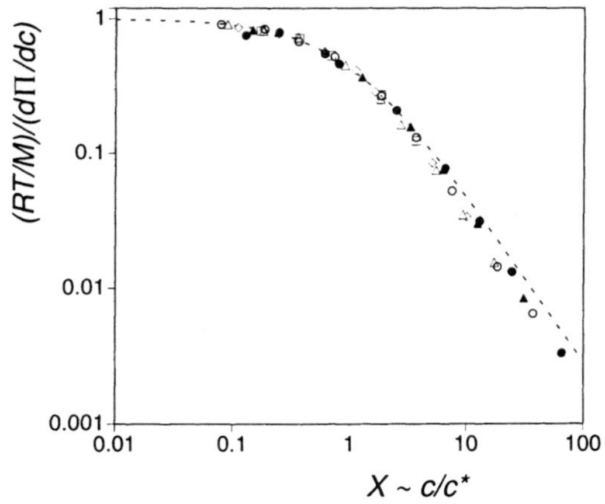

Fig. 23. Reduced osmotic compressibility $(N_A k_B T/M_w)/(\partial\Pi/\partial c)$ as a function of the reduced concentration c/c^* for polystyrene in toluene for several different values of M_w [3000 (■), 13,000 (◇), 28,000 (△), 68,000 (○), 120,000 (▲) and 350,000 (●)]. The data are taken from Pedersen and Schurtenberger(1999). The dashed line shows the result of the renormalization group calculation.

Eq. (71) provides us now with the complete background for the experimental determination of the molar mass of a polymer with light or neutron scattering. In particular we see that for not too high values of c we obtain the correct molar mass from a linear extrapolation of the concentration dependence of the osmotic compressibility, i.e., from a plot of $1/I(q)$ vs. c.

3.2. Semi-dilute solutions: Blob model and scaling laws

Polymer solutions are called semi-dilute if they obey two conditions: The polymer concentration is above the overlap threshold ($c > c^* = M/(N_A \overline{R_G^3})$), and the volume fraction of the polymer is small compared to that of the solvent ($\Phi \ll 1$). While we have found universal scaling functions for the osmotic pressure and the osmotic compressibility that are capable of reproducing these experimental quantities over an extended range of concentrations, this is much more difficult to achieve for the complete structure factor $S(q)$. In contrast to dilute solutions and the polymer melt, semi-dilute solutions are very difficult to treat experimentally due to an additional characteristic length scale, the so-called mesh size or correlation length ξ_s. The meaning of this quantity is schematically shown in Fig. 24. This figure shows that a semi-dilute polymer solution can be looked at as a transient network of entangled chains. The mesh size of this network, i.e., the average distance between entanglement points, is large compared to the monomer size. We can thus immediately design a blob model that captures the most essential features of such a solution: For distances small compared to ξ_s, the polymer chains will exhibit single coil behavior, i.e. possess a random walk or self avoiding walk conformation. At larger distances, we expect to find a behavior characteristic of a polymer melt as the sample space is densely packed with polymer segments of size ξ_s, the blobs. Therefore, for good solvent conditions the single chain static structure factor should show a different behavior for $q > 1/\xi_s$ (dilute solution behavior, $S(q) \sim q^{-5/3}$ provided that $\xi_s \gg b$) and $q < 1/\xi_s$ (melt behavior, $S(q) \sim q^{-2}$). Such a behavior can for example be seen for

Fig. 24. Semi-dilute polymer solution as an entanglement network with mesh size ξ_s. On length scales small compared to ξ_s we observe dilute solution behavior for the individual chains, whereas on length scales large compared to ξ_s the behavior is that of a polymer melt with the blobs playing the role of a coarse grained monomer.

mixtures of deuterated and non-deuterated chains with SANS, where a determination of the single chain structure factor is possible (Farnoux et al., 1978). Before attempting to derive the full structure factor we now first try to find simple scaling laws for the correlation length $\xi_s(c)$ and the radius of gyration $\overline{R}_G(c)$ of the single coil in the semi-dilute solution.

We can determine the correlation length experimentally from the low-q behavior of the static structure factor which is given by

$$S(\boldsymbol{q}) = \frac{S(0)}{\left(1+q^2\xi^2\right)}, \tag{72}$$

i.e., it follows a so-called Ornstein–Zernike law. Such a form arises from the modified form of the monomer correlation function

$$g(r) = n_m \frac{\xi}{r}\, e^{-r/\xi}, \tag{73}$$

where n_m is the monomer density (de Gennes, 1979). This form is based on $g(r) \sim 1/r$ for single coils on length scales $a < r < \overline{R}_G$ modified with an exponential screening term that describes the fact that excluded volume interactions are screened by entanglement effects on length scales $r > \xi_s$. $S(\boldsymbol{q})$ is then obtained through a Fourier transformation of $g(r)$.

The results for the concentration dependence of the correlation length for polystyrene in deuterated toluene at different values of the molar mass from an analysis of the SANS data shown in Fig. 20 using Eq. (72) is shown in Fig. 25 (Pedersen and Schurtenberger, 1999). Here we have plotted ξ_s/ξ_0 vs. c/c^*. ξ_0 is the correlation length in the limit of $c \to 0$ for which we find ${\xi_0}^2 = \overline{R}_G^2/3$ from a comparison of Eq. (72) and the Guinier approximation.

The data shown in Fig. 25 exhibit a behavior which is similar to the concentration dependence of the osmotic compressibility. At low concentrations $\xi_s(c)$ corresponds to the coil size given by the radius of gyration of the polymer. However, at higher values of $c > c^*$ we can resolve the structure of the many chain system only and $\xi_s(c)$ thus reflects the mesh size only, which on this length scale should be independent of M_w. We now try to obtain a universal scaling law for the concentration dependence of $\xi_s(c)$ (de Gennes, 1979). We start with the single chain for which we write $\overline{R}_G(0) = aN^{\nu}$. As we expect to find inside the blob

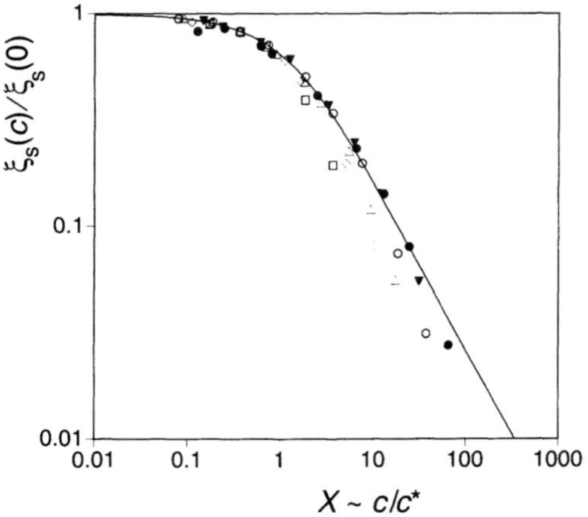

Fig. 25. Normalized correlation length ξ_s/ξ_0 vs. c/c^* determined from SANS experiments for polystyrene in toluene for several different values of M_w [3000 (□), 13,000 (◇), 28,000 (△), 68,000 (○), 120,000 (▲) and 350,000 (●)]. The data are taken from Pedersen and Schurtenberger (1999). The solid line shows the result of the renormalization group calculation.

the same conditions as for dilute polymer solutions we then obtain

$$\xi_s(c) = aN_\xi^\nu, \tag{74}$$

where N_ξ is the number of monomers per blob. Since the blobs are densely packed (melt of blobs) we write for the monomer density n_m

$$n_m = \frac{N_\xi}{\xi_s^3(c)} \tag{75}$$

and thus obtain for the mesh size

$$\xi_s(c) = a\left(n_m\xi_s^3(c)\right)^\nu = \left(an_m^\nu\right)^{1/(1-3\nu)}. \tag{76}$$

Together with $n_p = n_m/N$ and $\overline{R}_G(0) = aN^\nu$ we obtain $an_m^\nu = \overline{R}_G(0)n_p^\nu$ and thus

$$\xi_s(c) = \overline{R}_G(0)\left(\frac{c}{c^*}\right)^{-\nu/(\nu d-1)}. \tag{77}$$

For good solvents this leads to $\nu/(\nu d - 1) = 0.75$, i.e. we expect a scaling law of the form

$$\frac{\xi_s(c)}{\overline{R}_G(0)} \sim \left(\frac{c}{c^*}\right)^{-0.75}. \tag{78}$$

The scaling exponent of 0.75 is indeed in very good agreement with the results found for the concentration dependence of the correlation length in semidilute solutions (Fig. 25) (Pedersen and Schurtenberger, 1999).

An analogous treatment based on similar arguments can also be made for the concentration dependence of the radius of gyration of a chain in semi-dilute solutions. For good solvents this leads to

$$\overline{R}_G(c) = \overline{R}_G(0)\left(\frac{c}{c^*}\right)^{-1/8}. \tag{79}$$

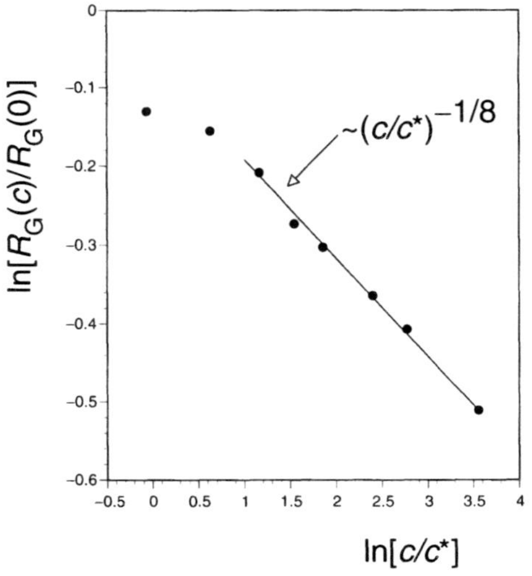

Fig. 26. The concentration dependence of the radius of gyration $\overline{R}_G(c)$ for polystyrene ($M_w = 1{,}114{,}000$ g/mol) in CS_2. Data taken from Daoud et al. (1975).

We thus expect that $\overline{R}_G(c)$ decreases with increasing polymer concentration due to the screening of intrachain excluded volume effects at distances larger than $\xi_s(c)$ by the presence of other chains until it approaches the value for the ideal random walk chain in the melt. The radius of gyration of a single chain in the entanglement network can be measured with SANS and mixtures of deuterated and non-deuterated chains (Daoud et al., 1975). Fig. 26 demonstrates that the agreement between the theoretical scaling law and the experimental data is indeed very good.

3.3. The structure factor in many chain systems

We have already seen that we can reproduce the static structure factor $S(q,c)$ for $q \to 0$ with a simple virial expansion (Eq. (71)) or with a scaling function of the form $S(0,c) \sim \tilde{F}(c/c^*)$. Moreover, we have found a simple Ornstein–Zernike law for the q-dependence of $S(\boldsymbol{q},c)$ in the Guinier regime which is characterized by a concentration dependent correlation length $\xi_s(c)$. $\xi_s(c)$ follows a scaling law that has been verified experimentally. However, while we are thus able to interpret the influence of interaction effects on $S(\boldsymbol{q})$ in the limit of small values of q, a complete treatment of $S(\boldsymbol{q})$ on all length has turned out to be extremely difficult in the semi-dilute regime. Several attempts to use a 'Random Phase Approximation' or integral equation theories in analogy to simple liquids have been made, but they were successful in concentrated solutions and in the melt only. However, a recent application of Monte Carlo simulations has turned out to yield very promising results. These Monte Carlo simulations have been performed on a system with a Kratky–Porod-type model for semiflexible chains, which has previously been introduced in the description of the single chain properties of worm-like polymers. The model has a fixed valence angle θ and free rotation around the bonds (Kratky and Porod, 1949; Pedersen et al., 1996). Hard spheres were placed along the chain to simulate the excluded volume of the polymer. Each chain has 6–8 spheres per Kuhn

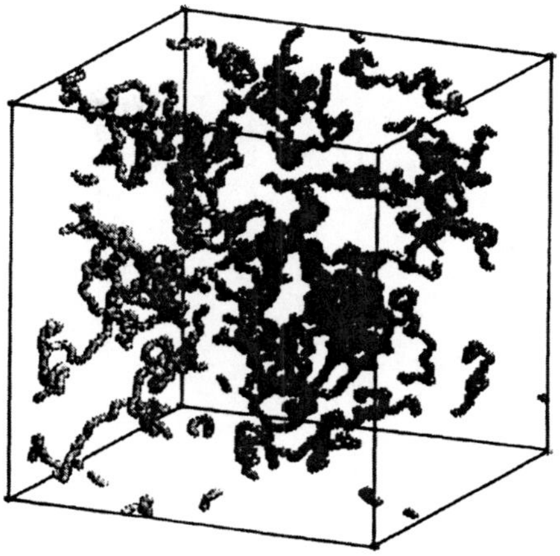

Fig. 27. Example of a snap shot from a MC many chain simulation with $L/b = 10.9$ and $\Phi = 0.0044$.

length b and the system typically contains 10,000–30,000 spheres of radius $R/b = 0.1$ in order to mimic properties of polystyrene. Simulations have been performed for different chain length using a box with periodic boundary conditions in order to limit the effort needed in the simulations of many chain systems at higher concentrations. For PS the contour lengths simulated correspond to molecular weights of about 2700, 9100, 25,000, 75,000 and 225,000 g/mol. For each L/b the concentration was varied from very dilute concentrations well below c^* up to concentrations with volume fractions $\eta = 0.1 - 0.25$ (Pedersen and Schurtenberger. 1999). For polystyrene the highest volume fractions are thus close to the melt density. An example of a snapshot from such a simulation is shown in Fig. 27.

The resulting structure factors as a function of concentration are shown in Fig. 28. We observe a behavior for $S(\boldsymbol{q},c)$ which resembles very closely that of polystyrene previously discussed in Fig. 20. These simulation results have then been analysed and compared to theoretical expressions for the static structure factor. The agreement of the scattering function with the classical RPA result (Daoud et al., 1975; Benoit and Benmouna, 1984)

$$S_{\mathrm{RPA}}(\boldsymbol{q}) = S_{sc}(\boldsymbol{q})/[1 + \beta S_{sc}(\boldsymbol{q})], \tag{80}$$

frequently used for analysing scattering data, was first investigated. In Eq. (80) β is related to the strength of the interaction potential. However, the function $S_{\mathrm{RPA}}(\boldsymbol{q})$ is not able to describe the simulation data and large deviations are found in particular for large concentrations at high q. An alternative approach is to use the polymer version of integral equation theory of liquids (Hansen and MacDonald, 1986), the polymer reference interaction site model (PRISM), which in the equivalent site approximation gives (Schweizer and Curro, 1994):

$$S_{\mathrm{PRISM}}(\boldsymbol{q}) = S_{sc}(\boldsymbol{q})/[1 + \beta c(q) S_{sc}(\boldsymbol{q})], \tag{81}$$

where $c(q)$ is the normalized Fourier transform of the direct correlation function for the spheres on the chains, and β is given by $\beta = 1/S(0) - 1$. In order to interpolate and fit the SANS data for PS, the simulated scattering functions were parameterized (Pedersen and Schurtenberger, 1999). The parameterization reproduces the experimentally measured $S(q)$ (Fig. 28) very well for all concentrations investigated. We have thus a functioning expression

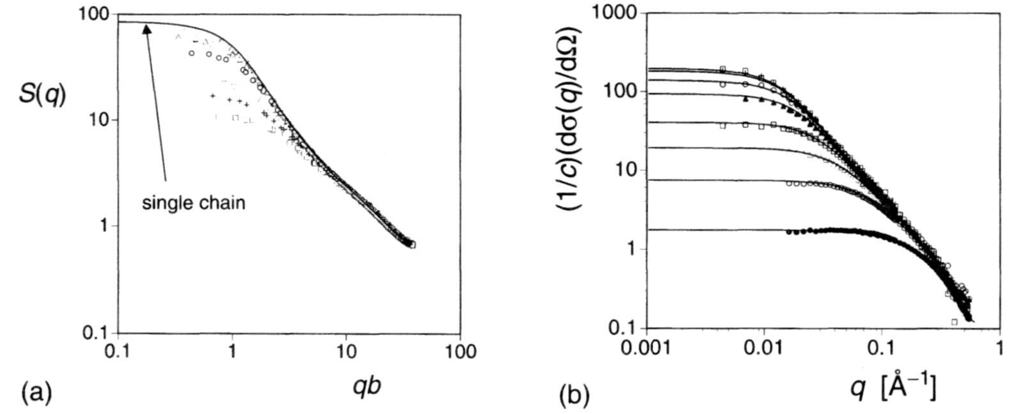

Fig. 28. (A) Scattering functions $S(\boldsymbol{q})$ for $L/b = 90$ and $\eta \geq 0.075$ from Monte Carlo simulations. The values of η are (from below) 0.075, 0.05, 0.03, 0.02, 0.015, 0.01, 0.007 0.005, 0.003, 0.002, 0.0015, 0.001, 0.0007, 0.0005, 0.0003, 0.0002, 0.00015, 0.0001, 0.00005, 0.000025. (B). Scattering data $S(q)$ for PS with $M = 120{,}000$ g/mol. The concentrations are (from below) $c = 0.27$, 0.11, 0.054, 0.028, 0.011, 0.0052, 0.0021 and 0.0013 g/ml. The curves are a fit using a PRISM-type parameterization of the Monte Carlo simulation results. Data has been taken from Pedersen and Schurtenberger (1999).

for the static structure factor $S(\boldsymbol{q})$ that is capable to reproduce the experimentally obtained data at all length scales and up to very high concentrations.

3.4. Concentrated solutions and melt

Dilute polymer solutions in a good solvent are more expanded than ideal chains under theta conditions due to excluded volume effects. We have found a scaling law of the form $\overline{R}_G \simeq M^{\nu}$, where $\nu = 3/5$ in good solvents. At increasing concentrations we intuitively expect that the chains become more compact again. Indeed one finds that a polymer in the melt behaves like an ideal random walk chain. This surprisingly simple behavior can be understood based on ideas proposed by Flory (Flory, 1953; de Gennes, 1979). He has been able to show that inter- and intrachain excluded volume effects cancel in the melt. A chain then behaves like a Gaussian chain with $\overline{R}_G = aN^{1/2}$. The asymptotic q-dependence of $S(q)$ is given by $S(q) \sim q^{-2}$. These predictions have indeed been verified with neutron scattering experiments on mixtures of deuterated and non-deuterated chains (Fig. 21).

4. Summary

We have seen that we can distinguish three different concentration regimes with distinctly different behavior for polymers in solution: Dilute, semidilute and concentrated solutions and the melt. The crossover from the dilute to the semidilute regime is given by the overlap concentration c^* where the individual coils start to overlap. A schematic overview has been given in Fig. 22. The different regimes can be summarized as follows:

- Dilute solutions: $c < c^*$. The individual coils are far apart from each other. With scattering experiments we can investigate the polymer conformation at all length scales from the local monomer structure to the overall size given by the radius of gyration. The theoretical treatment is relatively easy as interactions between different chains are of minor importance only. We have seen that we can incorporate the effects of the excluded volume interactions, flexibility, temperature and solvent quality and obtain a universal description of the global polymer properties that is chemistry independent. Analogous to an ideal gas we can treat a solution of polymer chains as a system of hard spheres, and deviations from ideal behavior can be incorporated with a simple virial expansion in the concentration.

- Semi-dilute polymer solutions: $c > c^*$ and $\phi \ll 1$. The polymer concentration is above the overlap threshold c^*. The different polymer chains interpenetrate, but the volume fraction of polymers ϕ is still small compared to the solvent. Interactions between the monomers can be treated relatively easily. However, the correlation function is already quite complicated as interactions need to be considered on different length scales. The correlation length or mesh size ξ_s has shown to be the important characteristic quantity. At distances below ξ, i.e., for $r \ll \xi$, the single chain behavior dominates as the influence from the neighbouring chains is not visible. Therefore a semi-dilute solution behaves analogous to a dilute solution on length scales smaller than ξ_s. At distances $r > \xi$ we now observe the influence of the other chains in the system. At these length scales we find a behavior that is characteristic for a polymer melt. The entire space is densely packed with polymer segments (blobs) of size ξ_s, and we model the system as a melt of blobs. We have found universal scaling functions that describe the concentration dependency of quantities such as the osmotic pressure, the osmotic compressibility, the correlation length or the radius of gyration as a function of a reduced concentration c/c^*.
- Concentrated solutions and melts: $c \gg c^*$ and $\phi \sim 1$. The polymer concentration is so high that the polymer volume fraction is almost one. Surprisingly such systems are once again very easy to describe theoretically. Concentration fluctuations are extremely reduced, and the polymer conformation exhibits a random walk statistics, i.e., polymers in the melt can be described as Gaussian chains.

Acknowledgements

I gratefully acknowledge the important contributions made by Jan Skov Pedersen, Hans Christian Öttinger, Götz Jerke, Luigi Cannavacciuolo and Anna Stradner. Without their help this chapter would not have been possible.

References

Arai, T., Abe, F., Yoshizaki, T., Einaga, Y. and Yamakawa, H. (1995). *Macromolecules*, **28**, 5458.

Ballard, D.G.H., Rayner, M.G. and Schelten, J. (1976). *Polymer*, **17**, 349.

Benoit, H. and Benmouna, M. (1984). *Polymer*, **25**, 1059.

Benoit, H. and Doty, P. (1953). *J. Phys. Chem.*, **57**, 958.

des Cloizeaux, J. and Jannink, G. (1990). *Polymers in Solutions: Their Modelling and Structure*. Claredon Press, Oxford.

Daoud, M., Cotton, J.P., Farnoux, B., Jannink, G., Sarma, G., Benoit, H., Duplessix, R., Picot, C. and de Gennes, P.-G. (1975). *Macromolecules*, **8**, 804.

Debye, P. (1947). *J. Phys. Coll. Chem.*, **51**, 18.

Doi, M. (1995). *Introduction to Polymer Physics*. Oxford University Press, Oxford.
Duplantier, B. (1985). *J. Phys.*, **47**, 1633.
Farnoux, B., Boue, F., Cotton, J.P., Daoud, M., Jannink, G., Nierlich, M. and deGennes, P.G. (1978). *J. Phys.*, **39**, 77.
Flory, P.J. (1953). *Principles of Polymer Chemistry*. Cornell University Press, Ithaca.
Flory, P.J. (1969). *Statistical Mechanics of Chain Molecules*. Interscience Publishers, New York.
Freed, K.F. (1987). *Renormalization Group Theory of Macromolecules*. Wiley, New York.
Fujita, H. (1990). *Polymer Solutions*. Elsevier, Amsterdam.
de Gennes, P.G. (1979). *Scaling Concepts in Polymer Physics*. Cornell University Press, Ithaca.
Hansen, J.P. and MacDonald, I.R. (1986). *Theory of Simple Liquids*. Academic Press, London.
Higgins, J.S. and Benoit, H.C. (1996). *Polymers and Neutron Scattering*. Oxford University Press, Oxford.
Huber, K., Bantle, S., Lutz, P. and Burchard, W. (1985). *Macromolecules*, **18**(7), 1461.
Kirste, R.G. and Oberthür, R.C. (1982). In: *Small Angle X-ray Scattering* (Glatter, O. and Kratky, O., eds.). Academic Press, London.
Kirste, R.G. and Wild, G. (1969). *Makromolekulare Chem.*, **121**, 174.
Kratky, O. and Porod, G. (1949). *Rec. Trav. Chim.*, **68**, 1106.
bibitemKuhn34Kuhn, W. (1934). *Kolloid Z.*, **68**, 2.
Kuhn, W. and Kuhn, H. (1943). *Helv. Chim. Acta*, **26**, 1394.
Meirowitch, H. (1991). In: *Computer Simulation of Polymers* (Roe, R.J., ed.). Prentice-Hall, Englewood Cliffs.
Miyaki, Y., Einaga, Y. and Fujita, H. (1978). *Macromolecules*, **11**(6), 1180.
Oono, Y. (1985). *Adv. Chem. Phys.*, **61**, 301.
Ohta, T. and Oono, Y. (1982). *Phys. Lett.*, **89A**, 460.
Pedersen, J.S. and Schurtenberger, P. (1996). *Macromolecules*, **29**, 7602.
Pedersen, J.S. and Schurtenberger, P. (1999). *Europhys. Lett.*, **45**, 666.
Pedersen, J.S., Laso, M. and Schurtenberger, P. (1996). *Phys. Rev. E*, **54**, R5917.
Rawiso, M., Duplessix, R. and Picot, C. (1980). *Macromolecules*, 13, 1518.
Schäfer, L. (1999). *Excluded Volume Effects in Polymer Solutions as Explained by the Renormalization Group*. Springer-Verlag, Berlin.
Schwahn, D. (1991). In: *Physik der Polymere, Proceedings of the 22. IFF-Ferienkurs*, Forschungszentrum Jülich GmbH, pp. 7.1.
Schweizer, K.S. and Curro, J.G. (1994). *Adv. Polymer Sci.*, **116**, 319.
Sharp, P. and Bloomfield, V.A. (1968). *Biopolymers*, **6**, 1201.
Strobl, G. (1996). *The Physics of Polymers*. Springer-Verlag, Berlin.
Utiyama, H., Tsunashima, Y. and Karate, M.J. (1971). *J. Chem. Phys.*, **55**, 3133.
Yamakawa, H. (1971). *Modern Theory of Polymer Solutions*. Harper & Row, New York.
Yoshizaki, T. and Yamakawa, H. (1980). *Macromolecules*, **13**, 1518.

Neutrons, X-Rays and Light
P. Lindner and Th. Zemb (Editors)

Chapter 12

Surfactant Micelles and Bilayers: Shapes and Interactions

Grégoire Porte

Groupe de Dynamique des Phases Condensées, UMR 5581 CNRS/UM.II, Case 026, Université Montpellier II, F-34095 Montpellier CEDEX 5, France

1. Introduction

One of the most distinctive properties of surfactant systems is certainly the impressive polymorphism of their mesoscopic structures in aqueous solutions (Ekwall, 1975). The selective solvent effect is of course at the origin of their self assembling properties in water: aggregates form spontaneously where the hydrophobic tails are confined into the core and so remain protected from contact with water by the layer of hydrophilic head groups which are located at the surface, wetted by the aqueous solvent. Such micro-partitioning of heads and tails is compatible with a large variety of different shapes and examples of morphologies currently encountered in dilute surfactant solutions are shown in Fig. 1. Starting from common globular micelles and tuning the experimental conditions (salinity, temperature, etc.) one can induce the growth of flexible cylindrical micelles (Mazer et al., 1976). When these wormlike objects are long enough, the dilute solutions exhibit remarkable viscoelastic properties. In other conditions, self assembling leads to very large 2D fluid bilayers (Benton and Miller, 1983). They show various topologies at large scale: singly connected infinite sheets (smectic lamellar phase), disconnected finite size vesicles (so-called L4 phase), multi-connected sheet (so-called 'sponge' phase). At higher concentrations, these morphologies often order in space (Ekwall, 1975) and form mesophases – nematic, hexagonal, smectic, cubic, . . . – which enrich even more the phase behavior of simple systems (even binary surfactant/water systems have complex phase behavior).

Such a rich set of structures induces an impressive variety of physical properties (in particular rheological properties) which are therefore equally sensitive to the experimental conditions. Reciprocally, characterizing the structures and morphologies in a given system is a prerequisite to understand and monitor its physical properties and scattering techniques have proved very appropriate for this purpose.

In the present chapter, we see how scattering methods are commonly used to characterize the various morphologies and structures schematized in Fig. 1. We take the point of view of an experimentalist who wants to make a first quick discrimination between possible shapes in a given sample or system: we therefore show how to use asymptotic data treatment and we leave more sophisticated fitting procedures for a later, more refined stage of characterization. By the way, we explain how the morphology sequence and the phase behavior is monitored by the spontaneous curvature of the hydrophilic/hydrophobic interface. Moreover, for two kinds

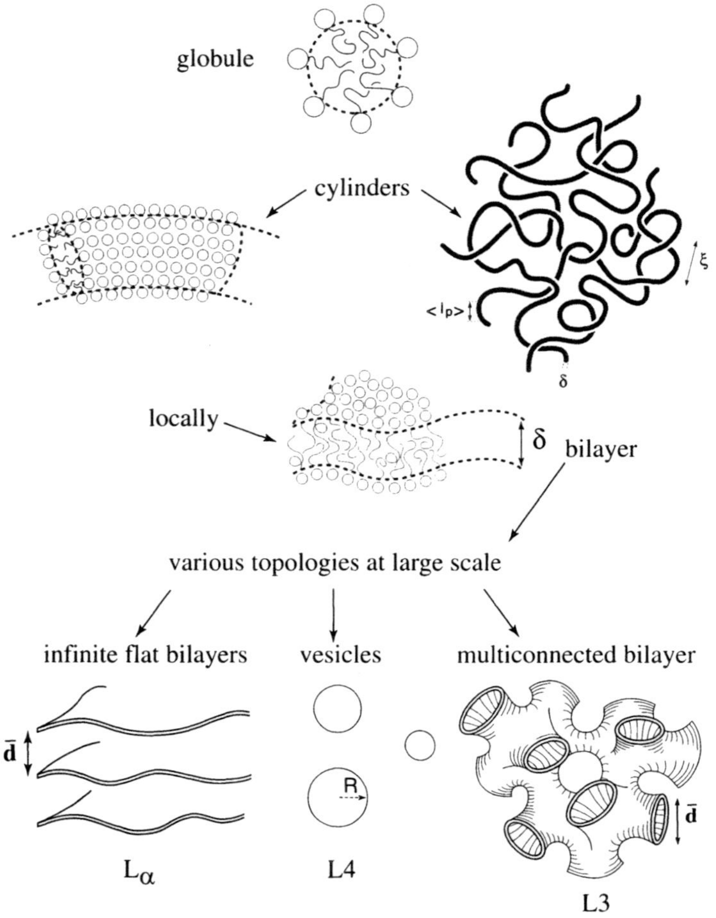

Fig. 1. Various morphologies and mesostructures in dilute surfactant aqueous solutions.

of microstructures only: globular micelles and lamellar phase of infinite bilayers, we shall examine how interactions affects the scattering pattern.

2. Phase behaviour

Even if we restrict to moderate concentrations (below 20% amphiphile), the phase behaviour is surprisingly rich. In Fig. 2, which represents the generic phase diagram of many surfactant systems, the horizontal axis is the total concentration in amphiphiles ϕ (volume fraction). The vertical axis is a control parameter noted η, the choice of which depends on the system considered: for a non-ionic surfactant binary solution η is simply the temperature (Olsson and Wennerström, 1994), for a ionic surfactant aqueous solutions η is the salinity, and for a more complex ternary system involving a surfactant and a cosurfactant in water η could be the cosurfactant to surfactant ratio (Porte et al., 1988). As we shall see below, η can be

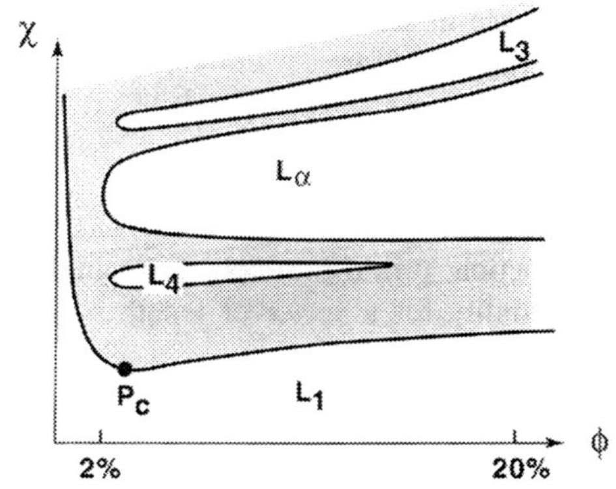

Fig. 2. Generic phase diagram of dilute surfactant solutions.

any experimentally tunable parameter which *induces a monotonic decrease of the spontaneous curvature of the amphiphilic interface*. Provided this appropriate choice for the axis η and ϕ, the 2D section of the phase diagram of many different systems has the same general geometry (Fig. 2).

The phase boundaries are roughly horizontal except at high dilutions. At the lowest values of η, we find the domain of the classical micellar phase (L1 phase): it is transparent (weak turbidity), shows no optical birefringence and exhibits no long range order. The viscosity in the L1 phase increases steadily with η up to the upper phase boundary.

The next one-phase domain met upon increasing η is the L4 domain (vesicle phase): it is more turbid than the L1 phase; it is fluid below $\phi = 5\%$, optically isotropic and shows no long range order.

Increasing further η, one then finds a birefringent phase showing 'focal conic textures' in polarized optical microscopy characteristic of *a smectic long range order*: this is the Lα lamellar phase.

At still higher η, we meet another fluid, transparent, moderately turbid, fluid, disordered phase: this is the L3 'sponge' phase.

Although the L4 phase is not always observed, the three other phases are ubiquitous for almost all systems: they always appear in the same order as described above.

3. Shape sequence

3.1. Evidenced by scattering

The morphological transformation underlying the above phase behaviour have been investigated by neutron and X-ray scattering. According to the basic principles, interactions mainly affect the scattering pattern in a wave vector range of the order or below the inverse average interaggregate distance. At sufficient dilution, the shape dominates the scattering over a large q-range comprising the inverse of the size of the aggregate. A first classification of the shapes can be proposed according to their dimensionality D: $D = 0$ for globular micelles, $D = 1$ for cylinders, and $D = 2$ for bilayers. Assuming simple contrast for simplicity, characterized by $\Delta\rho$ the difference in scattering length density between the aggregates and the solvent, we write for an isotropic solution:

$$I(q) = \frac{(\Delta\rho)^2}{V} \int_0^\infty 4\pi r^2 \langle\gamma(0)\gamma(r)\rangle \frac{\sin(qr)}{qr}\, \mathrm{d}r \tag{1}$$

where $\gamma(r)$ is a two values function: 0 in the solvent and 1 inside the aggregate. Therefore $\langle\gamma(0)\gamma(r)\rangle$ is simply the probability for a vector of length r to have both ends inside the aggregate. For a 1D cylinder of radius a:

$$\langle\gamma(0)\gamma(r)\rangle \approx \phi\frac{2\pi a^2}{4\pi r^2} = \frac{\phi}{2}\left(\frac{a}{r}\right)^2 \quad \text{at large } r. \tag{2}$$

Similarly for a 2D bilayer of thickness δ:

$$\langle\gamma(0)\gamma(r)\rangle \approx \phi\frac{2\pi r\delta}{4\pi r^2} = \phi\frac{\delta}{2r} \quad \text{at large } r. \tag{3}$$

Using these asymptotic expressions in (1), we get immediately:

$$I(q) \approx \pi^2(\Delta\rho)^2\phi a^2/q \propto q^{-1} \quad \text{for cylinders} \tag{4}$$

and:

$$I(q) \approx 2\pi(\Delta\rho)^2\phi\delta/q^2 \propto q^{-2} \quad \text{for bilayers} \tag{5}$$

More generally, we expect $I(q)$ to show a dominant q^{-D} dependence. Using asymptotic expressions for $\langle\gamma(0)\gamma(r)\rangle$ such as (2) and (3) amounts to neglecting the finite transverse dimension of the aggregate at small r. More appropriate expressions are given in Glatter and Kratky (1982) which include the well known Guinier approximation for globules:

$$I(q) \approx \phi V(\Delta\rho)^2 \exp(-q^2R_G^2/3) = \phi V(\Delta\rho)^2 \exp(-q^2R^2/5) \tag{6}$$

for globules of volume V and radius of gyration R_G (or spheres of radius R);

$$q \cdot I(q) \approx \pi^2\phi a^2(\Delta\rho)^2 \exp(-q^2a^2/4) \tag{7}$$

for cylinders;

$$q^2 \cdot I(q) \approx 2\pi\phi\delta(\Delta\rho)^2 \exp(-q^2\delta^2/12) \tag{8}$$

for bilayers. Of course these expressions only apply at small q's compared to transverse size R, a or δ.

Fig. 3 shows an example (Marignan et al., 1989) where expression (8) is used to characterize by neutron scattering the bilayers in a dilute L3-sponge phase [examples for the other morphologies can also be found in Marignan et al. (1989)]. Depending on the dimensionality of the aggregate a straight line is obtained in the appropriate semi-log plot ($\ln(I(q))$, $\ln(qI(q))$ or $\ln(q^2I(q))$ versus q^2 respectively for globules, cylinders and bilayers). Note that in expressions (6), (7), and (8), the transverse size (R, a or δ) appears not only in the exp term (that is in the slope of the semi-log plot) but also in the prefactor (that is in the extrapolation to the origin in the semi-log plot). It is therefore very important to collect the data in absolute units and check that the slope and the extrapolation are consistent with the

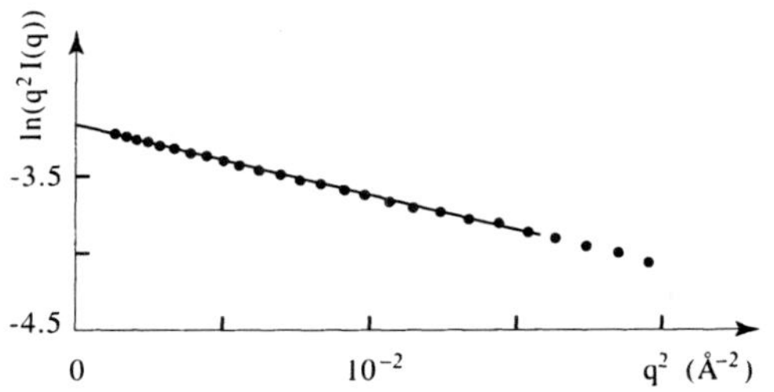

Fig. 3. $\ln(q^2 I(q))$ versus q^2 for a dilute phase of fluid bilayers: the L3 sponge phase. System CPCl/n-hexanol/brine.

expected value of the contrast $\Delta\rho$. If not, this means that interaggregates interactions cannot be neglected: they bias the scattering at low q's. When this happens, it is much safer to fit the data at high q's using the classical expressions for the form factors computed for the different morphologies:

$$I(q) = (\Delta\rho)^2 \phi V \left(3\frac{\sin qR - qR\cos qR}{(qR)^3}\right)^2 \quad \text{for spheres} \tag{9}$$

where $V = \frac{4}{3}\pi R^3$ is the volume of each sphere;

$$I(q) = \frac{\pi^2}{q}\phi a^2 (\Delta\rho)^2 \left(2\frac{J_1(qa)}{qa}\right)^2 \quad \text{for long cylinders} \tag{10}$$

$$I(q) = \frac{2\pi}{q^2}\phi\delta(\Delta\rho)^2 \left(\frac{\sin(q\delta/2)}{q\delta/2}\right)^2 \quad \text{for bilayers} \tag{11}$$

In order to emphasize the high q's, which are less affected by the interactions, it is recommended (Cabane et al., 1985) to use a so-called Porod's plot: that is $q^4 I(q)$ versus q. Wavy patterns are so obtained, as illustrated in Fig. 4. According to (9)–(11), the positions of the maxima and minima discriminate the morphology of the aggregates. On the other hand, the damping of the oscillations arises from the polydispersity of R, a or δ, which is neglected in (9)–(11) but could be easily incorporated. Note in Fig. 4 that data taken from concentrated

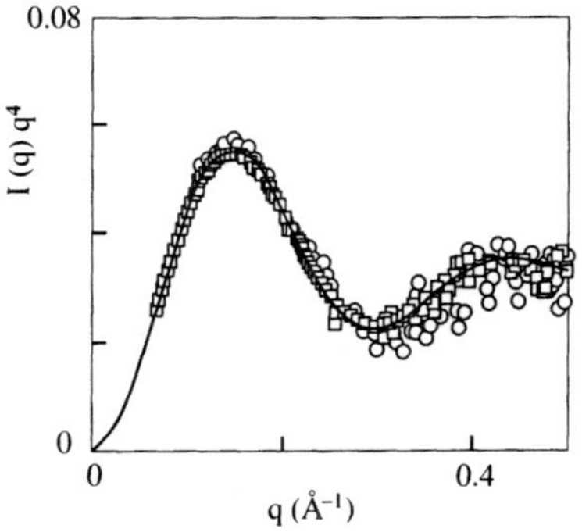

Fig. 4. Porod plot ($q^4 I(q)$ versus q) for 20% samples of the Lα (o) and L3 (□) phases of bilayers. System: CPCl/n-hexanol/brine.

(20%) Lα (lamellar) and L3 (sponge) samples of the same system superimpose almost exactly at high q's, indicating that the morphology at local scale (the thickness of the bilayer) is identical in the two phases although the structure at large scale is different. This confirms that using a Porod's plot is reliable for shape and size characterizations despite the interactions. Moreover, a leading q^{-4}-dependence at high q's is characteristic of a sharp interface for the aggregates. Diffuse boundaries would produce a scattering dropping down much faster than q^{-4}. Therefore Porod's plots such as the one in Fig. 4 provide strong evidence that, whatever their shape, the water/hydrocarbon interface of the aggregates is sharp and well defined.

Similar analysis of scattering data taken from various systems during the late 1980s have shown that the phase sequence is indeed related to a well defined morphological sequence: a progressive unidimensional growth of initially globular micelles into long cylinders is first observed upon increasing the control parameter η in the L1 micellar phase; on the other hand, the three other phases, successively observed at higher η's, correspond to the same local morphology as of a bilayer but arranged differently at large scales.

3.2. Shape sequence and curvature of the interface

The well known shape sequence: globules, cylinders and bilayers, observed at increasing η has been discussed at length in the literature [see, for instance, Israelachvili et al. (1976)]; it can be rationalized as follow. The respective requirements of the hydrophilic head group on the one side of the hydrocarbon water interface and those of the hydrophobic tails on the other side of the interface actually work as in a bimetallic strip and determine an optimum curvature for the interface which we denote c_0. Indeed, changing the experimental conditions (that is changing the control parameter η) will change the heads and tails requirement and so change the optimum curvature. As long as η is such that c_0 is of the order or larger than $2/l$ where l is the length of the surfactant molecules, a quasi-spherical shape of radius $R = 2/c_0$ will be preferred. If, on the other hand, c_0 decreases to values smaller than $2/l$, a spherical micelle of radius $R = 2/c_0 > l$ would exhibit a void at the center of the core which is indeed energetically not favorable. A first bifurcation occurs towards a cylindrical shape of radius $a = 1/c_0$ which will be favored as long as c_0 remains larger than $1/l$. When c_0 decreases further below $1/l$, cylinders become less and less favorable and a second morphological bifurcation ultimately happens towards bilayers for which the curvature of the interface is zero. So, the polymorphism arises from the competition between the preferred curvature c_0 and the constraint that, in one direction of space at least, the thickness of the aggregate may not exceed $2l$.

4. Globular micelles

Beyond the above characterization of their size and shape, scattering can be used further to investigate the internal structure of globular micelles. One of the most exhaustive, detailed studies of SDS globular micelles to date is reported in Cabane et al. (1985). Deuterating selectively the methyl end group of the SDS molecule, the authors showed that it is evenly distributed everywhere in the core of the micelles. Their results confirm the liquid structure of the core with only little (if any) penetration of water.

In absence of added salt, globular ionic micelles experience strong electrostatic repulsion.

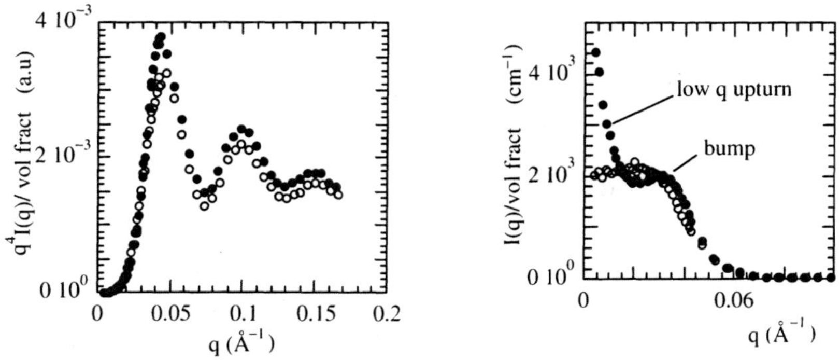

Fig. 5. (a) Oil swollen micelles: without (o) and with 8 telechelic polymers per micelle (•). (b) Same as (a).

Their distribution in space is far from random and a correlation peak arises in the q-range of the order of the inverse distance between the micelles. Expressing the scattered intensity as the product of the form and structure factors, Hayter and Penfold (1981) were the first to exploit the peak in terms of micellar interactions using the statistical methods initially worked out in the field of the physics of liquids [see also Cabane et al. (1985), Zemb and Charpin (1985)]. These methods are presented in detail in Chapter 14 by R. Klein.

We here only show an illustration of the effect of the interactions on the scattering pattern of spherical micelles in Fig. 5a,b. Two identical batches of a 5% dispersion in brine (0.2 M NaCl) of spherical micelles of mixed CPCl (surfactant) and n-octanol (cosurfactant) slightly swollen with decane are prepared. In one batch, a small amount of a hydrophobically end modified polyoxyethylene ($C_{18}H_{37}$ group grafted at both ends of the polymer) is incorporated (Filali et al., 1999). The end-groups of the polymer reversibly stick onto the micelles which so link to their neighbors. The scattering patterns of the two batches (with and without the polymer) are compared. The intensity scattered by the polymer itself is negligible. The above recommended Porod plot is used in Fig. 5a: maxima and minima at high q's for the two batches coincide remarkably: the size and shape of the micelles (spheres of 62 Å radius) are not affected by the addition of polymer. On the other hand, at low q's the scattering is very different (Fig. 5b). The upturn at very low q's, seen for the solution with polymer only, indicates that bridging induces an effective attractive interaction between the micelles. Moreover, the bump at moderate q's shows that there is a preferred distance between micelles in the transient clusters they form due to the bridging attraction.

5. Giant wormlike micelles: analogy with 'living' polymers

Very long wormlike micelles spontaneously grow when η (or c_0) favors locally a cylindrical shape. However, the end caps of a long but finite cylinder necessarily have a more or less hemispherical shape for which the curvature requirement is not satisfied: so, there is a positive excess curvature energy ΔE_{ec} in each end cap. ΔE_{ec} indeed increases monotonically with η (that is when c_0 decreases). When two giant micelles of respective length L' and L'' combine so to form a longer micelle of size $L = L' + L''$ there is an energetic gain of $2\Delta E_{ec}$ (four end caps in the initial state and only two in the merged final state). On the other hand, there is an entropy penalty to be paid due to restricted translational degrees of freedom: two micelles moving independently are replaced by only one. Balancing these opposite contributions, one can easily calculate the size distribution of the giant micelles (Israelachvili et al., 1976; Cates and Candau, 1990):

$$\langle L\rangle \propto \sqrt{\phi \exp\left(\frac{2\Delta E_{\mathrm{ec}}}{k_B T}\right)} \tag{12}$$

and so explain the progressive increase of the average size with η which is well documented in the current literature. For some systems under favorable conditions, ΔE_{ec} can reach very large values (above $30k_B T$) and the average contour length of the giant micelles is expected of the order of several micrometers!

One-dimensional cylindrical structures have finite bending rigidity: flexibility brings in an internal degree of freedom which plays its role if the object is long enough. The elastic energy $\mathrm{d}E_{\mathrm{el}}$ to be paid in order to bend a length element $\mathrm{d}L$ of the cylinder with a curvature c is:

$$\mathrm{d}E_{el} = \frac{1}{2}Kc^2\,\mathrm{d}L \tag{13}$$

where K is the rigidity modulus which has the dimension of an energy times a length. At finite temperature, a characteristic length is introduced naturally: namely the persistence length L_p:

$$L_p = \frac{K}{k_B T} \tag{14}$$

The persistence length separates the short scales over which the cylinder is essentially stiff and the large scales over which the giant micelle has the shape of a random walk of unit step of the order of L_p. The ensemble of the various bent conformations of a giant micelle is therefore very similar to that of a flexible polymer. But contrary to that of a polymer molecule, the size of a given micelle is not frozen but fluctuates in time around the average value specified in (12). To underline this essential difference, giant micelles are sometimes compared to 'living polymers'.

Of course, as long as static properties are considered, the frozen or annealed character of the size distribution makes no difference. And the analogy of the scattering patterns of giant micelles and polymers solutions in the semi-dilute regime is striking. An example is shown in Fig. 6 where the intensity scattered by a solution of extremely long flexible micelles is collected over three orders of magnitude in q's, combining static neutron and light scattering (Appell and Marignan, 1991). At high q's, we see the leading q^{-4}-dependence (sharp interface) modulated by the oscillations due to the well defined radius a of the cross-section of the cylinders. At lower q's in the range: $L_p^{-1} < q < a^{-1}$, we find a leading q^{-1}-dependence confirming the rod like shape of the micelles at scales shorter than the persistence length. At still larger scale ($\xi^{-1} < q < L_p^{-1}$), the structure is that of a self avoiding random walk the fractal D-dimension of which is 1.66: we so find a leading $q^{-1.7}$ dependence for $I(q)$ in this range. At the lowest q's, below the inverse of entanglement length ξ of the semi-dilute solution, $I(q)$ levels off as expected. Each q-range is treated separately in Appell and Marignan (1991) so to characterize quantitatively the different characteristic lengths of the semi-dilute solution: $a = 21$ Å, $L_p = 170$ Å and $\xi = 1150$ Å. More integrated and accurate procedures for the data treatment, with which the entire scattering pattern is fitted at a time,

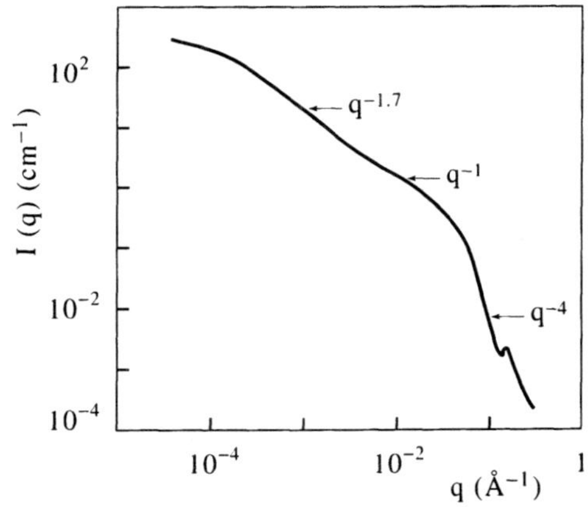

Fig. 6. Scattering pattern of a solution of giant micelles in the semi-dilute regime. $CPClO_3$ (0.25% in weight) in 1 M $NaBrO_3$. Data combine light and neutron scattering.

are presented and used in Pedersen et al. (1995). They provide more reliable quantitative values.

6. L4, Lα and L3: three phases of fluid bilayers

As shown above, the growth of giant wormlike micelles is driven by the excess energy in the end caps. Similarly, the formation of large bilayers can be considered as the result of the growth of disk like micelles driven by the excess energy in their hemicylindrical rims (Israelachvili et al., 1976). Contrary to the unidimensional case where the excess energy in the end cap does not depend on the size of the cylinder, the length of the rim, and so the excess energy attached to it, increases with the size of the disk. This feature makes the 2D reversible aggregation much more singular and sudden than the 1D growth: below some characteristic energy per unit length of the rim no noticeable growth is seen; but as soon as this characteristic value is exceeded, growth diverges abruptly and leads to the formation of infinite bilayers. There is one possibility however to avoid the divergence: this is to allow the bilayer to bend and form a closed shape such as a hollow vesicle which has no rim. Including this remark, we conclude that 2D reversible aggregation leads to the formation of *finite or infinite bilayers but with no rim*. The structures at large scales in the three phases of bilayers all agree with this prediction.

The bending elasticity of 2D fluid bilayers presents distinct features compared to that of 1D cylinders [expression (13)]. Helfrich (1973) was the first to show that the elastic energy dE_{el} to be paid to bend an area element dA of a fluid film can be written as:

$$dE_{el} = \left[\tfrac{1}{2}\kappa(c_1 + c_2 - c_0)^2 + \bar{\kappa}c_1c_2\right] dA \qquad (14)$$

where c_1 and c_2 are the principal curvatures of the film surface. This expression involves three quantities characteristic of the film:

- κ is the mean curvature bending modulus (the sum $c_1 + c_2$ is called the mean curvature);
- $\bar{\kappa}$ is the gaussian curvature bending modulus (the product c_1c_2 is called the gaussian curvature); and

- c_0 expresses the asymmetry of the film: this is the spontaneous curvature (which was introduced above for the specific case of the water hydrocarbon interface).

However, a bilayer consists of two identical monolayers stuck opposite to each other in a symmetrical manner: therefore its spontaneous curvature c_0 is identically zero. (This is indeed not true for the constituting monolayers which are in general asymmetric: asymmetry between the heads and tails of the surfactant molecules.) So for bilayers the elasticity reduces to:

$$\mathrm{d}E_{el} = \left[\tfrac{1}{2}\kappa(c_1 + c_2)^2 + \bar{\kappa}c_1c_2\right] \mathrm{d}A \tag{15}$$

We first note that according to (15), κ and $\bar{\kappa}$ have the dimension of an energy and *there is no direct way to specify a characteristic length.* Even more, integrating (15) over a sphere shows that the energy to be paid in order to form a vesicle of radius R is: $4\pi(2\kappa + \bar{\kappa})$ *independent of the size* R. In difference with the bending elasticity of 1D cylinders, the elasticity of symmetrical bilayers involves no notion of a persistence length and rather obeys a principle of *'scale invariance'*. On the other hand, a flat bilayer is mechanically stable only if any bending deformation is paid by a positive energy price: that is, Eq. (15) must be definite positive. The conditions of stability come out immediately:

$$\kappa > 0 \quad \text{and} \quad -2\kappa < \bar{\kappa} < 0 \tag{16}$$

κ and $\bar{\kappa}$ actually play very contrasting roles. The Gauss–Bonnet theorem states that the gaussian curvature integrated over a surface with no rim is a topological invariant: it only depends on the topology of the surface.

$$\int_A c_1c_2 \, \mathrm{d}A = 4\pi(n_d - n_h) \tag{17}$$

where n_d is the number of separated subunits and n_h is the number of 'handles' of the surface. For example, the integrated gaussian curvature is 4π for a sphere ($n_d = 1$, $n_h = 0$) and 0 for a torus ($n_d = 1$, $n_h = 1$). Therefore $\bar{\kappa}$ plays the role of the chemical potential for the topology of the surface: positive values of $\bar{\kappa}$ favor the formation of a multiconnected bilayer (as in the L3 sponge phase, see Fig. 1) whereas sufficiently negative values will rather favor a disconnected topology such as separated vesicles in the L4 phase (Fig.1). On the other hand, bending fluctuations which preserve the topology do not involve $\bar{\kappa}$. So, to sum up (Porte, 1992):

- $\bar{\kappa}$ controls the topology of the bilayer
- whereas κ controls the amplitude of the thermally excited bending fluctuations of the membrane at fixed topology.

6.1. Lα: a dilute phase with quasi-long range smectic order

The lamellar phase shows finite uniaxial birefringence in polarized microscopy: it bears long range orientational order. The neutron scattering pattern of an oriented sample (Porte, 1992) (obtained by suitable surface treatment) is shown in Fig. 7. There appears clearly a pair of Bragg peaks in the q_z direction parallel to the director. This feature as well as the focal conic textures and oily streaks observed in misoriented samples by optical microscopy (Fig. 8) is the signature of quasi-long range smectic order: infinite, flat on the average, bilayers regularly stack parallel to one another normal to the director of the phase.

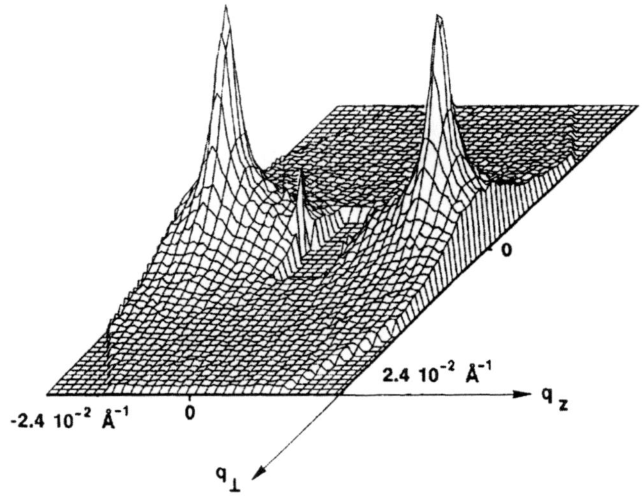

Fig. 7. 2D scattering pattern (ILL-D11) of a well oriented lamellar sample. The Bragg peaks stand in the q_z-direction normal to the bilayers (CPCl/n-hexanol/brine: $\overline{d} = 360$ Å).

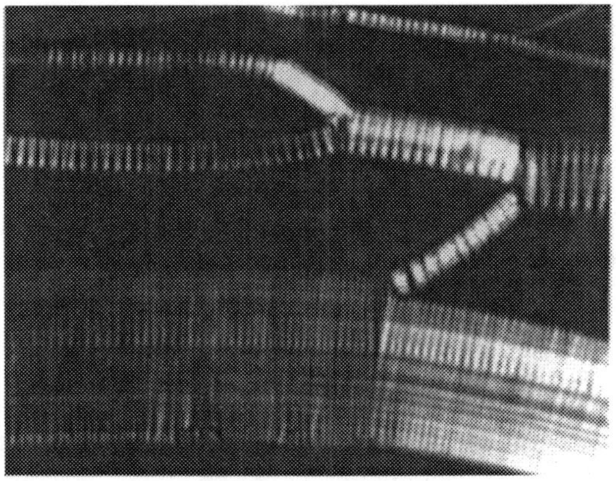

Fig. 8. Focal conic texture in polarized microscopy (same sample as in Fig. 7).

This is further confirmed by the swelling behaviour of the phase upon dilution. Assuming constant thickness for the bilayers at different dilutions, we indeed expect the simple relation between the concentration ϕ and the periodicity of the lamellar stacking $\overline{d}$: $\overline{d} = \delta/\phi$. Combining with the Bragg condition $q_B = 2\pi/\overline{d}$ (where q_B is the first-order Bragg wave vector), one immediately gets:

$$q_B = 2\pi\phi/\delta \tag{18}$$

Fig. 9 is an illustration of the linear swelling behaviour according to (18).

But the scattering pattern can be further exploited to get information on the interactions between bilayers. A given deformation of the smectic stacking can be represented by the local displacement $u_n(x, y)$ of the nth bilayer in the z-direction relative to its nominal position ($n\overline{d}$).

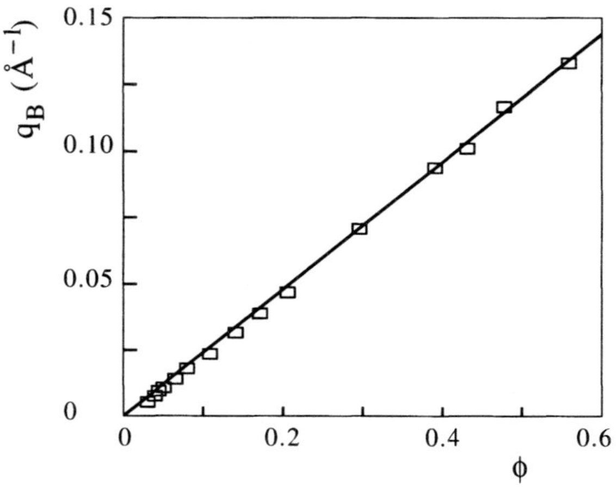

Fig. 9. Linear swelling behaviour of the lamellar phase. System: CPCL/n-hexanol/brine.

The energy associated with such a deformation writes:

$$E_{\text{smectic}} = \int \mathrm{d}x\ \mathrm{d}y \sum_n \left[\frac{1}{2} B(u_n(x,y) - u_{n-1}(x,y))^2 + \frac{1}{2}\kappa \left(\frac{\partial^2 u_n}{\partial x^2} + \frac{\partial^2 u_n}{\partial y^2} \right)^2 \right] \tag{19}$$

where B is the so-called smectic compression modulus which scales the energy associated with deviations of the interlayer distance from the equilibrium value $\bar{d}$ and κ is as before the mean curvature modulus of each bilayer. Note that we only consider smectic deformations at constant topology for all bilayers: $\bar{\kappa}$ does not appear in (19). The modulus B is simply related to the interlayer effective interaction per unit area $V(d)$ by the simple relation (Roux and Safinya, 1988):

$$B = d \frac{\partial^2 V(d)}{\partial d^2} \tag{20}$$

Taking the continuous limit and switching to the reciprocal space, the free energy density for a displacement mode of wave vector $\boldsymbol{q}$ and amplitude $u_{\boldsymbol{q}}$ is simply:

$$E_{\boldsymbol{q}}/\text{volume} = \tfrac{1}{2}[Bq_z^2 + Kq_{xy}^4]|u_{\boldsymbol{q}}|^2 \quad \text{with } K = \kappa/\bar{d} \tag{21}$$

The energy in (19) is harmonic; so the equipartion theorem applies to each $\boldsymbol{q}$-mode: their average square amplitude at temperature T is simply:

$$\langle |u_{\boldsymbol{q}}|^2 \rangle = \frac{k_B T}{Bq_z^2 + Kq_{xy}^4} \tag{22}$$

Within these premises, it is possible to calculate the correlation function of bilayers in the direct space and by Fourier transformation to derive the asymptotic form of the singularity at the Bragg peak position (Caille, 1972; Gunther et al., 1980). (This is rather technical.) The main predictions are:

$$I(q_{xy} = 0, q_z) \propto (q_z - q_B)^{2-\nu} \quad \text{for a well oriented single crystal} \tag{23}$$

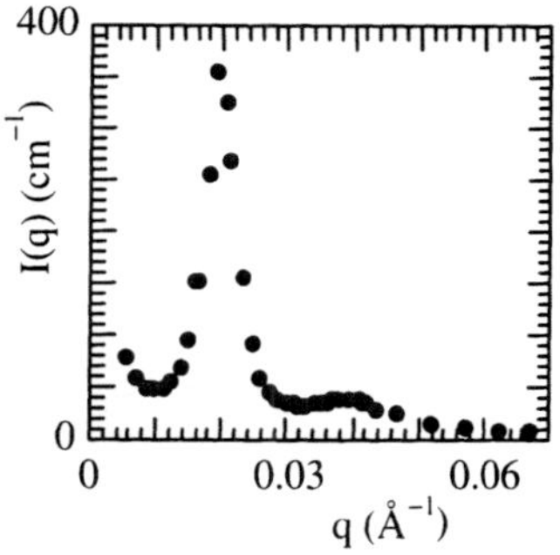

Fig. 10. Lamellar non-ionic sample (10% concentration) doped with a small amount of a cationic surfactant.

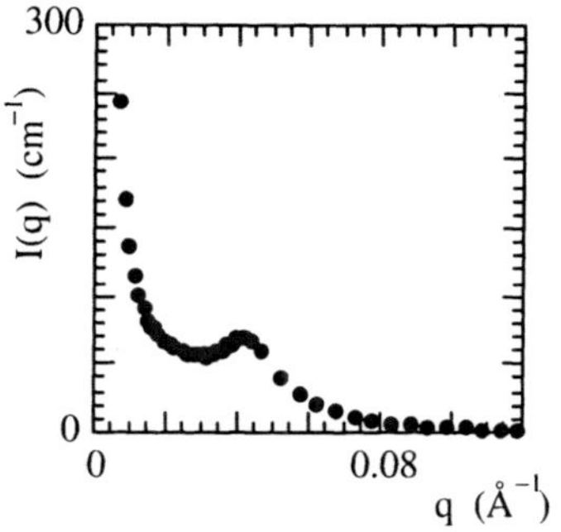

Fig. 11. Lamellar non-ionic sample (20% concentration); undoped.

and

$$I(q) \propto (q - q_B)^{1-\nu} \quad \text{for a powder averaged polycrystal} \tag{24}$$

where the exponent ν is determined by the moduli B and K:

$$\nu = \frac{q_B^2 k_B T}{8\pi\sqrt{KB}} \tag{25}$$

On the other hand for a powder polycrystal, the asymptotic form of the intensity scattered at very low q's is (Castro-Roman, 1999):

$$I(q \to 0) = \phi^2 \frac{k_B T}{B}\left(1 - \frac{\pi}{2}\sqrt{\frac{K}{B}}q + \cdots\right) \tag{26}$$

At given rigidity of the bilayer, a strong interaction potential determines a high value of B (20). This results in a low value for ν (24) and according to (25) the Bragg peak is sharp (Roux and Safinya, 1988). Correlatively (26), the intensity scattered at very low q's is very low. Of course just the opposite features are expected when the interaction potential is soft. An illustration is given in Figs. 10 and 11. Fig. 10 is taken from a lamellar sample at $\phi = 10\%$ involving non-ionic surfactant and cosurfactant (TX100 and TX35); the bilayers are weakly charged by small addition (1% in weight fraction of all surfactants) of a cationic surfactant

(CPCl). Strong electrostatic interactions arise: in spite of the dilution the peak is sharp and the scattering at the beam stop is low. In Fig. 11 the concentration is twice as large but no CPCl is added: there is no electrostatic interaction between the neutral bilayers. Only weak steric interactions remain. The peak is broad and soft and the intensity scattered at beam stop almost diverges. More quantitative procedures have been worked out along these line in order to derive quantitatively the interlayer interactions (Nallet et al., 1993; Zhang et al., 1996; Castro-Roman, 1999).

6.2. L4: an equilibrium dispersion of polydisperse vesicles

The scattering patterns of samples of the L4 phase show nothing specific and provide no direct evidence for vesicles. An example is shown in Fig. 12. Of course, at high q's we find the leading q^{-2} dependence characteristic for the locally bilayered morphology. But at lower q's, there is a progressive crossover towards a weaker dependence which depends in a non-straightforward manner on the system and the concentration of the sample. Therefore, the first evidence of vesicles in the L4 phase was obtained by the Bordeaux group (Herve et al., 1993) from a clever combination of viscosimetry, conductimetry and quasi-elastic light scattering and using 'direct visualization' by freeze fracture electron microscopy. The main reason for the non-specific character of the static scattering is the size polydispersity. Neglecting interactions, for a monodisperse dispersion of vesicles of internal radius R_i and external radius R_e (so $R_e - R_i = \delta$), we would expect for the scattered intensity:

$$I(q) = \phi V(\Delta\rho)^2 \left(3\,\frac{\sin qR_e - qR_e\cos qR_e - \sin qR_i + qR_i\cos qR_i}{q^3(R_e^3 - R_i^3)}\right)^2 \quad (27)$$

Eq. (27) indeed shows a series of maxima and minima when $q^2 I(q)$ is plotted versus q, characteristic of the two radii: R_e and R_i. But for the very polydisperse situation typical for the L4 phase, these maxima and minima are blurred out and nothing specific remains to be exploited in the scattering. Static scattering alone cannot characterize unambiguously phases of polydisperse vesicles.

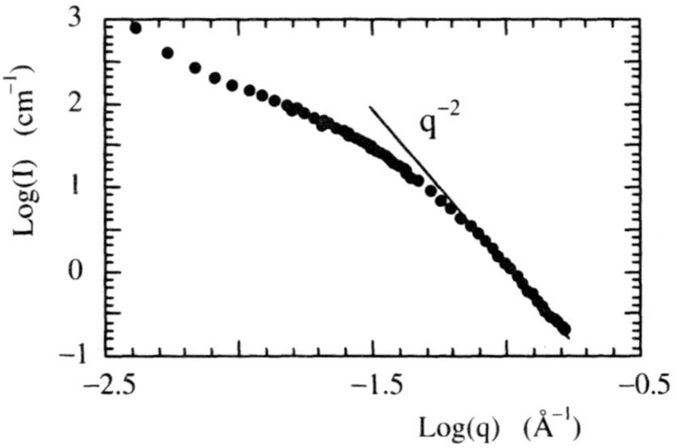

Fig. 12. L4 phase of a non-ionic (TX100, TX35) ternary system.

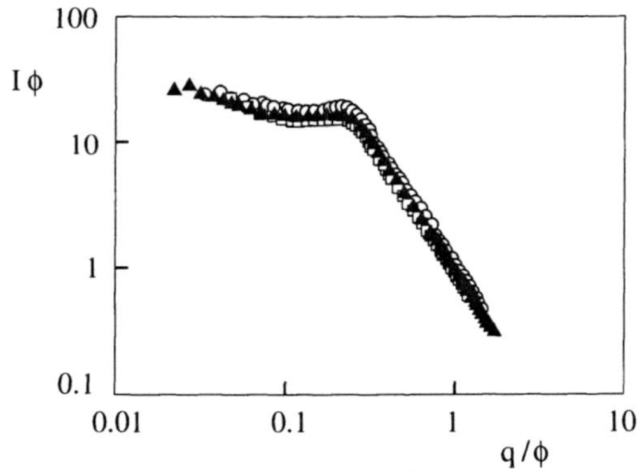

Fig. 13. Scattering pattern of L3 sponge samples. Reduced units are used so that data taken at various concentrations superimpose on the same master curve (5%, 7% and 10%, respectively). System: AOT/water/NaCl.

6.3. L3 sponge phase a multiconnected bilayer

Fig. 13 shows the typical scattering pattern of a sample in the L3 sponge phase. As for the L4 phase, the q-dependence is rather smooth: again at high q's the scattering is dominated by the leading q^{-2} dependence, and again a crossover to a weaker dependence is seen at lower q's. But this crossover is sharper and allows to point out a characteristic wave vector q_m. Plotting the evolution of q_m as function of the dilution, one reveals the linear swelling behaviour of the sponge phase:

$$q_m = 1.5\left(\frac{2\pi\phi}{\delta}\right) \tag{28}$$

Apart from the 1.5 prefactor, the swelling behaviour is very similar to that of the above lamellar phase. It is shown in Porte et al. (1988) that the 1.5 prefactor is actually consistent with the random multiconnected structure represented in Fig. 1. Of course this agreement alone does not make a proof, and other methods including freeze fracture electron microscopy (Strey et al., 1990) were used later on and provided further convincing evidence.

6.4. Topology sequence and optimum curvature of the monolayer

So, the phase sequence observed upon increasing the control parameter η above the domain of existence of the micellar phase corresponds to the same bilayer but organized at large scale according to *different topologies*: disconnected small finite subunits (vesicles in L4); singly connected infinite flat bilayers (in the lamellar phase); and finally bilayers multiconnected in the three directions of space (in the L3 sponge phase). Since the topologies always appear in the same order upon increasing η, it is tempting (Marignan et al., 1989) to relate their occurrence with the spontaneous curvature of the amphiphilic interface. As mentioned above, in difference with the bilayer, the constituting monolayers are not symmetrical structures. Their bending elasticity writes according to (14) and in general they have a finite optimum curvature $c_{0,\mathrm{mono}}$ (which is indeed identical to the spontaneous curvature of the amphiphilic interface). It is quite easy to compute the bending elasticity of the bilayer obtained after sticking two identical monolayers against one another in a symmetrical manner and one gets:

$$c_{0,\mathrm{bil}} \equiv 0 \tag{29}$$

$$\kappa_{\mathrm{bil}} = 2\kappa_{\mathrm{mono}} \tag{30}$$

$$\overline{\kappa}_{\mathrm{bil}} = 2\overline{\kappa}_{\mathrm{mono}} - 2\kappa_{\mathrm{mono}}\delta c_{0,\mathrm{mono}} \tag{31}$$

Eqs. (29) and (30) are indeed expected due to the symmetry. More unexpected is the contribution in (31) to the gaussian modulus $\overline{\kappa}_{\mathrm{bil}}$ arising from the optimum curvature of the monolayer $c_{0,\mathrm{mono}}$. This contribution is at the origin of the topology sequence. The lamellar phase occurs for values of η which make $c_{0,\mathrm{mono}}$ close to zero. Its contribution to $\overline{\kappa}_{\mathrm{bil}}$ remains moderate and the bilayers remain flat on the average ($-2\kappa < \overline{\kappa} < 0$). Increasing η one makes $c_{0,\mathrm{mono}}$ more and more negative until the point is reached where $\overline{\kappa}_{\mathrm{bil}}$ becomes positive. The flat bilayer becomes unstable against proliferation of 'handles' and so the multiconnected structure of the L3 sponge forms spontaneously. If, on the other hand, η is decreased, $c_{0,\mathrm{mono}}$ is made more and more positive until the point where $\overline{\kappa}_{\mathrm{bil}}$ becomes smaller than $-2\kappa_{\mathrm{bil}}$. At that point, the flat bilayer becomes unstable against the 'evaporation' of disconnected vesicles: and the disconnected L4 phase forms spontaneously.

7. Conclusion

So finally, the complex phase behaviour of diluted surfactant systems is driven by the shape sequence at local scale (globules, cylinders and bilayers) and also by the topological changes of the bilayers at larger scale (disconnected, singly connected, multiconnected). The shapes as well as the topologies are controlled by the spontaneous curvature c_0 of the amphiphilic interface alone. More precisely the rich behaviour results from the compromise between two competing requirements: the first one is the tendency for the interface curvature to keep closest to c_0 and the other is the constraint that, in one direction of space at least, the thickness of the aggregate does not exceed twice the surfactant molecular length. When dealing with one particular sample or system, the experimentalist faces a typical inverse problem. The scattering pattern in the appropriate q-range is generally easy to collect by X-rays or neutrons; the difficult point is to derive the structure. We have presented here how the asymptotic behaviors of the scattering profile in the relevant q ranges can be exploited for a first rough characterization of the structure. We must keep in mind however, that such quick evaluation procedures should always be complemented by the more detailed data treatments presented in the preceding chapters of the present book.

Acknowledgement

The author thanks Christian Ligoure and Francisco Castro-Roman for interesting discussions and for communicating their recent results. The figures and drawings are made by Jean Pierre Fluxench.

References

Appell, J. and Marignan, J. (1991). *J. Phys II France*, **1**, 1447.

Benton, W.J. and Miller, C.A. (1983). *J. Phys. Chem.*, **87**, 4981.
Cabane, B., Duplessix, R. and Zemb, T. (1985). *J. Phys. France*, **46**, 2161.
Caille, A. (1972). *C.R. Acad. Sci.*, **274**, 891.
Castro-Roman, F. (1999). Ph.D. Thesis, Université Montpellier II (calculation by F. Castro-Roman and C. Ligoure).
Cates, M.E. and Candau, S. (1990). *J. Phys.: Cond. Matter*, **2**, 6869.
Ekwall, P. (1975). *Adv. Liq. Cryst.*, **1**, 1.
Filali, M., Aznar, R., Svensson, M., Porte, G. and Appell, J. (1999). *J. Phys. Chem. B*, **103**, 7293.
Glatter, O. and Kratky, O. (1982). *Small Angle X-Rays Scattering*. Academic Press, New York.
Gunther, L., Imry, Y. and Lajzerowicz, J. (1980). *Phys. Rev. A*, **22**, 1733.
Hayter, J.B. and Penfold, J. (1981). *J. Chem. Soc. Faraday II*, **77**, 1851.
Helfrich, W. (1973). *Z. Naturf.*, **C28**, 693.
Herve, D., Roux, A.-M., Bellocq, F., Nallet, T. and Gulik-Krzywicki, T. (1993). *J. Phys II France*, **3**, 1255.
Israelachvili, J.N., Mitchell, D.J. and Ninham, B.W. (1976). *J. Chem. Soc. Faraday II*, **72**, 1525.
Marignan, J., Appell, J., Bassereau, P., Porte, G. and May, R.P. (1989). *J. Phys. France*, **50**, 3553.
Mazer, N.A., Benedeck, G.B. and Carey, M.C. (1976). *J. Phys. Chem.*, **80**, 1075.
Nallet, F., Laversanne, R. and Roux, D. (1993). *J. Phys II France*, **3**, 487.
Olsson, U. and Wennerström, H. (1994). *Adv. Colloid Interface Sci.*, **49**, 113.
Pedersen, P.J.S., Egelhaaf, S.U. and Schurtenberger, P. (1995). *J. Phys. Chem.*, **99**, 1299.
Porte, G. (1992). *J. Phys.: Cond. Matter*, **4**, 8649.
Porte, G., Marignan, J., Bassereau, P. and May, R. (1988). *J. Phys. France*, **49**, 511.
Roux, D. and Safinya, C.R. (1988). *J. Phys. France*, **49**, 307.
Strey, R., Jahn, W., Porte, G. and Bassereau, P. (1990). *Langmuir*, **6**, 1635.
Zemb, T. and Charpin, P. (1985). *J. Phys. France*, **46**, 249.
Zhang, R., Tristram-Nagle, S., Sun, W., Headrick, R.L., Irving, T.C., Suter, R.M. and Nagle, J.F. (1996). *Biophys. J.*, **70**, 349.

Neutrons, X-Rays and Light
P. Lindner and Th. Zemb (Editors)

Chapter 13

Scattering by Microemulsions

Thomas Zemb

DSM/DRECAM, CEA – Saclay, F-91191 Gif-sur-Yvette CEDEX, France

1. Introduction

Microemulsions are clear, optically homogeneous and thermodynamically stable fluids. They contain two non-miscible liquids and at least one surface active component. In practice, the two components to be mixed ('cosolubilized') are water and oil. At least one of the solutes reduces the surface tension between water and oil: the surfactant. Other components are sometimes present in microemulsions: salt or other water-soluble components in the polar part; oil-soluble components in the apolar part. Some components such as long-chain alcohols (called co-surfactants) partition between polar and apolar parts.

All fluids called 'reverse micelles' as well as 'oil swollen micelles' belong to the class of microemulsions. There is no clear obvious delimitation between emulsions and microemulsions, because some emulsions are optically clear and extremely stable, some can even form spontaneously from two liquids when they are mixed without stirring. The main difference between emulsions and microemulsions is the typical size of water or oil domains, which are invisible to the microscope in microemulsions, and visible since larger than one micron in the case of emulsions. Experimental methods described here are common to emulsions and microemulsions, only the relevant q-range of investigation is different: 10^{-3} Å^{-1} to 0.5 Å^{-1} will be sufficient to characterize a microemulsion, while one decade lower q (sometimes using classical light scattering) will be needed in order to characterize a typical emulsion.

Microstructure is the structure at the scale of the surfactant film, i.e. structural models of packing at the scale from 10 Å to 10,000 Å. They are three general methodologies used to determine a microstructure:

- directly determine some features of the microstructure, such as the specific area,
- use some fitting procedure, with a presumption that either spheres are present or some parametric expression holds for the scattering,
- test microstructural models, using dilution lines, since the presence of one microstructure cannot be ascertained by studying one sample only, dilution lines for several samples need to be confronted to explicit predictive models to proof the presence of one microstructure.

We will describe these three experimental approaches in this chapter. Note that scattering is still the most used method because it does not require freezing, nor etching or labelling of the sample, like electron microscopy. The next step after determination of the microstructure is the understanding of phase boundaries, solubility limits and all types of physical or chemical properties of microemulsions, such as conductivity, osmotic pressures, etc. This part is beyond the introductory material assembled here (see Chevalier and Zemb, 1990).

We do not consider in this chapter the case of liquid crystals, neither lyotropic nor thermotropic, since they exhibit quasi-Bragg peaks due to the long range order detected in the scattering. By far the three most encountered microstructures in liquid crystals are:

(a) lamellar or smectic (indexed 1, 2, 3, . . .)

(b) hexagonal (indexed 1, square root of three, 2, . . .)

(c) cubic (most of the time indexed 1, square root of 2, . . .)

Lamellar and hexagonal liquid crystals are easily distinguished from microemulsions, because they belong to the class of liquid crystals, and show optical birefringence. They appear clear at rest between crossed polarisers. Note that static birefringence should be clearly distinguished from flow birefringence, i.e., transient birefringence detectable while stirring he fluid, an optical property that can be easily observed with a large number of microemulsions when they are not made of spherical droplets.

Note that more than one sharp peak has to be observed to ascertain the presence of long range order as is the case of a liquid crystal.

For example, Fig. 1 shows a disordered microstructure, with equal volumes of water and oil (volume fractions 0.5) and a typical size. Since this structure has been generated with randomly interfering wavelets, the scattering of this microstructure only exhibits one sharp quasi-Bragg peak, but there is no long range order in the microstructure.

In practice, the sharp peaks corresponding to this microstructure have not been encountered yet while investigating microemulsions, but only in the case of mesoporous solid materials.

2. The oldest model: the inhomogeneous solid (Debye model)

Any microemulsion contains two media, 'polar' and 'non-polar', distributed in a volume and separated by a surfactant film. Therefore, there is a possibility to describe very generally this fluid at one instant by a model which was developed for solids (Debye and Bueche, 1949; Debye et al., 1957; Brumberger and Goodisman, 1983, Brumberger et al., 1986; Delaglio et al., 1986). This model was introduced by Debye and Bueche to describe the large scale fluctuations inducing scattering of light in glass. Debye defines the correlation function $g(r)$ of the scattering length density $\rho(r)$:

$$g(r) = \frac{\langle \rho(0)\rho(r)\rangle}{\langle(\rho(r)-\rho_{\mathrm{av}})^2\rangle} \tag{1}$$

where ρ_{av} is the average value of the scattering length density:

$$\rho_{\mathrm{av}} = \phi\rho_1 + (1-\phi)\rho_2 \tag{2}$$

where volume fractions ϕ and $(1-\phi)$ represent the two volume fractions of polar and apolar liquid on the microscopic scale. Volume fractions can always be derived from the known composition of the sample: for the oil and water part, volume fractions are obtained from weight fractions and densities. For the surfactant, the head-group belongs to the polar fraction, while the hydrophobic tail belongs to the apolar volume. Knowing the total molar volume (cm^3/mol) of the surfactant and the chemical group volumes given in tables (for example. 25 Å^3 for a CH_2 group and 54 Å^3 for an end methyl group), it is always possible then

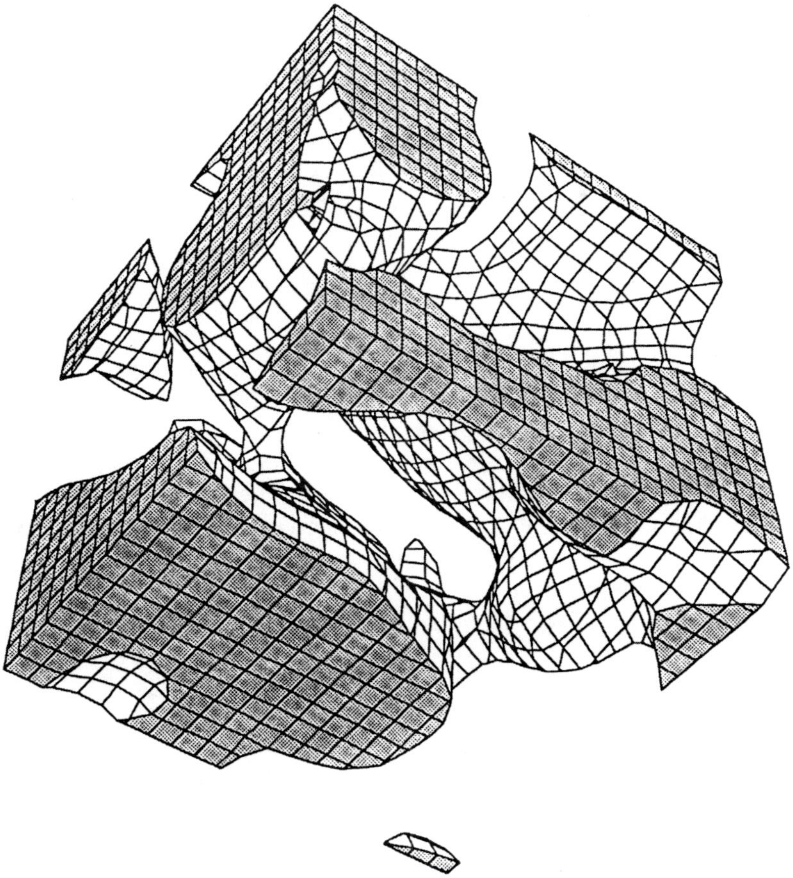

Fig. 1. An example of a bicontinuous structure generated using random wavelets. Volume fraction is one half and average curvature is zero. However, this microstructure, constructed with random infinite wavelength, produces one sharp 'quasi-Bragg' peak without any higher-order reflection. The sharp peak in the scattering corresponds to a typical length, but without long range order. For all these microstructures with equal polar and apolar volume fractions, the relation $\Sigma D^* \cong 1.5$ approximately holds. Experiments near symmetric case with volume fractions 0.5 cannot test validity of microstructural models.

to determine with a precision of 1% polar and apolar volume fractions. If a solute such as typically a short-chain alcohol is present, the balance of solubility between water and oil has to be known before calculating polar and apolar volume fractions of a given sample. This quantity is needed to check any microstructural model.

The correlation $g(r)$ function describes the loss of 'memory' of being located in medium 1 associated to a translation r from one 'origin point' located in medium 1. Debye considers an isotropic sample, so that the three-dimensional $g(\boldsymbol{r})$ is replaced by the scalar $g(r)$. Debye postulated that the correlation decay can be described using only *one* typical size ξ, the correlation length:

$$g(r) = e^{-r/\xi} \tag{3}$$

The corresponding scattered intensity has been calculated by Debye and Bueche:

$$I(q) = \frac{8\pi\xi^3\Delta\rho^2}{(1+\xi^2q^2)^2} \tag{4}$$

There is no peak in the scattering in the Debye model of porous material: most of the time, a maximum of scattered intensity is detected when microemulsions are investigated. Hence, the model of Debye and Bueche, with only one characteristic length, is no more applied in the case of microemulsions.

3. Fractals

A fractal structure is defined by the following relation: take a sphere of radius R, the average mass of medium 1 included inside the sphere varies like R^f, where f is the fractal dimension ($f < 3$). When f is close to 3, the medium 1 is very dense. A fractal dimension closer to 1 indicates a loosely connected network, as branches on a tree. Gels made of aggregating particles usually show fractal dimension between 1.5 and 2. The scattered intensity (in absolute units of cm^{-1}) of a fractal aggregate made of monodisperse spheres of radius r_e and where the fractality holds up to a maximum size κ is given by the product (Teixeira, 1986, 1988, 1990):

$$I(q) = n\Delta\rho^2 p_{sph}(q,r)\left\{1+\frac{f\Gamma(f-1)}{(qr)^f\left(1+\dfrac{1}{q^2\kappa^2}\right)}\right\}\sin\left(\frac{f-1}{tg(q\kappa)}\right) \tag{5}$$

where n is the density (spheres per cm^{-3}) and $p_{sph}(q,r)$ is the form factor of one sphere of radius r,

$$p_{sph}(q,r) = \left(3\frac{\sin(qr)-(qr)\cos(qr)}{(qr)^3}\right)^2 \tag{6}$$

Thus, scattering of a fractal aggregate cannot produce an interaction peak. If $\kappa \gg r$ holds, the structure factor behaves as $S(q) = q^{-f}$. A misleading procedure should be avoided: determination of the fractal dimension from the slope of $I(q)$ at low angles. This is very dangerous since particles of a given symmetry such as independent cylinders (q^{-1}) or lamellae (q^{-2}) or critical fluctuations (q^{-2}) all produce linear plots of $\log I(q)$ versus $\log q$ without having a fractal structure.

4. The Talmon–Prager cell model

The first model compatible with the continuous transformation from water droplets to oil droplets via a bicontinuous network has been developed by Talmon and Prager (1977). Bicontinuity over macroscopic distances exists at polar volume fractions between 18% and 82%.

The construction of a Talmon–Prager structure is as follows:

(1) choose of a set of random (Poisson) points, density n per cm^3 of sample and then construct the associated Voronoï tessellation of space using the bisector planes;

(2) fill the different Voronoï cells with water or oil according to the available volume fractions ϕ and $(1-\phi)$ of water and oil; then
(3) set the interface Σ between adjacent cells with different contents.

The scattering of this structure can be calculated analytically (Kaler, 1982). In any case, there is no peak in the scattering. Since nearly all micro-emulsions exhibit a peak, no scattering experiment yet made with microemulsions has been compatible with the Talmon-Prager model.

5. The cubic random cell (CRC) model

We consider now samples which show a scattering peak: this means that there is a cell size in the structure. The size of this cell is $D^* = 2\pi/q_{\text{peak}}$. Bragg relation is used because adjacent cells produce a peak at a distance equivalent to their diameter, which is also the typical size of one polar or apolar 'domain'.

First introduced by Jouffroy et al. (1982) as a 'decorated' Talmon–Prager model, this model assumes that the microstructure can be approximated by a set of cubes of size ξ filled at random with water or oil, according to volume fraction. This simple picture of microstructure has allowed to rationalise the phase diagram of microemulsions (Andelmann et al., 1987; Cates et al., 1988a,b; Safran et al., 1986; Safran, 1992), thus triggering a large number of studies of microstructure (Chandra and Safran, 1991). The 'persistence length' ξ is related to the specific area Σ (cm^2/cm^3). This specific area is set by the amount of surfactant:

$$\Sigma = c_s \sigma \tag{7}$$

where c_s is the concentration of surfactant (molecule/cm^{-3}) and σ the area in the surfactant film covered by one molecule (typically of the order of 100 Å^2/molecule). The value of σ can be obtained independently for any formulation by means of water/oil surface tension experiments if needed.

In the case of cubic random cells (CRC) covered by a surfactant film, the persistence length ξ is given by:

$$\xi = \frac{6\phi(1-\phi)}{\Sigma} \tag{8}$$

The corresponding scattering peak position is predicted to be at

$$D^* = \xi = 2\pi/q_{\max} \tag{9}$$

Indeed, in a large number of microemulsions, the peak is observed at a position corresponding to *twice* that distance in real space, i.e., it is observed that D^* is twice ξ. This constant ratio between prediction and observation has been taken as a success of the CRC model. As has been shown by Guinier, a distribution of random cells on a lattice can only give a peak at the position of the lattice, with excess scattering at small angles due to unoccupied positions in the lattice. However, this missing factor of about two in the observed periodicity is the *proof* of the existence of a local microstructure, and has been explained within the frame of the disordered open and connected (DOC) model which will be described below. Early models of microstructure of microemulsions are summarized in Table 1.

Table 1

Features of first structural models of microemulsions (adapted from Zemb et al., 1990).

	Talmon–Prager	CRC
Topology	Bicontinuous $0.18 < \phi < 0.82$	Bicontinuous $0.18 < \phi < 0.82$
Position of scattering peak	No peak	$D^*\Sigma = 6\phi(1-\phi)$
Average curvature $\langle H \rangle$	$1.57(1-2\phi)D^{*4/3}$	$(1-2\phi)/(12D^*)$
Gaussian curvature $\langle K \rangle$	$(3.65(2\phi-1)^2-1.5)D^{*2/3}$	$8/5((1-2\phi)^2)-\frac{1}{4})/D^{*2}$
Packing constraint	$p = 1 + Hl + Kl^2/3$	$p = 1 + Hl + Kl^2/3$
Problems	Incompatible with the observed conductivity	Incompatible with the observed conductivity
	No peak in the scattering	Scattering peak at twice the observed q-value
References	Talmon and Prager (1977), Kaler and Prager (1982)	Milner et al. (1988), Chandra and Safran (1991)

	Spheres w/o	Spheres o/w
Topology	Always water continuous	Always oil continuous
Position of scattering peak	$D^*\Sigma = 4.84\phi^{1/3}$	$D^*\Sigma = 4.84(1-\phi)^{1/3}$
Average curvature $\langle H \rangle$	$1/R$ with $R = 3\phi/\Sigma$	$1/R$ with $R = 3(1-\phi)/\Sigma$
Gaussian curvature $\langle K \rangle$	$1/R^2$	$1/R^2$
Packing constraint	$p = 1 + Hl + Kl^2/3$	$p = 1 - Hl + Kl^2/3$
Problems	Scattering peak position versus dilution	Scattering peak position versus dilution
	Strong curvature constraints	Strong curvature constraints
References	Pileni et al. (1985), Nave et al. (2000a,b)	Hayter and Penfold (1983), Huang et al. (1984), Billman and Kaler (1990), Chevalier and Zemb (1990)

Obtaining information about microstructure of a microemulsion requires acquisition of either small-angle scattering of X-rays or neutrons (SAXS or SANS). Static light scattering in most of the axes gives only osmotic compressibility or the value of a gyration radius, since q-range investigated corresponds to distances larger than typical size of a microemulsion. Only experimental results expressed in absolute scale are useful. Scattering vectors q are usually expressed in Å^{-1} or nm^{-1} while scattered intensity is expressed in scattering cross-section per unit thickness and is expressed in cm^{-1} (see Chapter 2 by P. Lindner in this book). Experimental results without scale, i.e. in arbitrary scale, even if present sometimes in the literature, are of no value for determining microstructure of a microemulsion.

As we will see, experimental data have to be acquired on the largest possible q range: the ratio between $q_{\min}$ and $q_{\max}$ is the key point which allows determination of structure, because fitting a curve on a limited q range will always succeed, but does not give information about the sample. Depending on the parameter to be determined, experimental results should always be shown (and used) on several scales, at least lin–lin, log–log and Iq^n versus q^n. Three plots are mainly used together in order to analyse experimental results: the linear plot, the logarithmic plot and the invariant plot. The linear plot is used to show the presence of peaks and the evolution at low angles; asymptotic limits and general decay can only be found in log–log plots. The relevance of the q-range investigated can only be demonstrated using a plot of $I(q)q^2$ versus q: the surface under the curve is 'the invariant', labelled Q

$$Q = \int_0^\infty I(q)q^2 \, dq = 2\pi^2\phi(1-\phi)(\rho_w - \rho_o)^2 = 2\pi^2\phi(1-\phi)\Delta\rho^2 \tag{10}$$

Experimental measurement of the invariant (in [length]$^{-4}$) allows either consistency checks of the known values of volume fraction and scattering contrast, or determination of one term, if the other one is not known. Comparing invariant plots of two samples allows immediate identification of increase or decrease of the global fluctuation, i.e. domain size at constant contrast $\Delta\rho$. An example of a non-ionic microemulsion, showing the same SANS results on several scales is shown in Fig. 2. Linear scale is shown in Fig. 2a.

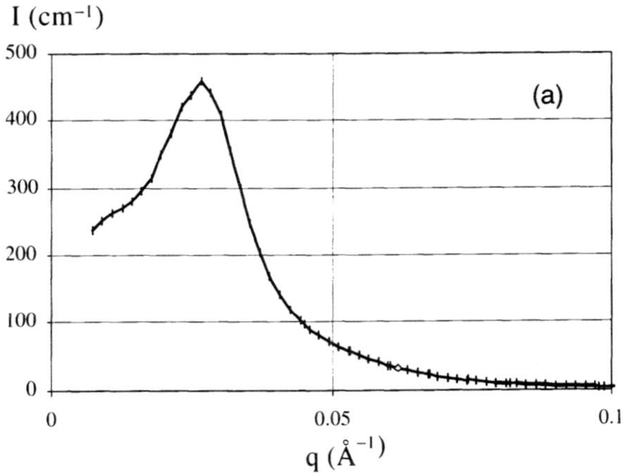

Fig. 2a. Small-angle neutron scattering produced by a solution containing 0.4 M/l on a non-ionic surfactant ($C_{12}E_5$), with an apolar volume fraction of 48.6%. Ordinate is in absolute scale, i.e., intensity is expressed in probability of scattering (cm^{-1}).

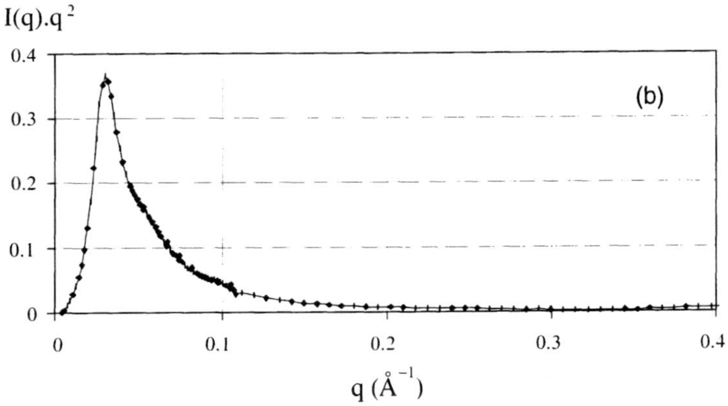

Fig. 2b. The plot used to measure the invariant by SANS, i.e. the second moment of the scattering curve: $I(q)q^2$. Taken in a sufficiently large interval, the experimental value of the invariant corresponds to the value expected from the known volume fractions and contrast. Here the value of the invariant is 1.59×10^6 cm^{-4}.

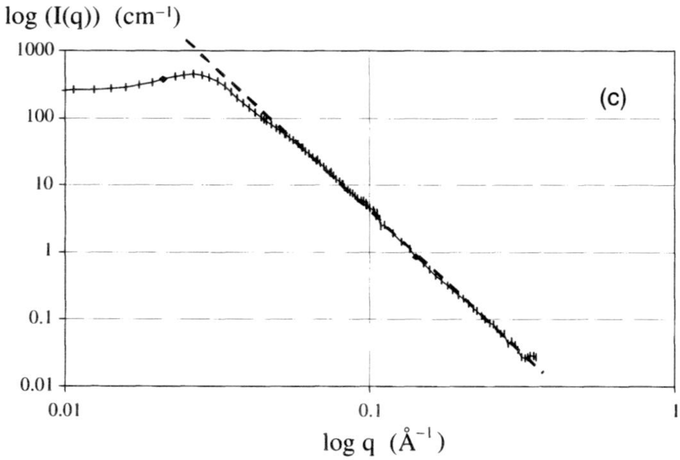

Fig. 2c. Double logarithmic plot of the scattered intensity (SANS, same sample as (a) on logarithmic scale, which shows the asymptotic behaviour due to the sharp water/oil interface, since water is deuterated. The specific area of the sample is 168 m^2/cm^3, corresponding to an area per molecule of $\sigma = 70$ Å^2/molecule, derived from the Porod asymptotic limit shown as the dotted line and the known contrast $\Delta\rho^2 = 5.57 \times 10^{10}$ cm^{-1}.

6. Direct determinations from the scattering curve

The two most important parameters the specific area Σ (usually expressed in cm^2 of interface/cm^3 of sample or m^2/g or Å^2/Å^3) and the volume fraction ϕ. In the common case of a sample made only of two homogenous media 1 and 2 characterised by a scattering length density ρ_1 and ρ_2 (cm^{-1}), the volume fraction has to be determined from the composition of the sample: no attempt of interpretation can be made if the volume fraction is not known independently from the scattering experiment.

6.1. Volume fraction and scattering lengths deduced from invariant

For any composition, the polar and apolar volume fraction can be trivially calculated from the known composition of the sample and molar volumes (or molar mass and density relative to water).

The scattering length density ρ in cm/cm^3 has to be calculated from the chemical formula of each component (Stuhrmann and Miller, 1978). For neutron scattering, a scattering length is associated to each nucleus. For X-rays, the elastic scattering length associated to each electron has to be added up for each molecule. In older literature, scattering length densities are expressed in electronic densities. For water, 10 electrons are dispersed over 0.3 nm^3 since the density is one; in scattering length densities, this corresponds to 9.38×10^{10} cm^{-2}.

Once scattering length densities and volume fraction are evaluated for a given microemulsion, the contrast $\Delta\rho^2 = (\rho_1 - \rho_2)^2$, usually expressed in cm^{-4} is known together with the volume fraction ϕ, this should be checked by comparing the experimental invariant Q^*_{exp} and the theoretical invariant Q^*_{th} (in cm^{-4}). Therefore, the value of the invariant expected should be checked with the value experimentally measured for the second moment of the scattered intensity. A mismatch is either an experimental error, or sometimes can demonstrate some unexpected high solubility of the surfactant in one of the phases present. This was the case in an experiment described by Warr et al. (1990): a long-chain cationic surfactant showed a high solubility as monomers in water at high temperature, thus reducing the apolar volume fraction, since only part of the amount of surfactant present in the sample contributed to form the surfactant film. The case when more than two homogenous media have to be considered has been discussed by Brumberger et al. (1986).

6.2. Determination of specific surface

If the scattering curve $I(q)$ has been measured towards angles large enough up to the 'Porod limit' asymptotic regime, it is possible to determine directly the area per unit volume, and its equivalent, the area per surfactant. General conditions and practical examples are given in Chapter 3 by O. Spalla.

In practice, in the case of particles of typical size R_t, measuring the 'Porod limit' means that the measurement range has been extended up to q values around $qR_t = 10$. In the case of connected networks, the typical size R_t of a 'particle' has no more meaning. This concept of 'typical size' has to be replaced by the average chord l_c which can be calculated from the general integral relation:

$$l_c = \frac{\pi}{Q}\int_0^\infty qI(q)\,\mathrm{d}q \qquad (11)$$

where Q is the invariant. Since scattering intensities appear twice in this formula, one should notice that average chord lengths can be evaluated even if the Intensity $I(q)$ is known on arbitrary units only.

If the sample produces measurable scattering in the right q-range, three experimentally useful methods of measuring the specific area Σ from a sample:

(a) when the scattered intensity is precisely known in absolute units (cm^{-1}), Σ is given by:

$$\Sigma = \frac{1}{2\pi\,\Delta\rho^2}\lim_{q\to\infty} I(q)q^4 \qquad (12)$$

(b) if the integral of the invariant converges to a better value, an alternative method is to use the invariant Q:

$$\Sigma = \frac{\pi\phi(1-\phi)}{Q}\lim_{q\to\infty} I(q)q^4 \qquad (13)$$

(c) if the correlation function has been calculated, a general theorem demonstrated by Porod gives (see Chapter 3 by O. Spalla in this book):

$$\Sigma = \frac{4\phi(1-\phi)}{\bar{l}} \qquad (14)$$

where the length $\bar{l}$, which is usually numerically close within 20% to the average chord length l_c, is the slope at the origin of the correlation function.

A standard graphical method used to separate the high-angle term in q^{-4} from an unavoidable constant parasitic 'background scattering' B_d is to plot $I(q)q^4$ versus q^4. Since $I(q) = Cq^{-4} + B_d$. Target value C is determined from best linear fit in the plot. Depending on the experimental situation, the incertitude interval associated to each of the three methods can be very different: for large objects, methods (a) and (b) are more precise than method (c). When l_c is of the order of 20–50 Å, method (c) is usually more precise. An example of derivation of an area per molecule of 63 ± 3 Å^2/molecule is shown in Fig. 3a. Careful

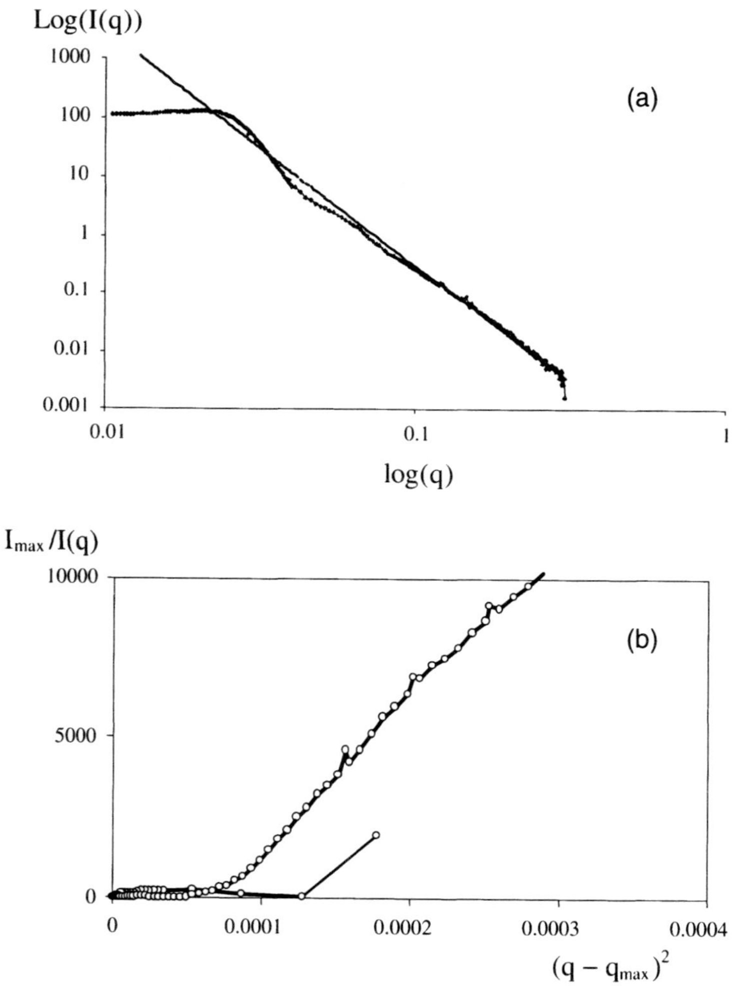

Fig. 3. (a) Small-angle X-ray scattering produced by an ionic microemulsion containing 25% polar volume and an area of 78.3 m^2/cm^3 with 0.19 M of a surfactant with $\sigma = 68$ Å^2/molecule, deduced from the Porod limit, which is the asymptotic q^{-4} behaviour. (b) The same SAXS data as (a), for an ionic microemulsion, in the coordinates $I_{max}/I(q)$ versus $(q - q_{max})^2$ which should give a parabola in the range where the parametric Teubner–Strey expression for the intensity could be used for extracting two distances from one broad peak.

determination of area per molecule using Porod plots have been given in the case of the popular surfactant AOT (Nave et al., 2000a,b).

Another useful method allowing direct determination is possible using SANS and contrast variation. If partial deuteration allows water and oil to have the same scattering length density, scattering is produced by the film only. The general expression in ‘film contrast’ is then given by (Auvray et al., 1984, 1986; Eastoe et al., 1997):

$$I_{\text{film}}(q) = 2\pi\,\Sigma\Delta\rho^2\frac{l^2}{q^2}(1 + O(q^2l^2)) \tag{15}$$

Fitting data to this expression allows determination of the thickness l of the surfactant film (Auvray et al., 1984). The specific surface Σ is then easily extracted from volume conservation since the product $l\Sigma$ should represent the volume fraction of the surfactant film, containing surfactant and part of the co-surfactants. Consistency can be checked using the fact that the next term contains $(\sin(ql)/(ql))^2$. Thus a plot of $I(q)q^2$ in the case of film contrast shows a distinct minimum at a value $q = \pi/l$.

6.3. Assuming the presence of spheres

The simplest microstructure known for a microemulsion is droplets of oil in water or the reverse. This situation is frequent, but not always the case, when volume fraction of the minority component comes to less than 10%. This situation is very easily checked. Since volume fraction and specific area are known, there is only one possible radius for the spherical droplets. That radius is set by surface to volume ratio of a sphere:

$$R_{\text{sph}} = 3\frac{\phi}{\Sigma} \tag{16}$$

Scattering of spheres of a given radius is given by a simple expression:

$$I(q) = n\Delta\rho^2\left(\tfrac{4}{3}\pi R_{\text{sph}}^3\right)^2 p_{\text{sph}}(q)S(q) \tag{17}$$

where $S(q)$ is the structure factor. Structure factors are given explicitly in literature for hard spheres, attractive potentials (‘Baxter’ model) and repulsive potentials by Hayter and Penfold. Therefore, the presence of disconnected spheres can be checked easily, with just magnitude of contact potential (an intensive quantity supposed independent from concentration). Close to a liquid–liquid transition, the structure factor $S(q)$ can be approximated by the Ornstein–Zernike expression: $S(q) = 1 + \chi_T/(1 + q^2\xi^2)$, where χ_T is the osmotic compressibility which is the most sensitive parameter to local attractive or repulsive interactions and ξ an arbitrary correlation length.

In practice, the value of $D^* = 2\pi/q_{\max}$ derived from the position of the maximum of $I(q)$ for repulsive spheres is given by an empirical approximate expression:

$$D^* = \frac{3}{\sqrt{2}}\left(\frac{8\pi}{3}\right)^{1/3}\Sigma^{-1}\phi^{2/3} \approx 4.32\Sigma^{-1}\phi^{2/3} \tag{18}$$

The presence of droplets in a microemulsion can therefore be easily checked, since the only possible radius is known. A very bad procedure, sometimes found in literature, is to use the radius as a free parameter, thus leading to inconsistent ‘fits’ to data, since radius cannot be

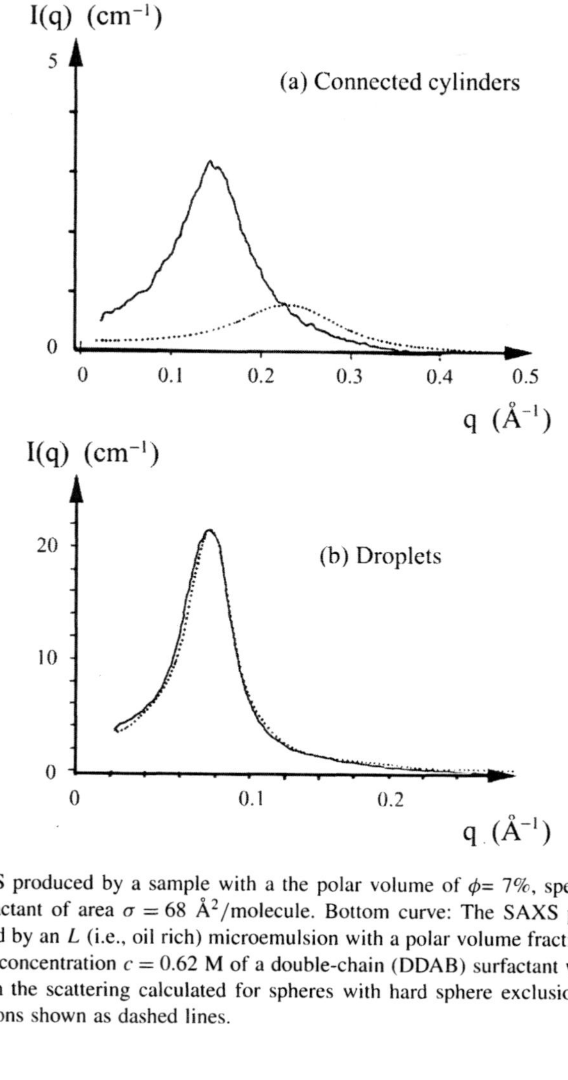

Fig. 4. Top curve: SAXS produced by a sample with a the polar volume of ϕ= 7%, specific area $\Sigma = 0.031$ Å^{-1} due to $c = 0.72$ M surfactant of area $\sigma = 68$ Å^2/molecule. Bottom curve: The SAXS pattern – shown on linear absolute scale – produced by an L (i.e., oil rich) microemulsion with a polar volume fraction $\phi = 24\%$, specific area $\Sigma = 0.025$ Å^{-1} due to a concentration $c = 0.62$ M of a double-chain (DDAB) surfactant with $\sigma = 68$ Å^2/molecule. There is no parameter in the scattering calculated for spheres with hard sphere exclusion calculated according to Hayter–Penfold expressions shown as dashed lines.

varied unless either volume or surface conservation is violated. An example of this is given in the Fig. 4.

At low volume fraction as shown in the top curve, a radius of 7 Å for the polar cores should produce after fitting the radius, a scattering pattern as shown by the full line. It would be possible to 'fit' the observed scattering, assuming a droplet radius of 12 Å, assuming the unphysical value of $\sigma = 42$ Å^2/molecule. The only microstructure consistent with the pattern observed – including the peak position – *and* the known area per molecule is a local structure as connected cylinders.

If microstructure is a monodisperse droplets, the only possible radius is $R = 3\phi/\Sigma = 28$ Å. The dotted line is the scattering calculated for a density of spheres per unit volume $n_s = S^3/36\pi\phi^2$. We will see that in this case, the DOC model allows analytic calculation of peak position as well as numeric calculation of the full scattering curve, shown with dotted lines, considering that progressive formation of a network of connected cylinders is formed when the polar volume decreases. In the case given in this example, at high volume fraction, dispersions of spheres of given radius give a perfect agreement with the measured scattering. At low volume fraction, the assumption of the presence of spheres breaks down. A 'fit' would be possible by varying the radius of the larger droplets assumed to be present, and thus the area per head-group forced 'ad hoc' to be varied by more than 30%. This wrong procedure, i.e., assuming the presence of spheres, and then fitting with radius as a free parameter is found sometimes in literature and obscures the problem of determination of microstructures.

If the complete scattering curve calculated for a dispersion of spherical interacting droplets is different from the observed scattering, there is an important consequence: either the microstructure is not made of droplets, or the particles are connected and there is a gradual change in local morphology from spheres to cylinders or locally flat bilayers.

6.4. Determination of two characteristic lengths

A general expression has been derived by Kaler and Vonk: they assume that any microemulsion structure is locally lamellar of periodicity D^*, with a persistence length ξ. The Patterson function of the structure is now given by (Vonk et al., 1988):

$$\Gamma_{\mathrm{KV}}(r) = \gamma_{D^*}(r)\, \mathrm{e}^{-r/\xi} \tag{19}$$

where γ_{D^*} represents the correlation function of a lamellar phase. Use of *two* characteristic length scales as in this expression has been very successful in describing semi-crystalline polymers. The fitting procedure consists in Fourier transforming either the correlation function to fit to data or the reverse. In the case of microemulsions, the corresponding microstructure – a lamellar phase with very small crystallites – is confined to relatively small domains in phase diagrams. If one considers the standard free energy of bending:

$$F_{\mathrm{bending}} = 2k_c(H - H_0)^2 + \kappa(K - K_0) \tag{20}$$

where the rigidity of the film is represented by the two bending constants k_c and k. H is the average of the surfactant film and K is the Gaussian curvature. Values of H_0 and K_0 are the values of the curvatures of the surfactant film which would be observed if no external constraint (such as packing imposed by concentration or fluctuations due to temperature) is present. Even if this expression has no self-consistency problems as been demonstrated by Fogden et al. (1991), it is very useful in distinguishing stiff microemulsions with $k_c \gg kT$ from flexible film when bending energy is of the order of kT. Also, low spontaneous curvature corresponds to surfactant films which spontaneously tend to be locally flat. Probably the most important relation in surfactant microstructure is the relation linking the packing parameter p_0 to spontaneous average and Gaussian curvature:

$$p_0 = 1 + H_0 l + \tfrac{1}{3} K_0 l^2 \tag{21}$$

This purely geometric 'coverage' relation links the scalar parameter p_0 to the spontaneous curvatures adopted by a hypothetic unconstrained surfactant film. The surfactant parameter p_0 is related to the average shape of surfactant molecules in a free surfactant film. This 'coverage' relation simply links molecular property such as shape to the curved films that un-teared continuous films can be formed without deforming the shape of molecules.

Cases where the Kaler–Vonk expression has been successful in describing microemulsion structure with a locally lamellar structure and a persistence length are cases with stiff interface and low spontaneous curvature, i.e., p_0 close to 1 (Vonk, 1978; Vonk et al., 1988)

Another way to introduce two length scales to describe microemulsions is to use a parametric expression of the intensity which merges at high q into the Porod limit. This has been proposed by Teubner and Strey (Teubner and Strey, 1987; Teubner, 1990):

$$I_{\mathrm{TS}}(q) = \frac{I_0}{a + bq^2 + cq^4} \tag{22}$$

This corresponds also to a correlation function with two lengths:

$$\Gamma_{\mathrm{TS}}(r) = e^{-r/\xi}\left[\frac{\sin(\pi r/D^*)}{(\pi r/D^*)}\right] \tag{23}$$

where the periodicity D^* and the correlation length ξ are given by:

$$D^* = 2\pi\left[\frac{1}{2}\sqrt{\frac{a}{c}} - \frac{1}{4}\frac{b}{c}\right]^{-1/2} \tag{24}$$

$$\xi = \left[\frac{1}{2}\sqrt{\frac{a}{c}} + \frac{1}{4}\frac{b}{c}\right]^{-1/2} \tag{25}$$

When results are adjusted to expression on an absolute scale, it has to be taken into account that the constant I_0 is given by: $8\pi c\langle\phi(1-\phi)\Delta\rho^2\rangle/\xi$.

The position of the maximum of the scattering is related to both the periodicity and the correlation length:

$$q_{\max} = \left[\left(\frac{2\pi}{D^*}\right)^2 - \frac{1}{\xi^2}\right]^{1/2} \tag{26}$$

The Teubner–Strey simple analytic expression, can be fitted easily on data, especially if they are plotted in $I_0/I(q)$ versus $(q - q_{\max})^2$, where a figure close to a parabola is found, as in Fig. 3b:

$$I_{\mathrm{TS}}(q) = \frac{I_0}{(1 - C/I_{\max})(q^2/q_{\max}^2 - 1)^2 + C/I_{\max}} \tag{27}$$

The Teubner–Strey expression allows to extract the peak position D^* and the peak width expressed in terms of a persistence length ξ from any scattering curve by fitting three parameters a_2, c_1 and c_2. Fitting an observed scattering $I(q)$ to these expressions succeeds as long as there is only one peak in the scattering, as in the case where microstructure is a dispersion on spheres or even on calculated scattering of interacting spheres using expression (17).

The Teubner–Strey expression has been applied successfully in a large collection of papers (Choi and Chen, 1997; Choi et al., 1998, 1999; Glatter et al., 1996; Gompper et al., 2001; Gradzielski et al., 1996; Komura et al., 1996; Sottmann et al., 1997; Strey, 1994) in the case of flexible films and low spontaneous curvature. This is the case for single-chain non-ionic surfactants where the packing parameter p_0 and hence the spontaneous curvature H_0 can be tuned by temperature and even set to zero at one peculiar point in the phase diagram. This is the inversion temperature, where a microemulsion with lowest amount of surfactant and equal volumes of water and oil can be prepared.

Important results of fitting data in the case of flexible films are:

- the persistence length is always close to three times the periodicity, which turns out to be a general property of Voronoï tessellation (Brumberger and Goodisman, 1983)
- the general relation found for CRC model holds with another value of the constant:

$$\xi = \frac{7,16\phi(1-\phi)}{\Sigma} \tag{28}$$

This general relation has been also verified by Abillon et al. (1988) using four component microemulsions. In this study, the presence of a short-chain co-surfactant lowers the film bending constants.

Even if fitting to Teubner–Strey expression has no predictive power and can be applied to any scattering curve with the danger of obscuring the microstructure determination, there are two important properties to be noticed: surface per molecule as well as surfactant film curvatures can be extracted from the scattering at each composition by using the Teubner–Strey expression.

6.5. Determination of curvature

The spontaneous curvature is a property of a surfactant film at a given temperature, salinity and chemical potential of water (osmotic pressure). Due to sterical and packing constraints, this spontaneous curvature cannot always correspond to the effective average curvature in a real sample. This is the physical origin of the emulsification failure: excess 'inside' fluid is expelled if there is a too large difference between the spontaneous and actual curvatures (Safran, 1992; Safran et al., 1986).

It is therefore extremely useful to have direct methods to evaluate the average and Gaussian curvatures in a given sample. This can be achieved using neutron scattering using three similar methods:

(a) Using contrast variation by partial deuteration of water, the signal of a microemulsion can be considered as the sum of three correlations: water–water, water–oil and oil–oil.

$$I(q) = C\left[(\Delta\rho_{wo})^2\chi_{ww}(q) + (\Delta\rho_{of})^2\chi_{ff}(q) + \Delta\rho_{wo}\Delta\rho_{fo}\chi_{wf}(q)\right] \tag{29}$$

By studying the behaviour of the three functions which can be separated by graphical constructions, Auvray has shown in that average curvature can be extracted from contrast variation experiments (Auvray, 1985; Auvray et al., 1984, 1986).

(b) Contrast variation experiments can be made in two different conditions: when water and film have the same contrast and when oil and film have the same contrast. Therefore, for a sample of the same chemical composition, two surface to volume ratios can be determined. One can measure separately the interfacial area along both sides of the surfactant film. Identical specific area is a direct proof of zero average curvature while differences can be converted in exact value of average curvature (Chen and Choi, 1997a,b; Chen et al., 1998).

(c) A general relation first demonstrated by Teubner (1991a,b) allows direct extraction of Gaussian curvature K at the point with equal volume of water and oil:

$$\langle K(\phi = 0.5)\rangle = -\frac{\pi^2}{8}\Sigma^2 \qquad (30)$$

In the more general case of anisometric volumes, the average and Gaussian curvatures have to be extracted from a third order polynomial expansion of the correlation function (Lee and Chen, 1994a,b). The third-order expansion of the correlation function has been calculated by Kirste and Porod [for corrections specific of X-rays, see Robertus et al. (1988)]:

$$\Gamma(r \to 0) \approx 1 - \frac{\Sigma}{4\phi(1-\phi)} r(1 - C^* r^2) + \cdots \qquad (31)$$

where the constant C^* is related to principal curvatures C_1 and C_2:

$$C^* = \left\langle \tfrac{1}{12} C_1 C_2 + \tfrac{1}{32}(C_1 - C_2)^2 \right\rangle \qquad (32)$$

It is equivalent to fit directly the scattering function to:

$$I(q) = 2\pi(\Delta\rho)^2 \frac{\Sigma}{q^4}\left\{1 + \frac{1}{q^2}\left[\tfrac{1}{4}\langle (C_1 + C_2)^2\rangle + \tfrac{1}{8}\langle (C_1 - C_2)^2\rangle\right]\right\} \qquad (33)$$

Value of average curvature and Gaussian curvature are then easily calculated since average curvature is the average of the principal curvatures and the Gaussian curvature is the product of the two principal curvatures.

Fitting intensity or correlation functions to higher-order polynomials has allowed for example the measurement of average curvature in linear non-ionic surfactant model systems and to show that curvature decreases when molecular length decreases, thus linking the rigidity to thickness of interfacial film (Chen et al., 1996; Chen and Choi, 1997a,b; Choi and Chen, 1997, Choi et al., 2002)

7. Scattering of connected systems: locally cylinders or lamellae

We consider now the very frequent case when assuming a priori the presence of spheres of known radius does not give agreement with the scattering observed. Specific area Σ for the samples to investigate is known. Since one spectra at one given composition is not sufficient to determine a microstructure, so the data available to disentangle the microstructure is along a dilution line, either diluting with water, oil a given concentrated sample. Fitting to Teubner–Strey formula will always give two characteristic lengths. If the film is very flexible, fluctuation in curvature dominates, and one cannot determine more than two characteristic lengths. In the case of stiff interfacial films, simple observations such as non-monotonic conductivity are however totally incompatible with the absence of a constraint local microstructure (Rushforth et al., 1986; Billman and Kaler, 1990).

Direct determination of local microstructure by fitting either the scattered intensity (Eastoe et al., 1994; Bardez et al., 1995) or the Fourier transformed Patterson function to a local shape succeeds at low volume fraction (Glatter et al., 1996; Brunner-Poppela et al., 1999) in order to determine if the local microstructure is connected cylinders or connected bilayers [also called 'sponge' structure (Gazeau et al., 1989; Safran, 1991)]. These two microstructures are the only types which have been shown to exist so far.

Connected cylinders can be represented 'molten' hexagonal or cubic phases who have lost long range order while connected lamellar are a disordered state of a smectic phase. There is a possibility of biliquid foam microstructure which exist in microemulsion and has been produced in theoretical modelling of anisometric microemulsions (Arleth et al., 2001). Experimentally, as can be seen by fitting any data to Teubner–Strey formula, the order is typically lost at about three times the peak position: $D^* = \xi/3$.

Asserting the presence of a local microstructure in an optically isotropic fluid requires first the geometrical construction of a complete set of possible microstructures, calculating their volume fraction, specific area Σ and curvatures via the packing parameter. Then compare the observed scattering along a dilution line, assuming fluctuations are dominated by the rigidity of the film, i.e. that he fluctuations due to temperature do not overcome packing constraints. An easy experimental check is to perform scattering at several temperatures for a given composition: if topological and bending constraints dominate, position of the peak, Porod limit as well as scattering at low angle does not move. This is the case with double-chain surfactants, surfactants with 16 or more carbons in one chain addition of long-chain alcohols, microemulsions with lipids, long co-surfactants and most microemulsions used in industry.

Disordered lamellae is the likely microstructure if spontaneous curvature is low compared to D^*, while connected cylinders is the likely microstructure at high surfactant content and high curvature. Before showing how data can discriminate between the different possible microstructures, we first describe shortly the construction of a complete set of microstructure.

At first order in terms of topology, any disordered open and connected (DOC) microstructure, is constructed on a set of Voronoï polyhedra constructed on seed points of density n (cm^{-3}). But instead of filling each Voronoï cell by either water or oil as in early microemulsion models, we now split each cell according to volume fraction and construct in each cell either cylinders perpendicular to cell faces or some flat pentagons or hexagons on the faces of cells. We use the simplest approximation of cell packing of known average nearest-neighbour distance. These are Voronoï cells, a complete tessellation of space with polyhedra. These polyhedra are constructed from a set of central seed points as shown in Fig. 5a and b. We choose the seed points, not at random, but with a distance of minimum approach d, which is fixed to twice the chain length to avoid edge effects introduced by the presence of too large curvatures (cusps). The density of these seed points in space introduces an underlying Voronoï cut-off volume fraction Φ_v, defined as:

$$\Phi_v = \tfrac{1}{6}\pi d^3 n \tag{34}$$

A value of Φ_v exceeding 0.7 should correspond to a divergence in the possibility of packing a DOC structure. This is the origin of one of the composition limits observed for microemulsions at low water content, i.e., the phase limit opposite to the emulsification failure. This density n of Poisson points or the 'periodicity' of the structure is fixed by the amount of available interface. When surfactant is added at constant 'interior' volume fraction, the peak position shifts to larger angles since $D^* = 1.22n^{-1/3} = 2\pi/q_{\max}$ increases.

The three constraints in each Voronoï cell are: conservation of volume fraction ϕ, specific area Σ, and packing parameter p that characterises a given surfactant at a given temperature, with a given oil. We apply in each cell conservation of volume, surface and curvature, which stands:

$$p = 1 + \langle H \rangle l + \tfrac{1}{3}\langle K \rangle l^2 \tag{35}$$

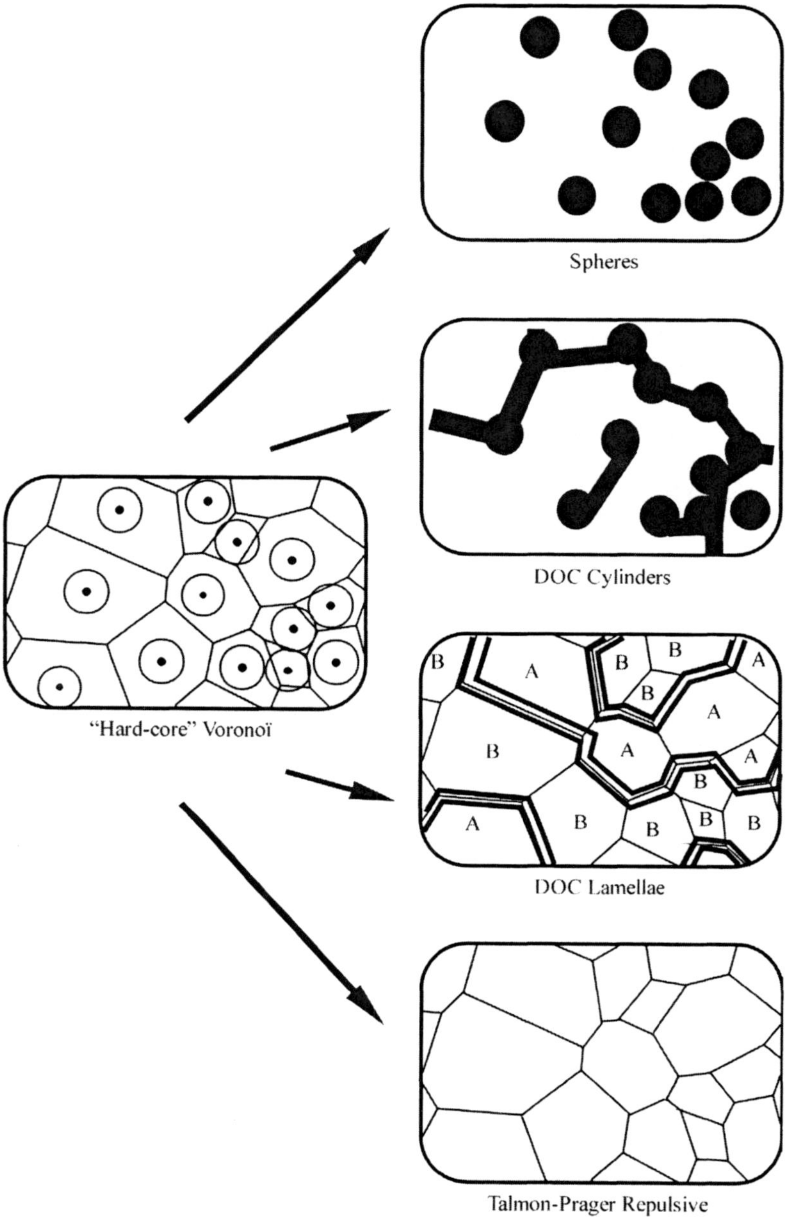

Fig. 5. (a) The same basic underlying lattice of seed points define a set of Voronoï cells. All microstructures constructed inside each cell using the same seed point distributions produce scattering patterns with the same position of maximum, but different absolute intensities and different asymptotic limits at zero and at large angles. Top structure: one sphere occupies the centre of each Voronoï cell. DOC cylinder: an average number of Z cylinders per cell connects neighbouring cells. DOC-lamellae: a locally flat swollen bilayer decorates all surfaces located between cells of type 'A' and type 'B'. This construction is the seed of all 'sponge' and locally lamellar microstructures. Talmon–Prager original model assumes random filling by water or oil of individual cells. This model does not produce peaks in the scattering.

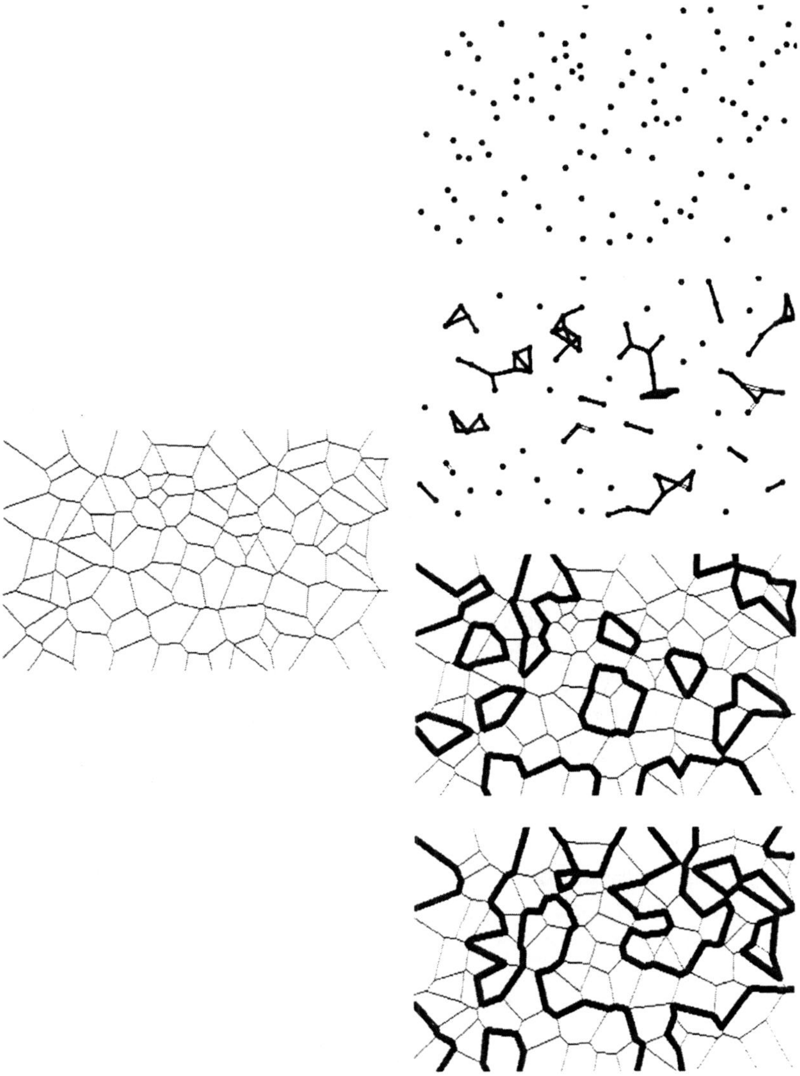

Fig. 5 (continued). (b) Two-dimensional examples of microstructures as repulsive spheres, connected cylinders with connectivity $Z = 2$ and asymmetric as well as symmetric connected lamellae (sponge) with the same scattering peak position, but different contact length λ.

where $\langle H \rangle$ and $\langle K \rangle$ are the curvatures averaged over one cell, including the lines at intersection of cylinders points. Any deviation from this coverage constraint would have an high energetic toll (Hyde et al., 1996)

Now consider a disordered surface of constant mean curvature, i.e., a disordered version of the infinite periodic minimal and H surfaces. This surface has to be constructed in random cells of size $D^* = 1.22n^{-1/3}$ and return the right value of the conserved quantities: volume, surface and curvature. Drawings of two-dimensional analogs of these structures have been made by Welberry and Zemb (1988) and Barnes and Zemb (1988).

7.1. The DOC-cylinder microstructure

If the spontaneous radius of curvature is small versus the density n of Poisson points, i.e., if $(1-p_0)l < n^{-1/3}$, where l is the typical surfactant tail length. We approximate the polar/apolar limiting surface in the surfactant film by a network of spheres of radius R connected by an average of Z cylinders per sphere to the nearest neighbouring spheres. Of course, the cusps between cylinders and spheres have no special meaning and can always been smoothed out according to the constant coverage condition. The average number of cylinders meeting each sphere is the coordination number Z which can vary from zero to 4, corresponding to molten cubic phase. In the limiting case were $Z=0$, the DOC structure is just the hard spheres model. A value of $Z>1.1$ ensures bicontinuity. $Z=2$ is the case of swollen giant cylindrical micelles. To maintain the same value of the surfactant parameter throughout the structure, it is necessary to fix the ratio between spheres and cylinders to satisfy the coverage condition. The packing parameter remains the same in the cylindrical parts as in the spherical parts of the structure.

To describe any connected cylinder structure, we have three parameters describing a structure (n, Z, and R or r) to be determined by the three geometrical constraints. A simplified version of the DOC-cylinder model, fixing connectivity to Z close to 2, has been described (Tlusty et al., 1997; Bernheim-Grosswasser et al., 1999).

For given values of n, Z and R, the DOC-cylinder structure is constructed as follows:

First, a random hard-sphere distribution of points is generated in space with minimum separation $2R+2l$, ensuring that the tails of the surfactants coating the water spheres do not intermesh. A sphere of radius R is then placed at each of these points. Then, we determine the nearest neighbours of each point by constructing the Voronoï polyhedra. Two points are neighbours if their Voronoï polyhedra share a face. Finally, we order the resulting pairs of neighbours by their separation distance, and connect neighbouring spheres by cylinders of radius $r=R/2$, i.e., approximating as $\rho=r/R\approx 2$, in order of proximity, until the average coordination number for the structure is Z. It is possible now to derive equations for such a random network relating the parameters n, R and Z to the system properties Φ_i, Σ, and p. We have the following expressions for Φ and Σ which are exact as long as the cylinders do not intersect (Ninham et al., 1987; Barnes et al., 1988a,b):

$$\phi=\left[\frac{4\pi}{3}R^3-\frac{1}{3}Zr^3(\Omega\rho^3-\pi\sqrt{\rho^2-1})\right]n+\pi R^2 Z\left[\frac{8.15n^{2/3}}{13.4}-nr\sqrt{\rho^2-1}\right] \tag{36}$$

$$\Sigma=\left[(4\pi-\Omega Z)R^2 n+2\pi Zr\left(\frac{8.15n^{2/3}}{13.4}-nr\sqrt{\rho^2-1}\right)\right] \tag{37}$$

where $\rho=R/r$ and $\Omega=2\pi\left(1-\sqrt{\rho^2-1}\right)$.

The constraint of average film curvature is given by an imposed value of p:

$$p=\frac{\phi(R+l,Z,n)-\phi(R,Z,n)}{l\Sigma(R,Z,n)} \tag{38}$$

This coverage relation expresses that a surfactant film of packing parameter p is contained in a volume defined by displacing the dividing surface by a distance l. This differential relation is equivalent to satisfying the coverage condition with the spontaneous packing parameter p.

We thus have a simple geometrical model with three geometrical parameters: n, R, and Z which are uniquely fixed by the two experimentally available quantities Φ, Σ, leaving only p as a unknown parameter. If the phase diagram is known and a boundary between droplets and expelled 'interior' (Winsor I or II equilibria) is located, it is even possible to derive experimentally p and thus to eliminate p. The trick is to use a linear phase boundary in a ternary phase diagram, when unconnected droplets of the largest stable radius R_{max} are in equilibrium with the pure internal liquid. The radius of these maximum swollen spheres gives at emulsification failure the surfactant parameter p for the given sample:

$$p = 1 + \left(\frac{l}{R_{max}}\right) + \frac{1}{3}\left(\frac{l}{R_{max}}\right)^2 \tag{39}$$

A detailed description of how to evaluate p from the emulsification failure boundary in ternary phase diagrams is available (Barnes et al., 1988b; Zemb et al., 1987).

The DOC-cylinder microstructures have been found in ternary microemulsions with a stiff interfacial film by Friberg and Burasczenka (1978), Ninham et al. (1987), Barnes et al. (1988b) and Hyde et al. (1989).

7.2. The DOC-lamellar microstructure

When the spontaneous curvature is larger than the typical cell size given by the peak position D^*, i.e., when $(1 - p_0)l > n^{-1/3}$, the general surface of constant curvature is approximated in a different way (Hyde et al., 1990, 1996). The structure more closely resembles a tightly folded thick sheet of bilayer. We now approximate the bicontinuous volume by a set of connected pieces of bilayer of known volume fraction.

Again a random hard-sphere distribution of points (density n) in space is constructed with a minimum approach distance of $2(l + t)$, where l is the surfactant tail length and t is the half-thickness of the polar layer. Voronoï polyhedra are then used to divide the volume into regions about each point. In this construction, the faces of the polyhedra contain a bilayer of surfactant separated by a water layer, and we again require the internal volume fraction and total interfacial area to meet the constraints.

If all faces are coated as in a biliquid foam, the resulting structure will be water continuous (hence electrically conducting) with the oil domains segregated and a high curvature energy involved at the vertices where four bilayers join. In order not to generate a foam, but infinitely folded bilayer without edges, one has to imagine the initial Voronoï tessellation to be composed of two types of oil, I and II, with pseudo-volume fractions Ψ and $1 - \Psi$. This is similar to constructing a microemulsion model with a set of adjacent oil- and water-filled cubes (Cates et al., 1988a,b). The polyhedra are then filled with oil 'type I' and 'type II' accordingly, and all the polyhedra faces between domains of like oil are removed. The remaining film of total area Σ contains neither joints nor unconnected edges and may be swelled to a thickness $2t$ with water so that its volume fraction is correct, and the area conservation condition is satisfied.

For a Voronoï tessellation without the hard sphere condition, the surface area of the Voronoï polyhedra per unit volume is $2.91n^{1/3}$ (Barnes et al., 1990). The probability that a given face actually lies between the two different oil regions is $2\Psi(1-\Psi)$. Assuming that the edge and vertex effects can be neglected, volume and surface constraints are expressed as:

$$\phi = 11.64t\psi(1-\psi)n^{1/3} \tag{40}$$

$$\Sigma = 11.64\psi(1-\psi)n^{1/3} \tag{41}$$

since $t = \Phi/\Sigma$ as for smectic liquid crystals. From this line of reasoning one would also derive a value of $p = V_{\mathrm{mol}}/\sigma l$ close to one for the curvature, which is independent of the pseudo-volume fraction Ψ. Note that if $\Psi = 0.5$, The DOC-lamella model reduces to the symmetric sponge microstructure (Gazeau et al., 1989). For different values, the asymmetric sponge phase is produced (Warr et al., 1988). If pseudo-volume is one half, a symmetric sponge microstructure shown in Fig. 5b is produced. In Fig. 5b, the case where the pseudo-volume is 0.25 is also shown. There is a continuous way to change a symmetric sponge into closed vesicles.

The coverage relation and curvature constraints are also analytic within the DOC model approximations. These are described in detail (Zemb, 1997). A robust approximation is:

$$p = 1 + \frac{tl + l^2/3}{\left(\frac{3}{4}\pi n\right)^{2/3} + t^2}\left(1 - 4\psi(1-\psi)\right) \tag{42}$$

As was the case in the DOC-cylinder model, the three constraints as imposed ϕ, Σ and p are satisfied by varying the density of Poisson points n, the local thickness t and the asymmetry ψ. The peak position of DOC-cylinder and DOC-lamellar models are also obtained from the density of the initial set of Poisson points, $D^* = 1.22n^{-1/3}$ (Welberry and Zemb, 1988; Zemb et al., 1993). The disordered lamellae microstructure has been proven to exist in the ternary DDAB/tetradecane/water system (Barnes et al., 1990) and confirmed in a four-component system by Haviqsson et al. (1994). A detailed description of conductivity and phase limits set by maximal curvature constraints in the case of connected bilayers has been given by Barnes et al. (1990).

7.3. Checking for the presence of connected microstructures

Once the three constraints ϕ, Σ and p are met, the density n of Poisson points equivalent to the scattering peak position D^* and the connectivity Z can be calculated by means solving the three implicit equations (36), (37) and (38) by an iterative procedure (a simple and short program written in BASIC, is available from the author). This has allowed the peak position shift versus composition to be predicted and used to evidence the existence of connected structures. The most general method is to use the dilution plot which also discriminates between microstructures. As can be seen on the two-dimensional model below, the surface to volume ratio at constant peak position (underlying lattice), is largely controlled by microstructure. In the three parts of Fig. 6, underlying lattice as well as part occupied (25%) are the same, black/white contact lines are different. This is the basis of the 'dilution plot'.

The existence of connected cylinders form stiff bilayers has been proven by means of peak

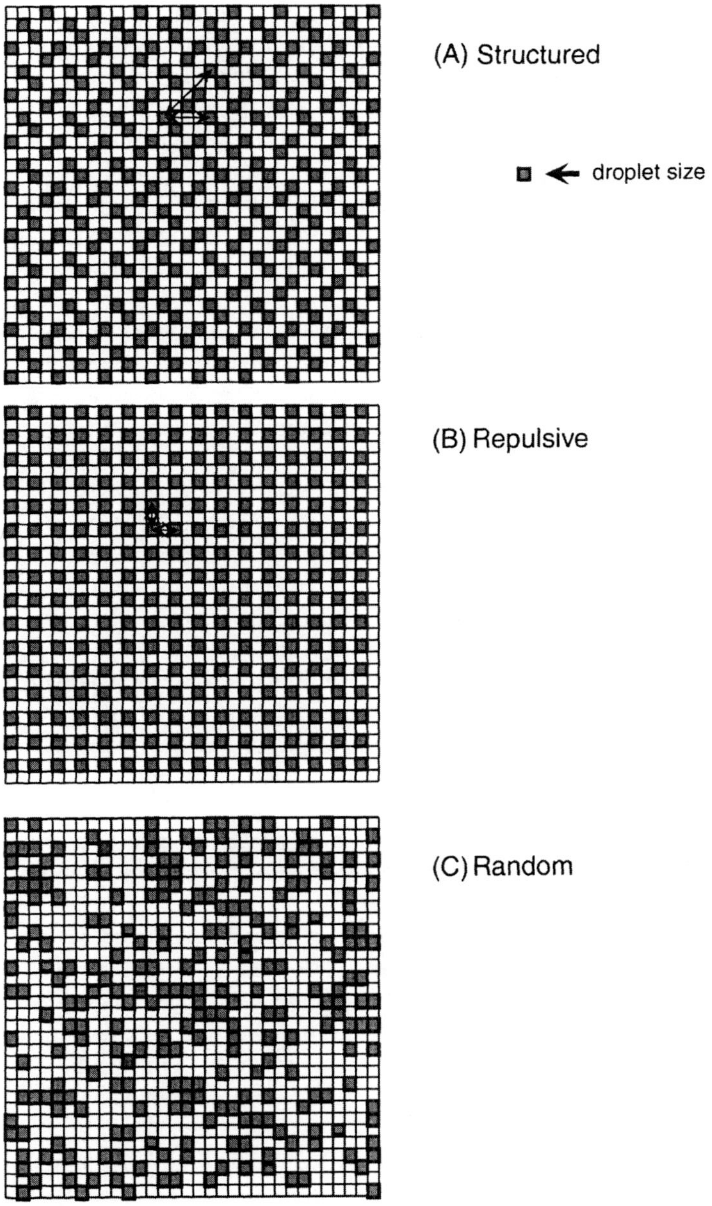

Fig. 6. How microstructure controls the surface to volume ratio. In this two-dimensional model, the ratio of scattering peak position to length per unit surface depends is controlled by local arrangements, a feature which cannot be included in mean-field approximations. 'Microdomains' of size ξ are juxtaposed on a lattice: (A) 'Structured' model considers a periodic with units vector as four lattice steps ξ. (B) Repulsive distributions minimize contacts between adjacent droplets of size ξ. (C) 'Random' distributions imply partial coalescence between droplets, thus decreasing the length of interface per unit surface. The two-dimensional analogue of ΣD^* is λD^*, where λ is the length of black/white contact lines per unit surface. λD^* is 4, 2 and 0.75 in cases A, B and C, respectively.

position shift versus concentration, in the case of highly curved and stiff interfaces with ionic double-chain surfactants, when a short alcane is used as the oil. In this case, the value of p depends of the oil. If a non-penetrating oil such as tetradecane is used, the spontaneous packing parameter p tends to 1, and a DOC-lamellar microstructure is present (Warr et al., 1990). If the two chains of the surfactant are shortened, ionic microemulsions forms a film with low spontaneous curvature. An asymmetric sponge phase, made of oil swollen locally flat bilayers, is present. The DOC model is also the only one compatible with the non-monotonic behaviour of conductivity (Rushforth et al., 1986). It has long been a mystery how conductivity can increase when polar content decreases. When water is added to this class of microemulsions, the conductivity is not monotonic. There is first an increase, followed by a sharp decrease of conductivity. This behaviour is also a consequence of microstructure, since connected cylinders filled with water transform progressively in isolated droplets at constant curvature. In the case of DOC cylinders, there is an anti-percolation when the connectivity Z is 1.1. This allows to predict the position of the percolation lines on ternary phase diagrams (Zemb, 1987; Hyde et al., 1989).

Use of DOC model is not restricted to stiff charged interfaces made of double-chain quaternary ammonium. Similar behaviour has been observed in other systems of so-called 'reverse micelles' (RM), i.e., microemulsion with polar volume fraction less than 50%. Usually, they are assumed (without direct experimental evidence) to be made of more or less monodisperse water in oil disconnected spherical droplets. However, counter-examples of non-spherical reverse micelles are abundant in the literature.

The first example of connected cylinder structure in the literature has been given by Friberg and Burasczenka (1978). In the water/potassium oleate/oil system, where the oil was the adequate mixture of benzene and pentanol, the extent of the reverse micelle isotropic domain increased when the cosurfactant (pentanol) content in the solvent was increased. When salt was added, the reverse micelle (L_2) domain extended up to the water corner. The direct (L_1) and reverse micelle domain then merged. At some point, the picture of a reverse micelle as spherical droplets of water has to break down, and a structural transition has to occur by means of connected cylinders.

To our knowledge, the first connected bilayer – i.e., the one which have been rediscovered under the term of 'sponge' microstructure – has been demonstrated in a ternary system where the reverse micelle domain could be diluted up to 95% with water in the sodium octanoate/octanoic acid/water system by Ekwall in the late sixties, by noticing that the scattering peak position shifts linearly with volume fraction, as in smectic phases. The ternary system where the locally lamellar microstructure is present at high water content is best known and described with all details, including activity of all components, involves a short-chain fatty acid and its soap (Ekwall and Mandell, 1967; Ekwall and Solyom, 1967; Ekwall and Stenius, 1967; Ekwall et al., 1967a,b; Ekwall, 1988, 1989; Fontell and Mandell, 1989a,b, 1993). The gradual transformation of droplets into connected cylinders and finally bilayers has been confirmed by Olsson and Schurtenberger (1993) and Olsen et al. (1993).

Locally lamellar or cylindrical connected disordered structures observed in microemulsions, are also observable with solid co-polymers, when parts of copolymers are not miscible. They are represented in Fig. 7 (Förster et al., 2001). The analogy between microemulsions and the closely related copolymers which obey also to volume, surface and curvature constraints has been described by Jinnai et al. (2000).

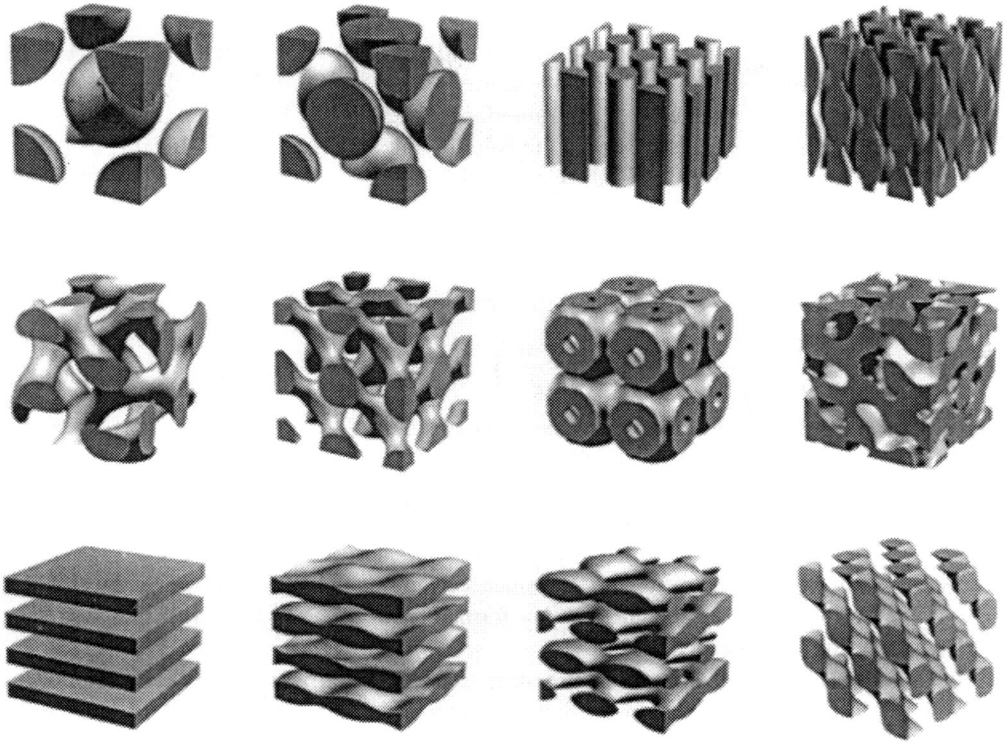

Fig. 7. A general family of microstructures as spheres, cylinders and planes on a regular lattice, as observed with solid copolymers (Smarsly et al., 2002). As microemulsions, copolymers and mixtures of polymers obey to surface, volume and curvature constraints: locally spherical, cylindrical or lamellar objects can be observed. However, the underlying network has no long range order in microemulsions. Microemulsions can be represented with the same morphologies than the family developed to describe polymers, but without long range order.

The DOC model is valid only when the curvature is constant in the amorphous dispersion: when local variation of the due to thermal fluctuation and spontaneous curvature co-exist, the microstructural constraints and the thermal fluctuations have to be taken into account. The only model with thermal equilibration of bending energy, fluctuations, non-zero spontaneous curvature and non-equivalent polar and apolar volumes is the random localized wavelet model as initiated by Marcelja (1990).

8. Mixing fluctuations and geometrical constraints: random wavelets models

This model has been proposed by Berk (1987). The idea is to produce a random structure of typical size D^* by superposing waves of wavelength D^* with random phases and random directions. The structure produced is random over long distances and the scattering has a peak at the desired position. No bending constraint or total interface were introduced at that stage. At a given, water/oil ratio, the total interface is fixed and cannot be varied. This model is useful in the case of large spontaneous radius of curvature (p close to 1) and low bending energy of the interfacial film ($k_c < kT$). This model allows precise calculation of the scattering in the conditions of film contrast only from the bulk contrast one. The first success of this random wavelet model was to explain how scattering is modified while shifting from bulk contrast to film contrast in the case of fluid interfaces (Trevino et al., 1998). Varying the level of the cut in the random wavelets controls the volume fraction (Teubner, 1991a,b).

An improved version of the random wave model has been introduced by Marcelja using localised random waves (Marcelja, 1990; Pieruschka and Marcelja, 1992; Pieruschka and Safran, 1993). The localized random interfering monochromatic waves are now replaced by a set of localized wavepackets. The envelope $u(r - r_0)$ of the wavelets used to generate the microstructure is characterized by the correlation length ξ:

$$U(r - r_0) = \frac{1}{r}\exp\left(\frac{-r}{\xi}\right) - \frac{1}{r}\exp\left(\frac{-ar}{\xi}\right) \tag{43}$$

where a is an arbitrary constant. When one level-cut is used, microstructures corresponding to equal volumes of water and oil are generated. Cutting asymmetrically a random field allows modelling of polar volumes different from 50% (Chen and Choi, 1997a,b; Chen et al., 1998; Choy and Chen, 2001). Using two level cuts, Arleth has developed a general model (ACGRF) which can take into account any type of local energy, and minimize it. For example, microemulsions where bending energy dominates over for instance long range electrostatic energy can be treated using the free energy formula given in Eq. (20) with the two usual bending constants k_c and κ.

This model allows local variation of curvature from the preferred value. If k_c is set to kT or lower, the general swelling behaviour which has been observed in all experiments is found as a consequence of surface, volume and bending constraints:

$$\Sigma\xi = T(k_c)\phi(1 - \phi) \tag{44}$$

The topological constant $T(k_c)$ is now dependent on k_c. If k_c is set smaller than one, a value of $T = 5.6$ is found. If k_c is larger than kT, a value $T = 7.1$ is found. At constant surface and volumes, increasing the stiffness consumes less surface through crumpling and the peak position shifts towards lower angles (D^* increases). In this case, the ACGRF model merges toward the DOC-cylinder and DOC-lamellar models, depending on spontaneous curvature (Arleth et al., 2001). Examples of connected microstructures generated with different combinations of spontaneous curvature and bending constants by the asymmetrically-cut are shown in Fig. 8. All structures are thermalised at room temperature for the case of 'stiff film' (bending constant $k_c = 2k_bT$ and the Gaussian bending constant set to $-k_bT$). In Fig. 8, (a) represents equal apolar and polar volume fractions, with a volume fraction of surfactant set to 20% and zero spontaneous curvature; (b) is the same as (a), but with a spontaneous curvature radius $1/\langle H\rangle$ pointing towards the white side of the interfacial film and set to 133 Å. (c) to (f) correspond to cases where 'inside' volume fraction represented in black as 25% and a volume fraction of surfactant 10% with the same 'stiff film' bending constants. In (c), the spontaneous curvature towards the black side favours closed 'droplets' of the material represented in black. (d) is the same, but with no spontaneous curvature, i.e., a generic bicontinuous structure which can be approximated by spheres connected by cylindrical domains. (e) is the case of frustrated curvature, but with a spontaneous curvature towards the white side. Frustration in this case produces the disordered open connected lamellar

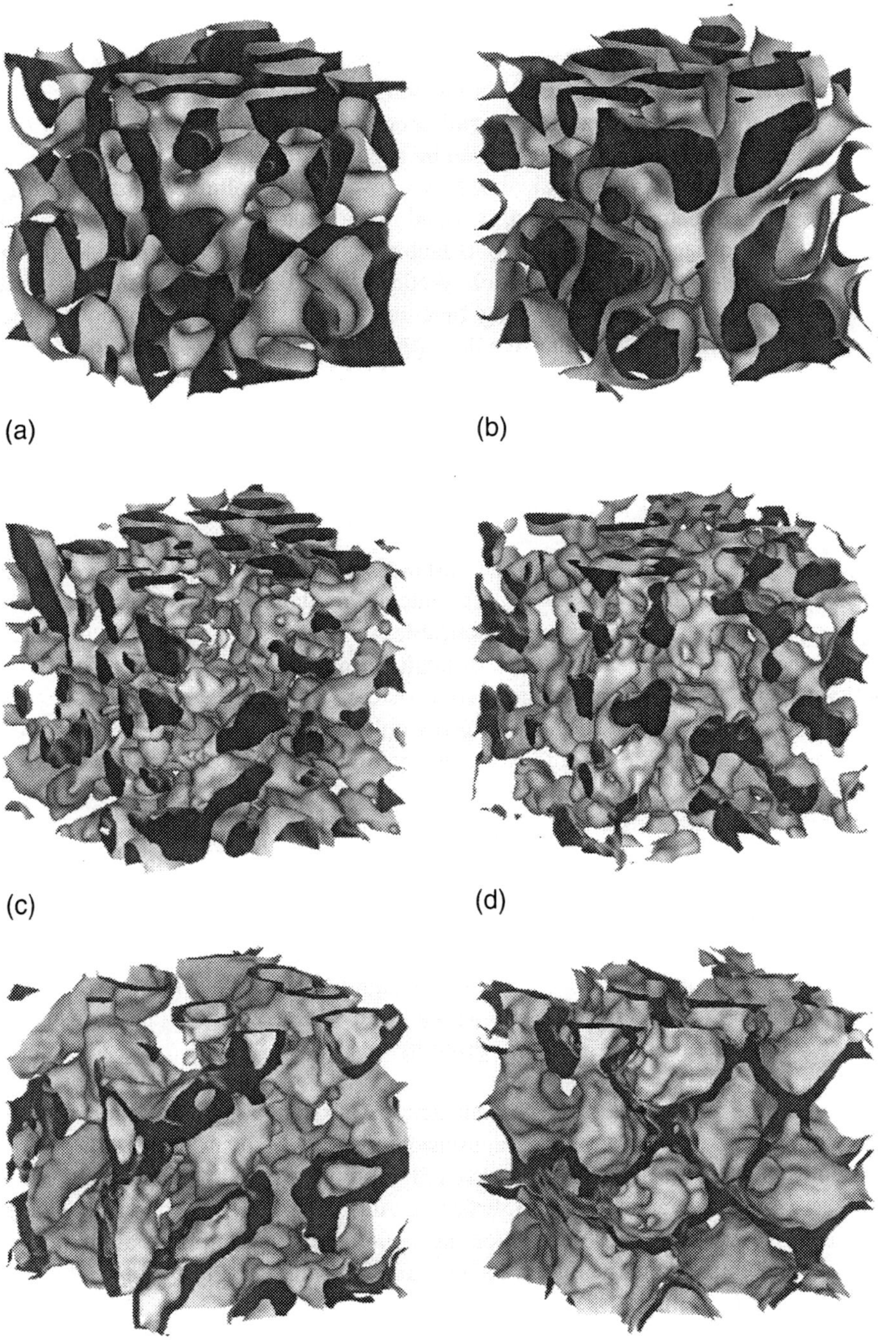

Fig. 8. Examples of microstructures of microemulsions generated using asymmetrically-cut thermalised Gaussian random fields, using different physical constraints in terms of spontaneous curvatures and a bending constant of $2kT$ corresponding to a slightly 'stiff' case (Arleth et al., 2001).

microstructure (asymmetric sponge). (f) is the same with stronger curvature constraint, producing the 'biliquid foam' topology which is known for emulsions. The side length for (a) and (b) represents 500 Å, while the scale for the others figures is 1000 Å.

The ACGRF model is the most powerful model available yet on microemulsion microstructure: microstructure and scattering can be generated for thermalised samples. Surface to volume ratio, and curvatures are calculated for a given set of volume as well as stiffness of the film. When *kc* tends to zero, the results found using Teubner–Strey expression or similar (Choi et al., 2002) are generated. When k_c is larger than one, the ACGRF model reproduces the locally cylindrical and lamellar models predicted by the DOC model. Both limits of flexible and stiff films, known since long time in experimental determinations (Chevalier and Zemb, 1990) are found using the ACGRF approach. Use of the ACGRF model requires thermalisation of random wavelets and is not analytic. Note however, that effective curvature as well as full scattering in film and in bulk contrast are generated: zero angle limit, peak position, Porod limit for any initial constraints.

9. Practical methods to determine microstructure

Due to the order of magnitude of the distances involved, typically between 100 and 1000 Å, neutron scattering is the easiest experimental method. Small-angle X-ray scattering using collimation at a synchrotron is also applicable. On a laboratory source, obtaining enough precision at the low-q edge is sometimes difficult. In any case, the spectra should be taken at least to $q = 0.4$ Å^{-1} to allow determination of the interfacial area per molecule. Static light scattering cannot be used, because it gives only osmotic compressibility and has only be used close to a demixtion in order to investigate attractive interactions, often using the Ornstein–Zernike expression or similar.

The first check should be for a microstructure as hard spheres or spheres interacting with a known potential (w/o or o/w). Since volume and surface are known, the only possible radius is known to be $3\phi/\Sigma$. If the analytically calculated intensity is close to the observed one, but not identical, the main reason is interaction between droplets (with a slight effect of polydispersity). The only unknown parameter is the contact potential. Therefore, the proof of the presence of spheres is simple and straightforward. Well developed examples of microemulsions well described as spheres with a radius linearly dependent on inside volume to surfactant ratio (Pileni et al., 1985) are given by Olsson et al. (1993), Bagger-Jorgensen et al. (1997) and Nave et al. (2000a,b)

Note that fitting to Teubner–Strey parametric expression would also converge even if spherical droplets are present. In this case, division of X-ray by neutron may allow to resolve the ambiguity (Barnes et al., 1993). A detailed example where microstructure as spheres works is the case of the largely studied surfactant AOT far from the boundary of the reverse micellar region (Nave et al., 2000a,b). For the same ternary system, the assumption of the presence of spheres breaks down and a locally lamellar microstructure can be demonstrated using a dilution plot (Zemb, 1997).

A very general and powerful method is to use the inversion method proposed by Glatter. Due to the presence of strong interactions, a peak is observed in scattering. If no electrostatic term is involved, this peak can be approximated as an effective volume fraction of hard spheres (ϕ_v). This effective excluded volume fraction appearing in $S(q)$ is the same which sets the seed point in Voronoï polyhedra. Leaving this parameter free, a Fourier inversion including a division by the structure factor involved $S(q)$ allows direct determination of local microstructure, if the inversion program converges. Using this methodology, Glatter et al. (1996) has shown that even short-chain non-ionic surfactants show the progressive transition from spheres to globules to locally planar geometry consistently with the DOC-model.

9.1. The dilution plot

We have seen that all microstructures, cubic random cells, interacting droplets correspond to a peculiar behaviour of the product ΣD^* along a dilution line. The DOC-lamellar and DOC-cylinder models also give peculiar peak shifts, depending on packing parameter p, a molecular property which should not vary along a dilution line. The ACGFR model also will give relation (27), where the topological constant depends on the stiffness of the film. These swelling behaviours predicted for different microstructural models in reduced units are compared in Fig. 9a and b.

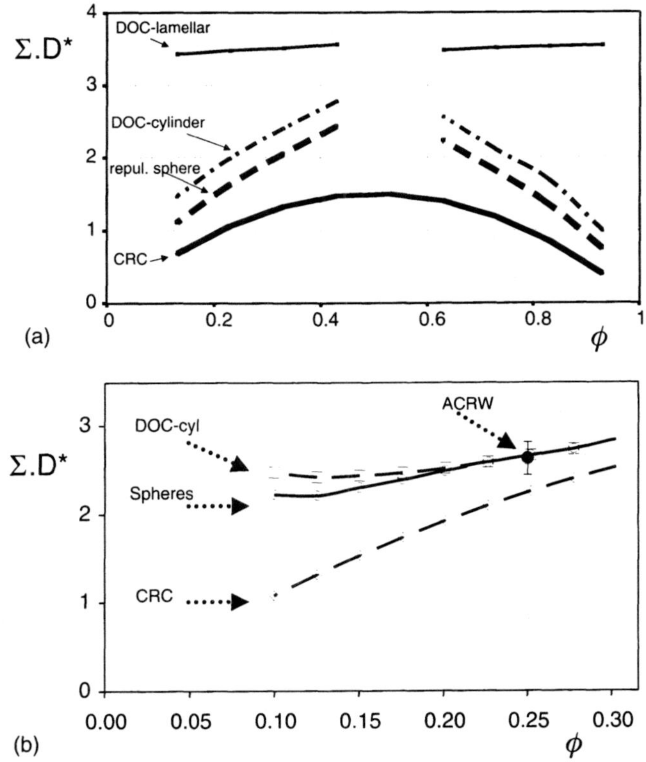

Fig. 9. (a) The general dilution plot, which allows determination of microstructure: ordinate is peak position in real space, multiplied by specific surface. For flexible films, the experimental result lies at twice the CRC swelling curve, which is the prediction of the ACGRF model. (b) Compared welling plots geometrical models of microstructure: CRC, monodisperse repulsive droplets, connected cylinders with the thermalised Gaussian random field with two level cuts developed ACGRF (Arleth et al., 2001).

The CRC models as developed by Jouffroy et al. (1982) predict a peak as shown on the black line. Repulsive spheres consume more interface and are represented by the dashed line. Dot dashed line represents the connected cylinder model with constant curvature. Continuous line with square represents the locally lamellar microstructure. The flexible case of cubic random cell, predicts a peak at about twice the CRC result (Chandra and Safran, 1991) and is not far from the swelling plot obtaining with the DOC model, where connected cylinders with variable connectivity at constant curvature are considered. The DOC-lamellar (asymmetric sponge) microstructure is the easiest evidence using a dilution plot, since the peak position is at lower angles (Warr et al., 1988). If the dilution plot are not sufficient to discriminate between possible microstructures, full numerical calculation of the scattering curve corresponding to connected microstructures or ACGRF is needed (Barnes et al, 1988a,b; Arleth et al., 2001).

10. An example of determining microemulsion microstructure using dilution lines

Microstructure can be tested experimentally using he largest dilution ratio possible: the idea is to make a series of samples with known interfacial area per unit volume (i.e., surfactant concentration). Area per unit volume is either deduced from the sample composition, or better, directly accessed from the scattering if the Porod limit is measured. Then, the maximum of the scattering ($q_{\max}$) is converted into real space peak position $D^* = 2\pi/q_{\max}$.

Experimental results obtained for a double-chain surfactant, using a non-penetrating solvent which sets low spontaneous curvature, are given in the Table 2.

We can now test the models and methods described in this chapter (Fig. 10):

(a) In the case of CRC models, one expects: $\Sigma D^* = 6\phi(1-\phi)$. Taking into account effective repulsion as described by Chandra and Safran (1991, 1992), we have to multiply the peak position D^* by a multiplicative factor of two. The microstructure is a random dispersion domains of size ξ with an effective repulsion due to not vanishing flexibility. As can be seen on figure 10, taking twice the CRC prediction gives reasonable agreement, but with a maximum of the dilution plot at 50%. Neglecting any type of more defined microstructure gives a fair prediction of the peak shift, but with a maximum due to symmetry of the peak position when ϕ is replaced by $(1-\phi)$. This property is not observed in the experiment when the polar volume is close to 50%.

Table 2

Experimental data for a dilution plot

ϕ	Σ	D^*	ΣD^* (DOC-lamella)	ΣD^* ACGRF
0.279	0.03	82	2.49	3.1
0.337	0.028	97	2.77	3.25
0.406	0.025	120	3.025	3.43
0.487	0.022	160	3.344	3.48
0.572	0.018	190	3.438	3.5

Polar volume fraction, specific surface in $Å^2/Å^3$. Peak position $D^* = 2\pi/q_{\max}$ and product ΣD^* as calculated according to the DOC-lamellae and with the ACGRF models in the case of diluted bicontinuous microstructure. Data for DDAB/water/tetradecane microemulsions (Barnes et al., 1990).

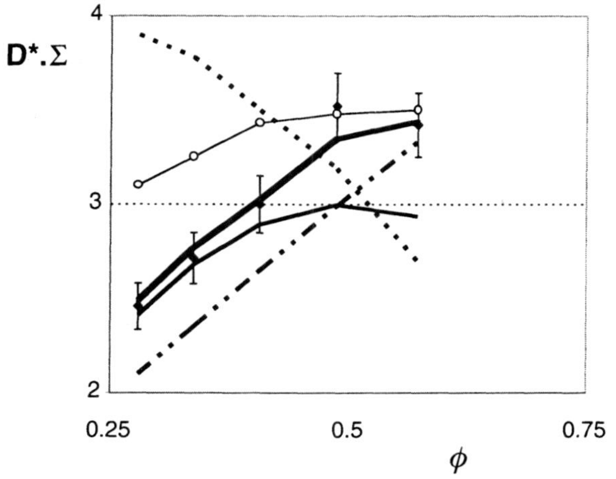

Fig. 10. Testing microemulsion microstructures: Dilution plot observed for a ternary microemulsion with a stiff film and low curvature, non-penetrating oil in the concentrated regime, near the centre of phase diagrams. Data points are shown with error bars. Experimental data points are compared to CRC models including the correction factor of 2 (thin line), spheres (w/o and o/w) shown as dashed lines, the DOC-lamellar model (thick line) as well as the ACGRF model in the case of diluted bicontinuous microstructure (line with circles).

(b) Assuming that microstructure is w/o spheres or o/w spheres. In this case, scattering can be calculated using the Hayter–Penfold expressions (Hansen and Hayter, 1982; Hayter, 1985; Hayter and Penfold, 1983; Hayter et al., 1984), or the empiric analytic approximation [Eq. (18)], valid for repulsive spheres. The results are shown with dashed lines. Examination of the corresponding dilution plots shows that peak shifts expected if only droplets are present are not compatible with the observed value. Note here that a common mistake is to allow the radius of droplet, and hence the area per surfactant to vary as an independent variable.

(c) The DOC-lamellar model, in the case where the spontaneous curvature radius is larger than typical size ξ of polar domains, reduces to the DOC-lamellar model. Peak positions D^* are simply calculated with a packing parameter of $p = 1.1$. The microstructure is then as a slightly asymmetric sponge microstructure, i.e., connected locally lamellar domains of oil and water. Close to 50% volume fraction, the microstructure corresponds to locally flat bilayers, which can be either considered as oil swollen normal bilayers or water swollen reverse bilayers. The DOC model prediction for the dilution plot, without any other constraint that constant curvature, is in very good agreement with experimental results.

(d) Assume finally that the two level cut model (Arleth et al., 2001), with flexible interface holds ($k_c = 2kT$). The predicted swelling plot is shifted towards higher ΣD^* value indicating that the stiffness of the locally lamellar microstructure corresponds to more than $2kT$. However, the microstructure should be close to the one shown in Fig. 8.

11. Conclusion

Microstructures as a dispersion of spheres can be shown without ambiguity. Other microstructures can be tested versus experimental results obtained by scattering using a dilution plot. For flexible interfaces, CRC models give a good approximation of the conformation of an infinitely folded film. If strong curvature constraints come into play, gradual transformation of a microstructure from droplets to locally connected cylinders and finally disordered lamella (sponge or vesicles) is the relevant microstructure.

Determination of microstructure of a microemulsion is useful in understanding other properties of microemulsions, such as electrical conductivity, macroscopic diffusion, or reactivity between species dissolved in microemulsions. An important case for application is to understand the large variation in solubilisation power of a given microemulsion for a slightly soluble compound. Relating solubilisation power to microstructure in the case of isolated droplets has been done by in the case of connected microstructures by Testard et al. (1997) and Testard and Zemb (1999, 2000).

References

Abillon, O., Binks, B.P., Langevin, D. and Meunier, J. (1988). *Prog. Colloid Polym. Sci.*, **76**, 84–89.

Andelman, D., Cates, M.E., Roux, D. and Safran, S.A. (1987). *J. Chem. Phys.*, **87** (12), 7229–7241.

Arleth, L., Marcelja, S. and Zemb, T. (2001). *J. Chem. Phys.*, **115** (8), 3923–3936.

Auvray, L. (1985). *J. Phys. Lett.*, **46** (4), 163–172.

Auvray, L., Cotton, J.P., Ober, R. and Taupin, C. (1984). *J. Phys. Chem.*, **88**, 4586–4598.

Auvray, L., Cotton, J.P., Ober, R. and Taupin, C. (1986). *Phys. B & C*, **136)(1–3**, 281–283.

Bagger-Jorgensen, H., Olsson, U. and Mortensen, K. (1997). *Langmuir*, **13** (6), 1413–1421.

Bardez, E., Vy, N.C. and Zemb, T. (1995). *Langmuir*, **11** (9), 3374–3381.

Barnes, I.S. and Zemb, T.N. (1988). *J. Appl. Crystallogr.*, **21** (4), 373–379.

Barnes, I.S., Hyde, S.T., Ninham, B.W., Derian, P.J., Drifford, M., Warr, G.G. and Zemb, T.N. (1988a). *Prog. Colloid Polym. Sci.*, **76**, 90–95.

Barnes, I.S., Hyde, S.T., Ninham, B.W., Derian, P.J., Drifford, M. and Zemb, T.N. (1988b). *J. Phys. Chem.*, **92** (8), 2286–2293.

Barnes, I.S., Derian, P.J., Hyde, S.T., Ninham, B.W. and Zemb, T.N. (1990). *J. Phys. (Paris)*, **51** (22), 2605–2628.

Barnes, I.S., Corti, M., Degiorgio, V. and Zemb, T. (1993). *Prog. Colloid Polym. Sci.*, **93**, 205–206.

Berk, N.F. (1987). *Phys. Rev. Lett.*, **58** (25), 2718–2721.

Bernheim-Groswasser, A., Tlusty, T., Safran, S.A. and Talmon, Y. (1999). *Langmuir*, **15** (17), 5448–5453.

Billman, J.F. and Kaler, E.W. (1990). *Langmuir*, **6** (3), 611–620.

Brumberger, H. and Goodisman, J. (1983). *J. Appl. Crystallogr.*, **16** (1), 83–88.

Brumberger, H., Delaglio, F., Goodisman, J. and Whitfield, M. (1986). *J. Appl. Crystallogr.*, **19** (5), 287–299.

Brunner-Popela, J., Mittelbach, R., Strey, R., Schubert, K.-V., Kaler, E.W. and Glatter, O. (1999). *J. Chem. Phys.*, **110** (21), 10623–10632.

Cates, M.E., Roux, D., Andelman, D., Milner, S.T. and Safran, S.A. (1988a). *Europhys. Lett.*, **5** (8), 733–739.

Cates, M.E., Andelman, D., Safran, S.A. and Roux, D. (1988b). *Langmuir*, **(4)**, 802–806.

Chandra, P. and Safran, S.A. (1991). *Langmuir*, **7** (9), 1849–1854.

Chandra, P. and Safran, S.A. (1992). *Europhys. Lett.*, **17** (8), 691–696.

Chen, S.H. and Choi, S.-M. (1997a). *Phys. A*, **236** (1–2), 38–51.

Chen, S.-H. and Choi, S.-M. (1997b). *J. Appl. Crystallogr.*, **30** (5, Pt. 2), 755–760.

Chen, S.H., Lee, D.D., Kimishima, K., Jinnai, H. and Hashimoto, T. (1996). *Phys. Rev. E*, **54** (6), 6525–6531.

Chen, S.H., Choi, S.M. and LoNostro, P. (1998). *Nuovo Cimento Soc. Ital. Fis., D*, **20D** (12bis), 1971–1988.

Chevalier, Y. and Zemb, T. (1990). *Rep. Prog. Phys.*, **53** (3), 279–371.

Chevalier, Y., Belloni, L., Hayter, J.B. and Zemb, T. (1985). *J. Phys.*, **46** (5), 749–759.

Choi, S.M. and Chen, S.H. (1997). *Prog. Colloid Polym. Sci.*, **106**, 14–23.

Choi, S.M., Chen, S.H., Sottmann, T. and Strey, R. (1998). *Phys. B*, **1997**, 241–243, 976–978.

Choi, S.-M., LoNostro, P. and Chen, S.-H. (1999). *Prog. Colloid Polym. Sci.*, **112**, 98–104.

Choi, S.M., Chen, S.H., Sottmann, T. and Strey, R. (2002). *Phys. A*, **304** (1–2), 85–92.

Choy, D. and Chen, S.-H. (2001). *Phys. Rev. E*, **63(2–1)**, 021401/1–021401/9.

Debye, P. and Bueche, A.M. (1949). *J. Appl. Phys.*, **20**, 518–525.
Debye, P., Anderson Jr., H.R. and Brumberger, H. (1957). *J. Appl. Phys.*, **28**, 679–683.
Delaglio, F., Goodisman, J. and Brumberger, H. (1986). *J. Catal.*, **99** (2), 383–390.
Eastoe, J., Steytier, D.C., David, C., Robinson, B.H., Heenan, R.K., North, A.N. and Dore, J.C. (1994). *J. Chem. Soc. Faraday Trans.*, **90** (17), 2497–24504.
Eastoe, J., Hetherington, K.J., Dalton, J.S., Sharpe, D., Lu, J.R. and Heenan, R.K. (1997). *J. Coll. Interface Sci.*, **190**, 449–455.
Ekwall, P. (1988). *Colloid Polym. Sci.*, **266** (3), 261–278.
Ekwall, P. (1989). *Colloid Polym. Sci.*, **267** (7), 607–621.
Ekwall, P. and Mandell, L. (1967). *Acta Chem. Scand.*, **21** (6), 1612–1618.
Ekwall, P. and Solyom, P. (1967). *Roy. Swed. Acad. Eng. Sci., Acta Chem. Scand.*, **21** (6), 1619–1629.
Ekwall, P. and Stenius, P. (1967). *Acta Chem. Scand.*, **21** (6), 1643–1646.
Ekwall, P., Eikrem, H. and Stenius, P. (1967a). *Roy. Swed. Acad. Eng. Sci., Acta Chem. Scand.*, **21** (6), 1639–1642.
Ekwall, P., Lemstroem, K.E., Eikrem, H. and Holmberg, P. (1967b). *Acta Chem. Scand.*, **21** (6), 1401–1407.
Ekwall, P., Mandell, L. and Fontell, K. (1986). *Colloid Polym. Sci.*, **264** (6), 542–551.
Fogden, A., Hyde, S.T. and Lundberg, G.J. (1991). *J. Chem. Soc. Faraday Trans.*, **87** (7), 949–955.
QQQ Fontell, K. and Mandell, L. (1989a). *Kolloid-Z. Z. Polym.*, **233** (1–2), 938–944.
Fontell, K. and Mandell, L. (1989b) *Kolloid-Z. Z. Polym.*, **233** (1–2), 945–968.
Fontell, K. and Mandell, L. (1993). *Colloid Polym. Sci.*, **271**, 974–991.
Förster, S., Berton, B., Hentze, H.-P., Kraemer, E., Antonietti, M. and Lindner, P. (2001). *Macromolecules*, **34** (13), 4610–4623.
Friberg, S. and Burasczenka, I. (1978). *Prog. Colloid Polym. Sci.*, **63**, 1–9.
Gazeau, D., Bellocq, A.M., Roux, D. and Zemb, T. (1989). *Europhys. Lett.*, **9**, 447–452.
Glatter, O., Strey, R., Schubert, K.-V. and Kaler, E.W. (1996). *Ber. Bunsen-Ges.*, **100** (3), 323–335.
Gompper, G., Endo, H., Mihailescu, M., Allgaier, J., Monkenbusch, M., Richter, D., Jakobs, B., Sottman, T. and Strey, R. (2001). *Europhys. Lett.*, **56** (5), 683–689.
Gradzielski, M., Langevin, D., Sottmann, T. and Strey, R. (1996). *J. Chem. Phys.*, **104** (10), 3782–3787.
Hansen, J.P. and Hayter, J.B. (1982). *Mol. Phys.*, **46** (3), 651–656.
Haviqsson, I., Soedermann, O. and Regev, O. (1994). *Langmuir*, **10**, 4731–4734.
Hayter, J.B. (1985). *Proc. Int. Sch. Phys. 'Enrico Fermi'* (Phys. Amphiphiles), **90**, 59–93.
Hayter, J.B. and Penfold, J. (1983). *Colloid Polym. Sci.*, **261** (12), 1022–1030.
Hayter, J.B., Hayoun, M. and Zemb, T. (1984). *Colloid Polym. Sci.*, **262** (10), 798–803.
Helsen, J.A. and Teixeira, J. (1986). *Colloid Polym. Sci.*, **264** (7), 619–622.
Huang, J.S., Safran, S.A., Kim, M.W., Grest, G.S., Kotlarchyk, M. and Quirke, N. (1984). *Phys. Rev. Lett.*, **53** (6), 592–595.
Hyde, S.T., Ninham, B.W. and Zemb, T. (1989). *J. Phys. Chem.*, **93** (4), 1464–1471.
Hyde, S.T., Barnes, I.S. and Ninham, B.W. (1990). *Langmuir*, **6** (6), 1055–1062.
Hyde, S., Andersson, S., Larsson, K., Blum, Z., Landh, T., Lidin, S. and Ninham, B.W. (1996). In: *The Language of Shape, The Role of Curvature on Condensed Matter: Physics, Chemistry and Biology*. Elsevier, Amsterdam.
Jinnai, H., Nishikawa, Y., Chen, S.H., Kiozumi, S. and Hashimoto, T. (2000). *Phys. Rev. E.*, **61** (6B), 6773–6780.
Jouffroy, J., Levinson, P. and De Gennes, P.G. (1982). *J. Phys.*, **43** (8), 1241–1248.
Kaler, E.W. and Prager, S. (1982). *J. Colloid Interface Sci.*, **86** (2), 359–369.
Komura, S., Deto, H., Takeda, T., Nagao, M., Ito, Y. and Imai, M. (1996). *J. Chem. Phys.*, **105** (8), 3264–3277.
Lee, D.D. and Chen, S.H. (1994a). *Nuovo Cimento Soc. Ital. Fis., D*, **16D** (9), 1357–1366.
Lee, D.D. and Chen, S.H. (1994b). *Phys. Rev. Lett.*, **73** (1), 106–109.
Marcelja, S. (1990). *J. Phys. Chem.*, **94** (18), 7259–7261.
Milner, S.T., Safran, S.A., Andelman, D., Cates, M.E. and Roux, D. (1988). *J. Phys.*, **49** (6), 1065–1076.
Nave, S., Eastoe, J. and Penfold, J. (2000a). *Langmuir*, **16** (23), 8733–8740.
Nave, S., Eastoe, J., Penfold, J. and Grillo, I. (2000b). *Langmuir*, **16** (23), 8733–8740.
Ninham, B.W., Barnes, I.S., Hyde, S.T., Derian, P.J. and Zemb, T.N. (1987). *Europhys. Lett.*, **4** (5), 561–568.
Olsson, U. and Schurtenberger, P. (1993). *Langmuir*, **9** (12), 3389–3394.
Olsson, U., Wuerz, U. and Strey, R. (1993). *J. Phys. Chem.*, **97** (17), 4535–4539.
Pieruschka, P. and Marcelja, S. (1992). *J. Phys. II*, **2** (2), 235–247.

Pieruschka, P. and Safran, S.A. (1993). *Europhys. Lett.*, **22** (8), 625–630.
Pileni, M.P., Zemb, T. and Petit, C. (1985). *Chem. Phys. Lett.*, **118 (4)**, 414–420.
Regev, O., Ezrahi, S., Aserin, A., Garti, N., Wachtel, E., Kaler, E.W., Khan, A. and Talmon, Y. (1996). *Langmuir*, **12** (3), 668–674.
Robertus, C., Joosten, J.G.H. and Levine, Y.K. (1988). *Prog. Colloid Polym. Sci.*, **77**, 115–119.
Rushforth, D.S., Sanchez-Rubio, M., Santos-Vidals, L.M., Wormuth, K.R., Kaler, E.W., Cuevas, R. and Puig, J.E. (1986). *J. Phys. Chem.*, **90** (25), 6668–6673.
Safran, S.A. (1991). *Langmuir*, **7** (9), 1864–1866.
Safran, S.A. (1992). *NATO ASI Ser., Ser. C*, **369**, 61–67.
Safran, S.A., Roux, D., Cates, M.E. and Andelman, D. (1986). *Phys. Rev. Lett.*, **57** (4), 491–494.
Smarsly, B., Antonietti, M. and Wolff, T. (2002). *J. Chem. Phys.*, **116** (6), 2618–2627.
Sottmann, T., Strey, R. and Chen, S.-H. (1997). *J. Chem. Phys.*, **106** (15), 6483–6491.
Strey, R. (1994). *Colloid Polym. Sci.*, **272** (8), 1005–1019.
Strey, R., Glatter, O., Schubert, K.-V. and Kaler, E.W. (1996). *J. Chem. Phys.*, **105** (3), 1175–1188.
Stuhrmann, H.B. and Miller, A. (1978). *J. Appl. Crystallogr.*, **11** (5), 325–345.
Talmon, Y. and Prager, S. (1977). *Nature* (London), **267** (5609), 333–335.
Teixeira, J. (1986). *NATO ASI Ser., Ser. E*, **100**, 145–162.
Teixeira, J. (1988). *J. Appl. Cryst.*, **21**, 781–793.
Teixeira, J. (1990). *New J. Chem.*, **14** (3), 217–219.
Testard, F. and Zemb, T. (1999). *J. Colloid Interface Sci.*, **219** (1), 11–19.
Testard, F. and Zemb, T. (2000). *Langmuir*, **16** (2), 332–339.
Testard, F., Strey, R. and Zemb, T. (1997). *Prog. Colloid Polym. Sci.*, **105**, 332–339.
Teubner, M. (1990). *J. Chem. Phys.*, **92** (7), 4501–4507.
Teubner, M. (1991a). *J. Chem. Phys.*, **95** (7), 5072–5081.
Teubner, M. (1991b). *Europhys. Lett.*, **14** (5), 403–408.
Teubner, M. and Strey, R. (1987). *J. Chem. Phys.*, **87** (5), 3195–3200.
Tlusty, T., Safran, S.A., Menes, R. and Strey, R. (1997). *Phys. Rev. Lett.*, **78** (13), 2616–2619.
Trevino, S.F., Joubran, R., Parris, N. and Berk, N.F. (1998). *J. Phys. Chem. B*, **102** (6), 953–960.
Vonk, C.G. (1978). *J. Appl. Crystallogr.*, **11** (5), 541–6.
Vonk, C.G., Billman, J.F. and Kaler, E.W. (1988). *J. Chem. Phys.*, **88** (6), 3970–3975.
Warr, G.G., Zemb, T. and Barnes, I.S. (1988). In: *Proc. 2nd World Surfactant Congr., Paris, May 1988*. Aspa, Paris, Vol. 1, pp. 368–376.
Warr, G.G., Zemb, T.N. and Drifford, M. (1990). *J. Phys. Chem.*, **94** (7), 3086–3092.
Welberry, T.R. and Zemb, T.N. (1988). *J. Colloid Interface Sci.*, **123** (2), 413–426.
Wuerz, U. (1988). *Prog. Colloid Polym. Sci.*, **76**, 153–158.
Zemb, T.N. (1997). *Colloids Surf., A*, **129–130**, 435–454.
Zemb, T.N. and Welberry, T.R. (1993). *Colloid Polym. Sci.*, **271** (2), 124–132.
Zemb, T.N., Hyde, S.T., Derian, P.J., Barnes, I.S. and Ninham, B.W. (1987). *J. Phys. Chem.*, **91** (14), 3814–3820.
Zemb, T.N., Barnes, I.S., Derian, P.J. and Ninham, B.W. (1990). *Prog. Colloid Polym. Sci.*, **81**, 20–29.

Neutrons, X-Rays and Light
P. Lindner and Th. Zemb (Editors)

Chapter 14

Interacting Colloidal Suspensions

Rudolf Klein

Fakultät für Physik, Universität Konstanz, Postfach 5560, D-78434 Konstanz, Germany

1. Introduction

The purpose of this contribution is to present an introduction to the statistical mechanical methods and tools to describe quantitatively scattering from particle suspensions, in particular when effects of the finite concentration of suspended particles give rise to interactions. It is assumed that the reader is familiar with some of the basic results of equilibrium statistical mechanics of classical systems (McQuarrie, 1976; Hansen and McDonald, 1986). A macroscopic or thermodynamic state, which is defined by a small set of independent thermodynamic variables, can be realized microscopically by a very large number of microstates. The set of systems of all possible microstates, which are compatible with the prescribed values of the few independent thermodynamic variables, is known as an ensemble. Among the most useful ensembles are the canonical and the grand canonical ensembles; in the first case, the independent variables are the system volume V, the number of particles N and the temperature T, which corresponds to a closed system in energy contact with a large bath. In the second case, particle exchange with the bath is allowed, so that the chemical potential μ together with V and T are the independent variables.

Thermodynamic properties of a system, such as the internal energy or the pressure, with prescribed values of either (N, V, T) or (μ, V, T) are obtained as suitable averages of these properties taken for all microstates. The weights, with which the different microstates enter the average, are different in each of the ensembles. They are known as phase space distribution functions. Starting from the distribution function for the canonical ensemble, it will be shown how the quantities most relevant to scattering experiments, such as the static structure factor $S(q)$ and the pair distribution function $g(r)$, are defined and how they are related to each other.

A second problem in relation to scattering experiments is to understand how $g(r)$ and $S(q)$ depend on the interactions between the suspended particles, their sizes and shapes and their concentration. This information can be obtained from the solution of integral equations; here, the Ornstein–Zernike equation will be introduced together with so-called closure relations. Typical results will be given for two extreme cases of particle interactions, the extremely short-ranged hard core or excluded volume interaction and the soft and long-ranged interaction between charge-stabilized particles.

Another aspect of scattering experiments from soft matter in general and colloidal particles in particular is that these systems are often unavoidably polydisperse. It will be shown how the scattering relations change when effects of polydispersity are properly taken into account (Klein and D'Aguanno, 1996).

The case of binary mixtures will be considered to give an introduction to depletion interactions, an area of active research in recent years. Depletion effects provide an additional mechanism to influence effective interactions by varying the concentrations and the diameters of the two species properly, resulting in an interesting and rich phase behavior.

2. Colloidal particles as superatoms

The complete statistical mechanical description of a colloidal suspension on a microscopic level is complicated since one has to deal with the big colloidal particles and the small molecules of the solvent. When charge-stabilized suspensions are considered there are in addition the counterions and possibly the ions of added salt. Since one is usually only interested in the properties of the colloids it is sufficient to know their distribution function. The transition from the microscopic level to this mesoscopic level is an example of a coarse-graining process. Because of the very large size ratio of colloids to solvent molecules the latter are fast degrees of freedom which may be integrated out. The result of this procedure is that the potentials of interaction between the colloids depend not only on properties of the colloidal particles (such as their excluded volumes and the number of charges carried by each of them), but also on solvent properties: The colloid–colloid interaction is in general state-dependent. Whereas the microscopic total potential energy $U(\boldsymbol{r}^N, \boldsymbol{R}^{N_0})$ of the suspension depends on the positions of the centers of N colloidal particles (which are assumed to be spherical), denoted collectively by $\boldsymbol{r}^N = (\boldsymbol{r}_1, \boldsymbol{r}_2, \ldots, \boldsymbol{r}_N)$, and on the positions of N_0 solvent molecules $\boldsymbol{R}^{N_0} = (\boldsymbol{R}_1, \boldsymbol{R}_2, \ldots, \boldsymbol{R}_{N_0})$, the coarse-graining procedure results in a potential energy $V(\boldsymbol{r}^N)$ which depends only on the positions $\boldsymbol{r}^N$ of the colloidal particles. It is in general not strictly pairwise additive. Approximating it as a sum of pair-potentials, one obtains

$$V(\boldsymbol{r}^N) = \frac{1}{2} \sum_{i \neq j} v(r_{ij}) \tag{1}$$

where $r_{ij} = |\boldsymbol{r}_i - \boldsymbol{r}_j|$. These (state-dependent) potentials are then equivalent to the interaction potentials between atoms (or small molecules) in liquids or solids. It is in this sense that a colloidal suspension may be considered as a collection of super-atoms. Methods developed for the statistical mechanical description of liquids and solids can then be applied to derive macroscopic properties of colloidal suspensions starting from the interaction potentials on the mesoscopic scale.

What are these potentials? What is the equivalent of, for instance, the Lennard-Jones potential between neutral atoms? Neutral atoms are known to interact by van der Waals potentials which vary at large distances proportional to $-r^{-6}$. When this interaction is summed over the atoms of two extended spheres of diameter σ the result is (Hamaker, 1937)

$$v_{\mathrm{vdW}}(r) = -\frac{A}{12}\left[\frac{\sigma^2}{r^2-\sigma^2} + \left(\frac{\sigma}{r}\right)^2 + 2\ln\left(1-\frac{\sigma^2}{r^2}\right)\right]; \qquad r > \sigma \tag{2}$$

where r is the distance between the centers of the spheres. The strength of this attractive van der Waals potential is measured by the Hamaker constant A, which depends on the dielectric functions of the spheres and the solvent. For typical cases A is of the order of the thermal energy, k_BT. In simple models the Hamaker constant can be expressed in terms of the difference $n_c - n_s$ of the refractive indices of colloids and solvent, so that $A \to 0$ for $n_c \to n_s$. Therefore, a change to a solvent with a different n_s will change the strength of the van der Waals attraction.

The potential in Eq. (2) varies at large distances as $-r^{-6}$ and at short distances as $-(r-\sigma)^{-1}$ where it has a deep minimum. Its effective range extends typically to a few nanometers from the surface of the particles. If this were the only interaction between colloidal particles, the suspension would be unstable against aggregation. There are essentially two ways to stabilize the suspension. One method is charge stabilization. The particles contain ionizable groups which dissociate in a polar solvent (like water). The result is a diffuse layer of small counterions (in the simplest case protons) surrounding the oppositely charged colloidal particle. The interaction between two such objects has the form of a screened Coulomb (or Yukawa) potential [1]

$$v_c(r) = \frac{Q^2}{\varepsilon\left(1+\frac{\kappa\sigma}{2}\right)^2}\frac{e^{-\kappa(r-\sigma)}}{r}; \qquad r > \sigma. \tag{3}$$

Here Q is the charge on the particle (which can vary from 20 to 50 elementary charges on small ionic micelles to several hundreds or even more on latex particles) and ε is the dielectric constant of the solvent. The screening parameter κ is given by

$$\kappa^2 = \frac{4\pi}{\epsilon k_B T}\sum_i \rho_i q_i^2 \tag{4}$$

where the sum runs over all species of small ions in the system; ρ_i being the number density and q_i the charge of the ions of species i. When there are only the counterions, dissociated from the colloids, electroneutrality gives $\kappa^2_{\text{counter}} = (4\pi e^2/\varepsilon k_B T)\rho|Z|$, where ρ is the number density of colloids and $Z = Q/e$ their valency. For the case of an added $1-1$-electrolyte of number density ρ_I

$$\kappa^2 = \frac{4\pi e^2}{\varepsilon k_B T}(\rho|Z| + 2\rho_I). \tag{5}$$

The range of the repulsion due to the screened Coulomb potential is given by the Debye screening length $\lambda = \kappa^{-1}$. It changes strongly with the salt concentration; in aqueous solution and for salt concentrations of 1 M, 1 mM and 1 μM the Debye lengths are roughly 0.3, 10 and 300 nm, respectively. Therefore, the competition between the repulsive and attractive interactions depends strongly on the amount of added salt. At high salt concentrations the Debye length is smaller than the range of the attractions, and as a result the suspension is unstable and shows irreversible flocculation. A suspension is therefore only stable, when λ is sufficiently large, meaning that the added salt content has to be sufficiently low. Then the range of the repulsive interactions can easily become larger than the separation between neighboring colloidal particles so that the system becomes highly correlated, although the actual volume

[1] This form of the interaction has been obtained by rather different methods and under various assumptions. For a selection of the extensive literature on this subject see Schmitz (1993), Löwen et al. (1993), Belloni (1998), van Roij et al. (2000) and von Grünberg et al. (2000).

occupied by the particles is rather low. This competition between double-layer repulsions and dispersive attractions is the content of the theory developed by Derjaguin, Landau, Verwey and Overbeek (DLVO; Derjaguin and Landau, 1941; Verwey and Overbeek, 1948). Direct measurements (Vondermassen et al., 1994; Crocker and Grier, 1994) and computer simulations (Löwen, et al., 1993) have shown good agreement under salt-free conditions where the repulsive screened Coulomb part of the DLVO effective interaction dominates.

Also the potential $v_c(r)$ of Eq. (3) is a case of a state-dependent potential mentioned before. It depends on the dielectric constant of the solvent and through the screening constant on temperature and number density of colloids. Moreover, $v_c(r)$ depends on the small ions only through κ; the small ions are not treated on the same level as the colloids. This model is known as the one-component macrofluid model which can be derived from the more basic primitive model under certain restrictive assumptions. The latter treats the colloids and the small ions as separate species on the same level, but considers the solvent as a structureless continuous medium of dielectric constant ε.

There has been a discussion in recent years about the general validity of the DLVO theory. Several experimental observations have been reported on deionized suspensions (Ise et al., 1983; Ito et al., 1994; Tata et al., 1997) which seem to indicate that a dense crystalline phase or a homogeneous liquid phase can coexist with a dilute (gas-like) phase. This would require the existence of long-ranged attractions between the colloidal particles. The van der Waals interaction can not provide this attraction, because the Hamaker constants that would qualitatively explain the observations are at least one order of magnitude too large. But it has been shown that the effective Hamiltonian for the colloidal particles obtained by integrating out the counterions does not only provide the DLVO expression for the effective pair potential but in addition a term which is independent of the particle positions but dependent on the macroion density (van Roij et al., 1999). When this term is used to deduce the phase diagram it is possible to have a spinodal instability and hence coexistence between dilute and dense phases without the presence of effective attractions between the colloidal particles.

An alternative method of stabilization is to adsorb short polymers to the surface of the colloidal particles. When the solvent is a good solvent for the polymers they form brushes and these brushes repel each other. In many cases, the thickness of the polymer layer can be made small compared to the diameter of the colloids so that the resulting interaction potential between two colloids is very short-ranged and strongly repulsive. This potential is then well approximated by that between hard spheres

$$v_{\mathrm{HS}}(r) = \begin{cases} \infty; & r < \sigma \\ 0; & r > \sigma. \end{cases} \tag{6}$$

Systems of hard spheres are the best understood systems of statistical mechanics.

This survey of the most common pair potentials shows the wide variability of interactions in colloidal systems, ranging from practically hard-sphere behavior to the long-ranged soft potentials in deionized charge-stabilized suspensions. Moreover, the potentials can easily be modified by changing certain system parameters so that the strength and the range of the interactions can be varied. This is in general not possible for conventional liquids and solids. Therefore, colloidal suspensions provide very valuable model systems for condensed matter in general. One can, for instance, study the phase behavior of a system as a function of the parameters, on which the interactions depend.

3. Some basic results from statistical mechanics

The aim of statistical mechanics is to calculate macroscopic quantities and structural properties of a system starting from the microscopic characteristics of the constituents of the system. Microscopically the system is fully characterized by its Hamiltonian, consisting of the sum of kinetic and potential energies,

$$H(\boldsymbol{p}^N,\boldsymbol{r}^N) = T(\boldsymbol{p}^N) + V(\boldsymbol{r}^N), \tag{7}$$

where $\boldsymbol{p}^N = (\boldsymbol{p}_1,\ldots,\boldsymbol{p}_N)$ and $\boldsymbol{r}^N = (\boldsymbol{r}_1,\ldots,\boldsymbol{r}_N)$ abbreviate the set of momenta $\boldsymbol{p}_i$ and positions $\boldsymbol{r}_i, i = 1,\ldots,N$, of N particles, respectively. A system of N particles in a volume V at temperature T can microscopically be represented in many ways, each one of them being a possible microstate; there are many different sets $\boldsymbol{p}^N$ and $\boldsymbol{r}^N$ compatible with identical macroscopic properties. A microstate is represented as a point in $6N$-dimensional phase space, a space spanned by the Cartesian components of the positions $\boldsymbol{r}^N$ and momenta $\boldsymbol{p}^N$. The set of all microstates of the system having the prescribed values of N, V and T forms the canonical ensemble, and the distribution of the microstates in phase space is given by

$$f(\boldsymbol{p}^N,\boldsymbol{r}^N) = \frac{1}{N!h^{3N}Q_N(V,T)}\exp\bigl[-\beta H(\boldsymbol{p}^N;\boldsymbol{r}^N)\bigr]. \tag{8}$$

The factor $N!$ takes care of the indistinguishability of the particles, $h = 2\pi\hbar$ is Planck's constant, $\beta = (k_B T)^{-1}$ where k_B is Boltzmann's constant, and the canonical partition function $Q_N(V,T)$ normalizes $f(\boldsymbol{p}^N,\boldsymbol{r}^N)$ to one,

$$\int \mathrm{d}\boldsymbol{p}^N \int \mathrm{d}\boldsymbol{r}^N f(\boldsymbol{p}^N,\boldsymbol{r}^N) = 1, \tag{9}$$

so that

$$Q_N(V,T) = \frac{1}{N!h^{3N}}\int \mathrm{d}\boldsymbol{p}^N \int \mathrm{d}\boldsymbol{r}^N \exp\bigl[-\beta H(\boldsymbol{p}^N;\boldsymbol{r}^N)\bigr]. \tag{10}$$

Therefore, $f\,\mathrm{d}\boldsymbol{p}^N\,\mathrm{d}\boldsymbol{r}^N$ is the probability that the system is in a microstate with values of $\boldsymbol{p}^N$ and $\boldsymbol{r}^N$ in the $6N$-dimensional volume element $\mathrm{d}\boldsymbol{p}^N\,\mathrm{d}\boldsymbol{r}^N = \mathrm{d}\boldsymbol{p}_1\ldots\mathrm{d}\boldsymbol{p}_N\,\mathrm{d}\boldsymbol{r}_1\ldots\mathrm{d}\boldsymbol{r}_N$.

Macroscopic properties of the system are obtained as averages of the corresponding microscopic quantities, performed with the distribution function $f(\boldsymbol{p}^N,\boldsymbol{r}^N)$. Taking the number density as an example, the corresponding microscopic density is

$$\hat{\rho}(\boldsymbol{r}) \equiv \hat{\rho}(\boldsymbol{r};\boldsymbol{r}^N) = \sum_{i=1}^{N}\delta(\boldsymbol{r}-\boldsymbol{r}_i); \tag{11}$$

it is a function of the phase space variables $\boldsymbol{r}^N$, which depends parametrically on the position $\boldsymbol{r}$ in physical space. The macroscopic density is then given by

$$\rho(\boldsymbol{r}) = \langle\hat{\rho}(\boldsymbol{r})\rangle \equiv \int \mathrm{d}\boldsymbol{p}^N \int \mathrm{d}\boldsymbol{r}^N \hat{\rho}(\boldsymbol{r};\boldsymbol{r}^N) f(\boldsymbol{p}^N,\boldsymbol{r}^N), \tag{12}$$

which defines the canonical average, denoted by $\langle\ldots\rangle$.

The link of the phase space description to thermodynamics is provided (in the present case of the canonical ensemble) by the expression for the Helmholtz free energy

$$A(N,V,T) = -k_B T \ln Q_N(V,T), \tag{13}$$

from which other thermodynamic properties can be obtained. Using, for example, the thermodynamic relation for the internal energy, $U = (\partial(\beta A)/\partial\beta)_V$, one obtains

$$U = -\frac{\partial}{\partial\beta}\ln Q_N(V,T) \tag{14}$$

or with (10) for Q_N

$$U(N,V,T) = \int \mathrm{d}\boldsymbol{p}^N \int \mathrm{d}\boldsymbol{r}^N H(\boldsymbol{p}^N;\boldsymbol{r}^N) f(\boldsymbol{p}^N,\boldsymbol{r}^N) = \langle H \rangle; \tag{15}$$

it is the canonical average of the Hamiltonian. Similarly, the pressure is obtained from $p = -(\partial A/\partial V)_T$.

For the Hamiltonian in the form of Eq. (7) the integration over the momenta in expression (10) for Q_N can readily be carried out, giving a factor $(2\pi m k_B T)^{1/2}$ for each degree of freedom with m denoting the mass of a particle. The partition function is then represented by

$$Q_N(V,T) = \frac{Z_N(V,T)}{N!\Lambda^{3N}}, \tag{16}$$

where

$$Z_N(V,T) = \int \mathrm{d}\boldsymbol{r}^N \exp[-\beta V(\boldsymbol{r}^N)] \tag{17}$$

is the configurational integral and $\Lambda = h/(2\pi m k_B T)^{1/2}$ denotes the thermal de Broglie wavelength.

For an ideal system, for which $V(\boldsymbol{r}^N) \equiv 0$, Eq. (17) gives $Z_N^{\mathrm{id}}(V,T) = V^N$, so that the partition function of ideal systems becomes $Q_N^{\mathrm{id}} = V^N/(N!\Lambda^{3N})$. From it the results

$$A^{\mathrm{id}} = k_B T N\left[\ln(\rho\Lambda^3) - 1\right] \tag{18}$$

$$U^{\mathrm{id}} = \tfrac{3}{2} N k_B T \tag{19}$$

$$p^{\mathrm{id}} = \rho k_B T \tag{20}$$

follow with $\rho = N/V$ as the equilibrium number density. It is convenient to rewrite (16) as

$$Q_N(V,T) = Q_N^{\mathrm{id}} Z_N(V,T)/V^N. \tag{21}$$

Then the excess contributions to the thermodynamic quantities, which arise from the interactions, are given in terms of the configurational integral. As an example, the internal energy (14) follows from (21) and (17) as

$$U = U^{\mathrm{id}} + U^{\mathrm{ex}} \tag{22}$$

where

$$U^{\mathrm{ex}} = -\frac{\partial}{\partial\beta}\ln Z_N(V,T) = \frac{1}{Z_N(V,T)}\int \mathrm{d}\boldsymbol{r}^N V(\boldsymbol{r}^N)\exp[-\beta V(\boldsymbol{r}^N)] \tag{23}$$

is the average of the potential energy taken over configuration space, which is the $3N$-dimensional space spanned by the Cartesian components of $\boldsymbol{r}^N = (\boldsymbol{r}_1,\ldots,\boldsymbol{r}_N)$. The distribution function in configuration space appearing in (23) is obtained from the full phase space distribution function $f(\boldsymbol{p}^N,\boldsymbol{r}^N)$ in Eq. (8) by integrating over the momenta and using (16) for Q_N

$$f(\boldsymbol{r}^N) = \int \mathrm{d}\boldsymbol{p}^N f(\boldsymbol{p}^N,\boldsymbol{r}^N) = \frac{1}{Z_N(V,T)} \exp[-\beta V(\boldsymbol{r}^N)]. \tag{24}$$

$f(\boldsymbol{r}^N)\mathrm{d}\boldsymbol{r}^N$ is probability of finding the system in a microstate with particle positions in the configuration space volume element $d\boldsymbol{r}^N$, irrespective of the momenta of the particles.

Even the function $f(\boldsymbol{r}^N)$ contains unnecessary detailed information about the system. It is for many purposes sufficient to know the distribution of sets of only a small number n of particles in the presence of the remaining $N-n$. Therefore, one introduces reduced distribution functions

$$\rho^{(n)}(\boldsymbol{r}^n) \equiv \rho^{(n)}(\boldsymbol{r}_1,\ldots,\boldsymbol{r}_n) = \frac{N!}{(N-n)!} \int \mathrm{d}\boldsymbol{r}^{N-n} f(\boldsymbol{r}^N), \tag{25}$$

where $\mathrm{d}\boldsymbol{r}^{N-n} = \mathrm{d}\boldsymbol{r}_{n+1}\ldots\mathrm{d}\boldsymbol{r}_N$. The integral in (25) is the probability density of finding any subset of n particles in the reduced configuration space volume element $\mathrm{d}\boldsymbol{r}^n = \mathrm{d}\boldsymbol{r}_1\ldots\mathrm{d}\boldsymbol{r}_n$, and the prefactor in (25) is equal to the number of ways of choosing a set of n particles from a total of N. The most important reduced distribution functions are $\rho^{(1)}(\boldsymbol{r}_1)$ and the two-particle or pair density $\rho^{(2)}(\boldsymbol{r}_1,\boldsymbol{r}_2)$. From (24) and (25) the latter is

$$\rho^{(2)}(\boldsymbol{r}_1,\boldsymbol{r}_2) = \frac{N(N-1)}{Z_N(V,T)} \int \mathrm{d}\boldsymbol{r}^{N-2} \exp[-\beta V(\boldsymbol{r}^N)]. \tag{26}$$

The normalizations of $\rho^{(1)}$ and $\rho^{(2)}$ follow from (9), (24) and (25) as

$$\int \mathrm{d}\boldsymbol{r}_1\, \rho^{(1)}(\boldsymbol{r}_1) = N \tag{27}$$

$$\int \mathrm{d}\boldsymbol{r}_1 \int \mathrm{d}\boldsymbol{r}_2\, \rho^{(2)}(\boldsymbol{r}_1,\boldsymbol{r}_2) = N(N-1). \tag{28}$$

For uncorrelated particles the distributions are completely random so that $\rho^{(2)}(\boldsymbol{r}_1,\boldsymbol{r}_2) = \rho^{(1)}(\boldsymbol{r}_1)\rho^{(1)}(\boldsymbol{r}_2)$ for this case. This suggests to introduce the function

$$g(\boldsymbol{r}_1,\boldsymbol{r}_2) = \frac{\rho^{(2)}(\boldsymbol{r}_1,\boldsymbol{r}_2)}{\rho^{(1)}(\boldsymbol{r}_1)\rho^{(1)}(\boldsymbol{r}_2)} \tag{29}$$

as a measure of correlations; this function is the pair distribution function. These distributions simplify under most circumstances of practical importance. The one-particle density $\rho^{(1)}(\boldsymbol{r}_1)$ cannot depend on $\boldsymbol{r}_1$ for homogeneous systems because of translational invariance; it has to be a constant, which can be taken out from under the integral in (27), giving $\rho^{(1)}(\boldsymbol{r}) = N/V = \rho$. For the same reason of translational invariance $\rho^{(2)}(\boldsymbol{r}_1,\boldsymbol{r}_2)$ and $g(\boldsymbol{r}_1,\boldsymbol{r}_2)$ can only depend on $\boldsymbol{r}_1 - \boldsymbol{r}_2 = \boldsymbol{r}_{12}$. When the system is also isotropic, $\rho^{(2)}$ and g are only functions of $r_{12} = |\boldsymbol{r}_{12}|$. Using these simplifications in (26) and (29), the statistical mechanical expression for the pair or radial distribution function of homogeneous and isotropic systems is

$$g(r_{12}) = \frac{V^2(1-\frac{1}{N})}{Z_N(V,T)} \int \mathrm{d}\boldsymbol{r}^{N-2} \exp[-\beta V(\boldsymbol{r}^N)]. \tag{30}$$

It is a measure of the correlations between two particles separated by r_{12} in the presence of the remaining $N-2$ particles. For ideal systems ($V(\boldsymbol{r}^N) \equiv 0, Z_N = V^N$), $g(r_{12}) = 1-(1/N) \approx 1$

for all r, so that deviations of $g(r_{12})$ for a non-ideal system from this value is a signature of correlations.

The interpretation of $g(r)$ as the most important quantity describing the microstructure of a correlated system, follows from the normalization condition (28). Using (29) it becomes, for a homogeneous and isotropic system,

$$\int \mathrm{d}\boldsymbol{r}_1 \int \mathrm{d}\boldsymbol{r}_2\, \rho^2 g(|\boldsymbol{r}_1 - \boldsymbol{r}_2|) = N(N-1). \tag{31}$$

Changing integration variables from $\boldsymbol{r}_1$ and $\boldsymbol{r}_2$ to $\boldsymbol{r}_1$ and $\boldsymbol{r}_1 - \boldsymbol{r}_2 = \boldsymbol{r}$ gives

$$\rho \int \mathrm{d}\boldsymbol{r}\; g(r) = \rho \int_0^\infty 4\pi r^2 \mathrm{d}r\; g(r) \approx N. \tag{32}$$

Therefore, the number of particles in a spherical shell of radius r and thickness $\mathrm{d}r$ around a particle at $r = 0$ is given by $4\pi\rho r^2 g(r)\mathrm{d}r$.

Besides describing the microstructure the pair distribution function is also very useful for the calculation of thermodynamic properties. Assuming the pairwise additivity of the total potential energy as in Eq. (1), the internal energy is given by

$$U = \frac{3}{2} N k_B T + 2\pi\rho N \int_0^\infty \mathrm{d}r\; r^2 v(r) g(r) \tag{33}$$

and the pressure by

$$p = \rho k_B T - \frac{2\pi}{3}\rho^2 \int_0^\infty \mathrm{d}r\; r^3 \frac{\mathrm{d}v}{\mathrm{d}r} g(r). \tag{34}$$

The latter result is known as the virial equation of state. These expressions look formally much simpler than (22) and (23) for the internal energy; the difficulties are, however, shifted to the determination of $g(r)$.

Starting from the grand canonical ensemble one can express the isothermal compressibility, defined by $\chi_T = \rho^{-1}(\partial\rho/\partial p)_T$ in terms of $g(r)$ as

$$\rho k_B T \chi_T = 1 + \rho \int \mathrm{d}\boldsymbol{r}\; (g(r) - 1). \tag{35}$$

The pair distributions can also be used to express the correlations of the local densities at two different points $\boldsymbol{r}$ and $\boldsymbol{r}'$. With (11) for the microscopic density one considers the density correlation function

$$\langle \hat{\rho}(\boldsymbol{r})\hat{\rho}(\boldsymbol{r}')\rangle = \Big\langle \sum_{i,j} \delta(\boldsymbol{r} - \boldsymbol{r}_i)\delta(\boldsymbol{r}' - \boldsymbol{r}_j) \Big\rangle. \tag{36}$$

The diagonal part ($i = j$) is $\delta(\boldsymbol{r} - \boldsymbol{r}')\sum_i \langle \delta(\boldsymbol{r} - \boldsymbol{r}_i)\rangle = \delta(\boldsymbol{r} - \boldsymbol{r}')\rho(\boldsymbol{r})$, where (12) has been used. Since the N particles of the system are statistically equivalent the macroscopic density $\rho(\boldsymbol{r})$ can also be written as

$$\rho(\boldsymbol{r}) = N \int \mathrm{d}\boldsymbol{r}^N\; \delta(\boldsymbol{r} - \boldsymbol{r}_1) = \rho^{(1)}(\boldsymbol{r}), \tag{37}$$

where the last step follows from (25) for $n = 1$. The non-diagonal contribution of (36) is, again using the statistical equivalence of the particles,

$$\begin{aligned}\left\langle \sum_{i\neq j} \delta(\boldsymbol{r}-\boldsymbol{r}_i)\delta(\boldsymbol{r}'-\boldsymbol{r}_j)\right\rangle &= N(N-1)\langle \delta(\boldsymbol{r}-\boldsymbol{r}_1)\delta(\boldsymbol{r}'-\boldsymbol{r}_2)\rangle \\ &= \frac{N(N-1)}{Z_N(V,T)}\int \mathrm{d}\boldsymbol{r}^N\, \delta(\boldsymbol{r}-\boldsymbol{r}_1)\delta(\boldsymbol{r}'-\boldsymbol{r}_2)\exp[-\beta V(\boldsymbol{r}^N)] \\ &= \rho^{(2)}(\boldsymbol{r},\boldsymbol{r}'). \end{aligned} \tag{38}$$

The second step uses the definition of the canonical average and the last step follows from (26). Therefore

$$\langle \hat{\rho}(\boldsymbol{r})\hat{\rho}(\boldsymbol{r}')\rangle = \rho^{(2)}(\boldsymbol{r},\boldsymbol{r}') + \delta(\boldsymbol{r}-\boldsymbol{r}')\rho^{(1)}(\boldsymbol{r}). \tag{39}$$

Correlations of density fluctuations $\delta\hat{\rho}(\boldsymbol{r}) = \hat{\rho}(\boldsymbol{r}) - \langle \hat{\rho}(\boldsymbol{r})\rangle$ are then described by

$$\begin{aligned} S_{\rho\rho}(\boldsymbol{r},\boldsymbol{r}') &\equiv \langle \delta\hat{\rho}(\boldsymbol{r})\delta\hat{\rho}(\boldsymbol{r}')\rangle \\ &= \rho^{(2)}(\boldsymbol{r},\boldsymbol{r}') + \rho^{(1)}(\boldsymbol{r})\delta(\boldsymbol{r}-\boldsymbol{r}') - \rho^{(1)}(\boldsymbol{r})\rho^{(1)}(\boldsymbol{r}'). \end{aligned} \tag{40}$$

For scattering experiments it is essential to consider the Fourier components of density fluctuations

$$\begin{aligned} \delta\hat{\rho}(\boldsymbol{q}) &= \int \mathrm{d}\boldsymbol{r}\; e^{i\boldsymbol{q}\cdot\boldsymbol{r}}\delta\hat{\rho}(\boldsymbol{r}) \\ &= \sum_i \mathrm{e}^{i\boldsymbol{q}\cdot\boldsymbol{r}_i} - \langle \hat{\rho}(\boldsymbol{q})\rangle. \end{aligned} \tag{41}$$

The static structure factor is defined as

$$\begin{aligned} S(q) &= \frac{1}{N}\langle |\delta\hat{\rho}(\boldsymbol{q})|^2\rangle \\ &= \frac{1}{N}\int \mathrm{d}\boldsymbol{r}\; \mathrm{e}^{i\boldsymbol{q}\cdot\boldsymbol{r}}\int \mathrm{d}\boldsymbol{r}'\; \mathrm{e}^{-i\boldsymbol{q}\cdot\boldsymbol{r}'} S_{\rho\rho}(\boldsymbol{r},\boldsymbol{r}') \\ &= \frac{1}{N}\int d\boldsymbol{r}\rho^{(1)}(\boldsymbol{r}) + \frac{1}{N}\int \mathrm{d}\boldsymbol{r}\int \mathrm{d}\boldsymbol{r}'\; \mathrm{e}^{i\boldsymbol{q}\cdot(\boldsymbol{r}-\boldsymbol{r}')}[\rho^{(2)}(\boldsymbol{r},\boldsymbol{r}') - \rho^{(1)}(\boldsymbol{r})\rho^{(1)}(\boldsymbol{r}')]. \end{aligned} \tag{42}$$

For a homogeneous and isotropic system this expression reduces to

$$S(q) = 1 + \rho\int \mathrm{d}\boldsymbol{r}\; \mathrm{e}^{i\boldsymbol{q}\cdot\boldsymbol{r}}(g(r) - 1), \tag{43}$$

which shows that the static structure factor is related to the Fourier transform of the total correlation function $h(r) \equiv g(r) - 1$. An alternative expression for $S(q)$ in terms of the positions of particles is obtained by using the second line of Eq. (41),

$$S(q) = \frac{1}{N}\left\langle \sum_{i,j} \mathrm{e}^{i\boldsymbol{q}\cdot(\boldsymbol{r}_i-\boldsymbol{r}_j)}\right\rangle - \frac{1}{N}\langle \hat{\rho}(\boldsymbol{q})\rangle^2. \tag{44}$$

The first term will be shown to be proportional to the intensity of scattered radiation. For homogeneous systems $\langle \hat{\rho}(\boldsymbol{q})\rangle = (2\pi)^3\rho\delta(\boldsymbol{q})$ so that $S(q)$ is proportional to the scattered radiation for $\boldsymbol{q} \neq 0$.

Comparing the expression (35) for the isothermal compressibility with Eq. (43) shows that

$$S(0) = \rho k_B T \chi_T. \tag{45}$$

4. Calculation of the microstructure

It has been shown in the previous section that the pair correlation function $g(r)$ describes the microstructure and thermodynamics of the system. To calculate $g(r)$ from a given pair potential one uses the Ornstein–Zernike (OZ) equation. It is an integral equation for the total correlation function $h(r_{12}) = g(r_{12}) - 1$,

$$h(r_{12}) = c(r_{12}) + \rho \int \mathrm{d}\boldsymbol{r}_3\, c(r_{13}) h(r_{23}). \tag{46}$$

Here, $c(r)$ is the direct correlation function. The basic idea behind the OZ equation is seen when (46) is iterated and written as an infinite geometric series. Then the total correlation between particles at $\boldsymbol{r}_1$ and $\boldsymbol{r}_2$ consists of the direct correlation between them, the direct correlation of the particle at $\boldsymbol{r}_1$ with an intermediate particle at $\boldsymbol{r}_3$ and its correlation with the particle at $\boldsymbol{r}_2$, etc. As it stands, Eq. (46) is not very useful, since $c(r)$ is unknown. Furthermore, Eq. (46) does not contain the specific pair potential $v(r)$ of the system under investigation. It is therefore necessary to have a further relation between $h(r)$, $c(r)$ and $v(r)$. This further relation is known as closure relation and it introduces approximations to the calculation of the correlation functions.

Closure relations can be obtained from systematic density expansions (Hansen and McDonald, 1986) or from functional derivative arguments [see appendix A in Klein and D'Aguanno (1996)]. To get an idea of this procedure one considers (46) in the limit of vanishing density, $\rho \to 0$. Then, the integral term is absent and $g(r) = \exp[-\beta v(r)]$, which is exact in this limit, so that (46) reduces to

$$c(r) = \mathrm{e}^{-\beta v(r)} - 1, \qquad (\rho \to 0) \tag{47}$$

or, if $\beta v(r) \ll 1$, to

$$c(r) = -\beta v(r). \qquad (\rho \to 0) \tag{48}$$

This condition is fulfilled asymptotically for large separations.

The asymptotic result (48) has been used as starting point of an approximation to solve the OZ equation, known as mean-spherical approximation (MSA). The direct correlation function is assumed to be given by (48) for all $r > \sigma$ and for finite ρ. Since $g(r) = 0$ for $r < \sigma$, which is exact for hard particles, the OZ equation is used to calculate $c(r)$ for $r < \sigma$ and $g(r)$ for $r > \sigma$.

When $c(r)$ is known it can be used to calculate the static structure factor. Taking the Fourier transform of the OZ equation, one obtains

$$\tilde{h}(q) = \frac{\tilde{c}(q)}{1 - \rho \tilde{c}(q)}. \tag{49}$$

When this is used in (43), $S(q)$ is given by

$$S(q) = \frac{1}{1 - \rho \tilde{c}(q)}. \tag{50}$$

The tilde indicates the Fourier transforms of $h(r)$ and $c(r)$. With (45) the compressibility is given by

$$(\rho k_B T \chi_T)^{-1} = 1 - \rho\tilde{c}(0) = 1 - 4\pi\rho \int_0^\infty \mathrm{d}r r^2 c(r). \tag{51}$$

There are further, improved, closure relations (Hansen and McDonald, 1986). The hypernetted chain (HNC) closure is

$$g(r) = \exp[-\beta v(r) + h(r) - c(r)] \tag{52}$$

or

$$c(r) = -\beta v(r) + h(r) - \ln(h(r) + 1). \tag{53}$$

Since $h(r) \to 0$ for $r \to \infty$, the HNC gives $c(r) = -\beta v(r)$, like in (48), but now for arbitrary ρ. The OZ equation (46) together with (52) or (53) can only be solved numerically. It is found by comparison with Monte Carlo simulations that the HNC scheme works especially well for long-ranged potentials.

The Percus–Yevick (PY) closure can formally be obtained from an expansion of the HNC closure (52),

$$g(r) = \mathrm{e}^{-\beta v(r)}(1 + h(r) - c(r)). \tag{54}$$

Solutions of the OZ equation together with (54) are known for several forms of the pair potential $v(r)$. The PY scheme works well for short-ranged potentials. For hard spheres, Eq. (6), the PY solution is[2]

$$c(r) = -\lambda_1(\phi) - 6\phi\lambda_2(\phi)\frac{r}{\sigma} - \frac{1}{2}\phi\lambda_1(\phi)\left(\frac{r}{\sigma}\right)^3; \qquad r < \sigma \tag{55}$$

and $c(r) = 0$ for $r > \sigma$. Here,

$$\lambda_1(\phi) = \frac{(1+2\phi)^2}{(1-\phi)^4}; \qquad \lambda_2(\phi) = -\frac{(1+\frac{1}{2}\phi)^2}{(1-\phi)^4} \tag{56}$$

and $\phi = (\pi/6)\rho\sigma^3$ denotes the packing fraction. Taking the Fourier transform of (55) and using it in (50) gives an expression for the static structure factor which is quite accurate for packing fractions up to about 0.4.

The pair distribution function $g(r)$ for the hard-sphere fluid can only be obtained from an inverse numerical Fourier transform of the function

$$\tilde{\gamma}(q) = \tilde{h}(q) - \tilde{c}(q) = \frac{\rho\tilde{c}^2(q)}{1 - \rho\tilde{c}(q)}. \tag{57}$$

One gets

$$g(r) = FT^{-1}(\tilde{\gamma}(q)) + c(r) + 1; \qquad r > \sigma \tag{58}$$

---

[2] Different methods to derive this result can be found in Balescu (1975) and in appendix B of Hansen and McDonald (1986).

and $g(r) = 0$ for $r < \sigma$. This procedure has to be used since $\gamma(r) = h(r) - c(r)$ is continuous at σ, whereas $c(r)$ and $h(r)$ are not. (The symbol FT^{-1} in (58) denotes inverse Fourier transformation.)

The equation of state of a system of hard spheres in the PY approximation is obtained by using $c(r)$ in the compressibility equation (51) and integrating it with respect to ϕ; the result is

$$\frac{\beta p^c}{\rho} = \frac{1+\phi+\phi^2}{(1-\phi)^3}. \tag{59}$$

Alternatively, the equation of state can also be obtained from the virial equation (34)

$$\frac{\beta p^v}{\rho} = \frac{1+2\phi+3\phi^2}{(1-\phi)^2}. \tag{60}$$

These two results for $\beta p/\rho$ are seen to be different; they show the internal thermodynamic inconsistency of the PY scheme, which results from its approximate character. The comparison with exact simulation data shows the inaccuracy of both expressions, in particular at densities close to the freezing transition of hard spheres at $\phi_f = 0.49$. An empirical equation of state, which fits nicely the simulation data for $\phi < \phi_f$, is due to Carnahan and Starling (1969); it is an interpolation of the PY virial and compressibility equations of state,

$$\frac{\beta p}{\rho} = \frac{\beta}{\rho}\left(\frac{2}{3}p^c + \frac{1}{3}p^v\right) = \frac{1+\phi+\phi^2-\phi^3}{(1-\phi)^3}; \tag{61}$$

it does, however, not reproduce the freezing transition.

The thermodynamic inconsistencies inherent in the HNC and PY closure schemes can be partially removed by the Rogers–Young (RY) closure (Rogers and Young, 1984). It mixes HNC and PY and the amount of mixing is determined by demanding that the compressibility calculated from Eq. (51) is identical to that obtained from taking the appropriate derivative of the virial equation of state, Eq. (34). Since this procedure removes the inconsistency from χ_T, the structure factor at $q = 0$ is determined correctly, and it can be expected that the calculation of $S(q)$ by the RY scheme will give accurate results. This is indeed the case, as shown in Fig. 1, where results for $g(r)$ obtained from PY, HNC and RY are compared to Monte Carlo simulation data for hard spheres and Yukawa potentials (Mendez-Alcaraz et al., 1992). Whereas HNC is seen to overestimate the contact value of $g(r)$ for hard spheres, it is the PY scheme which overestimates $g(r)$ for the soft Yukawa potential rather strongly.

Fig. 1 shows also the main differences of the pair distribution functions of hard spheres compared to those of charge-stabilized particles, interacting by the Yukawa potential. Whereas $g(r)$ has its maximum at contact ($r = \sigma$) for hard spheres given by $g(\sigma^+) = (1+\phi/2)/(1-\phi)^2$ in PY approximation, $g(r)$ exhibits a correlation hole for charge-stabilized particles; $g(r)$ remains to be zero for certain distances outside the central particle with its center at $r = 0$. This situation is illustrated further in Figs. 2 and 3. In Fig. 2 $g(r)$ is shown for hard spheres and in Fig. 3 for Yukawa particles.

It is instructive to compare the results for $\phi = 0.01$ for the two cases; whereas $g(r)$ is practically equal to 1 for hard spheres the system of Yukawa particles of valency $Z = 200$ is already strongly structured. There is a well-developed first shell of neighbors. Increasing ϕ reduces the extent of the correlation hole, sharpens the nearest neighbor shell and increases the main maximum of $g(r)$. The position of this maximum agrees rather precisely with the mean

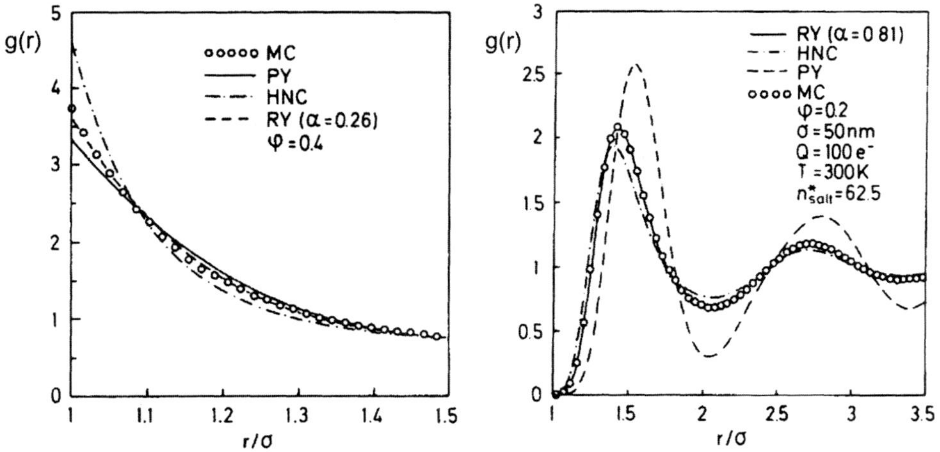

Fig. 1. Pair correlation functions for hard spheres (left) and Yukawa particles (right). The results of integral equation methods are compared to Monte Carlo simulations. For hard spheres (left) PY and HNC deviate somewhat from the simulation results near contact for a concentrated system of volume fraction $\phi = 0.4$. These deviations are more pronounced for charge-stabilized suspensions (right), but the RY scheme is seen to reproduce the Monte Carlo results very well.

interparticle distance $\bar{r} = (\sigma/2)(4\pi/3\phi)^{1/3}$ for deionized suspensions. When salt is added at a fixed volume fraction, $g(r)$ looses its structure as shown in Fig. 4.

5. Scattering from colloidal suspension

Most of the presently available knowledge about the microstructure and dynamics of colloidal suspensions is the result of scattering experiments. Depending on the length scales to be resolved and the contrast between suspended particles and solvent one uses the scattering of neutrons, X-rays or of light. The contrast and therefore the strength of the coupling of the radiation to the system depends in the first case on the difference of neutron scattering lengths, in the second case on the difference of electron densities and in the case of light on the dielectric properties of the particles and the solvent. The formalism is essentially identical in all three cases; here, the light scattering will be discussed (Berne and Pecora, 1976).

Light scattering arises from inhomogeneities of the dielectric properties of a medium by solving Maxwell's equations. An incoming wave excites dipoles which radiate in all directions and the superposition of all fields generates the scattering pattern of the sample. The sample consists of solvent and particles; the particles can have a core-shell or a crystalline structure, so that their dielectric properties are described by a position-dependent dielectric tensor $\varepsilon_p(\boldsymbol{r},t)$, whereas the solvent is characterized by a scalar dielectric constant ε_s. The electric displacement is written as $\boldsymbol{D} = \varepsilon_0 \boldsymbol{E} + 4\pi \boldsymbol{P} = \varepsilon(\boldsymbol{r}) \cdot \boldsymbol{E}$, where ε_0 is the averaged dielectric constant of the sample. The first part of $\boldsymbol{D}$ takes into account that the field, which gives rise to the polarization $\boldsymbol{P}$, propagates in an effective medium of dielectric constant ε_0.

Writing $4\pi \boldsymbol{P} = \delta\varepsilon(\boldsymbol{r}) \cdot \boldsymbol{E}$, the deviations of the dielectric properties are $\delta\varepsilon(\boldsymbol{r}) = \varepsilon_p(\boldsymbol{r}) - \varepsilon_0$

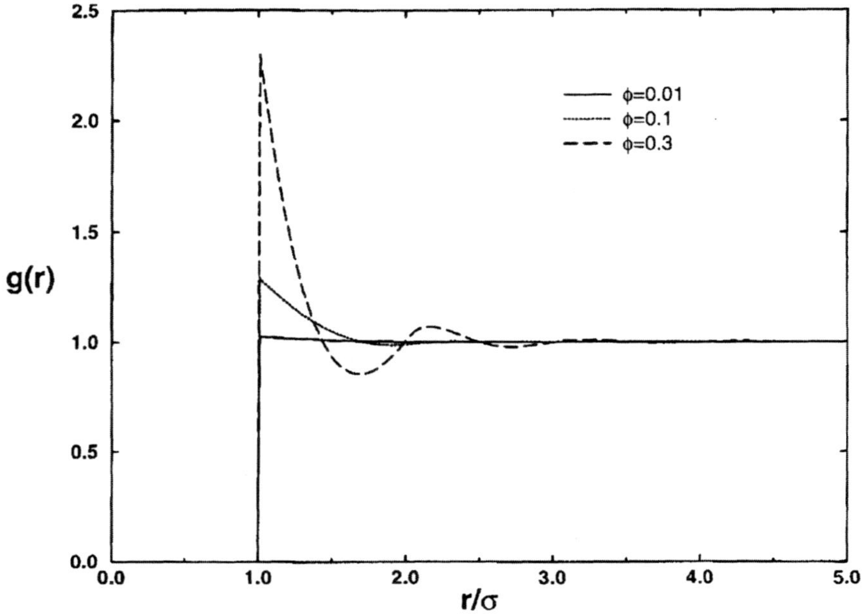

Fig. 2. Radial distribution functions for hard spheres using PY. At low volume fractions $g(r)$ is structureless except for the excluded volume ($r < \sigma$). Structure develops at volume fractions larger than 0.1.

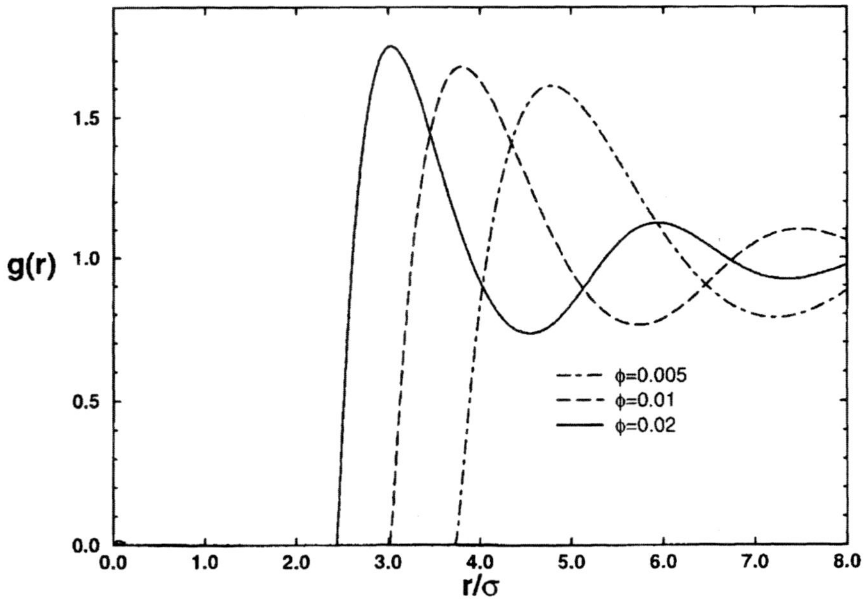

Fig. 3. Radial distribution functions for salt-free suspensions of charged particles with $Z = 200$. Because of the longer range of interactions there is already a well developed structure at rather low volume fractions.

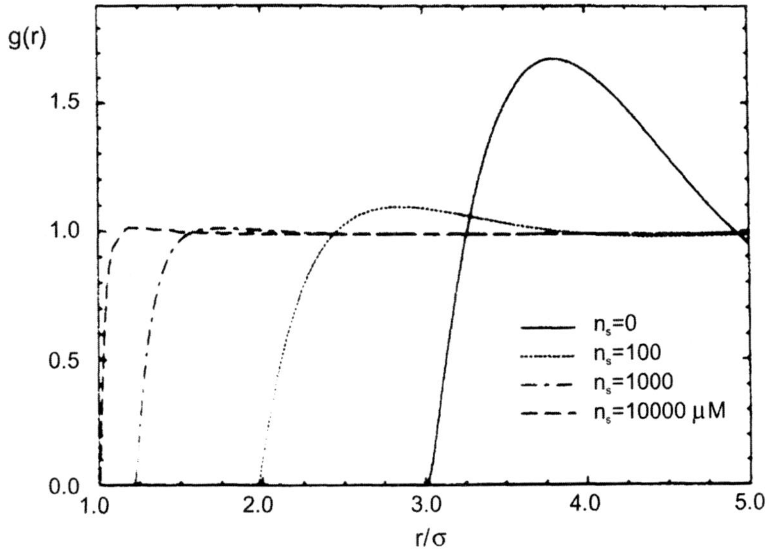

Fig. 4. Radial distribution functions for charge-stabilized systems as a function of salt content, n_s. The curve for $n_s = 0$ corresponds to the result shown in Fig. 3 for $\phi = 0.01$. The addition of salt reduces the structure of $g(r)$ and at high salt content the system is hard-sphere like.

inside the particles and $\delta\varepsilon = \varepsilon_s - \varepsilon_0$ outside. The contrast of the regions inside the particles can be rewritten as $\delta\varepsilon(\boldsymbol{r}) = (\varepsilon_p(\boldsymbol{r}) - \varepsilon_s) + (\varepsilon_s - \varepsilon_0)$. Assuming $\varepsilon_s - \varepsilon_0$ to be constant, which amounts to neglecting thermal fluctuations in the solvent, the only remaining spatially varying contrast arises from the particles. Therefore, $4\pi \boldsymbol{P} = \varepsilon_0 \delta\varepsilon_r(\boldsymbol{r})$, where $\delta\varepsilon_r(\boldsymbol{r}) = (\varepsilon_p(\boldsymbol{r}) - \varepsilon_s)/\varepsilon_0$.

From Maxwell's equations it is easy to derive a wave equation for $\boldsymbol{D}(\boldsymbol{r},t)$. Assuming a time dependence proportional to $\exp(-i\omega t)$ for all fields one obtains

$$(\nabla^2 + q^2)\boldsymbol{D} = -4\pi \nabla \times \nabla \times \boldsymbol{P} \tag{62}$$

where $q^2 = \mu_0 \varepsilon_0 \omega^2 / c^2 = (2\pi/\lambda)^2; \lambda$ is the wavelength of light in the effective medium. The solution of (62) is of the form $\boldsymbol{D} = \boldsymbol{D}_i + \boldsymbol{D}_s$, where $\boldsymbol{D}_i$ solves the homogeneous equation, which describes the unperturbed incoming wave. The scattered part $\boldsymbol{D}_s$ is obtained by using the Green function of the Helmholtz equation. Since one is interested in the scattered field $\boldsymbol{E}_s$ at the point $\boldsymbol{R}$ of the detector, the far-field approximation for the Green function can be used. The result is

$$\boldsymbol{E}_s(\boldsymbol{R}) = \frac{e^{iqR}}{R} \frac{q^2}{4\pi} \int d\boldsymbol{r}' \{ -\hat{\boldsymbol{q}}_f \times (\hat{\boldsymbol{q}}_f \times \delta\varepsilon_r(\boldsymbol{r}') \cdot \boldsymbol{E}(\boldsymbol{r}')) \} \, e^{-i\boldsymbol{q}_f \cdot \boldsymbol{r}'} \tag{63}$$

The first factor is the outgoing spherical wave and the rest the scattering amplitude which depends on the direction of observation $\hat{\boldsymbol{q}}_f$ and the direction of the total electric field $\boldsymbol{E}(\boldsymbol{r}')$. The integral is over the volumes of the suspended particles; $\boldsymbol{q}_f = q\hat{\boldsymbol{q}}_f$. The field $\boldsymbol{E}(\boldsymbol{r}')$ under the integral in Eq. (63) is the sum of the incident field and the field coming from other particles. Eq. (63) is an integral equation for the scattered field.

The simplest approximation to the general expression for the scattering amplitude in

Eq. (63) is to replace the total field $\boldsymbol{E}(\boldsymbol{r})$ by the incident field

$$\boldsymbol{E}_i(\boldsymbol{r}) = E_0 \boldsymbol{n}_i \, e^{i\boldsymbol{q}_i \cdot \boldsymbol{r}} \tag{64}$$

where $\boldsymbol{n}_i$ is the polarization direction and $\boldsymbol{q}_i$ the incident wavevector. This procedure, known as the first Born approximation, not only restricts the following results to single scattering but it also assumes that the phase of the wave traversing the scattering particle is the same as it would be if the particle were not present. This assumption is valid when

$$\frac{2\pi\sigma}{\lambda}\left|\frac{n_p}{n_s} - 1\right| \ll 1 \tag{65}$$

where n_p and n_s are the refractive indices of the particle and the solvent, respectively. This condition defines the Rayleigh–Debye–Gans approximation; it breaks down for large particles and/or if the refractive indices are rather different.

With these assumptions the component of the scattered field in a polarization direction $\boldsymbol{n}_f$ is given by

$$\boldsymbol{n}_f \cdot \boldsymbol{E}_s(\boldsymbol{R}) = E_0 \frac{e^{iqR}}{R} \frac{q^2}{4\pi} \int d\boldsymbol{r}' \delta\varepsilon_{if}(\boldsymbol{r}') \, e^{i\boldsymbol{q}_s \cdot \boldsymbol{r}'} \tag{66}$$

where $\delta\varepsilon_{if}(\boldsymbol{r}) = \boldsymbol{n}_f \cdot \delta\boldsymbol{\varepsilon}_r(\boldsymbol{r}) \cdot \boldsymbol{n}_i$, and $\boldsymbol{q}_s = \boldsymbol{q}_i - \boldsymbol{q}_f$ is the scattering vector. For elastic scattering the magnitude of $\boldsymbol{q}_s$ is $2q \sin(\vartheta/2)$ and ϑ is the scattering angle. When the suspension consists of N particles of m different species such that N_α particles belong to species α,

$$N = \sum_{\alpha=1}^{m} N_\alpha \tag{67}$$

and when $\boldsymbol{r}_{j\alpha}$ denotes the position of the center of mass of particle j of species α, Eq. (66) becomes

$$E_s(\boldsymbol{R}) = E_0 \frac{q^2 \, e^{iqR}}{4\pi R} \sum_{\alpha=1}^{m} \sum_{j=1}^{N_\alpha} e^{i\boldsymbol{q}_s \cdot \boldsymbol{r}_{j\alpha}} \int_{V_\alpha} d\boldsymbol{r} \; \delta\varepsilon_{if}^{\alpha}(\boldsymbol{r}) e^{i\boldsymbol{q}_s \cdot \boldsymbol{r}}; \tag{68}$$

the integral is over the volume of a particle of species α. This expression can be used to obtain the scattered intensity from a polydisperse suspension of particles of arbitrary shapes and of an arbitrary dependence of the dielectric properties inside the particles; it also allows to consider polarized and depolarized scattering.

For the simple case of spherical particles, which are homogeneous in their dielectric properties, the integral in Eq. (68), which is known as the form amplitude, becomes

$$f_\alpha(\boldsymbol{q}_s) = \delta\varepsilon_{if}^{\alpha} V_\alpha \frac{3 j_1(q_s \sigma_\alpha/2)}{q_s \sigma_\alpha/2}; \tag{69}$$

here, $j_1(x) = (\sin x - x \cos x)/x^2$ is the spherical Bessel function of index 1; σ_α is the diameter of a sphere of type α and V_α is its volume. For spherical particles f_α is seen to depend only on the magnitude q_s of the scattering vector. The situation is different for non-spherical particles; for a thin rod of length L with the rod axis in the direction of the unit vector $\boldsymbol{u}$, one has

$$f_{\text{rod}}(\boldsymbol{q}_s) = \delta\varepsilon_{if} L j_0(\boldsymbol{q}_s \cdot \boldsymbol{u}/2), \tag{70}$$

where $j_0(x) = \sin x/x$. There is now a dependence on the magnitude q_s and on the angle between $\boldsymbol{q}_s$ and $\boldsymbol{u}$.

When the Rayleigh–Debye–Gans condition (65) is not satisfied one has to consider the proper boundary conditions at the surface of the particles for the electromagnetic fields inside and outside of the particles. The corresponding formalism is Mie scattering, which results in a rather more complicated behavior of the scattering amplitude (Kerker, 1969).

If the condition (65) is satisfied the scattered intensity in the first Born approximation follows from (68) as

$$\langle |E_s(\boldsymbol{R})|^2 \rangle = \frac{E_0^2 q^4}{16\pi^4 R^2} \left\langle \left| \sum_{\alpha=1}^{m} \sum_{i=1}^{N_\alpha} f_\alpha(\boldsymbol{q}_s) e^{i\boldsymbol{q}_s \cdot \boldsymbol{r}_{i\alpha}} \right|^2 \right\rangle. \tag{71}$$

The bracket denotes an ensemble average, which is to be performed with the configuration space distribution function. The latter is essentially the Boltzmann factor of the total potential energy of interaction. For non-spherical particles the pair potential depends in general on the distance between their centers of mass and in addition on their relative orientation. For two rods the pair potential is of the form $v(\boldsymbol{r}, \boldsymbol{u}_1, \boldsymbol{u}_2)$, where $\boldsymbol{r}$ is the vector connecting their centers of mass and $\boldsymbol{u}_1$ and $\boldsymbol{u}_2$ denote the orientations of their axes. Therefore, the configurational integral in (71) contains the rod orientations both in the amplitudes f_{rod} and in the Boltzmann factor, so that a factorization of the ensemble average into an average over the amplitudes and an average over the phase factors is not possible.

The situation is much simpler for spherical particles, where the form amplitudes are given by Eq. (69). Then the ensemble average is only over the phase factors depending on the positions of the centers of the spheres. For monodisperse suspensions the index α is irrelevant so that the scattered intensity, apart from the prefactor in (71), becomes

$$I(q) = N f^2(q) S(q). \tag{72}$$

Here and in the following q_s is written as $q = (4\pi/\lambda)\sin(\vartheta/2)$, where λ is the wavelength of light in the solvent. The strength of the scattering is entirely in the quantity $f^2(q)$. It is convenient to factorize the form amplitude as $f(q) = f(0)B(q)$ and to express $\delta\varepsilon_{if}$ in terms of refractive indices n_p, n_s and n_0 of particles, solvent and the effective medium. Then

$$f(0) = \frac{\pi}{6} \frac{n_p + n_s}{n_0^2} (n_p - n_s)\sigma^3 \approx \frac{\pi}{3} \frac{n_p - n_s}{n_0} \sigma^3. \tag{73}$$

The quantity

$$P(q) = B^2(q) = \left(\frac{3 j_1(q\sigma/2)}{q\sigma/2} \right)^2 \tag{74}$$

is the normalized form factor ($P(0) = 1$). The last factor in (72) is

$$S(q) = \frac{1}{N} \left\langle \sum_{i,j} e^{i\boldsymbol{q} \cdot (\boldsymbol{r}_i - \boldsymbol{r}_j)} \right\rangle \tag{75}$$

which, apart from a term proportional to $\delta(\boldsymbol{q})$, is the static structure factor of Eq. (44). Discarding this term which describes scattering only in the forward direction, the scattered intensity for a system of N monodisperse spherical particles follows from (72) and (43) as

$$I(q) = Nf^2(0)P(q)\left\{1+\rho\int d\boldsymbol{r}\, e^{i\boldsymbol{q}\cdot\boldsymbol{r}}(g(r)-1)\right\}$$

$$= Nf^2(0)P(q)\left\{1+\frac{4\pi\rho}{q}\int_0^\infty dr\, r(g(r)-1)\sin qr\right\}. \tag{76}$$

This result establishes the link of the statistical mechanical description of the microstructure, given by the pair distribution function $g(r)$, to the static light scattering experiment. For the present case of spherical particles the scattered intensity separates into a product of single particle properties, such as the material constants determining $f^2(0)$, the distribution of scattering material inside the particle, described by the form factor $P(q)$, and the correlations between the particles as described by $S(q)$ or $g(r)$. This last factor is independent of the quantities which determine the coupling of light to the system and the geometrical shape. The factorization property makes the experimental determination of $S(q)$ very easy; by diluting the sample until all correlations have disappeared one measures the intensity $I_0(q) = N_0 f^2(0)P(q)$ so that the structure factor of the original system can be determined from the ratio

$$\frac{I(q)}{I_0(q)} = \frac{N}{N_0}S(q). \tag{77}$$

As an example of the relation between pair correlation functions and corresponding static structure factors Figs. 5 and 6 show results for systems of moderately charged particles ($Q = 107e$) of diameter $\sigma = 160$ nm, $\kappa\sigma = 3.4$ in an organic solvent ($\varepsilon = 10$). These parameters refer to suspensions of silica particles which have been experimentally investigated (Philipse and Vrij, 1988). Fig. 5 shows $g(r)$ for four volume fractions. For the lowest value,

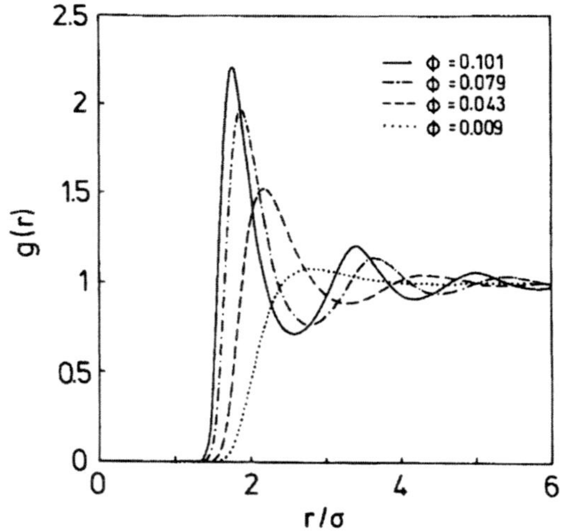

Fig. 5. Radial distribution functions for a series of samples of silica particles in an organic solvent at different volume fractions for a valency $Z = 107$.

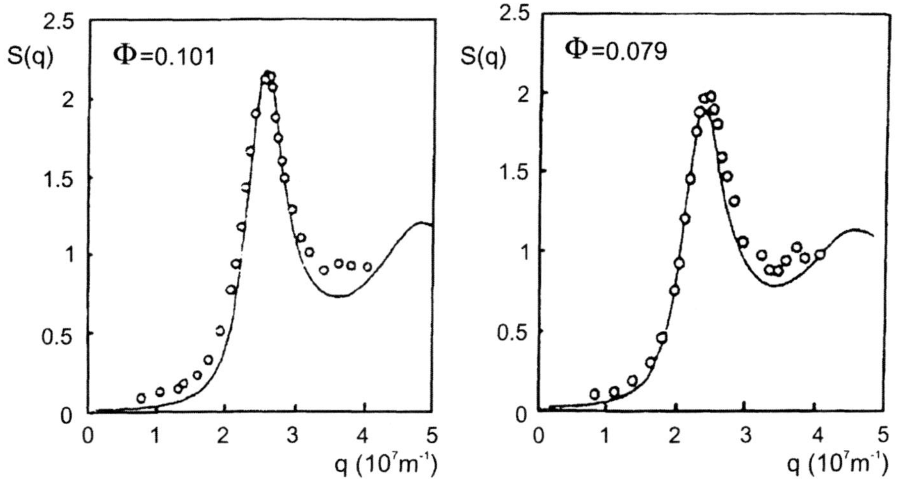

Fig. 6. Static structure factors corresponding to two of the radial distribution functions shown in Fig. 5, together with experimental results.

$\phi = 0.009$, $g(r)$ has very little structure, except that it exhibits a correlation hole; $g(r) = 0$ for $r < 1.6\sigma$, reflecting the fact that the particles avoid each other because of the electrostatic repulsion. When ϕ increases the local structure becomes more pronounced, rather well defined shells of nearest and next-to-nearest neighbors develop. The correlation hole gets somewhat smaller because of the decreasing mean interparticle distances. The peak positions of $g(r)$ are approximately equal to the interparticle distances (Nägele et al., 1993).

These results for $g(r)$ can now be used to calculate $S(q)$. Two examples are shown in Fig. 6 together with the experimental results. Since the value of the charge Q used to calculate $g(r)$ is not known experimentally, one proceeds as follows: one takes the experimentally determined values for $S(q)$ for one value of ϕ, for instance $\phi = 0.101$ in the present case, and varies Q until an acceptable fit is obtained. This procedure resulted in $Q = 107e$. The value of Q is then kept fixed in the calculations of $g(r)$ and $S(q)$ for the other volume fractions, assuming that the value of the charge is independent of the concentration (Nägele et al., 1993). Although the fits to the experimental results are rather good at and around the main maximum, the agreement at low values of q and around the minima following the main peak is not that good. This kind of disagreement of theory and experiment is often observed and has to be attributed to polydispersity effects.

6. Mixtures and polydispersity effects

For a mixture of m components the scattered intensity is obtained from (68), again apart from the prefactor,

$$I(q) = N \sum_{\alpha,\beta=1}^{m} f_\alpha(q) f_\beta(q) S_{\alpha\beta}(q). \tag{78}$$

The form amplitude of particles of species α is

$$f_\alpha(q) = \frac{\pi}{3} \frac{n_{p,\alpha} - n_s}{n_0} \sigma_\alpha^3 \frac{3 j_1(q\sigma_\alpha/2)}{q\sigma_\alpha/2}, \tag{79}$$

where $n_{p,\alpha}$ is the refractive index of a particle of type α and σ_α is its diameter. The $S_{\alpha\beta}(q)$ in (78) are the partial static structure factors

$$\begin{aligned} S_{\alpha\beta}(q) &= \frac{1}{N} \left\langle \sum_{i=1}^{N_\alpha} \sum_{j=i}^{N_\beta} e^{i\boldsymbol{q}\cdot(\boldsymbol{r}_{i\alpha} - \boldsymbol{r}_{j\beta})} \right\rangle \\ &= x_\alpha \delta_{\alpha\beta} + \rho x_\alpha x_\beta \tilde{h}_{\alpha\beta}(q) \end{aligned} \tag{80}$$

where $x_\alpha = N_\alpha/N$ is the fraction of particles of type α and $\tilde{h}_{\alpha\beta}(q)$ is the Fourier transform of $g_{\alpha\beta}(r) - 1$. The pair correlation functions $g_{\alpha\beta}(r)$ describe the microstructure of the mixture; $g_{\alpha\beta}(r)$ is proportional to the probability to find a particle of species β a distance r apart from one of species α.

When there are no correlations, $S_{\alpha\beta}(q) = x_\alpha \delta_{\alpha\beta}$, so that the scattered intensity becomes

$$I_0(q) = N \sum_{\alpha=1}^{m} x_\alpha f_\alpha^2(q) \equiv N\overline{f^2(q)}. \tag{81}$$

This suggests to rewrite (78) formally as the scattered intensity for monodisperse suspensions, Eq. (72),

$$I(q) = N\overline{f^2(q)} S^M(q) \tag{82}$$

where

$$S^M(q) = \frac{1}{\overline{f^2(q)}} \sum_{\alpha,\beta} f_\alpha(q) f_\beta(q) S_{\alpha\beta}(q) \tag{83}$$

is called the measurable structure factor, since it is this quantity which can be obtained from the experiment. Although Eq. (82) has the same form as Eq. (72) it should be noted that the factorization of $I(q)$ into a one-particle property and a genuine statistical mechanical quantity like $S(q)$, which does not depend on the coupling of the radiation to the system, is no longer possible; $S^M(q)$ depends on the form amplitudes $f_\alpha(q)$. The truly thermodynamic quantities are the $m(m+1)/2$ partial structure factors $S_{\alpha\beta}(q)$. They can be determined experimentally only by contrast matching so that $f_\alpha = 0$ for a certain α.

In order to calculate the $S_{\alpha\beta}(q)$ it is necessary to solve coupled Ornstein–Zernike equations

$$h_{\alpha\beta}(r) = c_{\alpha\beta}(r) + \rho \sum_{\gamma=1}^{m} x_\gamma \int d\boldsymbol{r}' c_{\alpha\gamma}(|\boldsymbol{r} - \boldsymbol{r}'|) h_{\gamma\beta}(r') \tag{84}$$

together with closure relations, appropriately generalized to mixtures. This has been done for mixtures of hard spheres, for mixtures of charge-stabilized particles and for mixtures of charged and uncharged hard spheres (Klein and D'Aguanno, 1996).

The theoretical methods for the description of mixtures can be used to study polydispersity effects. Realistic colloidal suspensions are in most cases polydisperse; there is a certain variation in particle diameters, which in the case of charged particles may be combined with a variation of charge per particle. If this is the case, the larger and more strongly charged particles will be mostly surrounded by less charged small particles so that the polydispersity will affect the local structure. Because of the factor σ_α^3 in the form amplitude (79) the larger particles in a size distribution $p(\sigma)$ scatter light more strongly than smaller ones. These are different effects of polydispersity on the scattering intensity; the first one is on the $S_{\alpha\beta}(q)$ and the second on $f_\alpha(q)$.

The polydisperse suspension, described by the continuous distribution function $p(\sigma)$ of diameters, is approximated by a m-component mixture by replacing $p(\sigma)$ by a histogram. The diameters σ_α, where $\alpha = 1,\ldots,m$, of particles of type α and their molar fractions x_α are determined by requiring the equality of the first $2m$ moments of both the histogram representation and the continuous distribution. For colloids the Schulz distribution

$$p(\sigma) = \left(\frac{t+1}{\langle\sigma\rangle}\right)^{t+1} \frac{\sigma^t}{\Gamma(t+1)} \exp\left[\frac{t+1}{\langle\sigma\rangle}\sigma\right] \tag{85}$$

is often a good representation. In (85) $\langle\sigma\rangle$ is the mean diameter and the parameter t is related to the standard deviation (or width of the distribution) by

$$s \equiv \frac{\left(\langle\sigma^2\rangle - \langle\sigma\rangle^2\right)^{1/2}}{\langle\sigma\rangle} = \frac{1}{(t+1)^{1/2}}. \tag{86}$$

The particle diameters σ_α and the molar fractions x_α for $\alpha = 1,\ldots,m$ are calculated from the $2m$ equations

$$\sum_{\alpha=1}^{m} x_\alpha \sigma_\alpha^n = \langle\sigma^n\rangle = \frac{(n+t)!}{t!(t+1)^n}\langle\sigma\rangle^n; \qquad n = 0, 1,\ldots, 2m-1. \tag{87}$$

This defines the m-component mixture, which represents the continuous distribution: the solution of the coupled system of Eqs. (84) gives the $g_{\alpha\beta}$ and from them one can obtain $S^M(q)$. This program has been used for polydisperse hard spheres and for Yukawa particles where the charge per particle is assumed to be proportional to the surface area ($Z_\alpha \propto \sigma_\alpha^2$). A typical result for $S^M(q)$ is shown in Fig. 7 for charged particles (D'Aguanno and Klein. 1991); the number density is kept constant, but the standard deviation s is increased from 0 to 0.4. Curve 1 corresponds to the monodisperse case, all particles have the same diameter $\sigma = \langle\sigma\rangle = 50$ nm and they carry charges $Q = 220e$. Curves 2 to 5 are the results for increasing s. Three important observations can be made.

(i) The peak position decreases with s, although the number density and therefore the mean interparticle distance remains constant.

(ii) The peak height of $S^M(q)$ decreases, although the average charge per particle increases with s. This is in contrast to the monodisperse case, where the peak of $S(q)$ increases when the charge increases at fixed concentration.

(iii) The value $S^M(0)$ increases with s. This is due to an increasing incoherent scattering contribution. It should also be emphasized that $S^M(0)$ is not simply related to the isothermal compressibility of the polydisperse system, which is in contrast to the relation $S(0) = \rho k_B T \chi_T$ for monodisperse systems. For the systems of Fig. 7 the compressibility is practically constant when s is varied from 0 to 0.4 (it actually decreases slightly); the values of $S^M(0)$ however.

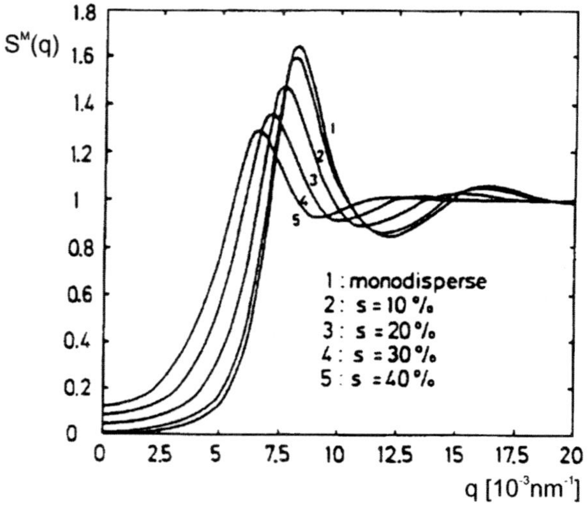

Fig. 7. The measurable structure factor $S^M(q)$ for a series of samples having the same number density but different degrees of polydispersity.

increase by a factor of 2.5 in going from $s = 0$ to $s = 0.1$ and by a factor of 16 in going from $s = 0$ to $s = 0.3$.

On the basis of these results it is evident that it can be quite misleading when experiments on systems with a known or unknown amount of polydispersity are interpreted in terms of theoretical results for monodisperse systems. There are many examples in the literature of experimentally determined structure factors which can be well represented by calculated structure factors for monodisperse systems for intermediate values of q, including the main peak. But it is often observed that the calculated $S(q)$ is too small at small q and that the secondary extrema at high q are also not well reproduced. It has been shown for several cases that the assumption of a certain amount of polydispersity leads to a remarkably better overall agreement, so that in these cases nominally monodisperse samples were really somewhat polydisperse (Klein and D'Aguanno, 1996).

The discrepancies at high q arise mainly from the fact that the structure factors of the various components of the mixture have their extrema at slightly different values of q so that increasing polydispersity gives rise to a progressive washing-out of the extrema. The increase of the structure factor of a polydisperse sample at small q is due to the incoherent scattering. This can already be demonstrated for the simple case of only optical polydispersity. One assumes that the correlations between all particles are identical regardless of their different diameters, but the form amplitudes are correctly taken into account according to Eq. (79). In this case the total correlations $\tilde{h}_{\alpha\beta}(q)$ in Eq. (80) are the same for all α and β, so that $\rho\tilde{h}_{\alpha\beta}(q) = \rho\tilde{h}(q) = S(q) - 1$. With this simplification Eq. (83) becomes

$$\begin{aligned} S^M(q) &= \frac{1}{\overline{f^2(q)}} \sum_{\alpha,\beta} f_\alpha(q) f_\beta(q) \left[x_\alpha \delta_{\alpha\beta} + x_\alpha x_\beta (S(q) - 1) \right] \\ &= R(q) + (1 - R(q)) S(q) \end{aligned} \tag{88}$$

where

$$R(q) = 1 - \frac{\overline{f(q)}^2}{\overline{f^2(q)}} \tag{89}$$

with

$$\overline{f(q)} = \sum_{\alpha=1}^{m} x_\alpha f_\alpha(q). \tag{90}$$

For a narrow size distribution of width s the incoherent contribution $R(q)$ is approximately equal to $9s^2$. For a strongly correlated suspension the static structure factor at $q \approx 0$ can be as small as 10^{-2} or even smaller, so that for a polydisperse sample with a standard deviation of $s = 0.1$ the first term in Eq. (88) is significantly larger than the second and the scattered intensity is therefore mostly due to incoherent contributions.

7. Depletion interaction

In 1954 Asakura and Oosawa considered two colloidal particles in a sea of smaller non-adsorbing polymers and showed that when the separation between the colloids becomes smaller than the size of a polymer so that the latter is squeezed out from the space between the particles, an attraction between the colloids results (Asakura and Oosawa, 1954). The attraction is induced by an imbalanced osmotic pressure which arises from the depletion of the polymers. The appearance of effective attractive interactions has several interesting aspects. One is that in the mixture, consisting of colloids and polymers, the pair interactions among all particles are repulsive, but when this system is mapped to an effective one-component system, consisting explicitly only of colloids, the original bare repulsive interaction between colloids is replaced by an effective one having an attractive part. This raises the question whether such mixtures can exhibit a phase behavior reminiscent of the gas–liquid coexistence in simple fluids. According to arguments first given more than 125 years ago by van der Waals it is the interplay between short-ranged (excluded volume) repulsions and long-ranged attractions between atoms which is responsible for gas-liquid coexistence. The depletion mechanism has qualitatively the ingredients for this kind of phenomenon. However, the existence of attractive interactions is only a necessary condition, it is not sufficient. The question is, therefore, how much attraction is needed? In contrast to atoms or small molecules colloids provide the possibilities for tuning interactions in strength and range, so that mixtures of big colloids with small ones or with non-adsorbing polymers serve as suitable model systems to study this question.

The statistical mechanical treatment of the depletion interaction starts from the partition function of a binary mixture. Species 1 consists of N_1 particles, having momenta $\boldsymbol{p}^{N_1} = (\boldsymbol{p}_1, \ldots, \boldsymbol{p}_{N_1})$ and positions $\boldsymbol{r}^{N_1} = (\boldsymbol{r}_1, \ldots, \boldsymbol{r}_{N_1})$ and species 2 has N_2 particles of momenta $\boldsymbol{P}^{N_2}$ at positions $\boldsymbol{R}^{N_2}$. The partition function is an obvious generalization of the expression (10) for a one-component system,

$$\begin{aligned} Q_{N_1 N_2}(V,T) &= \frac{1}{N_1! h^{3N_1}} \frac{1}{N_2! h^{3N_2}} \int \mathrm{d}\boldsymbol{p}^{N_1} \int \mathrm{d}\boldsymbol{r}^{N_1} \int \mathrm{d}\boldsymbol{P}^{N_2} \int \mathrm{d}\boldsymbol{R}^{N_2} \exp[-\beta H] \\ &\equiv Tr_1 Tr_2\, \mathrm{e}^{-\beta H} \end{aligned} \tag{91}$$

The operations Tr_i, $i = 1,2$, are a short-hand notation for the integrations over momenta and positions of particles of species i, including the prefactors $(N_i!h^{3N_i})^{-1}$. The Hamiltonian is now

$$H = H_{11} + H_{22} + V_{12}(\boldsymbol{r}^{N_1}, \boldsymbol{R}^{N_2}) \tag{92}$$

where

$$H_{11} = T_1 + V_{11}(\boldsymbol{r}^{N_1}) \tag{93}$$

consists of the kinetic energies of particles of species 1 and the potential of interactions between them; similarly

$$H_{22} = T_2 + V_{22}(\boldsymbol{R}^{N_2}) \tag{94}$$

for species 2. The interactions between particles of different species give rise to the last term in (92).

The kinetic energy contributions to $Q_{N_1N_2}$ can again be integrated, just like in obtaining the result (16) for the one component case,

$$Q_{N_1N_2}(V,T) = \frac{1}{N_1!\Lambda_1^{3N_1}} \frac{1}{N_2!\Lambda_2^{3N_2}} Z_{N_1N_2}(V,T) \tag{95}$$

where

$$Z_{N_1N_2}(V,T) = \int \mathrm{d}\boldsymbol{r}^{N_1} \exp[-\beta V_{11}(\boldsymbol{r}^{N_1})] \int \mathrm{d}\boldsymbol{R}^{N_2} \exp[-\beta(V_{22}(\boldsymbol{R}^{N_2}) + V_{12}(\boldsymbol{r}^{N_1}, \boldsymbol{R}^{N_2}))] \tag{96}$$

is the configurational integral of the mixture, and Λ_i denotes the de Broglie wavelength $h/(2\pi m_i k_B T)^{1/2}$, where m_i is the mass of a particles of species i.

The aim is now to map the mixture to an effective one-component system with effective interactions between particles of one species only. These effective interactions must include the effects of the other species and will of course be different from those in the original mixture. The mapping to the effective system, described by a Hamiltonian H_{eff}, has to be performed such that the thermodynamics of the effective system is exactly equal to that of the original mixture. Therefore, the partition function (91) must have the form

$$Q_{N_1N_2}(V,T) = Tr_1\, \mathrm{e}^{-\beta H_{\mathrm{eff}}}, \tag{97}$$

so that H_{eff} has to be determined by rewriting (91) as in (97).

Using (92) in (91) and observing that H_{11} depends only on the momenta and coordinates of particles of species 1, one obtains

$$Q_{N_1N_2}(V,T) = Tr_1[\mathrm{e}^{-\beta H_{11}} Q_{N_2}(V,T;\boldsymbol{r}^{N_1})], \tag{98}$$

where

$$\begin{aligned} Q_{N_2}(V,T;\boldsymbol{r}^{N_1}) &= Tr_2 \exp[-\beta(H_{22} + V_{12})] \\ &= \frac{1}{N_2!h^{3N_2}} \int \mathrm{d}\boldsymbol{P}^{N_2} \int \mathrm{d}\boldsymbol{R}^{N_2} \exp[-\beta(H_{22} + V_{12})] \\ &= \frac{V^{N_2}}{N_2!\Lambda_2^{3N_2}} \frac{Z_{N_2}(V,T;\boldsymbol{r}^{N_1})}{V^{N_2}} \end{aligned} \tag{99}$$

and

$$Z_{N_2}(V,T;\boldsymbol{r}^{N_1}) = \int \mathrm{d}\boldsymbol{R}^{N_2} \exp\left[-\beta\left(V_{22}(\boldsymbol{R}^{N_2}) + V_{12}(\boldsymbol{r}^{N_1},\boldsymbol{R}^{N_2})\right)\right] \tag{100}$$

Here, $Q_{N_2}(V,T;\boldsymbol{r}^{N_1}) \equiv Q_2(\boldsymbol{r}^{N_1})$ is the partition function and $Z_{N_2}(V,T;\boldsymbol{r}^{N_1}) \equiv Z_2(\boldsymbol{r}^{N_1})$ the configurational integral of the system of particles of species 2 in the external field created by the particles of species 1, sitting at positions $\boldsymbol{r}^{N_1}$. This dependence on $\boldsymbol{r}^{N_1}$ is explicitly noted in the arguments of Q_2 and Z_2.

The comparison of Eqs. (97) and (98) shows that the effective Hamiltonian is determined by

$$\mathrm{e}^{-\beta H_{\mathrm{eff}}} = \mathrm{e}^{-\beta H_{11}} Q_2(\boldsymbol{r}^{N_1}) \tag{101}$$

or

$$\begin{aligned} H_{\mathrm{eff}} &= H_{11} - k_B T \ln Q_2(\boldsymbol{r}^{N_1}) \\ &= H_{11} + A_2(\boldsymbol{r}^{N_1}). \end{aligned} \tag{102}$$

Here, $A_2(\boldsymbol{r}^{N_1}) \equiv A(N_2,V,T;\boldsymbol{r}^{N_1})$ is the free energy of species 2 in the presence of species 1, as can be seen from (13). The effective Hamiltonian consists of the sum of the Hamiltonian of the particles of species 1 in the original system and the free energy of species 2 in the external field created by the particles of species 1. In the applications to colloidal mixtures or colloids and non-adsorbing polymers species 1 refers to big colloidal particles and species 2 are either small colloids or polymers treated as small spheres.

Using (99) in (102) the effective Hamiltonian becomes

$$H_{\mathrm{eff}} = H_{11} + k_B T N_2(\ln(\rho_2 \Lambda_2^3) - 1) - k_B T \ln\left[Z_2(\boldsymbol{r}^{N_1})/V^{N_2}\right]. \tag{103}$$

The second term is the ideal gas part of the free energy of the small particles, similar to (18), with $\rho_2 = N_2/V$. It is independent of the momenta and coordinates. A further contribution of this type is also contained in the last term of (103), which can be expanded as a sum of contributions involving no coordinates, one coordinate, two coordinates, etc. (Dijkstra et al., 1999a). The parts of H_{eff} which do not involve momenta and positions of the big particles are irrelevant for the calculation of correlation functions such as the static structure factor of the big colloids, but these so-called volume terms are important for the calculation of thermodynamic properties, including the phase behavior.

The further discussion will be restricted to the simplest case, which is the pair contribution in (103). Then the potential part of H_{eff} consists of a sum of pair potentials. For two big particles separated by $r = |\boldsymbol{r}_1 - \boldsymbol{r}_2|$ the effective potential is

$$v_{\mathrm{eff}}(r;\rho_1,\rho_2,T) = v_{11}(r) - k_B T \ln \frac{Z_2(r)}{Z_2(\infty)}, \tag{104}$$

where $Z_2(\infty)$ is the pair contribution to (100), when the big particles are infinitely separated. This is a constant term which is included as a normalization. The simple case to be considered consists of a mixture of big hard spheres of diameter σ_1 (species 1) and small entities of diameter σ_2 which are excluded by the big spheres but which do not interact with each other. In this case the pair potentials of the mixture are

$$v_{11}(r) = \begin{cases} \infty; & r < \sigma_1 \\ 0; & r > \sigma_1; \end{cases} \tag{105}$$

$$v_{12}(|\boldsymbol{r}_i - \boldsymbol{R}_j|) = \begin{cases} \infty; & |\boldsymbol{r}_i - \boldsymbol{R}_j| < (\sigma_1 + \sigma_2)/2 \\ 0; & \text{otherwise} \end{cases} \tag{106}$$

and $v_{22} = 0$. To calculate $Z_2(r)$ one starts from (17), where now $V_{22} = 0$, so that

$$Z_2(r) = \int \mathrm{d}\boldsymbol{R}^{N_2} \prod_{j=1}^{N_2} \exp\bigl[-\beta v_{12}(|\boldsymbol{r}_1 - \boldsymbol{R}_j|)\bigr] \exp\bigl[-\beta v_{12}(|\boldsymbol{r}_2 - \boldsymbol{R}_j|)\bigr]. \tag{107}$$

The integrand vanishes when one or more of the exponentials vanishes, which is the case when $|\boldsymbol{r}_i - \boldsymbol{R}_j| < (\sigma_1 + \sigma_2)/2$, $i = 1,2$; otherwise the integrand is equal to one. Consider first the case when the distance $r = |\boldsymbol{r}_1 - \boldsymbol{r}_2|$ between the centers of the big spheres is larger than $\sigma_1 + \sigma_2$. Then $Z_2(r)$ is equal to the power N_2 of the free volume available to the small spheres, $V_{\text{free}}^{N_2}$, where

$$V_{\text{free}} = V - 2\frac{4\pi}{3}\left(\frac{\sigma_1 + \sigma_2}{2}\right)^3. \tag{108}$$

When the big spheres come closer, so that $\sigma_1 < r < \sigma_1 + \sigma_2$, the two depletion zones of thickness $\sigma_2/2$ around each sphere overlap; they have some volume in common, so that the volume available to the small spheres increases. This additional volume is equal to twice the volume of a spherical cap, V_{cap}, which is cut-off from a sphere of radius $a = (\sigma_1 + \sigma_2)/2$ by a plane which has a distance $b = r/2$ from its center:

$$V_{\text{cap}} = \frac{2\pi}{3}a^3\left(1 - \frac{3}{2}\frac{b}{a} + \frac{b^3}{2a^3}\right) \tag{109}$$

and $Z_2(r)$ is equal to $(V_{\text{free}} + 2V_{\text{cap}})^{N_2}$.

Therefore, $Z_2(\infty) = Z_2(r > \sigma_1 + \sigma_2) = V_{\text{free}}^{N_2}$ and $Z_2(r) = (V_{\text{free}} + 2V_{\text{cap}})^{N_2}$. Using the identifications of a and b, one obtains

$$v_{\text{eff}}(r; \rho_1, \rho_2, T) = \begin{cases} \infty; & r < \sigma_1 \\ v_{A0}(r); & \sigma_1 < r < \sigma_1 + \sigma_2 \\ 0; & r > \sigma_1 + \sigma_2 \end{cases} \tag{110}$$

where

$$v_{A0}(r) = -k_B T \rho_2 \frac{\pi \sigma_1^3 (1+q)^3}{6}\left[1 - \frac{3r}{2(1+q)\sigma_1} + \frac{r^3}{2(1+q)^3 \sigma_1^3}\right] \tag{111}$$

with $q = \sigma_2/\sigma_1$ the ratio of diameters. This expression for the effective potential was first derived by Asakura and Oosawa (1954) using a method rather different from the one given here. It was later independently rederived by Vrij (1976) by still another procedure. It should be noted that $v_{A0}(r)$ depends explicitly on the temperature and on the number density ρ_2 of the small spheres. It is a purely attractive potential of a range given by the diameter of the small spheres and a depth determined by ρ_2. Since the component 2 has been treated as an ideal gas, its osmotic pressure is given by $\pi = \rho_2 k_B T$ so that the potential strength can equally be expressed by π. The depletion potential has been measured directly in several experiments (Kaplan et al., 1994; Ohshima et al., 1997; Rudhardt et al., 1998, 1999).

The result (110) has been derived from (103) by neglecting the higher order potentials. For size ratios $q < 0.154$ it is known (Gast et al., 1983) that such contributions are identically zero. Therefore, the mapping of the two-component model with no interactions among the small particles onto the effective one-component Hamiltonian with pairwise Asakura–Oosawa potentials is exact.

This Hamiltonian, including the volume terms, has been used in simulations (Dijkstra et al., 1999a) to determine phase diagrams. For $\rho_2 = 0$ there are only hard sphere colloids, for which a single fluid phase exists up to a colloid volume fraction $\phi_c = 0.49$ and a solid phase above $\phi_c = 0.54$ with fluid–solid coexistence in between. The simulation results show that the addition of the second component widens the fluid–solid coexistence region substantially. For small size ratios, $q \leq 0.4$, and high concentrations ρ_2 a fluid phase of colloids exists only at fairly low values of ϕ_c. When q is increased ($q \approx 0.5$) a critical point develops between two fluid phases, qualitatively similar to the van der Waals theory of dense gases; a fluid phase dilute in colloids is in coexistence with a fluid phase more concentrated in colloids. This result is remarkable since the two-component system has only purely repulsive interactions, and it is only through the formal contraction procedure to the effective one-component system that the effect of the second component manifests itself by the appearance of the attractive interactions which are necessary for the van der Waals argument for the existence of fluid–fluid coexistence. Since the interactions in the original system, Eqs. (105) and (106), are only between hard spheres, there are no internal energy contributions to the free energy, so that the origin of these effects is purely entropic.

Dijkstra et al. (1999a) have, moreover, found isostructural solid–solid transitions at high colloid concentrations, a coexistence between two face-centered-cubic colloidal crystal phases. They also calculated pair distribution functions and static structure factors for the colloidal particles, and assessed the accuracy of earlier theories by Gast et al. (1983) and Lekkerkerker et al. (1992). There are some experimental data on structure factors for colloidal–polymer mixtures (Ye et al., 1996; Mossaid et al., 1999) which show rather similar characteristics to those obtained from the Asakura–Oosawa pair potential.

Effective potentials and the resulting phase behavior have also been studied on the basis of Eq. (104) for mixtures of big and small hard spheres. Now the particles of species 2 interact with each other through their excluded volumes. This gives rise to correlation effects which are reflected in the depletion interaction between the big spheres. As a result the effective potential deviates significantly from the Asakura–Oosawa form, in particular when the concentration of small spheres is increased. The effective potential develops repulsive contributions and has oscillatory behavior as a function of separation; it also becomes more long-ranged (Mao et al., 1995; Götzelmann et al., 1998). Concerning the phase behavior of hard-sphere binary mixtures it has been shown that although a fluid–fluid transition occurs for small size ratios, this remains metastable with respect to a broad fluid–solid transition (Dijkstra et al., 1998, 1999b,c).

As long as the interactions are of hard-sphere type there are no internal energy contributions to the free energy since the excess internal energy of hard-sphere systems is identically equal to zero. Therefore, the depletion effects and the resulting phase behavior are purely entropy driven. When the particles interact via soft potentials in addition to the excluded volume, there are energy-driven depletion forces in addition to those of entropic origin. A model of this kind has been studied where charges are put on the small particles whereas the big spheres are still uncharged. This investigation uses an approach to depletion effects different from the one described so far (Mendez-Alcaraz and Klein, 2000). It starts from the coupled OZ equations (84) for the binary mixture and rewrites them as a single OZ equation for the effective system. From the direct correlation function and the use of an approximate closure relation one obtains the effective interaction between the big hard spheres.

For charged small spheres the effective interaction between the uncharged big spheres is repulsive at contact in contrast to the case of uncharged small spheres considered above. This kind of behavior is to be expected since the big hard spheres represent a volume without charge, and the small charged spheres therefore try to be at contact with the big particles. This layer of charge around each of the big spheres is the origin of the repulsive interaction at contact. For separations such that one small sphere fits in between two large ones the interaction is attractive, and at larger separations it is oscillatory. The particular microstructure of the small spheres, due to their Yukawa repulsions, reflects itself in the effective interaction between the big uncharged hard spheres. A detailed understanding of the depletion effects has to consider the three-dimensional non-spherical distribution of the small charged spheres in the space between and around two big ones; this has been done by density-functional theory (von Grünberg and Klein, 1999).

8. Concluding remarks

As mentioned at the beginning, this chapter is meant to give an introduction to methods to treat the effects of interactions among colloidal particles on structural properties of suspensions. Because of the introductory character several important aspects of the effects of interactions have not been mentioned. One of them is the progress made by computer simulations. A review by Dijkstra (2001) discusses recent results on charge and sterically stabilized suspensions.

Another aspect is the dynamics of suspended particles, which can be measured by dynamic light scattering, diffusing-wave spectroscopy and neutron-spin echo experiments. The dynamics is also important to understand rheological properties and the glass transition. The direct potential interactions between particles, which determine the static structure factor, have of course also important consequences on collective and self-diffusion. But in addition thre are hydrodynamic interactions, which arise from the presence of the solvent. A diffusing particle creates a long-ranged velocity field in the solvent which influences the motion of other particles. The dynamics of one particle is therefore affected by the dynamics of another one, even when the distance between them is larger than the range of direct interactions. The dynamics of interacting suspensions of colloids and polymers can only be properly described by taking both kinds of interactions into account. The interested reader is referred to the book by Dhont (1996) and to the articles by Nägele (1996) and by Klein (2002).

References

Asakura, S. and Oosawa, F. (1954). *J. Chem. Phys.*, **22**, 1255.

Balescu, R., (1975). *Equilibrium and Nonequilibrium Statistical Mechanics*. Wiley, New York.

Belloni, L. (1998). *Colloids Surfaces A*, **140**, 227.

Berne, B.J. and Pecora, R. (1976). *Dynamic Light Scattering*. Wiley, New York.

Carnahan, N.F. and Starling, K.E. (1969). *J. Chem. Phys.*, **51**, 635.
Crocker, J.C. and Grier, D.G. (1994). *Phys. Rev. Lett.*, **73**, 352.
D'Aguanno, B. and Klein, R. (1991). *J. Chem. Soc. Faraday Trans.*, **87**, 379.
Derjaguin, B. and Landau, L. (1941). *Acta Phys. Chim. URSS*, **14**, 633.
Dhont, J.K.G. (1996). *An Introduction to Dynamics of Colloids*. Elsevier, Amsterdam.
Dijkstra, M. (2001). *Current Opin. Colloid Interface Sci.*, **6**, 372.
Dijkstra, M., van Roij, R. and Evans, R. (1998). *Phys. Rev. Lett.*, **81**, 2268.
Dijkstra, M., Brader, J.M. and Evans, R. (1999a). *J. Phys.: Condens. Matter*, **11**, 10079.
Dijkstra, M., van Roij, R. and Evans, R. (1999b). *Phys. Rev. Lett.*, **82**, 117.
Dijkstra, M., van Roij, R. and Evans, R. (1999c). *Phys. Rev. E*, **59**, 5744.
Gast, A.P., Hall, C.K. and Russel, W.B. (1983). *J. Colloid Interface Sci.*, **96**, 251.
Götzelmann, B., Evans, R. and Dietrich, S. (1998). *Phys. Rev. E*, **57**, 6785.
Hamaker, H.L. (1937). *Physica*, **4**, 1058.
Hansen, J.P. and McDonald, I.R. (1986). *Theory of Simple Liquids*. Academic Press, San Diego.
Ise, N., Okubo, T., Sugimura, M., Ito, K. and Nolte, H.J. (1983). *J. Chem. Phys.*, **78**, 536.
Ito, K., Hiroshi, H. and Ise, N. (1994). *Science*, **263**, 66.
Kaplan, P.D., Faucheux, L.P. and Libchaber, A.J. (1994). *Phys. Rev. Lett.*, **73**, 2793.
Kerker, M. (1969). *The Scattering of Light and Other Electromagnetic Radiation*. Academic Press, New York.
Klein, R. and D'Aguanno, B. (1996). In: *Light Scattering, Principles and Development* (Brown, W., ed). Oxford.
Lekkerkerker, H.N.W., Poon, W.C.K., Pusey, P.N., Stroobants, A. and Warren, P.B. (1992). *Europhys. Lett.*, **20**, 559.
Löwen, H., Hansen, J.P. and Madden, P.A. (1993). *J. Chem. Phys.*, **98**, 3275.
Mao, Y., Cates, M.E. and Lekkerkerker, H.N.W. (1995). *Phys. A*, **222**, 10.
McQuarrie, D.A. (1976). *Statistical Mechanics*. Harper Collins Publishers, New York.
Mendez-Alcaraz, J.M. and Klein, R. (2000). *Phys. Rev. E*, **61**, 4095.
Mendez-Alcaraz, J.M., D'Aguanno, B. and Klein, R. (1992). *Langmuir*, **8**, 2913.
Mossaid, A., Poon, W.C.K., Pusey, P.N. and Soliva, M.F. (1999). *Phys. Rev. Lett.*, **82**, 225.
Nägele, G. (1996). *Phys. Rep.*, **272**, 215.
Nägele, G., Kellerbauer, O., Steininger, B. and Klein, R. (1993). *Phys. Rev. E*, **47**, 2562.
Klein, R. (2002). In: *Structure and Dynamics of Polymer and Colloidal Systems* (Borsali, R. and Pecora, R., eds.). Kluwer, Dordrecht.
Ohshima, Y.N., Sakagami, H., Okumoto, K., Tokoyoda, A., Igarashi, T., Shintaku, K.B., Toride, S., Sekino, H., Kabuto, K. and Nishio, A.I. (1997). *Phys. Rev. Lett.*, **78**, 3963.
Philipse, A.P. and Vrij, A. (1988). *J. Chem. Phys.*, **88**, 6459.
Rudhardt, D., Bechinger, C. and Leiderer, P. (1998). *Phys. Rev. Lett.*, **81**, 1330.
Rudhardt, D., Bechinger, C. and Leiderer, P. (1999). *J. Phys.: Condens. Matter*, **11**, 10073.
Rogers, F.J. and Young, D.A. (1984). *Phys. Rev. A*, **30**, 999.
Schmitz, K.S. (1993). *Macroions in Solution and Colloidal Suspensions*. VCH, New York.
Tata, B.V.R., Yamahara, E., Rajamani, P.V. and Ise, N. (1997). *Phys. Rev. Lett.* **78**, 2660.
von Grünberg, H.H.and Belloni, L. (2000). *Phys. Rev. E*, **62**, 2493.
van Roij, R., Dijkstra, M. and Hansen, J.P. (1999). *Phys. Rev. E*, **59**, 2010.
Verwey, E.J.W. and Overbeek, J.T.G. (1948). *Theory of the Stability of Lyotropic Colloids*. Elsevier, Amsterdam.
von Grünberg, H.H. and Klein, R. (1999). *J. Chem. Phys.*, **110**, 5421.
Vondermassen, K., Bongers, J. Müller, A.and Versmold, H. (1994). *Langmuir*, **10**, 1351.
Vrij, A. (1976). *Pure Appl. Chem.*, **48**, 471.
Ye, X., Narayanan, T., Tong, P., Huang, J.S., Lin, M.Y., Carvalho, B.L. and Letters, L.J. (1996). *Phys. Rev. E*, **54**, 6500.

Neutrons, X-Rays and Light
P. Lindner and Th. Zemb (Editors)

Chapter 15

Monte Carlo Simulation Techniques Applied in the Analysis of Small-Angle Scattering Data from Colloids and Polymer Systems

Jan Skov Pedersen

Department of Chemistry, University of Aarhus, Langelandsgade 140, DK-8000 Aarhus C, Denmark

1. Introduction

Analysis of small-angle scattering data by the modelling approach (see Chapter 16 by J.S. Pedersen) is in principle straightforward if an analytical expression for the scattering cross-section is available. It is then possible to optimize the structural parameters using a standard least-squares approach (Pedersen, 1997) and obtain information on the size, structure and interactions of the objects in the sample. In many cases analytical expressions are not available and in these situations the Monte Carlo simulation technique can be of great help. This is the case for many polymer and colloid systems with interactions of varying complexity (Pedersen and Gerstenberg, 1996; Yoshisaki and Yamakawa, 1980; Pedersen and Schurtenberger, 1996, 1999; Cannavacciuolo et al., 2000, 2002; Svaneborg and Pedersen, 2000, 2002).

Scattering functions can be generated for the model structures and some structural parameters, such as radius, chain length, etc. can be varied. Subsequently the scattering functions obtained from simulations can be parameterized as a function of the parameters, so that numerical expressions and routines are available, which can be used for model fitting (Pedersen and Schurtenberger, 1996; Jerke et al., 1997, 1998; Svaneborg and Pedersen, 2000, 2002). Another very important application of Monte Carlo simulations is for checking the results of analytical calculations and for investigating the influence of various approximations made in the calculations (Pedersen and Gerstenberg, 1996; Svaneborg and Pedersen, 2000). Simulation techniques have also been used for investigating instrumental smearing effects in SANS (Pedersen, 1994) and simulations of the entire SANS experiment (Wignall et al., 1988). The most recent application within the field of small-angle scattering is the use of the closely related simulated annealing technique for determining the structure of biological macromolecular structures (Svergun, 1999).

In the following, the basic principles of the Monte Carlo simulation technique are outlined with emphasis on the determination of scattering functions. A few recent applications are presented.

2. Monte Carlo simulations

The Monte Carlo (MC) simulation technique is very helpful in studies of polymer systems with randomness due to the many internal degrees of freedom in such systems. Even for a system consisting of a single polymer chain with excluded interactions it is very difficult to obtain rigorous analytical results concerning the general statistics and the scattering functions. The MC simulation can be used for generating series of allowed configurations and for obtaining the configurational average values of various parameters, like radius of gyration and end-to-end distance, which are sampled during the simulation. In this way insight into the statistical behaviour is obtained. It is of course also possible to use the generated polymer configurations for calculating the scattering function.

Most MC algorithms have a dynamical character, which means that one performs an operation (a move) on a current polymer configuration to get to the next (new) configuration. The most usual move is one in which one end of the polymer is chosen at random and a certain fraction of segments is cut away. By generating a new random part of the same length at the opposite end, a new configuration is obtained. This is called a 'reptation', a 'cut-and-past', or a 'slithering snake' move. Another type of move is the pivot move in which a randomly selected bond is chosen and the torsion angle at this bond is changed, so that a part of the chain is rotated as a rigid body. Some background literature can be found in e.g., Stellman and Gans (1972), Baumgärtner, (1992), Kremer and Binder, (1988), Binder (1995), and Frenkel and Smit (1996).

The simplest MC simulation is based on a few basic principles:

(i) All configurations should be generated with equal probability.
(ii) The probability of going from state A to state B should be identical to the probability of going from state B to state A (the principle of detailed balance).
(iii) For two states, A and B, it should always be possible to come from state A to state B (the principle or ergodicity).
(iv) If a new state is forbidden, for example due to chain overlap, the old state is kept and counted once more when averages are calculated.

For systems with hard-sphere interactions it is straightforward to decide whether a new state is accepted: the overlap between all pairs of spheres are checked and the new state is accepted if no overlaps are detected. For systems with a finite interaction potential, like e.g., Lennard-Jones or Coulomb potentials, the so-called Metropolis criterion is used (Metropolis et al., 1953). It states that a new configuration is always accepted if it has a lower energy than the old one. If it has a higher energy, it is accepted with the probability $\exp(-\Delta E/kT)$, where ΔE is the energy difference between the new and the old state and kT is the Boltzmann constant multiplied by the absolute temperature. This approach gives configurations that are weighted by the Boltzmann factor $\exp(-E_A/kT)$, where E_A is the energy of state A, so that thermodynamic averages can be calculated as simple averages over the configurations. Note that there are several other methods which use biased configurations with corrections for the bias in the calculation of the averages (Kremer and Binder, 1988; Frenkel and Smit, 1996).

As the simulations involve repetitions of the same operations and calculations of typically 10^6 times or more, it is important to use efficient algorithms. Usually the most time consuming part of a simulation is the calculation of energies (or equivalently, the check for overlaps) as this, in principle requires N^2 operations for a systems with N particles or subparticles.

The simulation execution time can often be greatly reduced if some tricks are employed. For example for a single chain, one can use the fact that the chain is a connected structure. Let i and j be the indices of two spheres on the chain. If these have been found to be separated by the distance d, which is larger than the separation of the spheres along the contour, one does not have to check for overlap between, for example, i and $j+1$. One can make a jump from j along the contour of length $d-2R$, as this is the first sphere after j which can be close to the sphere with index i. This is known as the 'zippering' method and scales approximately as $N^{1.2}$, which has an enormous advantage for long chains (Stellman et al., 1971; Stellman and Gans, 1972; Pedersen et al., 1996). For many-chain and many-particle simulations in a box with periodic boundary conditions, the 'minimum image convention' gives a very convenient way of deciding between which particles interactions should be considered (see, for example, Allen and Tildesley, 1987). It is sometimes possible to reduce the computation time significantly by using neighbouring tables, which are constantly updated.

Often simulations are started from a biased configuration which is not representative of the equilibrium configurations. This is for example the case if one generates a polymer chain from one end allowing several attempts at each new step in order to get a self-avoiding configuration. With such a biased configuration it is necessary to run the simulation for a certain period in order to get an equilibrium configuration before the sampling is started. The extent of the equilibrium period depends on the relaxation times of the system and the moves that are used. For a single chain with N points, reptation moves with M points moved in each move it is of magnitude $(N/M)^2$, whereas it is only of magnitude N for the pivot moves.

As a rule of thumb, a MC simulation is most efficient in sampling the configuration space, if the fraction of accepted moves is not smaller than 50%. If the fraction is significantly lower, it means that an accepted configuration is sampled many times, which is of course inefficient. If the fraction is significantly larger, it could mean that the new configuration only differs marginally from the old one, which is also inefficient.

For most simulations it is important that a good random number generator is used. This usually means that the system supplied random number generator cannot be used. The quality of the generator is particular important if only a small fraction (typically $< 1\%$) of the attempted moves are accepted: The simulation can very easily turn into a test of the generator, rather than a real simulation. The generators described in *Numerical Recipes* (Press et al., 1989) have an acceptable quality for most purposes.

For a system consisting of independent objects (i.e., at infinite dilution) with random orientation, the scattering function can be calculated using the Debye equation:

$$S(q) = \frac{1}{N^2} \sum_{i,j} \frac{\sin qr_{ij}}{qr_{ij}}, \tag{1}$$

where N is the number of scattering points on the object and r_{ij} is the distance between the points with indices i and j. Note that the normalization $S(q=0)=1$ has been chosen. As the calculation of the sine term is computationally expensive the calculation can be speeded up by using a histogram $p(r_i)$ of distances. The scattering function can then be obtained after the simulation is done as

$$S(q) = \sum_{i} p(r_i) \frac{\sin qr_i}{qr_i}. \tag{2}$$

The calculation of $p(r_i)$ required the same number of calculations, namely N^2, as the number of terms is the same as in (1), however, a calculation of the sine function for each term is not required and this saves time.

For interacting particles, a super cell with periodic boundaries is used as this reduces the effects of the boundaries and the finite size of the system. For such a cell the scattering function should only be calculated in the reciprocal lattice points of the unit cell. In this way one does not try to derive information on the structure for distances that are larger than the length of unit cell. The 'density' of reciprocal lattice points becomes quite large for large q and therefore it is usual to calculate the function only in representative directions, like the 001, 110 and 111 directions and their equivalents (Frenkel et al., 1986). This gives in total 13 directions. For each of the directions (h,k,l) the scattering function is then calculated as

$$S(q_p) = \left| \sum_{j=1}^{N} \exp\left[-ip2\pi(hx_j + ky_j + lz_j)/L\right] \right|^2, \tag{3}$$

where L is the side length of the unit cell, (x_j, y_j, z_j) is the position of the jth point and $q_p = p(2\pi/L)(h^2+k^2+l^2)^{1/2}$ is the length of the scattering vector, with $p = 1,2,\ldots$. The scattering functions from the different equivalent directions are averaged and finally the functions from different directions are merged. Note that the use of Eq. (3) requires only $13N$ calculations as opposed to the N^2 for Eq. (1). For large structures and many particles it is therefore an advantage to use Eq. (3) and define a unit cell of sufficient size even for single particle simulations (Cannavacciuolo et al., 2000). The use of Eq. (3) can furthermore be combined with Fast Fourier Transform (FFT) routines (Frenkel et al., 1986) which also speeds up the calculations. However, in many cases it is found that the discretization involved in using FFT gives numerical errors which cannot be accepted (Pedersen, 1995). It is in stead recommendable to apply the addition formulas for sine and cosine in (3), which can also give a significant reduction in the computational times. Note that another efficient approach for single particle simulations also based on (3) has been described by Svaneborg and Pedersen (2000).

The dynamical Monte Carlo simulation algorithms give configurations that are correlated in the sense that a new configuration has some parts and properties which are identical to the old one. This means that some quantities like the radius of gyration and the scattering function need not to be calculated and sampled in each step. This is known as 'stroboscopic sampling' and often reduces the simulation times significantly. When calculating the errors on a sampled quantity, the possible presence of correlation has to be taken into account. The easiest way is to split a simulation into several blocks (10–20, typically) each having a length which is significantly larger than the relaxation time of the system (Flyvbjerg and Petersen, 1989). In this case the blocks are individual independent observations and the standard error on the average $\langle A \rangle$ of a quantity A is given by

$$\sigma\left(\langle A \rangle\right)^2 = \frac{1}{M(M-1)} \sum_{m=1}^{M} \left[\langle A \rangle_m - \langle A \rangle\right]^2, \tag{4}$$

where M is the number of blocks and $\langle A \rangle_m$ is the average of the quantity A in the mth block. This equation can also be used for the scattering function for each of the q values.

When applying the MC technique for a particular problem it is important that the model is chosen with great care. The model has to include as many details as required for describing the measurements with the resolution these have. At the same time it should be as simple as possible, so that it allows simulations to be performed with sufficient statistical accuracy within a reasonable time. The accuracy should be better than that of the experimental data. For most systems, a coarse-grained model with effective interactions is used, which means that the solvent molecules are not specifically included in the simulation. When the simulations have been done, the statistics of the system is known and the scattering functions have been determined. In order to use the functions for analysing experimental data a parameterization of the results has to be done. The actual choice of approach for the parameterization depends on the system. The usual procedure is to fit the data for each set of input parameters (concentration, chain length, etc.) individually using empirical expressions with some input from insight in the physics of the system. One then looks at the systematic variation of the parameters in the parameterization and describe again by empirical expressions how these vary with the input parameters. Once this goal is accomplished a numerical expression is available and the experimental data can be analysed.

In the following two examples of applications are summarized. Details can be found in the original references.

3. Examples

3.1. Semi-flexible chains with excluded volume interactions

Large flexible micelles with a tubular cross-section structure behave essentially as semi-flexible polymers with excluded volume interactions (Schurtenberger and Cavaco, 1994). Analysis of the combined data from static light scattering and small-angle neutron scattering (SANS) experiments requires that scattering functions for semi-flexible chains with excluded volume interactions are available. In order to obtain such functions a Kratky–Porod type model (Kratky and Porod, 1949) with hard spheres along the chain was studied (Pedersen et al., 1996; Pedersen and Schurtenberger, 1996). The model has a fixed angle between bonds and free rotation around the bonds. The fixed angle results in local stiffness of the chain. The micelles typically have a statistical segment length of $b = 300$ Å and a radius of $R = 30$ Å, and in agreement with this the spheres for describing the model were chosen to have a radius of $R/b = 0.1$. The simulations were done using pivot moves and the check for overlap was performed using the zippering method.

The scattering functions with and without excluded volume effects are shown in Fig. 1. They are calculated using Eq. (2). At low q values the scattering exhibits a Guinier-type behavior. For longer chains an intermediate power law behavior of $q^{-1/\nu}$ is observed, where $\nu = 1/2$ for chains without excluded volume and $\nu = (1+\epsilon)/2 = (1+0.176)/2 = 0.588$ for chains with excluded volume. At short length scales the local stiffness of the chain influences the chain conformation and it behaves more like a stiff rod and therefore a q^{-1} behavior as for rods is found is observed at large q values.

The functions were parameterized by different methods (Pedersen and Schurtenberger, 1996) and have successfully been applied in the analysis of scattering data from wormlike micelles (Jerke et al., 1997, 1998) (see Fig. 2). Due to the equilibrium nature of the micelles, a

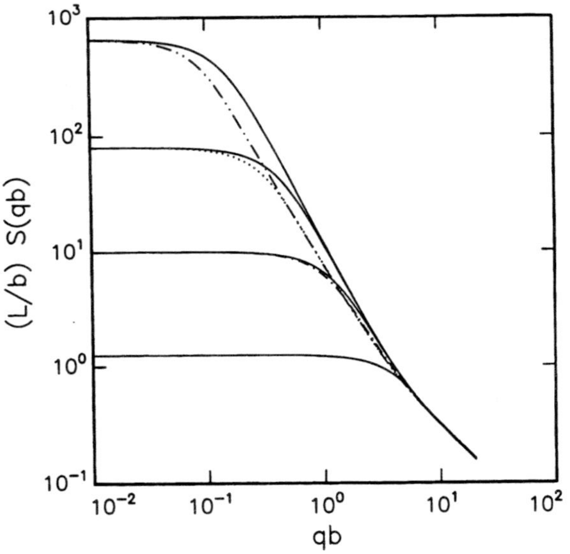

Fig. 1. Scattering functions for semi-flexible polymer chains for chains without excluded volume effects (full curve) and for chains with excluded volume effects (broken, dotted and dashed-dotted curves). From below: $L = 1.25b$, $L = 10b$, $L = 80b$, and $L = 640b$, where L is the contour length of the chains.

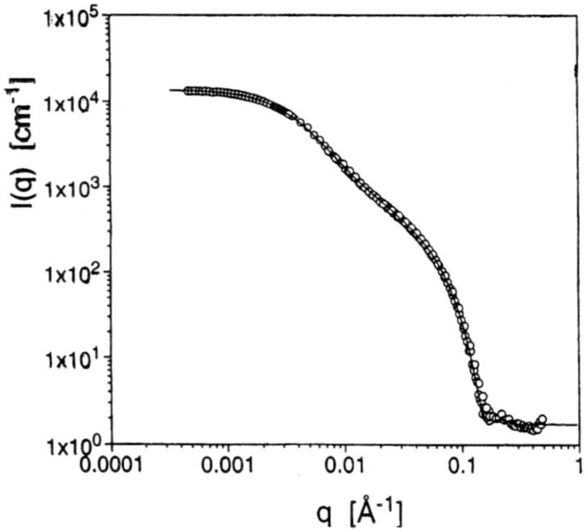

Fig. 2. Scattering data from inverse micelles of lecithin in deuterated isooctane. The micelles have a small water core. The curve is the fit using a model which includes micellar size polydispersity and a finite cross-section size (Jerke et al., 1997).

size polydispersity was included in the analysis using an exponentially decaying number size distribution. Finally a cross-section form factor was included. The fits provided information on the contour length of the micelles, the Kuhn length, and the cross-section radius. The good agreement between model and data furthermore confirms that the data is in agreement with an exponential number size distribution.

It has been shown that the chosen value of R/b gives rise to interactions which are similar to those found for polystyrene (PS) in a good solvent (Pedersen et al., 1996). This has been confirmed by the successful analysis (Pedersen et al., 1996; Pedersen and Schurtenberger, 1996) of SANS data for PS in carbon disulfide at infinite dilution (Rawiso et al., 1980). The applicability of the model to PS has been demonstrated further in a study of semi-dilute solutions of PS in toluene (Pedersen and Schurtenberger, 1999) and of partially labelled PS stars in tetrahydrofuran and cyclohexane (Pedersen et al., 1998). The simulations on semi-dilute solutions required use of a box with periodic boundary conditions and the use of reptation moves at high concentrations.

3.2. Block copolymer micelles

A block copolymer consists of two or more chemically distinct polymers which are linked together by a chemical bond. If the block copolymer is placed in a solvent, which is a good solvent for one block and a poor solvent for the other, micelles will form. These will usually have a compact core of the block for which the solvent is poor, surrounded by a corona of the blocks for which the solvent is good. The chains in the corona will have configurations that are similar to those for a polymer dissolved in a good solvent, however, with some perturbations due to the core-chain and chain-chain interactions.

The form factor of a micelle with a spherical homogeneous core with Gaussian chains attached to the surface has been calculated analytically (Pedersen and Gerstenberg, 1996). The analytical results were checked by performing Monte Carlo simulations on the same model. The simulations were done in such a way that a contrast variation series could be calculated in which the scattering of the chains relative to the core was varied. The simulated scattering data could be closely reproduced by calculations done with the analytical expression.

As there are no interactions between the chains and the core, the chains are free to penetrate into the core region. This is believed to be an unfavourable situation for a real system and therefore non-penetration of the chains should be included in the model. It is difficult, if not impossible, to do this in the analytical calculations and therefore simulations were performed. The chains that penetrated the core were simply rejected and chains were generated until the desired number of configurations was reached. A contrast variation series for the form factor was calculated (see Fig. 3). The data were analysed by the analytical expression with the extra freedom that the position of the chains relative to the core surface could be varied. The model reproduced the data nicely when the chains were moved a certain distance away from the surface which was equal to the radius of gyration of the chains.

Pedersen and Gerstenberg (1996) used the model for analysing SANS and SAXS data (Glatter et al., 1994) for the triblock copolymer polyethyleneoxide–polypropyleneoxide–polyethyleneoxide (PEO–PPO–PEO) in water at 50°C. A new analysis of the data on an absolute scale with the model expressions of Pedersen and Gerstenberg is presented in the chapter on modelling. Note that a more elaborate model with self- and mutually-avoiding semi-flexible chains with core expulsion is now available (Svaneborg and Pedersen, 2002).

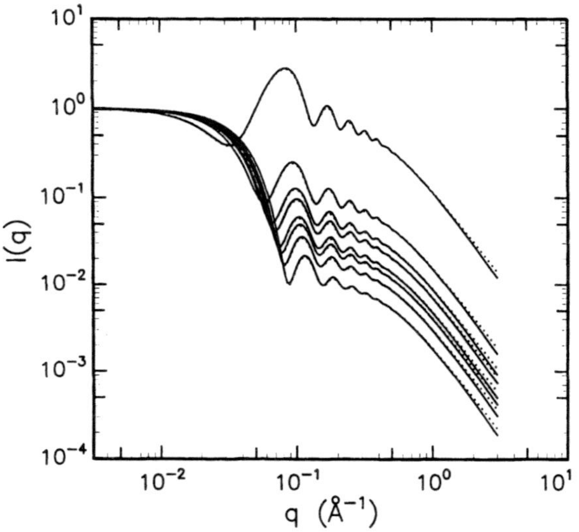

Fig. 3. Contrast variation for the form factor of spherical micelles with non-penetrating chains at the surface. The contrast of the chains in the corona relative to that of the core is varied. The simulation results are the full curves and the broken curves are fits with the analytical expression (Pedersen and Gerstenberg, 1996).

4. Summery and conclusions

Monte Carlo simulation techniques are well-suited for studying polymer and colloidal systems with complex structure and interactions. There are numerous studies in the literature [see, e.g., Binder (1995)], however, only few of them includes calculation of the scattering functions, which make their usefulness quite limited in connection with interpretation of small-angle scattering data. Another rather obvious problem can be that the published simulations are not for exactly the same system that has been investigated experimentally. There might be differences in the molecular weight, concentration, interactions, etc. All of these parameters can be controlled if the simulations are performed by the experimentalist himself. It is therefore important to realize that the implementation and execution of Monte Carlo simulations is not more complicated than it can actually be done by an experimentalist. (In fact some scientists like to call computer simulations for 'computer experiments'.) The experimentalist has in this connection the advantage of being already familiar with data analysis and modelling. It should be kept in mind that the simulations do not always solve a problem in a theoretical sense. However, the simulations provide results for a systems with well-defined objects and well-defined interactions and application of the results in analysis of experimental data can often give insight into the fundamental properties of the system. The application in analysis of experimental data is obviously greatly facilitated if numerical parameterizations and interpolations of the simulation results are available.

The basic principles of Monte Carlo simulations have been reviewed in this chapter and applications to polymer and colloidal systems have been discussed. Hopefully this can give a good starting point for readers that are interested in entering the field. One should not forget that the literature contains a huge amount of papers on simulations of polymer and colloidal systems and that these papers comprises many decades of developments and experiences with the technique. It is therefore worthwhile to consult the literature when choosing the appropriate models and simulation techniques for a particular problem.

5. Acknowledgements

The stimulating collaborations with Carsten Svaneborg, Peter Schurtenberger and Michael C. Gerstenberg in the work described in these notes are gratefully acknowledged. Comments from Carsten Svaneborg and Lise Arleth to the chapter are also acknowledged.

References

Allen, M.P. and Tildesley, D.J. (1987). *Computer Simulations of Liquids*. Clarendon Press, Oxford.

Baumgärtner, A. (1992). In: *Application of the Monte Carlo Simulation Method in Condensed Matter Physics* (Binder, K., ed). Springer, Berlin.

Binder, K. (ed.) (1995). *Monte Carlo and Molecular Dynamics Simulations in Polymer Science*. Oxford University Press, Oxford.

Cannavacciuolo, L., Sommer, C., Pedersen, J.S. and Schurtenberger, P. (2000). *Phys. Rev. E*, **62**, 5409–5419.

Cannavacciuolo, L., Pedersen, J.S. and Schurtenberger, P. (2002). *Langmuir*, **18**, 2922–2932.

Flyvbjerg, H. and Petersen, H. G. (1989). *J. Chem. Phys.*, **91**, 461–466.

Frenkel, D. and Smit, B. (1996). *Understanding Molecular Simulations*. Academic Press, London.

Frenkel, D., Vos, R.J., de Kruif, C.G. and Vrij, A. (1986). *J. Chem. Phys.* **84**, 4625–4630.

Glatter, O., Scherf, G., Schillén, K. and Brown, W. (1994). *Macromolecules, 27*, 6046–6054.

Jerke, G., Pedersen, J.S., Egelhaaf, S.U. and Schurtenberger, P. (1997). *Phys. Rev. E*, **56**, 5772–5788.

Jerke, G., Pedersen, J.S., Egelhaaf, S.U. and Schurtenberger, P. (1998). *Langmuir*, **14**, 6013–6024.

Kratky, O. and Porod, G. (1949). *Rec. Trav. Chim. Pays-Bas*, **68**, 1106–1122.

Kremer, K. and Binder, K. (1988). *Computer Phys. Rep.*, **7**, 259–310.

Metropolis, N, Rosenbluth, A.W., Rosenbluth, M.N., Teller, A.H. and Teller, E. (1953). *J. Chem. Phys.*, **21**, 1087–1092.

Pedersen, J.S. (1994). *J. Phys. IV, France, Coll. C8*, **3**, 491–498.

Pedersen, J.S. (1995). Numerical errors in the calculation of scattering functions from the use of Fast Fourier Transform methods. *Unpublished*.

Pedersen, J.S. (1997). *Adv. Colloid Interface Sci.*, **70**, 171–210.

Pedersen, J.S. and Gerstenberg, M.C. (1996). *Macromolecules*, **29**, 1363–1365.

Pedersen, J.S. and Schurtenberger, P. (1996). *Macromolecules*, **29**, 7602–7612.

Pedersen, J.S. and Schurtenberger, P. (1999). *Europhys. Lett.*, **45**, 666–672.

Pedersen, J.S., Laso, M. and Schurtenberger, P. (1996). *Phys. Rev. E*, **54**, R5917–R5920.

Pedersen, J.S., Almdal, K., Ndoni, S., Egelhaaf, S.U. and Schurtenberger, P. (1998). A small-angle neutron scattering study of partially labelled polystyrene star molecules in good and theta solvents. *Unpublished*.

Press, W.H., Flannery, B.P., Teukolsky, S.A. and Vetterling, W.T. (1989). *Numerical Recipes*. University Press, Cambridge.

Rawiso, M., Duplessix, R. and Picot, C. (1980). *Macromolecules*, **13**, 630–648.

Schurtenberger, P. and Cavaco, C. (1994). *J. Phys. Chem.*, **98**, 5481–5486.

Stellman, S.D. and Gans, P.J. (1972). *Macromolecules*, **5**, 516–526.

Stellman, G.D., Froimowitz, M. and Gans, P.J. (1971). *J. Comp. Phys.*, **7**, 178–181.

Svaneborg, C. and Pedersen, J.S. (2000). *J. Chem. Phys.*, **112**, 9661–9670.

Svaneborg, C. and Pedersen, J.S. (2002). *Macromolecules*, **35**, 1028–1037.

Svergun, D.I. (1999). *Biophys. J.* **76**, 2879–2886.

Wignall, G.D., Christen, D.K. and Ramakrishnan, V. (1988). *J. Appl. Cryst.*, **21**, 438–451.

Yoshizaki, T. and Yamakawa, H. (1980). *Macromolecules*, **13**, 1518–1525.

Neutrons, X-Rays and Light
P. Lindner and Th. Zemb (Editors)

Chapter 16

Modelling of Small-Angle Scattering Data from Colloids and Polymer Systems

Jan Skov Pedersen

Department of Chemistry, University of Aarhus, Langelandsgade 140, DK-8000 Aarhus C, Denmark

1. Introduction

The small-angle scattering (SAS) technique is a low-resolution method for determining structures on a length scale from a few thousand Ångströms down to 5–10 Å. SAS experiments give an intensity distribution in reciprocal space and usually considerable effort has to be invested in the data analysis in order to obtain the corresponding real-space structure. This effort should be reasonably matched to the effort and economical resources which are invested in obtaining the experimental data, which is not always the case, as one can see by going through the literature. The investments in time and money are often considerable as a large fraction of the experiments is performed at large scale facilities such as synchrotrons or nuclear reactors.

This chapter concerns modelling of small-angle scattering data from colloidal and polymer systems consisting of particles or molecules in solution. The systems are supposed to have short range order and isotropic scattering spectra, so that the scattering intensity is only a function of the modulus of the scattering vector.

Analysis of small-angle scattering data are usually done by either *model-independent* approaches or by *direct modelling* (Pedersen, 1997). Both approaches require the application of least-squares methods, which will only be briefly discussed in the present context. A detailed discussion of these methods based on the presentations by Bevington (1969) and by Press et al. (1989) and with emphasis on the application to small-angle scattering data can be found in Pedersen (1997).

The classical model-independent approaches are described in Chapter 5 by O. Glatter. The methods consist of an inverse Fourier transformation of the experimental scattering curve. which provides the pair distance distribution function $p(r)$. The inverse Fourier transformation is usually done by the Indirect Fourier Transformation (IFT) method introduced by Glatter (1977, 1980a). This method can be applied for all systems for which the correlations have a finite range. The $p(r)$ function provides real-space information and comparisons to model calculation may provide key information and give suggestions for the structure of the particles (Glatter, 1979, 1980a). After interpretation of the $p(r)$ function it may be possible to construct a model on an analytical form, which can be fitted to the data. For particles with centro symmetry it is possible to go one step further with the model-independent analysis and deconvolve $p(r)$, so that the radial scattering length density profile $\rho(r)$ is obtained (Glatter, 1981; Glatter and Hainish, 1984; Pedersen and Schurtenberger, 1996b and Chapter 5 by O. Glatter in this book). The information obtained by this procedure would of course also be incorporated in later attempts to perform model fits to the data.

For systems consisting of a collection of identical monodisperse particles, the shape of particles can be determined model independently by fitting directly to the scattering data. There exists a method which requires that the particles have a nearly homogeneous distribution of scattering length density. In this method a multipole expansion is used for the shape of the particles and the coefficients of this expansion are determined by non-linear least-squares methods (Stuhrmann, 1970; Svergun, 1994). The method has been modified so that it can be used for shape determination of the components of 'two-phase' particles (Svergun, 1994), and it has been applied successfully to the 50S subunit of the ribosome of *E. Coli* (Svergun et al., 1994a,b,c). In the latest development of the model-independent methods for structure and shape determination (Svergun, 1999), the particle is represented by a large collection of 'dummy atoms', which represent the various phases of the particles or solvent. Constraints on 'interconnectivity' and 'looseness' are employed to reduce the number of degrees of freedom. The fit to the experimental data is done using simulated annealing techniques.

For polydisperse systems, the aim of the analysis is to extract the size distribution of the particles when a particular shape of the particles is assumed. This can be done with a free-form size distribution (Glatter, 1980b; Pedersen, 1994), so that no assumptions on the distribution has to be invoked.

The present chapter is based on a previously published review (Pedersen, 1997) on analysis of small-angle scattering data, which also included a quite extensive review of least-squares methods. The part on the least-squares methods is not repeated here; however, the concepts related to the methods are so important that a brief discussion is given in the following section. The present chapter contains an update of the available model expressions, as well as an example of modelling of the scattering from block copolymer micelles in which the application of molecular constraints is demonstrated. However, it should be noted that a compilation of form and structure factors can never be complete. The expressions given in the present text therefore reflect the author's own interests and preferences to some extent.

2. Least-squares methods

The conventional approach for optimizing the agreement between a model expression and measured data is by means of least-squares methods. The *chi-squared* (χ^2) functional is used as a measure for the deviation between the model and the experimental data. Let $I^{\exp}(q_i)$, $i = 1, \ldots, N$ be the data points measured for the independent variable q_i. In a scattering experiment $I^{\exp}(q_i)$ is the measured intensities and q_i is the modulus of the scattering vector. The counting statistics will give rise to the statistical uncertainties σ_i on the data point $I^{\exp}(q_i)$. The chi-squared functional is defined as:

$$\chi^2 \equiv \sum_{i=1}^{N} \left(\frac{I^{\exp}(q_i) - I^{\mathrm{mod}}(q_i)}{\sigma_i} \right)^2, \tag{1}$$

where $I^{\mathrm{mod}}(q_i)$ is the model intensities which depend on the parameters a_i with $i = 1, \ldots, M$.

It is often convenient to consider the *reduced chi-squared* χ_r^2, which is given by:

$$\chi_r^2 = \frac{\chi^2}{N-M}, \tag{2}$$

where $N-M$ is the *number of degrees of freedom*. The optimum (most likely) set of parameter values for a model is determined by minimizing (1). A fit with $\chi_r^2 = 1$ is considered to be an ideal fit. Note that for $N \gg M$ a fit with $\chi_r^2 = 1$ has 'on average' $|I^{\exp}(q_i) - I^{\mathrm{mod}}(q_i)| = \sigma_i$, which means that the deviations are on average equal to the statistical uncertainties. A more rigorous discussion of the chi-square functional can be found in text books mentioned in the previous section.

Experimental scattering data are always influenced by instrumental smearing, and it may be necessary to include corrections for this in the data analysis, as already described in Chapter 6. Smearing correction can be included for both X-rays and neutrons. As the mathematical structure of the smearing equations is similar, the following description is restricted to neutron SAS.

Owing to the finite resolution of the instrument the scattering in a region around the nominal scattering vector $\langle q \rangle$ is probed. For small-angle neutron scattering, the distribution of scattering vectors q can be described by a resolution function $R(\langle q \rangle, q)$ (see, for example, Pedersen et al., 1990). Using this, the model function can be written as

$$I^{\mathrm{mod}}(\langle q \rangle) = \int R(\langle q \rangle, q) \frac{\mathrm{d}\sigma(q)}{\mathrm{d}\Omega}\, \mathrm{d}q, \tag{3}$$

where $\mathrm{d}\sigma(q)/\mathrm{d}\Omega$ is the ideal scattering intensity (equal to the cross section, see next section). The experimental data which enter the expression for chi-squared (1) should be written as $I(\langle q_i \rangle)$ as they are recorded for the nominal scattering vectors.

In most small-angle neutron scattering experiments, the data are recorded using more than one instrumental setting. Each setting corresponds to a set of values of the wavelength, wavelength resolution, collimation and sample-to-detector distance. Therefore each of the N_{set} settings has its own resolution function $R_j(\langle q \rangle, q)$, and it is therefore convenient to write

$$\chi^2 = \sum_{j=1}^{N_{\mathrm{set}}} \sum_{i=1}^{N(j)} \left(\frac{I_j^{\exp}(\langle q_i \rangle) - I_j^{\mathrm{mod}}(\langle q_i \rangle)}{\sigma_{ji}} \right)^2, \tag{4}$$

where the index j refers to which data set is considered. The model intensity is then

$$I_j^{\mathrm{mod}}(\langle q \rangle) = \int R_j(\langle q \rangle, q) \frac{\mathrm{d}\sigma(q)}{\mathrm{d}\Omega}\, \mathrm{d}q. \tag{5}$$

The actual calculation of the integral in (3) and (5) is done numerically. For small-angle neutron scattering experiments, the resolution function is well approximated by a Gaussian function (Pedersen et al., 1990; Pedersen, 1993; Barker and Pedersen, 1994) and only about 10 points are required for obtaining a good approximation of the integral. In an efficient implementation the resolution function is calculated once and for all in the beginning of the program, and stored for later use. In this way unnecessary repeated calculations are avoided. It should also be noted that although only about 10 points are required in the convolution, it means that the cross section should be evaluated for $10 \times N_m$ values, where N_m is the number of measured points. Therefore, if the expressions for the cross section are complex, the calculation times can be greatly reduced by making use of master curves and interpolations (Pedersen, 1993).

The least-squares problem for the indirect Fourier transformation is linear in the unknown parameters. This has the advantage that the optimum can be found without iterations by solving a set of N linear equations with N unknowns, which are the parameters describing the distribution. When determining free-form size distributions, the least-squares problem is also linear, however, the application of a non-negativity constraint requires that a limited number of iterations are performed (Lawson and Hanson, 1974; Pedersen, 1994).

Most other applications of the least-squares methods within the field of small-angle scattering are non-linear. This is the case for general modelling using the expressions given in the next section. There exists are large number of different iterative methods of varying complexity for minimizing chi-square for non-linear problems. All of these methods require that a set of starting values for the model parameters is provided. If these are not close enough to the true minimum values, the fit might get stuck in a local minimum. It is therefore advisable to try out different initial guesses. Furthermore, many of the advanced methods for minimizing chi-squared might not work reliable far from the minimum or if the parameters are strongly correlated. Therefore great care has to be taken to ensure that the fit has converged. It is always advisable to check this by performing a grid search in which the individual parameters are successively and repeatedly optimized one by one. Such a grid search is very time consuming and inefficient; however, it has the advantage that it will always be able to locate the closest local minimum if the search is repeated for many cycles even for correlated parameters.

3. Models

In this section some of the most common expressions for form factors and structure factors will be reviewed. It is impractical to give all the mathematical expressions, and for some of the models only the references to the original papers are given. The reader is encouraged always to look into the original papers in order to check the range of validity of the expressions (and for checking for possible typing errors in the present text). For a general introduction to scattering theory and definition, Chapter 1 by P.N. Pusey in this book should be consulted.

The differential scattering cross section $\mathrm{d}\sigma(q)/\mathrm{d}\Omega$ of a sample can be defined as the number of scattered neutrons or photons per unit time, relative to the incident flux of neutron or photons, per unit solid angle at q per unit volume of the sample. The flux is the number of neutron or photons per unit time and per unit area. This is a convenient way of expressing the scattering from a sample as it does not depend on the form or transmission of the sample. It has furthermore the advantage that molecular constraints can be used, as demonstrated in Section 4, which contains an example.

For a monodisperse collection of spherically symmetric particles the scattering cross section can be written as

$$\frac{\mathrm{d}\sigma(q)}{\mathrm{d}\Omega} = n\Delta\rho^2 V^2 P(q)S(q), \tag{6}$$

where n is the number density of particles, $\Delta\rho$ is the difference in scattering length density between the particles and the solvent/matrix, V is the volume of the particles, $P(q)$ is the particle *form factor* and $S(q)$ is the *structure factor*. The form factor describes the structure of the particles and fulfills $P(q) = 1$. The structure factor describes the interference of scattering from different particles and contains information on the interaction between the particles. For very dilute systems there is no position-position correlations and $S(q) = 1$ and the structure factor can be neglected. Chapter 14 by R. Klein in this book deals specifically with the determination of $S(q)$ by integral equation theory.

An alternative approach is to express the excess scattering contrast as excess scattering length per unit mass, $\Delta\rho_m$. Using this, (6) becomes

$$\frac{\mathrm{d}\sigma(q)}{\mathrm{d}\Omega} = \frac{c}{M}\Delta\rho_m^2 M^2 P(q)S(q) = c\Delta\rho_m^2 M\, P(q)S(q), \tag{7}$$

where M is the molecular mass of a particle and c is the concentration of solute. Note that $n = c/M$.

For dilute systems with $S(q) = 1$, it is straightforward to modify (6) and (7) to include size polydispersity. In this case (6) becomes

$$\frac{\mathrm{d}\sigma(q)}{\mathrm{d}\Omega} = \Delta\rho^2 \int_0^\infty D(R)\, V(R)^2 P(q,R)\, \mathrm{d}R. \tag{8}$$

where $D(R)$ is the number size distribution, $V(R)$ is the volume of a particle with radius R and form factor $P(q,R)$.

Polydispersity can be included for an arbitrary size distribution (as long as the integral is convergent) by performing the integration numerically. However, the normalized ($\int_0^\infty D(x)\mathrm{d}x = 1$) Schulz–Zimm distribution (Schulz, 1939; Zimm, 1948)

$$D(x) = \left(\frac{z+1}{x_{\mathrm{av}}}\right)^{z+1} \frac{x^z}{\Gamma(z+1)} \exp\left(-(z+1)\frac{x}{x_{\mathrm{av}}}\right) \tag{9}$$

allows polydispersity to be included analytically for many form factors (see, e.g., Sheu, 1992; Greschner, 1973). The distribution has the further advantage that it in many cases is a good approximation for theoretical size distributions as derived, for example, from thermodynamical theories. Note that the first moment of the distribution is x_{av} and the variance of the distribution is $\sigma^2/x_{\mathrm{av}}^2 = 1/(1+z)$. Higher moments of the distribution can easily be calculated.

The expressions (6) and (7) assumes spherical symmetry of the particle shape and the interactions. For anisotropic identical particles the cross section is

$$\frac{\mathrm{d}\sigma(q)}{\mathrm{d}\Omega} = \Delta\rho^2 V^2 \left[\sum_i F_i(q,\boldsymbol{e}_i)^2 + \frac{1}{N}\sum_{i,j} F_i(q,\boldsymbol{e}_i)F_j(q,\boldsymbol{e}_j)[S_{i,j}(q,\boldsymbol{e}_i,\boldsymbol{e}_j) - 1]\right] \tag{10}$$

where the sums are over all particles in the sample and $F_i(q,\boldsymbol{e}_i)$ is the amplitude of the form factor for the ith particle with orientation given by the unit vector $\boldsymbol{e}_i$. The $S_{i,j}(q,\boldsymbol{e}_i,\boldsymbol{e}_j)$ functions are the partial structure factors which depend on orientations and N is the number of particles. Note that the first term in (10) is proportional to the orientational averaged form factor: $P(q) \equiv \langle F^2(q)\rangle_o$.

For particles with only a small anisotropy, it can be assumed that the interactions are independent of the orientation and given by the average size of the particles. This leads to the decoupling approximation (Kotlarchyk and Chen, 1984)

$$\frac{\mathrm{d}\sigma(q)}{\mathrm{d}\Omega} = \Delta\rho^2 V^2 P(q)[1+\beta(q)(S(q)-1)], \tag{11}$$

where

$$\beta(q) = \langle F(q)\rangle_o^2 / \langle F^2(q)\rangle_o, \tag{12}$$

and $S(q)$ is the structure factor calculated for the average particle size defined as $R_{\mathrm{av}} = [3V/(4\pi)]^{1/3}$, where V is the particle volume.

For particles with a random character, like block copolymer micelles with a compact core surrounded by a corona of dissolved chains, an expression similar to (11) holds (Pedersen, 2001). Note that for particles with a spherical core, the potential is to a very good approximation independent of particle orientations and the structure factor then refers to the effective interaction potential between the particles. The expression for random structures is (Pedersen, 2001):

$$\frac{\mathrm{d}\sigma(q)}{\mathrm{d}\Omega} = n\Delta\rho^2 V^2 [P(q) + A(q)^2 (S(q)-1)], \tag{13}$$

where $P(q)$ is an ensemble average of the form factor over the conformations (and therefore also an average over orientations). $A(q)$ is the normalized ($A(q=0)=1$) Fourier transform of the ensemble-averaged radial excess scattering length density distribution. For details see form factor 35 in the following and the example in Section 4.

For polydisperse systems it is also not possible to write the scattering cross section as a product of a form factor and a structure factor. The scattering cross section has the form:

$$\frac{\mathrm{d}\sigma(q)}{\mathrm{d}\Omega} = \Delta\rho^2 \left[\int_0^\infty D(R)\,V(R)^2 F(q,R)^2 \,\mathrm{d}R + \frac{1}{N}\int_0^\infty\int_0^\infty D(R)V(R)D(R')V(R')F(q,R)F(q,R')[S(R,R',q)-1]\,\mathrm{d}R\,\mathrm{d}R' \right], \tag{14}$$

where $D(R)$ is the number size distribution, $V(R)$ is the volume of a particle with radius R and form factor amplitude $F(q,R)$, and $S(R,R',q)$ are partial structure factors. N is the number of particles and is given by:

$$N = \int D(R)\,\mathrm{d}R. \tag{15}$$

For systems with small polydispersities, a decoupling approach similar to the one for anisotropic particles (Kotlarchyk and Chen, 1984) can be used. It is assumed that interactions are independent of size. With this one obtains:

$$\frac{\mathrm{d}\sigma(q)}{\mathrm{d}\Omega} = \Delta\rho^2 \langle V^2 P(q)\rangle \big[1+\beta(q)(S(q)-1)\big], \tag{16}$$

where $\langle V^2P(q)\rangle = \int D(R)V(R)^2F(q,R)^2\,\mathrm{d}R$ and

$$\beta(q)=\left[\int D(R)V(R)F(q,R)\,\mathrm{d}R\right]^2\bigg/\left[\int D(R)V(R)^2F(q,R)^2\,\mathrm{d}R\right], \tag{17}$$

and $S(q)$ is the structure factor calculated for the average particle size given by

$$R_{\mathrm{av}}=\left(\int D(R)R^3\,\mathrm{d}R\bigg/\int D(R)\,\mathrm{d}R\right)^{1/3}. \tag{18}$$

Note that (14) and (16) can also be used for slightly anisotropic particles, if $F_i(q,R)$ is replaced by $\langle F_i(q,R)\rangle_o$ and $F_i(q,R)^2$ is replaced by $\langle F_i(q,R)^2\rangle_o$. They can also be used for particles with randomness due to internal degrees of freedom like block copolymer micelles with $F_i(q,R)$ replaced by $A(q,R)$ and $F_i(q,R)^2$ replaced by $P(q,R)$, where the dependence of R is explicitly written.

The opposite limit of the approximations as used for the decoupling approximation is used in the local monodisperse approximation (Pedersen, 1994). In this approach it is assumed that a particle of a certain size is always surrounded by particles with the same size. Following this the scattering is approximated by that of monodisperse sub-systems weighted by the size distribution:

$$\frac{\mathrm{d}\sigma(q)}{\mathrm{d}\Omega}=\Delta\rho^2\int D(R)V(R)^2F(q,R)^2S(q,R)\,\mathrm{d}R, \tag{19}$$

in which it has been indicated that the structure factor is for particles of size R. This approach works better than the decoupling approximation (16) for systems with larger polydispersities and higher concentrations. As compared to the decoupling approximation and the scaling approximation described below, it has the advantage that the cross section is linear in the size distribution.

A scaling approximation has recently been introduced by Gazzillo et al. (1999). It is assumed that the pair correlation functions are identical except for a scaling constant. This leads to the following expression:

$$\frac{\mathrm{d}\sigma(q)}{\mathrm{d}\Omega}=\Delta\rho^2\left[\int_0^\infty D(R)V(R)^2F(q,R)^2\,\mathrm{d}R+\frac{1}{N}\int_0^\infty\int_0^\infty D(R)V(R)D(R')V(R')\right.$$
$$\left.\times F(q,R)F(q,R')\left(\frac{R+R'}{2R_{\mathrm{av}}}\right)^3[S(q,R+R')-1]\,\mathrm{d}R\,\mathrm{d}R'\right], \tag{20}$$

where R_{av} is given by (18) and the (monodisperse) structure factor is evaluated for the radius $(R+R')/2$. Compared to the local monodisperse approximation, it gives a much better agreement for q values in the vicinity of the structure factor peak and at higher q, whereas it gives a worse agreement for $q\rightarrow 0$. Note that the expression is not linear in the size distribution and that it involves double integrations, which makes least-square fitting with this expression relatively slow. This latter problems has been overcome by Gazzillo et al. (1999) in their implementation by use of a master curve for the structure factor.

3.1. Form factors

In the following it will be assumed that the particles are randomly oriented in the sample so that the theoretical form factors for anisotropic particles have to be averaged over orientation. Note that for spherically symmetric objects the form factor can be written as $P(q) = F^2(q)$, where $F(q)$ is the amplitude of the form factor.

1. Homogeneous sphere: The form factor amplitude of a homogeneous sphere was calculated already in 1911 by Lord Rayleigh. For a sphere with radius R:

$$F_1(q,R) = \frac{3[\sin(qR) - qR\cos(qR)]}{(qR)^3}. \tag{21}$$

Sheu (1992) has also given analytical expressions for polydisperse spheres with the number distribution given by, respectively, a box function, a triangular function, an exponential distribution, a Gaussian distribution, a log-normal distribution, and a Schulz–Zimm distribution.

Bagger-Jörgensen et al. (1997) and Svaneborg and Pedersen (2002) [1] have given expressions for the form factor of a sphere with a scattering length density, which gradually decrease at the sphere surface following a half-Gaussian distribution (see form factor 39 and the example in Section 4 for an approximate expression).

2. Spherical shell: This form factor amplitude is obtained by subtracting the empty core with a proper weighting by the volumes:

$$F_2(q) = \frac{V(R_1)F_1(q,R_1) - V(R_2)F_1(q,R_2)}{V(R_1) - V(R_2)}, \tag{22}$$

where $V(R) = 4\pi R^3/3$ and R_1 and R_2 are the outer and inner radius of the shell, respectively. An infinitely thin shell with radius R has the form factor $F_2(q)' = \sin(qR)/(qR)$.

Gradzielski et al. (1995) have given the expressions for a shell with a Gaussian radial scattering length density profile.

3. Spherical concentric shells: This form factor amplitude is a generalization of the shell form factor. Let R_i, $i = 1, \ldots, N$ be the radii of the shells and ρ_i be their scattering densities. With this:

$$F_3(q) = \frac{1}{M_3}\left[\rho_1 V(R_1)F_1(q,R_1) + \sum_{i=2}^{N}(\rho_i - \rho_{i-1})V(R_i)F_1(q,R_i)\right], \tag{23}$$

where M_3 is the scattering mass or scattering volume of the particle, given by:

$$M_3 = \rho_1 V(R_1) + \sum_{i=2}^{N} V(R_i)(\rho_i - \rho_{i-1}). \tag{24}$$

Förster and Burger (1998) have recently considered core-shell particles with an algebraic radial scattering length density profile which is piecewise of the form $\rho(r) \propto r^\alpha$. The form factors are given in terms of hypergeometric functions.

[1] Some errors in the paper by Bagger-Jörgensen et al. (1997) are corrected.

4. Particles consisting of spherical subunits: The expression was derived by Debye in 1915. For a particle consisting of P subunits:

$$P_4(q) = \frac{1}{M_4^2} \sum_{i,j=1}^{P} M_3(i) M_3(j) F_3(q,i) F_3(q,j) \frac{\sin(q r_{ij})}{q r_{ij}}, \tag{25}$$

where $F_3(q,i)$ and $M_3(i)$ are the form factor and scattering mass of the ith particle, respectively. r_{ij} is the distance between the centers of the ith and the jth subunit, and

$$M_4 = \sum_{i=1}^{P} M_3(i). \tag{26}$$

5. Ellipsoid of revolution: This expression was determined by Guinier (1939). The averaging over orientations has to be done numerically. For the semi-axes $R, R, \epsilon R$:

$$P_5(q,R,\epsilon) = \int_0^{\pi/2} F_1\big[q, r(R,\epsilon,\alpha)\big]^2 \sin\alpha \, d\alpha, \tag{27}$$

where $r(R,\epsilon,\alpha) = R(\sin^2\alpha + \epsilon^2 \cos^2\alpha)^{1/2}$. It is straight forward to generalize (27) to particles consisting of concentric elliptical shells. $F_1(q,r)$ has to be replaced by $F_3(q)$ and the volumes by $V(R) = 4\pi\epsilon R^3/3$. Note, that the different shells can have different values of ϵ. The form factor of an infinitely thin elliptical shell is given by (27) with $F_1(q,r)$ replaced by $F_2(qr)'$.

6. Tri-axial ellipsoid: For this object two numerical integrations have to be performed in order to get the orientational average. For the semi-axes a, b, c (Mittelbach and Porod, 1962):

$$P_6(q,a,b,c) = \frac{2}{\pi} \int_0^{\pi/2} \int_0^{\pi/2} F_1\big[q, r(a,b,c,\alpha,\beta)\big]^2 \sin\alpha \, d\alpha \, d\beta, \tag{28}$$

where $r(a,b,c,\alpha,\beta) = \big[(a^2 \sin^2\beta + b^2 \cos^2\beta) \sin^2\alpha + c^2 \cos^2\alpha\big]^{1/2}$. The form factor for the tri-axial ellipsoid can be generalized in the same way as (27) for tri-axial ellipsoids consisting of concentric shells. The form factor of an infinitely thin elliptical shell is given by (28) with $F_1(q,r)$ replaced by $F_2(qr)'$.

7. Cube and rectangular parallelepipedons: Two orientational averages have to be performed. For the edge lengths a, b, c (Mittelbach and Porod, 1961a):

$$P_7(q,a,b,c) = \frac{2}{\pi} \int_0^{\pi/2} \int_0^{\pi/2} \left[\frac{\sin(qa \sin\alpha \cos\beta)}{qa \sin\alpha \cos\beta} \frac{\sin(qb \sin\alpha \sin\beta)}{qb \sin\alpha \sin\beta} \frac{\sin(qc \cos\alpha)}{qc \cos\alpha} \right]^2 \sin\alpha \, d\alpha \, d\beta. \tag{29}$$

8. Truncated octahedra: The equations for an oriented particle were given by Hendricks, Schelten and Schmatz (1974). The orientational average has to be done as in (29).

9. Faceted sphere: The equations for an oriented particle were given by Dubey (1990) (see also Dubey et al., 1991). The orientational average has to be done as in (29).

10. Cube with terraces: The equations for an oriented particle were given by Rodríguez et al., (1991). The orientational average has to be done as in (29).

11. Cylinder: The expression for a cylinder with radius R and length L was given by Fournet (1949):

$$P_{11}(q) = \int_0^{\pi/2} \left[\frac{2J_1(qR\sin\alpha)}{qR\sin\alpha} \frac{\sin(qL\cos\alpha/2)}{qL\cos\alpha/2} \right]^2 \sin\alpha \, d\alpha, \tag{30}$$

where $J_1(x)$ is the first order Bessel function. An expression for cylinders consisting of concentric shells can be constructed by an approach similar to the one used for spherical particles. The form factor of a infinitely thin cylindrical shell with closed ends is given by (30) with the terms in the square brackets replaced by (Pedersen, 2000)

$$\left[\frac{R}{(R+L)} \frac{2J_1(qR\sin\alpha)}{qR\sin\alpha} \cos(qL\cos\alpha/2) + \frac{L}{(R+L)} J_0(qR\sin\alpha) \frac{\sin(qL\cos\alpha/2)}{qL\cos\alpha/2} \right],$$

where $J_0(x)$ is the zeroth order Bessel function.

Note that the first term is from the ends of the cylinder and the last term is from the side (Pedersen, 2000). A cylinder with symmetric layers in the length direction can be obtained by replacing the $\sin(qL\cos\alpha/2)/(qL\cos\alpha/2)$ term by the appropriate difference terms as shown in the following for form factor 41 for a very anisotropic particle with local planer geometry.

12. Cylinder with elliptical cross section: This expression was given by Mittelbach and Porod (1961b) for a cylinder of length L and with cross section semi-axes a and b:

$$P_{12}(q) = \frac{2}{\pi} \int_0^{\pi/2} \int_0^{\pi/2} \left[\frac{2J_1(qr(a,b,\phi,\alpha))}{qr(a,b,\phi,\alpha)} \frac{\sin(qL\cos\alpha/2)}{qL\cos\alpha/2} \right]^2 d\phi \sin\alpha \, d\alpha, \tag{31}$$

where $r(a,b,\phi,\alpha) = [a^2\sin^2\phi + b^2\cos^2\phi]^{1/2}\sin\alpha$. An expression for cylinders consisting of concentric shells with elliptical cross section can be constructed by an approach similar to the one used for spherical particles and for tri-axial ellipsoids. The shells can have varying eccentricities (a/b).

As for cylinders with spherical cross section, cylinders with symmetric layers in the length direction can be obtained by replacing the $\sin(qL\cos\alpha/2)/(qL\cos\alpha/2)$ term by the appropriate difference terms as shown in the following for form factor 41 for a very anisotropic particle with local planer geometry.

13. Cylinder with hemi-spherical end caps: The equations were given by Cusack (1981). Expressions for a shell structure were also given.

14. Toroid: Förster et al. (1999) have given the form factor of a toroid with spherical cross section:

$$P_{14}(q) = \int_0^{\pi/2} F_{\mathrm{tor}}(q,\beta)^2 \sin\beta \, \mathrm{d}\beta, \tag{32}$$

where

$$F_{\mathrm{tor}}(q,\beta) = \frac{2\pi}{V} \int_{-R_c}^{R_c} \left[R_+ J_1(qR_+ \sin\beta) - R_- J_1(qR_- \sin\beta)\right] \frac{\cos(qz\cos\beta)}{(q\sin\beta)} \, \mathrm{d}z. \tag{33}$$

$V = 2\pi^2 R_c^2 R_t$ is the volume of the toroid with radius R_t and cross section radius R_c. Furthermore $R_+ = R_t + \sqrt{R_c - z}$ and $R_- = R_t - \sqrt{R_c - z}$.

15. Infinitely thin rod: The expression was determined by Neugebauer (1943):

$$P_{15}(q) = 2\,\mathrm{Si}(qL)/(q\,L) - 4\sin^2(qL/2)/(q^2L^2), \tag{34}$$

where

$$\mathrm{Si}(x) = \int_0^x t^{-1} \sin t \, \mathrm{d}t, \tag{35}$$

and L is the length.

16. Infinitely thin circular disk: The expression was determined by Kratky and Porod (1949a):

$$P_{16}(q) = \frac{2}{q^2R^2}\left[1 - \frac{J_1(2qR)}{qR}\right], \tag{36}$$

where R is the radius of the disk.

17. Fractal aggregates: An empirical expression for a mass fractals consisting of spheres with a radius R has been given by Teixeira (1988):

$$P_{17}(q) = \left[1 + \frac{1}{(qR)^D} \frac{D\Gamma(D-1)}{[1+1/(q^2\xi^2)]^{(D-1)/2}} \sin\left[(D-1)\tan^{-1}(q\xi)\right]\right] F_1^2(q), \tag{37}$$

where D is the fractal dimension, ξ is a cutoff length for the fractal correlations, and $\Gamma(x)$ is the gamma function. [Note that expressions for fractal surfaces have been given by Bale and Schmidt (1988).]

18. Flexible polymers with Gaussian statistics: Flexible polymer chains which are not self-avoiding and obey Gaussian statistics. Debye (1947) has calculated the form factor of such chains:

$$P_{18}(q) = 2[\exp(-u) + u - 1]/u^2, \tag{38}$$

with $u = \langle R_g^2 \rangle q^2$, where $\langle R_g^2 \rangle$ is the ensemble average radius of gyration squared: $\langle R_g^2 \rangle = (Lb)/6$, where L is the contour length and b is the statistical (Kuhn) segment length.

19. Polydisperse flexible polymers with Gaussian statistics: Polydispersity has been included in terms of a Schulz–Zimm mass distribution by Zimm (1948) and Greschner (1973):

$$P_{19}(q) = \frac{2\left[(1+Ux)^{-1/U} + x - 1\right]}{(1+U)x^2}, \tag{39}$$

with $x = \langle R_g^2 \rangle_z q^2/(1+2U)$, where $\langle R_g^2 \rangle_z$ is the z-averaged radius of gyration squared. In terms of the mass distribution $m(M)$ it is

$$\langle R_g^2 \rangle_z = \int Mm(M)\langle R_g(M)^2 \rangle \, \mathrm{d}M \Big/ \int Mm(M) \, \mathrm{d}M, \tag{40}$$

in which the mass dependence of the radius of gyration is written explicitly (note $L \propto M$). The parameter U is related to the polydispersity parameter z in (9) by $U = 1/z$. Furthermore $U = (M_w/M_n) - 1$, where M_w is the weight-averaged molecular mass (first moment of $m(M)$) and M_n is the number-averaged molecular mass.

20. Flexible ring polymers with Gaussian statistics: Zimm and Stockmayer (1949) have originally calculated the form factor:

$$P_{20}(q) = \left(\frac{2}{x^2}\right)^{1/2} D\left[\left(\frac{x^2}{2}\right)^{1/2}\right], \tag{41}$$

with $x = \langle R_g^2 \rangle q^2$, where $\langle R_g^2 \rangle$ is the ensemble average radius of gyration squared: $\langle R_g^2 \rangle = (Lb)/12$, where L is the contour length and b is the statistical (Kuhn) segment length. $D(x) = \exp(-x^2)\int_0^x \exp(t^2)\,\mathrm{d}t$ is the Dawson integral.

The result has been extended by Burchard, Michel and Trappe (1996) to connected ring structures: 'Twisted' ring structures and 'daisy-like' structures with an arbitrary number of rings.

21. Flexible self-avoiding polymers: Empirical expressions have been given by Utiyama et al. (1971). The parameters should be taken as $\epsilon = 0.176$, $t = 2/(1-\epsilon)$, and $s = 2.90$ (see Pedersen and Schurtenberger (1996a) which contain a simple approximation).

22. Polydisperse flexible self-avoiding polymers: The results given by Utiyama et al. (1971) have been extended to the polydisperse case by Förster and Burger (1998). They used a Schulz-Zimm number distribution. As for the monodisperse case, the parameters should be taken as $\epsilon = 0.176$, $t = 2/(1-\epsilon)$, and $s = 2.90$ (Pedersen and Schurtenberger, 1996a). Note that Förster and Burger use a notation in which $\theta = s - 4$. The result for the form factor is given as two series expansions for, respectively low q in powers of q and for high q in powers of $1/q$.

23. Semi-flexible polymers without self-avoidance: Numerical interpolation formula for the Kratky–Porod model (1949b) have been given by Yoshizaki and Yamakawa (1981). These have recently been corrected using results from Monte Carlo simulations (Pedersen and Schurtenberger, 1996a).

24. Semi-flexible polymers with self-avoidance: Numerical interpolation formula have been given by Pedersen and Schurtenberger (1996a). The results are given for $R/b = 0.1$, where R is the cross section radius and b is the Kuhn length. This corresponds to a reduced binary cluster integral of 0.3, which is similar to the value found for polystyrene in a good solvent.

25. Star polymer with Gaussian statistics: The expression was given by Benoit (1953). For a star with f arms:

$$P_{25}(q) = \frac{2}{fv^2}\left[v - [1 - \exp(-v)] + \frac{f-1}{2}[1 - \exp(-v)]^2\right], \tag{42}$$

with $v = u^2 f/(3f - 2)$, and $u^2 = \langle R_g^2 \rangle q^2$, where $\langle R_g^2 \rangle$ is the ensemble average radius of gyration squared of an arm.

26. Polydisperse star polymer with Gaussian statistics: For a Schulz–Flory (most probable) distribution (Schulz–Zimm distribution with $z = 1$) for the mass distribution of the arms, Burchard (1974) has given the form factor:

$$P_{26}(q) = \frac{1 + u^2/(3f)}{(1 + u^2(f+1)/(6f))^2}, \tag{43}$$

where f is the number of arms and $u^2 = \langle R_g^2 \rangle_z q^2$, where $\langle R_g^2 \rangle_z$ is the z-average radius of gyration squared of an arm.

27. Regular star-burst polymer (dendrimer) with Gaussian statistics: Expressions have been given by Burchard and Kajiwara (1982) and by Hammouda (1992). The results are obtained by summation of geometrical series. However, it cannot be recommended to use these as the performance of the sums introduces singularities. It is better to use the expressions *without* performing the sums (Pedersen and Gerstenberg, 1995). See also form factor 32 in the following, which can be used for calculating the same form factor.

28. Polycondensates of A_f monomers: Kajiwara et al. (1970) have calculated the form factors of structures obtained by random condensation of f-functional monomers (A_f) in which the resulting statistics of the subchains is Gaussian. For equal reaction probability of the f functional groups:

$$P_{28}(q) = \frac{1}{(1 + u^2/3)}, \tag{44}$$

where $u^2 = \langle R_g^2 \rangle_z q^2$.

29. Polycondensates of AB_f monomers: Burchard (1977) have calculated the form factors of structures obtained by random condensation of $f+1$-functional monomers (AB_f) in which the resulting statistics of the subchains is Gaussian. AB_f means that a monomer has one A group and f B groups. There is the further constraint that an A group can react only with a B group (with a certain probability), whereas a B group does not react with a B group. In this case:

$$P_{29}(q) = \frac{1}{(1 + u^2/6)^2}, \tag{45}$$

where $u^2 = \langle R_g^2 \rangle_z q^2$.

30. Polycondensates of ABC monomers: Burchard (1972) have calculated the form factors of structures obtained by random condensation of monomers with three different reactive groups, A, B and C (ABC). The only allowed reactions are A with B group with probability β and A with C group with probability γ. In this case:

$$P_{30}(q) = \frac{1 + Cu^2/3}{\left(1 + (1+C)u^2/6\right)^2}, \tag{46}$$

where

$$C = (\beta^2 + \gamma^2)\left[(\beta + \gamma) + \frac{2\beta\gamma}{1 - \beta - \gamma}\right]^{-1} \tag{47}$$

and $u^2 = \langle R_g^2 \rangle_z q^2$.

31. Regular comb polymer with Gaussian statistics: Expressions have been given by Casassa and Berry (1966). In the same paper Casassa and Berry treat the case of random coupling of the side chains for a monodisperse backbone and monodisperse sidechains, as well as the case of polydisperse sidechains.

32. Arbitrarily branched polymers with Gaussian statistics: The form factor for P subchains can be written as (Pedersen and Gerstenberg, 1995)

$$P_{32}(q) = \frac{1}{M_{32}^2}\left[\sum_{i=1}^{P} \rho_i^2 P_{18}(qR_{g,i}) + 2\sum_{i>j}^{P} \rho_i \rho_j \psi(qR_{g,i})\psi(qR_{g,j})\exp(-q^2 d_{ij}^2)\right], \tag{48}$$

where $R_{g,i}$ is the root-mean-square radius of gyration of the ith subchain, and ρ_i is the total excess scattering length of the ith subchain. Furthermore,

$$\psi(qR_{g,i}) = [1 - \exp(-u)]/u, \tag{49}$$

where $u = R_{g,i}^2 q^2$. The parameter $d_{ij}^2 = L_{ij} b$, where L_{ij} is the separation in contour length between the starting points of the ith and the jth subchains. Note that for Gaussian chains $d^2 = 6R_g^2$. The scattering mass M_{32} is given by $M_{32} = \sum_{i=1}^{P} \rho_i$.

Note that the form factor 25 of a star given by Benoit is a special case of the expression (48). Burchard et al. (1984) have given the results for a regular star polymer with f arms and two blocks per arm. Although these results look very much different from (48), it is just a special case of (48).

33. Arbitrarily branched semi-flexible polymers: An expression similar to (48) for semi-flexible chain has been given recently by Svaneborg and Pedersen (2000a). The semi-flexibility is treated in the second Daniels approximation, which means that the results are valid for subchains with $L/b > 10$ and $qb < 3.1$.

34. Arbitrarily branched self-avoiding polymers: An expression similar to (48) for self-avoiding chain has been given recently by Svaneborg and Pedersen (2000a). The self-avoidance was treated on two different levels: (i) within each subchain as for a linear polymer; (ii) for all paths as for a linear polymer. This means that higher order interactions are neglected for both cases, although (ii) takes excluded volume interactions into account between pair of subchains. Note that the parameters in the probability distributions should be taken as $\epsilon = 0.176$, $t = 2/(1-\epsilon)$, and $s = 2.90$ (Pedersen and Schurtenberger, 1996a). The result is given as two series expansions for, respectively low q in powers of q and for high q in powers of $1/q$.

35. Sphere with Gaussian chains attached: The expressions have been derived by Pedersen and Gerstenberg (1996) (see also Pedersen, 2000, 2001). For a sphere with radius R and total excess scattering length ρ_s with N_c attached chains

$$P_{\text{mic}}(q) = \frac{1}{M_{\text{mic}}^2}\left[\rho_s^2 F_1^2(q,R) + N_c\rho_c^2 P_{18}(q,L) + N_c(N_c-1)\rho_c^2 S_{cc}(q) + 2N_c\rho_s\rho_c S_{sc}(q)\right], \tag{50}$$

with:

$$S_{sc}(q) = F_1(q,R)\psi(qR_g)\frac{\sin\bigl(q(R+dR_g)\bigr)}{q(R+dR_g)} \tag{51}$$

and

$$S_{cc}(q) = \psi(qR_g)^2\left[\frac{\sin\bigl(q(R+dR_g)\bigr)}{q(R+dR_g)}\right]^2. \tag{52}$$

where R_g is the root-mean-square radius of gyration of a chain. The scattering mass is: $M_{\text{mic}} = \rho_s + N_c\rho_c$, where ρ_c is the total excess scattering length of a single chain. Non-penetrating of the chains into the core region is mimicked by $d \approx 1$ for $R \gg R_g$.

The expressions have been modified by Pedersen et al. (2000) to account for the fluctuation scattering caused by swelling of the core.

Svaneborg and Pedersen (2000b) have recently published a version of the model in which the corona chains starts as straight radial rods and continue as Gaussian chains. The form factor of a micelle with straight radial rods and that of a star polymer with straight arms are special cases of the expressions.

Plestil (2000) has recently calculated analytically the form factor of polydisperse micelles with spherical cores and gaussian chains attached. The number distribution of micellar radii was described by a Schulz–Zimm distribution and an approximation with fixed $R_{\text{chains}} = R_{\text{av}} + R_g$, where R_{av} is the average core radius was used.

36. Ellipsoid with Gaussian chains attached: An ellipsoid of revolution is considered with radius R, eccentricity ϵ, and total excess scattering length ρ_s with N_c chains with radius of gyration R_g and total excess scattering length ρ_c. The form factor is (Pedersen and Gerstenberg, 1995; Pedersen, 2000) given by (50) with

$$S_{sc}(q) = \psi(qR_g)\int_0^{\pi/2} F_1\bigl[q, r(R,\epsilon,\alpha)\bigr]\frac{\sin\bigl[q\bigl(r(R,\epsilon,\alpha)+dR_g\bigr)\bigr]}{q\bigl(r(R,\epsilon,\alpha)+dR_g\bigr)}\sin\alpha\,\mathrm{d}\alpha \tag{53}$$

and

$$S_{cc}(q) = \psi(qR_g)^2 \int_0^{\pi/2} \left[\frac{\sin[q(r(R,\epsilon,\alpha)+dR_g)]}{q(r(R,\epsilon,\alpha)+dR_g)} \right]^2 \sin\alpha \, d\alpha. \tag{54}$$

where $r(R,\epsilon,\alpha) = R(\sin^2\alpha + \epsilon^2\cos^2\alpha)^{1/2}$. For R and $R\epsilon \gg R_g$ non-penetration is mimicked by $d \approx 1$.

The expressions have been modified by Pedersen et al. (2000) to account for the fluctuation scattering caused by swelling of the core.

37. Cylinder with Gaussian chains attached: A cylinder is considered with radius R, length L_{cyl}, and total excess scattering length ρ_s with N_c chains with a radius of gyration R_g and total excess scattering length ρ_c. The form factor is (Pedersen, 2000) [2] given by (50) with:

$$S_{sc}(q) = \psi(qR_g) \int_0^{\pi/2} \Psi(q,R,L,\alpha)\Xi(q,R+dR_g,L+2dR_g,\alpha)\sin\alpha \, d\alpha, \tag{55}$$

where $\Xi(q,R,L,\alpha)$ is the form factor amplitude of an infinitely thin cylindrical shell with circular end pieces:

$$\Xi(q,R,L,\alpha) = \left[\frac{R}{(R+L)} \frac{2J_1(qR\sin\alpha)}{qR\sin\alpha} \cos(qL\cos\alpha/2) + \frac{L}{(R+L)} J_0(qR\sin\alpha) \frac{\sin(qL\cos\alpha/2)}{qL\cos\alpha/2} \right]. \tag{56}$$

Furthermore

$$S_{cc}(q) = \psi^2(qR_g) \int_0^{\pi/2} \Xi(q,R+dR_g,L+2dR_g,\alpha)^2 \sin\alpha \, d\alpha. \tag{57}$$

$\psi(q,R,L,\alpha)$ is the expression in the square brackets of Eq. (30).

Non-penetrating chains are for R and $L \gg R_g$ mimicked by $d \approx 1$.

Cylindrical cores with chains only on the side can be obtained by omitting the first term in (56), whereas cylinders with chains only at the circular end pieces are obtained by omitting the last term in (56). To ensure proper normalization the prefactors $R/(R+L)$ and $L/(R+L)$ should be omitted. Note that the expressions can easily be modified to a cylindrical core with elliptical cross section using the expression for form factor 12.

The expressions have been modified by Pedersen et al. (2000) to account for the fluctuation scattering caused by swelling of the core.

38. Polydisperse thin cylinder with polydisperse Gaussian chains attached to the ends: Burchard and Vogel (2000) have calculated the form factor of triblock copolymer micelle, for which the middle block is stiff. The polydispersity of the thin cylindrical middle part and of the chains at the ends are described by the most probable (mass) distribution [$z = 1$ in (9)].

---

[2] Note that there are errors in the expressions given in Pedersen (1997).

39. Sphere with corona of semi-flexible interacting self-avoiding chains: The expressions have been derived by Svaneborg and Pedersen (2002) and are base on a Monte Carlo simulation study. The average radial profile of the corona is reasonably well described by a box and half of a Gaussian function: $\rho(r) = 0$ for $r < R$, $\rho(r) = 1$ for $R \leq r < R_c$, and $\rho(r) = \exp\left[-(r - R_c)/(2s^2)\right]$ for $r \geq R_c$.

The form factor is:

$$P_{39}(q) = \frac{1}{M_{39}^2}\left[\rho_s^2 F_1^2(q,R) + N_c\rho_c^2 P_{\text{chain}}(q) + N_c\left[N_c - P_{\text{chain}}(q=0)\right]\rho_c^2 S_{cc}(q) + 2N_c\rho_s\rho_c S_{sc}(q)\right], \tag{58}$$

with:

$$P_{\text{chain}}(q) = \frac{P_{24}(q)}{1 + \nu P_{21}(q)}, \tag{59}$$

where $P_{24}(q)$ is the form factor of semi-flexible self-avoiding chains and $P_{21}(q)$ is the form factor of flexible self-avoiding chains. It is here used that $P_{21}(q) \approx c(q)P_{24}(q)$, where $c(q)$ is the direct correlation function between the sites on the chains. Furthermore, for the particular strength of the excluded volume ($R/b = 0.1$) one has

$$\nu \approx 1.4\sigma^{1.04}, \tag{60}$$

where $\sigma = N_c R_g^2/(4[(R + R_g)]^2)$ is the reduced surface coverage in terms of the number of chains N_c, the radius of gyration of the chains R_g and the core radius R.

$$S_{sc}(q) = F_1(q,R)A_c(q) \quad \text{and} \quad S_{cc}(q) = A_c(q)^2 \tag{61}$$

with:

$$A_c(q) = \frac{V_0 S_g(q) + V(R_c)F_1(q,R_c) - V(R)F_1(q,R)}{V_0 + V(R_c) - V(R)} \tag{62}$$

where $V(R) = 4\pi R^3/3$, $V_0 = 2\pi\left(4R_c s^2 + \sqrt{2\pi}(R_c^2 + s^2)s\right)$ and:

$$S_g(q) = \left\{qR_c\left(4R_c s + \sqrt{2\pi}(R_c^2 + s^2)\right)\right\}^{-1}\left\{2R_c s\sin(qR_c) + \sqrt{2\pi}\exp\left(-\frac{(qs)^2}{2}\right) \times \left(qR_c s^2\cos(qR_c) + R_c^2\sin(qR_c)\right) + 2\sqrt{2}D\left(\frac{qs}{\sqrt{2}}\right)\left(R_c^2\cos(qR_c) - qR_c s^2\sin(qR_c)\right)\right\}. \tag{63}$$

$D(y) = \exp(-y^2)\int_0^y \exp(t^2)\,dt$ is the Dawson integral. From the simulation one finds that $|R - R_c| \approx 0$ and $s \approx R_g$. The scattering mass in (58) is: $M_{39} = \rho_s + N_c\rho_c$, where ρ_s is the total excess scattering length of the core and ρ_c is the total excess scattering length of a corona chain.

40. Very anisotropic particles with local planar geometry: The form factor of such particles can be shown to factorize (see, e.g., Porod, 1948) into a cross section form factor $P_{cs}(q)$ for the shorter dimensions and a shape factor $P(q)'$ for the larger dimension:

$$P_{40}(q) = P(q)'\,P_{cs}(q) \tag{64}$$

where the cross section form factor for symmetric cross sections is:

$$P_{cs}(q) = \frac{1}{M_{cs}^2}\left[2\int_0^\infty \Delta\rho_{cs}(r)\cos(qr)\,\mathrm{d}r\right]^2 \tag{65}$$

where $\Delta\rho_{cs}(r)$ is scattering length density distribution function and $M_{cs} = 2\int_0^\infty \Delta\rho_{cs}(r)\mathrm{d}r$. For a homogeneous cross section (65) reduces to:

$$P_{cs}(q) = \left(\frac{\sin(qL/2)}{qL/2}\right)^2, \tag{66}$$

and for a cross section with two infinitely thin planes separated by L:

$$P_{cs}(q) = \left[\cos(qL/2)\right]^2. \tag{67}$$

A layered centro symmetric cross section structure with outer thickness L_1 and a core of thickness L_2, where the core and the outer part have the excess scattering lengths density ρ_2 and ρ_1, respectively, has

$$P_{cs}(q) = \frac{1}{M^2}\left(\frac{\rho_1 L_1 \sin(qL_1/2)}{qL_1/2} - \frac{L_2(\rho_1-\rho_2)\sin(qL_2/2)}{qL_2/2}\right)^2, \tag{68}$$

where $M = \rho_1 L_1 - L_2(\rho_1 - \rho_2)$. This expression is easily generalized to symmetric structures with many layers using an approach similar to the one used for spherical particles with concentric shells.

The expression corresponding to (65) for particles with Gaussian chains attached to the surface has recently been given (Pedersen, 2000):

$$P_{cs}(q) = \frac{1}{M_{cs}^2}\left[\rho_s^2\left(\frac{\sin(qL/2)}{qL/2}\right)^2 + \rho_c^2 N_c P_{18}(q,L) + N_c(N_c-1)\rho_c^2 S_{cc}(q) + 2N_c\rho_s\rho_c S_{sc}(q)\right], \tag{69}$$

where

$$S_{sc}(q) = \psi(qR_g)\frac{\sin(qL/2)}{qL/2}\cos\left[q(L/2+dR_g)\right] \tag{70}$$

and

$$S_{cc}(q) = \psi(qR_g)^2\cos\left[q(L/2+dR_g)\right]^2. \tag{71}$$

L is the thickness of the central core, $d \approx 1$, and $M_{cs} = (\rho_s + N_c\rho_c)$, where ρ_s is the total excess scattering length of the core and ρ_c is that of a single chain.

The shape form factor $P(q)'$ can be that of an infinitely thin spherical shell, elliptical shell, cylindrical shell, or disk (see form factors 2, 5, 6, 11, 12 and 16).

41. Very anisotropic particles with local cylindrical geometry: The form factor of such particles can also be shown to factorize (see, e.g., Porod 1948) into a cross section form factor $P_{cs}(q)$ for the short dimension and a shape factor $P(q)'$ for the larger dimensions.

For particles with a centro-symmetric scattering length density distribution:

$$P_{cs}(q) = \frac{1}{M_{cs}^2}\left(2\pi\int_0^\infty \Delta\rho_{cs}(r)J_0(qr)r\,\mathrm{d}r\right)^2 \tag{72}$$

where $\Delta\rho_{cs}(r)$ is the excess scattering length density distribution function and $M_{cs} = 2\pi\int_0^\infty \Delta\rho(r)r\,dr$. For a homogeneous particle with cross section radius R (72) reduces to:

$$P_{cs}(q) = \left(\frac{2J_1(qR)}{qR}\right)^2. \tag{73}$$

An expression for cross section structures consisting of concentric shells can be constructed by an approach similar to the one used for spherical particles.

The expression can easily be generalized to an elliptical cross section. For a homogeneous particle with semi-axes a and b (cf. form factor 12):

$$P_{cs}(q) = \int_0^{\pi/2}\left[\frac{2J_1(qr(a,b,\phi))}{qr(a,b,\phi)}\right]^2 \mathrm{d}\phi, \tag{74}$$

where $r(a,b,\phi) = [a^2\sin^2\phi + b^2\cos^2\phi]^{1/2}$. An expression for cross section structures consisting of concentric shells with elliptical cross section can be constructed by an approach similar to the one used for spherical particles and for tri-axial ellipsoids. The shells can have varying eccentricities (a/b).

The expression corresponding to (72) for particles with Gaussian chains attached to the surface has recently been given (Pedersen, 2000):

$$P_{cs}(q) = \frac{1}{M_{cs}^2}\left[\rho_s^2\left(\frac{2J_1(qR)}{qR}\right)^2 + \rho_c^2 N_c P_{18}(q,L) + N_c(N_c-1)\rho_c^2 S_{cc}(q) + 2N_c\rho_s\rho_c S_{sc}(q)\right], \tag{75}$$

where

$$S_{sc}(q) = \psi(qR_g)\frac{2J_1(qR)}{qR}J_0[q(R+dR_g)] \tag{76}$$

and

$$S_{cc}(q) = \psi(qR_g)^2 J_0[q(R+dR_g)]^2. \tag{77}$$

R is the radius of the central core, $d \approx 1$, and $M_{cs} = (\rho_s + N_c\rho_c)$, where ρ_s is the total excess scattering length of the core and ρ_c is that of a single chain.

The shape form factor $P(q)'$ can be the form factor of an infinitely thin rod, a semi-flexible polymer chain with or without excluded volume interactions (see form factors 15, 23, 24, and 34).

3.2. Structure factors

There are only very few cases for which the structure factor can be calculated analytically. Most of the available results have been obtained from liquid state theory for particles with spherical symmetry interacting with a spherically symmetric potential. The liquid state theory combines the Ornstein–Zernike integral equation with an approximate closure relation that relates the interaction potential to the direct correlation function (see e.g., D'Aguanno and Klein, 1991 and Chapter 14 by R. Klein in this book). If the equations cannot be solved analytically, it is possible to obtain numerical results for the structure factor. In this case a closure relation can be chosen which gives thermodynamically self-consistent results (Rogers and Young, 1984; Zerah and Hansen, 1986).

1. Hard-sphere potential: The particles interact with the hard-sphere radius R_{HS} and have a hard-sphere volume fraction η. The expressions for $S(q)$ have been calculated with the Percus-Yevick approximation for the closure relation (see, e.g., Kinning and Thomas, 1984):

$$S_1(q) = \frac{1}{1 + 24\eta G(2R_{HS}\,q)/(2R_{HS}\,q)}. \tag{78}$$

In this equation:

$$G(A) = \frac{\alpha(\sin A - A\cos A)}{A^2} + \frac{\beta\big(2A\sin A + (2 - A^2)\cos A - 2\big)}{A^3} + \frac{\gamma\big[-A^4\cos A + 4\{(3A^2 - 6)\cos A + (A^3 - 6A)\sin A + 6\}\big]}{A^5} \tag{79}$$

and

$$\alpha = \frac{(1+2\eta)^2}{(1-\eta)^4}, \qquad \beta = \frac{-6\eta(1+\eta/2)^2}{(1-\eta)^4}, \qquad \gamma = \frac{\eta\alpha}{2}. \tag{80}$$

2. Sticky hard-sphere potential: The model was introduced by Baxter (1968). The particles have a hard-sphere radius R_{HS} and a delta function attractive potential at the surface. The stickiness of the particles is given by the parameter τ. The expressions for the structure factor can be calculated with the Percus–Yevick approximation for the closure relation (see for example, Regnaut and Ravey (1989) and Menon et al. (1991)). For a hard-sphere volume fraction η and $\kappa = 2qR_{HS}$

$$S_2(q) = \frac{1}{A^2 + B^2}, \tag{81}$$

where

$$A = 1 + 12\eta\left\{\alpha\frac{\sin\kappa - \kappa\cos\kappa}{\kappa^3} + \beta\frac{1-\cos\kappa}{\kappa^2} - \frac{\lambda}{12}\frac{\sin\kappa}{\kappa}\right\},$$

and

$$B = 12\eta\left\{\alpha\frac{\kappa^2/2 - \kappa\sin\kappa + 1 - \cos\kappa}{\kappa^3} + \beta\frac{\kappa - \sin\kappa}{\kappa^2} - \frac{\lambda}{12}\frac{1-\cos\kappa}{\kappa}\right\}.$$

Furthermore,

$$\lambda = \text{Min}\left\{6\left[\frac{\tau}{\eta}+\frac{1}{1-\eta}\right] \pm \left\{36\left[\frac{\tau}{\eta}+\frac{1}{1-\eta}\right]\right\}^{1/2}\right\}, \qquad \mu = \lambda\eta(1-\eta),$$

$$\alpha = \frac{1+2\eta-\mu}{(1-\eta)^2}, \qquad \beta = \frac{-3\eta+\mu}{2(1-\eta)^2}. \tag{82}$$

3. Screened Coulomb potential: The particle interaction is described by the hard-sphere radius R_{HS} and an interaction potential given by:

$$V(r) = \frac{C_B Z^2}{(1+\kappa R_{\text{HS}})^2}\frac{\exp[-\kappa(r-2R_{\text{HS}})]}{r}, \tag{83}$$

for $r > 2R_{\text{HS}}$. $C_B = e^2/(4\pi\epsilon)$, where ϵ is the permittivity of the solvent, e the elementary charge, Z the number of charges per particle, and κ is the inverse Debye–Hückel length.

The structure factor have been calculated in the Mean-Spherical Approximation (MSA) and the expressions can be found in Hayter and Penfold (1981). The results work well for high to medium concentrations, but the pair correlation function becomes unphysical for low concentrations. A better result is obtained by a rescaling of the hard-sphere radius as described by Hansen and Hayter (1982). This procedure is known as the Rescaled MSA (RMSA).

It is also possible to obtain structure factors from thermodynamically self-consistent approaches (Rogers and Young, 1984; Rogers, 1980; D'Aguanno and Klein, 1991). However, these have the problem that the integral equations have to be solved numerically (and iteratively) and this makes the applications very time consuming.

4. Hard-sphere potential, polydisperse system: Vrij (1979) has given a solution for polydisperse hard spheres for an arbitrary size distribution. It is relatively straight forward to implement it in a computer program, as it only requires the calculation of certain averages of products of form factors and trigonometric functions weighted by the size distribution. The expressions can be found in the paper by Vrij.

5. Sticky hard-sphere potential, polydisperse system: A solution within the Percus–Yevick approximation has been given by Robertus et al. (1989). This solution is for arbitrary size distributions. In order to obtain the solution a set of equations quadratic in the parameters have to be solved numerically.

6. Screened Coulomb potential, polydisperse system: The solution in the MSA has been given by Blum and Hoye (1978) and Blum (1980): Ruiz-Estrada et al. (1990) have given a procedure for making a computer implementation using the rescaled mean-spherical approximation.

The implementation of a thermodynamically self-consistent (numerical) approach based on the closure relation of Rogers and Young (1984) has been described by D'Aguanno and Klein (1991, 1992).

7. Cylinders, RPA: The scattering can be calculated in the random phase approximation (RPA) (Edwards, 1966). The measured intensity for cylinders of length L and radius R, with $L \gg R$

is (Shimada et al., 1988)

$$\left(\frac{d\sigma(q)}{d\Omega}\right)_7 = n\rho_{\text{cyl}}^2 \frac{P_{11}(q)}{1+\nu P_{11}(q)}, \tag{84}$$

where ν is proportional to the concentration of the cylinders, ρ_{cyl} is the total scattering length of a cylinder, and n is the number density.

ν can be calculated from the expression given by Cotter and Martire (1970) for the osmotic compressibility:

$$\nu = \frac{\left(1+2(B+C)^2\right)+2D\left[1+B+\frac{5}{4}C\right]}{(1-B-C)^4} - 1, \tag{85}$$

where $B = \pi R^2 Ln$, $C = 4\pi R^3 n/3$, and $D = \pi RL^2/2$.

8. Cylinders, 'PRISM': Based on recent Monte Carlo simulation studies (Bergmann et al., 2001), it is better to suggest an expression which is in line with the polymer reference interaction site model (PRISM) (see, e.g., Schweizer and Curro, 1994):

$$\left(\frac{d\sigma(q)}{d\Omega}\right)_8 = n\rho_{\text{cyl}}^2 \frac{P_{11}(q)}{1+\nu c(q)P_{15}(q,L-2R)}, \tag{86}$$

where $P_{15}(q, L-2R)$ is the form factor of an infinitely thin rod of length $L-2R$ and $c(q)$ is a 'direct' correlation function which can be approximated by the Fourier transform of the correlation hole around each site:

$$c(q) = F_1(q, 2R) = \frac{3\left[\sin(q2R) - q2R\cos(q2R)\right]}{(q2R)^3}. \tag{87}$$

The parameter ν should be chosen according to (85).

9. Solutions of flexible polymers, RPA: The simplest approach for taking interactions into account is when using the random phase approximation (RPA). The measured scattering intensity is (Edwards, 1966; Benoit and Benmouna, 1984)

$$\left(\frac{d\sigma(q)}{d\Omega}\right)_9 = n\rho_c^2 \frac{P_{18}(q)}{1+\nu P_{18}(q)}, \tag{88}$$

where ν is proportional to the polymer mass concentration, to the contour length and to the strength of the excluded volume interaction. ρ_c is the total scattering length of a chain. The RPA is only a good approximation for high polymer concentrations. At lower concentrations, it can be recommended to use (88) with $P_{18}(q)$ replaced by $P_{21}(q)$, and with the parameters normalized, so that (88) gives the forward scattering (intensity at $q = 0$) predicted by renormalization group theory (Nakanishi and Ohta, 1985). One has according to Ohta and Oono (1982):

$$\nu = \frac{1}{8}\left(9X - 2 + \frac{2\ln(1+X)}{X}\right)\exp\left(\frac{1}{4}\left[\frac{1}{X} + \left(1 - \frac{1}{X^2}\right)\ln(1+X)\right]\right), \tag{89}$$

with $X = 16A_2cM$, where c is the polymer concentration and M is the polymer mass. The second virial coefficient A_2 is given by

$$A_2 = 4\pi^{3/2} N_A \frac{R_g^3}{M^2} \Psi, \tag{90}$$

where N_A is Avogadro's number, R_g is the ensemble average radius of gyration of a single chain and $\Psi \approx 0.23$ is the degree of interpenetration (Oono and Freed, 1982).

10. Solutions of semi-flexible polymers, 'PRISM': The following expressions are based on the simulations by Pedersen and Schurtenberger (1999) for semi-flexible polymers with $R/b = 0.1$, where R is the cross section radius and b is the Kuhn length. The expression for the cross section is:

$$\left(\frac{\mathrm{d}\sigma(q)}{\mathrm{d}\Omega}\right)_{10} = n\rho_c^2 \frac{P_{24}(q)}{1 + \nu P_{21}(q)}, \tag{91}$$

where $P_{24}(q)$ is the form factor of semi-flexible self-avoiding chains and $P_{21}(q)$ is the form factor of flexible self-avoiding chains. It is here used that $P_{21}(q) \approx c(q)P_{24}(q)$, where $c(q)$ is the direct correlation function between the sites. ν is given by (89), however, in order to get agreement with simulations the factor $1/4$ is replaced by $1/2.56$. The simulations furthermore suggest that $\Psi \approx 0.21$.

4. An application: block copolymer micelles

The example described in this section concerns a block copolymer in a selective solvent. The polymers is a tri-block copolymer and consists of two end blocks of poly(ethylene oxide) (PEO) and a central block of poly(propylene oxide) (PPO). The composition is $EO_{25}PO_{40}EO_{25}$ and the molecular weight is 4500 Da. The molecular weights are 2300 Da for the PPO part and 1100 Da for each of the PEO parts. The polymer is commercially available from Serva AG, Heidelberg, Germany, under the name P85.

The aggregation behavior of P85 in water has been studied by SANS (Mortensen and Pedersen, 1993; Pedersen and Gerstenberg, 1996) and SAXS (Glatter et al., 1995). The P85 is present as single chains at low temperatures ($T < 5$°C) and forms aggregates/micelles at higher temperatures. This behavior is ascribed to the properties of PPO, which is hydrophilic at room temperatures and becomes hydrophobic at higher temperatures. The PEO is reasonably soluble for temperatures up to $T = 70$°C. The present discussion is limited to the dilute region with well-defined micelles, where the structure factor effects are relatively small.

Measurements on a 1 %wt solution at $T = 50$°C were performed at the SANS instrument at Risø National Laboratory (Pedersen, 1995). In order to enhance the contrast and reduce the incoherent background, D_2O was used as the solvent. SAXS data have been recorded for a 1% wt solution $T = 50$°C by a conventional Kratky camera (see, e.g., Pedersen, 1995) by Glatter et al. (1995). The SAXS data were corrected for instrumental smearing effects in connection with an indirect Fourier transformation. The desmeared SAXS scattering curve is shown in Fig. 1 together with the SANS data. Note that the SANS data have been recorded using three different instrumental settings. Due to the different smearing effects for the different settings, the data do not coincide in the overlap regions.

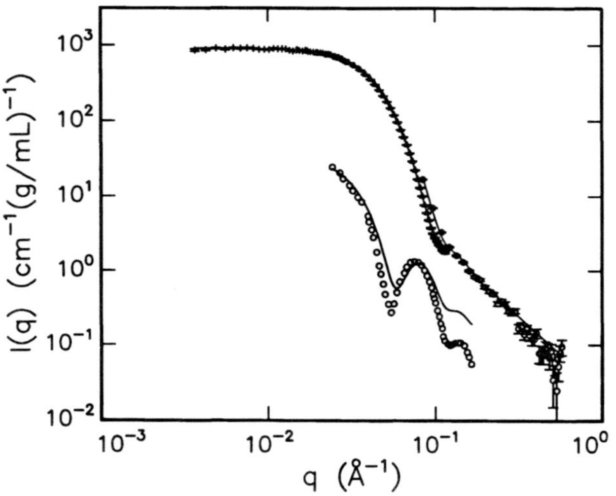

Fig. 1. The small-angle neutron scattering data of 1.0 wt% P85 in D_2O (upper data) (Pedersen and Gerstenberg, 1996). The curves are the fit by the analytical model including polydispersity of the micelles. Note that the fitted curves are discontinuous where the data from different settings overlap due to the instrumental smearing. The lower data are SAXS results on similar micelles (Glatter et al., 1995). The curve is calculated for the results determined by fitting the SANS data.

The SANS and SAXS scattering curves are remarkably different. The SANS data have a smooth decrease with increasing q and resemble the scattering form factor of a sphere. However, no oscillations are observed (cf. (21)). A q^{-2} scattering is observed at large q, which is similar to the scattering from flexible chains obeying Gaussian statistics (cf. (38)). In contrast to this, the SAXS data have large oscillations. The range of scattering vectors is not large enough to identify power law scattering at large q.

Accurate density measurements (Armstrong et al., 1993) as a function of temperature for a series of PEO–PPO–PEO block copolymers can be used for calculating the specific volumes of the PEO and PPO parts. Such calculations (Pedersen and Gerstenberg, 1996) show that the contrast of the PPO is small and negative, whereas the contrast of the PEO is larger and positive. This shows that it is mainly the PEO and thus the corona which is observed in SAXS. If a model should be constructed from the available information, a first guess would consist of a homogeneous and spherical PPO core and a shell of dissolved PEO chains.

Information obtained from the model-independent approaches (Pedersen, 1997) confirms that the form factor given by (50), for a particle with spherical core and Gaussian chains attached to the surface, is appropriate to use for fitting the data. The number of fitting parameters in the model is quite large and in order to reduce it, molecular constraints have been employed. Note that the analysis presented in the present text which is done on absolute scale is different from the analysis previously reported (Pedersen and Gerstenberg, 1996; Pedersen, 1997).

The density measurements by Armstrong et al. (1993) were used for estimating the specific volumes (v_{EO} and v_{PO}, respectively) of the EO and PO monomers at 50°C. Note that these vary as much as 10–15% with temperature. The volume of the PEO and PPO blocks are then, respectively: $V_{\mathrm{PEO}} = 25v_{\mathrm{EO}}$ and $V_{\mathrm{PPO}} = 40v_{\mathrm{PO}}$, where the polymerization indices of, respectively, 25 and 40 are used. The total excess scattering length of a PEO chain is

$$\rho_c = 25 \sum_{i \in \mathrm{EO}} b_i - \rho_{\mathrm{D_2O}} V_{\mathrm{PEO}}, \tag{92}$$

where the sum over i is over the atoms in the EO monomer, b_i is the scattering length of the ith atom, and $\rho_{\mathrm{D_2O}}$ is the scattering length density of D_2O. The corresponding total excess scattering length of the core is with similar notation:

$$\rho_s = N_{\mathrm{agg}}\left(40 \sum_{i \in PO} b_i - \rho_{\mathrm{D_2O}} V_{\mathrm{PPO}}\right), \tag{93}$$

where N_{agg} is the aggregation number of the micelle. As there are two PEO chains for each PPO block one has $N_c = 2N_{\mathrm{agg}}$.

The volume of the core is

$$V = \frac{V_{\mathrm{PPO}} N_{\mathrm{agg}}}{1-f}, \tag{94}$$

where f is the volume fraction of solvent in the core. The core radius can be calculated from the relation $V = 4\pi R^3/3$.

The number density of micelles is:

$$n = \frac{c}{M_{\mathrm{P85}} N_{\mathrm{agg}}}, \tag{95}$$

where M_{P85} is the total mass of a triblock copolymer. In the analysis of the SANS data we use the expression (13), however, with the data normalized by the concentration c of polymer. This leads to

$$I(q) = \frac{1}{c}\frac{\mathrm{d}\sigma(q)}{\mathrm{d}\Omega} = \frac{1}{M_{P85} N_{\mathrm{agg}}} M_{\mathrm{mic}}^2 \left(P(q) + A(q)^2\left[S(q) - 1\right]\right), \tag{96}$$

where $P(q)$ is form factor 35 and:

$$A(q) = \frac{1}{M_{\mathrm{mic}}}\left[\rho_s F_1(q,R) + N_c \rho_c \psi(qR_g) \frac{\sin(q(R + dR_g))}{q(R + dR_g)}\right] \tag{97}$$

is the normalized Fourier transform of the ensemble averaged radial scattering length density distribution.

It turned out that two additional effect had to be considered: (i) Diffuseness of the core interface, (ii) Polydispersity of the aggregation number of the micelles. The first effect was included by substituting $F_1(q,R)$ by $F_1(q,R)\exp(-s^2q^2/2)$, where s describes the width of the core surface. Polydispersity was described by a Gaussian function $n(N_{\mathrm{agg}}) = \exp\left(-(N_{\mathrm{agg}} - N_{\mathrm{av}})^2/\sigma(N_{\mathrm{agg}})^2/2\right)/\sqrt{2\pi\sigma(N_{\mathrm{agg}})^2}$, where N_{av} is the average aggregation number and $\sigma(N_{\mathrm{agg}})$ gives the width of the distribution. Using the local monodisperse approximation for the inclusion of the structure factor effects (96) becomes:

$$I(q) = \frac{\int M_{\mathrm{mic}}(N_{\mathrm{agg}})^2 \left(P(q,N_{\mathrm{agg}}) + A(q,N_{\mathrm{agg}})^2\left[S(q,N_{\mathrm{agg}}) - 1\right]\right) n(N_{\mathrm{agg}})\,\mathrm{d}N_{\mathrm{agg}}}{M_{P85}\int N_{\mathrm{agg}} n(N_{\mathrm{agg}})\,\mathrm{d}N_{\mathrm{agg}}}, \tag{98}$$

where the dependence on the aggregation number has been expresses specifically. A hard-sphere structure factor (78) was used for $S(q, N_{agg})$ with a hard-sphere radius of $R_{HS} = R + \Delta R_{HS}$.

The conditions for the SANS and SAXS measurements are not identical and it was therefore decided first to fit only the SANS data. The results were then used for calculating the SAXS spectra by substituting neutron scattering length densities by electron densities.

When the SANS data were fitted, the theoretical scattering curve was smeared by the instrumental resolution function (Pedersen et al., 1990). For each setting the appropriate resolution function was used. The fit parameters were N_{av}, $\sigma(N_{agg})$, f, s, R_g, d, ΔR_{HS}, and η_{HS}, where η_{HS} is the volume fraction of the hard spheres. The polydisperse model fits the scattering data very well as can be seen in Fig. 1, which also shows the SAXS curve calculated for the same set of parameters. The agreement with the SAXS data is reasonable, but not perfect, however, it is satisfactory considering that the two sets of data have been obtained on two very different instruments, on different batches of polymers in two different laboratories.

The results from the fit are $N_{av} = 46$, $\sigma(N_{agg})/N_{av} = 0.33$, $f = 0.35$, $s = 3.8$ Å, $R_g = 13$ Å, $d = 0.77$, $\Delta R_{HS} = 33$ Å, and $\eta_{HS} = 0.028$. Note that the polydispersity in aggregation number corresponds to a polydispersity in core radius of only 11%, which demonstrates that the micelles are quite monodisperse. The average core radius is 40 Å. There is a significant fraction of solvent in the core. The width of the core interface is about 4 Å which is reasonable considering that this corresponds to about the size of a monomer. The radius of gyration is similar to that expected for free Gaussian chains for which $R_g^2 = Lb/6$, where $L = 90$ Å is the chain contour length of the PEO block and $b = 10$ Å is the Kuhn length (Aharoni, 1983). This give $R_g = 12$ Å. Note furthermore that $d \approx 1$ as expected.

The width of the polymer corona is about $2R_g = 26$ Å and therefore the interaction parameter $\Delta R_{HS} = 33$ Å, which describes the interaction range of the corona, is reasonable. The hard-sphere volume fraction of 2.8%, also seems reasonable for a 1 wt% solution considering the large amount of solvent present in the core and corona.

It is somewhat surprising that the present model, which neglects chain self-avoidance and chain-chain interactions work so well. As the chains are relatively short with $L \approx 9b$, the excluded volume interactions within the chains are expected to have only a small influence on the polymer configuration. Furthermore, water is not really a good solvent for PEO at 50°C, and therefore excluded volume and inter-chain interactions are relatively weak. The last point to be considered is the degree of coverage of the core surface. It can be characterized by the reduced surface coverage $\sigma = N_c R_g^2/(4[(R + R_g)]^2)$, which is the area of a chain on the surface at a radius $R + R_g$ relative to the available area per chain. In the present case the value is only 1.4 and therefore the inter-chain interactions are relatively small.

5. Summary and conclusions

Modelling of small-angle scattering data from systems consisting of colloidal particles or polymers in solution has been described. Only systems with short range correlations have been considered. The basic principles of the least-squares methods, applied for optimizing the parameters in the model when fitting to the experimental data have also been discussed. The most common analytical and numerical expressions for model form factors and structure factors have been summarized with emphasis on practical applications. It has been described how size polydispersity can be taken into account in the form factor expressions and some attention was given to the construction of form factors for very anisotropic particles. Also approximate expressions for inclusion of structure factor, for example, for anisotropic and polydisperse particles have been described.

The chapters deals, as already stated, with modelling of small-angle scattering data. However, before the data can be fitted a model cross section has to be constructed and this may be a tedious job. First the most important features of the structure have to be identified, so that a model structure can be formed. Model independent methods, like indirect Fourier transformation (Glatter, 1977, 1980a) and square-root deconvolution (Glatter, 1981) might be very useful in this connection. But occasionally, analytical expressions are not available for the scattering cross section of the model structure and it is necessary to construct them or calculate them. This was the case for the P85 micelles described in the previous section. When analytical expressions are derived, it is often convenient to use Monte Carlo simulations (Hansen, 1990) for checking the analytical expressions as it was done for the P85 micelles (Pedersen and Gerstenberg, 1996). In some cases it may be impossible to obtain analytical expressions as it is for semi-flexible polymers. If scattering functions are available, for example, from computer simulations for a large range of parameters, the scattering functions can be parameterized and interpolated, so that they can be used for least-squares fitting (Yoshizaki and Yamakawa, 1981; Pedersen and Schurtenberger, 1996b; Jerke et al., 1997, 1998).

When fitting experimental data, it is not unusual that the information content of a data set is quite low. The parameters of the model are then very correlated and when physical parameters, like the mass density, are calculated from the fit parameters it may give unphysical values. Such problems can be greatly reduced if the data are on an absolute scale and the molecular properties of the constituting molecules are taken into account in the model as constraints. If the densities are known from other types of measurements this information can be included and unphysical solutions can be avoided. An example with application of molecular constraints was given in the previous section. The information content of the data can also be enhanced by measuring for several different contrasts, or by measuring the samples by both X-ray and neutron scattering. For neutrons the contrast variation can be done by varying the deuterated fraction of the solvent. If the particles have components with different composition, the contrast variation will provide additional information. The combination of X-ray and neutron scattering is in some cases very efficient (Arleth et al., 1997; Iampietro et al., 1998; Cantù et al., 1998; Pedersen et al., 2000, see also previous section). This is the case for surfactant systems with heavy atoms in the head group, for which the X-ray scattering is almost entirely from the head group region, whereas the micelles appear to be homogeneous for neutron scattering in D_2O. The extra information from contrast variation is most efficiently taken into account if the data from different contrasts are fitted simultaneously. In this case the same parameters for the geometry can be used and only the scattering length densities depend on which contrast is considered.

Acknowledgements

I thank Carsten Svaneborg for numerous useful discussions in connection with the writing of the present text and for his corrections to the text.

References

Aharoni, S.M. (1983). *Macromolecules*, **16**, 1722–1728.
Arleth, L., Posselt, D., Gazeau, D., Lampent, C., Zemb, T., Mortensen K. and Pedersen, J.S. (1997). *Langmuir*, **13**, 1887–1896.
Armstrong, J.K., Parsonage, J., Chowdhry, B., Leharne, S., Mitchell, J., Beezer, A., Löhner, K. and Laggner, P. (1993). *J. Phys. Chem.*, **97**, 3904–3909.
Barker, J.G. and Pedersen, J.S. (1995). *J. Appl. Cryst.*, **28**, 105–114.
Bagger-Jörgensen, H., Olsson, U. and Mortensen, K. (1997). *Langmuir*, **13**, 1413–1421.
Bale, H.D. and Schmidt, P.W. (1984). *Phys. Rev. Lett.* **53**, 596–599.
Baxter, R.J. (1968). *J. Chem. Phys.*, **49**, 2770-2774.
Benoit, H. (1953). *J. Polym. Sci*, **11**, 507–510.
Benoit, H. and Benmouna, M. (1984). *Polymer*, **25**, 1059–1067.
Bergmann, A., Glatter, O. and Pedersen, J.S. (2001). Flux increase of an optimized 'high-volume' version of the NanoSTAR small-angle x-ray scattering instrument (*unpublished*).
Bevington, B.R. (1969) *Data Reduction and Error Analysis for the Physical Sciences*. McGraw-Hill, New York.
Blum, L. and Hoye, J.S. (1978). *J. Stat. Phys.* **19**, 317–324.
Blum, L. (1980). *J. Stat. Phys.*, **22**, 661–672.
Burchard, W. (1972). *Macromolecules*, **5**, 604–610.
Burchard, W. (1974). *Macromolecules*, **7**, 835–841.
Burchard, W. (1977). *Macromolecules*, **10**, 919–927.
Burchard, W. and Kajiwara, K. (1970). *Proc. Roy. Soc. London, Ser. A*, **316**, 185–199.
Burchard, W., Kajiwara, K., Nerger, D. and Stockmayer W.H. (1984). *Macromolecules*, **17**, 222–230.
Burchard, W., Michel, E. and Trappe, V. (1996). *Macromolecules*, **29**, 5934–5939.
Burchard, W. and Vogel, H.J. (2000). *Comp. Theor. Pol. Sci.*, **10**, 133–148.
Cantù, L., Corti, M., Del Favero, E., Dubois, M. and Zemb, T.N. (1998). *J. Chem. Phys. B*, **102**, 5737–5743.
Casassa, E.F. and Berry, G.C. (1966). *J. Polymer Sci. Part A-2*, **4**, 881–897.
Cotter, M. and Martire, D.E. (1968). *J. Chem. Phys.*, **4**, 1909–1919.
Cusack, S. (1981). *J. Mol. Biol.*, **145**, 541–543.
D'Aguanno, B. and Klein, R. (1991). *J. Chem. Soc. Faraday Trans.*, **87**, 379–390.
D'Aguanno, B. and Klein, R. (1992). *Phys. Rev. A*, **46**, 7652–7656.
Debye, P. (1915). *Ann. Phys. Leipzig*, **46**, 809–823.
Debye, P. (1947). *J. Phys. Colloid Chem.*, **51**, 18–32.
Dubey, P.A. (1990). Röntgenkleinwinkelstreuung an Al-Ag-Einkristallen: Frühe stadien der entmischung. *Dr. sc. nat. Dissertation No. 9077*, ETH Zürich.
Dubey, P.A. , Schönfeld, B. and Kostorz, G. (1991). *Acta Metall. Mater.*, **39**, 1161–1170.
Edwards, S.F. (1966). *Proc. Soc. London*, **88**, 265–280.
Fournet, G. (1951) *Bull. Soc. Fr. Minéral. Crist.*, **74**, 39–113.
Förster, S. and Burger, C. (1998). *Macromolecules*, **31**, 879–891.
Förster, S., Hermsdorf, N., Leube, W., Schnablegger, H., Regenbrecht, M., Akari, S, Lindner, P. and Böttcher, C. (1999). *J. Phys. Chem.*, **B103**, 6657–6668.
Gazzillo, D., Giacometti, A., Valle, R.G.D., Venuti, E. and Carsughi, F. (1999). *J. Chem. Phys.* **111**, 7636–7645.
Glatter, O. (1977). *J. Appl. Cryst.*, **10**, 415–421.
Glatter, O. (1979). *J. Appl. Cryst.*, **12**, 166–175.
Glatter, O. (1980a). *J. Appl. Cryst.*, **13**, 577–584.
Glatter, O. (1980b). *J. Appl. Cryst.*, **13**, 7–11.
Glatter, O. (1981). *J. Appl. Cryst.*, **14**, 101–108.

Glatter, O. and Hainish, B. (1984). *J. Appl. Cryst.*, **17**, 535–441.
Glatter, O., Scherf, G., Schillén, K. and Brown, W. (1994). *Macromolecules*, **27**, 6046–6054.
Gradzielski, M., Langevin, D., Magid, L. and Strey, R. (1995). *J. Phys. Chem.* **99**, 13234–13238.
Greschner, G.S. (1973). *Makromol. Chem.*, **170**, 203–229.
Guinier, A. (1939). *Ann. Phys.*, **12**, 161–237.
Hammouda, B. (1992). *J. Polym. Sci. B: Polym Phys.*, **30**, 1387–1390.
Hansen, J.-P. and Hayter, J.B. (1982). *Mol. Phys.*, **46**, 651–656.
Hansen, S. (1990). *J. Appl. Cryst.*, **23**, 344–346.
Hayter, J.B. and Penfold, J. (1981). *Mol. Phys.*, **42**, 109–118.
Hendricks, R.W., Schelten, J. and Schmatz, W. (1974). *Phil. Mag.*, **30**, 819–837.
Iampietro, D., Brasher, L.L., Kaler, E.W., Stradner, A. and Glatter, O. (1998). *J. Chem. Phys. B*, **102**, 3105–3113.
Jerke, G., Pedersen, J.S., Egelhaaf, S.U. and Schurtenberger, P. (1997). *Phys. Rev. E*, **56**, 5772–5788.
Jerke, G., Pedersen, J.S., Egelhaaf, S.U. and Schurtenberger, P. (1998). *Langmuir*, **14**, 6013–6024.
Kajiwara, K., Burchard, W. and Gordon, M. (1970). *Br. Polym. J.*, **2**, 110–115.
Kinning, D.J. and Thomas, E.L. (1984). *Macromolecules*, **17**, 1712–1718.
Kotlarchyk, M. and Chen, S.-H. (1983). *J. Chem. Phys.* **79**, 2461–2469.
Kratky, O. and Porod, G. (1949a). *J. Colloid Sci.*, **4**, 35–70.
Kratky, O. and Porod, G. (1949b). *Rec. Trav. Chim. Pays-Bas*, **68**, 1106–1122.
Lawson C.L. and Hanson, R.J. (1974). *Solving Least Squares Problems*. Prentice-Hall, Englewood Cliffs.
Menon, S.V.G., Manohar, C. and Srinivasa Rao, K. (1991). *J. Chem. Phys.* **95**, 9186–9190.
Mittelbach, P. and Porod, G. (1961a). *Acta Phys. Austriaca*, **14**, 185–211.
Mittelbach, P. and Porod, G. (1961b). *Acta Phys. Austriaca*, **14**, 405–439.
Mittelbach, P. and Porod, G. (1962). *Acta Phys. Austriaca*, **15**, 122–147.
Mortensen, K. and Pedersen, J.S. (1993). *Macromolecules*, **26**, 805–812.
Nakanishi, A. and Ohta. T. (1985). *J. Phys. A*, **18**, 127–139.
Neugebauer, T. (1943). *Ann. Phys. (Leipzig)*, **42**, 509–533.
Oono, Y. and Freed, K.F. (1982). *J. Phys. A*, **15**, 1931–1950.
Ohta, T. and Oono, Y. (1982). *Phys. Lett. A* **89**, 460–464.
Pedersen, J.S. (1993). *J. de Phys. IV France, Coll. C8*, **3**, 491–498.
Pedersen, J.S. (1994). *J. Appl. Cryst.*, **27**, 595–608.
Pedersen, J.S. (1995). In: *Modern Aspects of Small-angle Scattering* (Brumberger, H., ed.). Kluwer Academic Press, Dordrecht, pp. 57–91.
Pedersen, J.S. (1997). *Adv. Colloid Interface Sci.*, **70**, 171-210.
Pedersen, J.S. (2000). *J. Appl. Cryst.*, **33**, 637–640.
Pedersen, J.S. (2001). *J. Chem. Phys.*, **114**, 2839–2846.
Pedersen, J.S., Posselt, D. and Mortensen, K. (1990). *J. Appl. Cryst.*, **23**, 321–333.
Pedersen, J.S. and Gerstenberg, M.C. (1995). Calculation of form factors of branched molecules and block copolymer micelles (*unpublished*).
Pedersen, J.S. and Gerstenberg, M.C. (1996). *Macromolecules*, **29**, 1363–1365.
Pedersen, J.S. and Schurtenberger, P. (1996a). *Macromolecules*, **29**, 7602–7612.
Pedersen, J.S. and Schurtenberger, P. (1996b). *J. Appl. Cryst.* **29**, 646–661.
Pedersen, J.S., Hamley, I.W. , Ryu, C.Y. and Lodge, T.P. (2000). *Macromolecules*, **33**, 542–550.
Pedersen, J.S. and Schurtenberger, P. (1999). *Europhys. Lett.*, **45**, 666–672.
Plestil, J. (2000). *J. Appl. Cryst.*, **33**, 600–604.
Porod, G. (1948). *Acta Phys. Austriaca*, **2**, 255–292.
Press, W.H., Flannery, B.P., Teukolsky, S.A. and Vetterling, W.T. (1989). *Numerical Recipies*. University Press. Cambridge.
Rayleigh, Lord (1911). *Proc. Roy. Soc. London, Ser. A*, **84**, 25–46.
Regnaut, C. and Ravey, J.C. (1989). *J. Chem. Phys.*, **91**, 1211–1221.
Robertus, C., Phillipse, W.H., Joosten, J.G.H and Levine, Y.K. (1989). *J. Chem. Phys.*, **90**, 4482–4490.
Rodríguez, F., Gómez Sal, J.C., Moreno, M., de Geyer, A. and Janot, C. (1991). *Phys. Rev. B*, **43**, 7519–7526.
Rogers, F.J. (1980). *J. Chem. Phys.*, **73**, 6272–6278.
Rogers, F.J. and Young, D.A. (1984). *Phys. Rev. A*, **30**, 999–1007.

Ruiz-Estrada, H., Medina-Noyola, M. and Nägele, G. (1990). *Phys. A*, **168**, 919–941.
Schweizer, K.S. and Curro, J.G. (1994) *Adv. Polym. Sci.*, **116**, 319–377.
Schulz, G.V. (1939). *Z. Physik. Chem. B*, **43**, 25–46.
Sheu, E.Y. (1992). *Pys. Rev. A* **45**, 2328–2438.
Shimada, T., Doi, M. and Okano, K. (1988). *J. Chem. Phys.*, **88**, 2815–2821.
Stuhrmann, H.B. (1970). *Z. Phys. Chem. (Frankfurt am Main)*, **72**, 177–184, 185–198.
Svaneborg, C. and Pedersen, J.S. (2000a). Form factors of branched semiflexible polymers and branched self-avoiding polymers (*unpublished*).
Svaneborg, C. and Pedersen, J.S. (2000b). *J. Chem. Phys.*, **112**, 9661–9670.
Svaneborg, C. and Pedersen, J.S. (2002). *Macromolecules*, **35**, 1028–1037.
Svergun, D.I. (1994). *Acta Cryst. A*, **50**, 391–402.
Svergun, D.I. (1999). *Biophys. J.*, **76**, 2879–2886, **77**, 2896.
Svergun, D.I., Koch, M.H.J. and Serdyuk, I.N. (1994a). *J. Mol. Biol.*, **240**, 66–77.
Svergun, D.I., Koch, M.H.J., Pedersen, J.S. and Serdyuk, I.N. (1994b). *J. Mol. Biol.*, **240**, 78–96.
Svergun, D.I., Koch, M.H.J., Pedersen, J.S. and Serdyuk, I.N. (1994c). *Proc. Natl. Acad. Sci. USA*, **91**, 11826–11830.
Teixeira, J. (1988). *J. Appl. Cryst.*, **21**, 781–785.
Utiyama, H., Tsunashima, Y. and Kurate, M. (1971). *J. Chem. Phys.*, **55**, 3133–3145.
Vrij, A. (1979). *J. Chem. Phys.*, **71**, 3267–3270.
Yoshizaki, T. and Yamakawa, H. (1980). *Macromolecules*, **13**, 1518–1525.
Zerah, G. and Hansen, J.P. (1986). *J. Chem. Phys.* **84**, 2336–2343.
Zimm, B.H. (1948). *J. Chem. Phys.*, **16**, 1093–1116.
Zimm, B.H. and Stockmayer (1949). *J. Chem. Phys.*, **17**, 1301–1314.

Part IV
Special applications

Neutrons, X-Rays and Light
P. Lindner and Th. Zemb (Editors)

Chapter 17

Scattering Experiments under External Constraints: SANS and Shear Flow

Peter Lindner

Institut Laue-Langevin, B.P. 156, F-38042 Grenoble CEDEX 9, France

1. Introduction

During the past 30 years the application of the neutron scattering technique, in particular at *small angles* (SANS), has considerably enhanced the understanding of colloidal and polymeric systems with respect to their structure as well as their dynamics. Variable wavelengths and sample-to-detector distances at modern SANS instruments provide a range of momentum transfer Q which is suitable for studying molecular structures that are between some 1 and 100 nm across. Studies under *equilibrium conditions* allow characterization of single molecule parameters and molecular interaction parameters in dilute and concentrated solutions. Other examples are studies of critical phenomena in phase separation, of chain morphology in aggregated systems such as networks and crystalline polymers and characterization of deformation and orientation processes on a molecular level. The outstanding advantage of this technique is the possibility of an isotopic labelling of hydrogen rich compounds, such as synthetic or biological polymers. The difference in neutron scattering lengths of both hydrogen isotopes allows for a specific 'colouring' of molecules by replacement of ^{1}H against ^{2}H. Thus an increased contrast is provided in a neutron scattering experiment but the polymer system remains chemically (almost) unchanged. Furthermore, in a neutron experiment there are fewer restrictions with respect to optical cleanness of the solution (dust problem in light scattering, LS), transparency of container material and absorption (a problem in small-angle X-ray scattering, SAXS). However, all three methods (SANS, SAXS, LS) are in principle complementary for structure determination.

This chapter illustrates the application of the 'classical' SANS technique to systems under *non-equilibrium conditions* (Oberthür, 1983, 1984) such as flow. *Flow fields* are often encountered in process engineering and application of polymeric or colloidal material. There is a fundamental interest in understanding the microscopic structure and dynamics of flowing complex fluids as the macroscopic material properties might change considerably with application of an external perturbation (e.g. non-Newtonian flow behaviour). The technological and practical implications of such studies become immediately evident when we look for instance at injection moulding with polymeric melts, polymer fibre- and foil-extrusion, modern dispersion paints, flocculation, coating techniques or enhanced oil recovery. It is obvious that an 'on-line' scattering study with *flowing* systems gives insight into a very realistic state of matter since those processes involve *non-equilibrium* rather than equilibrium states.

2. Non-equilibrium sample environment

In the following we are concerned with SANS experiments using 'cold' neutrons, whose wavelengths are in the range between 0.45 and 2.0 nm. The high penetrating power of neutrons into matter allows for a typical sample thickness of the order of 1 mm, with a typical beam cross-section of the order of 0.5 to 1 cm^2. These macroscopic dimensions hence facilitate handling of the sample and its alignment with respect to the beam path. If the sample is a liquid, it will be confined in a quartz cell with defined path length or another material transparent to neutrons (e.g. Vanadium). In a *standard SANS* experiment the sample will usually be in thermodynamic equilibrium with its environment, at given chemical composition (concentration), pressure and temperature. *Without* external constraints, the scattered intensity (e.g. from a dissolved macromolecule) is a result of the space and time average of all molecular conformations and orientations. An (azimuthally) isotropic scattering pattern is observed on the two-dimensional multidetector, typically used at SANS instruments such as D11 of the ILL (Schmatz, Springer, Schelten and Ibel, 1974; Ibel, 1976; Lindner, May and Timmins, 1992).

Under *non-equilibrium conditions* in general, an external field is applied to the sample during the scattering experiment. By submitting the sample to external constraints such as *magnetic*, *electric*, or *hydrodynamic fields* (cf. Fig. 1) structural changes can be induced. Furthermore, various fields can be imposed in different modes of operation as shown schematically in Fig. 2. Thus, in a *kinetic* experiment (cf. Fig. 2a), an activation process of the sample can be studied after a short perturbation pulse (dashed line), or it is allowed to relax back to equilibrium after an externally imposed perturbation of its equilibrium configuration (relaxation). In a *cyclic* experiment (cf. Fig. 2b) the sample is periodically distorted around its equilibrium state. A third type of non-equilibrium experiment is the *steady-state* experiment with keeping the external constraint constant during the measuring time t (cf. Fig. 2c).

Studies under *steady-state flow* are particularly interesting because of their technological implications. When the solute particles of a liquid sample are exposed to a hydrodynamic field they experience forces due to viscous drag in the streaming fluid which tend to orient, to deform or to order them. The scattering intensity distribution on a two-dimensional multidetector might in this case become anisotropic with respect to the direction of flow. Complex features, such as viscoelasticity, thixotropy, rheopexy or drag reduction, are well

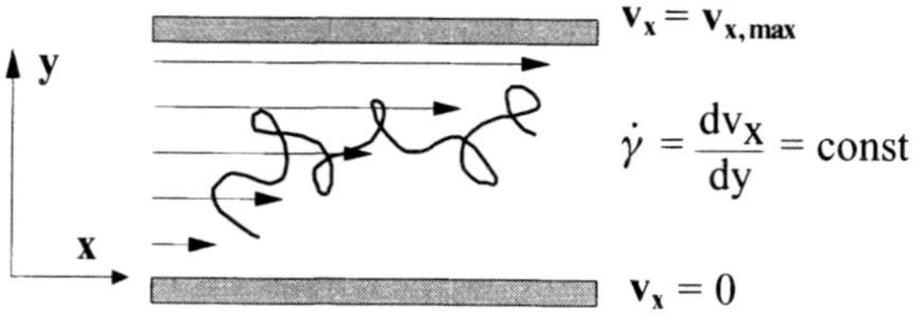

Fig. 1. Example for a hydrodynamic field (plane Couette flow): a constant, transverse (shear) gradient $\dot{\gamma}$ is acting between two parallel plates on a sheared fluid containing polymer molecules.

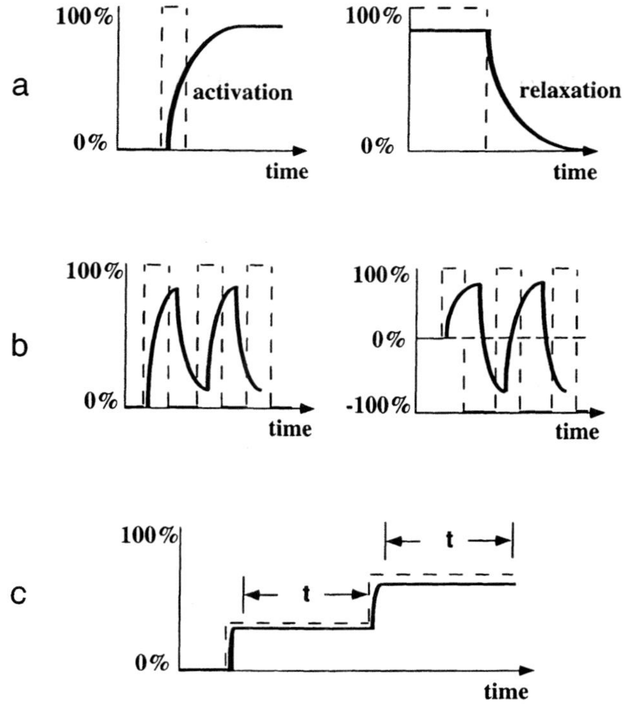

Fig. 2. Possibilities for kinetic (a), periodically cyclic (b) and steady state modes (c) in non-equilibrium experiments. The dashed line represents the perturbation of the system by an externally applied field. Note that in a steady-state experiment the system is usually exposed to a constant perturbation during the measuring time t (e.g. measurement at constant flow velocity in a hydrodynamic field).

known from (macroscopic) rheological measurements. However, the underlying (microscopic) mechanisms on the *molecular scale* have until now not been completely understood and are a matter of intensive ongoing research activity both from the experimental and the theoretical point of view (Bird et al., 1977).

3. Flow apparatuses for small-angle neutron scattering

From theoretical aspects as well as for experimental reasons, the linear laminar *Couette flow* is of great importance because of its simplicity. Couette shear flow is given between two parallel plates, one at rest and the other plate moving with constant velocity v_x due to the action of a constant external force (cf. Fig. 1). In the case of the *plane* Couette flow the transverse (shear) gradient $\dot{\gamma}$ can be written as:

$$\dot{\gamma} = \frac{v_x}{d} = \text{const} \tag{1}$$

where d is the distance between the two plates. A practical example is the Couette-type shear apparatus constructed for neutron scattering experiments with liquid systems in a constant shear gradient at the SANS instruments of the Institut Laue-Langevin (Lindner and Oberthür.

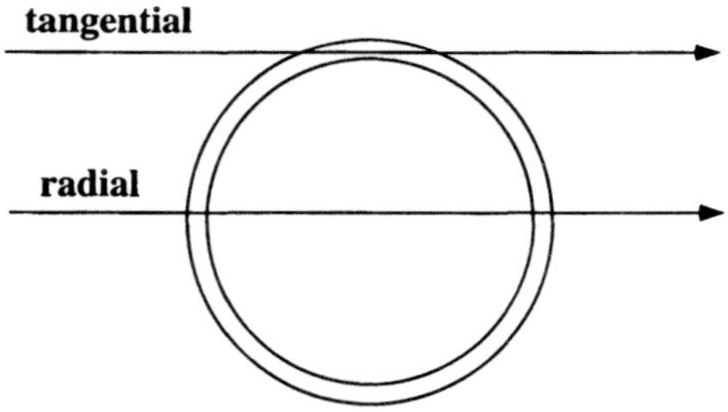

Fig. 3. Schematic sketch of a SANS shear cell (top view) with the radial and tangential beam configuration. The sample is confined in the gap between the two concentric cylinders.

1984). The sample container consists of an inner fixed piston (stator) and a concentric outer rotating beaker (rotor), both made of quartz glass, which is highly transparent for thermal and cold neutrons and shows a very low small-angle scattering background. The sample is confined in the annular gap between rotor and stator. The outer cylinder is rotating at a constant speed. The inner static cylinder is under temperature control. The gap width is sufficiently small ($d \approx 0.5$ mm) compared to the cylinder diameter ($d \approx 48$ mm) and to a good approximation a *plane* Couette flow with a *constant transverse* (or shear) gradient [cf. Eq. (1)] up to $\dot{\gamma} \approx 12{,}500\ \mathrm{s}^{-1}$ can be realized in the annulus (sample volume ≈ 4.5 ml).

The incident primary neutron beam hits the sample normal to the axis of rotation (z-direction) of the outer cylinder. Depending on the precisely adjustable horizontal position of the shear apparatus, two different measurement configurations can be realized (cf. Fig. 3). In the *radial* configuration, the neutron beam passes through the middle of the cell, along the direction of the shear gradient $\dot{\gamma} = |\boldsymbol{v}|/d$. It is thus *perpendicular* to the direction of flow $\boldsymbol{v}$ (cf. Fig. 1). In the *tangential* configuration (by using an appropriate slit diaphragm) the neutron beam passes through the gap of the shear cell along the flow direction and perpendicular with respect to the gradient direction. These two experimental possibilities are an important tool for studying shear-induced orientation and structural transitions with lyotropic lamellar phases (cf. Section 4.2).

The Couette type shear apparatus has become routine equipment for SANS experiments at ILL with sheared complex fluids. Similar flow cells for special requirements (such as a variation of the shear gradient direction with respect to the direction of the primary neutron beam or an air-tight flow cell for investigation of volatile liquid systems) have been constructed by other groups and successfully used at the ILL (Thurn et al., 1984; Johnson et al., 1988; Cummins et al., 1990; Kalus et al. 1990) or elsewhere (Nakatani et al., 1990a; Baker et al., 1994; Straty et al., 1998).

A recent development (rheo-SANS) is the installation of a commercial rheometer (Bohlin CVO 120HR) at the SANS instruments of the ILL, in order to correlate directly rheological properties of complex fluids with the underlying microstructure (Lindner et al., 1999). For this purpose a temperature controlled Searle type shear cell has been developed. The cell consists of two concentric cylinders (inner cylinder rotating) and is made of high quality quartz glass. Its geometry is similar to the Couette cell described above. The outer cylinder is fixed in a specially designed steel jacket for controlling the temperature in the gap of the shear cell by a water flow circuit around the outer cylinder. The equipment allows to perform many different rheological tests like simple shear flow either at constant shear stress or constant shear rate, creep and creep recovery as well as low and large amplitude oscillatory shear deformation, together with on-line SANS experiments in radial and tangential beam configuration (cf. Fig. 3).

Other types of flow can be distinguished from Couette- or Searle flow and might be of experimental relevance either in pure form or in combination. For instance, *elongational* flow with a *longitudinal* gradient $\dot{\varepsilon}$ is best illustrated with a liquid flowing from an orifice in the bottom of a vessel, due to the action of a gravitational field. On the other hand, pure *shear* flow with a *transverse* velocity gradient is realized when a fluid streams through a tube at constant flow rate. The velocity at the tube wall is zero and increases towards the centre of the tube. The parabolic velocity profile (*Poiseuille flow*), causes a non-constant, transverse flow gradient which linearly decreases from its maximum value at the tube wall to zero in the central tube axis. Poiseuille-type experiments have been performed both with a quartz tube flow apparatus for SANS measurements with drag reducing rod-like micellar solutions under laminar and turbulent flow conditions (Lindner et al., 1990), as well as with a quartz slit die apparatus for measurements with flowing concentrated polymer dispersions (Laun et al., 1992).

4. Examples

The investigation of polymer solutions and of lyotropic lamellar phases in laminar (steady state) shear flow has been chosen here as two examples for illustrating the various possibilities which are offered by the SANS technique under non-equilibrium conditions and which are complementary to macroscopic rheological measurements.

4.1. Sheared polymer solutions

The behaviour of dilute and semi-dilute polymer solutions in the *quiescent* state has been extensively studied during the past decades by scattering techniques, and in particular with SANS (Higgins and Benoît, 1994; Chapter 11 by P. Schurtenberger in this book). The technique is resolving the range of momentum transfer Q in the polymer scattering curve from length scales of the order of the overall coil size to the scattering of the flexible segment. It is therefore a particularly suitable method for obtaining information (i) on the statistical conformation of the single coil molecule in dilute solution, as well as (ii) on the structure of an entangled system in the semi-dilute concentration range. This section reviews some basic experimental results of shear-induced structural changes obtained with *flowing* polymer solutions. In particular, the method of small-angle neutron scattering in conjunction with laminar Couette shear flow in a scattering geometry as sketched in Fig. 1 has been applied.

4.1.1. Polymer conformation in sheared dilute solution (good solvent)

Under laminar shear flow conditions, information about the morphological changes on the molecular level are of fundamental importance for an understanding of the macroscopic properties of the flowing system. To what extent is the conformation of a single and more or less flexible polymer chain altered by the velocity gradient in the streaming solution, as sketched for instance in Fig. 1? A distortion of the polymer coil conformation in flow has long been predicted by theory. In a longitudinal gradient $\dot{\varepsilon}$, a sharp coil-stretch transition should occur for $\dot{\varepsilon}\tau \geq 1$, where τ is the longest relaxation time of the intra-molecular chain motion (DeGennes, 1974). In a transverse (or constant shear) gradient $\dot{\gamma}$ a gradual transition towards an anisotropic intersegmental distance distribution is expected with increasing shear gradient (Peterlin et al., 1958). A common feature of the different theoretical treatments of the ideal flexible chain molecule in shear flow is to consider the balance of (i) frictional forces, (ii) contractile forces, and (iii) Brownian motion of the chain segments. The friction forces acting on the polymer chain segments in the shear gradient result in a combined rotation and deformation of the coil beside the translative motion in flow direction, whereas the elastic properties of the chain resist a change of the shape to less probable chain conformations. According to the theories, the coil shape is deformed in shear flow and changes to an anisotropic conformation with the symmetry of an ellipsoid of rotation. A dynamic effect, however, limits the deformation of the single coil in shear flow: shape changes require changes in chain conformation, e.g. intra-molecular bond rotation, which is hindered by local energy barriers. With *scattering* techniques it is possible in principle to observe the overall molecular dimensions *directly* and to probe the chain extension of a single macromolecule under flow conditions.

The shear-induced deformation of polystyrene in dilute solution has been systematically studied by SANS experiments in laminar shear flow at the ILL. Considering the Q-dependent dynamics of a single polymer coil in solution, experimental conditions for shear-induced deformation can be roughly estimated (Lindner and Oberthür, 1985): the interaction between shear gradient $\dot{\gamma}$ and the longest segmental relaxation time $\tau_{\max}$ of the polymer chain implies the condition

$$\dot{\gamma}\tau_{\max} \geq 1 \qquad (2)$$

in order to permanently distort the chain conformation in shear flow. It turns out that by *increasing* the solvent viscosity the segmental relaxation can be slowed down such that the relationship [2] is fulfilled and a shear deformation can be observed under given experimental conditions with SANS measurements (Lindner and Oberthür, 1985).

Fig. 4 shows as an example the uncorrected two-dimensional SANS spectra of a dilute polystyrene solution (molecular mass 280,000 g/mol, mass concentration of polymer $c = 8$ mg cm^{-3}) in a viscous ('good') solvent and measured in laminar Couette shear flow with radial beam configuration (cf. Section 3). The SANS intensity of the solution at rest (data not shown) is azimuthally isotropic, whereas the sheared polymer solution (Fig. 4) shows an anisotropic intensity distribution on the two-dimensional multidetector with elliptical symmetry (long axis perpendicular with respect to flow direction). Moreover, the magnitude of the effect depends upon molecular mass, shear gradient, solvent viscosity and Q-range. From the normalized scattering intensity and after extrapolation to zero concentration the scattering curve (form factor) of the single polystyrene chain in sheared dilute solution is obtained and shown in Fig. 5.

The distinct anisotropy, already visible in the uncorrected, two-dimensional detector pattern (cf. Fig. 4) is observed in the region of small values of the momentum transfer Q, on the length scale of the radius of gyration (a measure of the overall molecular size of the polymer coil). The anisotropy of the form factor decreases with increasing momentum transfer Q. A

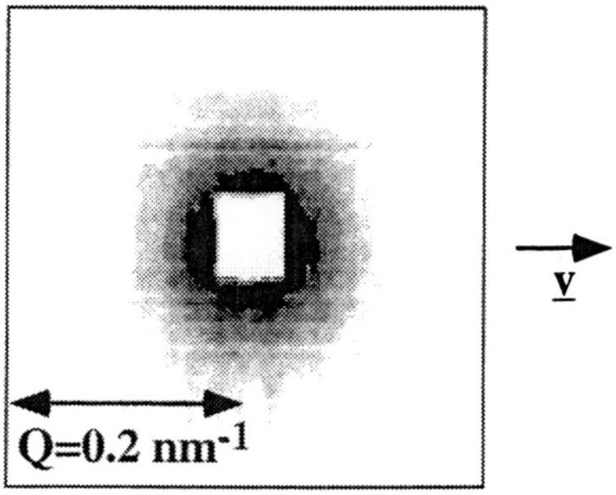

Fig. 4. SANS pattern (raw data) of a dilute polymer solution in a good solvent (polystyrene, $M_w = 280{,}000$ g/mol, solvent oligostyrene + toluene, solvent viscosity $\eta = 50$ mPa s, polymer concentration $c = 8$ mg cm^{-3}) measured at D11 of the ILL, Grenoble. Solution in laminar shear flow at a gradient of $\dot{\gamma} = 6000$ s^{-1} (the light rectangle in the middle of the detector is due to the primary neutron beam stop for detector protection, the arrow on the right side indicates the flow direction $\underline{v}$).

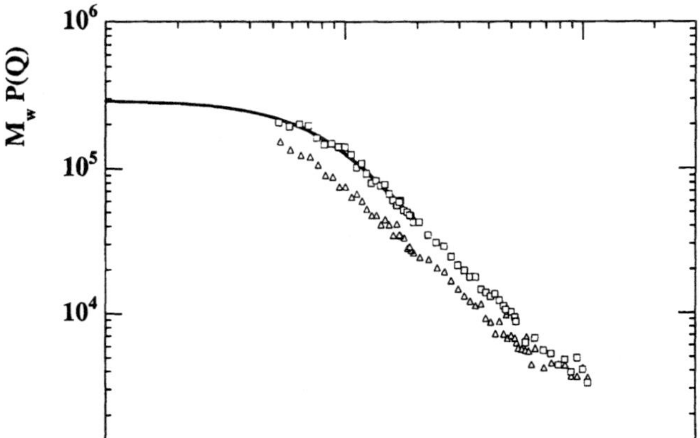

Fig. 5. Normalized scattering curve $M_w \cdot P(q) = f(q)$ of polystyrene ($M_w = 280{,}000$ g/mol, good solvent) in a shear gradient of $\dot{\gamma} = 6000$ s^{-1}, evaluated in the directions perpendicular ($\perp$, squares) and parallel ($\parallel$, triangles) to the flow direction (data from Fig. 4). The drawn line corresponds to the scattering curve of the polystyrene in the quiescent state (the momentum transfer q is given in units of Å^{-1}, 1 Å^{-1} = 10 nm^{-1}).

crossover to isotropic scattering with respect to the directions parallel and perpendicular is observed at large Q-values. It indicates that the *short range* distribution of chain segments is not affected by the flow field on the length scale of the statistical chain element. Fig. 6 shows the experimental radii of gyration, resulting from the low Q region, as a function of the various shear gradients $\dot{\gamma}$ applied to the polystyrene solution. The value $\langle R_\parallel \rangle$ (obtained in direction parallel to the flow) increases significantly with increasing shear gradient $\dot{\gamma}$, whereas

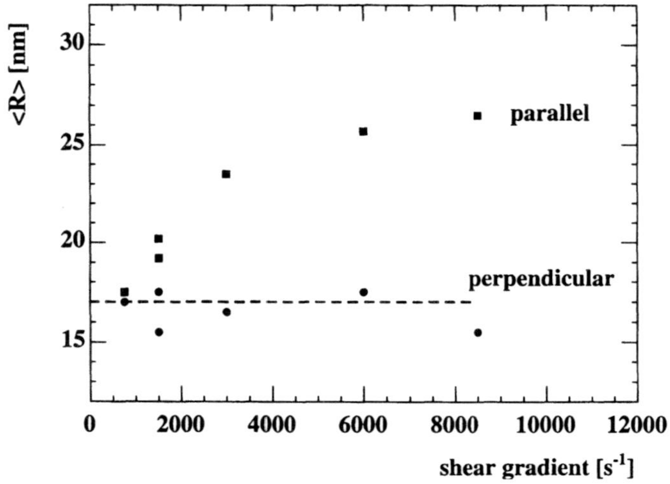

Fig. 6. Experimental radii of gyration as determined from the solution at rest ($\langle R_{\rm iso}\rangle$, dashed line) and from the sheared solutions ($\langle R_\perp\rangle$: filled circles, $\langle R_\parallel\rangle$: filled squares) as a function of shear gradient $\dot\gamma$.

the value $\langle R_\perp\rangle$ (obtained in direction perpendicular to the flow) is within experimental error identical to the radius of gyration $\langle R_{\rm iso}\rangle$ of polystyrene in the quiescent state and independent of $\dot\gamma$.

Based on the experimental radii of gyration (cf. Fig. 6), the deformation ratio of the polystyrene chain,

$$\frac{\langle R^2\rangle}{\langle R^2_{\rm iso}\rangle} - 1 \tag{3}$$

can be calculated as a function of the reduced shear gradient

$$\beta = \frac{[\eta]\eta_{\rm s} M_w \dot\gamma}{RT} \tag{4}$$

$\langle R^2\rangle$ and $\langle R^2_{\rm iso}\rangle$ are the averaged squares of the radii of gyration of the deformed coil and of the coil in solution at rest respectively; $[\eta]$ is the intrinsic viscosity, $\eta_{\rm s}$ is the solvent viscosity and R is the gas constant (the other symbols have the usual meaning). The reduced gradient β is a dimensionless orientation variable for re-scaling the different experimental conditions (e.g., molecular mass M_w, solvent viscosity $\eta_{\rm s}$ and shear gradient $\dot\gamma$). The gradient dependence of the deformation ratio (cf. Fig. 7) allows quantitative comparison with theoretical predictions for shear deformation of a single polymer coil, including the effect of a limitation of dynamic flexibility.

An *ideal flexible* behaviour with a β^2 dependence is only found at low gradients whereas with increasing β, the deformation of the polymer chain is found to be smaller than for a dynamically infinitely flexible molecule, in qualitative agreement with theory (Lindner and Oberthür, 1988). Recently, a N-bead chain composed of N monomers ($N \le 50$) in a solvent subjected to shear flow has been treated by non-equilibrium molecular dynamics (NEMD) in

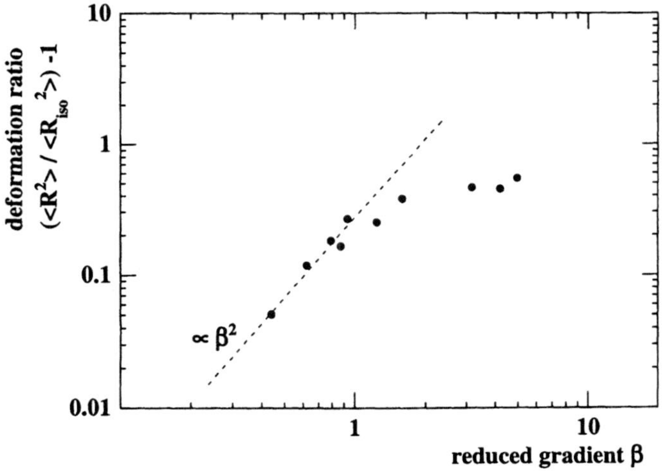

Fig. 7. Experimental deformation ratio $(\langle R^2\rangle/\langle R^2_{iso}\rangle) - 1$ as a function of the reduced gradient β. The β^2-dependence (continuous line) is predicted for an infinitely flexible coil for the Rouse as well as for the Zimm model (Lindner and Oberthür, 1988).

a computer simulation (Pierleoni and Ryckaert, 1995). Results are consistent with the above described SANS results for polystyrene in good solvent conditions.

4.1.2. Sheared dilute polymer solutions: variation of solvent quality

The example discussed so far was focussed on the case of sheared *dilute* polymer solutions under *good* solvent conditions, i.e. the study of a deformed single polymer chain with excluded volume. An investigation of the polymer-solvent system under shear and close to the theta point (second osmotic virial coefficient $A_2 = 0$) is not only interesting with respect to a check of the theoretical predictions concerning the overall coil deformation. The thermodynamic response under non-equilibrium conditions is a further point of interest. An increasing theta temperature under shear, due to the externally applied shear force, has been predicted theoretically in terms of a decrease of excluded volume due to intersegmental interaction (orientation and deformation of the chain) (Sariban, 1986). Shearing of the polymer system is, however, not equivalent to a simple effective temperature shift but involves a more complicated interplay between hydrodynamic and thermodynamic effects. For instance, shear-induced mixing has been observed in polymer blends with an apparent shift of the phase boundary only in direction parallel to the flow (Nakatani et al., 1990b). Recent SANS experiments at D11 under shear (Lindner, 1996) confirmed earlier findings with a solution of perdeuterated polystyrene in the solvent di-(octylphthalate) [DOP, which is a convenient and non-volatile solvent for polystyrene and a theta solvent for *protonated* polystyrene at a temperature of $T \approx 22°C$ (Park and Berry, 1989)]. At a temperature of $T \approx 22°C$ and at shear gradients below $\dot{\gamma} \approx 5000\ s^{-1}$, an anisotropic SANS pattern is observed, indicating a similar gradient-dependent chain deformation as in the good solvent case. At higher shear gradients, however, above a critical shear gradient of the order of $\dot{\gamma} \approx 5000\ s^{-1}$, a (reversible) increase

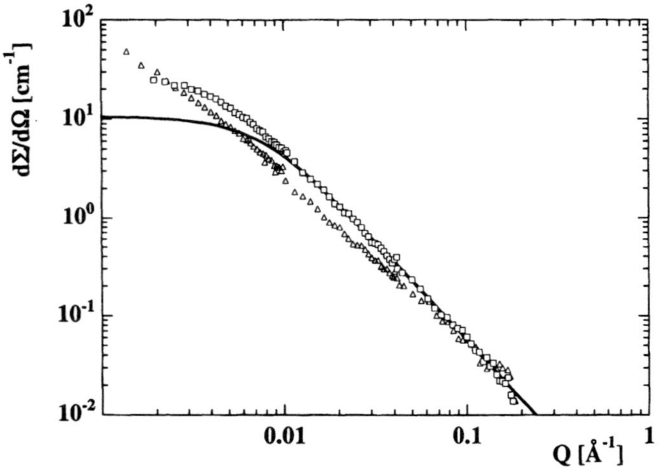

Fig. 8. Differential scattering cross section of polystyrene-d8 ($M_w = 672{,}000$ g/mol) in the theta solvent DOP at $T = 22$°C and at a shear gradient of $\dot{\gamma} = 7500\ \mathrm{s}^{-1}$, evaluated in the directions perpendicular (⊥, squares) and parallel (‖, triangles) to the flow direction. The drawn line corresponds to the scattering curve of the polystyrene in the quiescent state (the momentum transfer Q is given in units of Å^{-1}, 1 Å^{-1} = 10 nm^{-1}).

of the scattering intensity is observed in the low Q-range, in addition to the anisotropy in the intermediate Q-range.

Fig. 8 shows the resulting normalized scattering curve for a sheared polystyrene of molecular mass $M_w = 672{,}000$ g/mol in the theta solvent DOP (to be compared with the good solvent case shown in Fig. 5). In both directions, *perpendicular* and *parallel* with respect to flow, the intensities in the sheared solution are at low Q higher, compared to the solution measured at rest. A similar result, but without observation of a scattering anisotropy had been found by Hammouda et al. (1992). Further experiments are necessary to elucidate the physical origin of these results, either in terms of shear enhanced density fluctuations (Helfand et al., 1989, Milner, 1991) or in terms of shear-induced phase separation based on thermodynamic arguments (Rangel-Nafaile et al., 1984).

4.1.3. Sheared semi-dilute polymer solutions

A recent subject of theoretical and experimental interest is scattering studies of polymers in sheared *semi-dilute* solution (Larson, 1992). It is intuitively clear that the situation is more complex than for the case of dilute solutions, where single chain behaviour is observed: with increasing concentration the deformation of the polymer coils in flowing solution is increasingly influenced by topological constraints, due to molecular entanglements. Above the overlap concentration c^* the chains form a three-dimensional transient network that extends throughout the solution. During flow, the network is stretched but also continuously regenerated. Using light scattering (LS) and SANS with various scattering geometries (different beam paths with respect to the shear gradient direction) and shear flow geometries (Couette or cone-and-plate), it had been observed by several groups that the scattering intensity of a semi-dilute solution is strongly increased by shearing the system. It should be

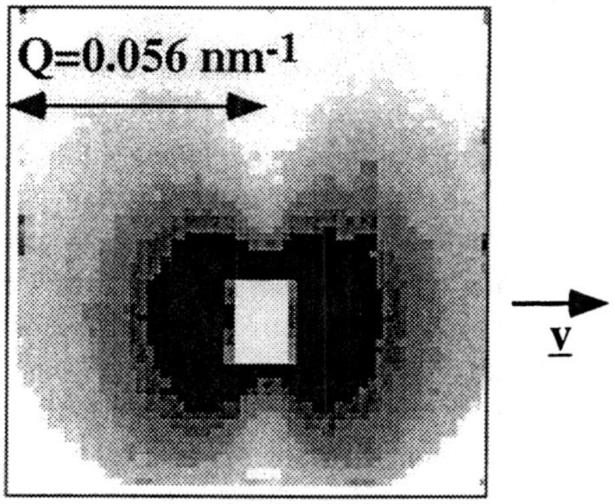

Fig. 9. SANS 'butterfly' pattern (raw data) of a semi-dilute polymer solution (polystyrene, $M_w = 514{,}000$ g/mol, solvent Di-octylphthalate (DOP), concentration 97 g/l $\approx$ 9 wt%, temperature $T = 22°C$), measured at D11 of the ILL, Grenoble. Solution in laminar shear flow at a gradient of $\dot{\gamma} = 150\ s^{-1}$ (the light rectangular spot in the middle of the detector is due to the primary neutron beam stop, the arrow on the right side indicates the flow direction $\underline{v}$).

stressed at this point that LS and SANS differ by about an order of magnitude in the range of momentum transfer Q of the experiment. SANS, on the higher Q side, explores shorter length scales in the system (ranging from $Q\xi < 1$ to $Q\xi \gg 1$, ξ being the correlation length in the quiescent solution) as compared to LS. Flow enhanced concentration fluctuations have been found to be either anisotropic with respect to the flow direction (Wu et al., 1991; van Egmond et al., 1992; Hashimoto et al., 1991, 1992) or to be isotropic (Hammouda et al., 1992). An ongoing debate concerns the question whether shear induces a shift in the cloud point of the solution or whether the concentration fluctuations in the entangled polymer system are amplified and distorted by other mechanisms, *without* a shift in the critical temperature for phase separation (Rangel-Nafaile et al., 1984; Helfand and Fredrickson, 1989; Onuki, 1990; Bastide et al., 1990; Milner, 1991).

In recent SANS studies, semi-dilute solutions of *perdeuterated* polystyrene in DOP were investigated under steady state shear flow over wide ranges of concentration, temperature and shear rate (Boué and Lindner, 1994, 1995, Morfin et al., 1999). The experiments have been performed at different shear gradients $\dot{\gamma}$ in a regime $Wi \geq 1$, where Wi is the Weissenberg number $Wi = \dot{\gamma}\tau_c$, with τ_c being the longest relaxation time of the system. Variation of the temperature allowed tuning of the solvent quality. Hence, different regions of the phase diagram of the polymer solution could be explored, ranging from poor solvent conditions, in the vicinity of the phase boundary, across the theta region to the good solvent regime.

Fig. 9 shows a typical example of a SANS pattern, obtained with a semi-dilute polystyrene solution (9 wt%) under steady state Couette shear flow in radial beam configuration at ambient temperature. At rest (not shown), an azimuthally isotropic scattering pattern is observed. Under shear and at low momentum transfer Q the pattern shown in Fig. 9 reveals a significantly different feature compared to the scattering of the sheared *dilute* solution (cf. Fig. 4). The pronounced intensity increase *parallel* to the flow direction leads to a double winged shape of the pattern which resembles closely the 'butterfly patterns' observed in deformed gels, rubbers and in polymer melts (Mendès et al., 1991; Bastide et al., 1988; Boué et al., 1991). The effect is increasing with increasing shear. Compared to the measurements in the theta region, it is smoother but still visible in the good solvent regime ($T = 50°C$). At temperatures below $T = 22°C$, however, the scattering becomes stronger in all directions and contours tend to have an elliptical shape which superimposes on the butterfly pattern. At even lower temperatures, when approaching the two-phase region, the scattering is finally dominated by an increase of intensity along all directions (characteristic for shear-induced demixing). Globally speaking, a structure coarsening with different stages can be deduced from the scattering patterns observed under different conditions (solution at rest, low shear gradient/medium–high temperature, large shear gradient/room temperature or low shear gradient/low temperature, highest shear gradient/lowest temperature in the vicinity of the cloud point). Interestingly, it has been also observed that the mechanical properties of the semi-dilute solution under shear flow follow a time-temperature superposition: the measured shear viscosity does only depend on the Weissenberg number *Wi*, in contrast to the flow-induced structures as seen with SANS (Morfin et al., 1999).

Even more of the rich behaviour of sheared semi-dilute polymer systems has been revealed in a recent work, combining SANS with small-angle light scattering (SALS) experiments (Saito et al., 2002). The shear-induced structures in a higher molecular mass polystyrene solution were analyzed in a range covering 3 orders of magnitude in momentum transfer Q and 7 orders of magnitude in intensity. At a given temperature, shear flow induces a structural change in the solution at shear rates larger than a critical shear rate $\dot{\gamma}$. It gives rise to a butterfly pattern (see example of Fig. 9) in the low Q-region of SANS, but the same pattern as that in the quiescent solution is observed with SALS. With increasing $\dot{\gamma}$, the butterfly pattern also evolves in the Q region of SALS, indicating that the shear-induced structures responsible for the butterfly pattern are developing over a wide Q-range with increasing $\dot{\gamma}$. At a given shear rate $\dot{\gamma}$ larger than a critical shear rate $\dot{\gamma}_c$, both SALS and SANS intensities increase with a decrease of temperature, indicating that the shear-induced structures are more pronounced at lower temperature. Especially at the lowest temperature covered in the experiment the low-Q SANS pattern is developing an ellipical shape with a major axis perpendicular to the flow direction while the butterfly pattern is maintained in the SALS Q-region.

4.2. Shear-induced transitions in lamellar surfactant phases

Amphiphilic molecules can form a variety of different structures in water, due to self-assembly. These mesophases are composed of aggregates with different morphology such as spheres, rods or disks, depending on the concentration, temperature and salinity. At high concentrations, the aggregates can form ordered phases of nematic, hexagonal, smectic or other symmetry. Surfactant phases as well as systems composed of amphiphilic polymers display in general a complex rheological behaviour and they are the subject of intense research because of their great technological relevance. Again, as for polymer solutions, different techniques are applied in order to relate macroscopic flow properties to the underlying micro- and mesoscopic structures of these complex fluids. For learning about more details of this research field the reader is referred to recent reviews (Butler, 1999; Hamley, 2000; Mortensen, 2001; Richtering, 2002) and references given therein.

In the following we briefly discuss the case of sheared *lyotropic* lamellar phases, i.e., stacks of surfactant bilayers (repeat distance d), separated by a solvent (usually water). Two main different processes can be distinguished when studying shear-induced structural changes in

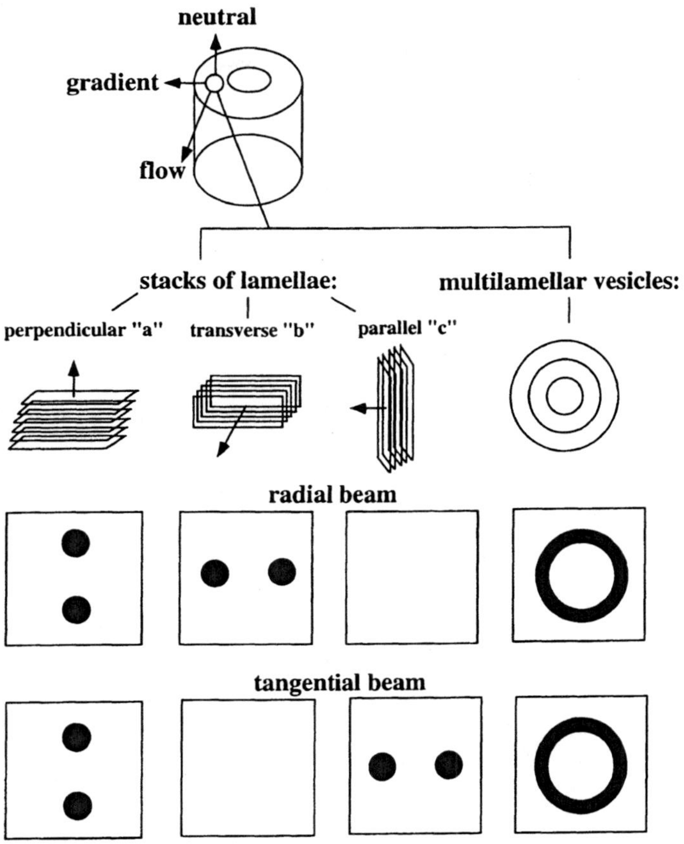

Fig. 10. Schematic view of oriented lamellae and of spherical multilamellar vesicles in Couette shear flow together with the corresponding scattering patterns on a two-dimensional detector with the beam in radial configuration (the detector plane corresponds to the plane defined by the neutral direction and the flow direction) and tangential configuration (the detector plane corresponds to the plane defined by the neutral direction and the gradient direction). Lamellar layers can have three possible orientations with respect to the wall of the shear cell: perpendicular ("a"): layer normal parallel to the neutral direction, transverse ("b"): layer normal parallel to the flow direction and parallel ("c"): layer normal parallel to the shear gradient direction. Spherically symmetric multilamellar vesicles have an isotropic orientation of layer normals, leading to an azimuthally isotropic diffraction pattern for both beam configurations.

these systems: (i) lamellae can align in a flow field with different orientations, and (ii) they can form shear-induced multilamellar vesicles (MLV's or 'onions'), as sketched in Fig. 10 (Zipfel, 2000).

For the three possible orientations of lamellae and for MLV's, the corresponding scattering patterns (Bragg scattering of the inter-lamellar distance) are schematically shown in the lower part of Fig. 10 for the radial (beam along the velocity gradient) and for the tangential (beam along the flow direction) configuration (see Section 3). The radial beam experiment is sensitive for the perpendicular ("a") and the transverse ("b") orientation of lamellae, whereas the tangential beam experiment can distinguish between parallel ("c") and perpendicular ("a") orientation. No scattering is expected when the normal of the lamellae is oriented along the beam. As the transverse ("b") orientation is almost never observed experimentally, the tangential beam is crucial when shear-induced transitions occur. The spherical symmetry of multilamellar vesicles with an isotropic distribution of membranes in the shear cell gives rise to an isotropic distribution of layer normals with an isotropic diffraction pattern in the radial as well as in the tangential configuration. An increase of the scattering intensity in the flow direction (radial beam) is a hint to the onset of shear-induced vesicle formation since the transverse ("b") orientation is not expected. A diffraction ring in the tangential configuration could also be caused by an isotropic distribution of the directors of the lamellae oriented in the gradient-vorticity plane. Therefore, the radial beam experiment is important when the shear-induced formation of MLV's is investigated.

One aspect of recent research on surfactant membranes is to reveal the influence of macromolecules as guest components and the response of the system upon mechanical deformation, such as shear flow (shear-induced formation of multilamellar vesicles). Studying the addition of water-soluble polymers to surfactant systems is of considerable practical interest because polymer–surfactant mixtures are widely applied in pharmaceutical (drug encapsulation) and cosmetic applications and it is therefore important to know how polymers affect the vesicle formation. The influence of added, water soluble polymer on the structure and flow behaviour of a lyotropic lamellar phase SDS/decanol/water is illustrated with the example shown in Fig. 11 (Zipfel et al., 1999; Berghausen et al., 2001). Obviously, the flow curve of the sample was extremely altered when 0.5 wt% of poly(N-isopropylacrylamide) (PNIPAM) is added to the system. *Without* added polymer, the sample was shear thinning, whereas shear thickening *and* shear thinning as observed *with* the polymer-doped sample. The micostructure of the sample was investigated simultaneously by SANS at the instrument D11 of the ILL as a function of increasing shear stress, using the Bohlin CVO120HR rheometer in tangential beam configuration (rheo-SANS, see Section 3). For the sample without polymer (lower curve in Fig. 11, filled squares), a shear-induced alignment of the lamellae was found. At low shear stress, a Bragg peak is observed along the velocity gradient direction, thus the lamellae are aligned parallel to the walls of the shear cell. With increasing shear, the layers flip to the perpendicular orientation, which is characterized by a scattering peak along the neutral (vorticity) direction. The data for the sample without polymer clearly show that shear thinning is correlated with a re-orientation process of the surfactant double layers.

A different behaviour was observed when a small amount (0.5 wt%) of water-soluble PNIPAM was added to the aqueous surfactant solution (upper curve in Fig. 11, filled circles). The SANS data obtained at low and high shear stress were similar to those obtained from the sample without polymer. At *intermediate* shear stress, however, the Bragg peak was observed along the entire azimuthal trace on the two-dimensional multidetector in the tangential beam configuration. From the isotropic Bragg scattering one can deduce the shear-induced formation of multilamellar vesicles (MLVs). Obviously, the vesicles are destroyed at high shear and a perpendicular orientation of the planar layers was found. Again, the simultaneous detection of rheological and scattering data allows for a direct correlation of flow properties with the shear-induced microstructure. For the surfactant system discussed above, the transition from parallel to perpendicular layer alignment was hardly influenced by addition of polymer. The added polymer does, however, strongly affect the vesicle formation. Presence of other water soluble polymers such as hydroxyethyl starch or N-vinyl caprolactame can even result in

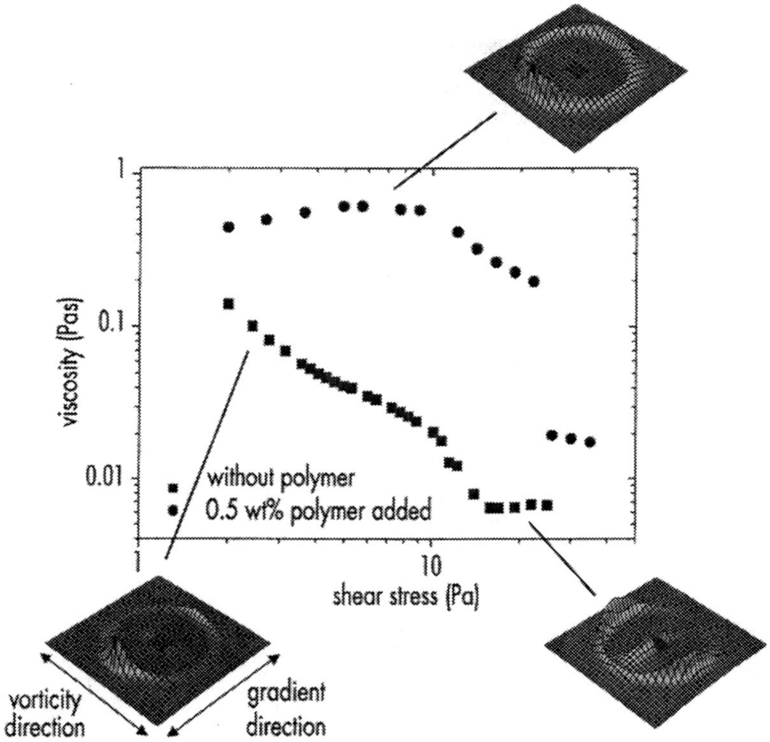

Fig. 11. Flow curves and SANS spectra measured in tangential beam configuration for a lyotropic lamellar phase SDS/decanol/water. Filled squares: no polymer added. Filled circles: after addition of 0.5 wt% PNIPAM (poly(N-isopropylacrylamide)).

a complete suppression of vesicle formation (Berghausen et al., 2001), in contrast to the surfactant/PNIPAM system shown as an example in Fig. 11.

5. Summary

Neutron scattering is a convenient technique for studying changes of the microscopic structure of complex fluids in a non-equilibrium state such as shear flow. In soft condensed matter research this type of experiments is attracting increasing attention. Combining the classical scattering method with an externally applied field, for instance shear flow, important information is obtained, about structural changes of deformable or anisotropic particles with respect to shape and size, on dynamic properties as well as on thermodynamic behaviour.

References

Baker, S.M., Smith G., Pynn, R., Butler, P., Hayter, J., Hamilton, W. and Magid, L. (1994). *Rev. Sci. Instr.*, **65**, 412.
Bastide, J., Buzier, M. and Boué, F. (1988). *Springer Proc. Phys.*, **29**, 112–120.

Bastide, J., Leibler, L. and Prost, J. (1990). *Macromolecules*, **23**, 1821.
Berghausen, J., Zipfel, J., Lindner, P. and Richtering W. (2001). *J. Phys. Chem. B*, **105**, 11081–11088.
Bird, R.B., Armstrong, R.C. and Hassager, O. (1977). *Dynamics of Polymeric Liquids*. John Wiley, New York.
Boué, F. and Lindner, P. (1994). *Europhys. Lett.*, **25** (6), 421.
Boué, F. and Lindner, P. (1995). In: *Flow Induced Structure in Polymers, ACS Symposium Series 597* (Nakatani, A.I. and Dadmun, M.D., eds.). ACS, Washington.
Boué, F., Bastide, J., Buzier, M., Lapp, A., Herz, J. and Vilgis, T.A. (1991). *Colloid Polym. Sci.*, **269**, 195–216.
Butler, P. (1999). *Current Opinion Colloid Interface Sci.*, **4**, 214–221.
Cummins, P.G., Staples, E., Millen, B. and Penfold, J. (1990). *J. Meas. Sci. Technol.*, **1**, 179.
DeGennes, P.G. (1974). *J. Chem. Phys.*, **60**, 5030.
Hamley, I.W. (2000). *Current Opinion Colloid Interface Sci.*, **5**, 342–350.
Hammouda, B., Nakatani, A.I., Waldow, D.A. and Han, C.C. (1992). *Macromolecules*, **25**, 2903–2906.
Hashimoto, T. and Fujioka, K. (1991). *J. Phys. Soc. Japan*, **60**, 356.
Hashimoto, T. and Kume, T. (1992). *J. Phys. Soc. (Japan)*, **61**, 1839.
Helfand, E. and Fredrickson, G.H. (1989). *Phys. Rev. Lett.*, **62**, 2468.
Higgins, J.S. and Benoît, H.C. (eds.) (1994). *Polymers and Neutron Scattering*. Clarendon Press, Oxford.
Ibel, K. (1976). *J. Appl. Cryst.*, **9**, 296–309.
Johnson, S.J., deKruiff, C.G. and May, R.P. (1988). *J. Chem. Phys.*, **89**, 5909.
Kalus, J., Neubauer, G. and Schmelzer U. (1990). *Rev. Sci. Instrum.*, **61**, 3384.
Larson, R.G. (1992). *Rheol. Acta*, **31**, 497–520.
Laun, H.M., Bung, R., Hess, S., Loose, W., Hess, O., Hahn, K., Hädicke, E., Hingmann, R., Schmidt, F. and Lindner, P. (1992). *J. Rheol.*, **36** (4), 743–787.
Lindner, P. (1996). unpublished results.
Lindner, P. and Oberthür, R.C. (1984). *Rev. Phys. Appl.*, **19**, 759.
Lindner, P. and Oberthür, R. (1985). *Colloid Polym. Sci.*, **263**, 443–453.
Lindner, P. and Oberthür, R. (1988). *Colloid Polym. Sci.*, **266**, 886–897.
Lindner, P., Bewersdorff, H.W., Heen, R., Sittart, P., Thiel, H., Langowski, J. and Oberthür, R. (1990). *Progr. Colloid Polym. Sci.*, **81**, 107–112.
Lindner, P., May, R.P. and Timmins, P.A. (1992). *Phys. B*, **180 & 181**, 967–972.
Lindner, P., Zipfel, J., Berghausen, J., Nettesheim, F. and Richtering, W. (1999). *Ann. Rep. Inst. Laue-Langevin*, 84–85.
Mendès, E., Lindner, P., Buzier, M., Boué, F. and Bastide, J. (1991). *Phys. Rev. Lett.*, **66**, 1595.
Milner, S. (1991). *Phys. Rev. Lett.*, **66**, 1477.
Morfin, I., Lindner, P. and Boue, F. (1999). *Macromolecules*, **32**, 7208–7223.
Mortensen, K. (2001). *Current Opinion Colloid Interface Sci.*, **6**, 140–145.
Nakatani, A.I. and Dadmun, M. (1995). *Flow-Induced Structure in Polymers*. ACS, Washington.
Nakatani, A.I., Kim, H. and Han, C.C. (1990a). *J. Res. Natl. Inst. Stand. Technol.*, **95**, 7.
Nakatani, A.I., Kim, H., Takahashi, Y., Matsuhita, Y., Takano, A., Bauer, B.J. and Han, C.C. (1990b). *J. Chem. Phys.*, **93**, 795.
Oberthür, R.C. (1983). *Inst. Phys. Conf. Ser.*, **64**, 321.
Oberthür, R.C. (1984). *Rev. Phys. Appl.*, **19**, 663.
Onuki, A. (1990). *J. Phys. Soc. (Japan)*, **59**, 3423; **59**, 3427.
Park, J.O. and Berry, G.C. (1989). *Macromolecules*, **22**, 3022.
Peterlin, A., Heller, W. and Nakagaki, M. (1958). *J. Chem. Phys.*, **28**, 470.
Pierleoni, C. and Ryckaert, J.P. (1995). *Macromolecules*, **28**, 5097–5108.
Rangel-Nafaile, C., Metzner, A.B. and Wissbrun, A.F. (1984). *Macromolecules*, **17**, 1187.
Richtering, W. (2001). *Current Opinion Colloid Interface Sci.*, **6**, 446–450.
Saito, S., Hashimoto, T., Morfin, I., Lindner, P. and Boué, F. (2002). *Macromolecules*, **35**, 445–459.
Sariban, A. (1986). *Macromolecules*, **19**, 843.
Schmatz, W., Springer, T., Schelten, J. and Ibel, K. (1974). *J. Appl. Cryst.*, **7**, 96–116.
Straty, G.C., Muzny, C.D., Butler, B.D., Lin, M.Y., Slawecki, T.M., Glinka, C.J. and Hanley, H.J.M. (1998). *Nucl. Instrum. Methods Phys. Res.*, **A 408**, 511.
Thurn, H., Kalus, J. and Hoffmann, H. (1984). *J. Chem. Phys.*, **80**, 3440.

van Egmond, J.W., Werner, D.E. and Fuller, G.G. (1992). *J. Chem. Phys.*, **96**, 7742.
Wu, X.L., Pine, D.J. and Dixon, P.K. (1991). *Phys. Rev. Lett.*, **66**, 2408.
Zipfel, J. (2000). *Ph.D. thesis*, Univ. of Freiburg.
Zipfel, J., Berghausen, J., Lindner, P. and Richtering, W. (1999). *J. Phys. Chem. B*, **103**, 2841–2849.

Neutrons, X-Rays and Light
P. Lindner and Th. Zemb (Editors)

Chapter 18

Using Synchrotron Radiation to Study Structure Development in Polymer Processing

Anthony J. Ryan

Department of Chemistry, University of Sheffield, Sheffield S3 7HF, UK

1. Introduction

To make a useful artefact from a polymer, most polymer processes form a liquid polymer or a monomer into a useful shape using a device such as a die or a mould. Polymer processes may be broadly divided into the continuous (extrusion, coating) and discontinuous (moulding). If the liquid is not to lose its shape it must be solidified either in the mould or shortly after exit from the die. The primary solidification mechanisms are vitrification, crystallization, phase separation and crosslinking (Wilkinson and Ryan, 1998). Synchrotron radiation may be most readily used in the study of the crystallization and phase separation processes whereas spectroscopy (FT–IR, UV and Raman) and rheology are more appropriate for crosslinking and vitrification (Stanford et al., 1997).

Much of the knowledge concerning the crystallization of polymers comes from the application of programmed heating and cooling, and isothermal crystallization studies by differential scanning calorimetry (DSC) or optical microscopy combined with post-mortem assessment of morphology by either X-ray diffraction or electron microscopy (Barham, 1993). Similarly, the existence of phase separation in polymer blends or microphase separation in block copolymers is often assessed by DSC with confirmation sought by scattering, microscopy and dynamic mechanical thermal analysis. The kinetics of (micro)phase separation can be followed by DSC but the thermal response is weak and these experiments are often far from quantitative.

There is a reciprocity between the scattering angle and the size scale of the structure probed. The technique of wide-angle X-ray diffraction (WAXS) probes atomic distances (1–10 Å) and may be used to solve the crystal structure (establish the unit cell and atomic positions) if the full diffraction pattern of a single crystal or fibre is available (Baltá-Calleja and Vonk, 1989). WAXS is based on the scattering of radiation by atoms and may be used to calculated atomic positions. Due to the polycrystalline nature of most polymers it is more common to obtain the radially symmetric one-dimensional powder diffraction pattern and, wherever possible, index the structure from this. For example, the two crystalline forms of polypropylene can be readily distinguished from their powder diffraction patterns (Baltá-Calleja and Vonk, 1989).

Small-angle X-ray scattering (SAXS) is a well established technique for studying the morphology of multi-phase polymers (Baltá-Calleja and Vonk, 1989) and probes larger length scales (50–1000 Å). In SAXS, X-rays are scattered by regions with different electron densities. It is often used in tandem with DSC to study polymer crystallization and microphase separation in block copolymers (Ryan et al., 1995). Information is obtained in the form of a scattering pattern; as with WAXS unoriented materials have radially symmetrical one-dimensional patterns which can be analysed using Bragg's law leading to information on the structural features with size scales from 50 to 1000 Å. For liquid crystalline and semi-crystalline polymers this corresponds to the crystallite size, for block copolymers the unit cell. In some cases correlation function analyses can yield further spatial information such the thickness of the interface between microphases in a block copolymer or between the crystalline and amorphous regions of a semi-crystalline polymer (Baltá-Calleja and Vonk, 1989).

Conventional SAXS and WAXS experiments, that is those utilising sealed-tube or rotating anode X-ray generators, are limited to stable materials due to the long times (hours) required to obtain patterns of sufficient statistical quality or spatial resolution. To obtain statistically significant SAXS and WAXS patterns at heating rates used in DSC experiments (time resolution of less than one minute and preferably less than one second) the high flux of synchrotron radiation and fast, position-sensitive, electronic detectors must be used (Bras and Ryan, 1998). Synchrotron radiation is produced on storage rings by maintaining electrons in a relativistic orbit, the radiation is taken off at ports which are at a tangent to the electron trajectory. The radiation diverges slightly but can be easily monochromated and focussed using X-ray mirrors and monochromators. It is thus possible to obtain low-divergence, point-collimated, monochromatic X-ray sources. There are around twenty suitable synchrotrons worldwide each containing up to sixty beamlines. The use of X-ray scattering combined with other techniques has been reviewed by Bras and Ryan (1998).

Continuous processes, such as the extrusion of fibres, do not need time resolution because in a steady-state process distance from the die is directly proportional to time, however the article being produced is often very small (<50 μm) so the intensity available from synchrotron radiation is required to obtain X-ray data in reasonable times. A recent application of synchrotron radiation was to the continuous spinning of silk by a spider and in this situation the fibre was only 5 μm in diameter (Riekel et al., 1999). Discontinuous processes, such as moulding, require time resolution to follow the development of structure during quiescent solidification. The thickness of the structure (≈ 1 mm) is more readily matched to diffraction experiments but there is often a high absorption from the sample environment. Reaction injection moulding of a copolyurethane has been studied by small-angle X-ray scattering (SAXS) (Ryan et al., 1991) where evolution of (micro)phase separation was probed. A small reaction injection molding machine was used to meter, mix and inject stoichiometric amounts of reactants into a thermostatted cell and a scattering pattern was collected every 0.7 seconds. Generally, higher polymerization temperatures led to a faster rate of (micro)phase separation but a lower ultimate degree of (micro)phase separation. An induction time was observed which decreased with increasing temperature. The scattered intensity at fixed angle increased with the square of time and the length scale in the material increased with temperature. However, it was not possible to distinguish whether this structure was due to hard segment crystallization or non-crystalline microphase separation in this system. Extrusion of polymer fibres and film has been studied by synchrotron SAXS and WAXS (Cakmak et al., 1993) by building a film extrusion line above the SR beamline. Areal detectors were used in the small- and wide-angle regions to follow the crystallization process along the spin line. Mechanisms of solidification were proposed for the different draw ratios (Cakmak et al., 1993).

2. The early stages of crystallization

In this section we describe experimental evidence for precrystallization density fluctuations in a range of polymer melts. In order to separate nucleation from growth, two types of experiments have been performed. Rapid crystallizations have been studied by melt extrusion and indicate the development of well resolved oriented SAXS patterns associated with large-scale order before the development of crystalline peaks in the WAXS region. Crystallizations with long induction times, studied by small- and wide-angle X-ray scattering (SAXS and WAXS), reveal the onset of large-scale ordering prior to crystal growth.

2.1. Experimental methods

2.1.1. Materials

The polypropylene used was a commercial grade, S-30-S (ex. DSM), which was free from any additives. The number-average molar mass (GPC) and the polydispersity were 52 kg/mol and 2, respectively, and the melting point by differential scanning calorimetry (DSC) was 165 ± 2°C.

2.1.2. Simultaneous SAXS/WAXS/extrusion

Measurements were made on beamline 16.1 of the SRS at the CCLRC Daresbury Laboratory, Warrington, UK. The details of the storage ring, radiation and camera geometry and data collection electronics have been given in detail elsewhere (Bras and Ryan, 1998). An extruder above the X-ray position was used to provide a steady stream of crystallizing polymer past the X-ray beam. Tape extrusion is a steady-state process which shows post-die plug flow, therefore the distance down the spin-line where the observation was made correlates with the time since the material left the extruder die. The material in the X-ray beam is continuously replaced by material with the same shear and temperature history. A tape of polymer melt was extruded from a die (of dimensions 0.5 × 3 mm) at 205°C and collected via a wind up mechanism below the X-ray beam. The extruder used was an AXON BX18 that operated in starve-feed mode to minimize the time the polymer spent in the melt. The experimental set up is shown schematically in Fig. 1. The distance from the die head to the beam could be varied between 0.3 and 1.8 m.

2.1.3. Simultaneous SAXS/WAXS/DSC

Measurements were made on beamline 8.2 of the SRS at the CCLRC Daresbury Laboratory, Warrington, UK. The details of the storage ring, radiation and camera geometry and data collection electronics have been given in detail elsewhere (Bras and Ryan, 1998). Disk specimens of polymer (thickness ~1 mm, diameter ~8 mm) were cut from pre-molded sheet and encapsulated in a DSC pan fitted with mica windows (thickness ~25 μm, diameter ~7 mm), and the pan was inserted into a Linkam DSC apparatus of the single-pan design. In the present study, a multi-stage temperature program was used as follows: (i) heat to 200°C at 50°C min^{-1}; (ii) hold at 200°C for 1 min; (iii) cool at 60°C min^{-1} to the crystallization tem-

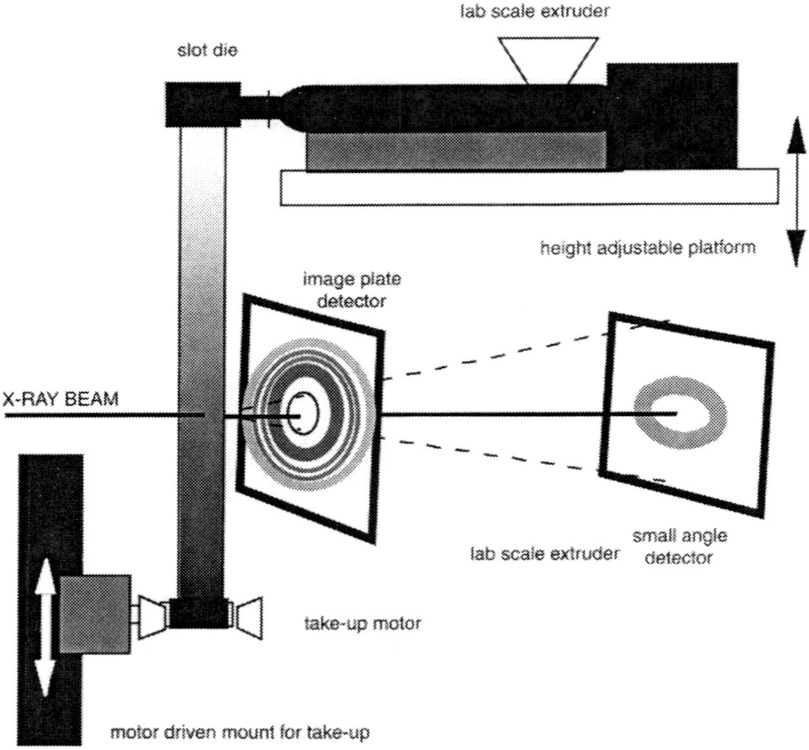

Fig. 1. A schematic diagram of the SAXS/WAXS/extrusion experimental set-up with 2 image plates as the detectors.

perature and hold for 10–120 minutes depending on the temperature. The crystallization time at each temperature was chosen so that at least half the primary crystallization kinetics were observed. To limit beam damage on longer crystallizations the samples were only exposed to the X-ray beam for 10% of the time after crystallization had started. For the SAXS data a Gaussian peak (whose position was a variable) was fitted on top of a Porod background. For the WAXS data Gaussian peaks were fitted on top of a polynomial background, the positions of the 4 Gaussian peaks were set according to the positions of the iPP reflections at *hkl* values of 110, 040, 130 and (111, 131, 041).

2.2. Early stages of crystallization in extrusion

It is difficult to separate nucleation from growth in due to the low concentration of nuclei, which gives rise to poor counting statistics in a scattering experiment. One potential method is to borrow from elementary chemical kinetics and use a flow apparatus. An extruder operating at steady state provides such a set-up (see Fig. 1). Polymer above the melting point is extruded from a die, the tape or fiber cools in the air (in our case in a column of chilled nitrogen) prior to being wound up as a solid. Extrusion of tape or fiber is a steady-state process where

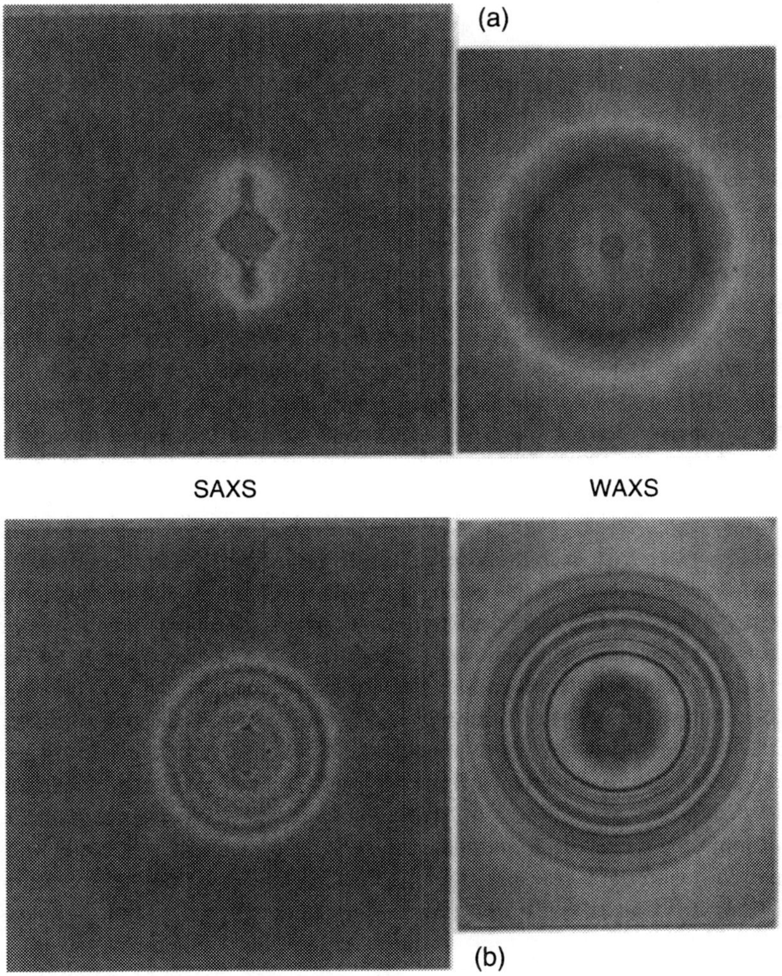

Fig. 2. Data taken during extrusion of iPP with a low wind-up speed using a two-dimensional (2D) wire chamber to collect the SAXS data and an image plate to collect the WAXS data. (a) Close to the die head tear shaped scattering features are observed in 2D SAXS with only an amorphous halo in 2D WAXS. (b) At the furthest position from the die head, when crystallization starts, isotropic rings from crystals can be seen in SAXS and in the WAXS.

the crystallization time increases down the spin-line. This allowed long data collection times (minutes) for very early stages of crystallization (milliseconds). Prior to the development of crystals, well resolved, oriented small-angle patterns could be observed with length scales (50–200 Å) and intensities that grew down the spin-line. The corresponding WAXS showed no Bragg peaks due to crystals.

Fig. 2 shows the scattering patterns collected during extrusion of iPP. At short times these early patterns have two SAXS peaks at finite q and no WAXS peaks. We interpret this as a signature of density fluctuations. The orientation observed in the scattering is caused by coupling of density fluctuations with the slight elongational flow-field (the take-up speed was approximately twice the extrusion speed). Once crystallization had been observed in the wide-angle region, the shape of the small-angle pattern changed to that typical of lamellar crystals. Since the elongational flow was weak the crystallization process dominated and only weakly anisotropic crystals were produced. The SAXS peaks shown at early times in Fig. 2a were approximately 100 times weaker than the diffraction ring observed in SAXS once spherulitic crystallization had occurred (Fig. 2b).

There are a number of other possible interpretations of the data. The SAXS peak could be due to the formation of oriented nuclei, the precursor to the 'shish' in 'shish-kebab' crystals and row nucleation (Strobl, 1996), but this should also lead to orientation in the WAXS which is not observed. Furthermore, the elongation is less than a factor of 2 which is very low for formation of such structures (Wilkinson and Ryan, 1998). Nucleation could have formed poorly ordered crystals that do not diffract. This would account for the lack of Bragg peaks in the WAXS (peak broadening due to small crystallites), but does not account for the peak in SAXS, the scattering from a low concentration of randomly oriented objects would give a peak at $q = 0$ from the shape factor of the scatterers (see Chapter 4 by O. Glatter in this book). Systems with conserved order parameters have nuclei surrounded by a depletion layer and these would give a weak peak in SAXS at low concentration due to the shape factor. However, in this system with a non-conserved order parameter, the nuclei are not surrounded by a depletion layer. Regions of high-density grow from a background of low density with an overall increase in density and the electron density profile does not have a peak at finite q. The most obvious alternative explanation of the observation is that there is an outer skin of crystalline material, due to the temperature gradient across the tape, which is giving rise to the SAXS. In this situation the crystals formed would be well ordered and one would expect them to diffract at wide angles, which is obviously not the case. Under the current experimental conditions, it is possible to observe diffraction from 1% by volume of a crystalline olefin dispersed in oil (Wang et al., 1999). In order to check the temperature of the melt in the scattering volume a number of techniques were tested, an optical pyrometer indicated that the data taken close to the X-ray position (in Fig. 2a) had a surface temperature in excess of 120°C. Whilst it is undoubtedly true that at long times crystallization proceeds from the exterior of the tape, is unlikely that it has occurred in the data presented here.

Previous SAXS/WAXS studies on polymer extrusion have concentrated on the growth and orientation of crystals (Cakmak et al., 1993; Katayama et al., 1968). Interestingly both studies, Cakmak et al. (1993) on PVDF tape (using synchrotron radiation) and Katayama et al. (1968) on PET fibers (using a sealed tube source and film as a detector) showed SAXS before WAXS down the spin-line, but made no comment on its significance.

2.3. Early stages of slow quiescent crystallization

Slow crystallizations, with long induction times, have been studied by simultaneous SAXS and WAXS. These experiments, on quiescent samples, show a clear development of a SAXS peak prior to the presence of crystals identified by WAXS, this is not obvious in the three-dimensional (intensity versus scattering vector versus time) SAXS/WAXS patterns in Fig. 3 but the peak area versus time data in Fig. 4 for iPP at 137°C show, unequivocally, that the SAXS peak grows before the WAXS peak. In the first 200 s of the experiment there is no scattering above the background intensity. Between 200 and 400 s there is a measurable SAXS

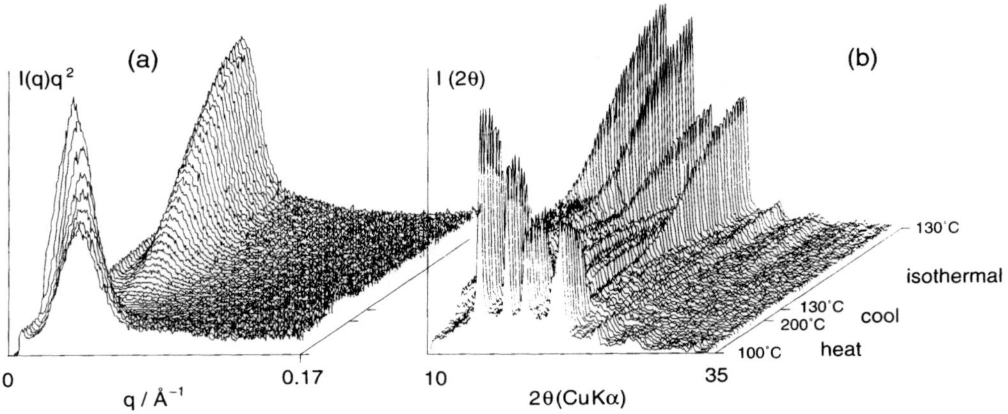

Fig. 3. Melting and recrystallization of iPP. (a) Three-dimensional plot of Lorentz corrected SAXS intensity, $I(q,t)q^2$, versus scattering vector, q, versus temperature, T. (b) Three-dimensional plot of WAXS intensity, $I(2\theta,t)$, versus scattering vector, q, versus temperature, T.

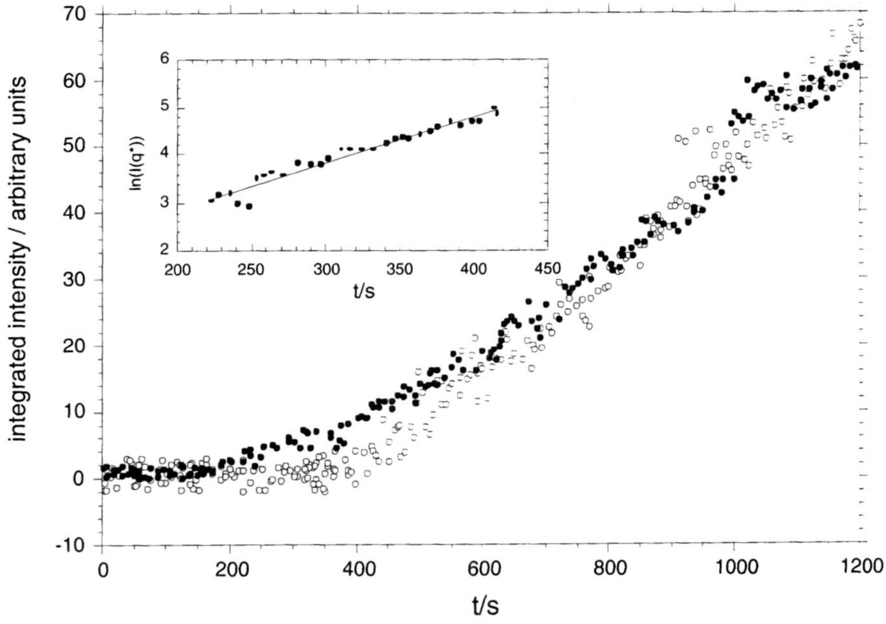

Fig. 4. Integrated intensity data for a crystallization of iPP, the inset to the data is the logarithm of the peak intensity with time in the region where there is SAXS but no WAXS peaks.

intensity with no WAXS above the background. After 400 s Bragg peaks are observed and after 800 s the growth in SAXS and WAXS map onto each other. The crystallinity at 1200 s is ≈ 0.1. The logarithm of the peak intensity versus time, for the period where there is SAXS without WAXS, gives a good straight line. Similar behavior has been reported previously for semi-rigid polymers crystallized by devitrifying a glass (Ezquerra et al., 1996) and the kinetics of crystallization after devitrification were analysed in terms of the Cahn–Hilliard (Imai et al., 1992a,b, 1994) theory for spinodal decomposition.

It has been shown (Imai et al., 1992a,b, 1994) that the general form of the variation in scattered intensity, $I(q,t)$, following a quench is given by:

$$I(q,t) = I(q,0)\exp[2R(q)t] \tag{1}$$

The variation in $I(q)$ at a given time interval is determined by the scattering law for the objects and $R(q)$ is termed the growth rate constant and is given by:

$$R(q) = -Mq^2\left(\frac{\partial^2 G}{\partial \rho^2} + 2\kappa q^2\right) \tag{2}$$

Here M is the mobility term, G is the Gibbs free-energy, and κ is a gradient free-energy term. Modifications to Eq. (2) have been discussed previously by Cook (1970) which take into account random thermal fluctuations (inclusion of a Brownian motion term). In employing Eq. (2) to analyse the data, the extrema are not strictly correct. The q dependence of the Onsager coefficient relating the diffusive flux of polymer molecules to the local chemical potential has been neglected. This may be valid for the early stages of phase separation and a shallow quench. It should be noted that the Onsager coefficient generally does have a q dependence, and according to Pincus (1981) it is proportional to q^2 for a polymer blend. $R(q)/q^2$ can be taken as a measure of the dynamic driving force for the growth of the concentration fluctuation with wave vector $q/2\pi$. There is a region of q in which $R(q)$, and thus $R(q)/q^2$, is positive and the concentration fluctuations do not decay but grow and give rise to phase separation. These growing concentration fluctuations have upper and lower critical boundaries to their wave numbers. Outside these limits, the concentration fluctuations decay and do not contribute to the phase separation dynamics.

The thermodynamic driving force for the growth of the concentration fluctuation with wave vector $q/2\pi$, $R(q)/q^2$, becomes a maximum at $q = q_\mathrm{m} = \sqrt{G''/\kappa})$. Thus, the wavelength, $q/2\pi$, of the dominant Fourier component of the growing fluctuations in the early stages of phase separation is determined by the maximum dynamic driving force. q_m is time independent in the early stages of phase separation and is controlled by thermodynamics (Cook, 1970; Pincus, 1981). $R(q)$ is further controlled by the transport properties and is related to the flux of molecules where the effective diffusion coefficient, D_eff, can be determined from an extrapolation to $q = 0$ of the straight-line portion of $R(q)/q^2$ during phase separation using

$$D_\mathrm{eff} = -\frac{2R(q)}{q^2} \tag{3}$$

and the linearity holds for $q_\mathrm{m} < q < \sqrt{2}q_m$. Values of the amplification factor $R(q)$, for the early stage of crystallization where we observe SAXS but no WAXS, were determined by plotting $\ln I$ versus t for discrete wave vectors and finding the slope (Bates and Wiltzius, 1989). $R(q)/q^2$ versus q^2 plots were constructed at a range of temperatures and are shown in Fig. 5, the data have been truncated at low q. As the crystallization temperature is increased it should be noted that the data get less noisy (as the kinetics slow down and counting statistics improve) and the linear part of the graph is reduced, this is because q_m moves to lower values.

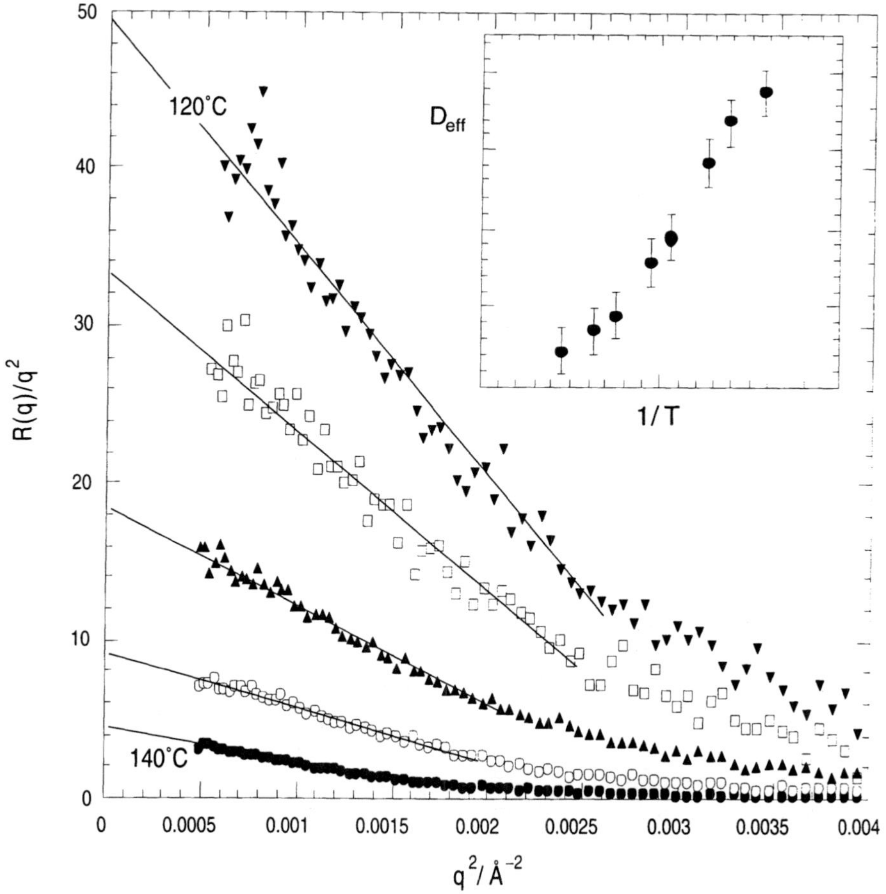

Fig. 5. A Cahn–Hilliard plot to estimate the effective diffusion coefficient, D_{eff}, from the $q = 0$ intercept of $R(q)/q^2$ versus q^2 for iPP at 120, 125, 130, 135 and 140°C. The solid lines are a linear fit to the data and are extrapolated to obtain D at $q = 0$. The inset shows D_{eff} versus $1/T$ (for iPP) to allow calculation of the spinodal temperature from extrapolation to $D_{\mathrm{eff}} = 0$.

The inset to Fig. 5 shows a plot of D_{eff} versus $1/T$, the spinodal temperature is determined by the $D_{\mathrm{eff}} = 0$ intercept. For example, in polypropylene at 410 K, we could estimate both the dominant length scale $L \approx 175$ Å and the effective co-operative diffusion coefficient $D_{\mathrm{eff}} = -4.5$ Å^2 s^{-1}. By conducting these experiments at a series of temperatures the stability limit could be found at 415 ± 5 K (Ryan et al., 1999) by extrapolation of D_{eff} to zero. The stability limits obtained in this way are compared to the melting point of an infinite crystal in Table 1 for iPP, PE and PET.

The stability limit is the temperature below which the polymer spontaneously separates into two phases. One of the phases is rich in polymer segments of the appropriate chain conformation to crystallize (*trans-gauche* arrangement of the carbon backbone in isotactic polypropylene) and the other is concentrated in sequences near entanglements and other

Table 1

Stability limits and thermodynamic melting points

	T_s	T_m^0	T_s/T_m^0
HDPE	408	417	0.98
iPP	415	459	0.90
PET	499	573	0.87

defects which cannot crystallize. The stability limit is 7 K below the measured melting point for polypropylene with a long spacing of 175 Å and 35 K below the thermodynamic melting point of isotactic polypropylene (Ryan et al., 1999). Once WAXS from crystals (atomic order on the 1 Å scale) was observed, the kinetics reverted to those of nucleation and growth, that is Avrami kinetics with an exponent $n \approx 3$ (see Fig. 2). Similar behavior has also been observed in devitrified glasses of PET, by Imai et al. (1992a,b, 1994), and PEEK, by Ezquerra et al. (1996) however, as the measurements are made close to the glass transition and the dynamics are dominated by the viscosity, estimation of the stability limit is not possible as D_{eff} increases with temperature.

2.4. Summary of crystallization studies

The quiescent time-resolved SAXS/WAXS and extrusion suggest that a process that strongly resembles spinodal decomposition of chain segments with different average conformations is the nucleation step in polymer crystallization. That polymer crystallization occurs with phase separation is in no doubt, since at the end of the process regions of well ordered crystalline polymer coexist with regions of disordered polymer in a layered morphology (lamellae) with a spherulitic super-structure. Sequences that can be oriented with the right conformation and incorporated into the crystal separate from sequences near entanglements and other defects that can not crystallize and can only be part of the amorphous regions. The transformation from the disordered phase to the better ordered partially crystalline phase proceeds continuously passing through a sequence of slightly more ordered states rather than building up a crystalline state instantaneously. This is consistent with the evolution of SAXS before WAXS. At some stage secondary nucleation must form crystals directly from the melt and a mechanism of continuous transformation could be consistent with a fast homogeneous nucleation process. It is difficult, however, to make a clear distinction between spinodal decomposition and nucleation and growth with nucleation barriers smaller than $k_B T$ (Olmsted et al., 1998). Polymer crystallization, like any other phase separation, is kinetically controlled. The structure formed is the one with the highest growth rate. Once a crystallite is formed, its lateral growth rate is much higher than that of the fluctuations and so dominates. In this case the growth mechanism of semi-crystalline polymer lamellae, in the form of spherulites, takes over because the lateral growth rate of crystals (typically $\mu\text{m s}^{-1}$) is much faster than the growth rate of the fluctuations (typically $\text{Å}^2\ \text{s}^{-1}$). Thus the combination of the steady-state extrusion and the high-intensity, synchrotron X-ray source allows nucleation to be observed.

The early stages of crystallization, in polymers and other systems, has been the subject of great debate recently and the field has become rather polarised whist relying on evidence from the same experimental techniques. Strobl and co-workers (Strobl, 2000; Heck et al., 2000; Iijima and Strobl, 2000) have developed a model based on the formation of crystalline blocks which self assemble into lamellar stacks, showing that the formation of the semi-crystalline state is a continuous process. This model also deals with some of the problems inherent in the conventional method of establishing the melting point of an infinite crystal, T_m^0, by recognising that there are separate melting and annealing curves of lamellar thickness versus temperature. Like the experimental work described above the block model is not universally accepted and a review by Strobl (2000) was followed (in the same issue of the journal) by three commentaries based on electron and optical microscopy (Lotz, 2000), analysis of length scales (Cheng et al., 2000) and theoretical and modelling considerations (Muthukumar, 2000). There is, however, good evidence (Matsuba et al., 1999a,b) for the formation of mesostructure prior to crystallization in syndiotactic and isotactic polystyrene where conformation changes and orientational fluctuations have been observed, by vibrational spectroscopy, that are consistent with the formation of helices in the melt. The conventional view of nucleation and growth has been invoked to explain the observation of SAXS before WAXS during the quiescent crystallization of homopolymers, Akpalu and Amis (1999, 2000) have analysed the crystallization of controlled molecular weight distributions of polyethylene and found that isolated lamellae are formed which lead to secondary nucleation process whereas Hsiao and co-workers (Wang et al., 2000) insist that the observation of SAXS before WAXS is an artefact of the detector systems and that the whole of the crystallization follows an Avrami curve. SAXS/WAXS studies of crystallization during melt spinning where the first to point out the early development of SAXS structure (Cakmak et al., 1993), and recent work (Schultz et al., 2000) analyses the data in the framework of a nucleation and growth mechanism. There is, however, widespread agreement that flow affects crystallization through the orientation of the high molecular weight end of the molecular weight distribution (Nogales et al., 2001) and this will most likely be the focus of many future efforts in the field of crystallization in polymers.

3. Reaction-induced microphase separation in polyurethanes

The reactive processing of water-blown flexible polyurethane foam from liquid monomers and oligomers involves a complex combination of both chemical and physical events. In less than five minutes, a liquid mixture of relatively low molecular weight components is transformed into the supramolecular architecture of solid foam as shown in Fig. 6. Information regarding both the reaction kinetics and development of morphology during processing are essential, such that an objective description of the events taking place and ultimately, selective control of the process can be achieved. Flexible polyurethane foam is formed by the simultaneous reaction between a diisocyanate with polyether polyol and water. Combination of these two exothermic reactions leads to the formation of a segmented block copoly(urethane-urea), of the $-(H_m S)_n-$ type (Elwell et al., 1995). This is blown into a foam by the co-generation of carbon dioxide gas evolved from the water–isocyanate reaction. As the polymerization proceeds, the core of the rising foam bun becomes self-insulated by the surrounding polymer and this results in the process occurring under quasi-adiabatic conditions. The unique combination of physical properties provided by segmented block copolymers is related to their microphase-separated morphologies. The development of

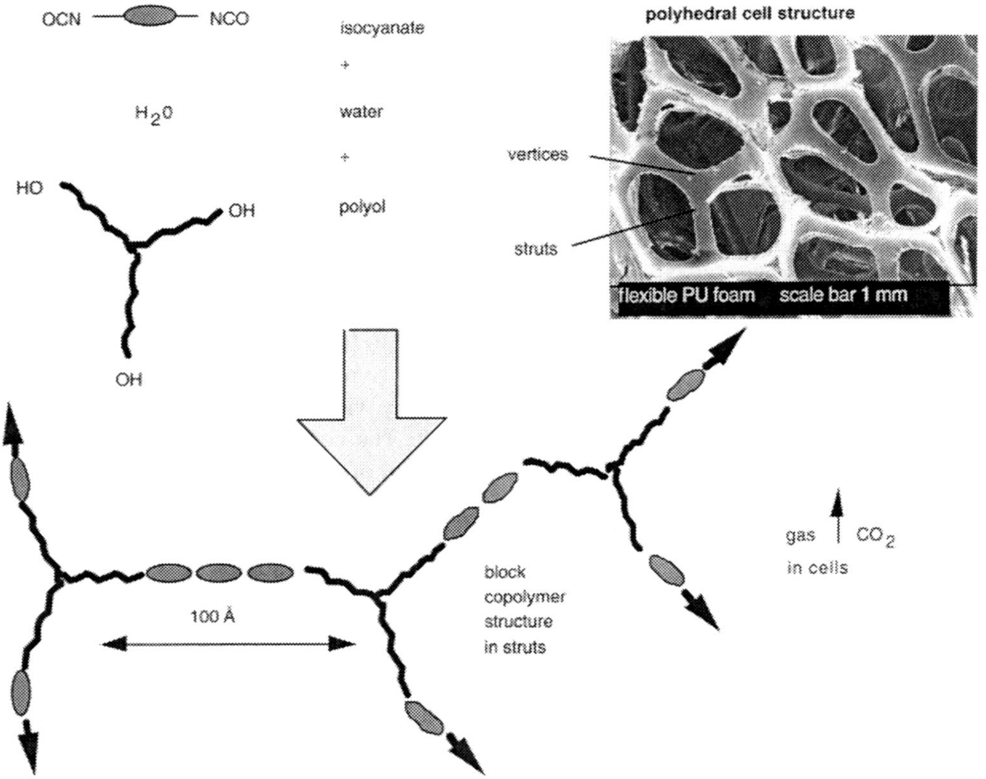

Fig. 6. An SEM micrograph of polyurethane foam and a schematic representation of the block copolymer formed by random copolymerization.

a microphase-separated morphology during polymerization is complex. As the different chemical reactions proceed, chain lengths increase giving an increase in the overall degree of polymerization, N. Additionally, the interaction parameter (χ), characterising the miscibilities between the different segments also increases as a result of polymerization. In the specific case of segmented block copolymerization, such changes in χ and N (or more strictly the product χN) drive the system across thermodynamic boundaries and result in a transition from an initial homogeneous (disordered) state to a microphase-separated (ordered) state. The resultant morphology is determined by the competition between the kinetics of polymerization and microphase separation (Elwell et al., 1995). Furthermore, homopolymeric and oligomeric species may undergo macrophase separation and the copolymer morphology that eventually evolves may reflect phase separation and/or vitrification, crosslinking or crystallization and is determined by the competition between the kinetics of these processes.

Understanding the complex competition between polymerization and microphase separation requires *in situ*, time-resolved analytical techniques. In multi-phase systems, FT–IR spectroscopy and rheometry are used to obtain chemical and macroscopic property information during polymer formation. In this section we describe investigations in which FT–IR spectroscopy, synchrotron small-angle X-ray scattering (SAXS) and dynamic rheometry have been used, under forced-adiabatic conditions, to observe how the multi-phase polymer structure evolves during foam formation in model, MDI based flexible polyurethane foam systems.

3.1. Experimental methods

3.1.1. Materials

A flexible foam formulations based on MDI and either 3.8 or 2.2 parts water per hundred polyether triol (5000 g/mol) was used at an isocyanate index of 105. Further details regarding the reactants are described elsewhere (Elwell et al., 1995).

3.1.2. FT–IR spectroscopy

Van Lieshout and Grünbauer have developed a forced adiabatic FT–IR cell where sampling takes place outside the spectrophotometer. Details of the experimental arrangement and procedures adopted are documented elsewhere (Elwell et al., 1996).

3.1.3. Synchrotron SAXS

The SAXS measurements were made on beam line 8.2 at the Synchrotron Radiation Source (SRS), Daresbury laboratory, Warrington, UK. Reaction injection moulding (RIM) was employed to remotely meter, mix and inject the reaction mixture into the SAXS reaction cell. A temperature-controlled reaction cell has been designed which can be positioned in the optical bench assembly of a synchrotron beam line and fed with reaction mixture from a micro-RIM machine. The specifications of the SAXS sampling cell, the experimental arrangement and the procedures adopted are described elsewhere (Elwell et al., 1994).

3.1.4. Dynamic rheometry

The application of electrical heating devices built into/around the disposable fixtures/plates for rheometers have been reported previously (Mora et al., 1991). A forced-adiabatic rheometer plate that can be attached to a modified lower fixture for the Rheometrics RMS 800 series rheometer was constructed in our laboratory; the specifications of the temperature-controlled rheometer plate, the experimental arrangement and procedures adopted are described elsewhere (Elwell, 1995).

3.2. Reaction kinetics and hydrogen bonding from FT–IR

The basic reaction chemistry taking place during foam formation is shown in Fig. 6. The isocyanate absorption band occurs at approximately 2300–2270 cm^{-1} in the mid infra-red spectrum and the decay in the intensity of this absorbance can be used to monitor the conversion of isocyanate functional groups during the reaction. Fig. 7a is a three-dimensional plot of the isocyanate absorbance, versus frequency, versus time for the foaming reaction. It shows clearly the decay in the intensity of the isocyanate absorbance with time. Each spectrum was recorded at 6 second intervals. Fig. 7b shows a similar three-dimensional plot

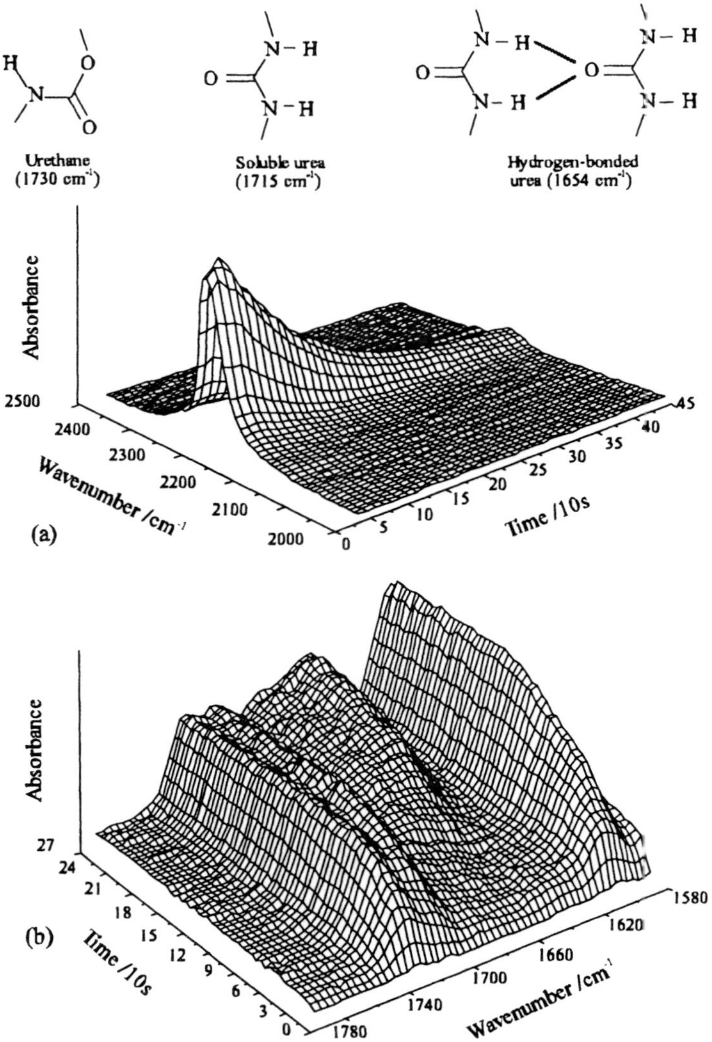

Fig. 7. (a) A three-dimensional plot of the isocyanate absorbance, versus frequency, versus time for the foaming reaction. Each spectrum was recorded at 6 second intervals. (b) A three-dimensional plot for the corresponding carbonyl region illustrating the evolution of urethane, soluble urea and hydrogen-bonded urea.

for the corresponding carbonyl region illustrating the evolution of urethane, soluble urea and hydrogen-bonded urea. It is apparent that urethane ($\approx$ 1730 cm^{-1}) and soluble urea ($\approx$ 1715 cm^{-1}) evolve early in the reaction. These two reactions take place simultaneously: not sequentially, as early research had suggested. It is also apparent that there is an induction time prior to the evolution of hydrogen-bonded urea ($\approx$ 1654 cm^{-1}).

The only source of heat during the foaming reaction is that arising from the consumption of isocyanate functional groups. It is thus possible to correlate the isocyanate conversion

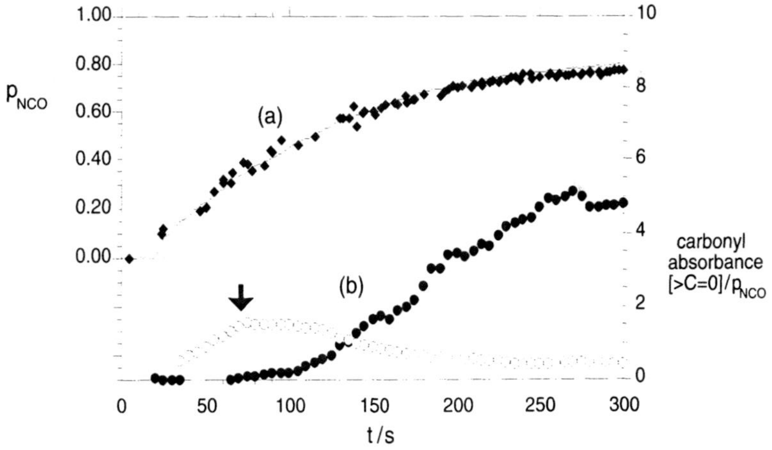

Fig. 8. (a) Comparison of the reaction kinetics data from the depletion of the isocyanate absorbance by FT–IR (solid symbols) and the adiabatic temperature rise (line). (b) Evolution of the carbonyl absorbances during reaction showing the initial growth and depletion of the free urea (open symbols) and the rapid growth of the H-bonded urea (solid symbols) after microphase separation has occurred.

calculated from the decay in the isocyanate absorption with time, and that calculated from normalization of the reaction exotherm. Fig. 8 compares isocyanate conversion calculated from the forced-adiabatic spectroscopy data (open symbols) with that calculated from the normalization of the foam reaction exotherm (solid line). The correlation between the isocyanate conversion calculated from adiabatic temperature rise and infra-red spectroscopy data is within ±3% after approximately 40 seconds of the reaction. It should be considered that the first 15 seconds depicted on the aforementioned curves are associated with mixing of the components and loading of the reaction mixture into the infra-red cell arrangement.

Representative plots of normalised soluble urea (1715 cm^{-1}, open symbols) and hydrogen-bonded urea (1654 cm^{-1}, closed symbols) versus reaction time are also shown in Fig. 8. The soluble urea and hydrogen-bonded urea absorbance bands were normalised against isocyanate conversion because the probe (carbonyl) concentration is directly proportional to conversion. The onset of microphase separation transition (MST) is indicated with an arrow and is taken as the point at which there is an acceleration in the hydrogen-bonded urea and a depletion in the free urea. At the MST, the system becomes both chemically and spatially heterogeneous, and this separates functional groups.

It has been observed for systems that cross a broad range of water concentrations, the conversion of isocyanate functional groups at the MST remains approximately constant at $p_{NCO} = 0.55 \pm 0.05$. It has been observed for systems that cross a broad range of water concentrations, the conversion of isocyanate functional groups at the MST remains approximately constant at $p_{NCO} = 0.55 \pm 0.05$ where the local number average hard segment sequence length, N_H, is approximately 1.5.

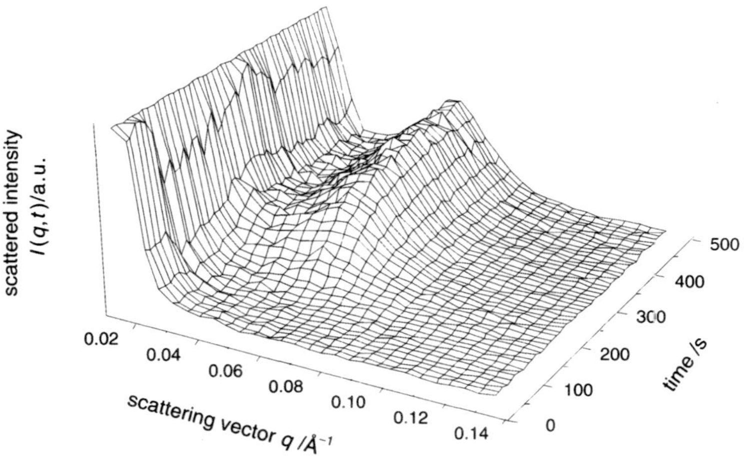

Fig. 9. Three-dimensional relief diagram of scattered intensity $I(q,t)$, versus scattering vector, q, versus time, t, for SAXS data obtained with a time resolution of 2 seconds during formation of polyurethane foam.

3.3. Microphase separation kinetics from SAXS

Time-resolved SAXS data that have been collected during foam formation (at 2 second intervals) are presented in Fig. 9 as three-dimensional surface plots of intensity, $I(q,t)$, versus scattering vector, q, versus time, t. The patterns illustrate that in the early stage of the reaction, there is a homogeneous liquid present. At the microphase separation transition (MST), there is the first appearance of a scattering maxima and the intensity of this peak continuously increases. Eventually, the growth slows down and the peak intensity becomes approximately constant where microphase separation is intercepted and arrested by vitrification of the phase that is richer in hard segment (Elwell et al., 1994, 1995, 1996). The maximum in $I(q)$ suggests the presence of structure with periodic electron density within the sample and the peak in $I(q)$ occurs at $q \approx 0.06$ Å^{-1} giving an interdomain spacing of 105 Å. The d spacing does not change during the structuring process for PU foam and this indicates that the process is dominated by the length scale of the polymer chains.

Fig. 10 shows representative plots of the invariant, Q' (a measure of the degree of microphase separation), versus reaction time. The onset of microphase separation is taken as 60 ± 2 seconds which corresponds to an isocyanate conversion, $p_{NCO} = 0.52 \pm 0.01$. For multiple runs the onset of microphase separation occurred at $p_{NCO} = 0.54 \pm 0.02$. The driving force for microphase separation is the thermodynamic incompatibility between the hard and soft-segment blocks. As polymerization proceeds, the degree of polymerization of the hard segment, N_H, increases and reaches a critical value where the product χN_H (where χ is the Flory–Huggins interaction parameter) is such that the system is no longer thermodynamically stable [i.e., $\Delta G \geq 0$ and $(\partial^2 \Delta G/\partial \phi^2) \leq 0$]. The polymerization has acted as a large thermodynamic quench. The depth of this quench determines the mechanism by which microphase separation occurs.

The material that starts to form structure is a combination of homopolymer, block copolymers and monomers and is discussed in terms of a mixture. The final structured

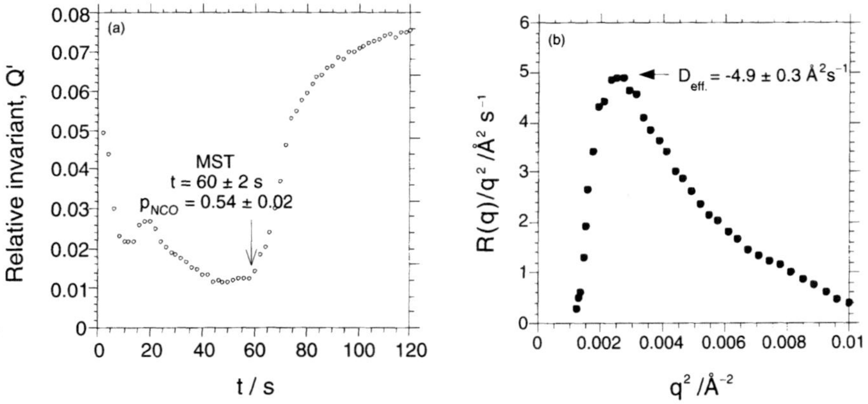

Fig. 10. (a) The integrated intensity (invariant), $Q'(t)$ and the peak intensity $I(q^*,t)$, versus time, t, showing the three-stage phase separation process. (b) A TGDL plot to estimate the effective diffusion coefficient, D_{eff}, from the peak in $R(q)/q^2$ versus q^2 for two polyurethane foam formulations.

material is predominantly a block copolymer; and thus the data is also analysed in terms of the TDGL theory for microphase separation (Elwell et al., 1994). There appears to be three distinct regions of phase behaviour.

Region I. Homogeneous liquid mixture comprised of unreacted monomer, homopolymer, urea hard segment sequences and isocyanate tipped, polyether oligomers.

Region II. This is the microphase separation transition (MST). At the MST there is observed a rapid increase in the scattered intensity. There is a good fit of the data to an exponential increase in intensity with time. This is a characteristic of phase separation via spinodal decomposition. The slope of the line provides a measure of the amplification rate, $R(q)$, of the composition fluctuations.

Region III. The expanding foam has reached the Berghmans point (onset of vitrification) and beyond this, the scattered intensity remains approximately constant.

Vitrification of the hard segments freezes in the morphology at that time and results in the evolution of a foam with an internal polymer morphology that comprises an interconnecting physical network of hydrogen-bonded urea hard segment sequences within a crosslinked polyether–urethane

To confirm mechanism of phase separation, the scattering patterns taken during the reaction have been analysed to obtain the amplification rate at discreet wave vectors, $R(q)$. For microphase separation via spinodal decomposition, a plot of $R(q)/q^2$ (a measure of the thermodynamic driving force for phase separation) versus q^2 should exhibit a maximum at a finite value of q as described in Section 2.3. Fig. 10b shows a plot of $R(q)/q^2$ versus q^2 for PU foam. From the peak in the curve the effective diffusion coefficient, D_{eff}, of the polyether soft-segment was calculated to be -4.9 ± 0.3 Å^2 s^{-1}. The negative diffusion coefficients indicate diffusion against the composition gradient (i.e., from a region of low concentration to a region of high concentration) as found in spinodal decomposition of a mixture.

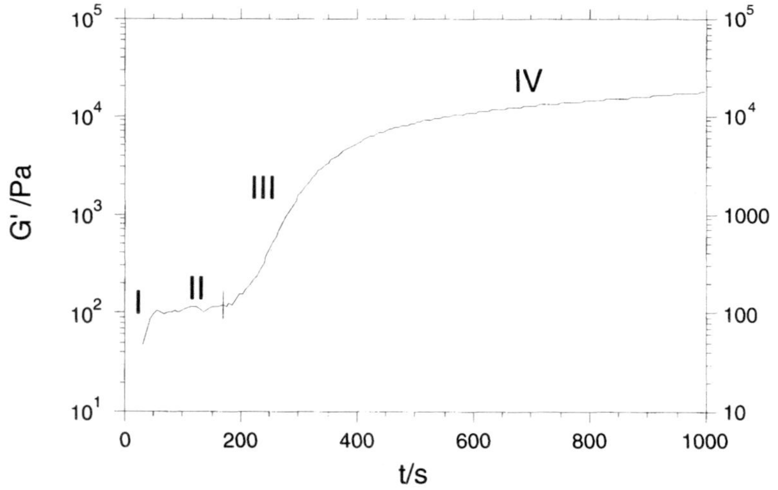

Fig. 11. The elastic shear modulus, G', versus reaction time for PU foam.

3.4. Solidification mechanism from rheometry

Macosko and co-workers (Mora et al., 1991) and Elwell (1995) have identified four main regions of rheological development during the formation of polyurethane foam the related to the behaviour seen in FT–IR and SAXS. Fig. 11 shows the elastic shear modulus, G', versus reaction time and indicated in the figure are the four regions of rheological development.

3.4.1. Region I: Bubble nucleation

In this stage, the dissolved bubbles of air that were 'whipped' into the system during mechanical agitation act as nucleation sites for the carbon dioxide gas that is generated from the water–isocyanate reaction. These bubbles then develop into a space filling network where the initial viscosity is determined by the monomer viscosity and the interfacial tension.

3.4.2. Region II: Liquid foam and microphase separation

At the end of region I the reacting mixture has an elastic shear modulus of approximately 100 Pa ($G' > G''$) and shows solid-like behaviour which is a characteristic of liquid foams (Mora et al., 1991; Elwell, 1995). The reacting mixture is still a low modulus foam that continues to expand as a result of the continued generation of carbon dioxide. The liquid cell walls still have both low molecular weight and viscosity. At approximately 60 seconds (isocyanate conversion is of the order of $54 \pm 2\%$) microphase separation of urea hard segment sequences takes place as evidenced from SAXS and infra-red measurements (see Figs. 7–10]. Upto and beyond the MST, the foam modulus still remains approximately constant. It has been reported, however, that the change in modulus associated with the MST of polyolefins (Almdal et al., 1992; Forster et al., 1994), and in the one reported case for polyurethane (Ryan et al., 1992), is of the order of a factor of two. There are three contributions to there being no apparent change in the modulus at the MST. Firstly, the material temperature is continuously changing (quasi-adiabatic). Secondly, the phase transition is not a particularly sharp one and is smeared even further by the polydisperse nature of the hard segment sequences. Finally, the polymer is at a volume fraction of ≈ 0.1 at the MST and it is estimated that the increase in modulus of the polymer at the MST would give rise to a change in the foam modulus of $\approx 2\%$.

3.4.3. Region III: Physical gelation

At the end of region II the isocyanate conversion is of the order of approximately $70 \pm 1\%$. At the start of region III there is a large increase the elastic storage modulus, G', of approximately 2 decades. This increase in modulus results from microphase separation being intercepted and arrested by vitrification of the phase that is richer in urea hard segment. Towards the end of region II, cell opening occurs. At the start of region III there is an increase in the elastic storage modulus, G', by a factor of 25.

3.4.4. Region IV: Foamed elastomer

In polyurethane foam, after cell opening has occurred, the reactions still continue but do so in a medium of increasing viscosity and stiffness. Further increases in modulus and elasticity of the polymer due to vitrification of the hard segments and formation of covalent crosslinks will severely restrict the escape of carbon dioxide from the foam. The cell membranes that have not previously ruptured will undergo reversible expansion, not rupture and remain *closed cells*. In this region, the foamed polymer reaches maximum temperature and modulus. The modulus of the foam in this region is dominated by the polymer modulus. Dynamic rheological measurements probe the macroscopic behaviour of the sample, not the molecular level (FT–IR, SAXS).

From an early stage in the reaction there are isocyanate tipped polyether oligomers and urea hard-segment sequences present in the system (refer to Fig. 6). The degree of polymerization of hard segment, N_H, increases steadily and at $p_{NCO} = 0.54 \pm 0.02$, N_H reaches N_{H^*} ($N_{H^*} \approx 1.5$ units) (Mora et al., 1991) and microphase separation takes place. Polymerization has provided the thermodynamic quench passing from the homogeneous (disordered) one-phase region to the heterogeneous (ordered) two-phase region. At $p_{NCO} = 0.71 \pm 0.02$ (refer to Fig. 11), microphase separation is intercepted and quickly arrested by vitrification of the phase that is richer in hard segment. This phase has attained a composition with a T_g equal to the temperature of the reacting system and is the composition where liquid–liquid phase separation is intercepted and arrested by glass transition. In general foam terminology it is the physical gel point or end of rise time. Vitrification of the hard segments 'freezes in' the morphology at that time and results in the evolution of a foam with an internal polymer morphology that comprises an interconnecting physical network of hydrogen-bonded urea hard-segment sequences within the crosslinked polyether–urethane.

3.5. Overview of PU foam formation

We have described how the techniques of FT–IR spectroscopy, synchrotron SAXS and dynamic rheometry can be used, to investigate *in situ*, the development of polymer morphology

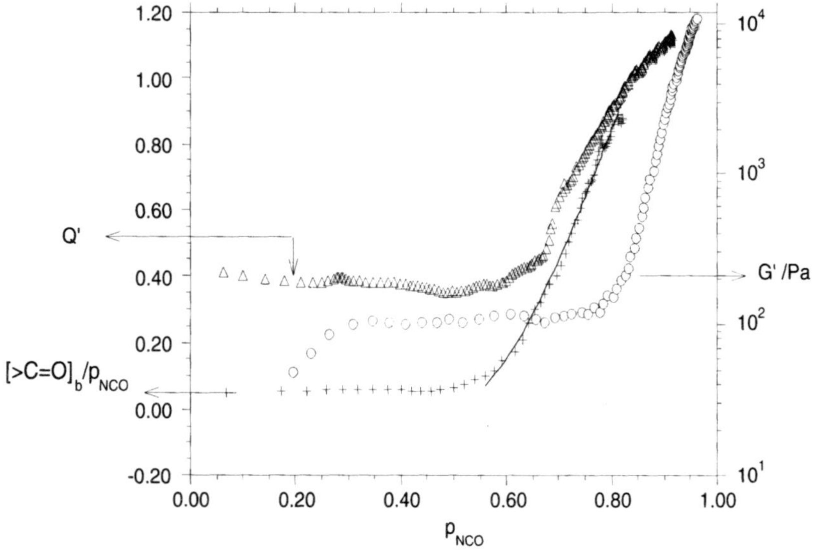

Fig. 12 Comparison of the changes in the key structuring parameters as isocyanate conversion increases during the foaming reaction. SAXS relative invariant, Q' (Δ), hydrogen-bonded urea carbonyl, $[{>}C{=}O]_b/p_{NCO}$ (+) and elastic storage modulus, G' (o). The curve through the hydrogen-bonded urea carbonyl data is a fit to: $[{>}C{=}O]_b/p_{NCO} = 1 - a[\exp(p_{NCO} - b)^c]$ according to a pseudo-Avrami analysis; where a, b and c are fitting parameters: a is a rate constant, b is the conversion at the onset of hydrogen bonding, and c is the phase-growth exponent.

during the reactive processing of flexible polyurethane foam. Each technique yields structural information of a different nature. The changes in the key structure parameters as isocyanate conversion increases can be summarized in a single diagram; and these are illustrated in Fig. 12. It will is apparent that the SAXS relative invariant, Q', *appears* (within the limits of experimental error) to start to increase prior to the rapid growth in the concentration of hydrogen-bonded urea carbonyl species, $[{>}C{=}O]_b/p_{NCO}$. The growth of foam's elastic shear modulus, G', lags the growth of the relative invariant and the hydrogen-bonded urea carbonyls. The structuring sequence of phase separation followed by hydrogen bonding and completed by vitrification only becomes clear after substantial experimental work.

In order that an objective description of both the chemical and physical events taking place and ultimately, selective control of the reactive processing of flexible polyurethane foam be achieved; *in situ* investigations into both the reaction kinetics and the development of morphology *under processing conditions* are essential. The use of model systems has demonstrated that the presence of covalent crosslinks delay the onset of microphase separation of the urea hard-segment sequence lengths; however, in terms of dimensional stability and the mechanical/physical properties of the foam, molecular connectivity between the microphases via urethane covalent crosslinks is essential. The use of Cahn–Hilliard kinetics to model the scattering data has led to the application of cell-dynamics simulations to predict the morphology of polyurethanes (Hamley et al., 2000) and there are overwhelming similarities between the TEM micrographs and the model leading to the confirmation of the mechanism proposed.

4. Conclusions

Time resolved experiments, especially spectroscopies and scattering with synchrotron radiation, have a lot to offer in the study of the fundamental processes occurring during solidification in polymer processing. The two examples given here are typical of the level of understanding that can be gained and subsequently used to design and optimise future experiments. The development of in-situ spectroscopic probes is possible in the factory and processing may be studied directly. An alternative approach is to record the temperatures and pressures experienced in a process and play them back to the material inside an analytical instrument. The limitations of the synchrotron radiation technology is that the process must be taken to the instrument and factory based measurements of structure in real-time are not possible at the moment.

References

Akpalu, Y.A. and Amis, E.J. (2000). *J. Chem. Phys.*, **113**, 392–403.
Almdal, K., Koppi, K.A., Bates, F.S. and Mortensen, K. (1992). *Macromolecules*, **25**, 1743–1751.
Akpalu, Y.A. and Amis, E.J. (1999). *J. Chem. Phys.*, **111**, 8686–8695.
Baltá-Calleja, F.J. and Vonk, C.G. (1989). *X-ray Scattering of Polymers*. Elsevier, Amsterdam.
Barham, P. (1993). In: Thomas, E.L. (ed.). *Structure Property Relations in Polymers*, Materials Science and Technology. VCH, Weinheim, Vol. 12 pp. 153–165.
Bates, F.S. and Wiltzius, P. (1989). *J. Chem. Phys.*, **91**, 3258–3274.
Bras, W. and Ryan, A.J. (1998). *Adv. Colloid Interface Sci.*, **75**, 1–43.
Cakmak, M., Teitge, A., Zachmann, H.G. and White, J.L. (1993) *J. Polym. Sci., Part B: Polym. Phys.*, **31**, 371–381.
Cheng, S.Z.D., Li, C.Y. and Zhu, L. (2000). *Eur. Phys. J. E*, **3**, 195–197.
Cook, H.E. (1970). *Acta Met.*, **18**, 297–306.
Elwell, M.J. (1995). *Thermochim. Acta*, **269/270**, 145–157.
Elwell, M.J., Mortimer, S. and Ryan, A.J. (1994). *Macromolecules*, **27**, 5428–5439.
Elwell, M.J., Ryan, A.J., Gruenbauer, H.J., Van Lieshout, H.C. and Lidy, W. (1995). *Plast., Rubber Compos. Process. Appl.*, **23**, 265–276.
Elwell, M.J., Ryan, A.J., Gruenbauer, H.J.M. and Van Lieshout, H.C. (1996). *Macromolecules*, **29**, 2960–2968.
Ezquerra, T.A., Lopez-Cabarcos, E., Hsiao, B.S. and Balta-Calleja, F.J. (1996). *Phys. Rev. E.*, **54**, 989–992.
Forster, S., Khandpur, A.K., Zhao, J., Bates, F.S., Hamley, I.W., Ryan, A.J. and Bras, W. (1994). *Macromolecules*, **27**, 6922–6935.
Hamley, I.W., Stanford, J.L., Wilkinson, A.N., Elwell, M.J. and Ryan, A.J. (2000). *Polymer*, **41**, 2569–2576.
Heck, B., Hugel, T., Iijima, M. and Strobl, G. (2000). *Polymer*, **41**, 8839–8848.
Iijima, M. and Strobl, G. (2000). *Macromolecules*, **33**, 5204–5214.
Imai, M., Mori, K., Mizukami, T., Kaji, K. and Kanaya, T. (1992a). *Polymer*, **33**, 4451–4456.
Imai, M., Mori, K., Mizukami, T., Kaji, K. and Kanaya, T. (1992b). *Polymer*, **33**, 4457–4462.
Imai, M., Kaji, K. and Kanaya, T. (1994). *Macromolecules*, **27**, 7103–7108.
Katayama, K., Amano, T. and Nakamura, K. (1968). *Kolloid-Z. Z. Polym.*, **226**, 125–134.
Lotz, B. (2000). *Eur. Phys. J. E*, **3**, 185–194.
Matsuba, G., Kaji, K., Nishida, K., Kanaya, T. and Imai, M. (1999a). *Macromolecules*, **32**, 8932–8937.
Matsuba, Go, Kaji, Keisuke, Nishida, Koji, Kanaya, Toshiji, Imai, Masayuki (1999b). *Polym. J.* (Tokyo), **31**. 5204–5214.
Mora, E., Artavia, L.D. and Macosko, C.W. (1991). *J. Rheol.* (N.Y.), **35**, 921–940.
Muthukumar, M. (2000). *Eur. Phys. J. E*, **3**, 199–202.
Nogales, A., Hsiao, B.S., Somani, R.H., Srinivas, S., Tsou, A.H., Balta-Calleja, F.J. and Ezquerra, T.A. (2001). *Polymer*, **42**, 5247–5256.
Olmsted, P.D., Poon, W.C.K., McLeish, T.C.B., Terrill, N.J. and Ryan, A.J. (1998) *Phys. Rev. Lett.*, **81**, 373–376.
Pincus, P. (1981). *J. Chem. Phys.*, **75**, 1996–2000.

Riekel, C., Mueller, M. and Vollrath, F. (1999). *Macromolecules*, **32**, 4464–4466.

Ryan, A.J., Macosko, C.W. and Bras, W. (1992). *Macromolecules*, **25**, 6277–6283.

Ryan, A.J., Willkomm, W.R., Bergstrom, T.B., Macosko, C.W., Koberstein, J.T., Yu, Chung, C. and Russell, T.P. (1991). *Macromolecules*, **24**, 2883–2889.

Ryan, A.J., Elwell, M.J. and Bras, W. (1995). *Nucl. Instrum. Methods Phys. Res., Sect. B*, **97**, 216–223.

Ryan, A.J., Fairclough, P.A., Terrill, N.J., Olmsted, P.D. and Poon, W.C.K. (1999). *Faraday Discuss.*, **112**, 13–29.

Schultz, J.M., Hsiao, B.S. and Samon, J.M. (2000). *Polymer*, **41**, 8887–8895.

Stanford, J.L., Ryan, A.J. and Elwell, M.J. (1997). In: Meijer, H.E.H. (ed.). *Processing of Polymers*, Materials Science and Technology. VCH, Weinheim, Vol. 18, pp. 466–512.

Strobl, G. (1996). *The Physics of Polymers*. Springer-Verlag, Berlin.

Strobl, G. (2000). *Eur. Phys. J. E*, **3**, 165–183.

Wang, Z.-G., Hsiao, B.S., Kopp, C., Sirota, E.B., Agarwal, P. and Srinivas, S. (1999). *Polym. Mater. Sci. Eng.*, **81**, 355–356.

Wang, Z.-G., Hsiao, B.S., Sirota, E.B. and Srinivas, S. (2000). *Polymer*, **41**, 8825–8832.

Wilkinson, A.N. and Ryan, A.J. (1998). *Polymer Processing and Structure Development*. Kluwer, Dordrecht.

Neutrons, X-Rays and Light
P. Lindner and Th. Zemb (Editors)

Chapter 19

Biological Applications of Small-Angle Neutron Scattering

Roland P. May

Institut Laue-Langevin, B.P. 156, F-38042 Grenoble CEDEX 9, France

1. Introduction

The knowledge of the structures of biological macromolecules is considered to be of prime importance for understanding their function. Powerful techniques have been developed to investigate biological structures. These include X-ray crystallography and, more recently, multi-dimensional NMR that allow one to obtain three-dimensional structure information, often at atomic resolution. However, X-ray crystallography requires the availability of single crystals of the molecule to be studied, and this is less frequently the case than one would like. With a few notable recent exceptions, in particular highly symmetrical objects like viruses, but also the structures of the small (Wimberly et al., 2000) and large (Ban et al., 1999) ribosomal subunits, most X-ray crystallographic structure determinations deal with objects of limited molecular mass, say 100 kDa, or less. In the case of NMR the size limit is even lower, at present about 30 kDa.

The resolution of small-angle scattering (for an introduction, see, e.g., Brumberger, 1995) cannot compete with these techniques, but it has continued to be a very useful tool for studying the *native* structures of biological macromolecules and of their complexes in solution. The scattering power of the different kinds of biologically relevant matter, i.e., proteins, nucleic acids, lipids, and sugars, for neutrons varies strongly, unlike for X-rays, mainly due to the different fraction of hydrogen compared to other atoms that they contain. Therefore, the scattering contributions from different matter can be separated by exposing the molecules to buffer solutions containing adequate mixtures of light (natural abundance) and heavy water; this is called 'solvent contrast variation' (cf. Chapter 7 by P. Schurtenberger in this book).

While deuteration of organic compounds like polymers requires sometimes complicated chemical techniques, deuterated biological molecules can be obtained in a relatively simple way by growing micro-organisms like bacteria or yeast in culture media containing heavy water (cf. Lederer et al., 1986). With the advancement of the overexpression of protein genes in foreign hosts, the amounts of D_2O needed have become affordable. Complete deuteration also requires feeding the microorganism with deuterated carbon sources.

Deuteration opens an opportunity that is unique to neutrons: single components (e.g., a single protein subunit) within a complex macromolecule can be highlighted by replacing all or part of their natural hydrogens by deuterium ('specific deuteration'). This technique requires that the complex in question can be taken apart and reconstituted from its components using biochemical and biophysical methods.

More recently, spin-contrast variation has appeared as a promising new technique which, however, requires important technical expenditures (Stuhrmann, 1994).

A number of biological investigations using classical techniques have been published (for more general reviews, see, e.g., Jacrot, 1976; Timmins and Zaccai, 1988). A few selected highlights are presented in this chapter.

2. Small-angle neutron scattering

Small-angle neutron scattering is introduced in the first chapter of this book. Other aspects can be found, e.g., in Williams et al. (1993). Here, the reader is only reminded of concepts and expressions that are necessary for understanding the context of this contribution.

2.1. Momentum transfer

The symbol for the momentum transfer used here is $\boldsymbol{q}$; $q = |\boldsymbol{q}| = (4\pi \sin\theta)/\lambda$, with λ, the wavelength and 2θ, the full scattering angle. $\boldsymbol{q}$ is often called $\boldsymbol{K}$, $\boldsymbol{h}$, $\boldsymbol{Q}$ by other authors, exceptionally even $\boldsymbol{s}$, which is usually defined as $s = (2\sin\theta)/\lambda$.

2.2. Scattering lengths and scattering length density of biological macromolecules

Only a few atoms are abundant in biological macromolecules. The coherent scattering lengths of the respective isotopes are listed in Table 1. They have quite similar (positive) values, with the exception of the light (normal abundance) isotope (^{1}H) of hydrogen with a negative scattering length; the heavy isotope deuterium (^{2}H, D) behaves 'normally'.

2.3. Scattering-length density and contrast variation

As described in Chapter 7 by P. Schurtenberger, one defines a scattering length density ρ, the sum of all scattering lengths b_i within a volume element V of a linear dimension corresponding to the resolution of SANS in real space, i.e., of the order of 10 Å. Because of the negative contribution of the scattering length of natural light hydrogen, the scattering length densities of the different kinds of biological matter are largely determined by the

Table 1

Coherent neutron (b_{coh}) and X-ray ($f_{X\text{-ray}}$) scattering lengths and incoherent cross-sections (σ_{inc}) for biologically relevant nuclei (after Jacrot, 1976)

Atom	Nucleus	b_{coh} (10^{-12} cm)	σ_{inc} (10^{-24} cm^2)	$f_{X\text{-ray}}$ for $q = 0$ (10^{-12} cm)
Hydrogen	^{1}H	−0.3742	80	0.28
Deuterium	^{2}H	0.6671	2	0.28
Carbon	^{12}C	0.6651	0	1.69
Nitrogen	^{14}N	0.940	0	1.97
Oxygen	^{16}O	0.5804	0	2.25
Phosphorus	^{31}P	0.517	0.3	4.23
Sulphur	mostly ^{32}S	0.2847	0	4.5

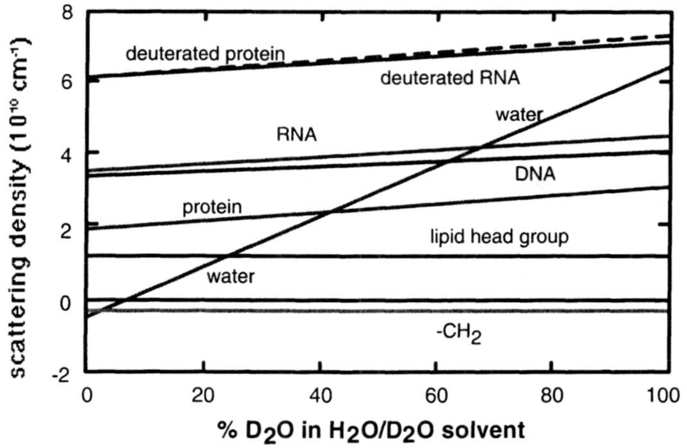

Fig. 1. The average scattering length densities ρ for some of the most common molecules found in biological material are plotted versus the fraction of D_2O in the solvent. Exchange of labile (non-covalently linked, solvent accessible) hydrogens accounts for the inclination of the protein and nucleic acid lines.

different fraction of hydrogen atoms present in them. Nucleic acids are less rich in hydrogen than proteins and thus have a lower average scattering length density. That of natural water is even negative. In 'heavy water' ^1H is replaced by deuterium, ^2H; its scattering length density is higher than that of any naturally occurring biological macromolecule. Mixtures of light and heavy water allow one to produce any intermediate scattering length density. The scattering densities of 100% deuterated proteins and nucleic acid exceed that of pure ^2H_2O.

The scattering intensity of a particle depends on the contrast $\Delta\rho = \rho_p - \rho_s$, the difference between the average scattering density ρ_p of the particle itself and that of the solvent around it, ρ_s. ρ_p depends on the ^2H_2O content of the solvent: labile (non-covalently bound and solvent-exposed) hydrogens in amino acids or nucleic acids in aqueous solution exchange with the solvent hydrogens until there is an equilibrium. Most of this exchange happens within fractions of seconds to a few hours.

At a certain fraction of ^2H_2O in an $^1H_2O/^2H_2O$ mixture (see Fig. 1), the average scattering length density of a given particle is identical to that of the solvent. At this so-called (scattering length density) match point or – semantically wrong – 'contrast match point' the zero-angle scattering of a homogeneous particle or of the corresponding part of a complex vanishes. The match point for proteins is at about 40 to 42% ^2H_2O, for RNA at about 68% ^2H_2O. This match-point difference of about 25% on the $^1H_2O/^2H_2O$ scale is sufficient to discriminate quite well between protein and RNA, and many biological applications of small-angle neutron scattering have been using exclusively this difference.

By partially replacing non-labile, covalently bound hydrogens with deuterium, the protein and nucleic acid lines in Fig. 1 can in fact take all intermediate values. The scattering of the nucleic acid in the ribosome (see Section 3.2), for example, can be matched by that of partially deuterated protein. The whole complex can then be made 'invisible' for neutrons in a buffer solution of the corresponding scattering density, and single components can be 'highlighted' by incorporating them in fully deuterated or fully protonated form.

2.4. Zero-angle scattering and radius of gyration

Biological molecules usually are identical particles of well-defined size and shape (at least in the limits of the resolution of SANS). In fact, there is no man-made polymer of similarly well-defined composition and mass as that of biological molecules like a particular protein or a virus. The relative molecular mass of such a particle can be determined from its zero-angle scattering known on an absolute scale (Jacrot and Zaccai, 1981). The zero-angle scattering needs to be extrapolated from the scattering curve that is known only from some minimal value $q_{\min}$ on, since the direct beam is hidden by a beam-stop.

Gross information on a particle's shape can be derived from the radius of gyration, R_G, which is an analogue to that in classical mechanics. It is usually obtained from a fit of a straight line to the innermost scattering points in a plot of the logarithm of the intensity versus q^2 ['Guinier plot': Guinier and Fournet (1955)]. The Zimm approximation is rarely used in biological applications of SANS.

Both the zero-angle scattering and R_G can also be calculated as the integral and the second moment of the pair distance distribution function, respectively (Glatter, 1977; see also Chapters 4 and 5 by O. Glatter in this book).

2.5. Indirect Fourier transformation

For particles with finite dimensions and which are sufficiently diluted not to show excluded-volume effects or other (e.g., electrostatic) interactions, the scattering intensity $I(q)$ can be expressed as a finite Fourier integral

$$I(q) = 4\pi \int_0^\infty p(r) \frac{\sin(qr)}{qr} \, \mathrm{d}r \tag{1}$$

The pair–distance distribution function $p(r)$ describes the probability of finding two volume elements at a given distance between r and $r + \Delta r$, weighted by their scattering length densities. Since the sign of the scattering densities of the two volume elements can be different, $p(r)$ can take negative values. Often, the pair distance distribution function bears more visual information than the shape of the scattering curve; practically only the scattering curves from symmetrical particles do not fit in this rule.

The Fourier integral needed to obtain $p(r)$ directly from the scattering curve $I(q)$ is infinite over q. The scattering curve is, however, only known over a limited q range. Therefore, $p(r)$ is calculated by an indirect Fourier transform: the coefficients of the Fourier transform of a set of functions in real space are fitted to the scattering data; convolutions due to geometrical and wavelength smearing can be taken into account. Glatter's (1979) method uses equidistant B-splines as the function set. Applying the fitted coefficients, which correspond to the heights of the splines, in real space results in the smeared and deconvoluted pair–distance distribution functions.

2.6. Triangulation

Biological macromolecules often are complexes that are built of many well-defined components. The sizes of those complexes can be so large that even if they can be crystallized

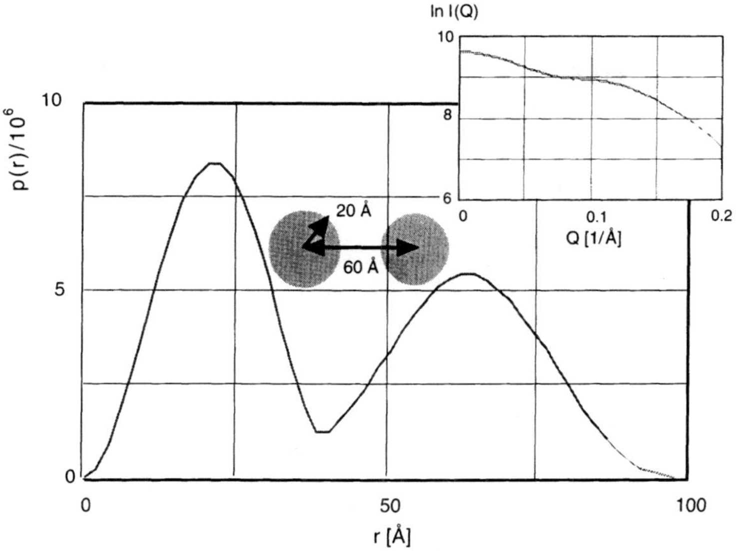

Fig. 2. The distance between two spheres appears as a peak in the pair–distance distribution function, but only as a slight oscillation in the scattering curve (insert).

– which is rather rarely the case – it is far from easy to reveal their structure by X-ray crystallography. By determining a sufficient number of distances between the components of the complex, however, the three-dimensional arrangement, i.e., the quaternary structure, of the complex can be reconstructed. Ribosomes (see Section 3.2) are an example of such complexes which can be taken apart and reconstituted from their constituents, proteins and nucleic acids, to form biochemically active entities again.

At least in principle, see Fig. 2, one can obtain the distance between two components of a complex by fitting a small number of splines with variable location, height and width to the scattering data (May and Nowotny, 1989).

In the case of a multi-component complex, determining the distance between two components is much more complicated due to the interference terms from the entire molecule that normally dominate the scattering.

The Label Triangulation Method (May, 1991), schematically shown in Fig. 3, allows one to measure distances between the components within a large complex and to determine their three-dimensional arrangement. Hoppe (1972) originally devised it for X-rays. Engelman and Moore (1972) first proposed to use neutrons instead. The method goes back to work by Kratky and Worthmann (1947) who studied the conformation of organic compounds by labelling their ends with heavy-metal atoms (iodine) and by measuring the X-ray scattering curves of these molecules containing two and one heavy atoms, respectively. The label distances were deduced from the periodicity of the difference between the scattering curves, which were dominated by the heavy-atom contributions. Later on, Vainshtein, Sosfenov and Feigin (1970) used a similar procedure to look at heavy-atom distances in gramicidin C. Even larger proteins (haemoglobin, histidine decarboxylase) were studied by Vainshtein et al. (1980) with an organo-mercury compound containing four Hg atoms.

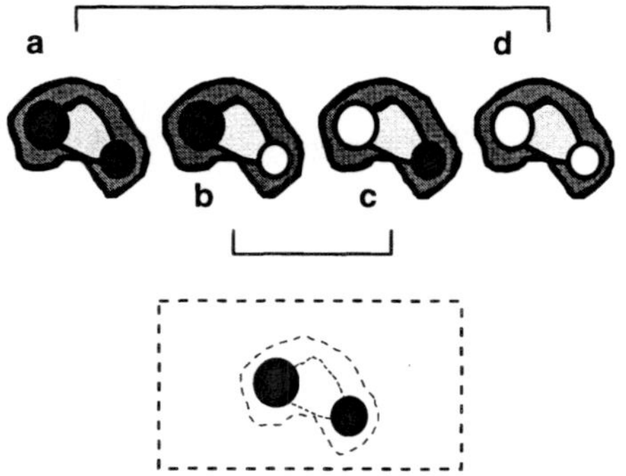

Fig. 3. Isomorphous replacement. Two single components of a multi-component complex are labelled specifically. Four different samples are prepared: (a) particles with two labelled components, (b) and (c) particles with either of the two components labelled, and (d) a (reference) particle that is not labelled at all. When the small-angle scattering from (b) + (c) is subtracted from that of (a) + (d), one obtains information on the scattering which is due exclusively to the vectors combining volume elements in one component with volume elements in the other component (bottom).

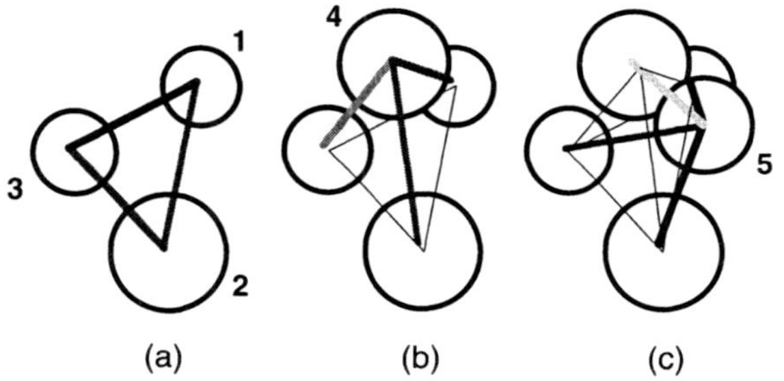

Fig. 4. Triangulation. Three distances define a triangle (a), another three a tetrahedron of undefined handedness (b). Each further four distances add another component in space (c).

Hoppe (1973) showed that in the case of high concentrations and/or heterogeneous complexes (consisting of different classes of matter), preparations (a) plus (d) and (b) plus (c) have to mixed. The scattering curves of these mixtures are subtracted from each other for obtaining the desired scattering difference curve, rather than measuring their scattering curves separately.

A direct result of a multitude of such scattering measurements is a table of inter-component distances. This is used to obtain a model of the spatial arrangement of the components: reconstructing the arrangement of n components in a complex (see Fig. 4) requires at least $n \cdot (n - 10)$ distance values: three distances define a basic triangle, three more yield a basic tetrahedron. The handedness of this tetrahedron is arbitrary; other techniques (like electron microscopy) are needed to define it. Fixing every further component in space requires at least four more distances. More than four distances may be needed if the resulting tetrahedron is too flat.

May et al. (1991) used an alternative approach rendering the naturally heterogeneous ribosomal particles homogeneous for neutrons by specific partial deuteration to a scattering length density level close to that of D_2O. This technique reduces undesired scattering contributions by measuring the scattering curves at the match point of the reference particle, making it quasi-invisible ('glassy ribosome'). The distance information is obtained by fitting the Fourier transform of 'moving splines' to the scattering curves from pairs of proteins according to a procedure developed by May and Nowotny (1989).

3. Some examples of biological applications

The examples that are presented in this chapter were chosen to show the role that neutrons play in biological structure investigation, but also to give scientists working in other fields some ideas on how they may use similar ideas in their research. Many, but not all of the examples, deal with complexes and assemblies of proteins and nucleic acid.

Viruses were amongst the first biological samples to be studied with neutrons. Besides the obvious interest in their structures they appeared to be ideal objects just for their large size and the strong scattering effects they provided, there is a basic reason for studying such particles: nucleic acids are at the origin of life. They store the genetic information and serve in the process of its translation into peptides and proteins. A multitude of proteins and complexes made of protein and nucleic acid is involved in these processes ruled by the central dogma of gene expression: "DNA makes RNA; RNA makes protein".

The first step of gene expression is the transcription. The enzyme DNA-dependent RNA polymerase produces a 'blueprint' of the genetic coded stored as a sequence of four different bases in the double-helical DNA; several SANS studies dealt with this molecule (Stöckel et al., 1979, 1980a,b). The duplicate consists of a thread of complementary single-stranded RNA. In a second step, an organelle called ribosome (see Section 3.2) reads this molecule and produces peptides and proteins, built from 20 different amino acids coded for by triplets of RNA bases.

Some proteins help to organize the nucleic acids or perform functions turning round the 'normal' course of the replication process, like in viruses. Either these molecules themselves are complexes of nucleic acid and protein, like the ribosome, or they form such complexes during their action. A few examples of structural studies using neutron scattering of such protein-nucleic acid complexes in solution follow below. In this chapter, we restrict ourselves to solution scattering. However, it is worthwhile to mention a few low-resolution crystallographic studies like those of nucleosomes, DNA storage particles (Finch et al., 1980; Bentley et al., 1981, 1984), of the satellite tobacco necrosis virus (Bentley et al., 1987), where the nucleic acid was located by neutron diffraction, or of detergents in Porin (Penel et al., 1998).

3.1. Viruses

Viruses are parasitic particles that always consist of nucleic acid and protein and sometimes of lipid; they convert the replication mechanism of infected ('host') cells in favour of their own reproduction. One piece or several pieces of RNA or DNA are encapsulated in a protein coat serving to protect the virus from being degraded. The nucleic acid contains genes that carry the code for the capsid protein(s) and for functional proteins involved in virus replication or cell recognition. Sometimes, e.g., in the case of the influenza virus, a lipid membrane derived from that of the host cell surrounds the virus.

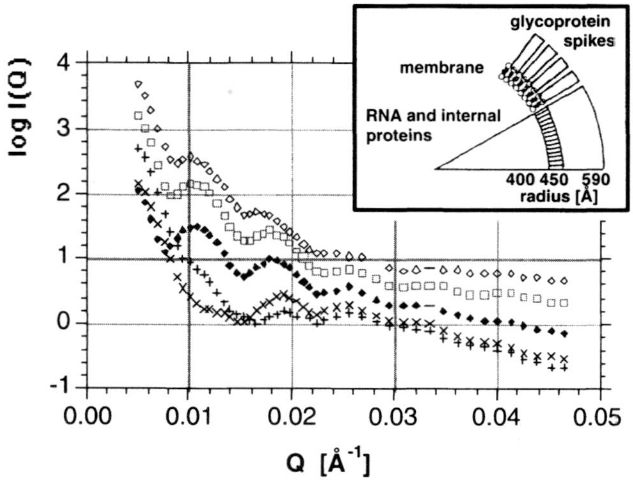

Fig. 5. Influenza virus. Neutron scattering curves from influenza virus in solutions of varying D_2O content (◇, 100%; □, 68%; ♦, 41%; +, 12%; ×, 22%). Insert: A low-resolution model derived from the neutron scattering data. (Drawn after Cusack, 1982.)

A review of structural studies of viruses with X-ray synchrotron and neutron radiation can be found in Timmins (1993). By electron microscopy, two kinds of structure are observed: rod-like and 'spherical' particles, which in fact are of icosahedral symmetry. Typical radii of the spherical viruses are of the order of 100 to 500 Å. These are dimensions that can be ideally studied by SANS.

Moreover, the virus components, nucleic acid, proteins and lipids can be easily distinguished due to their scattering length density differences (Fig. 1).

Due to the different weights of the scattering contributions from proteins, nucleic acids and lipids, the scattering curves show significant changes for the different contrasts. This is demonstrated in Fig. 5, taken from a study of the influenza virus (Cusack, 1982).

The extrapolated zero-angle scattering reveals the molecular mass of the virus as a function of the contrast. In 41% ^2H_2O, for example, only the proteins are 'visible'. The value of their molecular mass, which cannot be obtained easily with other methods, allows one to constrain the number of protein subunits within each particle; this can be of immense importance.

A model describing the virus as composed of a small number of spherical shells with different scattering length densities was fitted to the data. The model scattering curves were convoluted with parameters for the experimental smearing (wavelength dispersion, slit sizes). Viruses are one of the 'classical' cases where solvent scattering length contrast variation leads to a determination of the radial distribution of matter on an absolute scale. This is shown schematically in Fig. 6.

3.2. Ribosomes

Ribosomes are organelles that act as the protein factories in all living cells (e.g., Hardesty and Kramer, 1985). They are always composed of two subunits of unequal size, with a mass

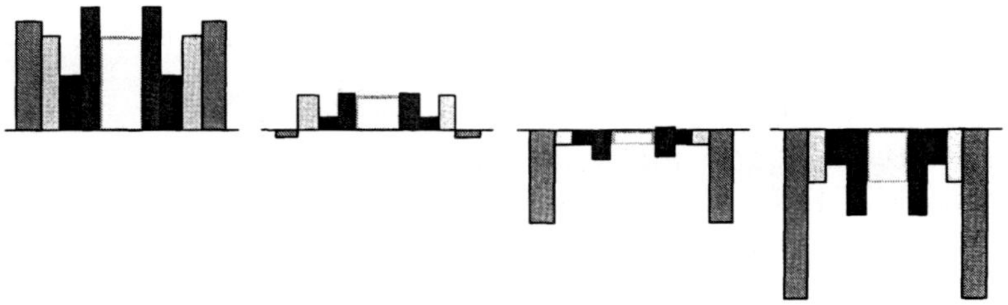

Fig. 6. Virus contrast variation. Schematic cut through a spherical model of SBMV in solutions of varying D_2O content (from left to right: 0%; 40%, 70%; 100%)

ratio of about 2 : 1, called after their Svedberg sedimentation coefficient, 30S and 50S in the case of the most extensively studied ribosome (70S), that of *E. coli*.

3.2.1. Overall shape of ribosomes

Already in the early 70s, both ribosomal subunits from *E. coli* were investigated with neutron small-angle scattering. The aim was to find out whether the protein and rRNA moieties are equally distributed over the volume of the subunits (30S: Moore et al., 1975; 50S: Stuhrmann et al., 1976; Serdyuk and Grenader, 1975).

The negative slope that is observed in a 'Stuhrmann plot' for 50S subunits isolated from *E. coli* grown in natural H_2O (Fig. 7) reveals that the protonated RNA (which is denser for neutrons than protonated proteins) is located in the centre of the particle. Since the data points in the Stuhrmann plot lie on a straight line rather than on a parabola one can conclude that within errors the mass centres of RNA and proteins coincide in both subunits. Several years later Ramakrishnan (1986) showed that salt washing the small subunit modified the scattering behaviour as compared to native particles because it removes certain proteins.

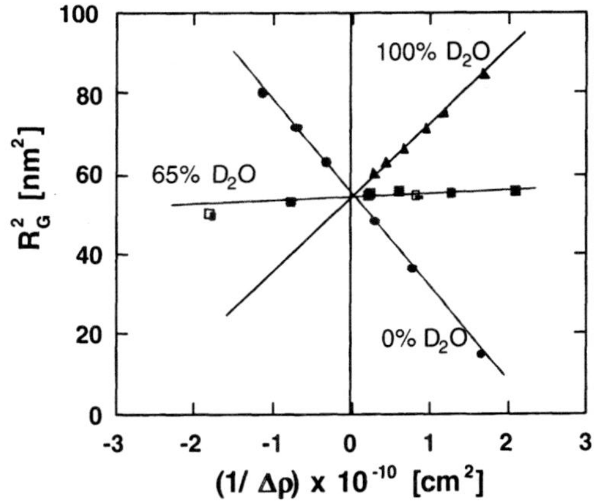

Fig. 7. Stuhrmann plot (square of the radius of gyration, R_G^2 versus the reciprocal contrast, $1/\Delta\rho$) for ribosomal 50S subunits from *E. coli* grown in media containing protonated glucose and nucleotides and different D_2O fractions as indicated. (Drawn after Koch and Stuhrmann, 1979.)

Stuhrmann et al. (1977) also gave a description of the shape of the 50S subunit by analysing the shape scattering function in terms of spherical harmonics.

More recently, a structural model of the 50S subunit was presented on the basis of X-ray (Svergun et al., 1994a) and neutron scattering (Svergun et al., 1994b). The authors used contrast variation and the decomposition of the scattering function in spherical harmonics in order to derive the shapes of the subunit and of its RNA-rich core.

3.2.2. Triple isotope substitution method (TIM)

TIM (Pavlov and Serdyuk, 1987) is a special case of specific partial deuteration. It requires the preparation of three samples: the component to be examined (e.g., a single protein, or the RNA moiety of a complex) is produced at three different levels of deuteration, in general protonated, highly deuterated and deuterated at an intermediate level, and incorporated into a matrix particle of identical composition in all three samples. The first two samples are mixed in such a way that the overall fraction of deuteration of the component is the same as that in the third sample; all concentrations are supposed to be equal. The neutron scattering curves of the mixture and that of the third sample are measured, and the latter subtracted from the former. The resulting intensity-difference curve is exclusively due to the nuclei in the component with varying contrast, whereas the rest of the complex vanishes. Pavlov et al. (1991) showed this for the complex of the ribosomal elongation factor EF-TUGDP with tRNA. The scattering contributions from the 30S protein and RNA moieties and of protein S4 *in situ* were determined by means of TIM by Harrison et al. (1993).

3.2.3. Triangulation

In 1987, Capel et al. published the complete map of all 21 protein positions within the small subunit from *E. coli* ribosomes, which was obtained with this method after more than ten years of experiments.

The 'glassy ribosome' approach of May et al. (1991), see Section 2.6, allowed them to report on 50 distances within the 50S subunit, and to present a model containing 7 protein positions (Fig. 8). As a 'by-product', also the scattering from single proteins in situ could be observed (Nowotny et al., 1994).

Finally, Willumeit et al. (2001) presented a map of the positions of 14 proteins in the large ribosomal subunit from *E. coli* as obtained by spin contrast variation.

For two decades, label triangulation remained the only method to provide a three-dimensional structure of this complicated object. Only very recently, some of the protagonists of the 30S triangulation studies and their collaborators succeeded in X-ray crystallographic determinations of the structures of the 30S (Wimberly et al., 2000) and 50S (Ban et al., 1999) subunits. The neutron data of the 30S subunit helped to establish the protein positions in the X-ray map.

3.3. Proliferating cell nuclear antigen (PCNA)

PCNA is a protein that acts in DNA replication and repair; it forms a sliding clamp for the DNA replication machinery. Schurtenberger et al. (1998) studied this protein by neutron

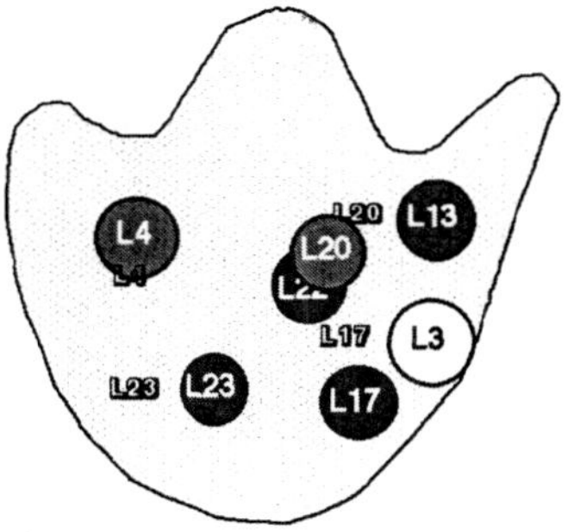

Fig. 8. 50S ribosomal subunit. The positions of 7 proteins determined by SANS are represented by spheres whose volume is the same as that of the protein. Framed numbers indicate immuno electron microscopy positions.

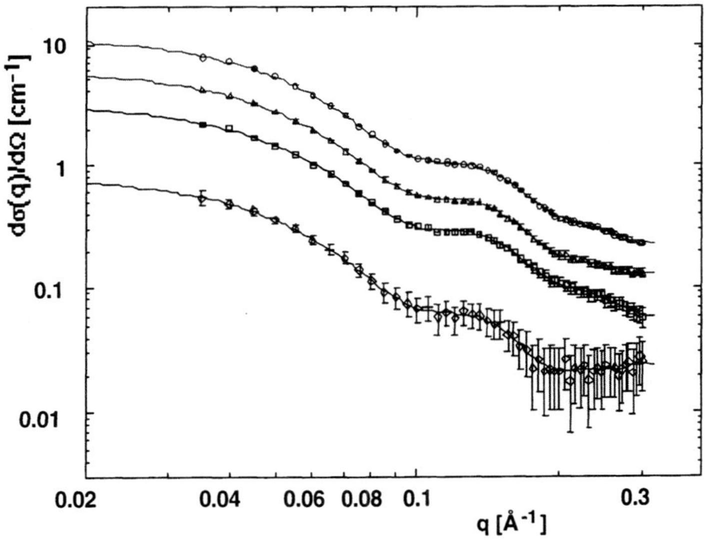

Fig. 9. Proliferating cell nuclear antigen. Scattering curves of PCNA at different concentrations, from top to bottom: 1.5 mg/ml, 0.8 mg/ml, 0.4 mg/ml, and 0.1 mg/ml. The solid lines are fits obtained by indirect Fourier transformation (after Schurtenberger et al., 1998).

small-angle scattering at D22 of the ILL. Decent scattering curves from a protein in D_2O buffer can be obtained with a concentration of only 0.1 mg/ml (Fig. 9).

As can be seen from Fig. 10, quite a few of the features that are visible in the pair distance distribution function calculated for a model based on the known X-ray crystallographic structure from yeast PCNA do not appear in the experimentally obtained $p(r)$ curve. This is true even if one takes into account the instrumental smearing of the scattering curve and the fact that the Glatter's (1977) indirect Fourier transformation procedure searches for the smoothest $p(r)$ function that can fit the scattering data.

The simplest model that can still describe the main features of the data is a hexagonal arrangement of spheres. Surprisingly, this very simple model approximates the scattering data and the $p(r)$ function (Fig. 10) extremely well.

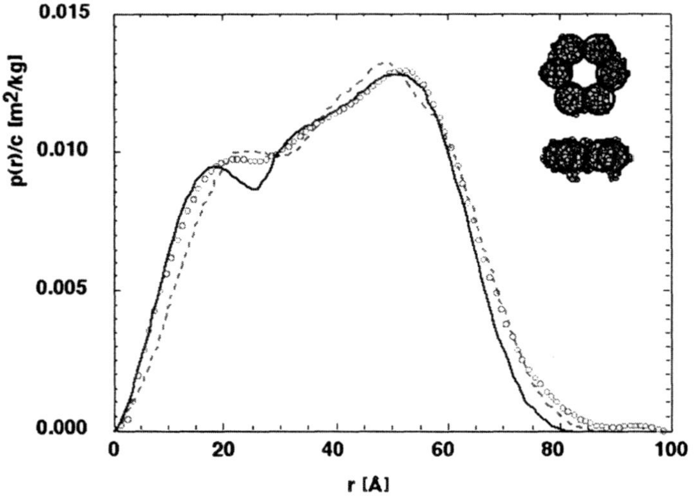

Fig. 10. Proliferating cell nuclear antigen. Pair–distance distribution functions of PCNA. Circles, calculated from neutron scattering data, dashed line, calculated from yeast PCNA X-ray crystallographic data (Krishna et al., 1994), using spheres of 3 Å radius at amino acid positions, solid line, simple six-sphere model (after Schurtenberger et al., 1998).

3.4. Chaperones

For proteins to function correctly, they must fold into their correct three-dimensional shape. This is carried out with the assistance of molecules called chaperonins – double-ring structures with a central cavity that holds the protein as it folds up. One of the best known chaperones is GroEL from *E. coli.* It belongs to the class of chaperonins/hsp60 proteins and has the form of a cylinder composed of two 7-rings. GroEL is assisted by its co-chaperonin GroES. Current models propose that in presence of ATP or ADP, 1:1-complexes of GroEL and GroES are formed that bind denatured proteins. These 'trans'-complexes, where the substrate protein is bound on the opposite side of GroES, are supposed to be transformed to *cis*-complexes. Current models propose a refolding of the proteins in the *cis*-complexes. After this, GroES and substrate protein are released. In certain conditions, GroEL-GroES 1 : 2 ('football') complexes and complexes with two substrate proteins have been observed.

3.4.1. Thermosomes

In a primitive group of organisms, the archaea, in particular in *Thermoplasma acidophilum* which survives in hot sulphur springs at temperatures between 55 and 59°C, but also in eukaryotic cells (including mammalian) the same process functions without a cofactor. The chaperonin involved is called a thermosome.

It was known that the thermosome, the archetype of group II chaperonins, has two different structures. It appears cylindrical when viewed, trapped in ice, with an electron microscope (Nitsch et al., 1998) but spherical in the crystalline state as observed with scattered X-rays (Ditzel et al., 1998). The interpretation is that the thermosome has a built-in lid instead of a

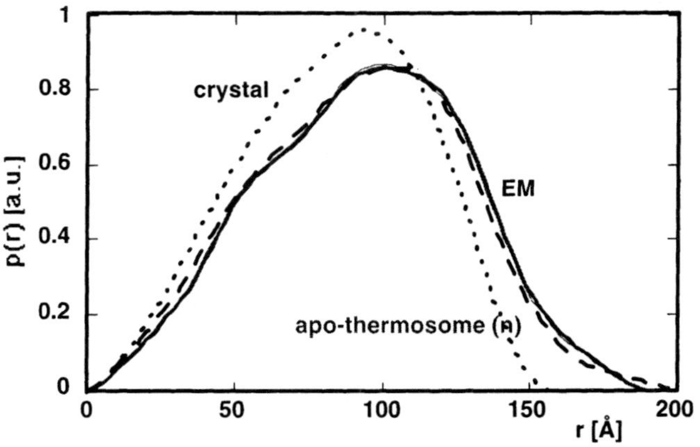

Fig. 11. Pair–distance–pair distribution functions of the thermosome obtained in solution by SANS (solid line), and calculated from X-ray crystallographic (dotted line; Ditzel et al., 1998), and electron microscopic data (dashed line; Nitsch et al., 1998).

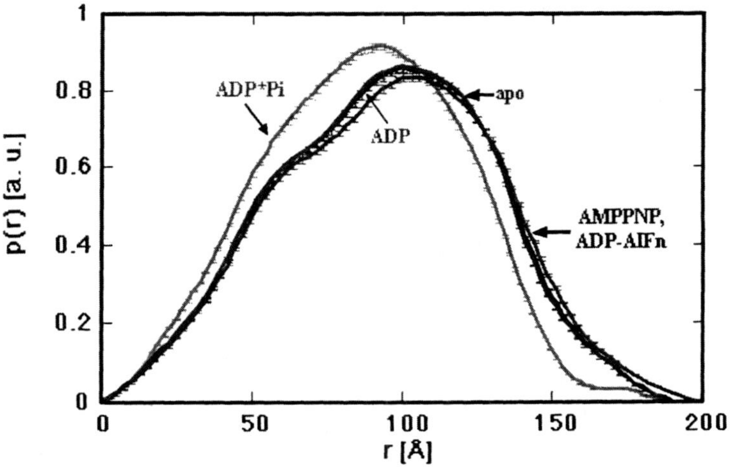

Fig. 12. Thermosome. SANS pair–distance distribution functions of the thermosome in solution with and without addition of nucleotides: apo, apo-thermosome; ADP, with ADP; AMPPNP, ADP-AlFn, with AMPPNP or ADP and AlF; ADP + Pi, with ADP and inorganic phosphate.

detachable cofactor, and that the cylindrical shape represents the open thermosome ready for ATP binding while the spherical shape indicates the closed, activated folding state.

Gutsche et al. (2000) tested these ideas using small-angle neutron scattering to examine the overall structure of the thermosome and its behaviour. The results confirmed that the unbound thermosome was indeed an open cylinder (Fig. 11). They then examined the structural changes occurring in solution close to real physiological conditions – in the presence of ATP, ADP and phosphate ions, and also analogues of ATP such as aluminium fluoride which do not hydrolyse and could thus mimic the intermediate bound states formed. The resulting data (Fig. 12) enabled them to unravel the reaction cycle.

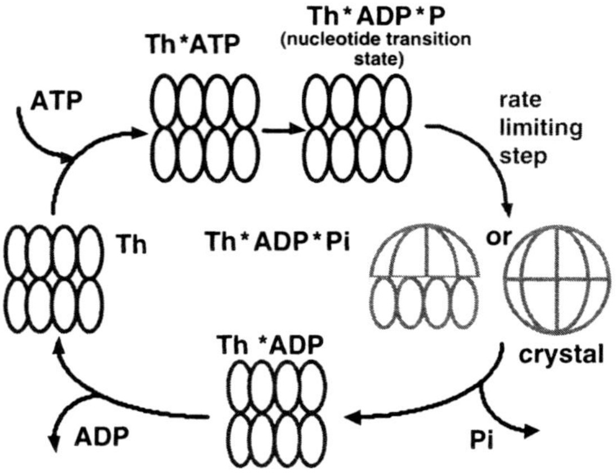

Fig. 13. Thermosome ATPase cycle. The open apo-chaperonin (left) expands slightly on binding ATP, hydrolyses it, and slowly closes on relaxation from the tight ADP-P state into a loose ADP-Pi state. Phosphate release triggers re-opening and ADP dissociates to complete the cycle. Adapted from Gutsche et al. (2000).

The ATPase cycle (Fig. 13) turned out to be quite complex. Surprisingly, at room temperature ATP binding does not induce closure of the chaperonin as in the case of the GroEL/GroES system.

In a later experiment that included a precise control of the sample temperature during the scattering measurement it was found (Gutsche et al., 2001) that ATP binding to the thermosome is sufficient to trigger its closure at physiologically relevant high temperatures (55°C). This reveals an analogy between the ATPase cycles of the two groups of chaperonins.

3.4.2. *GroEL/GroES*

Using deuterium labeling Stegmann et al. (1998) obtained information about the domain structures of GroEL and GroES as part of the chaperone complex and in the isolated state. The conformations of GroEL and GroES change upon complex formation. The central domains of GroES in the complex as seen by neutrons have a somewhat flatter appearance and display a less dome-like shape than in the isolated state. This finding is in contrast to the X-ray crystal structure published after completion of the neutron study (Xu et al., 1997). The apical domains of GroEL rotate away from the equatorial plane. These movements may facilitate dissociation by removing interacting surfaces from the substrate.

The process of substrate binding and release is necessarily highly dynamic. Therefore, it is important to develop methods that allow one to study the reactions that occur in the substrate cycle in real time. The high neutron flux available on the instrument D22 at the Institut Laue-Langevin in Grenoble (up to $>10^8$ neutrons/cm^2 s) rendered it possible to revive investigations of the association of macromolecules with each other and/or their behaviour after the modification of the solution conditions. This is achieved by using a stopped-flow apparatus in which two reagents in different syringes are rapidly mixed together in a mixing cell; the product (about 400 μl) is injected into a quartz observation cell. Simultaneously with

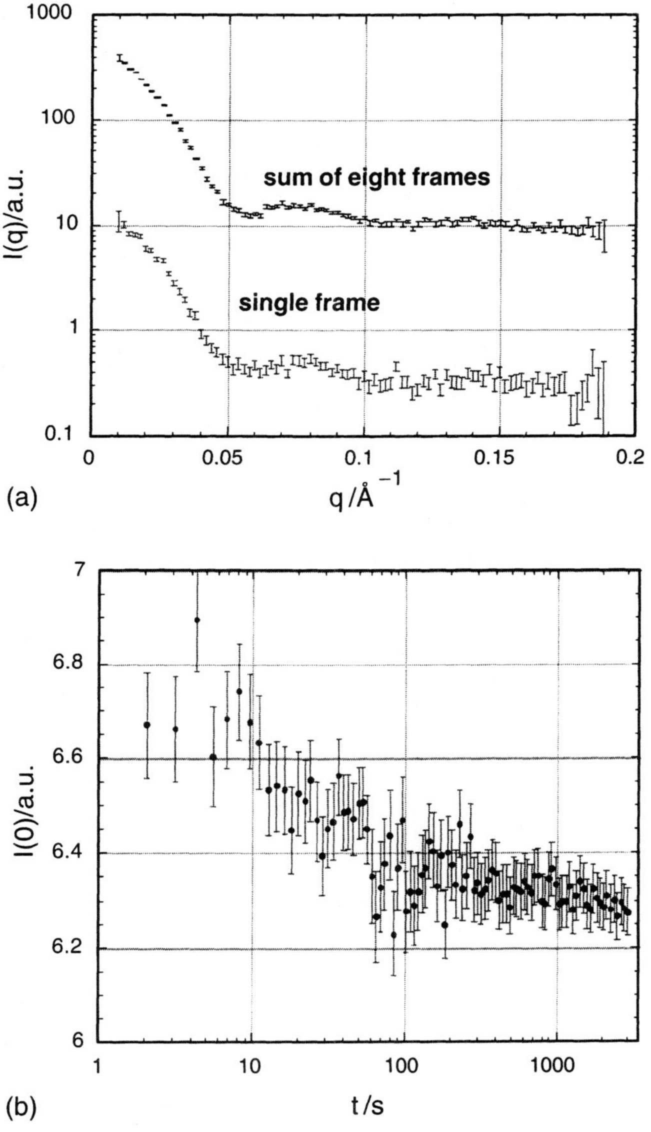

Fig. 14. (a) Second frame of a scattering experiment in which GroEL was mixed with GroES. Bottom, a single shot; top, sum of eight shots, shifted additionally for clarity (Holzinger et al., 2000). (b) Extrapolated zero-angle intensity after mixing deuterated GroEL with H-GroES in a D_2O solution containing ATP, as a function of time. In the presence of the ATP analogue AMP-PNP, GroEL forms a 'football' complex with GroES that is bound to both ends. In the particular contrast conditions used, GroES reduces the scattering mass of the complex upon binding. Therefore the scattering at zero angle decreases with time (Rössle et al., 2000).

the mixing, a control computer starts the acquisition of neutron patterns. At predefined times, the acquisition system switches to the next 'frame', corresponding to a higher address in the memory.

Fig. 14a shows that with the GroEL/GroES system integral counting times of about 10 seconds are sufficient to obtain useful signals. Smaller proteins, where the background (even with D_2O buffer solutions) dominates the scattering, longer integral times become necessary.

In the experiment shown in Fig. 14b, normal-abundance GroES was mixed with deuterated GroEL in the presence of the ATP analogue AMP-PNP in a D_2O solution. 'Football' complexes of GroEL and two GroES form, and because of the different signs of the scattering densities of the two molecules, the contrast of the whole complex at zero angle is smaller than that of GroEL alone. For this reason, $I(0)$ decreases with time.

4. Conclusion

Small-angle neutron scattering allows one to investigate the structures of biological macromolecules in solution, under native conditions identical to those used in biochemical studies and it does not depend on the availability of crystals. SANS is a relatively simple method, and there is practically no restriction on the maximum size of the molecules to be studied. The resolution of SANS data is limited compared to crystallography or NMR. On the other hand, information on parts of a molecule can be obtained by contrast variation and by employing specific deuteration. These techniques allow one to get more precise results by restricting the information to be gained in a single experiment. The contrast of proteins with respect to D_2O is high, and thus low concentrations are sufficient to obtain valid data. Time-resolved measurements are beginning to evolve, and they promise to enrich our knowledge of biological reactions on the macromolecular level.

General reading on small-angle and other scattering methods

Modern Aspects of Small-Angle Scattering. NATO ASI Series C, Vol. 451, edited by H. Brumberger. Kluwer Academic Publishers, Dordrecht, 1995.

Neutron, X-Ray and Light Scattering: Introduction to an Investigative Tool for Colloidal and Polymeric Systems, edited by P. Lindner and T. Zemb. North-Holland, Amsterdam, 1991.

Neutron and Synchrotron Radiation for Condensed Matter Studies, Vols. I to III, edited by J. Baruchel, J.-L. Hodeau, M. S. Lehmann, J.-R. Regnard and C. Schlenker. Les Editions de Physique/Springer Verlag, Les Ulis, 1993.

References

Arnold, E., Jacobo-Molina, A., Nanni, R.G., Williams, R.L., Lu, X., Ding, J., Clark Jr., A.D., Zhang, A., Ferris, A.L., Clark, P., Hizi, A. and Hughes, S.H. (1992). *Nature*, **357**, 85–89.

Ban, N., Nissen. P., Hansen. J., Capel, M., Moore, P.B., Steitz, T.A. (1999). *Nature*, **400**, 811–812.

Bentley G.A., Finch, J.T. and Lewit-Bentley, A. (1981). *J. Mol. Biol.*, **145**, 771–784.

Bentley G.A., Lewit-Bentley, A., Finch, J.T., Podjarny, A.D. and Roth, M. (1984). *J. Mol. Biol.*, **176**, 55–76.

Bentley G.A., Lewit-Bentley, A., Liljas, A., Skoglund, U., Roth, M. and Unge, T. (1987). *J. Mol. Biol.*, **194**, 129.

Brumberger, H., ed. (1995). *Modern Aspects of Small-Angle Scattering. NATO ASI Series C, Vol. 451*. Kluwer Academic Publishers, Dordrecht.

Capel, M.S., Engelman, D.M., Freeborn, B.R., Kjeldgaard, M., Langer, J.A., Ramakrishnan, V., Schindler, D.G., Schneider, D.K., Schoenborn, B.P., Sillers, I.Y., Yabuki, S. and Moore, P.B. (1987). *Science*, **238**, 1403–1406.

Cusack. S. (1982). In: *The Neutron and its Applications 1982* (Schofield, P., ed.). The Institute of Physics, Bristol and London, pp. 351–355.

Engelman, D.M. and Moore, P.B. (1972). *Proc. Natl. Acad. Sci. USA*, **69**, 1997–1999.
Finch, J.T., Lewit-Bentley, A., Bentley, G.A., Roth, M. and Timmins, P.A. (1980). *Philos. Trans. R. Soc. (London)*, **B290**, 635–638.
Glatter, O. (1977). *J. Appl. Crystallogr.*, **10**, 415–421.
Glatter, O. (1979). *J. Appl. Crystallogr.*, **12**, 166–175.
Guinier, A. and Fournet, G. (1955). *Small-Angle Scattering of X-Rays*. Wiley, New York.
Gutsche, I., Holzinger, J., Rößle, M., Heumann, H., Baumeister, W. and May, R.P. (2000). *Current Biol.*, **10**, 405–408.
Gutsche, I., Holzinger, J., Rauh, N., Baumeister, W. and May, R.P. (2001). *J. Struct. Biol.*, **134**, in press.
Hardesty, B. and Kramer, G., eds. (1985). *Springer Series in Molecular Biology*. Springer, New York.
Harrison, D.H., May, R.P. and Moore, P.B. (1993). *J. Appl. Crystallogr.*, **26**, 198–206.
Holzinger, J., Rößle, M., Heumann, H. and May, R.P. (2000). Unpublished.
Hoppe, W. (1972). *Isr. J. Chem.*, **10**, 321–333.
Hoppe, W. (1973). *J. Mol. Biol.*, **78**, 581–585.
Jacrot, B. (1976). *Rep. Prog. Phys.*, **10**, 911–953.
Jacrot, B. and Zaccai, G. (1981). *Biopolymers*, **20**, 2413–2426.
Koch, M.H.J. and Stuhrmann, H.B. (1979). *Methods Enzymol.*, **59**, 670–706.
Kratky, O. and Worthmann, W. (1947). *Mh. Chem.*, **76**, 263–281.
Krüse, J., Timmins, P.A. and Witz, J. (1982). *Virology*, **119**, 42–50.
Lederer, H., May, R.P., Kjems, J.K., Schaefer, W., Crespi, H.L. and Heumann, H. (1986). *Eur. J. Biochem.*, **156**, 655–659.
May, R.P. (1991). In: *Neutron, X-Ray and Light Scattering: Introduction to an Investigative Tool for Colloidal and Polymeric Systems*, (Lindner, P. and Zemb, T., eds.). North Holland, Amsterdam, pp. 119–133.
May, R.P. and Nowotny, V. (1989). *J. Appl. Crystallogr.*, **22**, 231–237.
May, R.P., Nowotny, V., Nowotny, P., Voß, H. and Nierhaus, K.H. (1991). *EMBO J.*, **11**, 373–378.
Moore, P.B., Engelman, D.M. and Schoenborn, B.P. (1975). *J. Mol. Biol.*, **91**, 101–120.
Nowotny, P., Rühl, M., Nowotny, V., May, R.P., Burkhardt, N., Voß, H. and Nierhaus, K.H. (1994). *Biophys. Chem.*, **53**, 115–122.
Pavlov, M.Y. and Serdyuk, I.N. (1987). *J. Appl. Crystallogr.*, **20**, 105–110.
Pavlov, M.Y., Rublevskaya, I.N., Serdyuk, I.N., Zaccai, G., Leberman, R. and Ostanevich, Y.M. (1991). *J. Appl. Crystallogr.*, **24**, 243–254.
Penel, S., Pebay-Peyroula, E., Rosenbusch, J., Rummel, G., Schirmer, T., Timmins, P.A. (1998). *Biochimie*, **80**, 443–451.
Ramakrishnan, V. (1986). *Science*, **231**, 1562–1564.
Rössle, M., Manakova, E., Holzinger, J., Vanatalu, K., May, R.P. and Heumann, H. (2000). *Phys. B*, **276–278**, 532–533.
Serdyuk, I.N. and Grenader, A.K. (1975). *FEBS Lett.*, **59**, 153.
Stöckel, P., May, R.P., Strell, I., Cejka, Z., Hoppe, W., Heumann, H., Zillig, W., Crespi, H.L., Katz, J.J. and Ibel, K. (1979). *J. Appl. Crystallogr.*, **12**, 176–185.
Stöckel, P., May, R.P., Strell, I., Cejka, Z., Hoppe, W., Heumann, H., Zillig, W. and Crespi, H.L. (1980a). *Eur. J. Biochem.*, **112**, 411–417.
Stöckel, P., May, R.P., Strell, I., Cejka, Z., Hoppe, W., Heumann, H., Zillig, W. and Crespi, H.L. (1980b). *Eur. J. Biochem.*, **112**, 419–423.
Schurtenberger, P., Egelhaaf, S.U., Hindges, R., Maga, G., Jonsson, Z.O., May, R.P., Glatter, O. and Hübscher, U. (1998). *J. Mol. Biol.*, **275**, 123–132.
Stegmann, R., Nieba-Axmann, S., Manakova, E., Rößle, M., Hermann, T., May, R.P., Wiedenmann, A., Plückthun, A. and Heumann, H. (1998). *J. Struct. Biol.*, **121**, 30–40.
Stuhrmann, H.B. (1994). *Neutron News*, **5**, 23–27.
Stuhrmann, H.B., Haas, J., Ibel, K., De Wolf, B., Koch, M.H.J., Parfait, R. and Crichton, R.R. (1976). *Proc. Natl. Acad. Sci. USA*, **73**, 2379–2383.
Stuhrmann, H.B., Koch, M.H.J., Parfait, R., Haas, J., Ibel, K. and Crichton, R.R. (1977). *Proc. Natl. Acad. Sci. USA*, **74**, 2316–2320.
Svergun, D.I., Koch, M.H.J. and Serdyuk, I.N. (1994a). *J. Mol. Biol.*, **240**, 66–77.

Svergun, D.I., Koch, M.H.J., Pedersen, J.S. and Serdyuk, I.N. (1994b). *J. Mol. Biol.*, **240**, 78–86.

Timmins, P.A. (1993). In: *Neutron and Synchrotron Radiation for Condensed Matter Studies Vol. III, Applications to Soft Condensed Matter and Biology* (Baruchel, J., Hodeau, J.-L., Lehmann, M.S., Regnard, J.-R. and Schlenker, C., eds.). Les Editions de Physique/Springer Verlag, Les Ulis, pp. 239–255.

Timmins, P.A. and Zaccai, G. (1988). *Eur. Biophysics J.*, **15**, 257–268.

Vainshtein, B.K., Sosfenov, N.I. and Feigin, L.A. (1970). *Dokl. Akad. Nauk USSR*, **190**, 574–577.

Vainshtein, B.K., Feigin, L.A., Lvov, Y.M., Gvozdev, R.I., Marakushev, S.A. and Likhtenshtein, G.I. (1980). *FEBS Lett.*, **116**, 107–110.

Williams, C., May, R.P. and Guinier, A. (1993). In: *Characterization of Materials. Materials Science and Technology, Vol. 2B* (Lifshin, E., ed.). VCH Verlagsgesellschaft, Weinheim, pp. 611–656.

Willumeit, R., Diedrich, G., Forthmann, S., Beckmann, J., May, R.P., Stuhrmann, H.B. and Nierhaus, K.H. (2001). *Biochim. Biophys. Acta*, **1520**, 7–20.

Wimberly, B.T., Brodersen, D.E., Clemons Jr., W.M., Morgan-Warren, R.J., Carter, A.P., Vonrhein, C., Hartsch, Th. and Ramakrishnan, V. (2000). *Nature*, **407**, 327–339.

Xu, Z., Horwich, A.L. and Sigler, P.B. (1997). *Nature*, **388**, 741–750.

Neutrons, X-Rays and Light
P. Lindner and Th. Zemb (Editors)

Chapter 20

Light Scattering in Turbid Suspensions, a Tutorial

Jaro Rička

Institute of Applied Physics, Sidlerstraße 5, CH-3014 Bern, Switzerland

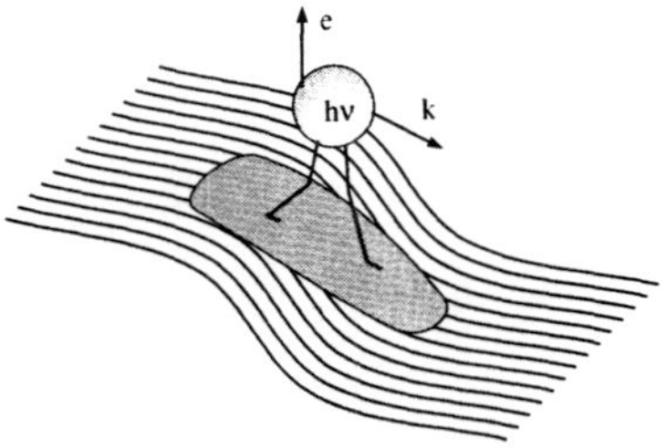

This cartoon combines three aspects of light: photons, waves, and *path*. We shall work entirely in the path model. I hope to please Georg Maret, one of the pioneers of the modern development of multiple scattering. He likes photon paths very much....

1. Introduction

1.1. What is multiple scattering?

Imagine four glasses:

0. The first glass is blank, filled with pure water. The other three glasses contain suspensions of oil in water at three different oil concentrations:
1. Water with a drop of milk added, roughly 0.01% volume of fat, in the form of droplets with about 100 nm radius.
2. Low fat skimmed milk, roughly 1% volume of the droplet phase.
3. Full cream milk from Swiss alps or Normandy, more than 10% of fat droplets.

The four glasses represent four different classes of light scattering problems. Case 0 is out of the scope of the present book, I only invoke one of the associated keywords: Brillouin scattering. Glass 1 represents a single scattering situation, such as has been discussed in most of the contributions in the present book. Characteristic for glasses 2 and 3 is the presence of *multiple scattering* in two different degrees. Let's be quantitative: we refill the suspensions in rectangular cells of 1 cm thickness and measure the light transmission with the setup sketched in Fig. 1. In case 1 we observe only 3% loss of intensity upon the transmission through the cell. In case 2 about 95% of the illuminating beam is lost because of the strong scattering. Light transmission decreases according to Beer's law $J = J_e \mathrm{e}^{-d/\ell}$, where d is

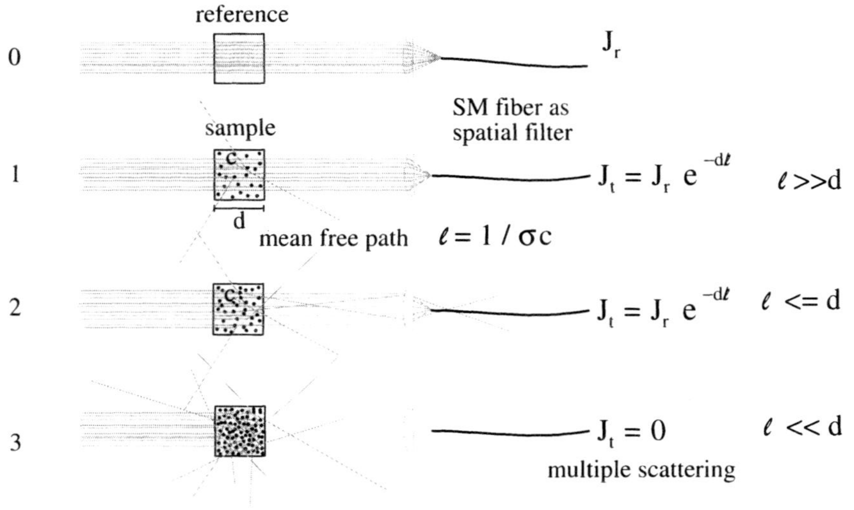

Fig. 1. Measuring mean free path ℓ. Single-mode fiber is good spatial filter selecting the transmitted beam so that the scattered light does not enter the detector.

the cell thickness. The *extinction length* can be regarded as the *mean free path* of photons traversing the sample. For a dilute particulate suspension, $\ell = 1/\sigma c$ can be calculated from the concentration c and the *scattering cross-section* σ of a scatterer. Our milk fat droplets with 50 nm radius exhibit a scattering cross-section of roughly 2 nm^2. Correspondingly the mean free path ℓ amounts to 350 mm, 3.5 mm and 0.35 mm for the three milk concentrations.

A scattering experiment is said to be in the single scattering regime if the mean free path ℓ is much larger than the size d of the sample cell. The other extreme with $\ell \ll d$ is the multiple scattering regime.

Case 1 is well in the *single scattering regime*, $\ell \gg d$. Case 2, skimmed milk in a 1 cm cell, is an intermediate *few scattering* regime with $\ell \approx d$. Case 3, $\ell \ll d$, is in the *multiple scattering* regime. Beer's law fails, the transmitted beam is gone completely. The sample is no more transparent, only translucent. Light propagates trough the concentrated suspension in a highly irregular fashion resembling diffusion. Note that both, mean free path ℓ *and* sample size d, determine the regime. (A fog layer is transparent over few meters, but the outlook over kilometers is lost.)

What happens if we keep increasing the fat concentration still further? Well, we obtain cream, then butter (milk fat with enclosed water droplets) and finally pure transparent milk fat. Thereby ℓ would first decrease to reach a minimum somewhere in the cream state. Then it will *increase* again, until we are back to single scattering! This, however, is a coarse macroscopic picture. The refractive index in the cell has changed as the result of a great number of microscopic single scattering processes. The scattered waves interfere, so that *macroscopic* scattering is suppressed in samples that are homogeneous on the length scale of light wavelength.

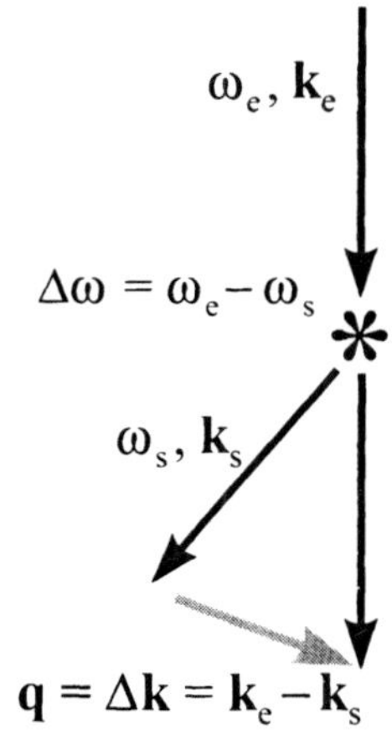

Fig. 2

1.2. Scattering in particle picture and diffusing wave spectroscopy

The star in Fig. 2 does not mean an individual scatterer, rather it designates a *single scattering event*. Because of the wave nature of the probing field, many scatterers (e.g., fat droplets) interfere in a single scattering event. In each event there occurs an energy exchange $\Delta u = \hbar\Delta\omega$ and momentum exchange $\Delta \boldsymbol{p} = \hbar\Delta\boldsymbol{k}$ between the scattering system and the probing field. Wave vector difference $\Delta\boldsymbol{k} = \boldsymbol{k}_s - \boldsymbol{k}_e$ is abbreviated with $\boldsymbol{q}$ (scattering vector) and the difference $\Delta\omega = \omega_s - \omega_e$ with ν. We assume *quasi-elastic* scattering. Only the orientation of the wave vector $\boldsymbol{k}$ changes in an event, i.e., $\nu/c_\ell \ll |\boldsymbol{q}|$. Momentum transfer $\boldsymbol{q}$ and energy transfer ν are random quantities. The probability density $S(\boldsymbol{q},\nu)$ for finding certain $\boldsymbol{q}$ and ν is known as the *dynamic structure factor*. Integrating $S(\boldsymbol{q},\nu)$ with respect to ν one obtains the familiar static *structure factor* $S(\boldsymbol{q}) = \int S(\boldsymbol{q},\nu)\mathrm{d}\nu$. The normalized spectral density $s(\boldsymbol{q},\nu) = S(\boldsymbol{q},\nu)/S(\boldsymbol{q})$ is closely related with the *first-order correlation function* $f(\boldsymbol{q},\tau)$ from P. Pusey's contribution on dynamic light scattering (DLS) (Chapter 9). $s(\boldsymbol{q},\nu)$ and $f(\boldsymbol{q},\tau)$ are a pair of Fourier transforms: $f(\boldsymbol{q},\tau) = \int s(\boldsymbol{q},\nu)\,\mathrm{e}^{i\nu\tau}\,\mathrm{d}\nu$. Recall that with DLS we measure the normalized signal correlation function $g(\tau,\boldsymbol{q})$ which is related to $f(\boldsymbol{q},\tau)$ by the so-called Siegert relation $g(\tau) = 1 + |f(\boldsymbol{q},\tau)|^2$.

Multiple scattering involves sequences of single scattering events. Fig. 3 shows only one of the infinite number of possible chains of scattering events, i.e., of paths of photons through a multiply scattering suspension. Momentum and energy are exchanged in each scattering event along a path. The total momentum transfer along a path of n single scattering events, $\boldsymbol{q}^{(p)} = \sum_{i=1}^{n} \boldsymbol{q}_i$, amounts to $\boldsymbol{k}_{\mathrm{out}} - \boldsymbol{k}_{\mathrm{in}}$. After a few scattering events the orientation of the resulting wave vector $\boldsymbol{k}_{\mathrm{out}}$ assumes essentially an isotropic distribution. Thus, in Case 3, strong multiple scattering, the information on the structure of the suspension is practically lost.

Dynamics, however, survives. The total energy transfer along a path is the sum $\nu^{(p)} = \sum_{i=1}^{n} \nu_i$ of independent random variables ν_i. Therefore, the spectral density $s^{(p)}(\nu)$ is the multiple convolution of single scattering spectral densities: $s^{(p)}(\nu,n) = s(\nu_1) * s(\nu_2) * \cdots * s(\nu_n)$. By virtue of the convolution theorem, the first-order correlation function $f^{(p)}(\tau,n)$ is the

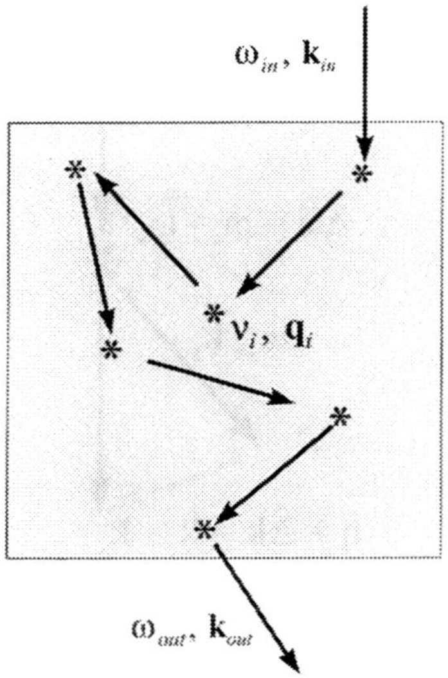

Fig. 3.

product of partial correlation functions:

$$f^{(p)}(\tau,n)=\prod_{1}^{n} f(\tau,q_i). \tag{1}$$

For illustration, we insert $f(\tau,q_i)=\mathrm{e}^{-q_i^2 D\tau}$ for Brownian motion and obtain: $f^{(p)}(\tau,n)=\mathrm{e}^{-n\overline{q^2}D\tau}$, $\overline{q^2}$ is the mean square momentum transfer along the path. Note in the exponent the number n of scattering events: multiple scattering greatly accelerates the decorrelation. The total spectral density $s_{\mathrm{tot}}(\nu)$ measured with a multiply scattering suspension is the average over a great number of possible paths with a broad distribution of n. Assuming that n is the only relevant parameter of a path (applies for Case 3 only), we may write

$$f(\tau)=\sum_{n=1}^{\infty} W_n \bar{f}_n(\tau). \tag{2}$$

In order to determine the path length distribution W_n we note that the chain of scattering events in Fig. 3 resembles a random flight. One may think of the photons as performing a kind of Brownian motion, which, in turn, can be approximated as diffusion. Density C of single scattering events obeys the diffusion equation $\partial C(\boldsymbol{r},t)/\partial t - D_s\nabla^2 C(\boldsymbol{r},t)=Q(\boldsymbol{r},t)$, where $D_s=c_\ell \ell^*/3$ The source term $Q(\boldsymbol{r},t)$ is given by the distribution of the *first* scattering events (e.g., the first star in the path shown in Fig. 3). In the diffusion coefficient D_s, c_ℓ is the light propagation velocity and ℓ^* is the *transport mean free path*, related but not identical with ℓ. From the diffusion equation one can deduce the distribution of times τ_t the photons need to traverse the sample and thus also the distribution W_n of $n\approx c_\ell\tau_t/\ell^*$. Note that these distributions depend strongly on the boundary conditions, i.e., on the size and shape of the sample cell.

This was, in a nutshell, the essence of the multiple scattering variant of DLS, namely diffusing wave spectroscopy (DWS) as originally introduced by Maret and Wolf (1987). We pictured here photons as classical particles, like billiard balls bouncing through the suspension. Wave nature of the probing field is only needed for single scattering. This is a highly naive, but highly useful approach. Justification will be provided in Section 2. In Section 3 we develop the DWS technique. We work out the path statistics and investigate how much of the suspension dynamics actually survives in the strong multiple scattering regime.

2. Fundamentals

2.1. Scattering and Feynman's rules

A quite general scattering experiment is sketched in Fig. 4. The probing particles may be photons or neutrons, the scattering system may be a turbid suspension or a piece of metal, single scattering or multiple scattering. One feature is common to all such experiments: there is a certain probability $\mathcal{P}(t,t_e)$ that a probing particle generated in the source at time t_e eventually reaches the detector at time t. If we know $\mathcal{P}$, then we can calculate the received scattered power simply as $J(t) = \int \mathcal{P}(t,t_e)J_e(t_e)\,\mathrm{d}t_e$. We take for simplicity $J_e = \text{const}$ and neglect propagation delays, so that $J(t) = \mathcal{P}(t)J_e$. Even so, the exact calculation of the probability $\mathcal{P}(t)$ is a formidable task, we shall be content with reasonable approximations. Thereby, a helpful guidance is a set of general rules formulated by Richard Feynman (1985):

1: Identify all possible sequences of consecutive events that lead a particle from the source into the detector. These sequences are called *paths*.
2: Calculate the *amplitudes* $\mathcal{A}_i$ of the events in a path. The complex square $\mathcal{A}_i\mathcal{A}_i^*$ is the probability for the event.
3: Multiply the amplitudes $\mathcal{A}_i$ of consecutive events in each path to get their *path amplitudes* $\mathcal{A}_p = \prod_i \mathcal{A}_i$.
4: Add the partial amplitudes $\mathcal{A}_p$ of all paths, in order to obtain the total amplitude $\mathcal{A}$ of the event 'photon is scattered from the laser into the detector': $\mathcal{A} = \sum_p \mathcal{A}_p$.
5: Compute the *probability* $\mathcal{P} = \mathcal{A}^*\mathcal{A}$ for the event 'particle is scattered from the source into the detector'.

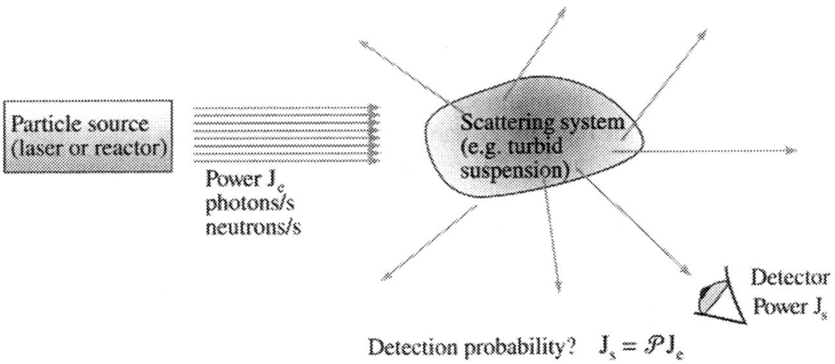

Fig. 4

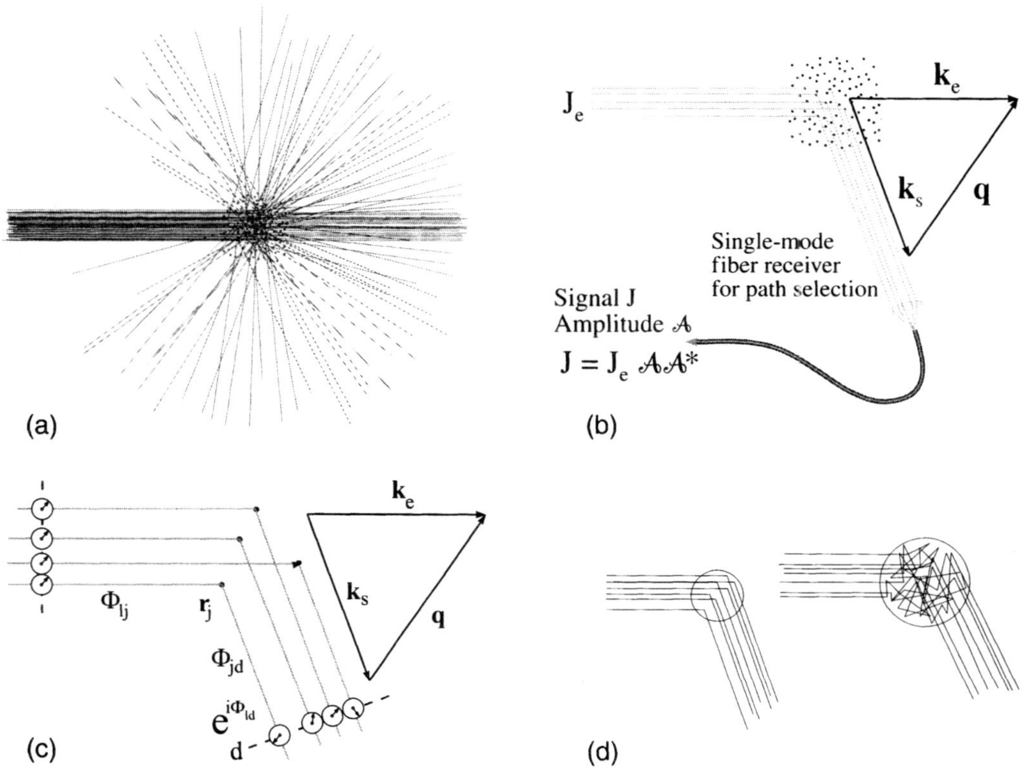

Fig. 5

6: Sometimes you find that the set of paths can be subdivided into n bundles that are in principle *distinguishable*. Then the total probability $\mathcal{P}_{\mathrm{tot}}$ is the sum $\mathcal{P}_{\mathrm{tot}} = \sum_{j=1}^{n} \mathcal{P}_j$. We also define the average probability $P = \langle \mathcal{P}_j \rangle = \mathcal{P}_{\mathrm{tot}}/n$.

In order to illustrate the application of Feynman's rules we consider a single scattering experiment in Fig. 5. The lines, or rays, represent a selection from an infinitely many possible paths of photons. Most of the photons of the laser beam pass through the sample, but some are scattered in all directions.

Rule 1, find the paths: From the multitude of paths we select a bundle with a well defined wave vector $\boldsymbol{k}_s$, as shown in Fig. 5b. A particularly simple situation arises when the path selection is done by a single-mode fiber receiver, such as are used in contemporary DLS instruments.

Rule 2, get event amplitudes: A single scattering path is actually a chain of at least 5 events: (i) generation, (ii) propagation from laser to a scatterer j, (iii) scattering event, (iv) propagation from the scatterer j to the detector, (v) detection:

$$\mathcal{A}_{\mathrm{single}} = \mathrm{generation} \cdot \mathrm{e}^{i\Phi_{lj}} \cdot b_j \cdot \mathrm{e}^{i\Phi_{jd}} \cdot \mathrm{detection} \tag{3}$$

Here the scattering event is represented by the amplitude b_j. We are not interested in details of photon generation and detection, and we only include the essential factors of the propagation amplitude in the phase factors $\mathrm{e}^{i\Phi_{lj}}$ and $\mathrm{e}^{i\Phi_{jd}}$. The sum $\Phi_{lj} + \Phi_{jd}$ represents the optical path length from a reference plane in the laser beam into a reference plane in the receiver beam, as sketched in Fig. 5c.

Rule 3, multiply event amplitudes: A simple geometrical calculation yields

$$\mathcal{A}_j \propto \mathrm{e}^{i\boldsymbol{k}_e\cdot\boldsymbol{r}_j} b_j \, \mathrm{e}^{-i\boldsymbol{k}_s\cdot\boldsymbol{r}_j} = b_j \, \mathrm{e}^{i\boldsymbol{q}\cdot\boldsymbol{r}_j}. \tag{4}$$

Rule 4, add path amplitudes: We collect all paths in Fig. 5b in a single path bundle characterized by the corresponding overall scattering amplitude $\mathcal{A}^{(s)}$

$$\mathcal{A}^{(s)} = \sum_j \mathcal{A}_j \approx \sum_j b_j \, \mathrm{e}^{i\boldsymbol{q}\cdot\boldsymbol{r}_j}. \tag{5}$$

A yet closer look reveals that each line through a scattering particle represents actually a whole bundle of paths, as indicated in Fig. 5d. The smaller scatterer illustrates the Rayleigh–Debye approximation, whereas the large scatterer is in the Mie regime, with microscopic multiple scattering paths within the particle. By virtue of Rule 4, we included all these partial paths into the particle amplitude $b_j(\boldsymbol{q})$ that depends on $\boldsymbol{q}$ and on the structure and orientation of the particle j.

Rule 5, square the total amplitude:

$$\mathcal{P} = |\mathcal{A}^{(s)}|^2 \approx \sum_j \sum_i b_j b_i \, \mathrm{e}^{i\boldsymbol{q}\cdot(\boldsymbol{r}_j - \boldsymbol{r}_i)}. \tag{6}$$

The right-hand sides of Eqs. (5) and (6) are identical with expressions encountered in various contributions to the present book. Obviously, the amplitude $\mathcal{A}$ is akin to the field E, and the probability $\mathcal{P}$ is akin to the intensity I.

Rule 6, average probabilities if necessary: This rule corresponds to incoherent addition of intensities. The paths selected with a single-mode receiver are indistinguishable, in other words the received light exhibits perfect *transversal* coherence. However, there are many *longitudinal* modes: the amplitude of a scattering suspension fluctuates in time. The measured quantities are time averages, experimental estimates of ensemble averages. We note that $\langle \mathcal{A}^{(s)} \rangle = 0$ (except for $\boldsymbol{q} = \boldsymbol{0}$), but this average is not measurable. More interesting are second order averages. From $P = \langle \mathcal{P} \rangle$ we obtain the structure factor $S(\boldsymbol{q})$

$$P = \langle \mathcal{A}^{(s)}(\boldsymbol{q}) \mathcal{A}^{(s)*}(\boldsymbol{q}) \rangle \propto S(\boldsymbol{q}), \tag{7}$$

whereas from the two-time average $\langle \mathcal{A}^{(s)}(\boldsymbol{q},0) \mathcal{A}^{(s)*}(\boldsymbol{q},\tau) \rangle$ we obtain the first-order correlation function $f(\tau,\boldsymbol{q})$:

$$f(\tau,\boldsymbol{q}) = \frac{\langle \mathcal{A}^{(s)}(\boldsymbol{q},0) \mathcal{A}^{(s)*}(\boldsymbol{q},\tau) \rangle}{\langle |\mathcal{A}^{(s)}(\boldsymbol{q})|^2(\boldsymbol{q}) \rangle}. \tag{8}$$

Thus, in the context of single scattering, the path amplitude approach does not bring anything new with respect to the conventional theory.

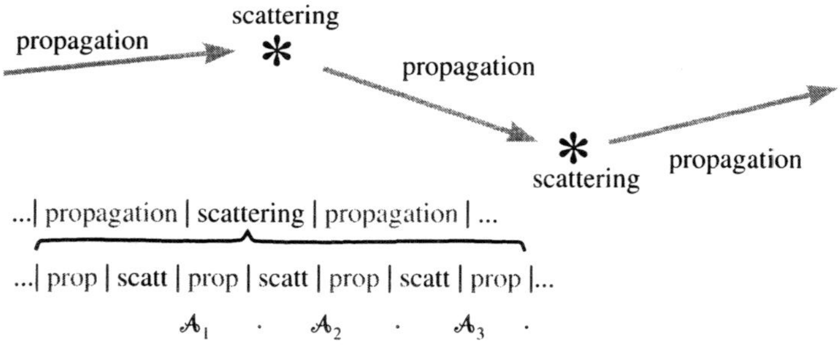

Fig. 6

Analysis of multiple scattering, however, depends on the path concept, as it permits a relatively simple yet reasonably rigorous treatment of the phenomena. The reason is Feynman's Rule 3: *amplitudes of consecutive events are multiplied.* (You can't do this with the field E!) In the presence of multiple scattering, we simply consider a chain of scattering and propagation events, multiplying the amplitudes as indicated in Fig. 6. Note that the stars do not mean individual suspension particles but rather single scattering events. The arrows are not the paths of individual photons, but rather 'path bundles', i.e., incoming and outgoing photon states.

2.2. Revisiting single scattering

In Eq. (5) for the single scattering amplitude $\mathcal{A}^{(s)}$ we generously neglected a number of features that are irrelevant for the basics of a single scattering experiment. However, for a consistent multiple scattering theory we need more realistic approximations. We take advantage of using a single-mode receiver for the selection of a 'well defined bundle of photon paths', which is actually a *photon state*. The classical electromagnetic field is a photon state filled with a certain number of photons. The amplitude $\mathcal{A}^{(s)}$ of the event 'photon is scattered from the laser into the detector', is the amplitude for the transition of a photon from the state 'laser beam' into the state 'observation beam selected by a single-mode optical receiver'. In a contemporary light scattering setup the single-mode selection is achieved by a single-mode optical waveguide (Rička, 1993, 1994; Gisler et al., 1995), as indicated in Fig. 7.

The oscillating electric field $\boldsymbol{E}_e(\boldsymbol{r},t) = \boldsymbol{E}_e(\boldsymbol{r})\mathrm{e}^{i\omega t}$ of the laser beam excites charge oscillations in the scatterers, which, in turn, couple to the field $\boldsymbol{E}_o(\boldsymbol{r})$ of the *observation beam.* This is a virtual light beam: a propagation reversed replica of the 'real' beam that would emerge from the lense when the fiber is used for illumination. Photons travel into the receiver on the same paths, as they would come out, if the device would be used as a transmitter. This virtual receiver field has all properties of a real electromagnetic field, but is 'empty' as long as we do not aim the receiver at a light source. Combining this *reciprocity* concept with the machinery of electromagnetic theory one can derive a simple yet rigorous expression for $\mathcal{A}^{(s)}$ (Rička, 1994). First we consider a single scatterer centered at the position $\boldsymbol{r}$. The amplitude $\mathcal{A}^{(1)}(\boldsymbol{r})$ for the transition of a photon from the laser into the receiver beam through this scatterer is

$$\mathcal{A}^{(1)}(\boldsymbol{r}) = -i\frac{ik}{2\epsilon}\tilde{\boldsymbol{E}}_o^*(\boldsymbol{r}) \cdot \hat{\boldsymbol{a}}_{oe} \cdot \tilde{\boldsymbol{E}}_e(\boldsymbol{r}). \tag{9}$$

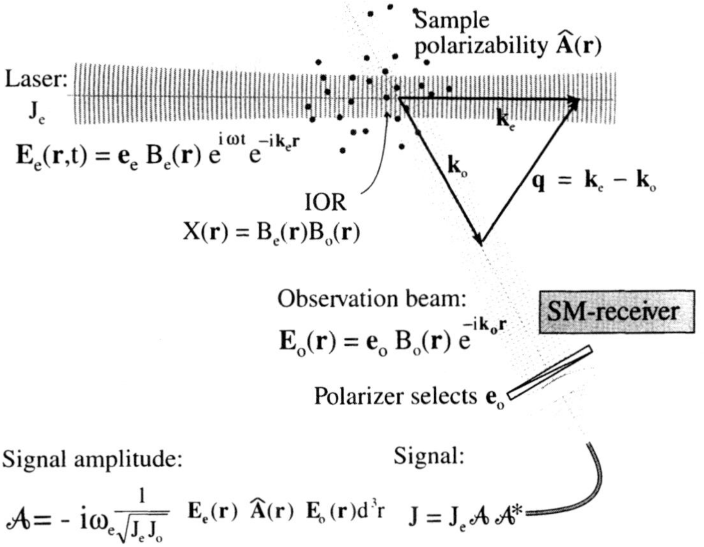

$$\mathcal{A} = -i\omega_e \frac{1}{\sqrt{J_e J_o}} \int E_e(\mathbf{r})\, \hat{A}(\mathbf{r})\, E_o(\mathbf{r})\, d^3r$$

Fig. 7

Here the scatterer is represented by its effective polarizability tensor $\hat{\boldsymbol{a}}_{oe}$. The fields with *tilde* are normalized so that $\oint |\tilde{\boldsymbol{E}}(\boldsymbol{r})|^2 \, d^2r = 1$ for an integration surface enclosing the source or the receiver. The only assumption in Eq. (9) is that the fields can be approximated by quasi-monochromatic quasi-plane waves; the scatterer is small compared with the profiles of $\boldsymbol{E}_o(\boldsymbol{r})$ and $\boldsymbol{E}_e(\boldsymbol{r})$. For example, choosing $\boldsymbol{E}_e(\boldsymbol{r})$ as plane wave and $\boldsymbol{E}_o(\boldsymbol{r})$ as the far field of a dipole radiator, one would end up with conventional light scattering analysis.

More realistic forms of $\tilde{\boldsymbol{E}}_o(\boldsymbol{r})$ and $\tilde{\boldsymbol{E}}_e(\boldsymbol{r})$ are the so-called beam waves. These are synchronized superpositions of plane waves with narrow distributions of wave vectors centered around the propagation direction $\boldsymbol{k}_o$ or $\boldsymbol{k}_e$ of the beam. A good approximation in the so-called Rayleigh range close to the focal point is:

$$\tilde{\boldsymbol{E}}_e(t,\boldsymbol{r}) = \frac{1}{\sqrt{\pi a_e^2}} B_e(\boldsymbol{r}) \boldsymbol{e}_e \, e^{i\psi(t,\boldsymbol{r})} \, e^{i(\omega t - \boldsymbol{k}_e \cdot \boldsymbol{r})}$$
$$\tilde{\boldsymbol{E}}_o(t,\boldsymbol{r}) = \frac{1}{\sqrt{\pi a_o^2}} B_o(\boldsymbol{r}) \boldsymbol{e}_o \, e^{i(\omega t - \boldsymbol{k}_o \cdot \boldsymbol{r})} \tag{10}$$

Here $B_e(\boldsymbol{r})$ and $B_o(\boldsymbol{r})$ are dimensionless profiles that are smooth compared with the wave structure $\exp(-i\boldsymbol{k} \cdot \boldsymbol{r})$ and with the size of the scatterer. πa^2 is an effective cross-section of the beam, defined through $\pi a^2 = \int B(\boldsymbol{r})^2 \, d^2r$. The effective radius a determines the divergence angle of the beam $\beta = 1/ka$ (uncertainty relation). The unit vectors $\boldsymbol{e}_o$ and $\boldsymbol{e}_e$ specify the polarizations. Note the factor $\exp i\psi(t,\boldsymbol{r})$ that accounts for phase fluctuations $\psi(t)$ of the laser: a laser exhibits a finite *coherence length* $l_{\text{coh}} = \tau_f c_\ell$, where τ_f is a characteristic fluctuation time. In a single scattering experiment l_{coh} is usually much larger than the size of scattering volume, and the coherence factor $\exp i\psi(t,\boldsymbol{r})$ can be neglected. When working with DWS, beware of coherence! l_{coh} must be large compared with the length of multiple scattering paths through the sample. After this warning we will assume a perfectly coherent laser.

In term of the beam waves, Eq. (9) can be recast in the following form

$$\mathcal{A}^{(1)}(\boldsymbol{r}) = -i\frac{1}{\sqrt{\pi a_e^2}}\frac{\lambda}{\sqrt{\pi a_o^2}}X_{eo}(\boldsymbol{r})b_{eo}\,\mathrm{e}^{i(\boldsymbol{k}_o-\boldsymbol{k}_e)\cdot\boldsymbol{r}}, \quad \text{where } b_{eo} = \frac{\pi}{\epsilon\lambda^2}\boldsymbol{e}_o\cdot\hat{\boldsymbol{a}}_{oe}\cdot\boldsymbol{e}_e \tag{11}$$

where the profile $X_{eo}(\boldsymbol{r}) = B_e(\boldsymbol{r})B_o(\boldsymbol{r})$ accounts for the size and shape of the illuminated and observed region (IOR). The meanings of the quantity b_{eo} and of the pre-factors become apparent in the following expression for the transition probability $\mathcal{P}^{(1)} = |\mathcal{A}^{(1)}|^2$:

$$\mathcal{P}^{(1)}(\boldsymbol{r}) = X_{eo}^2(\boldsymbol{r})\frac{\sigma_{eo}}{\pi a_e^2}\Omega_o \quad \text{where } \sigma_{eo} = |b_{eo}|^2. \tag{12}$$

Here $\Omega_o = \lambda^2/\pi a_o^2$ is the effective solid angle of observation and πa_e^2 is the effective cross-section of the laser beam. Obviously, σ_{eo} is the differential cross-section for the transition from the laser into the receiver. A simple explicit expression is available for the ensemble averaged cross-section in Rayleigh–Debye approximation (multiple scattering within the particle neglected, recall Fig. 5):

$$\langle\sigma_{eo}\rangle = \left(\frac{\pi\alpha}{\epsilon\lambda^2}\right)^2(\boldsymbol{e}_e\cdot\boldsymbol{e}_o)^2\,P(\boldsymbol{k}_o,\boldsymbol{k}_e), \tag{13}$$

where α is the polarizability of the scatterer and $P(\boldsymbol{k}_o,\boldsymbol{k}_e)$ the particle form factor that depends on $q_{eo} = |\boldsymbol{k}_o - \boldsymbol{k}_e|$; large particles scatter predominantly at low $\boldsymbol{q}$. Notice also the polarization factor $(\boldsymbol{e}_e\cdot\boldsymbol{e}_o)^2$. In a standard single scattering experiment the polarization states are aligned so that $\boldsymbol{e}_e\cdot\boldsymbol{e}_o = 1$. However, in the multiple scattering regime we find all possible polarizations.

Single scattering amplitude $\mathcal{A}^{(s)}$ of a particle suspension is the sum of particle contributions: $\mathcal{A}^{(s)} = \sum_j \mathcal{A}_j^{(1)}$. The summation extends over all particles in the sample cell. For simplicity, we assume monodisperse scatterers with spherical symmetry. This gives a particularly concise expression for $\mathcal{A}^{(s)}$:

$$\mathcal{A}^{(s)} = -i\frac{b_{eo}}{\sqrt{\pi a_e^2}}\frac{\lambda}{\sqrt{\pi a_o^2}}\int X_{eo}(\boldsymbol{r})c(\boldsymbol{r})\,\mathrm{e}^{i\boldsymbol{k}_o-\boldsymbol{k}_e\cdot\boldsymbol{r}}\,\mathrm{d}^3r, \quad \text{where } c(\boldsymbol{r}) = \sum_j \delta(\boldsymbol{r}-\boldsymbol{r}_j). \tag{14}$$

Here $c(\boldsymbol{r})$ is the microscopic concentration of the scatterers. The single scattering amplitude is related with $\tilde{c}(\boldsymbol{q}_{eo})$, the Fourier transform of the concentration at the point $\boldsymbol{q}_{eo} = \boldsymbol{k}_o - \boldsymbol{k}_e$ in the reciprocal space. The profile $X_{eo}(\boldsymbol{r})$ takes the finite size and shape of IOR into account. However, for simplicity we shall assume that the size a_e or a_o of $X_{eo}(\boldsymbol{r})$ is much larger than the extent of structural correlation in the suspension. With this assumption the ensemble averaged probability $P^{(s)} = \langle|\mathcal{A}^{(s)}|\rangle$ results as

$$P^{(s)} = \frac{1}{\pi a_e^2}\sigma_{eo}S_s(\boldsymbol{q}_{eo})cV_{eo}\Omega_o \tag{15}$$

where $S_s(\boldsymbol{q}_{eo})$ is the familiar suspension structure factor, c is the ensemble averaged number concentration and $V_{eo} = \int X_{eo}^2(\boldsymbol{r})\mathrm{d}^3r$ the effective volume of the IOR.

2.2.1. *Forward scattering and extinction*

Note that Eq. (15) predicts strong forward scattering. There is the well known singularity in the structure factor: $S_s(\boldsymbol{q}_{eo}) = 1 + c\tilde{h}(\boldsymbol{q}_{eo}) + (2\pi)^3c\delta(\boldsymbol{q}_{eo})$. In addition to the structure term

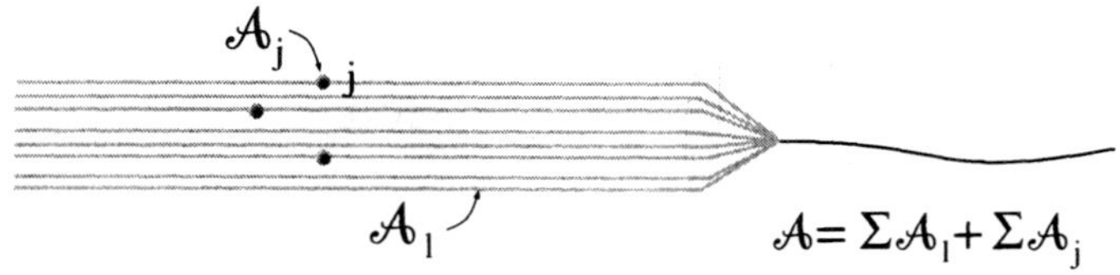

Fig. 8

$c\tilde{h}(\boldsymbol{q}_{eo})$ (Fourier transform of the pair correlation function), there is also a δ-peak at $\boldsymbol{q}_{eo} = \boldsymbol{0}$, i.e. at $\boldsymbol{k}_e = \boldsymbol{k}_o$. This term is usually neglected, because in standard scattering experiments $\boldsymbol{q}_{eo} = \boldsymbol{0}$ is not accessible. However, with multiple scattering we must consider all values of $\boldsymbol{q}_{eo}$. Strangely enough, forward scattering causes extinction: in the forward direction one receives both, photons that passed the sample without scattering, *and* photons that were scattered into the observation beam (viz. Fig. 8). Extinction results from the interference of forward scattered light with the original laser beam. Extinction length is given by $\ell = 1/\sigma c$, where σ is an effective total cross-section of a scatterer. This σ includes inter-particle interferences:

$$\sigma = \int \sigma_{eo}(\boldsymbol{e}_o, \boldsymbol{k}_o; \boldsymbol{e}_e, \boldsymbol{k}_e) S_s(\boldsymbol{k}_o, \boldsymbol{k}_e)\, \mathrm{d}\mathcal{O} \tag{16}$$

Here the symbol $\mathcal{O}$ symbolizes the set of Euler angles of the relative orientation of the vectors $\boldsymbol{e}_e, \boldsymbol{k}_e$ and $\boldsymbol{e}_o, \boldsymbol{k}_o$ that represent the incoming and outgoing beam waves.

2.3. Amplitude cross-correlations

The receiver in Fig. 9 selects two bundles of single scattering paths, passing through two distinct scattering volumes A and B. (Nevertheless, the receiver is still single-mode, there is no way to distinguish whether a photon comes from A or from B.) The total amplitude is the sum $\mathcal{A} = \mathcal{A}_A + \mathcal{A}_B$. The received signal, i.e., the total probability $\mathcal{P} = |\mathcal{A}|^2 = |\mathcal{A}|^2_A + \mathcal{A}^*_A \mathcal{A}_B + \mathcal{A}_A \mathcal{A}^*_B + |\mathcal{A}|^2_B$, contains two interference terms. The measured quantity, however, is the averaged signal $P = \langle \mathcal{P} \rangle$:

$$P = P_A + \langle \mathcal{A}^*_A \mathcal{A}_B \rangle + \langle \mathcal{A}_A \mathcal{A}^*_B \rangle + P_B \tag{17}$$

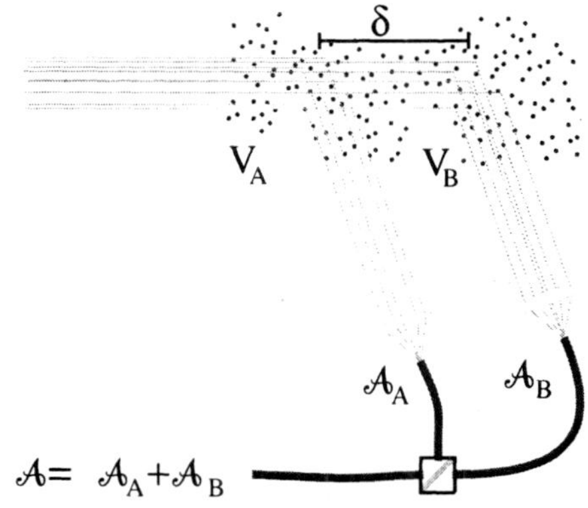

Fig. 9

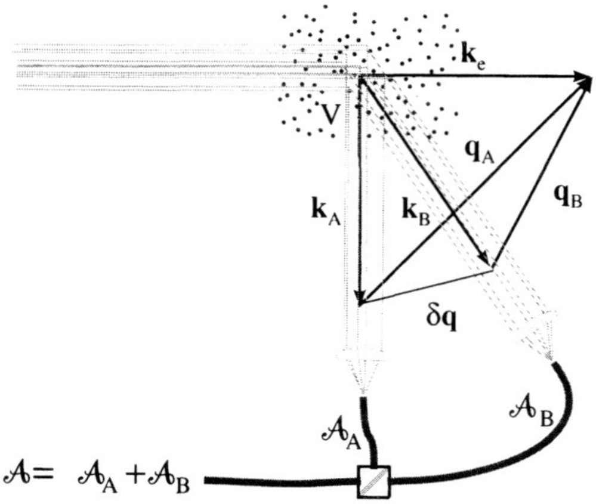

Fig. 10

The two observation volumes are well separated, and therefore the averaging can be factored to $\langle\mathcal{A}_A^*\rangle\langle\mathcal{A}_B\rangle$. Thus, the two *cross-correlation* terms vanish because $\langle\mathcal{A}\rangle = 0$. We end up with incoherent probability addition $P = P_A + P_B$. A rigorous analysis starting from Eq. (14) shows (Rička, 1994), that amplitude cross-correlation decreases with decreasing overlap of the two IOR profiles X_A, X_B:

$$\langle\mathcal{A}_A^*\mathcal{A}_B\rangle \propto \int X_A(\boldsymbol{r})X_B(\boldsymbol{r})\,\mathrm{d}^3r \propto \mathrm{e}^{-\delta^2/a_o^2} \tag{18}$$

The two receivers in Fig. 10 observe essentially the same volume, but under two different angles so that $\boldsymbol{q}_A \neq \boldsymbol{q}_B$. This situation is slightly less trivial, but one can show that the amplitudes of two receivers are uncorrelated as soon as the angle between the two observation beams exceeds the divergence angle β of the observation beam. This can be expressed in terms of an overlap integral of the spatial Fourier transforms of the two beams (Rička, 1994):

$$\langle\mathcal{A}_A^*\mathcal{A}_B\rangle \propto \int \tilde{X}_A(\boldsymbol{q})\tilde{X}_B(\boldsymbol{q})\,\mathrm{d}^3q \propto \mathrm{e}^{-\delta q^2 a_o^2/2} \tag{19}$$

Using Eqs. (18) and (19), one can estimate the number of *uncorrelated* path n_u for a photon scattered in a certain volume of characteristic size w: $n_u \approx w^2/\lambda^2$.

Vanishing amplitude cross-correlations are the deeper reason for the success of the naive particle approach to multiple scattering, as was outlined in the introduction. We shall practice this important topic in the section that follows.

2.4. Multiple scattering *gedanken* experiments

2.4.1. Double scattering

Consider a combination of two single scattering experiments (see Fig. 11). The task is to determine the first order correlation function $f(\tau) = \langle\mathcal{A}(0)\mathcal{A}^*(\tau)\rangle/\langle\mathcal{A}\mathcal{A}^*\rangle$ of such a serial DLS experiment. The amplitude $\mathcal{A}$ results from a chain of two events: (i) in the cell A photon is scattered from the state 0 (laser beam) to the state 1 (fiber mode 1); (ii) in the cell B photon

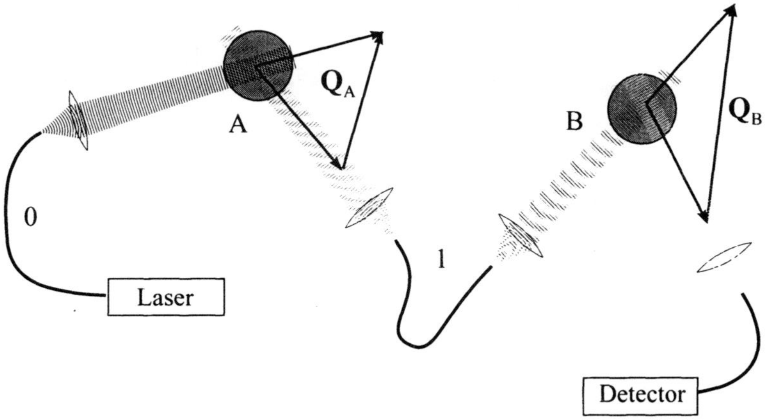

Fig. 11

is scattered from the state 1 (fiber mode 1) into the state 2 (observation beam). Feynman's multiplication rule gives:

$$\mathcal{A}(t) = \mathcal{A}_{0A1}(\boldsymbol{q}_A, t)\mathcal{A}_{1B2}(\boldsymbol{q}_B, t) = (0|A|1)_t(1|B|2)_t \tag{20}$$

The QT-like notation $(||)$ helps to save paper and ink. For $f(\tau)$ we need the average $\langle(0|A|1)_0(0|A|1)^*_\tau(1|B|2)_0(1|B|2)^*_\tau\rangle$. The partial amplitudes $(0|A|1)$ and $(1|B|2)$ concern two distinct scattering volumes, therefore they are uncorrelated. Factorizing the average one obtains

$$f(\tau) \propto \langle(0|A|1)_0(0|A|1)^*_\tau\rangle\langle(1|B|2)_0(1|B|2)^*_\tau\rangle. \tag{21}$$

Thus, the double scattering correlation function is simply the product $f(\tau) = f_A(Q_A, \tau)f_B(Q_B, \tau)$, where $f_A(Q_A, \tau)$ is the single scattering first-order correlation function one would measure with the output of fiber 1 plugged into the detector, and $f_B(Q_B, \tau)$ the corresponding quantity measured in the second sample cell B with fiber 1 plugged into the laser.

2.4.2. Loops

Note that there is a flaw in the argumentation above: we neglected an infinite number of photon paths. A photon coming through 1 into B is scattered backwards into 1 [amplitude $(1|B|-1)$], in the cell A it suffers a third scattering process, which brings it into 1 again [amplitude $(-1|A|1)$], and so on, until it manages to go into 2, or to escape unobserved. All such loop-like path must be added up in the correct amplitude $\mathcal{A}$ for the event 'photon goes from 0 to 1':

$$\begin{aligned}
\mathcal{A}^{(0)} &= (0|A|1)(1|B|2)\\
\mathcal{A}^{(1)} &= (0|A|1)(1|B|-1)(-1|A|1)(1|B|2)\\
\mathcal{A}^{(2)} &= (0|A|1)(1|B|-1)(-1|A|1)(1|B|-1)(-1|A|1)(1|B|2)\\
\ldots &= \ldots
\end{aligned} \tag{22}$$

The trouble with the loops are strong correlations, which would make the averaging difficult. For example, $\langle[(1|B|-1)(-1|A|1)]^n\rangle \neq \langle(1|B|-1)(-1|A|1)\rangle^n$. Anyway, we carry out the summation of the geometrical series formally (neglecting mathematical details) and obtain:

$$\mathcal{A} = \sum_{n=0}^{\infty} \mathcal{A}^{(n)} = \frac{(0|A|1)(1|B|2)}{1-(1|B|-1)(-1|A|1)} \tag{23}$$

Thus, we don't need to be concerned about the loops, as long as $|(1|B|-1)(-1|A|1)| \ll 1$. The single scattering amplitude, Eq. (14), is proportional to (λ/a_o), where a_o is the size of the IOR: loops can be neglected as long as $|(1|B|-1)(-1|A|1)| \propto (\lambda/a_o)^2$ is a small number. In the multiply scattering regime, the characteristic size is the mean free path ℓ. We neglect the loops as long as $\ell \gg \lambda$.

2.4.3. Double and single scattering

The setup in Fig. 12 opens a new path from the laser into the detector. The total amplitude reads $\mathcal{A}(t) = (0|A|1)_t(1|B|2)_t + (0|A|2)_t$. The correlation $\langle\mathcal{A}(0)\mathcal{A}^*(\tau)\rangle$ is the sum of four terms, but two cross-terms are zero, there are no correlations between $(0|A|1)$, $(0|A|2)$ and $(1|B|2)$. The first order correlation function from this experiment is the sum of two path correlation functions f_1 and f_2 weighted by the corresponding path probabilities P_1 and P_2:

$$f(\tau) = \frac{1}{P_1+P_2}\left[P_1 f_1(\tau) + P_2 f_2(\tau)\right], \tag{24}$$

where $f_1(\tau) = f_A(Q_{02},\tau)$, $f_2(\tau) = f_A(Q_{01},\tau)f_B(Q_{12},\tau)$, and $P_1 = P_{0A2} = \langle|(0|A|2)|^2\rangle$, $P_2 = P_{0A1}P_{1B2} = \langle|(0|A|1)|^2\rangle\langle|(1|B|2)|^2\rangle$.

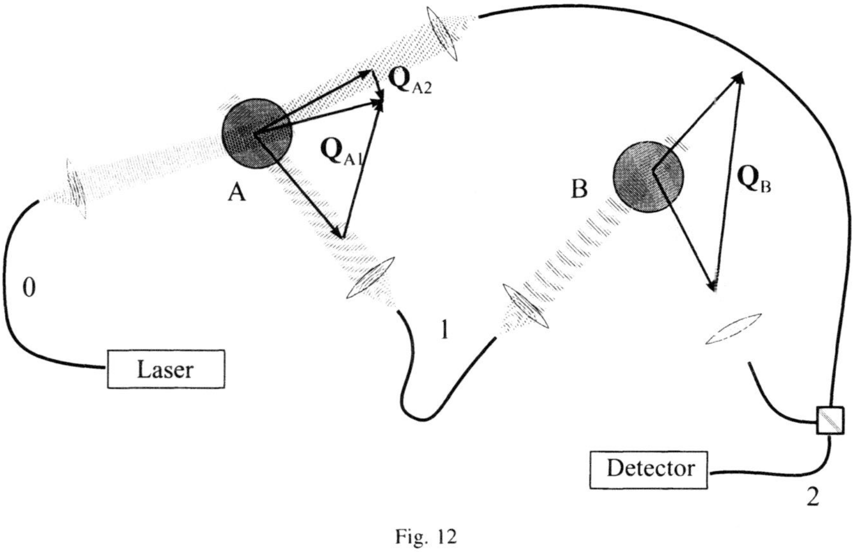

Fig. 12

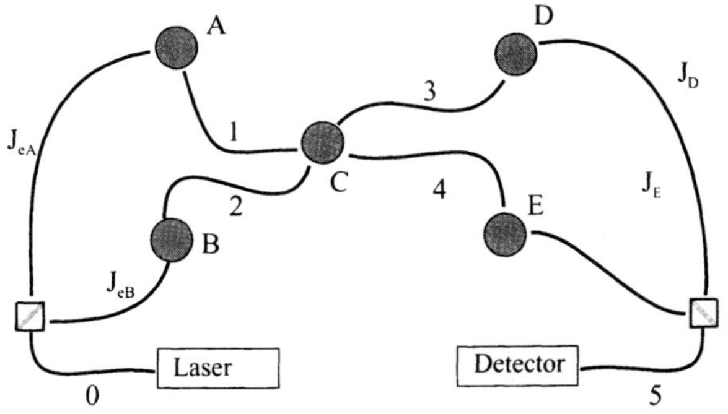

Fig. 13

2.4.4. Path crossings

The last path game, sketched in Fig. 13, contains a new feature: there is a path crossing in the cell C. One finds four different path bundles:

$$\begin{aligned}\mathcal{A} &= (0|A|1)(1|C|3)(3|D|5) + (0|A|1)(1|C|4)(4|E|5)\\ &\quad + (0|B|2)(2|C|3)(3|D|5) + (0|B|2)(2|C|4)(4|E|5)\end{aligned} \tag{25}$$

Notice that all but the C-type partial amplitudes occur in two terms of the sum. The four paths are not entirely different, they contain mutually correlated segments. Nevertheless, all cross-terms vanish in the total $f(\tau)$, which is the sum of four paths:

$$\begin{aligned}f(\tau) \propto\ & P_{0A1}P_{1C3}P_{3D5}f_A(Q_{01},\tau)f_C(Q_{13},\tau)f_D(Q_{35},\tau)\\ &+ P_{0A1}P_{1C4}P_{4E5}f_A(Q_{01},\tau)f_C(Q_{14},\tau)f_E(Q_{45},\tau)\\ &+ P_{0B2}P_{2C3}P_{3D5}f_B(Q_{02},\tau)f_C(Q_{23},\tau)f_D(Q_{35},\tau)\\ &+ P_{0B2}P_{2C4}P_{4E5}f_B(Q_{02},\tau)f_C(Q_{24},\tau)f_E(Q_{45},\tau)\end{aligned} \tag{26}$$

Though crossings do not play any role in $f(\tau)$, they contribute a so-called non-Gaussian term to the second order correlation: $g(\tau) = 1 + |f(\tau)|^2 + \text{cross}(\tau)$. The magnitude of crossings(τ) is usually negligibly small compared with $|f(\tau)|^2$, and therefore the crossings are of an academical interest only (Genack et al., 1990; Scheffold and Maret, 1998).

2.5. Conclusions

(1) Loops in photon path (bundles) can be neglected as long as $\ell \gg \lambda$.
(2) Hence, the first order correlation function f_p of a path of n_p single scattering events is the product of n_p-correlation functions of these events:

$$f_p(\tau) = \prod_{i=1}^{n_p} f_i^{(p)}(\tau). \tag{27}$$

(3) Two paths (bundles) are uncorrelated if they contain at least one independent segment, either an unpaired $\boldsymbol{q}$ or an unpaired cell.

(4) Hence, the total first-order correlation function $f(\tau)$ is the sum of path correlation functions weighted with the path probabilities P_p:

$$f(\tau) = \sum_p P_p f_p(\tau) \Big/ \sum_p P_p, \tag{28}$$

where P_p is the product of segment probabilities:

$$P_p = \prod_{i=1}^{n_p} P_i^{(p)}. \tag{29}$$

(5) Combining rules 2 and 4 we can write quite generally

$$f(\tau) = \sum_p \prod_{i=1}^{n_p} P_i^{(p)} f_i^{(p)}(\tau) \Big/ \sum_p \prod_{i=1}^{n_p} P_i^{(p)}, \tag{30}$$

where n_p is the number of scattering events in a path p.

2.6. Photons, scatterons, random flights

Our findings suggest to handle the multiple scattering problem as follows: we subdivide the sample cell in a number of independent scattering volumes and find all path through these compartments, their path probabilities and path correlation functions. Then we calculate the total correlation function according to Eq. (30). However, instead of bothering how to make the subdivision, we let the photons to do the job themselves (see Fig. 14): photons come riding the laser beam wave, which is characterized by a certain $\boldsymbol{k}_e$, $\boldsymbol{e}_e$ (the vector, $\boldsymbol{h}$, is unimportant with dielectrics). Upon entering the suspension the photons start suffering scattering events. Their number in the beam decreases exponentially $\propto \exp(-d/\ell)$. In each small volume element, whose size Δd is much smaller than ℓ, a certain number of photons are diverted to new paths, for example into a beam wave $\boldsymbol{k}_1$, $\boldsymbol{e}_1$. They surf the new wave, until becoming scattered into $\boldsymbol{k}_2$, $\boldsymbol{e}_2$. And so on, only few photons can make it into the receiver. The picture

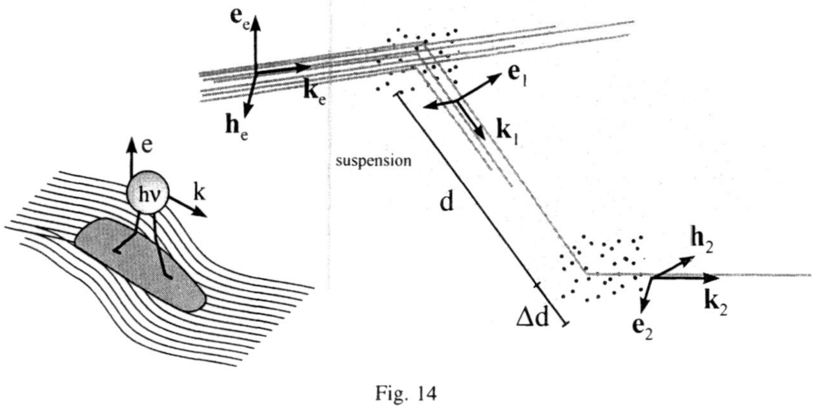

Fig. 14

of individual distinguishable photons, whose path can be traced from the laser to the receiver, is of course only a cartoon. However, there are chains of consecutive scattering events which can be more or less localized in different small volumes of the suspension. We shall call these path finding objects "scatterons". Theo Niewenhuisen uses the name 'diffusons', referring to the highly irregular path those objects travel: the scatterons perform a *random flight*.

This random flight of scatterons from one independent scattering volume to another, is a rough approximation. It will break down as soon as the 'characteristic size of the scattering volume', namely ℓ, approaches λ. Strange things, not accounted for in our model, may happen when $\ell \to \lambda$. As soon as $\ell \ll \lambda$, than we are in the realm of near-fields, ending up with transparent suspension with a new effective refractive index. However, for the intermediate range $\ell \approx \lambda$, the theoreticians predict the occurrence of the so-called *strong localization* of light: the light should stop to propagate, being stuck in the "loops" (recall Section 2.4). The experimental evidence for the strong localization is yet scarce. We shall stay in the safe regime $\lambda \ll \ell \ll L$, where L is the size of the sample cell.

3. Diffusing wave spectroscopy

Consider, for example, the DWS experiment outlined in Fig. 15: a laser beam illuminates a slab-like cell filled with strongly scattering suspension, (case 3, full milk), the scattered light is observed with a single-mode receiver. Mean free path ℓ of the scatterons (photons, if you wish) is short compared with L, i.e., $\ell \ll L$. The figure indicates a selection of possible path (bundles) through the cell. Some of the path (dark gray) lead all the way from the laser into the receiver, contributing to the signal $J(t)$. The measured quantity is the signal correlation function $g(\tau) = \langle J(t)J(t+\tau)\rangle / \langle J \rangle^2$. Because of the large number of possible path, we may safely expect the Gaussian property of the total signal amplitude $\mathcal{A}$, and thus the validity of the 'Siegert relation' $g(\tau) = 1 + |f(\tau)|^2$.

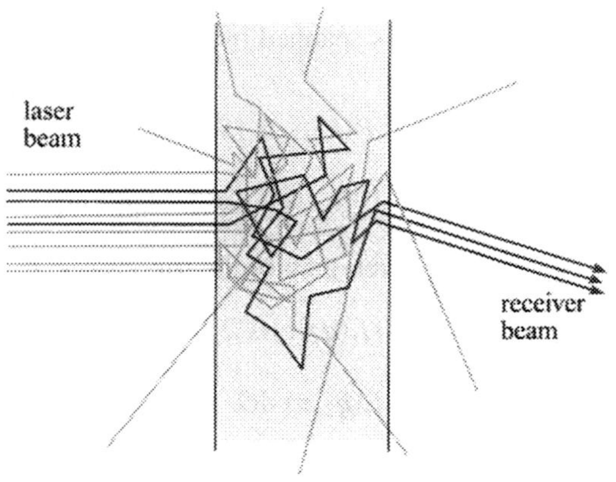

Fig. 15

3.1. Path averaging

Our general expression for $f(\tau)$, Eq. (30), requires that we find all possible path p from the laser to the receiver, as well as the probabilities $P_i^{(p)}$ and correlations $f_i^{(p)}(\tau)$ of their segments. This is an impossible task. We are forced to make an additional assumption. We assume that for each number of segments n_p the receiver samples *independently and uniformly* all realizations of the segments, i.e., all segment length and all orientations of the vectors $\boldsymbol{k}_i$ and $\boldsymbol{e}_i$, according to their distribution, Eq. (32). This is clearly a serious approximation! The segments can't be entirely independent, because we are considering only selected paths, which lead from the laser beam into the receiver beam. In particular, we must require that $\sum_i \boldsymbol{q}_i = \sum_i (\boldsymbol{k}_i - \boldsymbol{k}_{i-1}) = \boldsymbol{k}_o - \boldsymbol{k}_e$, where $\boldsymbol{k}_o$ and $\boldsymbol{k}_e$ are the wave vectors of the observation and excitation beam. This imposes a restriction on the available orientations of $\boldsymbol{k}_i$ and $\boldsymbol{e}_i$. However, the restriction relaxes with increasing n_p.

By virtue of the independence assumption we may replace Eq. (30) by a considerably simpler expression:

$$f(\tau) \approx \frac{\sum_{n=1}^{\infty} W_n \langle f_i(\tau, q_i) \rangle^n}{\sum_{n=1}^{\infty} W_n}, \tag{31}$$

where W_n is the number of paths consisting of n segments, and $\langle f_i(\tau, \boldsymbol{q}_i) \rangle$ the segment correlation function, pre-averaged according to segment probabilities $P_i^{(p)}$, Eq. (15). The expression in Eq. (15) can be re-cast into the following form:

$$P_i^{(p)} = P_{\boldsymbol{k}_i, \boldsymbol{e}_i \to \boldsymbol{k}_{i+1}, \boldsymbol{e}_{i+1}} = \frac{\Delta d}{\ell} \mathrm{e}^{-d/\ell} \cdot p(\boldsymbol{e}_{i+1}, \boldsymbol{k}_{i+1}; \boldsymbol{e}_i, \boldsymbol{k}_i) \Delta\Omega. \tag{32}$$

Here, $\exp(-d/\ell)$ accounts for the extinction of the beam wave between two scattering events separated by distance d. The factor $\Delta d/\ell$ has a simple interpretation: it is the total probability that a photon is scattered in a partial volume of thickness Δd. The distribution $p(\boldsymbol{e}_{i+1}, \boldsymbol{k}_{i+1}; \boldsymbol{e}_i, \boldsymbol{k}_i)$ reflects the path branching: $p(\boldsymbol{e}_{i+1}, \boldsymbol{k}_{i+1}; \boldsymbol{e}_i, \boldsymbol{k}_i)\Delta\Omega$ represents the probability that the scattering occurs from the state $\boldsymbol{e}_i, \boldsymbol{k}_i$ into a state $\boldsymbol{e}_{i+1}, \boldsymbol{k}_{i+i}$ (more precisely: into the range of $\boldsymbol{k}$ states that is spanned by the solid angle of observation $\Delta\Omega_{i+1}$). For Rayleigh–Debye particles

$$p(\boldsymbol{e}_{i+1}, \boldsymbol{k}_{i+1}; \boldsymbol{e}_i, \boldsymbol{k}_i) = \frac{(\boldsymbol{e}_i \cdot \boldsymbol{e}_{i+1})^2 P(\boldsymbol{k}_{i+1}, \boldsymbol{k}_i) S_s(\boldsymbol{k}_{i+1}, \boldsymbol{k}_i)}{\int (\boldsymbol{e}_i \cdot \boldsymbol{e}_{i+1})^2 P(\boldsymbol{k}_{i+1}, \boldsymbol{k}_i) S_s(\boldsymbol{k}_{i+1}, \boldsymbol{k}_i) \, \mathrm{d}\mathcal{O}}, \tag{33}$$

where $\int \ldots \mathrm{d}\mathcal{O}$ represents orientational averaging of the tripods $\boldsymbol{e}, \boldsymbol{k}, \boldsymbol{h}$. Thus:

$$\begin{aligned} \langle f_i(\tau, \boldsymbol{q}_i) \rangle &= \sum p(\boldsymbol{e}_{i+1}, \boldsymbol{k}_{i+1}; \boldsymbol{e}_i, \boldsymbol{k}_i) f_i(\boldsymbol{q}_i, \tau) \Delta\Omega_i \\ &\equiv \int p(\boldsymbol{e}_{i+1}, \boldsymbol{k}_{i+1}; \boldsymbol{e}_i, \boldsymbol{k}_i) f_i(\boldsymbol{q}_i, \tau) \, \mathrm{d}\mathcal{O} \end{aligned} \tag{34}$$

The summation on the left-hand side involves two polarization states and a finite number of uncorrelated paths through a partial scattering volume. It can be shown that the integral expression on the right is fully equivalent. Notice that the extinction factor $\frac{1}{\ell}e^{-d/\ell}\Delta d$, in Eq. (32), does not appear any more in Eq. (34), because the single scattering correlation function f_i does not depend on d and $\int \frac{1}{\ell}e^{-d/\ell}\Delta d = 1$.

A convenient set of integration variables for the explicit evaluation of the averages are the Euler angles θ, ϕ and ψ of the relative orientations of the consecutive tripods $\boldsymbol{k}_i, \boldsymbol{e}_i, \boldsymbol{h}_i$ and $\boldsymbol{k}_{i+1}, \boldsymbol{e}_{i+1}, \boldsymbol{h}_{i+1}$. We take θ to be the usual scattering vector spanned by $\boldsymbol{k}_{i+1}$ and $\boldsymbol{k}_i$, so that

$$q^2 = [\boldsymbol{k}_{i+1} - \boldsymbol{k}_i]^2 = 2k^2[1 - \cos\theta] \tag{35}$$

We restrict ourselves to the small particle approximation, Eq. (33). The particle form factor $P(q)$ and the structure factor $S(q)$ depend on θ only. The other two angles appear in the polarization factor $[\boldsymbol{e}_{i+1} \cdot \boldsymbol{e}_i]^2$, but this is readily pre-averaged to

$$\langle [\boldsymbol{e}_{i+1} \cdot \boldsymbol{e}_i]^2 \rangle_{\phi,\psi} = \tfrac{1}{4}[1 + \cos\theta^2] \tag{36}$$

Thus, the angle brackets actually mean

$$\langle \ldots \rangle = \frac{\displaystyle\int \ldots [1 + \cos^2\theta] P(\theta) S_s(\theta) \sin(\theta)\, \mathrm{d}\theta}{\displaystyle\int [1 + \cos\theta^2] P(\theta) S_s(\theta) \sin(\theta)\, \mathrm{d}\theta}. \tag{37}$$

Note that this only applies in the small particle approximation! For Mie-type of particles the averaging would be considerably more difficult.

3.1.1. Final result, including absorption

Our development suffers still from an unrealistic assumption: we only consider scattering, neglecting inelastic processes that cause absorption of the original lightwave. Absorption can be taken into account by weighing each path of length $n\ell$ with the factor $e^{-n\ell/\ell_a} = e^{-nA}$, where ℓ_a is the absorption extinction length. Introducing, for convenience, $\Gamma(\tau) = -\log\langle f_i(\tau, q_i) \rangle$ we write finally

$$f(\tau) \approx \frac{\displaystyle\sum_{n=1}^{\infty} W_n\, e^{-n[A+\Gamma(\tau)]}}{\displaystyle\sum_{n=1}^{\infty} W_n\, e^{-nA}}, \tag{38}$$

Note that the numerator and denominator each represent a discrete Laplace transform of the path length distribution W_n.

3.2. Averaging f

Since we were forced to assume that n is large, we take advantage of this assumption. We compute directly the average product

$$e^{-n\Gamma} = \langle f_i \rangle^n = \left\langle \prod_{i=1}^{n} f_i \right\rangle. \tag{39}$$

The aim is to determine Γ to be plugged into Eq. (38). We write the single scattering correlation function in terms of its logarithm $\gamma(\boldsymbol{q}_i,\tau) = -\log f(\boldsymbol{q}_i,\tau)$, so that

$$e^{-n\Gamma} = \left\langle e^{\sum_{i=1}^{n}\gamma_i(\boldsymbol{q}_i,\tau)}\right\rangle = \langle e^{-\mathcal{G}}\rangle. \tag{40}$$

Note that the exponential average $\langle\exp(-\mathcal{G})\rangle$ is the so-called characteristic function of the random variable $\mathcal{G}$ [see, e.g., van Kampen (1983)]. Our assumptions imply that $\mathcal{G} = \sum_i^n \gamma_i$ is the sum of a large number n of independent random variables. Thus, $\mathcal{G}$ is a Gaussian variable, whose characteristic function can be written in terms of the mean $\langle\mathcal{G}\rangle = n\langle\gamma\rangle$ and variance $\langle\mathcal{G}^2\rangle - \langle\mathcal{G}\rangle^2 = n(\langle\gamma^2\rangle - \langle\gamma\rangle^2)$. Thus, we arrive at

$$\Gamma = \langle\gamma\rangle - \frac{\langle\gamma^2\rangle - \langle\gamma\rangle^2}{2}, \qquad \gamma(\boldsymbol{q}_i,\tau) = -\log f(\boldsymbol{q}_i,\tau). \tag{41}$$

3.2.1. Examples

Brownian motion of monodisperse scatterers non-interacting scatterers is characterized by the well known single scattering correlation function $f_i(\tau,\boldsymbol{q}) = e^{-q^2 D\tau}$. This yields

$$\Gamma = \langle q^2\rangle D\tau - \frac{\langle q^4\rangle - \langle q^2\rangle^2}{2} D^2\tau^2. \tag{42}$$

The q averages can be expressed in terms of $\cos\theta$ averages (recall, however, that this only applies in the small particle approximation, see Section 3.1):

$$\Gamma = 2k^2(1 - \langle\cos\theta\rangle)D\tau - 4k^4\frac{\langle\cos^2\theta\rangle - \langle\cos\theta\rangle^2}{2} D^2\tau^2, \tag{43}$$

Here the angle bracket are defined through Eq. (37), where we set $S_s(\theta) = 1$, since we are neglecting interactions. We consider for illustration only the case $P(q) = 1$, i.e. point-like scatterers, where $\langle\cos\theta\rangle = 0$ and $\langle\cos^2\theta\rangle = 2/5$:

$$\Gamma \approx 2k^2 D\tau - \tfrac{1}{5}4k^4 D^2\tau^2. \tag{44}$$

On the other hand, an explicit formula for the average $\langle f_i(\tau)\rangle$ can be obtained with the Guinier approximation $P(q) = \exp(-q^2 R_g^2/3)$, where R_g^2 is the mean square radius of gyration of the particle (Flammer et al., 1998):

$$\langle f_i(\tau)\rangle = N(R_g)\left[\frac{u^2+u+1}{2u^3}\left(1-e^{-2u}\right) - \frac{1}{u^2}\right], \quad u = 2k^2(R_g^2/3 + D\tau) \tag{45}$$

where $N(R_g)$ is an ugly looking normalization factor enforcing $\langle f_i(\tau)\rangle = 1$. In Fig. 16 we compare three expressions for $\langle f_i(\tau)\rangle^n$. For $n = 1$, the agreement is miserable, but for $n = 10$ it becomes quite good in the τ range where $\langle f_i(\tau)\rangle^n$ differs from zero. This suggests that even the first cumulant $\langle\gamma\rangle = 2k^2 D\tau$ is sufficient, when the paths are sufficiently long.

Convective motion of the scatterers results in a complex single scattering correlation $f_i(\tau,\boldsymbol{q}_i) = e^{i\boldsymbol{q}_i\cdot\boldsymbol{v}_i\tau}$. The calculation of the product average $\langle\prod_{i=1}^n e^{i\boldsymbol{q}_i\cdot\boldsymbol{v}_i\tau}\rangle$ [viz. Eq. (39)] requires caution, because we must satisfy the condition $\sum\boldsymbol{q}_i = \boldsymbol{k}_o - \boldsymbol{k}_e$. Convective motion in a sample cell is never uniform and therefore we write the velocity field $\boldsymbol{v}_i(\boldsymbol{r})$ in terms

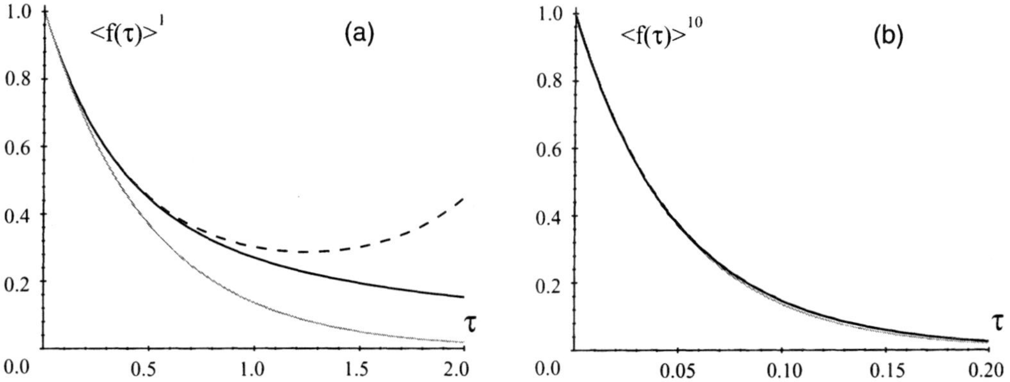

Fig. 16. Solid line: exact result using Eq. (45). Dashed line: cumulant expansion $\Gamma = 2k^2 D\tau - \frac{1}{5}4k^4 D^2\tau^2$. Lightly shaded line cumulant expansion $\Gamma = 2k^2 D\tau$, second cumulant neglected. Left: $n = 1$. Right: $n = 10$ (note the change in time scale). Parameters are $k^2 R_g = 0$, $k^2 D = 1$.

of its mean and deviation as $\boldsymbol{v}_i(\boldsymbol{r}) = \overline{\boldsymbol{v}} + \delta\boldsymbol{v}_i(\boldsymbol{r})$, where $\overline{\delta\boldsymbol{v}_i(\boldsymbol{r})} = \boldsymbol{0}$. Overline means position averaging. Thus, we write

$$\left\langle \prod_{i=1}^{n} \mathrm{e}^{i\boldsymbol{q}_i\cdot\boldsymbol{v}_i\tau} \right\rangle = \mathrm{e}^{i(\boldsymbol{k}_o-\boldsymbol{k}_e)\cdot\overline{\boldsymbol{v}}\tau}\,\overline{\left\langle \mathrm{e}^{i\sum_i \boldsymbol{q}_i\cdot\delta\boldsymbol{v}_i\tau} \right\rangle} \tag{46}$$

The first factor, which accounts for uniform translation of the sample, cancels in the measured $g(\tau) = 1 + |f(\tau)|^2$ [1]. Thus, we obtain

$$\Gamma = \tfrac{1}{6}\langle q^2\rangle\overline{\delta v^2}\tau^2 = \tfrac{1}{3}k^2\overline{\delta v^2}\tau^2(1 - \langle\cos\theta\rangle), \tag{47}$$

where $\overline{\delta v^2}$ is the variance of $\delta\boldsymbol{v}(\boldsymbol{r})$. Calculation of the averages $\overline{\boldsymbol{v}_i(\boldsymbol{r})}$ and $\overline{\delta v^2}$ is not trivial. We must take the spatial distribution $C_o(\boldsymbol{r}, n)$ of scattering events in the paths selected by the receiver into account (Bicout and Maret, 1994). Formally, we may write, for example, $\overline{\boldsymbol{v}_i(\boldsymbol{r})}_n = \int \boldsymbol{v}(\boldsymbol{r})C_o(\boldsymbol{r}, n)\,\mathrm{d}^3r$. These averages depend on n, spoiling thus the convenient Laplace transforms in Eq. (38). This puts the topic beyond the scope of the present paper, subsequently we assume uniformity of the suspension.

Interacting Brownian scatterers are characterized by a single-scattering first order correlation function $f_i(\boldsymbol{q}, \tau) = \exp\big(-\gamma_i(\tau) + \cdots\big)$ whose initial behavior is be

$$\gamma_i = q^2 H(q)/S(q)D\tau, \tag{48}$$

where $S(q)$ is the structure factor, and $H(q)$ is the hydrodynamic factor. How much does DWS tell about interactions? Hint 1: insert γ_i into Eq. (37). Hint 2: return to here after reading the next section, $S(q)$ also occurs in the transport mean free path ℓ^*. See also Fraden and Maret (1999).

[1] An exact analysis including the beam profile $B(\boldsymbol{r})$ would show that uniform translation causes decorrelation $\propto \exp(-\bar{v}^2\tau^2/a^2)$, where a is the size of the focus of the observation or excitation beam.

3.3. Random flights, diffusion and W_n

Statistics of multiple scattering paths depends strongly on the geometry of the problem. Figs. 20, 21 and 23 show three experimental geometries that are of importance in DWS practice. Common to all such experiments is the random flight of scatterons (or photons, if you wish) from the source to the receiver beam. In terms of random flyers the problem of finding W_n, the distribution of the number of received scattering events, can be formulated as follows: we start with a large number of random flyers at their initial positions, for example in the shaded strip of thickness Z in the slab. After some time (measured in the number of flyer's jumps) their number in the slab will have declined: in each step some photons are scattered out of the slab. How many of the original number of flyers are leaving the slab on the right hand side and on the left-hand side? This is a so-called *first passage problem*: as soon as a flyer crosses a sample boundary for the first time, he is out of the game.

We can't solve the problem of general random flight, but we can at least obtain good approximations in the limit of large n and small ℓ (which we anyway require): we replace the random flight with diffusion. In the same spirit, Brownian motion of a colloidal particle can be described as a diffusion process with diffusion coefficient D (Chandrasekhar, 1943). Density C of single scattering events obeys the diffusion equation

$$\frac{\partial C(\boldsymbol{r},t)}{\partial t} - D_\ell \nabla^2 C(\boldsymbol{r},t) = Q(\boldsymbol{r},t). \tag{49}$$

The *boundary condition* that is appropriate for solving first passage problems is absorption at the wall of the sample cell (Chandrasekhar, 1943). This implies:

$$C = 0\big|_{\text{boundary}} \qquad E(t,\rho) = -D_\ell \hat{\boldsymbol{n}} \cdot \nabla C(\boldsymbol{r},\tau)\big|_{\text{boundary}} \tag{50}$$

$E(t,\rho)\mathrm{d}t\delta a$ is the number of scatterons exiting through an element δa of the boundary in time interval $\mathrm{d}t$, $\hat{\boldsymbol{n}}$ is the outward normal.

Diffusion coefficient: D_ℓ is defined by requiring that, after large number $n \to \infty$ of steps, the mean square displacement $\langle L_n^2 \rangle$ equals the diffusive displacement $6D_\ell\tau$. Here τ is the time spent flying the distance $n\ell$ with some velocity c_ℓ. (*Could* be light velocity; it does not really matter, because c_ℓ does not enter the result for W_n.) Thus:

$$\langle L_{n\to\infty}^2 \rangle = 6D_\ell n\ell / c_\ell, \tag{51}$$

The average $\langle L_{n\to\infty}^2 \rangle$ is not quite trivial because we have a random flight with *orientational persistence:* the orientation of $\boldsymbol{e}_{i+1}, \boldsymbol{k}_{i+1}$ depends on the orientation of $\boldsymbol{e}_i, \boldsymbol{k}_i$, as expressed by the distribution $p(\boldsymbol{e}_{i+1}, \boldsymbol{k}_{i+1}; \boldsymbol{e}_i, \boldsymbol{k}_i)$. Large particles scatter predominantly forward, i.e., form factor $P(\theta)$ favors small θ. There are also the structure factor $S(\theta)$ and polarization factor $[1+\cos\theta^2]$. The situation is akin to a persistent polymer chain with fixed bond angles, except that we have exponential distribution $e^{-d/\ell}/\ell$ of the segment length d. In P. Flory's book on polymer statistics (Flory, 1969) one finds the recipe for calculating $\langle L_{n\to\infty}^2 \rangle$ in terms of averaging of products of rotational matrices. A final summation of a geometric series yields

$$\langle L_{n\to\infty}^2 \rangle = 2n\ell^2 \frac{1}{1 - \langle \cos\theta \rangle}. \tag{52}$$

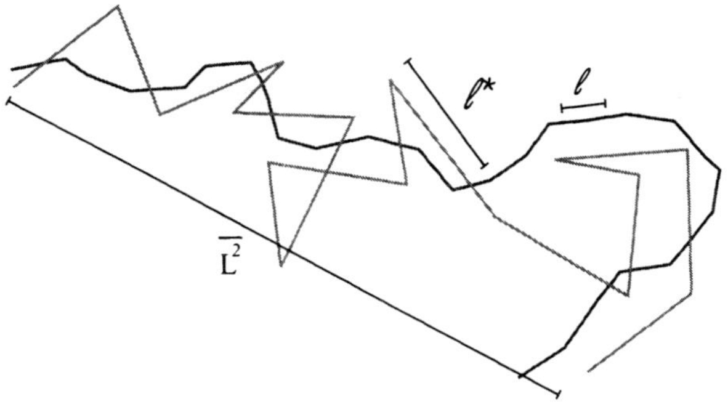

Fig. 17

Introducing this result into Eq. (51) we obtain the diffusion coefficient D_ℓ as

$$D_\ell = \frac{1}{3} c_\ell \ell^*, \quad \text{where } \ell^* = \frac{\ell}{1 - \langle \cos\theta \rangle}. \tag{53}$$

ℓ^* is known as the *transport mean free path*. The factor $1/(1 - \langle \cos\theta \rangle)$ measures the persistence of the flight. Fig. 17 illustrates the geometrical meaning of ℓ^*. We replace the persistent chain with a freely jointed chain (FJC). In a FJC the orientation of the segments is entirely independent of the orientation of its neighbors. Number n and mean length $\overline{d} = \ell$ in the FJC segments are the same as in the persistent chain. However, the mean square segment length $\overline{d^2}$ is different. The original flight with exponential distribution gives $\overline{d^2} = 2\ell^2$, whereas the FJC must satisfy $\overline{d^2} = 2\ell^2/(1 - \langle \cos\theta \rangle)$. Thus $\ell^* = \overline{d^2}/2\overline{d}$.

Source term $Q(\boldsymbol{r},t)$*:* Time profile that is adequate for the first passage problem is a δ-function: $Q(\boldsymbol{r},t) = Q(\boldsymbol{r})\delta(t)$. This can be deduced from an experiment: instead of CW illumination, we employ a femtosecond laser, $J_e(t) \approx \delta(t)$. Upon the passage through the multiply scattering medium we will find the pulse broadened: this pulse shape $J(t)$ reflects the distribution of the paths that are selected by the receiver. The space profile $Q(\boldsymbol{r})$ is given by the distribution of the *first* scattering events (e.g. the first kinks in the paths in Fig. 18). In the slab geometry, for example, we should take something like $Q(\boldsymbol{r}) \propto e^{-z/\ell^*}$. For simplicity, we place the source in a sharp distance Z from the plane of laser incidence, i.e. $Q(\boldsymbol{r}) \propto \delta(z - Z)$. A plausible choice is $Z = \ell^*$, but the topic is still subject to discussions. We take $Q(\boldsymbol{r},t) \propto \delta(z - \ell^*)\delta(t)$.

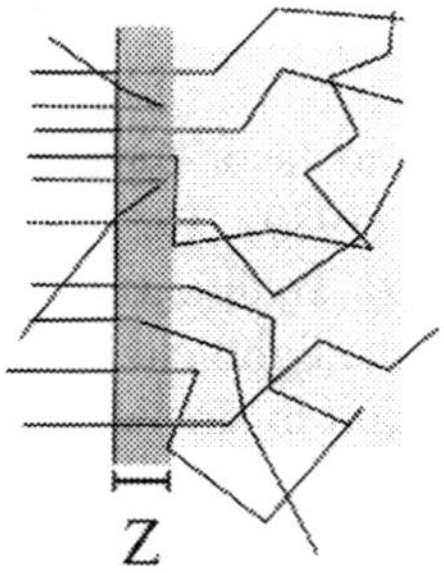

Fig. 18

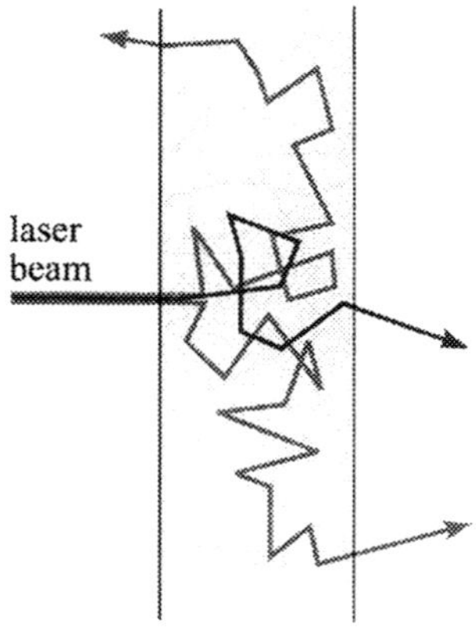

Fig. 19

Selection of paths: After solving Eqs. (49) for $C(\boldsymbol{r},t)$ and $E(t,\rho)$, we must take paths selection by the receiver into account. See Fig. 19, where the slab is illuminated by a focused narrow beam. Obviously, paths exiting far off axis are on average much longer than the paths close to the axis. On the other hand, the directions of exiting paths are essentially randomized, at each point the receiver samples uniformly the available population. Thus, we only need to compute the overlap integral

$$W(\tau) \propto \int B_o^2(\rho) E(\rho,\tau)\, \delta a, \tag{54}$$

where $B_o(\rho)$ is the profile of the observation beam at its intersection with the sample boundary. (This only applies for uniform samples. Otherwise one would have to extract also $C_o(\boldsymbol{r},\tau)$, the time space distribution of the *observed* scattering events. See discussion of the convective motion in Section 3.2.) Using the variable transformation $t = n\ell/c_\ell$, we obtain $W(n)dn$, a continuous version of the discrete path length distribution W_n. The final task is to find the Laplace transform of $W(n)$, recall Eq. (38). The outlined program, consisting in solution of the diffusion equation, path selection and Laplace transformation, is not easy. In practice we employ simple geometries, a selection of which is discussed in the next section. Most of the solved diffusion problems can be found, for example, in Crank (1992).

Comments on the diffusion approach. It has been argued that absorbing boundary condition, Eq. (50), is unphysical, because it violates a more general transport equation (Pine et al., 1989). An improved boundary condition should read

$$C - \tfrac{2}{3}\ell^* \hat{\boldsymbol{n}} \cdot \nabla C = 0\big|_{\text{boundary}}. \tag{55}$$

DWS experiments, however, seem to favor the simpler condition, Eq. (50), as already observed and discussed in Pine et al. (1989). This discrepancy may be due to a misunderstanding in the interpretation of Eq. (49). Traditionally, diffusion equation has been applied to *photon density*, i.e., energy density in the lightfield. However, there is a serious problem with combining all features of photon transport in simple diffusion. For example, diffusion implies infinite light speed (Ishimaru, 1978). For DWS we only need the distribution of *scattering events*, not caring what's going on between them. We regard the diffusion equation merely as a convenient mathematical tool for solving certain random flight problems.

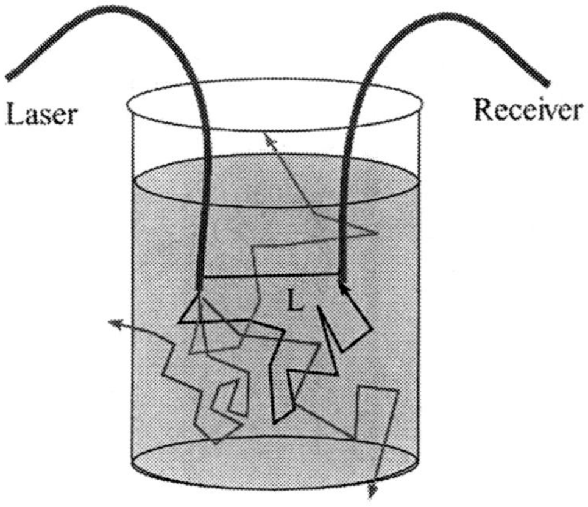

Fig. 20

3.4. Results for simple geometries

3.4.1. Point source and point receiver

The setup in Fig. 20 was suggested by Van Keuren et al. (1993): use two single-mode fibers immersed into a large beaker of the suspension. One fiber delivers the laser beam and the other is the receiver. We neglect, for simplicity, light losses into the fibers and assume the fiber distance L much smaller than the size of the beaker. This gives a particularly simple type of random flight: point source in an infinite medium. This is an excellent example for practicing the concepts of random flight, and the transition from random flight to diffusion. I recommend Chandrasekhars classical review (Chandrasekhar, 1943). There we take the standard 3D Gaussian for the random walk probability [eq. 107 in Chandrasekhar (1943)] insert the diffusion coefficient from Eq. (53) and apply the variable transformation $t = n\ell/c_t$. Then we plug the resulting W_n into Eq. (38), and look up the results in any table of Laplace transforms:

$$f(\tau) \approx \mathrm{e}^{-L\sqrt{3[A+\Gamma(\tau)]/(\ell\ell^*)}}/\mathrm{e}^{-\mathrm{e}^{-L\sqrt{3A/(\ell\ell^*)}}}, \tag{56}$$

where $\Gamma(\tau) = -\log\langle f(\tau)\rangle$, and $A = \ell/\ell_a$, where ℓ_a is the absorption extinction length. For illustration, we neglect absorption and insert into Eq. (56) the simplest approximation for $\Gamma(\tau)$, namely $\Gamma(\tau) \approx 2k^2 D[1 - \langle\cos\theta\rangle]\tau = 2k^2 D\tau\ell/\ell^*$:

$$f(\tau) \approx \mathrm{e}^{-(L/\ell^*)\sqrt{6k^2 D\tau}}. \tag{57}$$

3.4.2. Slab with homogeneous illumination, transmission

The slab in Fig. 21 is homogeneously illuminated by widened laser beam, so that the beam spot is much larger than thickness L. The receiver selects a path bundle somewhere from the middle of the broad spot of the transmitted light. (By virtue of reciprocity one could also exchange the receiver and the laser beam.) This geometry renders the problem 1-dimensional: we select with *equal probability* all types of paths that trans-cross the slab, disregarding the

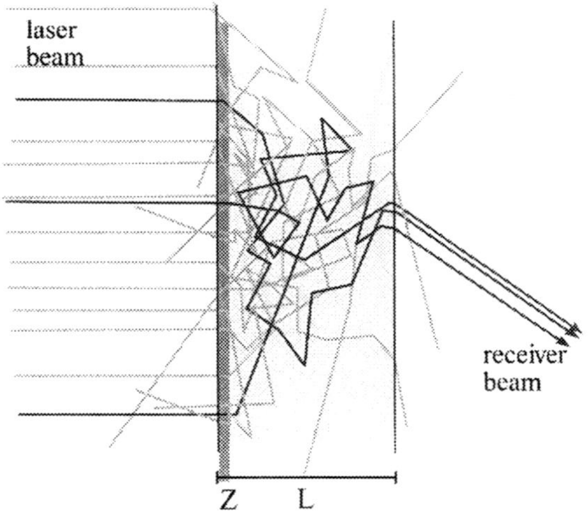

Fig. 21

lateral steps a scatteron takes. We take advantage of the one-dimensional character of this geometry in order to elucidate the relation between random flight and solutions by diffusion approximation, as given, e.g., in Pine at al. (1988, 1989). We map the persistent random flight on something even more simple than FJC: one-dimensional random walk on a lattice with fixed step length ℓ^w, and with equal probability 1/2 for left and right step. We require the mean square distance covered by a scatteron across the slab in L-direction to be the same for the scatteron as for the 1D random walker, this in the limit $n \to \infty$. This condition reads $\langle L^2_{n\to\infty}\rangle/3 = n\ell^{w2}$. The factor $1/3$ accounts for the projection on a single dimension. Inserting $\langle L^2_{n\to\infty}\rangle$ from Eq. (52), we obtain the walkers step length:

$$\ell^w = \ell\sqrt{\frac{2}{3}}\sqrt{\frac{1}{1-\langle\cos\theta\rangle}} = \ell^*\sqrt{\frac{2}{3}(1-\langle\cos\theta\rangle)}. \tag{58}$$

The random walk begins at the distance Z from the boundary. This problem can be solved exactly. The technique used is the so-called renewal equation (van Kampen, 1983) in combination with discrete Laplace transform. The technique yields directly the Laplace transforms $\sum W_n x^{-n}$. The results are:

$$f_T(\tau;\mathcal{Z},\mathcal{L}) = \frac{F_T(\tau;\mathcal{Z},\mathcal{L})}{T(\mathcal{Z},\mathcal{L})} \qquad F_T = \frac{G(\tau)^{-(\mathcal{Z}+1)} - G(\tau)^{(\mathcal{Z}+1)}}{G(\tau)^{-(\mathcal{L}+2)} - G(\tau)^{(\mathcal{L}+2)}}. \tag{59}$$

In the normalization, $T = F_T(0;\mathcal{Z},\mathcal{L})$ is the coefficient for the *total transmission* through the slab. $\mathcal{L} = L/\ell^w$ is the thickness of the slab and $\mathcal{Z} = Z/\ell^w$ is the walkers initial position, both measured in units of ℓ^w. A plausible choice is $Z = 1\ell^w$, i.e., $\mathcal{Z} = 1$. The function $G(\tau)$ contains dynamics $\Gamma(\tau) = -\log\langle f_i(\tau)\rangle$, and absorption $A = \ell/\ell_a$:

$$G(\tau) = \frac{\mathrm{e}^{-[A+\Gamma(\tau)]}}{1+\sqrt{1-\mathrm{e}^{-2[A+\Gamma(\tau)]}}} = \mathrm{e}^{-\sqrt{2[A+\Gamma(\tau)]}\{1+[A+\Gamma(\tau)]/6+[A+\Gamma(\tau)]^2/120+\cdots\}}. \tag{60}$$

We assume that $A \ll 1$ and $\Gamma(\tau) \ll 1$ in the relevant τ-range, and use the approximation $G(\tau) \approx \exp(-\sqrt{2[A+\Gamma(\tau)]})$. For $\mathcal{L} \gg 1$, $f_T(\tau;\mathcal{Z},\mathcal{L})$ will anyway decorrelate before the higher terms could appear. Thus, for $A \ll 1$ and $\mathcal{L} \gg 1$:

$$f_T(\tau) = \frac{1}{T} \cdot \frac{\sinh\big[(\mathcal{Z}+1)\sqrt{2[A+\Gamma(\tau)]}\big]}{\sinh\big[(\mathcal{L}+2)\sqrt{2[A+\Gamma(\tau)]}\big]}, \qquad T = \frac{\sinh\big[(\mathcal{Z}+1)\sqrt{2A}\big]}{\sinh\big[(\mathcal{L}+2)\sqrt{2A}\big]} \tag{61}$$

Exactly the same result is obtained in the diffusion approximation with the absorbing boundary condition, Eq. (50). For illustration, we insert into Eq. (61) the simplest approximation for $\Gamma(\tau)$, namely $\Gamma(\tau) \approx 2k^2D[1-\langle\cos\theta\rangle]\tau + \mathcal{O}(\tau^2)$. Recalling that $\mathcal{Z} = 1$ and using Eq. (58) for ℓ^w and ℓ^* we obtain

$$f(\tau) = \frac{1}{T} \cdot \frac{\sinh\big[\gamma\sqrt{3a^* + 6k^2D\tau}\big]}{\sinh\big[(L/\ell^* + \gamma)\sqrt{3a^* + 6k^2D\tau}\big]}, \qquad T = \frac{\sinh\big[\gamma\sqrt{3a^*}\big]}{\sinh\big[(L/\ell^* + \gamma)\sqrt{3a^*}\big]} \tag{62}$$

where

$$\gamma = 2\frac{\ell^w}{\ell^*} = 2\sqrt{\tfrac{2}{3}}\sqrt{1-\langle\cos\theta\rangle}, \qquad a^* = A\frac{8}{3\gamma^2} = \frac{\ell^*}{\ell_a}. \tag{63}$$

Random walk model yields $\gamma = 1.633$ for small, non-interacting particles, whereas experimental results suggest $\gamma \approx 1.9$. There is a certain arbitrariness in choosing the initial position Z of the random walkers. Using, e.g, $Z = \ell^*$ instead of $Z = \ell^w$, we would get $\gamma = (\ell^* + \ell^w)/\ell^* = 1 + \sqrt{2/3}\sqrt{1-\langle\cos\theta\rangle}$, yielding $\gamma = 1.82$ for small particles. The issue of the parameter γ is not yet fully clarified (Zhang et al., 1999).

Eq. (62) is almost identical with the result by Pine at al. (1988), that was obtained with the absorbing boundary condition in the diffusion problem. Later, this equation has been modified using the 'rigorous' boundary, Eq. (55). Our result, obtained without the diffusion approximation, suggests that this modification may not be entirely justified. However, a little calculation shows, that the 'rigorous' result, eq. 17 in Pine et al. (1989), is practically identical with $f(\tau) \propto \sinh\big[\gamma\sqrt{6k^2D\tau}\big]/\sinh\big[(L/\ell^* + \gamma/2)\sqrt{6k^2D\tau}\big]$. The only difference with our Eq. (62) is $\gamma/2$, which is easily compensated by a slight overestimation of ℓ^*. We will not dwell on this point, because the boundary apparently does not play any role when $L \gg \ell^*$. In Fig. 22 we compare the theoretical result with experimental data $g(\tau) = 1 + f(\tau)^2$.

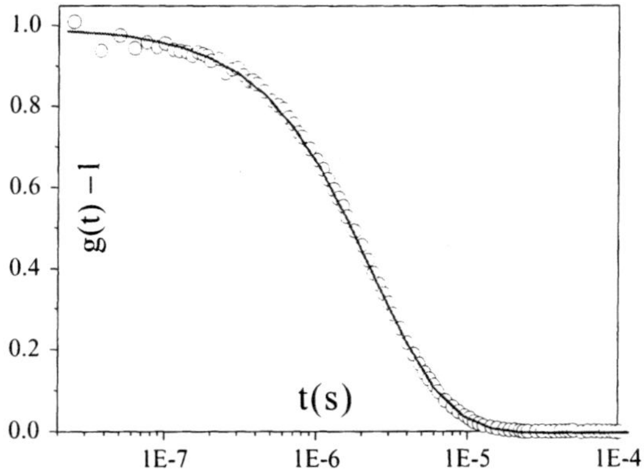

Fig. 22. Figure courtesy Frank Scheffold.

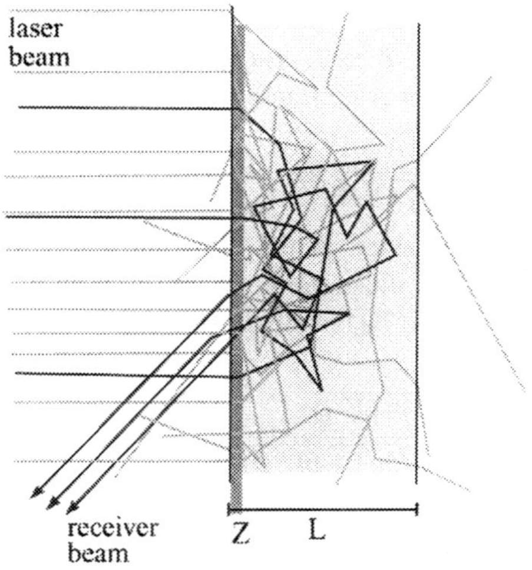

Fig. 23

3.4.3. Slab with homogeneous illumination, backscattering

The results for backscattering geometry, Fig. 23, are obtained by exchanging left and right, i.e., by replacing in Eq. (59) $\mathcal{Z}$ with $(\mathcal{L}-\mathcal{Z})$:

$$f_B(\tau;\mathcal{Z},\mathcal{L}) = \frac{1}{B(\mathcal{Z},\mathcal{L})}\cdot\frac{G(\tau)^{-(\mathcal{L}-\mathcal{Z}+1)} - G(\tau)^{(\mathcal{L}-\mathcal{Z}+1)}}{G(\tau)^{-(\mathcal{L}+2)} - G(\tau)^{(\mathcal{L}+2)}}. \tag{64}$$

where $G(\tau)$ is from Eq. (60), and B is the *total albedo*, the total fraction of light diffusely reflected from the slab.

$$B(\mathcal{Z},\mathcal{L}) = \frac{\sinh\left[(\mathcal{L}-\mathcal{Z}+1)\sqrt{2A}\right]}{\sinh\left[(\mathcal{L}+2)\sqrt{2A}\right]} \tag{65}$$

Here we assumed $A \ll 1$ [viz. Eq. (60)]. Letting $\mathcal{L}\to\infty$ (semi-infinite medium) we obtain:

$$f(\tau) = \frac{1}{B}\left\{\frac{e^{-[A+\Gamma(\tau)]}}{1+\sqrt{1-e^{-2[A+\Gamma(\tau)]}}}\right\}^{\mathcal{Z}+1} \approx \frac{1}{B}e^{-(\mathcal{Z}+1)\sqrt{2[A+\Gamma(\tau)]}}, \quad B = e^{-(\mathcal{Z}+1)\sqrt{2A}} \tag{66}$$

The right-hand approximation applies for $\Gamma(\tau)\ll 1$, which is less well satisfied than in the transmission case, because short paths are involved in back reflection (recall that $\mathcal{Z}\approx 1$). However, the approximate result is the same as the diffusion result with boundary condition Eq. (50). For illustration, we insert $\Gamma(\tau) = 2k^2D\left[1-\langle\cos\theta\rangle\right]\tau$ and use Eq. (58) to obtain:

$$f(\tau) \approx \frac{1}{B}e^{-\gamma\sqrt{3a^*+6k^2D\tau}}, \qquad B = e^{-\gamma\sqrt{3a^*}}. \tag{67}$$

The parameter γ is subject to further discussion in Section 3.5.

In Fig. 24 compare the exact result (full line) and the approximation for $\Gamma\ll 1$ (dashed line) with real data (squares) measured by Frank Scheffold. Slight absorption is included. Even the

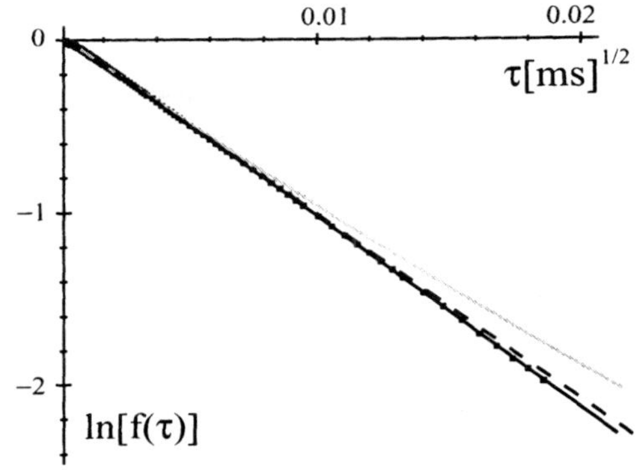

Fig. 24

simple expression fits experimental data well over a large dynamic range. For comparison, the lightly shaded curve results from the 'rigorous' boundary condition Eq. (55).

3.4.4. Tandem

Fig. 25 shows a tandem geometry: two different multiple scattering media are in a close contact. Scatterons may loop back and forth from one slab to the other, before exiting to either side. Results are easily deduced by applying Feynman's rules to these loops [viz. Eq. (23)]:

$$f_{T12} = \frac{1}{T_{\text{tot}}} \frac{T_1^1 f_{T1}^1 T_2 f_{T2}}{1 - B_1 f_{B1} B_2 f_{B2}}, \qquad T_{\text{tot}} = \frac{T_1^1 T_2}{1 - B_1 B_2}, \tag{68}$$

$$f_{B12} = \frac{1}{B_{\text{tot}}} \{ B_1^1 f_{B1}^1 + B_2 f_{B2} T_{\text{tot}} f_{T12} \}, \qquad B_{\text{tot}} = B_1^1 + B_2 T_{\text{tot}}, \tag{69}$$

The partial quantities are defined in Eqs. (59), (64) and (65). In the quantities denoted with superscript [1] set $\mathcal{Z} = 1$, but in all other functions $\mathcal{Z} = 0$, because the illumination of the

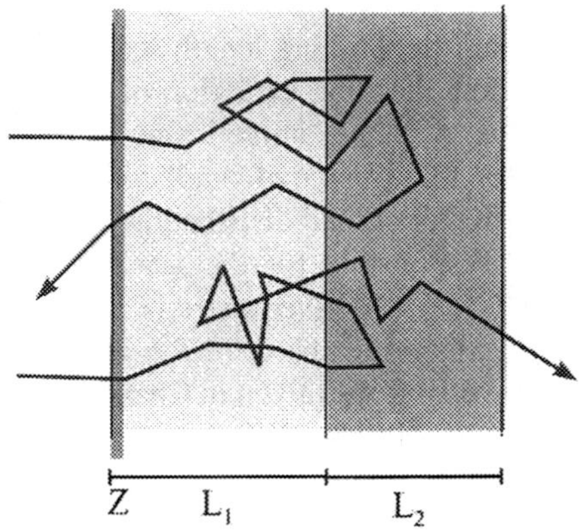

Fig. 25

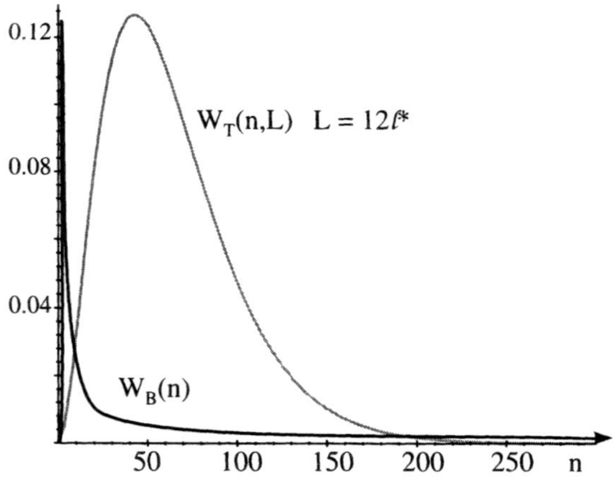

Fig. 26

interface is diffuse. Remember that ℓ^* and other parameters are different in the two media. See also Skipetrov and Maynard (1996).

3.5. Comments and practical considerations

The two most widely used DWS configurations are slab transmission and back reflection on semi-infinite medium (Sections 3.4.2 and 3.4.3). A fundamental difference between those two geometries results from their path length distributions $W_T(n, \mathcal{L})$ and $W_B(n)$, shown in Fig. 26.

Transmission $W_T(n, \mathcal{L})$ is a well behaved distribution peaked around the mean $\overline{n} = (\mathcal{L}^2 + 4\mathcal{L})/3$. Taking, for example $\mathcal{L} = 12$, we find that most of the contributions to $f(\tau)$ come from paths of about $\overline{n} = 64$ scattering events. This is a sufficiently large number to fulfill all the approximations of the DWS theory. Eq. (61) or, if appropriate, (62), can be applied with confidence. However, the parameter ℓ^* must be determined in an independent measurement. (The second parameter, $\gamma \approx 2$, plays only a negligible role when $L/\ell^* \gg 1$.) A concise survey of the methods for ℓ^* determination is given in Leutz and Rička (1996). Once ℓ^* is established, then one can vary the probing length scale, simply by varying $\overline{n}$. Choosing sufficiently large $\overline{n}$, one can follow the time evolution of mean square displacement of the scatterers down to few Ångstrøms. A nice example is due to Weitz et al. (1989), who studied the transition from ballistic motion to diffusion of small colloidal particles.

Back reflection exhibits an extremely broad distribution $W_B(n, \mathcal{Z})$. There is a tail, decaying slowly as $\propto n^{-3/2}$. This tail is responsible for the $\sqrt{\tau}$ behavior of $\log f(\tau)$, i.e., for the divergence of the initial slope of $f(\tau)$. However, there is also a large contribution from short path $n \approx 1$. This is not only an artefact of the diffusion approximation (obviously, on can't have $n < 1$). Even in the discrete version W_n [given in Chandrasekhar(43)] one finds $W_1 \approx 0.5$. With $n \approx 1$ the approximations of the simple DWS theory are not at all justified. In particular, much of the back reflection signal is due to single scattering. A rigorous treatment of this effect is difficult, but we can have a qualitative estimate: The bent curve in Fig. 27 results from replacing in $f_B(\tau)$ the pre-averaged path contributions corresponding to $n = 1$, with the

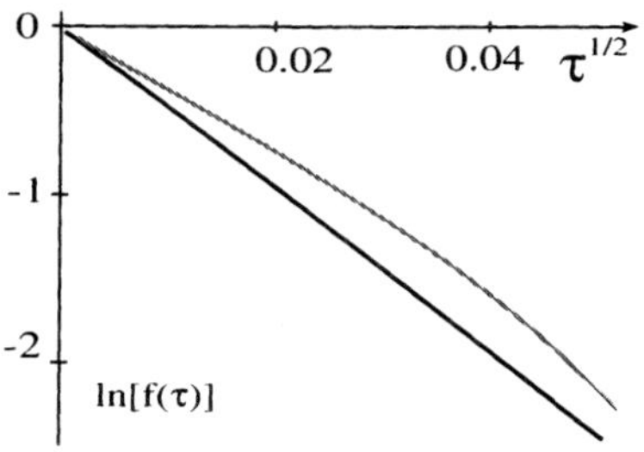

Fig. 27

single-scattering at $\approx 3\pi/4$ scattering angle, $f(q,\tau) \approx \exp(-3q^2 D\tau)$. Remarkably, the initial $\sqrt{\tau}$ behavior of $\log f(\tau)$ is essentially preserved, only the slope has changed. Thus, one may use Eqs. (66), or (67), despite problems with the theory.

However, the parameter γ that co-determines the slope, must be understood as an effective γ_{eff}, modified by the selection of short paths. This γ_{eff} depends on ℓ^* [as discussed in Pine et al. (1990)], as well as on the observation direction $\boldsymbol{k}_o$ and on the polarization $\boldsymbol{e}_o$. The effect of polarization is illustrated in Fig. 28 (data by Frank Scheffold). (Note that in VH geometry, most of the short paths are excluded, giving the good fit in Fig. 24.) In any case, accurate measurements require calibration of γ. However, γ_{eff} for *unpolarized observation* seems to vary in a relatively narrow range around 1.9, and this value can be used for rough evaluation of the data. This is an advantage of the back reflection with respect to the transmission: an estimate of suspension dynamics can be obtained from a single measurement, no need to measure ℓ^*.

A second advantage of back reflection geometry is the signal strength; compare the transmission and backreflection coefficients T and B from Eqs. (61) and (64). The effect of absorption is to cut the path lengths: the distributions in Fig. 27 are multiplied with the exponential $\exp(-nA)$. This suppresses the $n^{-3/2}$ tail in $W_B(n)$. Fig. 29 shows the corresponding change in the initial behavior of $f_B(\tau)$. On the other hand, the form of $W_T(n)$ remains more or less unchanged, only the mean $\bar{n}$ is shifted to a lower value. Correspondingly,

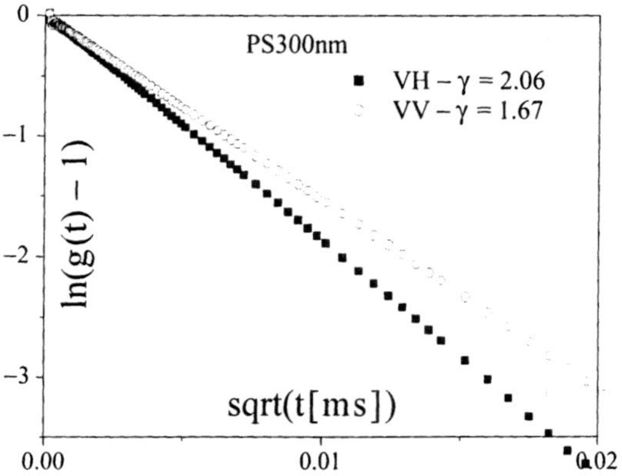

Fig. 28

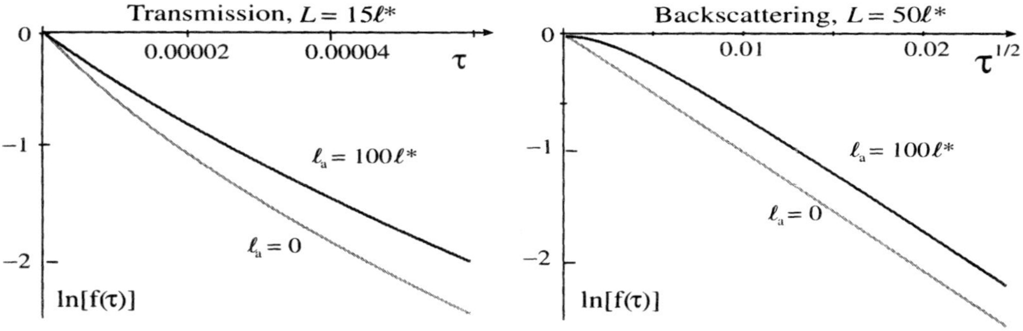

Fig. 29. $D = 1\ \mu^2\text{m/s}$. Absorption extinction length $\ell_a = 100\ell^*$.

$f_T(\tau)$ exhibits a smaller slope, but the form does not change much. Note that in transmission geometry one may easily overlook absorption. Back reflection is recommended for strongly absorbing samples.

Convection: Unlike conventional dynamic light scattering, DWS is highly sensitive to convective motion of the scatterers. This is illustrated in Fig. 30. Left-hand side is from a transmission experiment with $L = 15\ell^*$, right hand is for back-reflection. Transmission is much less sensitive to convection because the path lengths are long: the decay of the $f(\tau)$ curve is dominated by the first cumulant term $\propto k^2 D\tau$. The velocity term $\propto k^2\overline{\delta v^2}\tau^2$ simply does not have time to appear. Recall, however, that mean root square velocity in our example is only 50 μm/s!

Construction of the sample cell is an important issue with both techniques (Kaplan et al., 1993). The thickness must be large enough to guarantee the validity of the used approximations. For transmission, $L/\ell^* > 12$ is sufficient, but for back reflection the requirement is more strict. Eq. (66), that is valid for semi-infinite medium, $\mathcal{L} \to \infty$ requires a reasonably thick slab. This is illustrated in Fig. 31. See the effect of cutting the path length: the curvature at small τ is similar to the effect of absorption.

When constructing the sample cell, one should also keep in mind that the absorbing boundary condition is not fulfilled unless one takes precautions: there is reflection at the air–suspension interface. In particular, there is *total internal reflection* as soon as the angle

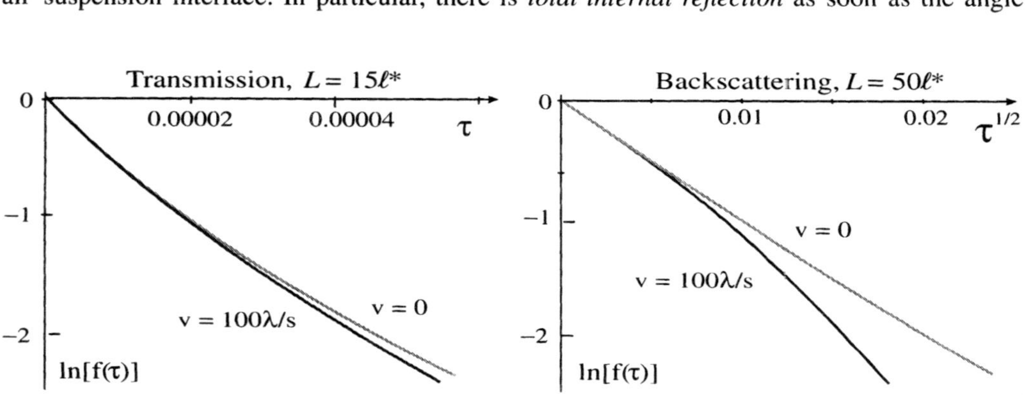

Fig. 30. $D = 1\ \mu^2\text{m/s}$. RMS speed: $v = 100\lambda/\text{s}$, i.e., only 50 μm/s.

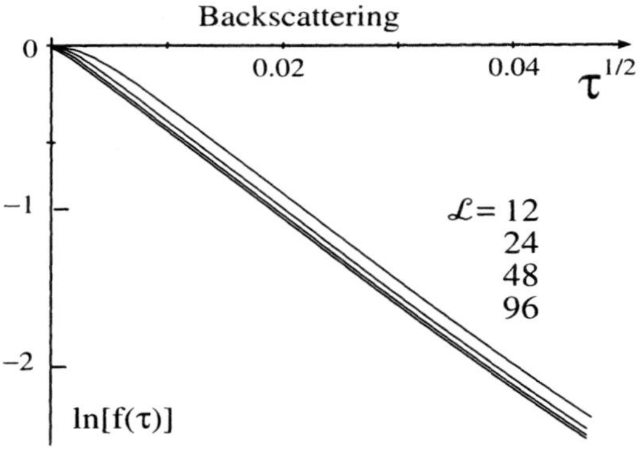

Fig. 31

of incidence exceeds the critical angle. Estimates of the effect of boundary reflection can be found in Langedijl et al. (1989) and Kaplan et al. (1993). However, the best solution is to insert the sample cell in an index-matching bath, a thermostatting bath is anyway required in order to prevent convection.

3.5.1. Non-standard geometries

Fiber optic setup (Section 3.4.1) is more or less equivalent to the transmission geometry, even though the resulting $f(\tau)$, Eq. (56), resembles back reflection, Eq. (66). The $\sqrt{\tau}$ dependence of $\log f(\tau)$ indicates an $n^{-3/2}$-tail in the distribution $W_F(n, \mathcal{L})$, which is almost identical with $W_B(n, \mathcal{Z})$, only the parameter $\mathcal{Z}$ is replaced by $\mathcal{L}$. However, the divergence of $W_F(n, \mathcal{L})$ at $n \to 0$ is only an artifact of diffusion approximation. The dangerous short paths are absent in the fiber setup with sufficiently large $\mathcal{L}$, and thus the theory for $f(\tau)$, Eq. (56), is well founded. A major advantage is the simplicity of varying $\mathcal{L}$. Also the possibility to measure *in situ*, in any sufficiently large container, makes the technique interesting. Note that the technique has been developed in an industrial environment (Keuren et al., 1993). The drawbacks are the necessity to determine ℓ^* separately, and the sensitivity of the delicate fibers to damage or to clogging by adsorbed material.

Tandem geometry (Section 3.4.4). An interesting practical application of the tandem has been suggested by Frank Scheffold et al. from Fribourg, Switzerland (Romer et al., 2000): the problem are frozen, so-called *non-ergodic* paths in nearly rigid media at sol–gel transition: a tandem cell with slow dynamics is a simple tool for efficient path averaging in such a sample. A particularly simple limiting case arises when $B_1 B_2 \ll 1$. Then $f_{12} = f_1 \cdot f_2$. This condition can be enforced for example be inserting an absorbing layer between the two samples, but the transmitted intensity is then low. Otherwise, the tandem geometry is still of mainly of academical interest: Eqs. (69) and (68) can be used as the starting point for the analysis of multiple scattering in a general layered medium (Skipetrov and Maynard, 1996).

3.6. Conclusion

I exhausted the allocated space. More information on DWS and related topics can be found in a special issue of *JOSA A*, **14**, 136 (1997). The fundamental aspects are reviewed in the recent 'tutorial' by Van Rossum and Niewenhuizen, *Rev. Mod. Phys*, **71**, 313–371 (1999).

References

Bicout, D. and Maret, G. (1994). *Phys. A*, **210**, 87.
Chandrasekhar, S. (1943). *Rev. Modern Phys.*, **15**, 1.
Crank, J. (1992). *The Mathematics of Diffusion*. Clarendon Press, Oxford.
Feynman, R.P. (1985). *QED: the Strange Theory of Light and Matter*. Princeton University Press, Princeton.
Flammer, I., Bucher, G. and Rička, J. (1998). *J. Opt. Soc. Am.*, **15**, 2066.
Flory, P.J. (1969). *Statistical Mechanics of Chain Molecules*. John Willey & Sons, New York.
Fraden, S. and Maret, G. (1999). *Phys. Rev. E*, **60**, 4843.
Genack, A., Garcia, N. and Polkosnik, W. (1990). *Phys. Rev. Lett.*, **65**, 2129.
Gisler, T., Rüger, H., Egelhaaf, S., Tschumi, J., Schurtenberger, P. and Rička, J. (1995). *Appl. Optics*, **34**, 3546.
Ishimaru, A. (1978). *Wave Propagation and Scattering in Random Media*. Academic Press, New York.
Kaplan, P.D., Kao, M.H., Yodh, A.G. and Pine, D.J. (1993). *Appl. Optics*, **32**, 3828.
Keuren, E.V., Wiese, H. and Horn, D. (1993). *Colloids Surf. A*, **77**, 29.
Langedijl, A., Vreeker, R. and Vries, P.D. (1989). *Phys. Lett. A*, **136**, 81.
Leutz, W. and Rička, J. (1996). *Optics Comm.*, **126**, 260.
Maret, G. and Wolf, P.E. (1987). *Z. Phys. B*, **65**, 409.
Pine, D.J., Weitz, D.A., Chaikin, P.M. and Herbolzheimer, E. (1988). *Phys. Rev. Lett.*, **60**, 1134.
Pine, D.J., Weitz, D.A., Maret, G., Wolf, P.E., Herbolzheimer, E. and Chaikin, P.M. (1989). *Dynamical Correlations of Multiply Scattered Light*. World Scientific, London and Singapore.
Pine, D.J., Weitz, D.A., Zhu, J.X. and Herbolzheimer, E. (1990). *J. Phys. (France)*, **51**, 2101.
Rička, J. (1993). *Appl. Optics*, **32**, 2860.
Rička, J. (1994). *Fluctuations, Correlations: Dynamic Light Scattering*. Internet publication www.fluor.unibe.ch, work in progress, Bern.
Romer, S., Scheffold, F. and Schurtenberger, P. (2000). *Phys. Rev. Lett.*, **85**.
Scheffold, F. and Maret, G. (1998). *Phys. Rev. Lett*, **81**, 5800.
Skipetrov, S. and Maynard, R. (1996). *Phys. Lett. A*, **217**, 181.
van Kampen, N.G. (1983). *Stochastic Processes in Physics and Chemistry*. North-Holland, Amsterdam.
Weitz, D.A., Pine, D.J., Pusey, P.N. and Tough, R.J.A. (1989). *Phys. Rev. Lett.*, **63**, 1747.
Zhang, Z., Jones, I., Schriemer, H., Page, J., Weitz, D. and Sheng, P. (1999). *Phys. Rev. E*, **60**, 4843.

Neutrons, X-Rays and Light
P. Lindner and Th. Zemb (Editors)

Chapter 21

Use of Scattering Methods in Chemical Industry – SAXS and SANS from Fibers and Films

Jens Rieger

BASF Aktiengesellschaft, Polymer Physics/Solid State Physics, D-67056 Ludwigshafen, Germany

1. Introduction

Scattering techniques are used in industry for the same reasons as in fundamental research: detecting and describing structures on length scales from 0.1 nm to some 100 nm. But although the physics of scattering of light, X-ray, neutrons, or electrons is the same in any application there are interesting differences as regards the systems under consideration and the philosophy of experimentation. These points are the subjects of this contribution.

First, the role of the various scattering methods in industrial research is shortly exemplified. In the following two sections two examples from our research at BASF have been chosen to stress the mutuality and differences concerning fundamental and applied research. As an example for X-ray scattering in classical polymer physics it will be shown how one can study the development and change of crystalline structures during the melt-spinning of polymer fibers. In the second example the structure of films that were obtained from colloidal polymer dispersions is studied with small-angle neutron scattering. Lastly, an outlook will be given as to present and future developments in the use of scattering techniques in industrial applications. Throughout this article emphasis will be put on applications in soft matter as produced, processed, and sold by chemical industry.

A comprehensive collection of industrial applications of X-ray scattering – both in classical solid-state physics as well as for soft matter systems – can be found in a voluminous book (Chung and Smith, 2000). Some recent achievements in the field of soft matter are presented in a symposium volume (Cebe et al., 2000). The list of references is quite extensive in order to provide an impression about the background of the topics discussed here and possibly as a starting point for one's own research.

2. Scattering techniques used and applications

In all major chemical companies scattering methods are well-established tools for the characterization of raw, intermediate, or final products. Methods and typical applications are the following:

- Wide-angle X-ray scattering (WAXS): crystal modification, crystallinity, crystal size. texture, i.e., orientation of crystallites in, e.g., polymeric fibers and films (Kakudo and

Kasai, 1972; Alexander, 1979; Murthy et al., 1995; Prazur and Prud'homme, 1996). WAXS is mainly used on a routine basis with existing commercial software and equipment. A special problem with the interpretation of WAXS data of polymers will be discussed below.

- Small-angle X-ray scattering (SAXS): periodicity of stacked lamellar crystals in semicrystalline polymers, such as polyethylene, polypropylene, polyamide, etc. (Crist, 1979; Koberstein and Stein, 1983; Kifle et al., 1986; Wilke and Bratrich, 1991; Stribeck, 1993; Rule et al., 1995; Hauser et al., 1998; Murthy et al., 1999; Goderis et al., 1999), particle size, particle structure and interaction between particles in disperse systems, such as in polymer blends, polymer dispersions, emulsions, micellar systems, etc. (Lindner and Zemb, 1991; Slagi and Rajagopalan, 1993; Chu et al., 1994; Pedersen, 1994; Brunner-Popela and Glatter, 1997; Dingenouts et al., 1999). SAXS is more demanding than WAXS since experimentation needs more care (correction of the raw data), software must be developed for the specific problem, and input is needed concerning the structural model.
- Small-angle neutron scattering (SANS): this method is related to SAXS and uses neutrons instead of photons (Lindner and Zemb, 1991; Higgins and Benoit, 1996). Due to the specific differences in the neutron scattering lengths of, e.g., hydrogen and deuterium SANS allows for unique contrast variation (Lindner and Zemb, 1991; Fischer et al., 1991; Cotton, 1996; Murthy et al., 1998; Leube et al., 2000) an example of which is given below. Major applications are the determination of the interaction parameter in polymer blends (Hammouda, 1993; Hellmann and Hellmann, 1997; Maurer et al., 1997; Crist, 1998; Reichart et al., 1998) and the structural investigation of colloidal systems (Chen, 1986; Jerke et al., 1998; Cosgrove et al., 1998; Zipfel et al., 1998; Lutterbach et al., 1999; Egelhaaf and Schurtenberger, 1999). Due to the higher expenditures this method is used in industry mainly for special problems that cannot be tackled with other methods or for fundamental studies.
- Static light scattering (LS) and dynamic light scattering (DLS): routine application for the determination of the molar mass of polymers and the particle size in colloidal systems (Kratochvil, 1987; Chu, 1991). For special applications such as particle size determination in turbid dispersions these techniques are adapted to the system, see, e.g., Wiese and Horn (1993), where a device for fiber-optic quasielastic light scattering is described that was developed especially for the monitoring of particle growth in concentrated latex dispersions.
- There are more specialized methods like, e.g., electron diffraction (Spence and Carpenter, 1986) and fluorescence correlation spectroscopy (Nörenberg et al., 1999) that will not be discussed here.

This list might give the impression that the scientific literature reflects the respective scattering activities in industry. But SAXS has considerably lost ground to electron microscopy and scanning force microscopy as regards characterization of the structure of semi-crystalline polymers, polymer blends, and colloidal systems, as, e.g., polymer dispersions and emulsions. The reason is that microscopy yields a real space image, which seems to be self-explaining to the customer.

However, there are many cases where scattering techniques are superior to competing methods for several reasons: standard scattering methods yield information that is averaged over some μm^3 to cm^3. This is an advantage when bulk properties are considered and a disadvantage when structural gradients are dominating the sample. The latter point applies to, e.g., component parts produced by injection molding of semicrystalline polymers. Scattering requires less sample preparation than, e.g., electron microscopy thus being less prone to artifacts. In some cases small-angle scattering is more sensitive to density differences between different phases than staining techniques that are required for rendering structures visible in transmission electron microscopy (TEM). Also, small-angle scattering allows for contrast variation – especially in SANS.

Having appreciated that scattering is useful for applied research a further point deserves attention: The system itself might not be well-defined in a scientific sense, i.e., it may contain several components that interact with each other – a typical example being shampoo containing different surfactants, polymers and further ingredients. Further examples are polymer samples containing residue catalysts, stabilizers, antioxidants, etc. In those cases it is necessary to clarify to which extent the essential information about structure–property relationships can be extracted from the complete formulation/system or whether it is indispensable to study the sole components and their mutual interactions. The guiding principle for learning the essential structural mechanisms relevant for application reads: design the experiments as close to application as possible and as idealized as necessary.

3. Structural investigations of polyamide 6 fibers

3.1. The structure of polyamide 6 fibers

The properties of polymeric fibers are determined to a large extent by their nano- and microstructure (Fig. 1), i.e., by the fraction and size of crystallites and their structural arrangement within the fiber (Murthy, 1997). This structure is by no means fixed during crystallization from the melt but changes during the spinning process and in the subsequent treatment steps.

In the following it shall be exemplified which structural changes may occur in the spinning process. The appropriate method for studying such structural changes is X-ray scattering. Wide-angle X-ray scattering permits to determine the crystallinity, modification, degree of crystal perfection and orientation. Small-angle X-ray scattering yields information about structural aspects on the nanometer scale, i.e., about the arrangement of the crystallites. As will be shown below it is also possible to derive from SAXS data information about the density of the amorphous phase.

The structure of polyamide 6 – in the bulk as well as in the form of fibers – has been described extensively in the literature (Heuvel and Huisman, 1981; Gianchandani et al., 1982; Koyama et al., 1986; Weyhe and Pietzsch, 1993; Murthy et al., 1995). But still, there are many open questions as regards the relationship between the morphology of the fibers (on all length scales) and application related properties as, e.g., shrinkage, dyeability, bleeding, etc.

3.2. Online detection of crystallization during fiber spinning

Polyamide 6 fibers are usually produced by melt-spinning: the polymer melt is extruded through spinnerets with orifices with a diameter of some 100 μm and drawn onto a winding-up machine with speeds up to 6000 m/min. The process is quasi-stationary, i.e., crystallization of

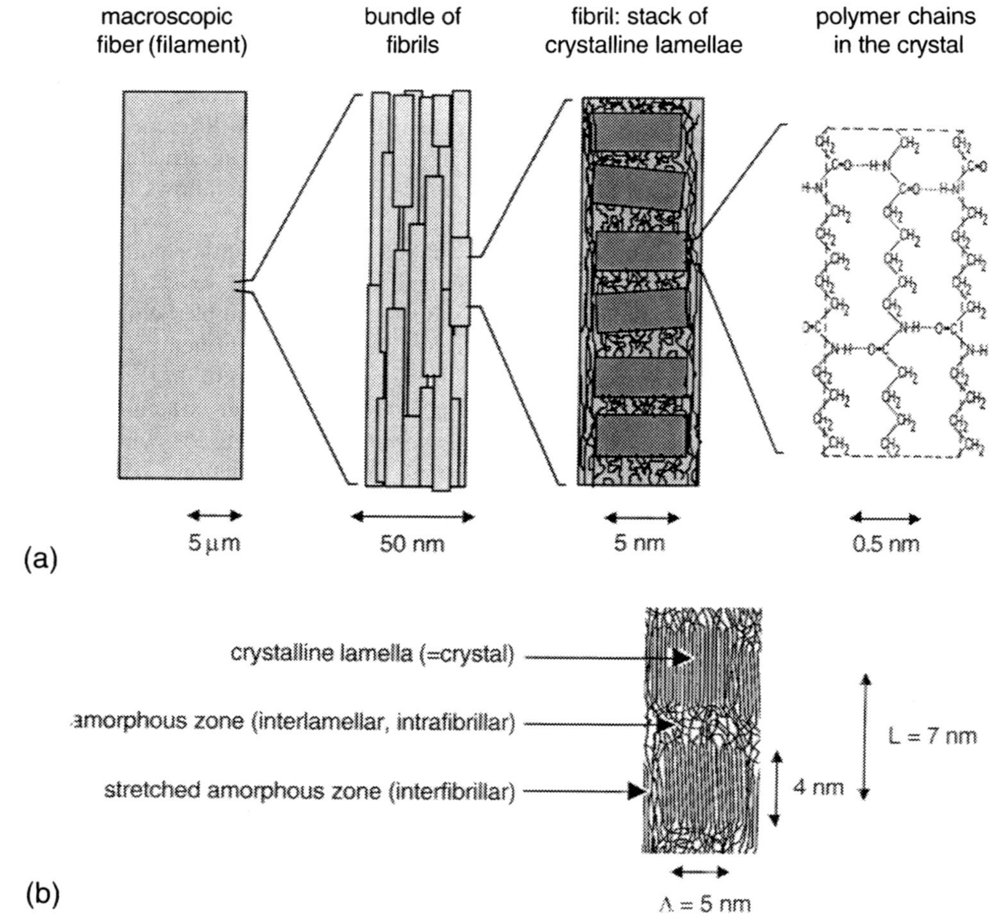

Fig. 1. (a) Hierarchy of structures in a polyamide 6 fiber, (b) structural organization of crystalline and amorphous zones on the nanometer level. The structural model is derived from SAXS and WAXS experiments as described in the text (see also Murthy, 1997).

the drawn fiber occurs at a fixed distance below the spinneret. It is of interest to characterize the crystallization process to obtain a better understanding of the spinning process and the resulting properties of the fibers. To this aim WAXS has been used for a long time either directly in large-scale spinning facilities using standard X-ray tubes or in experimental spinning apparatus using high-intensity synchrotron radiation (Chappel et al., 1964; Katayama et al., 1968; Bankar et al., 1977; Haberkorn et al., 1993; Hirahata et al., 1996; Samon et al., 1999; Kolb et al., 2000). The first approach has been followed since the crystallization processes are known to be sensitive to the actual spinning speed (Murase et al., 1997) and high spinning speeds require sophisticated winding-up machines. On the other hand, maximum information about crystallization is only obtained with the combination of WAXS with SAXS. This in turn is only possible with very high intensities which are beyond the limit of sealed tubes.

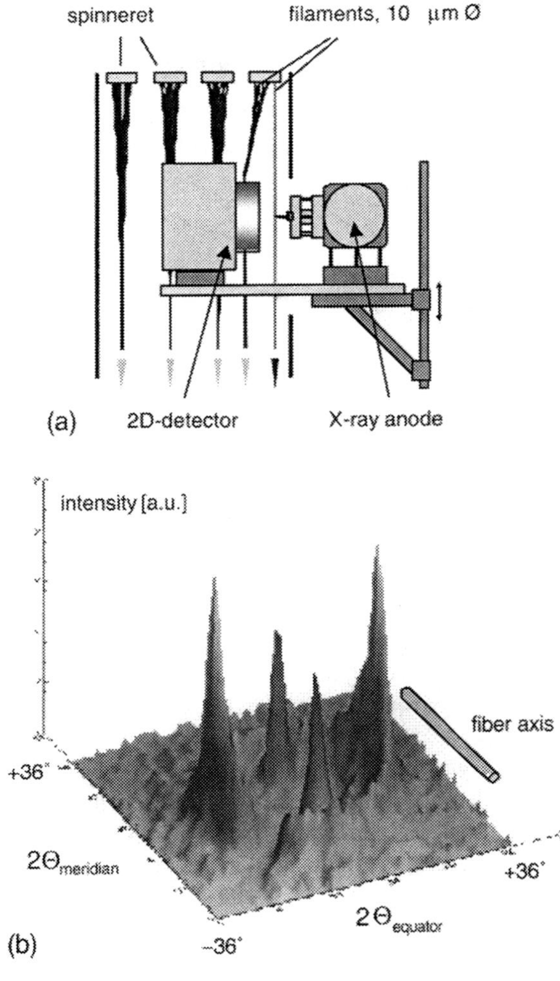

Fig. 2. (a) WAXS equipment for measuring the crystallinity online during melt-spinning. (b) WAXS diffractogram taken online with the equipment shown in (a). Air-scattering has already been subtracted.

We followed the first approach namely investigating crystallization in a large scale spinning-line at BASF with WAXS. The equipment is shown schematically in Fig. 2a. An X-ray tube is mounted on a platform together with the two-dimensional detector. This platform can be moved vertically up and down along the spinning path. A single filament with a diameter of about 10 μm is led past the collimator causing the scattered intensity detected. An example of a two-dimensional diffractogram of a filament at a distance where it is already solidified is given in Fig. 2b (the measurement takes about 10 minutes). After performing azimuthal averaging of the data it is possible to extract a crystallinity index by comparing the area of the Bragg peaks to the area of the amorphous halo. It is not possible to derive absolute crystallinities because of the inappropriate weighting of the meridian and equator peaks with

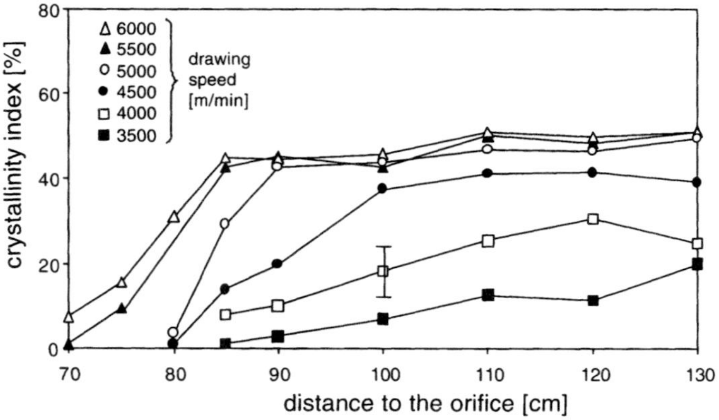

Fig. 3. Crystallinity index of melt-spun polyamide 6 fibers as a function of the distance to the spinneret and in dependence of the drawing speed.

the actual setup (Ruland and Dewaelheyns, 1967). The crystallinity indices as a function of the distance to the spinneret and of the spinning speed are shown in Fig. 3. As can be seen the crystallization mechanism changes when going from a low spinning speed such as 3500 m/min to high-speed spinning with 6000 m/min. In the first case there is a gradual increase of the crystallinity along the spinning line whereas in the latter case the crystallization takes place along a relatively short distance of 0.1 m. From the knowledge of this distance and of the value of the spinning speed one derives for the duration of the crystallization 10^{-3} seconds.

Further research is necessary in order to judge whether the fiber crystallizes directly into the final modification or whether crystalline precursors occur as has been claimed (Samon et al., 1999). And furthermore, it will be interesting to see by SAXS whether density fluctuations (Ryan et al., 2000) precede the actual crystallization in the present case.

3.3. Structural changes in polyamide 6 fibers during drawing and heating

In commercial high-speed processes for the production of carpet fibers the yarn undergoes further thermomechanical treatment steps after the initial crystallization before being wound up. In order to study micro-structural changes in the fiber during these steps we sampled polyamide 6 fibers at different stages of the spinning process. Thus we investigated by means of offline-WAXS and SAXS the original fiber, the drawn fiber (draw ratio 1 : 1.5), and the drawn and relaxed fiber (about 150 milliseconds at 180°C, i.e.. about 40 K below the melting point). Drawing and heating is applied in order to improve the final properties such as, e.g., strength and shrinkage. All these treatment steps are implemented in one spinning machine.

Typical WAXS and SAXS data are shown in Fig. 4. The X-rays were generated with a rotating Cu-anode, monochromatized by a Ni-filter and applied to the sample via a pin-hole camera. The scattered intensity was measured with a two-dimensional wire detector. The sample to detector distance was set to 4 cm (WAXS) and 64.8 cm (SAXS). The WAXS diffractogram is a typical fiber pattern of polyamide 6 containing crystals in the γ-modification [as evidenced by the meridional peak at about 11° (Murthy et al., 1995)] and

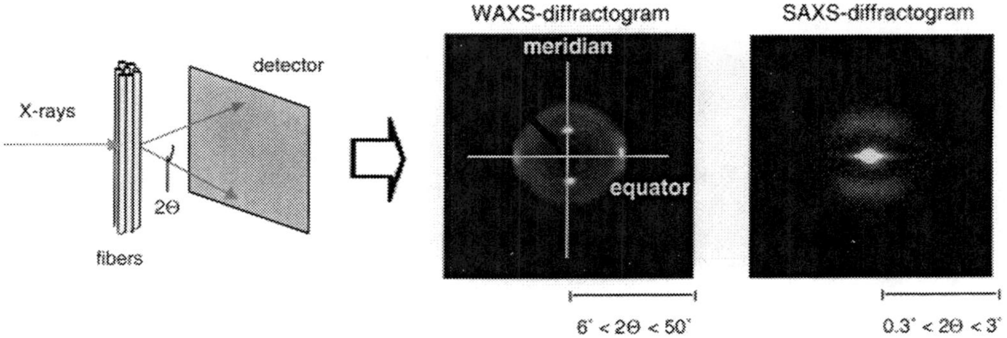

Fig. 4. Typical WAXS and SAXS data taken on polyamide 6 fiber bundles.

possibly additional α-crystals. The SAXS diffractogram exhibits two broad meridional peaks, which are slightly bent, and equatorial streaks at very small angles. The meridional peaks are due to the stacking of crystalline lamellae along the fiber axis as shown in Fig. 1. The broadening is due to the finite lateral size of these lamellae (Crist, 1979). The bending is caused by a misalignment of the stacked lamellae with respect to the fiber axis (Rule et al., 1995). The streaks are attributed to the fibrils, i.e., the stacks of crystalline lamellae as shown above (Murthy et al., 1996). The WAXS and SAXS data will now be discussed in a rather qualitative manner.

For the sake of a better resolution the WAXS data have also been taken with a Bruker D5005 diffractometer (CuK_α), equipped with a primary monochromator and a position sensitive detector. Fiber bundles with 500 to 1000 single parallel filaments were scanned in the equatorial mode and in rotation mode – in both cases data were taken in transmission. The resulting data contain information about the ratio of crystalline contents in different modifications (α, γ) and about crystal perfection. A θ–2θ geometry was chosen such that the meridional scattering peaks ($(00l)$ reflexes) lie on the surface of the Ewald sphere (Kakudo and Asai, 1972). The equatorial WAXS diffractograms of the three samples are shown in Fig. 5. The single peak at about $2\theta = 22°$ is the sum of six underlying peaks: four Bragg peaks due to crystals in the α and γ modification and the amorphous halo (Murthy et al., 1995). The Bragg peaks are comparatively broad because of the smallness of the crystallites. Obviously, the fitting approach is mathematically ill-defined because of the large number of parameters (peak position, width, shape). With knowledge about the influence of the thermal history on the crystalline state of the material it is possible to impose restrictions onto the fitting procedure thus achieving sensible results (Heuvel and Huisman, 1981). But for the present purpose it suffices to extract information from the data shown in Fig. 5 in a qualitative manner. First, the WAXS-scattering behavior does not change much under the conditions chosen here: The fraction of γ-crystals dominates. The crystallinity as estimated by the area of the peak hardly changes. The only obvious effect is the narrowing of the main peak when going from the drawn to the drawn and relaxed fiber. In this case narrowing is caused by an increase in crystal size as described by the Scherrer equation (Alexander, 1979). Quantitative evaluation yields lateral crystal sizes in the range of 7–10 nm – increasing when going from the untreated fiber to the relaxed fiber.

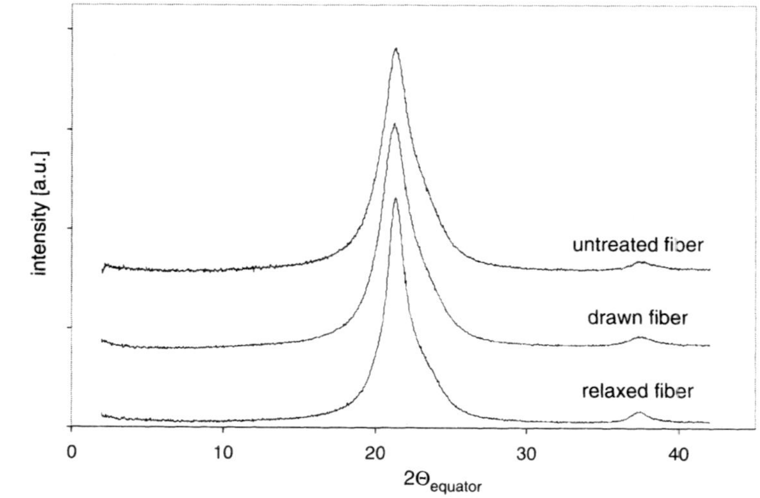

Fig. 5. WAXS scans taken perpendicular to the fiber axis, i.e., along the equator.

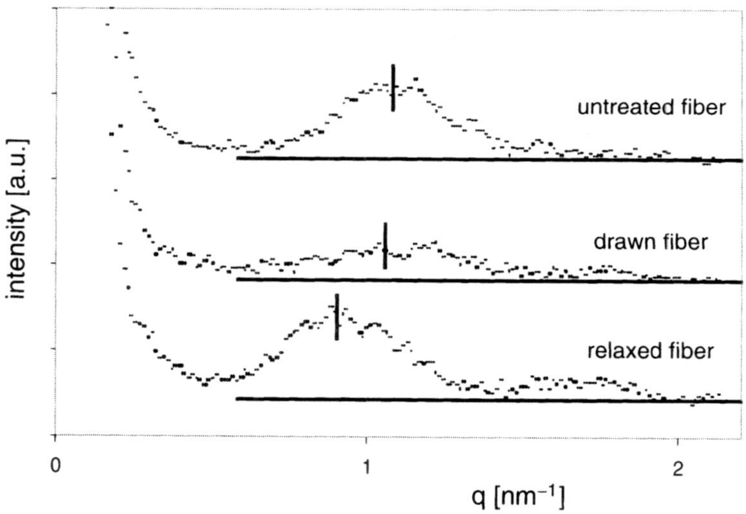

Fig. 6. SAXS scans taken along the fiber axis, i.e., along the meridian. $q = 4\pi \sin(\theta)/\lambda$. The straight lines are drawn as a guide for the eye.

Whereas the differences between the fibers of this set is small with respect to the properties of the crystallites in the fiber (as detected by WAXS) Fig. 6 shows that the respective SAXS diffractograms (measured along the fiber axis) differ significantly with respect to the peak position and intensity.

The position of the peaks is related in a first approximation to the longitudinal periodicity of the crystalline lamellae, cf. Fig. 1, via $L = 2\pi/q$. A more thorough approach is possible (Crist, 1979; Koberstein and Stein, 1983; Kifle et al., 1986; Stribeck, 1993; Rule et al., 1995;

Wilke and Bratrich, 1996; Hauser et al., 1998; Murthy et al., 1999; Goderis et al., 1999), but has not been used since the present simplistic evaluation yielded enough information for our purposes. The following values have been obtained for the long period L, i.e., the distance between the centers of gravity of two adjacent crystalline lamellae measured along the fiber axis: $L_{\text{untreated}} = 5.8$ nm, $L_{\text{drawn}} = 5.9$ nm, $L_{\text{relaxed}} = 6.9$ nm.

Assuming a two-phase structure of the fibers the two-dimensional integrated SAXS intensity is due to (a) the difference between the densities of the amorphous and the crystalline phase, and (b) the product of the fraction of each phase. Since we know from WAXS that the crystallinity and the crystal structure hardly changes during the process of drawing and relaxing it can be concluded that the observed change in SAXS intensity is due to a change in the density of the amorphous phase. Upon drawing the density of the amorphous phase approaches the density of the crystalline phase. During exposition to 180°C the density relaxes back. The decrease of the SAXS intensity upon drawing has also been reported before (Hsiao et al., 1997) but was interpreted in a different way.

Generally it is possible to extract more information from the two-dimensional SAXS patterns (Murthy et al., 1996). The corresponding data (e.g., meridional scans) are not shown for the sake of brevity. But it should be reminded that it is often quite difficult to obtain unequivocally reliable data about microscopic structures in real space (Rule et al., 1995; Goderis et al., 1999) as seen, e.g., by TEM.

Summarizing the above we arrive at the following statements for the structural changes occurring in the fibers during drawing and subsequent heating.

- *Drawing the fiber:* The long period hardly changes though the fibers are drawn by a factor of 1.5. This implies that the stacks of lamellae (fibrils) remain intact. The elongation is achieved by the sliding of fibrils past each other. This in turn means that polymer segments that connect adjacent fibrils are either drawn taut or broken during drawing. In the first case material may be supplied by unreeling chain segments from the crystalline lamellae. The scattering power strongly decreases. This means that either the crystallinity or the density difference between the crystalline and amorphous phase is strongly diminished. Since WAXS and calorimetry show no considerable change in crystallinity we must indeed assume that the latter point applies: the amorphous phase – and the network of crystalline lamellae – rearrange in such a way that $\rho_a \to \rho_c$ during drawing.
- *Heating ('relaxing') the drawn fiber:* The distance between adjacent lamellae *increases* slightly upon relaxation. The scattering power increases, i.e., the density difference between the crystalline and amorphous phase increases: the amorphous phase relaxes. The fibrils/crystals grow laterally. Thus, the relaxation of the amorphous phase and the rearrangement of the crystalline lamellae are coupled in a non-intuitive way. Relaxation of a stretched amorphous domain means shrinkage and one might assume that crystals which are coupled to the amorphous phase via tie chains and loops are drawn closer towards each other by this mechanism. The contrary is true: the distance between the lamellae increases. This can be explained by assuming that thin (and therefore less stable) crystals melt at the elevated temperatures of the relaxation procedure; the molten polymer segments crystallize onto existing crystals.

Summarizing the information from analysis of the online and offline X-ray data we conclude that the crystallites are predefined during the crystallization and that various relaxational processes are active subsequently on several length scales ranging from 0.1 nm up to about 100 nm. The structural data can be correlated with application-related properties thus allowing for a goal-directed development of new products and processes.

4. Films made out of ordered polymeric spheres

4.1. Film formation of polymer dispersions

Latices are colloidal dispersions consisting of discrete compact spherical polymer particles in a water phase. The size of the spheres is typically of the order of hundred nanometers. Usually, the polymeric spheres carry surface charges and/or steric stabilizers in order to prevent the particles from coagulation. These dispersions represent interesting model systems, which offer the possibility to investigate experimentally many-body effects more conveniently than in the case of gases or liquids consisting of single atoms or low molecular weight molecules. The advantage stems from the fact that both the typical time scale as well as the respective length scale lie in a window which is easily accessible by experiment (Pieranski, 1983; Stevens et al., 1991; Würth et al., 1995; Harland and van Megen, 1997; Pusey et al., 1997; Russel et al., 1997; Kegel and Dhont, 2000). Many investigations are dealing with colloidal crystals as found in liquid dispersions. These systems can be realized with a volume ratio of spheres down to 10^{-3} because of the repelling Coulomb forces acting among the charged spheres. In recent years there has been increasing interest in the structure of close-packed spheres for two reasons. First, polymer dispersions are applied as films in paints for buildings and cars, in paper coating, etc. Second, there is a lot of recent interest in colloidal crystals because of potential applications as photonic band gap materials (Joannopulos et al., 1997; Dinsmore et al., 1998; Cheng et al., 1999; Rogach et al., 2000). The aspect of film formation and structure is treated in several reviews (Dobler and Holl, 1996; Chevalier, 1996; Winnik, 1997; Pekcan, 1997; Keddie, 1997; Stewart et al., 2000) – specifically with SANS and SAXS (Joanicot et al., 1990; Chevalier et al., 1992; Joanicot et al., 1993; Richard and Wong, 1995; Rharbi et al., 1996; Joanicot et al., 1996; Dingenouts and Ballauff, 1998; Dingenouts and Ballauff, 1999). The following presentation is based on two BASF publications (Rieger et al., 1992; Rieger et al., 1995a).

Latex films are prepared from the dispersion by evaporation of the water. During the process of film formation three subsequent processes can be discerned, as is schematically illustrated in Fig. 7: Increasing concentration of the polymeric spheres during evaporation of the fluid resulting in dense packing of the particles (Fig. 7a and b), deformation of the particles to polyhedra and beginning of coalescence because of surface- and capillary-forces (Fig. 7c), further coalescence caused by the interdiffusion of polymeric chains emanating from contiguous particles (Fig. 7d).

The three processes sketched in Fig. 7 have been examined experimentally as well as theoretically, see references in Rieger et al. (1995a). But, still there are many open questions, e.g.: Which structure has evolved during film formation? How do different film formation temperatures affect this structure? How far has the coalescence of the particles already proceeded in a film with a given thermal history? To which degree is the process of coalescence homogeneous on spatial length scales?

In order to obtain more knowledge about the structure of latex films we performed SANS experiments on latex films. The films prepared differed with respect to their thermal history

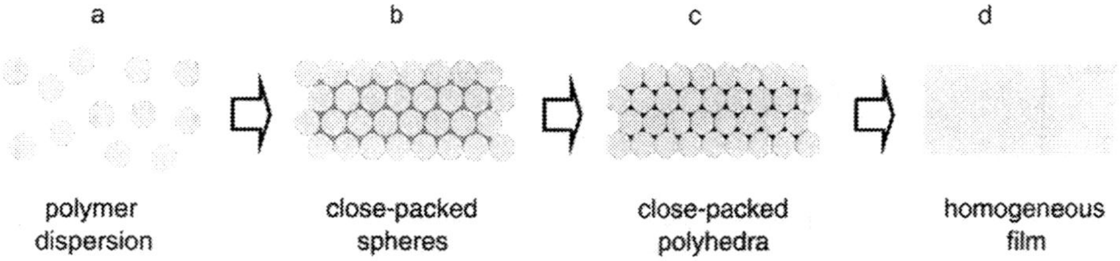

Fig. 7. Stages during the process of film formation from polymeric dispersions.

and with respect to the amount of water, which was brought into the films after drying. Our investigations differ from published experiments (Joanicot et al., 1990) insofar as we brought the water (D_2O) into the respective samples *after* completed formation of the film.

4.2. Samples and experimental setup

The dispersion from which the films were prepared is a styrene/n-butylacrylate copolymer latex with about 50% by weight of each monomer and a polymer content of about 50% by weight. The system is stabilized against coagulation by negative surface charges on the latex particles originating from sulfate groups. The dispersion contains additional emulsifier and electrolyte. The particle size distribution was quite narrow with an average diameter of 130 nm. Films were prepared by the evaporation of water at room temperature. The glass temperature of the films is at about room temperature. Film pieces of known mass were packed into polyethylene bags together with the amount of D_2O wanted and the bag was sealed off. D_2O instead of H_2O was used because of the enhanced scattering intensity due to the large differences in the respective scattering lengths of deuterium of D_2O and hydrogen of the polymer.

The SANS experiments were performed at the D11 instrument at the Institut Laue–Langevin (Grenoble). The sample to detector distance was chosen to be 20 m. The wavelength of the neutrons was 1nm with a wavelength distribution of $\Delta\lambda/\lambda = 0.09$. The data were sampled and manipulated in the standard way in order to obtain the scattering curves. On the two-dimensional detector of the D11 instrument diffraction of the powder-like (Debye–Scherrer) type is observed. The symmetry of the data is exploited for averaging and the determination of the radial distribution function $I(q)$. $I(q)$ denotes the intensity of neutrons scattered at an angle 2θ with respect to the beam sent onto the sample. q is defined by $q = 4\pi \sin(\theta)/\lambda$.

4.3. Interpretation of the scattering data

A typical set of data is shown in Fig. 8a. The scattering data exhibit two pronounced peaks, which indicates that the film possesses structural order. This type of scattering behavior is found independently of the actual D_2O content of the films – as long as this content is below ~10%. It is a priori not possible to derive the type of structure in an unequivocal way from such sets of scattering data. But, in the present case we know from electron microscopic pictures that the latex spheres are packed regularly. Thus, we show in the following how

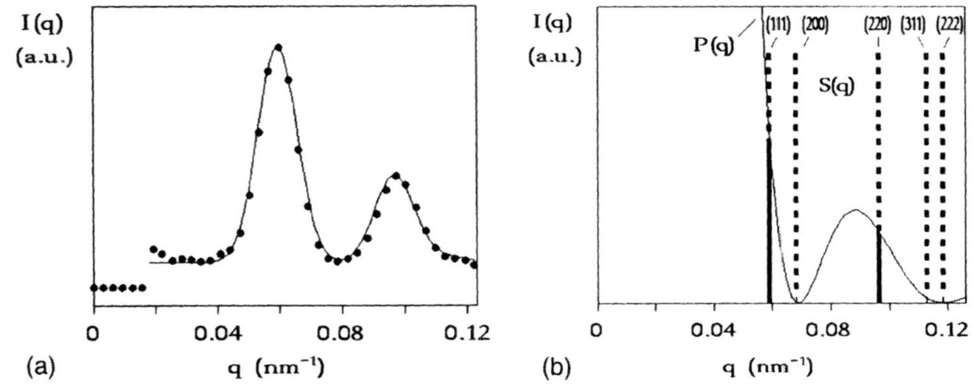

Fig. 8. (a) Scattering data obtained from SANS on a latex film prepared at room temperature which contained ~2% D_2O. The six points on the left are points of zero intensity (beamstop). Solid line: see text. (b) Thin line: q-dependence of the particle form factor $P(q)$ of a homogeneous sphere with radius 65 nm. Bold lines (dashed and solid part): positions of the Bragg-peaks $S(q)$ as expected in the case of an ideal face-centered cubic lattice made from close-packed spheres of radius 65 nm. The dashed parts indicate the degree of suppression of these peaks when calculating the total scattered intensity $I(q) \sim P(q)S(q)$; only two peaks are left (solid bold lines).

the observed scattering behavior can be explained by assuming that the latex spheres are close-packed in extended domains with a face centered cubic (fcc) lattice structure. The contrast which is necessary in order to obtain the observed scattering behaviour is due to the difference between the scattering length of the polymeric material in the latex spheres, on the one hand, and the scattering lengths of the components in the interstitial phase, on the other hand. The interstitial phase consists of material that is not miscible with the polymers, as, e.g., salt, surfactants, D_2O, etc., and possibly of micro-voids. Here, it must be stressed again that the uptake of D_2O (below ~10%) by the films does not lead to a qualitative change in the scattering curves. Hence, we can be sure that the assignment just given is correct. Making use of Babinet's theorem (Guinier and Fournet, 1955), we treat the data in a first-order approach as if all scattering effects were solely due to the scattering by the latex particles. Since the particles are fairly monodisperse we write the scattered intensity $I(q)$ as being proportional to the product of the particle form factor $P(q)$ and the lattice structure factor $S(q)$:

$$I(q) \propto P(q)S(q). \tag{1}$$

The particle form factor that is due to intraparticle interference effects is given by (Pedersen, 1997):

$$P(q) \propto \left([\sin(qr) - qr \cdot \cos(qr)]/(qr)^3\right)^2, \tag{2}$$

where r is the radius of the polymer spheres. The structure factor of an ideal, infinitely extended lattice is the sum of sharp Bragg peaks at q-values corresponding to the respective reciprocal lattice vectors q_{hkl} where hkl are the Miller indices of the lattice planes (Jenkins and Snyder, 1996). In the case of cubic symmetry the q_{hkl} are determined by

$$q_{hkl} = 2\pi \left(h^2 + k^2 + l^2\right)^{1/2}/a. \tag{3}$$

a is the lattice constant. For fcc-lattices only those q_{hkl} are non-vanishing where all indices are even or uneven, respectively: (111), (200), (220), (311), (222), etc. (Guinier, 1956). In Fig. 8b are shown $P(q)$ and $S(q)$ as calculated for the case of homogeneous spheres with radius r which are close-packed in a fcc structure. r is chosen to be the radius of the particles under investigation, namely 65 nm.

When calculating $I(q) \sim S(q)P(q)$ in order to obtain the theoretical scattering curves of an ideal fcc-lattice made of close-packed spheres one learns from Fig. 8b that some $S(q)$ peaks are strongly suppressed by the $P(q)$ term. Of the five fcc peaks expected in the chosen q range only two are left when taking into account $P(q)$. The sharp peaks discussed so far on a theoretical basis are broadened in an experiment because of apparatus effects (finite wavelength distribution, collimation effects) and deviations of the real system from ideality (finite domain size, lattice defects, polydispersity, non-sphericity of the particles). Taking into account all these points by convolution of the delta peaks with a single Gaussian we arrive at the curve shown in Fig. 8a which is depicted together with measured data taken at a film that was prepared at room temperature. Using the known value of the diameter of the spheres ($d = 130$ nm) and the respective lattice constant of 184nm, the only relevant fit parameter – besides shifting and scaling in the ordinate-direction – is the width of the Gaussian.

The coincidence between the fitted curve and the measured data is better than could be expected especially with regard to the peak intensities when noting that we did not take into account the following points: multiplicities of the reflecting planes, preferred deposition of latex spheres in (111)-planes along the plane of the film (Joanicot et al., 1990), the q-dependence of the scattered intensity because of instrumental effects (Pedersen et al., 1990). and deformation of the latex spheres. Concerning the latter point, we showed by simulating the scattering of deformed spheres via Monte Carlo techniques that the present interpretation of the scattering data is not affected by deformation effects (Rieger et al., 1995a).

Taking into account collimation effects and the wavelength distribution of the D11 instrument (Pedersen et al., 1990) we come to the conclusion that in the case of films prepared at room temperature the deconvoluted data should exhibit sharp Bragg peaks indicating that the crystalline order is well defined with a significant domain size. This is in agreement with data from TEM investigations that reveal crystalline domains exceeding the size of some dozen particles.

A similar analysis as the one leading to Fig. 8 was performed for body-centered cubic and hexagonal close-packed lattice structures. The respective predictions of the scattering behavior differ strongly from the $I(q)$ data that were found in our experiments. We conclude that indeed a fcc structure is realized.

As another kind of information that can be obtained by this type of experiment we show in Fig. 9 a set of SANS data taken on latex films that were annealed for 120 minutes at different temperatures T_a and then exposed to D_2O. For $T_a < 150$°C one observes the familiar two-peak structure which indicates that the water diffuses preferentially into the interstices. The picture drastically changes once the film has been tempered for two hours at 180°C: instead of the two peaks, which have almost completely disappeared, one observes an increasing SANS signal at very small q-values. We interpret this effect as arising from the existence of aggregates of water: all the tempered films were prepared at room temperature thus possessing essentially the same morphology. We conclude that the water no longer has access to the interstices for $T_a \geq 180$°C and is thus forced to aggregate in other locations of the film. This, in turn, might be explained by the assumption that the percolation structure of the network of interstices is lost when tempering for a sufficient time at $T_a \geq 180$°C. This effect is due to relaxation and

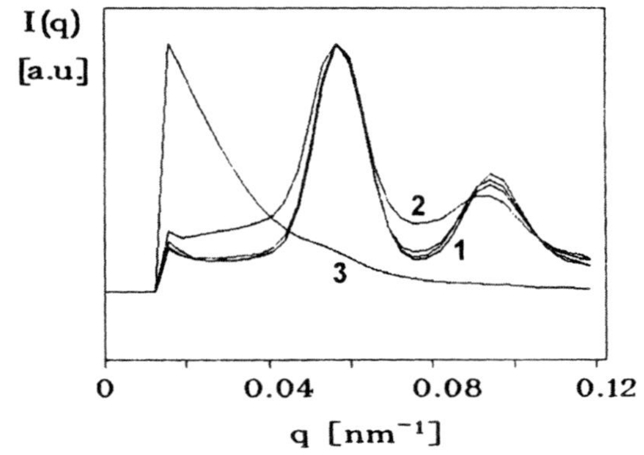

Fig. 9. SANS data obtained on films that were annealed for two hours at different temperatures T_a. D_2O content: 5%; $T_a = 70$, 100, 120 (data 1), 150 (data 2), 180°C (data 3). The intensities have been normalized to the maximal value of each data set.

further deformation of the close-packed particles. Since the water is likely to diffuse along the percolation network of the interstices, a loss of a certain fraction of bonds in this network makes part of the interstices inaccessible to water.

A discussion of the influence of the film formation temperature, of the amount of D_2O taken up by the film, and of annealing on the film structure has been given (Rieger et al., 1995a). Additionally, we studied by the same approach the 'fate' of surfactants in the latex films, i.e., whether the surfactant molecules are distributed homogeneously in the film or located preferentially in the interstices (Rieger et al., 1995b). In those studies the contrast variation approach has been employed by comparing the SANS scattering behavior of films with hydrogenated and deuterated surfactants, respectively.

5. Conclusion and outlook

The two examples presented here were chosen to highlight several aspects of scattering experiments as applied in industrial research:

- The technique that is selected for the investigation of a given system is chosen according to its potential value as regards the structural information to be expected – and its costs. This point is stressed because the situation often is different at institutes where methods are used because they are available. Thus, in the case of latex films SANS has been partly replaced meanwhile by atomic force microscopy and transmission electron microscopy.
- If necessary the equipment is adapted to the conditions of interest. During the investigation of fibers it was realized that it is essential to understand the crystallization in the spinning-line and at the high spinning speeds given in industrial production processes. It is reminded that it is an ongoing trend to establish physical methods for the characterization of structural aspects *online* in production lines.
- Research in industry often is confronted with 'dirty' samples, i.e., material that contains further ingredients besides the main components. For example, the dispersion used for the SANS investigations on latex films was taken directly from a production lot and contained surfactants and salt.

- When dealing with complex, i.e., multi-component systems one tries to reveal underlying structure–property relationships as directly as possible. Only if it is not possible to dismantle the various interactions being active–or if certain aspects are relevant for the understanding of other systems as well–more fundamental investigations are performed.
- Another point to be stressed is that the correlation between model systems and application-related systems is often very difficult if not impossible to establish.
- In many cases qualitative or semi-quantitative data are sufficient for understanding the essential mechanisms. As an example it was shown above that in the case of polyamide 6 fibers the relevant process during thermomechanical treatment is the rearrangement of structural units (the fibrils) rather than recrystallization processes of the single crystals.
- Frequently it happens that seemingly well-known topics such as fiber spinning or latex film formation become fashionable under new headings and/or with view to new applications. In the present case:
  – fiber spinning: phase transitions in confined systems under the influence of external fields,
  – latex films: self-organized photonic crystals.
- Increased use of synchrotron radiation for WAXS and SAXS is to be expected because of the opportunity to perform spatially resolved experiments with a micro-focus (Riekel et al., 1999), to characterize a large number of samples quickly (Dagani, 1999), and to study reactions and transitions in real time (Bras and Ryan, 1998; Butler et al., 1998; Bras et al., 1995; Rieger et al., 1998). From such studies new exciting insight into dynamic phenomena are to be expected–both in applied as well as in fundamental research.

Acknowledgements

I thank Peter Lindner (ILL) and Paul Bever (BASF) for their contributions to the work presented here.

References

Alexander, L.E. (1979). *X-Ray Diffraction Methods in Polymer Science*. Wiley-Interscience, New York.

Bankar, V., Spruiell, J.E. and White, J.L. (1977). *J. Appl. Polym. Sci.*, **21**, 2341–2358.

Bras, W. and Ryan, A.J. (1998). *Adv. Colloids Interface Sci.*, **75**, 1–43.

Bras, W., Derbyshire, G.E., Bogg, D., Cooke, J., Elwell, M.J., Komanschek, B.U., Naylor, S. and Ryan, A.J. (1995). *Science*, **267**, 996–999.

Brunner-Popela, J. and Glatter, O. (1997). *J. Appl. Cryst.*, **30**, 431–442.

Butler, M.F., Donald, A.M. and Ryan, A.J. (1998). *Polymer*, **39**, 781–792.

Cebe, P., Hsiao, B.S. and Lohse, D.J., eds. (2000). *Scattering from Polymers. ACS Symposium Series 739*. American Chemical Society, Washington.

Chappel, F.P., Culpin, M.F., Gosden, R.G. and Tranter, T.C. (1964). *J. Appl. Chem.*, **14**, 12–19.

Chen, S.H. (1986). *Ann. Rev. Phys. Chem.*, **37**, 351–399.

Cheng, Z.D., Russell, W.B. and Chaikin, P.M. (1999). *Nature*, **401**, 893–895.

Chevalier, Y. (1996). *Trends Polym. Sci.*, **4**, 197–203.

Chevalier, Y., Pichot, C., Graillat, C., Joanicot, M., Wong, K., Maquet, J., Lindner, P. and Cabane, B. (1992). *Colloids Polym. Sci.*, **270**, 806–821.

Chu, B. (1991). *Laser Light Scattering*. Academic Press, Boston.

Chu, B., Wu, G. and Schneider, D.K. (1994). *J. Polym. Sci., Polym. Phys.*, **32**, 2605–2614.

Chung, F.H. and Smith, D.K., eds. (2000). *Industrial Applications of X-Ray Diffraction*. Marcel Dekker, New York.

Cosgrove, T., Hone, J.H.E., Howe, A.M. and Heenan, R.K. (1998). *Langmuir* **14**, 5376–5382.
Cotton, J.P. (1996). *Adv. Colloids Interface Sci.*, **69**, 1–29.
Crist, B. (1979). *J. Appl. Cryst.*, **12**, 27–33.
Crist, B. (1998). *Macromolecules*, **31**, 5853–5860.
Dagani, R. (1999). *Chem. Eng. News*, **3**, 51–60.
Dingenouts, N. and Ballauff, M. (1998). *Macromolecules*, **31**, 7423–7439.
Dingenouts, N. and Ballauff, M. (1999). *Langmuir*, **15**, 3283–3288.
Dingenouts, N., Bolze, J., Pötschke, D. and Ballauff, M. (1999). *Adv. Polym. Sci.*, **144**, 1–47.
Dinsmore, A.D., Crocker, J.C. and Yodh, A.G. (1998). *Curr. Opinion Colloids Interface Sci.*, **3**, 5–11.
Dobler, F. and Holl, Y. (1996). *Trends Polym. Sci.*, **4**, 145–151.
Egelhaaf, S.U. and Schurtenberger, P. (1999). *Phys. Rev. Lett.*, **82**, 2804–2807.
Fischer, E.W., Stamm, M. and Voigt-Martin, I.G. (1985). In: *Chemistry and Physics of Macromolecules* (Fischer, E.W., Schulz R.C. and Sillescu, H., eds.). VCH, Weinheim, pp. 445–475.
Gianchandani, J., Spruiell, J.E. and Clark, E.S. (1982). *J. Appl. Polym. Sci.*, **27**, 3527–3551.
Goderis, B., Reynaers, H., Koch, M.H.J. and Mathot, V.B.F. (1999). *J. Polym. Sci., Polym. Phys.*, **37**, 1715–1738.
Guinier, A. (1956). *Théorie et Technique de la Radiocrystallographie*. Dunod, Paris, p. 116.
Guinier, A. and Fournet, G. (1955). *Small Angle Scattering of X-Rays*. Wiley, New York.
Haberkorn, H., Hahn, K., Breuer, H., Dorrer, H.-D. and Matthies, P. (1993). *J. Appl. Polym. Sci.*, **47**, 1551–1579.
Hammouda, B. (1993). *Adv. Polym. Sci.*, **106**, 87–133.
Harland, J.L. and van Megen, W. (1997). *Phys. Rev. A*, **55**, 3054–3067.
Hauser, G., Schmidtke, J. and Strobl, G. (1998). *Macromolecules*, **31**, 6250–6258.
Hellmann, E.W. and Hellmann, G.P. (1997). *Macromol. Chem. Phys.*, **198**, 329–340.
Heuvel, H.M. and Huisman, R. (1981). *J. Polym. Sci., Phys. Ed.*, **19**, 121–134.
Higgins, J. and Benoit, H. (1996). *Polymers and Neutron Scattering*. Clarendon Press, Oxford.
Hirahata, H., Seifert, S., Zachmann, H.G. and Yabuki, K. (1996). *Polymer*, **37**, 5131–5137.
Hsiao, B.S., Kennedy, A.D., Leach, R.A., Chu, B. and Harney, P. (1997). *J. Appl. Cryst.*, **30**, 1084–1095.
Jenkins, R. and Snyder, R.L. (1996). *X-Ray Powder Diffractometry*. Wiley-Interscience, New York.
Jerke, G., Pedersen, J.S., Egelhaaf, S.U. and Schurtenberger, P. (1998). *Langmuir*, **14**, 6013–6024.
Joanicot, M., Wong, K., Maquet, J., Chevalier, Y., Pichot, C., Graillat, C., Lindner, P., Rios, L. and Cabane, B. (1990). *Prog. Colloids Polym. Sci.*, **81**, 175–183.
Joanicot, M., Wong, K., Richard, J., Maquet, J. and Cabane, B. (1993). *Macromolecules*, **26**, 3168–3175.
Joanicot, M., Wong, K. and Cabane, B. (1996). *Macromolecules*, **29**, 4976–4984.
Joannopoulos, J.D., Villeneuve, P.R. and Fan, S. (1997). *Nature* **386**, 143–145.
Kakudo, M. and Kasai, N. (1972). *X-Ray Diffraction by Polymers*. Elsevier, Amsterdam.
Katayama, K., Amano, T. and Nakamura, K. (1968). *Kolloid Z., Z. Polym.* **226**, 125–134.
Keddie, J.L. (1997). *Mat. Sci. Eng.*, **21**, 101–170.
Kegel, W.K. and Dhont, J.K.G. (2000). *J. Chem. Phys.*, **112**, 3431–3436.
Kifle, Z., Harrison, I.R. and Herman, R.M. (1986). *J. Polym. Sci., Polym. Phys. Ed.*, **24**, 633–655.
Koberstein, J.T. and Stein, R.S. (1983). *J. Polym. Sci., Polym. Phys. Ed.*, **21**, 1439–1472.
Kolb, R., Seifert, S., Stribeck, N. and Zachmann, H.G. (2000). *Polymer*, **41**, 1497–1505.
Koyama, K., Suryadevara, J. and Spruiell, J.E. (1986). *J. Appl. Polym. Sci.*, **31**, 2203–2229.
Kratochvil, P. (1987). *Classical Light Scattering from Solutions*. Elsevier, Amsterdam.
Leube, W., Monkenbusch, M., Schneiders, D., Richter, D., Adamson, D., Fetters, L., Dounis, P. and Lovegrove, R. (2000). *Energy Fuels*, **14**, 419–430.
Lindner, P. and Zemb, Th., eds. (1991). *Neutron, X-Ray and Light Scattering: Introduction to an Investigative Tool for Colloidal and Polymeric Systems*. North-Holland, Amsterdam.
Lutterbach, N., Versmold, H., Reus, V., Belloni, L., Zemb, Th. and Lindner, P. (1999). *Langmuir*, **15**, 345–352.
Maurer, W.W., Bates, F.S., Lodge, T.P., Almdal, K., Mortensen, K. and Fredrickson, G.H. (1997). *J. Chem. Phys.*, **108**, 2989–3000.
Murase, S., Kashima, M., Kudo, K. and Hirami, M. (1997). *Macromol. Chem. Phys.*, **198**, 561–572.
Murthy, N.S. (1997). *Textile Res. J.*, **67**, 511–520.
Murthy, N.S., Bray, R.G., Correale, S.T. and Moore, R.A.F. (1995). *Polymer*, **36**, 3863–3873.
Murthy, N.S., Bednarczyk, C., Moore, R.A.F. and Grubb, D.T. (1996). *J. Polym. Sci., Polym. Phys.*, **34**, 821–835.

Murthy, N.S., Akkapeddi, M.K. and Orts, W.J. (1998). *Macromolecules*, **31**, 142–152.
Murthy, N.S., Wang, Z.-G. and Hsiao, B.S. (1999). *Macromolecules*, **32**, 5594–5599.
Nörenberg, R., Klingler, J. and Horn, D. (1999). *Angew. Chem. Int. Ed.*, **38**, 1626–1629.
Pedersen, J.S. (1997). *Adv. Colloids Interface Sci.*, **70**, 171–210.
Pedersen, J.S., Posselt, D. and Mortensen, K. (1990). *J. Appl. Cryst.*, **23**, 321–333.
Pekcan, Ö. (1997). *Trends Polym. Sci.*, **5**, 177.
Pieranski, P. (1983). *Contemp. Phys.*, **24**, 25–73.
Prazur, R.J. and Prud'homme, R.E. (1996). *Macromolecules*, **29**, 119–128.
Pusey, P.N., Segre, P.N., Behrend, O.P., Meeker, S.P. and Poon, W.C.K. (1997). *Phys. A*, **235**, 1–8.
Reichart, G.C., Graessley, W.W., Register, R.A. and Lohse, D.J. (1998). *Macromolecules*, **31**, 7886–7894.
Rharbi, Y., Boué, F., Joanicot, M.and Cabane, B. (1996). *Macromolecules*, **29**, 4346–4359.
Richard, J. and Wong, K. (1995). *J. Polym. Sci., Polym. Phys.*, **33**, 1395–1407.
Rieger, J., Hädicke, E., Ley, G. and Lindner, P. (1992). *Phys. Rev. Lett.*, **68**, 2782–2785.
Rieger, J., Dippel, O., Hädicke, E., Ley, G. and Lindner, P. (1995a). In: *Colloidal Polymer Particles* (Goodwin, J.W. and Buscall, R., eds.). Academic Press, London, pp. 29–48. *"Crystals Made of Close-Packed Polymeric Spheres: a Neutron Scattering Study on Latex Films"*
Rieger, J., Ley, G. and Lindner, P. (1995b). *ILL Experimental Reports*. Institut Laue–Langevin, Grenoble, p. 366.
Rieger, J., Bever, P., Gurke, I. and Stribeck, N. (1998). *HASYLAB Ann. Rep.*, HASYLAB, Hamburg, p. 757.
Riekel, C., Dieing, T., Engström, P., Vincze, L., Martin, C. and Mahendrasingam, A. (1999). *Macromolecules*, **32**, 7859–7865.
Rogach, A., Susha, A., Caruso, F., Sukhorukov, G., Kornowski, A., Kershaw, S., Möhwald, H., Eychmuller, A. and Weller, H. (2000). *Adv. Mat.*, **12**, 333–337.
Ruland, W. and Dewaelheyns, A. (1967). *J. Sci. Instrum.*, **44**, 236–238.
Rule, R.J., MacKerron, D.H. Mahendrasingam, A., Martin, C. and Nye, T.M.W. (1995). *Macromolecules*, **28**, 8517–8522.
Russel, W.B., Chaikin, P.M., Zhu, J., Meyer, W.V. and Rogers, R. (1997). *Langmuir*, **13**, 3871–3881.
Ryan, A.J., Terrill, N.J. and Fairclough, J.P.A. (2000). In: *Scattering from Polymers. ACS Symposium Series 739* (Cebe, P., Hsiao, B.S. and Lohse, D.J., eds.). American Chemical Society, Washington, p. 201.
Salgi, P. and Rajagopalan, R. (1993). *Adv. Colloids Interface. Sci.*, **43**, 169–288.
Samon, J.M., Schultz, J.M., Wu, J., Hsiao, B. and Kolb, R. (1999). *J. Polym. Sci., Polym. Phys.*, **37**, 1277–1287.
Spence, J.C.H. and Carpenter, R.W. (1986). In: *Principles of Analytical Electron Microscopy* (Joy, D.C., Romig, A.D. and Goldstein, J.I., eds.). Plenum Press, New York, p. 301.
Stevens, M.J., Robbins, M.O. and Belak, J.F. (1991). *Phys. Rev. Lett.*, **66**, 3004–3007.
Steward, P.A., Hearn, J. and Wilkinson, M.C. (2000). *Adv. Colloids Interface Sci.*, **86**, 195–267.
Stribeck, N. (1993). *Colloids Polym. Sci.*, **271**, 1007–1023.
Weyhe, G. and Pietzsch, H.-R. (1993). *Plaste Kautschuk*, **40**, 73–79.
Wiese, H. and Horn, D. (1993). *Ber. Bunsenges. Phys. Chem.*, **96**, 1818–1828.
Wilke, W. and Bratrich, M., J. (1991). *J. Appl. Cryst.*, **24**, 645–650.
Winnik, M.A. (1997). *Curr. Opinion Colloids Interface Sci.*, **2**, 192–199.
Würth, M., Schwarz, J., Culis, F., Leiderer, P. and Palberg, T. (1995). *Phys. Rev. E*, **52**, 6415–6423.
Zipfel, J., Lindner, P. and Richtering, W. (1998). *Prog. Colloids Polym. Sci.*, **110**, 139–143.

Author Index

Subject Index

D

E

N

O

P

R

S

Printed in the United Kingdom
by Lightning Source UK Ltd.
131577UK00001B/6/A

9 780444 511225